QUANTUM MECHANICS
With a SPIN On It
Fundamentals Of Quantum And Spin Physics

Kwaku Eason

First Edition 1.0

January 2019

Copyright © 2019 Kwaku Eason

This book contains information obtained from authentic and highly regarded sources. Reasonable efforts have been made to publish reliable data and information, however the Author and Publisher do not assume responsibility for the validity of all materials or the consequences of their use. The Author and Publisher have attempted to trace the copyright holders of all material reproduced in this publication, where applicable, and apologize to copyright holders if permission to publish in this form has not been clearly obtained. If any copyright material has not been properly acknowledged, please write and inform us so we may correct this in any future reprint.

All rights reserved. No part of this publication may be reproduced, distributed, or transmitted in any form or by any means, including photocopying, recording, or other electronic or mechanical methods, without the prior written permission of the publisher and/or author, except in the case of brief quotations embodied in critical reviews and certain other noncommercial uses permitted by copyright law. For permission requests, write to the publisher, addressed "Attention: SP Permissions Coordinator," at the email address below:

ISBN: 978-0-692-15979-8

Library of Congress Control Number: 2018908877

Front cover image by: Symphonious Publishers and istockphoto.com
Book design by: Symphonious Publishers
Published by: Symphonious Publishers
Website: http://www.symphoniouspub.com
Contact: qmechwspin@symphoniouspub.com

First printing, January 2019

Dedications

To my greatest inspirers Beryl, Symphony, Tala, Kojo, and Ash

...and to one of the greatest Professors the world never knew, Prof. Henry Gore, Morehouse College, Atlanta, GA.

...and to Professor Boris Luk'Yanchuk, Nanyang Technological University, Singapore, an early motivating force for the creation of this text.

Acknowledgements

The author gratefully acknowledges invaluable revision contributions from Professor Simon Greaves of Tohoku University in Sendai, Japan and Dr. Roger Woods, formerly of HGST and currently adjunct faculty at Washington State University.

Preface To The First Edition

Quantum Mechanics With A SPIN On It: Fundamentals of Quantum and Spin Physics, First Edition, we hope you'll find to be a unique textbook on the *fundamentals of quantum mechanics*. In addition to fundamental principles, it deeply emphasizes the often missing consequences associated with the electron spin. This flavor of quantum mechanics is sometimes known as *spin physics*. Today, this is an extremely relevant aspect of quantum mechanics being utilized in numerous novel technologies, both existing and on the horizon. These technologies include several forms of data storage devices such as spin-transfer torque magnetic random access memory (STT-MRAM), spin-orbit torque magnetic random access memory, *radically* novel quantum data processors with more advanced forms being based on Josephson junctions, utilizing superconductors, to novel types of spin-based field effect transistors (SPIN-FET), magnetoresistance sensors, and much more. Technology development is serious business. It's a risky business, as well. Part of the risk lies in how well things are understood. But *if* carried out rigorously, but efficiently, the rewards can greatly justify the risk, returning enormous value towards advancing our capabilities and knowledge. However, to contribute the kind of research leading to such success, among many other truths, a solid understanding of the relevant physics proves infinitely valuable. In fact, it often makes the difference between being the leader in the technology field and being the follower. Even more extreme, it can make the difference between success and failure. Amazingly, these are matters of choice. It lies in the acceptable level of knowledge and understanding to answering challenging technological questions compulsory for a successful technology design and manufacturing. This text aims to make an impact in this area, specifically to arm students and researchers with accessible, but sufficiently deep content that can be utilized in research, development, and the classroom.

The fundamentals of quantum mechanics are presented in combination with the relevant historical context as this enables a more grounded connection to the content. Further, the text provides uniquely complete and thorough derivations of both well-known fundamental results as well as *new and useful results* that are only contained in this text, at present. The reader will find nearly full derivations of all the important quantum mechanical results ranging from the fundamental topics to the newer spin physics results, as well as a novel presentation of superconductivity. This painstaking feature is intended for the *serious* student and/or researcher of quantum mechanics and spin physics. Having said that, be prepared for a level of rigor that is unavoidable in quantum mechanics to achieve correct and meaningful results. Because quantum mechanics utilizes tools from several branches of mathematics and physics, it is common in some texts to conveniently omit important details in derivations due to the uncharacteristically broad scope that may be needed to arrive at some results. This text resists such temptation *to facilitate a better learning experience*. Additionally, the author would rather place the reader in a position to be able to (1) more deeply understand content to confidently make use of it and (2) pick up from where this text brings the reader. Once understood *well enough*, there is greater value in extending the content and application to *new problems* rather than repeating the same problems robotically. *Quantum Mechanics With A SPIN On It* strategically guides the reader to all important conclusions, developing all the necessary tools successively, along the way. The text is therefore, predominantly self-contained.

As alluded, the text contains new and useful content that cannot be found currently in any other text on the subjects. Among the new content, the text features one of the first derivations of spin-dependent energy dissipation, which is implied from relativistic equations of motion. This is translated into the language of quantum mechanics to extend the reach of quantum mechanics and answer several technical questions, some of which have been bristling with thorns. This includes full quantum mechanical derivations of spin dependent phenomena including spin transfer torque and spin diffusion equations of motion. The dissipation theory also leads to more general dynamical equations of motion for electron spins, in particular, completely derivable in quantum mechanics using an extended form of the Generalized Ehrenfest Theorem given in the text. As an example, it is demonstrated how to derive the so-called Landau-Lifshitz-like equation of motion. Moreover, the exact solutions to such nonlinear equations are obtained and made available for the first time. Contemporary topics such as spintronics, relativistic quantum mechanics, spin-orbit coupling, and superconductors are analyzed sufficiently deeply. In the case of superconductors, the text leverages the overwhelming experimental evidence that suggests the BCS potential function is likely spin-dependent. From this premise, more precise statesments can be made. Thus, a slightly modified theory becomes possible to aid in explaining the remarkable phenomenon of superconductivity in an easy to understand manner.

The text begins in Chapter 1 with the history of humans' studying and analyzing light. This sets the stage for our first contemplation of wave-particle duality. Then, we discuss the historically relevant topic of thermodynamics to set the historical stage for deeper digestion of the discovery of Planck's constant and subsequent content. The formal introduction of quantum mechanic begins in Chapter 3. We evolve the topics from the initial descriptions containing wave-particle duality towards the inclusion of angular momentum. Once this topic is completed in Chapter 9, extensions into multi-electrons systems, relativistic quantum mechanics, and other more advanced topics are discussed, including the uniquely treated topic of quantum mechanical dissipation. We postpone the topic of Bra-Ket algebra until closer to the end of the text, intentionally, in order that the learning of the associated notation does not interfere with the learning of the physics. The text is crowned with the topic of superconductivity, introducing consequences of spin-dependent potentials to the well-known BCS theory. With this, we hope that the reader finds the content both educational and useful in its presentation. And lastly, but not least, thank you truly for your interest in this text. It was, without a doubt, written for you.

Sincerely,

Kwaku Eason, Ph.D.

Contents

Chapter 1 In the Beginning, There was Light ——— Page 1

1.1 Early Recorded Measurements of Light Bending 4

1.2 Dressing Early Light Observations in Mathematics 6

1.3 Thomas Young's Double Hole Experiment 12

1.4 Revelation That Light Is an Electromagnetic Wave 14

1.5 First Generation of EM Waves in a Laboratory 20

1.6 Electron Revealed By Electromagnetic Waves 22

1.7 Chapter Summary 29

1.8 Chapter Problems 30

1.9 Suggested Readings & References 31

Chapter 2 Blackbody Radiation & The Discovery Of $\hbar$ —— Page 33

2.1 Blackbody Radiation Dilemma 34

2.2 Classical Thermodynamics of Ideal Gas 36

2.3 Statistical Interpretation Of Entropy 37

2.4	Ideal Gas Law and Kinetic Energy	39
2.5	Statistical Thermodynamics and State Occupations	46
2.6	Maxwell Equations Give Blackbody Modal Density	55
2.7	Rayleigh-Jeans Distribution	59
2.8	Wiens Distribution	60
2.9	Planck's Distribution Law	62
2.10	Distribution From Pure Thermodynamics	68
2.11	Photoelectric Effect Needs Planck's Constant	70
2.12	Rydberg Formula And The De Broglie Relation	72
2.13	Chapter Summary	77
2.14	Chapter Problems	77
2.15	Suggested Readings & References	79

CHAPTER 3 — A Masterful Equation With Duality — Page 81

3.1	An Equation Describing Particles And Waves	82
3.2	The Schrödinger Equation	88
3.3	Introduction to Dirac delta function	89
3.4	Dirac Delta Function and Fourier Transforms	98
3.5	From Fourier Transform To Parseval's Theorem	104
3.6	Generalization of Operators and Operations	107
3.7	The Free Particle Problem	109
3.8	Expectation Values and Uncertainties	115
3.9	The Uncertainty Principle	118

xiii

3.10	The Generalized-Generalized Ehrenfest Theorem	122
3.11	Fundamental Particle Through A Single Slit	125
3.12	Revisiting Young's Double Slit Experiment	130
3.13	Chapter Summary	134
3.14	Chapter Problems	136
3.15	Suggested Readings & References	137

CHAPTER 4 — INFINITELY DEEP POTENTIAL WELL — PAGE 139

4.1	Infinitely Deep 1D Potential Well	140
4.2	Extending The IDPW To 3D	148
4.3	Introduction to Density of States	154
4.4	Orthogonality Of Schrödinger's Eigensolutions	161
4.5	Real Eigenvalues Of Schrödinger's Equation	164
4.6	Current Density And A Continuity Equation	166
4.7	Meaning of Divergence	170
4.8	Continuity Links to Gauss' Theorem	172
4.9	Chapter Summary	174
4.10	Chapter Problems	175

CHAPTER 5 — THE HARMONIC OSCILLATOR — PAGE 179

5.1	The Classical Harmonic Oscillator	180
5.2	Quantum Mechanical Harmonic Oscillator	184
5.3	Spatial Dependence Solvable By Frobenius Method	185

5.4	Symmetry & Annihilation/Creation Operators	206
5.5	Introduction To The Coherent State	213
5.6	Extending To 3D Harmonic Oscillator	216
5.7	Chapter Summary	219
5.8	Chapter Problems	222
5.9	Suggested Readings & References	223

CHAPTER 6 — EVANESCENCE, SCATTERING, & TUNNELING — Page 225

6.1	Creation Of Evanescent wave functions	226
6.2	Transmission and Reflection Probabilities	233
6.3	Electrons Across A Rectangular Potential Barrier	237
6.4	The Finite Potential Well Problem	250
6.5	Chapter Summary	258
6.6	Chapter Problems	259

CHAPTER 7 — ORBITAL ANGULAR MOMENTUM — Page 263

7.1	Orbital Angular Momentum Of A Rotating System	264
7.2	Canonical Problem Of The Magnetic Moment	267
7.3	Classical Orbital Magnetic Moment	272
7.4	Quantum Mechanical Angular Momentum $\widehat{L}$	274
7.5	The Laplacian and Legendre Equations	280
7.6	Extending To Associated Legendre Functions	287
7.7	Assoc. Legendre Funcs & Orb. Ang. Momentum	292

7.8	Special Commutation Relations 300
7.9	Extended Momentum Leads to Lorentz Force 308
7.10	Angular Momentum Ladder Operators 311
7.11	Chapter Summary. 322
7.12	Chapter Problems 325

CHAPTER 8 — THE HYDROGEN ATOM — PAGE 327

8.1	Atomic Spectra of Hydrogen. 328
8.2	Bohr's Atomic Hydrogen Model 330
8.3	Schrödinger's Hydrogen Atom Model 334
8.4	Schrödinger Equation Radial Solution 340
8.5	Conventional Shells and Orbitals 350
8.6	Observing the Zeeman Effect 354
8.7	Quantum Mechanics Of Ordinary Zeeman Effect 358
8.8	Chapter Summary. 360
8.9	Chapter Problems 363
8.10	Suggested Readings & References. 365

CHAPTER 9 — THE UNVEILING OF ELECTRON SPIN — PAGE 367

9.1	Evidence of Spin Existence 368
9.2	Understanding the Stern-Gerlach Experiment. 372
9.3	Anamolous Zeeman Effect 376
9.4	Principle Spin Matrices and Spinors. 380

9.5	Generalized Spin Matrix	387
9.6	The Pauli Equation and Spin Precession	390
9.7	The Spin Density Matrix	398
9.8	Chapter Summary	401
9.9	Chapter Problems	403
9.10	Suggested Readings & References	404

CHAPTER 10 — MULTI-ELECTRON SYSTEMS — PAGE 405

10.1	Consequences of Electron Indistinguishability	405
10.2	Two-Electron wave functions Reveal Exchange	407
10.3	Symmetric Spatial & Anti-symmetric Spin State	418
10.4	He Spectra Supports Asymmetry	422
10.5	Molecular Wave Functions And Bonds	427
10.6	The Simplest Diatomic Molecule: H_2^+	430
10.7	Vibrational-Rotational Spectra Of Molecules	436
10.8	Exchange Interaction With Spin	440
10.9	Multi-electron Total Angular Momentum	445
10.10	Clebsch-Gordan Coefficients	450
10.11	Multi-electron Generalized g-Factor	455
10.12	Spin Expectation Using Boltzmann Distribution	458
10.13	Chapter Summary	464
10.14	Chapter Problems	465
10.15	Suggested Readings & References	466

xvii

CHAPTER 11 — RELATIVISTIC QUANTUM MECHANICS — PAGE 469

- 11.1 Special Theory Of Relativity & Momentum 470
- 11.2 Relativistic Energy 478
- 11.3 The Klein-Gordon Equation 482
- 11.4 The Dirac Equation 484
- 11.5 Spin-Orbit Interaction 492
- 11.6 Cyclotron Orbits And Landau Levels 502
- 11.7 Spin Hall Spin-Orbit Effects 515
- 11.8 Rashba Spin-Orbit Effect 521
- 11.9 Chapter Summary 528
- 11.10 Chapter Problems 529
- 11.11 Suggested Readings & References 531

CHAPTER 12 — NON-HERMITIAN OPERATORS & DYNAMICS — PAGE 533

- 12.1 Lenz' Law Correction For Dynamical Effects 535
- 12.2 Non-Hermition Hamiltonian Including Transient Effects . . 542
- 12.3 Dissipative Continuity Equation 545
- 12.4 Dynamical Spin Angular Momentum 547
- 12.5 Dynamical Equation From G^2 Ehrenfest Theorem 554
- 12.6 Asymmetric Spin Flow Of Itinerant Electrons 562
- 12.7 Introduction To Magnetoresistance 578
- 12.8 Spin Diffusion Equations 585

12.9	Spin Transfer Torque On Localized Electrons	590
12.10	Chapter Summary.	596
12.11	Chapter Problems	598
12.12	Suggested Readings & References.	599

CHAPTER 13 METHODS OF APPROXIMATION — PAGE 601

13.1	Airy Functions: Prelude To JWKB Approximation	601
13.2	Approximation To Airy Functions In Region $x < 0$	604
13.3	Approximation To Airy Functions In Region $x > 0$	608
13.4	Zeroth Order JWKB Approximation.	611
13.5	First Order JWKB Approximation	613
13.6	The JWKB Eigenvalue Equation	615
13.7	First Order Perturbation Theory	619
13.8	Chapter Summary.	622
13.9	Chapter Problems	623
13.10	Suggested Readings & References.	625

CHAPTER 14 INTRODUCTION TO BRA & KET ALGEBRA — PAGE 627

14.1	Bra-Ket Preliminaries	627
14.2	Operators With Bras and Kets	632
14.3	Linear Harmonic Oscillator Revisited	634
14.4	Annihilation and Creation Operators Revisited	639

14.5	Ground And Above-Ground States of LHO	645
14.6	Coherent State Of The Oscillator Revisited	649
14.7	Chapter Summary	655
14.8	Chapter Problems	656

CHAPTER 15 — FUNDAMENTALS OF SUPERCONDUCTIVITY — PAGE 659

15.1	Kwik Nagenoeg Nul	661
15.2	The Critical Field Of A Superconductor	667
15.3	Type I & Type II Superconductors	670
15.4	London Equation Leads To Meissner Effect	674
15.5	Ginzburg & Landau Theory	677
15.6	Diamagnetism From Circulating Surface Currents	695
15.7	Correlating Cooper Pairs	699
15.8	Small Symmetric Splitting With Spin Matrices	703
15.9	Discontinuities In Electronic Properties Near T_c	719
15.10	Connecting Cooper, Ginzburg, and Landau	724
15.11	Higher Order Relativistic Correction	727
15.12	An Energy Gap To Excite Cooper Pairs	731
15.13	Chapter Summary	734
15.14	Chapter Problems	736
15.15	Suggested Readings & References	737

CHAPTER A	SUPPLEMENTAL: REAL ELECTRIC POTENTIAL	PAGE 739
A.1	Potential And Discrete Electric Charges	740
A.2	Continuous Charge In Volumes	744
A.3	Summary	747

Relevant Physical Constants In SI Units

Quantity	Symbol	Value
Avogadro constant	N_a	6.0221415×10^{23} mol^{-1}
Bohr radius	a_0	$0.5291772108 \times 10^{-10}$ m
Bohr magneton	μ_B	$9.27400949 \times 10^{-24}$ JT^{-1}
Boltzmann's constant	k_B	$1.3806505 \times 10^{-23}$ J·K^{-1}
Compton wavelength	λ_C	$2.426310238 \times 10^{-12}$ m
elementary charge	e	$1.60217662 \times 10^{-19}$ C
electron g-factor	g	2.0023193043718
electron mass	m_e	$9.1093826 \times 10^{-31}$ kg
gyromagnetic ratio	$|\gamma|$	$8.7941001136 \times 10^{-10}$ HzT^{-1}
light speed in vacuum	c	$299,792,458$ ms^{-1}
Planck constant	h	$6.6260693 \times 10^{-34}$ J·s
Planck reduced constant	$\hbar$	$1.05457168 \times 10^{-34}$ J·s
proton mass	m_p	$1.67262171 \times 10^{-27}$ kg
quantum of flux	Φ_0	$2.06783372 \times 10^{-15}$ Wb
Rydberg constant	R_d	$109,737,31.568525$ m^{-1}
universal gas constant	R	8.3144 JK^{-1}mol^{-1}
vacuum permittivity	ϵ_0	$8.854187817 \times 10^{-12}$ Fm^{-1}
vacuum permeability	μ_0	$4 \times \pi \times 10^{-7}$ N·A^{-2}

CHAPTER 1

In the Beginning, There was Light

In the year 1924, a German physicist named Max Born published a scientific paper bearing the novel name *Über Quantenmechanik*. Shortly thereafter, several publications by others would appear having in its title, part of that same German phrase, particularly, *quantenmechanik*. The title of the publication by Born translates to English as "About Quantum Mechanics" or "On Quantum Mechanics". It is from these early published works that the formal name of what became a new branch of physics was *Born*. Almost 100 years later, the name remains a keepsake from those early contributors for one of the most advanced, accurate, and fundamental branches of physics known today. Nowadays, it's even found its way into popular culture and language. At the time of Born's publication, a better name probably could not have been contrived because the most advanced physics models available to them describing newly discovered forms of matter were indeed mechanical, involving angular momentum, inertia, velocities, etc. However, there was an undeniable need for other branches of physics including fields like thermodynamics and electromagnetics. Under certain assumptions, these novel quantum mechanical models yielded unexpected *discrete* or *quantum* behavior. Ordinarily, it would have been easy to take these strange new models lightly, however, there was a difficulty with being dismissive with them. This is because these quantum models were spot-on in describing some very puzzling experiments at the time. We'll be getting into some of these experiments and models shortly. At the time, they were dealing with newly discovered forms of extremely tiny matter, as well as learning about new behaviors of light. It is from this distilled history that the subject known as *quantum mechanics* has come to traditionally focus on the physics of tiny "particles" such as atoms, as well as subatomic particles like electrons, protons and neutrons. Needless to say that before this time, no general physics framework existed that could be used to describe these tiny forms of matter that were being discovered. The mounting experimental evidence would motivate new questions for physics that were bristling with thorns. To address this satisfactorily, a new branch of physics would ultimately emerge.

Today, without exception, all matter encountered in everyday life is understood to form from such building blocks of atoms. It has not always been that way. The understanding nowadays is good enough that it is almost trivial for scientists to find clever ways to generate free subatomic particles such as electrons, neutrons, atoms, and electromagnetic waves. But, opposite to the large bridges and skyscrapers marvelously built by humans, this knowledge enables new forms of construction of matter or materials on length scales that sneak well below the resolution of the human eye. Now we understand that electrons, protons, and neutrons along with special forces form atoms, while atoms form molecules, atoms and molecules can form gases, liquids, solid state thin films or bulk materials. It is a remarkable fact that the same building blocks of matter give rise to an extraordinarily rich physical world for our experience. We are ever learning that so much is possible. With this understanding, this unique branch of physics regarding these tiny fundamental particles is formally known as **quantum mechanics** or **quantum physics**. Because it describes the building blocks, it necessarily becomes the study of nearly all forms of matter. An example of these building blocks is illustrated in Fig. 1.1. Using advanced equipment, scientists can now construct films on length scales smaller than one nanometer (nm), or one-billionth of a meter. These capabilities have played an important role in enabling the creation of technologies, some of which, we have come to take for granted, like computers, mobile phones, satellites, and more. Without question, the field of quantum mechanics has widened the doors of technological opportunities, and this fact continues today. From 1924 to now, the subject has come to be known as the canonical or standard theory of atomic and electronic motion, yet of the same consistency as classical mechanics, electrodynamics, and all other branches of physics.

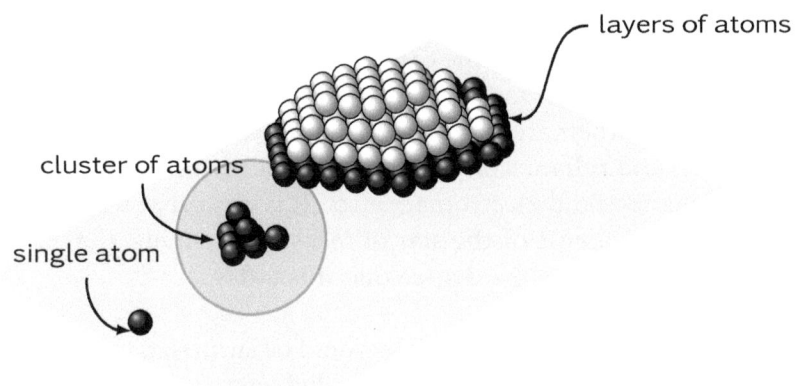

Figure 1.1: Illustration of the range of matter, building from atoms, to clusters of atoms, to layers of atoms, to multi-layer films, etc., deposited on a substrate (shown as gray).

As suggested by all of this, its content as well as history is rich and deep. Because of this, it is not easy to treat *all of* quantum mechanics in a single text (for it too, has many layers), and we will not attempt to do this. Instead, we shall devote our attention to more of the technically relevant *fundamentals of quantum mechanics* that are applicable to several mature as well as new technologies. It will be done with the distinction of emphasizing *both* properties of the electron, being *charge and spin*, as they are inseparable. For this reason, the fundamentals of spin-physics and magnetism and the inter-dependence with quantum mechanics is a definite part of the objective of this text. All stable quantum particles posses a property known as *intrinsic spin*. Don't worry if you are not familiar with this concept at this point, as we will formally introduce these ideas and come back to the topic over and over again in this text. It turns out to play an important role in much of the spin-physics being exploited in novel applications today.

The subject of *quantum mechanics* can be approached from many different directions. In this text, the emphasis will be placed, predominately, on the *electron*, its characteristics and its influence on atomic and/or molecular properties. For most people's everyday experience with materials, electrons are usually intimately involved. The behavior and characteristics of electrons turn out to provide a sufficiently deep understanding of the fundamentals of *spin-physics*. Hence, the *electron* is the main character of this show. However, no tale is justly conveyed without the proper historical context of its supporting cast and additional main characters all serving to edify our understanding of the main character, the *electron*. For the electron, the cast involves a combination of brilliant minds making key observations, contriving clever experiments, along with powerful mathematics, new levels of abstractness stretching our imaginations, and reliance on other rich predecessors of physics such as thermodynamics and electromagnetics. It is clear that without these *cast members*, the talents of the star of *this* show, namely, the electron, would not be revealed to the degree that it is today.

Where should we begin? Many people would be surprised to learn that if it were not for our interest in that stuff called *light*, quantum mechanics and the deeper understanding of electrons, would likely not exist as we know it. Some of the fundamental ideas that form pillars of quantum mechanics were taught to us, historically, out of the parallel evolution of the understanding of light behaving as a wave, and yet, as a particle. For centuries, light has intrigued humans, motivating many to work to understand what it *is*, and what is its *true* nature. As the story told in these pages unfolds, we will begin to see that this luminous curiosity ultimately gave rise to a conceptual framework enabling the foundations of *quantum mechanics*. This is where the concept known as *wave-particle duality* began. For centuries, it was believed by many of the prominent scientists of the time, that light behaved solely as a particle. At the same time, there was a growing belief that light behaved as a wave. Until the advent of quantum mechanics, it never occurred to us that something could, in fact, be either both or, more aptly, neither in a strict sense! Let us begin by tracing the developments of our recorded knowledge of light, and its role in the path to the discovery of the electron as well as the atomic nucleus. Perhaps then, we may come to understand just how the concept of *wave-particle duality* came about.

1.1 EARLY RECORDED MEASUREMENTS OF LIGHT BENDING

You are probably using *light* to read this page. Light has obviously not just enabled deeds like this, but it practically enables life itself! It cannot be emphasized enough that light has played an important role in the development of *many* areas of physics, and it is no less true for quantum mechanics. There is no wonder why man and woman would eventually take a keen interest in light. Therefore, this is where our tale begins...with humans starting to document the observed behavior of that remarkable stuff called *light*.

1.1 Early Recorded Measurements of Light Bending

The bending of light that can be observed when light passes from one material or medium into another, is called **refraction**. One person known to have measured bending or refraction angles, dating as far back as c.170 A.D., is the Egyptian-Greek polymath Claudius Ptolemy, of Alexandria, Egypt. He considered a system consisting of air interfaced with water and observed the light passing between them. He measured the incidence angle, denoted by θ_1, of light entering the water from the air, versus the bending angle, or refraction angle θ_2 in the water medium, as illustrated in Fig. 1.2.

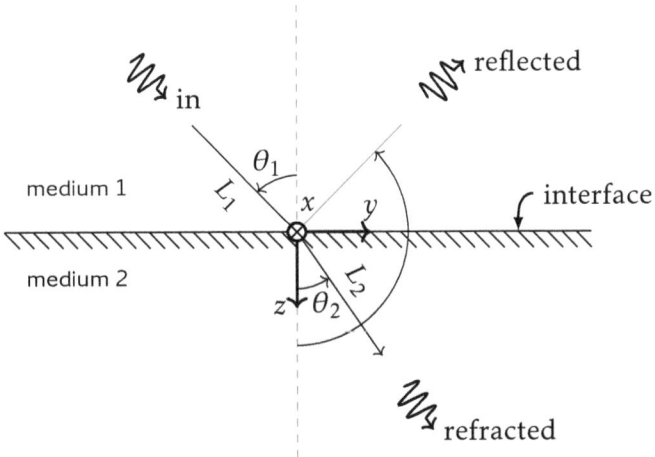

Figure 1.2: Illustration of reflection and refraction (or bending) of light. The light starts out propagating in medium 1 approaching the interface between it and medium 2. When crossing the interface, light partially reflects, while the rest penetrates, partially bending at the same time.

At a later time c.984 A.D., a Persian polymath named Ibn Sahl, from the time of the Islamic Golden Age, advanced this probing well beyond Ptolemy's records and insight of light refraction. His insights were published in a book called *On Burning Mirrors and Lenses*, which is the translation from Arabic. Sahl lived in modern day Baghdad, Iraq. Sahl not only understood the mathematical law describing light refraction, but also how curved surfaces bend and can focus light. He used it to design lenses during his lifetime. And while Ptolemy and Sahl are examples of people who clearly documented their work with light, recent archaeological findings have even uncovered perfectly spherical lenses originating in ancient Egypt (in northern Africa) suggesting a curiosity with light even predating both Ptolemy and Sahl. Therefore, history suggests that light has, indeed, been an interest of humanity for a long time. To better understand and describe these light observations, mathematics would inevitably prove to be an edifying companion.

1.2 Dressing Early Light Observations in Mathematics

Many centuries later, in 1621, the Dutch mathematician, Willebrord Snell noticed a mathematical relationship governing the kind of data documented by Ptolemy and others. Snell expressed the empirical relationship as

Snell's Law

$$\frac{\sin\theta_1}{\sin\theta_2} = \text{constant} \tag{1.1}$$

Eq. (1.1) is known as **Snell's law**. Initially an empirical law, it states that the ratio of the velocity vector components parallel to the interface, proportional to $\sin\theta_i$, remains constant. Only the component normal to the interface is affected from penetrating into a different material. Thus, in the observed measurements of Ptolemy, the ratio was found to be independent of the incidence angle. This relation provides a useful way to describe the way light bends or refracts when passing from one medium into another, provided the constant is known.

An equation of the form of Eq. (1.1) can be derived in a number of ways. One of the first known derivations was done in the 17th century by a French polymath named René Descartes (1596 - 1650), after Snell's empirical law was introduced. In Descartes' derivation, he regarded the light as a collection of particles. This particle-like concept for light became a popular notion known as the **corpuscular model of light**. A *corpuscle* is a basic minute particle or constituent of matter and/or light. For a time, it was the more common belief among prominent physicists until the early 19th century, but we shall come to that. The analyses done at the time of Descartes and other physicists can be framed with the later work of Swiss polymath Leonhard Euler (1707 - 1783), which among his many contributions, introduced the following integral known as the *integral of action*:

$$I_A = \int A(y, dy/dx)dx \tag{1.2}$$

1.2 Dressing Early Light Observations in Mathematics

In Eq. (1.2), one is interested in the function y^* that *minimizes* the integral I_A. Or what is the function such that if it were plugged into I_A, will always be less than the value of the integral with any other continuous function $y(x)$. This problem can be posed as an optimization problem, where the function y^* satisfies

$$\delta I_A = \int \delta A(y, dy/dx) dx = 0 \tag{1.3}$$

Since the integral of action I_A is completely described with two variables, the total variation of the *action A* can be expressed as

$$\delta A\left(y, \frac{dy}{dx}\right) = A\left(y + dy, \frac{dy}{dx} + \delta\left(\frac{dy}{dx}\right)\right) - A(y, \frac{dy}{dx}) \tag{1.4}$$

$$= \frac{\partial A}{\partial y}\delta y + \frac{\partial A}{\partial (dy/dx)}\delta(dy/dx) \tag{1.5}$$

After substitution of $\delta A(y, dy/dx)$ into Eq. (1.3), δI_A becomes

$$\delta I_A = \int \left[\frac{\partial A}{\partial y}\delta y + \frac{\partial A}{\partial (dy/dx)}\delta(dy/dx)\right] dx = 0 \tag{1.6}$$

For the second integral on the RHS, particularly, we can use integration by parts, which states $\int u\, dv = uv - \int v\, du$, where we have

$$\int \underbrace{\frac{\partial A}{\partial (dy/dx)}}_{u} \underbrace{\delta(dy/dx) dx}_{dv} =$$

$$\frac{\partial A}{\partial (dy/dx)} \cdot \delta y \bigg|_{x_i}^{x_f} - \int \delta y \frac{d}{dx}\frac{\partial A}{\partial (dy/dx)} dx \tag{1.7}$$

Note that $v = \delta y$ and dy have identical meaning. Different symbols δ and d are used, primarily when they appear together, to denote distinct variations (or a variation of a variation). In the first term on the RHS, Euler assumed that the initial and final conditions are the same. Therefore,

$$\delta y \bigg|_{x_i}^{x_f} = \delta y(x_i) - \delta y(x_f) = 0$$

This is valid because all the different *possible* functions y must be the same at x_i and x_f, so $\delta y(x_i) = 0$ and $\delta y(x_f) = 0$. Then, the first term vanishes, and we are left with

$$\int \frac{\partial A}{\partial (dy/dx)}\delta(dy/dx) dx = -\int \delta y \frac{d}{dx}\frac{\partial A}{\partial (dy/dx)} dx \tag{1.8}$$

Substituting this back into Eq. (1.6), we have

$$\delta I_A = \int \left[\frac{\partial A}{\partial y} \delta y - \delta y \frac{d}{dx} \frac{\partial A}{\partial (dy/dx)} \right] dx = 0 \tag{1.9}$$

To be true for all x, the integrand must equal zero, or

Euler's Equation

$$\frac{dA}{dy} - \frac{d}{dx} \frac{\partial A}{(\partial y/\partial x)} = 0 \tag{1.10}$$

The solution to **Euler's equation** given by Eq. (1.10) is y^*, or the function that minimizes Eq. (1.2). Euler's work was the basis for the introduction of several well-known concepts in physics, such as the *Hamiltonian* $= K + V$ named after Irish physicist Sir William Rowan Hamilton (1805-1865), as well as the *Lagrangian* $= K - V$, named after the Italian-French mathematician Joseph-Louis Lagrange (1736 - 1813). K is the kinetic energy and V is the potential energy. In quantum mechanics, we'll see that the Hamiltonian has the same meaning with a more general interpretation. It was understood that the laws of physics tend to correspond to the minimization of some sort of *action* $A(y, dy/dx)$.

One of the earliest *action functions* minimized was *time*. Minimizing *time* had been suggested by French mathematician (and lawyer) Pierre de Fermat (1607-1665). This principle of least time is known as *Fermat's Principle*. To derive Snell's law, Descartes computed the amount of time T_{ab} it takes for corpuscular light to travel from a point a in medium one, to a point b in medium two, using time as the action A. It can be expressed as a function of a single spatial variable x. These conditions reduce Euler's result in Eq. (1.10) to $dA/dy^* = dT_{ab}/dx = 0$.

The light traveling in medium one is taken to have an associated *constant* velocity v_1, while the light traveling in medium two has an associated *constant* velocity v_2, as illustrated in Fig. 1.2. The only things that changes are their respective angles. The total time T_{ab} can be written as

$$T_{ab} = \frac{L_1}{v_1} + \frac{L_2}{v_2} \tag{1.11}$$

After determining L_1 and L_2 in terms of θ_1, θ_2, L, and x, (using standard

1.2 Dressing Early Light Observations in Mathematics

trigonometry), Eq. (1.11) becomes

$$T_{ab} = \frac{\sqrt{((L-x)^2(1+\cot^2\theta_1))}}{v_1} + \frac{\sqrt{(x^2(1+\cot^2\theta_2))}}{v_2} \qquad (1.12)$$

The equation that minimizes the total time T_{ab} is, therefore, given by

$$\frac{dT_{ab}}{dx} = \frac{1}{v_1}\frac{dL_1}{dx} + \frac{1}{v_2}\frac{dL_2}{dx} = 0 \qquad (1.13)$$

Evaluating the derivatives in Eq. (1.11) leads to the following result:

$$\frac{dT_{ab}}{dx} = \frac{\sin\theta_1}{v_1} - \frac{\sin\theta_2}{v_2} = 0 \qquad (1.14)$$

Expressing this result in a form similar to Snell's empirical law, we have

$$\frac{\sin\theta_1}{\sin\theta_2} = \frac{v_1}{v_2} \quad \text{(Snell's Law from theory)} \qquad (1.15)$$

By inspection, Descartes' result would be identical to Snell's law, given by Eq. (1.1), only if

$$\frac{v_1}{v_2} = \text{constant}$$

When $v_1 = c$, where c is the speed of light in vacuum, it is known as the **refractive index**, defined as $n = c/v$. Since c is just a constant, Eq. (1.15) can also be written as

$$\frac{\sin\theta_1}{\sin\theta_2} = \frac{c}{v_2}\frac{v_1}{c} = \frac{n_2}{n_1} \qquad (1.16)$$

A derivation in the same corpuscular spirit as Descartes was carried out by Isaac Newton (1643-1727), the English polymath. Newton and Descartes were able to use corpuscular models of light to explain the concepts of *reflection* and *refraction*. Newton included his insights on light in the classic book entitled *Opticks*, published in 1704. However, the corpuscular model could not explain **diffraction**, which is the *spreading out* of light when passing through an aperture or opening. We will see later that *diffraction could not be explained without invoking characteristics of waves*. It appeared as though the success of a corpuscular model for light was taking root...that is, until c.1678. At that time, the Dutch mathematician and scientist Christiaan Huygens (1629-1695), surprisingly, proposed that light behaves as a wave. His proposed principle is known as **Huygens' Principle** and, like the corpuscular model, it also predicts the phenomena of *reflection* and *refraction*. Huygens principle states:

Every point on a known wave front in a given medium can be treated as a point source of secondary wavelets which spread out in all directions with a wave speed characteristic of that medium. The new wavefront at any subsequent time is an envelope of these secondary wavelets.
C. Huygens

The words of Huygens translate to a mathematical model that leads to Snell's empirical law, the same result of the corpuscular model. To see this, consider an interface between two media'with velocities v_1 and v_2, respectively. Light is propagating from medium one into medium two, as shown in Fig. 1.3. The gray lines are the wave-fronts with spacing determined by a fixed time interval Δt, which is the period of the wave. The circles in Fig. 1.3 represent expanding spherical waves from a point source along the wave-front. From the perspective of the point source, source light continues in the same direction. However, because the velocity reduces, for example, when crossing the interface into medium two, the point source wave travels a shorter distance in the same time span Δt. The result is an apparent rotation of the wave-front because the left side sees the interface first, and is forced to slow down, whilst the right side still propagates in the faster medium. It is akin to driving an automobile on a highway, and partially crossing one side of the vehicle into the shoulder on the side of the road. In this situation, the wheels on the shoulder are slowed down, then the car responds by turning towards the shoulder. A key difference, however, is that the car's velocity parallel to the road/shoulder interface *is* changed. Light behaves this way anytime it crosses over an interface. From this, two congruent triangles can be formed in the respective medium. The two congruent triangles are illustrated in Fig. 1.3. They can be used to relate the angles θ_1 (angles between wave-front and interface) and θ_2 as follows:

$$\sin\theta_1 = v_1 \Delta t / h_1$$

$$\sin\theta_2 = v_2 \Delta t / h_2$$

We can choose to construct the triangles such that $h_1 = h_2$. Thus just means we are extending our wave-fronts equal lengths along the interface in forming the two congruent triangles. By solving for Δt and equating the two, an identical result to the corpuscular light model is

1.2 Dressing Early Light Observations in Mathematics

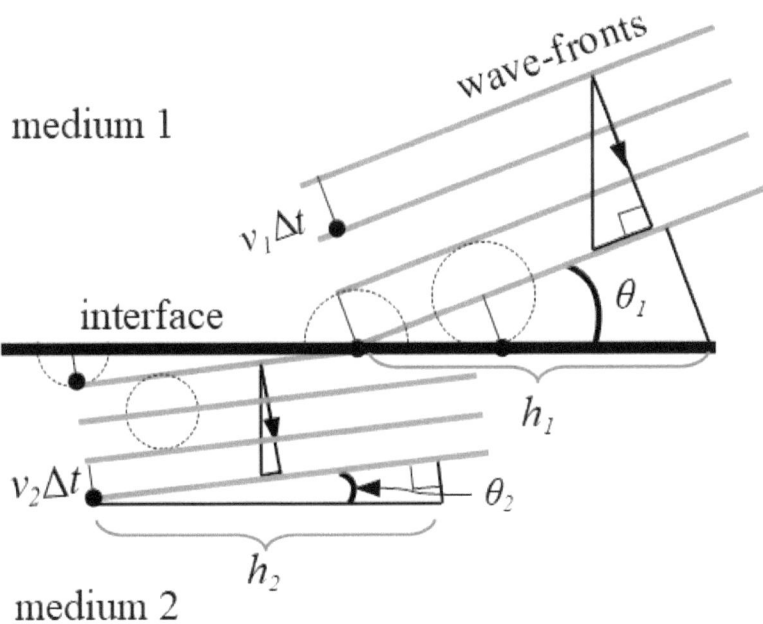

Figure 1.3: Illustration of Huygen's principle, which leads to the mathematical law of refraction.

obtained, where

$$\frac{\sin\theta_1}{v_1} = \frac{\sin\theta_2}{v_2} \Rightarrow \frac{\sin\theta_1}{\sin\theta_2} = \frac{v_1}{v_2}$$

As mentioned earlier, though Huygen's model can predict refraction (and reflection), it does not predict the phenomenon of diffraction, or the spreading of light passing through an aperture. However, unlike the corpuscular model, Huygen's wave-model turns out to be extendable in such a way to successfully predict diffraction, by applying the principle of superposition to the secondary light sources. Initially, one of the challenges many had in accepting Huygen's principle was that Newton's authority in the laws of physics were quite compelling at the time. For a time, few believed in the principle put forth by Huygens...in the beginning. Fortunately, that would begin to change with a simple yet unambiguous experiment with light, demonstrated by a Frenchman who firmly believed in the wave-model. His experiment is the subject of the next section.

1.3 Thomas Young's Double Hole Experiment

The belief in a corpuscular model of light persisted until around 1801. A paradigm shift towards a wave theory of light began when a French polymath named Thomas Young (1773-1829) performed his world famous interference experiments. The key was that his experimental observations could only be explained on the basis of a wave model of light. The phenomenon of wave interference was already known and understood to be based on a principle known as the **superposition principle**. Fig. 1.4 illustrates the idea behind Young's experiment.

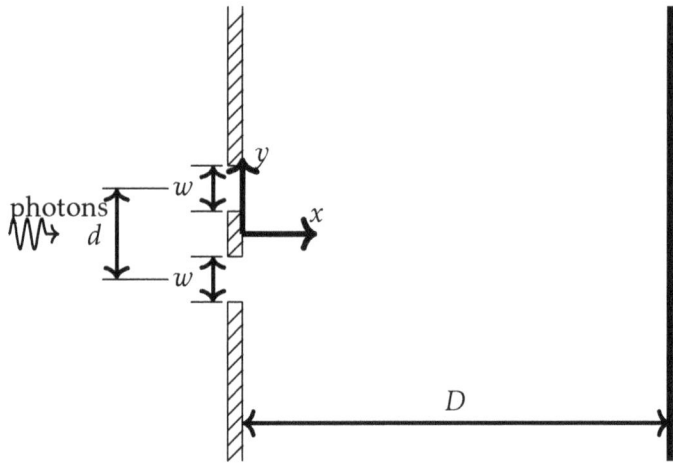

Figure 1.4: Illustration of Young's double hole experiment. Light enters from the left, later hitting a screen-wall a distance D away from the slits. The screen-wall 'captures' the light interference pattern.

Two holes at a distance d apart (from center-to-center), were pierced through a screen. On his first try, the light source was sunlight, though later, it was changed to a sodium chloride NaCl flame. The light enters from the left side of the pierced screen having the two holes. From this, one observes an interference pattern on a wall a distance D from the pierced screen, which was immediately understood to be caused by the *superposition* of two or more waves. The **superposition principle** states:

1.3 Thomas Young's Double Hole Experiment

 Superposition principle: The resultant displacement at any point, produced by a number of wave disturbances, is the vector sum of the displacements produced by each one of the disturbances separately.

For example, consider a wave displacement at a point O, produced by a wave that is given by

$$w_1 = A_1 \sin(\omega t) \tag{1.17}$$

If a second wave displacement, w_2 comes along to point O at the same instant of time, given by

$$w_2 = A_2 \sin(\omega t) \tag{1.18}$$

The resulting net displacement w_n, at point O, is given by

$$w_n = w_1 + w_2 = (A_1 + A_2)\sin(\omega t) \tag{1.19}$$

With superposition, a range of results become possible. If we have a displacement at a point O, produced by a wave that is given by

$$w_1 = A\sin(\omega t) \tag{1.20}$$

Then, if a second disturbance, w_2 is given by

$$w_2 = -A\sin(\omega t) \tag{1.21}$$

The resulting displacement at point O is given by

$$w_n = w_1 + w_2 = 0 \tag{1.22}$$

Thus, the superposition of waves can lead to a strengthening of the resultant disturbance due to both displacements, but it can also lead to a weakening, or even vanishing of the net disturbance. This behavior, specifically, is *not* a behavior that could come from a corpuscular model of light. When the interference cancels the disturbance amplitude in this way, it is known as **destructive interference**.

Thomas Young's double hole interference experiment was carried out the first time in 1801. As mentioned in the Nobel Prize speech from Denis Gabor, salt (NaCl) produces a bright yellow light in this situation because of its wavelength, which is around 600 nanometers (nm). It is in the middle of the visible light spectrum, and thus, was better to observe what Thomas was interested to observe. This observation by Young certainly created rightful doubt in the *corpuscular model* belief at the time, which was strong, due to the reverence of Newton, who continued believing in the corpuscular model of light.

To further strengthen the wave model of light, Augustine Fresnel (1788-1837) and others further developed Huygens' theory by modifying it so that secondary sources also mutually interfere and this was found to correctly predict the phenomenon of *diffraction*. Up until this time, only reflection and refraction were predicted by Huygens' theory, which was also what the corpuscular model of light was capable of deducing. After this extension to the wave-model of light, reflection, refraction, as well as diffraction were all predictable. This achievement was described well by the Hungarian-British engineer and physicist Dennis Gabor in his 1971 Nobel Lecture, where he stated:

the wave nature of light was demonstrated convincingly for the first time in 1801 by Thomas Young by a wonderfully simple experiment...he let a ray of sunlight into a dark room, placed a dark screen in front of it, pierced with two small pinholes, and beyond this, at some distance, a white screen. He then saw two darkish lines at both sides of a bright line, which gave him sufficient encouragement to repeat the experiment,...Thomas Young had expected it because he believed in the wave theory of light!
D. Gabor

1.4 Revelation That Light Is an Electromagnetic Wave

At this time, other significant developments in physics had also taken place. One was a discovery by English scientist Michael Faraday (1791 - 1867). He found that by inserting magnetic bars into electrical current loops, he could observe a direct relationship between the electrical current and the changing magnetic flux density **B**, created by moving the magnetic bar. It was as though the changing magnetic flux density **B** was giving rise to a current-driving-electric field **E**. Today, this relationship is commonly written as

Faraday's Law

1.4 Revelation That Light Is an Electromagnetic Wave

$$\nabla \times \mathbf{E} = -\frac{\partial \mathbf{B}}{\partial t} \tag{1.23}$$

Eq. (1.23) is known as **Faraday's Law Of Induction** or just **Faraday's Law**. Around the same time, French physicist and mathematician André-Marie Ampére (1775-1836), also observed another relationship between electrical current, or current density **J** (current per unit area), and magnetic field intensity, **H**. Ampére's experiments were motivated by the fact that ships would often see their compass needles violently rattle whenever they were struck by lightning. He sought to reproduce this phenomenon in his lab by bringing an electrical wire close to a compass needle. He found that the needle, in fact, deflected in response to the electrical current in the wire. This lead Ampére to propose a mathematical law that describes his observations. Today, it is commonly written as

$$\nabla \times \mathbf{H} = \mathbf{J} \tag{1.24}$$

Eq. (1.24) is known as **Ampére's Law**. Because his experiments only involved conducting materials carrying electricity, it later became better understood that Ampére's Law only applied to conducting materials. This relationship was later modified by British physicist James Clerk Maxwell (1831-1879), who added a displacement current term, $\partial \mathbf{D}/\partial t$ to the flowing current **J** on the RHS, giving the more complete law as

Ampére-Maxwell Equation

$$\nabla \times \mathbf{H} = \mathbf{J} + \partial \mathbf{D}/\partial t \tag{1.25}$$

D is the **electric displacement field**. The modification made by Maxwell may not appear to be much, however, it lead to a radically new prediction at the time, which turned out to provide a *quantum leap* in our understanding of *electricity* and *magnetism*. Maxwell considered the case of electric and magnetic fields in vacuum, which implies that there was no electrical current **J** to worry about. In this condition, using the Ampére-Maxwell equation, we then have

$$\nabla \times \mathbf{H} = \epsilon_0 \partial \mathbf{E}/\partial t \tag{1.26}$$

ϵ_0 is the **electrical permittivity of vacuum**, which has a measured value, in SI units, of

> **Electrical Permittivity In Vacuum**
>
> $$\epsilon_0 = 8.854187816 \times 10^{-12} \mathrm{C}^2 \cdot \mathrm{J}^{-1} \cdot \mathrm{m}^{-1} \tag{1.27}$$

In Eq. (1.27), C denotes Coulomb, the SI unit of charge. J denotes Joules, the SI unit of energy, and m for meters, the SI unit of length. Additionally, *Faraday's law* leads to

$$\nabla \times \mathbf{E} = -\mu_0 \frac{\partial \mathbf{H}}{\partial t} \tag{1.28}$$

μ_0 is known as the **magnetic permeability in vacuum** whose measured value is

> **Magnetic Permeability In Vacuum**
>
> $$\mu_0 = 4 \times \pi \times 10^{-7} \mathrm{N} \cdot \mathrm{A}^{-2} \tag{1.29}$$

In Eq. (1.29), N is the SI force unit Newton and A is the current unit Ampére. By taking the curl of Eq. (1.28), then substituting Ampére-Maxwell's law on the right-hand side (RHS), leads to

$$\nabla \times \nabla \times \mathbf{E} = -\epsilon_0 \mu_0 \frac{\partial^2 \mathbf{E}}{\partial t^2} \tag{1.30}$$

The left hand side (LHS) of Eq. (1.30) can also be written as

$$\nabla \times \nabla \times \mathbf{E} = \nabla(\nabla \cdot \mathbf{E}) - \nabla^2 \mathbf{E} \tag{1.31}$$

Since we are considering the case of fields in vacuum, we can assume $\nabla \cdot \mathbf{E} = 0$, since there is no charge density ρ in vacuum. This simplification lead Maxwell to the following important result:

$$\frac{\partial^2 \mathbf{E}}{\partial t^2} = \frac{1}{\epsilon_0 \mu_0} \nabla^2 \mathbf{E} \tag{1.32}$$

Eq. (1.32), obtained by Maxwell, is in the form of a well-known equation, the *wave equation*. The form of wave equations were already known due

1.4 Revelation That Light Is an Electromagnetic Wave

to the works of Jean le Rond d'Alembert (1717-1783), Leonhard Euler and others. Another known fact about this equation was that the speed of the wave is evident from the equation. The general wave equation is given by

---
Wave/Propagating Disturbance Equation

$$\frac{\partial^2 w}{\partial t^2} = v^2 \nabla^2 w \tag{1.33}$$

In Eq. (1.33), the parameter v is the speed of the wave disturbance $w(x)$. The general solution to Eq. (1.33) is given by

$$w(x) = f(x - vt) + g(x + vt) \tag{1.34}$$

So, a wave can be described, more generally, as a propagation of a disturbance described by the variation in x. In the wave solution, $f(x)$ represents *any* (determined by initial conditions) disturbance depending on x, and it propagates with speed v in the $+x$ direction, while $g(x)$ represents a disturbance propagating, however, with speed v, but in the $-x$ direction. Thus, c.1864, Maxwell was able to postulate that light was not only an electromagnetic wave, but also, that it was a wave that propagated with a speed $v = c$ given by

$$c = \frac{1}{\sqrt{\epsilon_0 \mu_0}} \quad \text{(speed of light in vacuum)} \tag{1.35}$$

Since ϵ_0 and μ_0 both had known measured values, it allowed Maxwell to predict the value of c to be

---
Speed Of Light In Vacuum

$$c = \frac{1}{\sqrt{\epsilon_0 \mu_0}} = 299{,}792{,}457 \ \frac{\text{m}}{\text{s}} \tag{1.36}$$

At the time Maxwell made this prediction, French scientist Hippolyte Fizeau (1819-1896) had already made approximate measurements of the speed of light and had found it to be 3×10^8 m/s. Thus, it was a provocative result from Maxwell, having contrived a theory correcting,

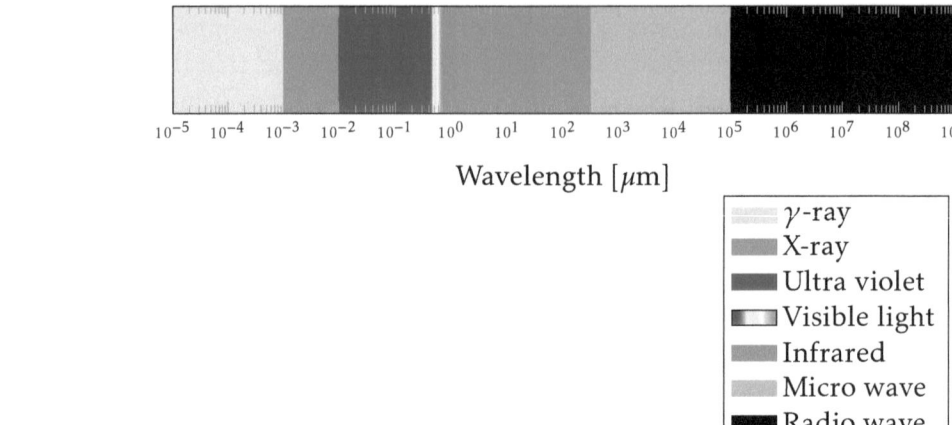

Figure 1.5: Electromagnetic spectrum. Note that although colors are used outside the visible light region in this figure, these parts of the spectrum are not visible to the human eye.

as well as combining electricity and magnetism, that predicted the remarkable speed of light measured by Fizeau.

From such developments, the wave model of light finally gained considerable traction, overshadowing the alternative belief that light was best described by a corpuscular model of light. The wave model was in complete agreement with most experimental observations concerning light, electrical, magnetic, and optical. Light was not only a wave, but a wave that was a part of a broader *electromagnetic spectrum*. We refer to the particular portion of the electromagnetic spectrum that can be seen with the naked eye as the *visible light* region, with wavelengths λ ranging from around 400 nm (blue in color) to 700 nm (red in color), spanning from blue to green to yellow to red. The spectrum of electromagnetic waves is illustrated in Fig. 1.5. We now know of many more types of electromagnetic waves, most being invisible to humans. There are x-rays, which have wavelength λ around 0.1 nm, or 1 angstrom (Å). Some technologies like televisions and FM radio have much longer wavelengths with λ around 1 m. Cellular phones operate using electromagnetic waves with λ around 0.01 m, or on the order of millimeters (e.g. 5G technology). Without any of these discoveries and insights taking place, it's quite possible that none of these technologies would exist today.

Finally, the beginning of a more lucid picture was emerging for the

1.4 Revelation That Light Is an Electromagnetic Wave

luminous waves, which according to the electromagnetic equations, are composed of fields **E** and **H**. In vacuum, for example, if we have the following propagating electric field **E**, given by

$$\mathbf{E} = E_0 \hat{x} \sin(kz - \omega t) \tag{1.37}$$

Then, we also know there will be an associated magnetic field **B** given by

$$\mathbf{B} = B_0 \hat{y} \sin(kz - \omega t) \tag{1.38}$$

The parameter k is the known as the **wave-number** defined as

> **Wavenumber k**
>
> $$k = 2\pi/\lambda \tag{1.39}$$

The angular rate is $\omega = 2\pi f$. Since $v = \lambda f$, the wavenumber k can also be expressed as $k = \omega/v$. The Maxwell electromagnetic equations, therefore, describe an electromagnetic wave as a pair of conjugated waves (**E** and **H**) moving along a propagation direction z, with speed $v = c$, summarized in Fig. 1.6.

Reaching this point was pivotal in the development of quantum mechanics. It lead to the first insight that electricity is intimately connected to magnetism. The idea that light behaves like a wave would give rise to additional questions. One significant question that arose was that, if light was a wave, how could it propagate through vacuum? Light was known to propagate through vacuum, however, it was also common experience that waves required a medium to propagate. For example, sound waves and water waves had already demonstrated this behavior. To further study electromagnetic waves, the ability to generate them became compulsory.

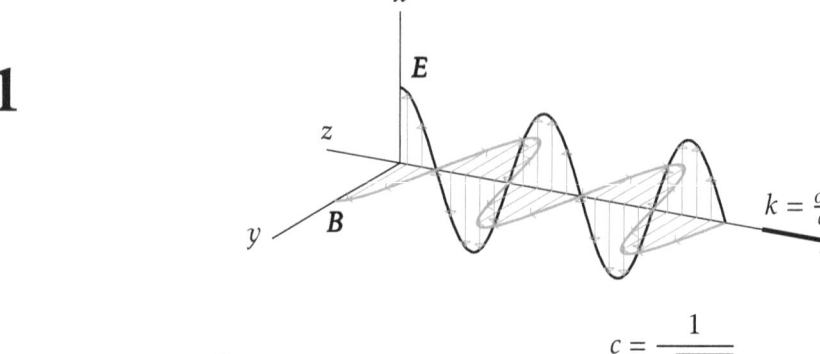

$$c = \frac{E}{B}$$

E = electric field amplitude V/m
B = magnetic field amplitude T
c = speed of light (3×10^8 m/s)

$$c = \frac{1}{\sqrt{\mu_0 \varepsilon_0}}$$

μ_0 = magnetic permeability in a vacuum, $\mu_0 = 1.3 \times 10^{-6}$ N/A^2
ε_0 = electric permeability in a vacuum, $\varepsilon_0 = 8.9 \times 10^{-12}$ C^2/Nm^2

Figure 1.6: Summary of electromagnetic waves described by the Maxwell equations. The illustration depicts propagation in vacuum.

1.5 First Generation of EM Waves in a Laboratory

Maxwell's prediction led the world of physics along a profoundly new direction in the understanding of light and electromagnetism. However, Maxwell's work was theoretical, and sufficient experimental confirmation was still lacking for years after Maxwell published his equations around 1865. It was not until 1887 that German physicist Heinrich Rudolph Hertz (1857-1894) generated the first electromagnetic waves in a laboratory by triggering electrical sparks between the gaps of a specially designed system shown in Fig. 1.7. With this setup, Hertz conducted a series of experiments that would, for the first time, and at will, produce the very thing that Maxwell's equations described. The

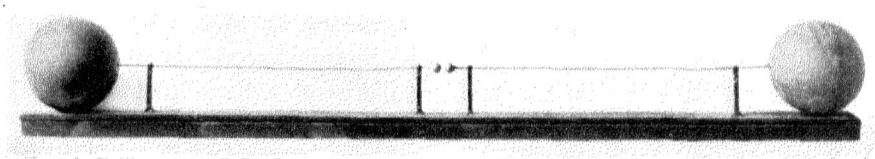

Figure 1.7: Heinrich Hertz first apparatus for generating EM waves in a laboratory.

1.5 First Generation of EM Waves in a Laboratory

two larger 30cm zinc spheres at the exterior ends are attached to two copper wires directed towards the center, terminated with a small gap. An induction coil was placed across the copper wires with a high voltage between them, causing sparks across the center gap with the smaller spheres. This generated standing waves of radio frequency current in the wires, which ultimately radiated *radio* waves of ∼ 50 MHz. Hertz also detected the radio waves with a ring resonator This experimental setup provided the first evidence of an ability to generate the electromagnetic waves described by Maxwell's theory. This represented a significant event for the future development of any technology utilizing electromagentic waves, of which there are thousands known today. It would

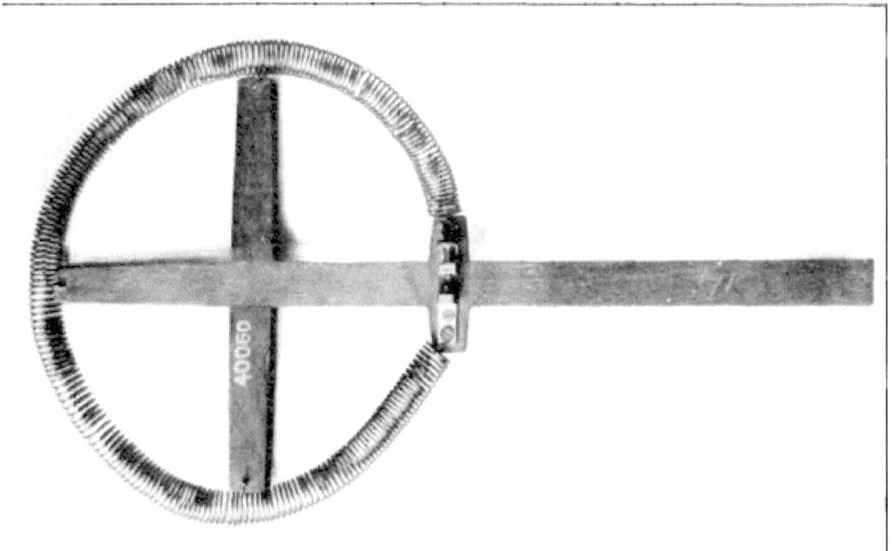

Figure 1.8: Radio-wave receiver used by Hertz. The reciever is in the shape of a ring, equipped with an adjustable gap (at the bottom center of ring) for tuning to resonance, thereby better enabling the waves to be picked up.

turn out that the electromagnetic waves generated by Hertz could be generated by nature using other means. In fact, it was a trail of visible light that ultimately lead to the first tangible proof in a laboratory, of the existence of a certain subatomic particle. These particles would eventually be manipulated to undergo transitions in matter that would also produce light.

1.6 Electron Revealed By Electromagnetic Waves

Despite the success in understanding and manipulating light and electromagnetic waves at the time, the notion of corpuscles or particles would not disappear entirely. In fact, strangely enough, it only better enabled an ability to detect both atomic and subatomic entities. Although observations relating to *magnetic* properties had been known for centuries from materials such as *lodestones*, it was only after the experimental discovery of the electron that a more fruitful pursuit of a deeper understanding could begin. Understanding the electron and its properties, as well as, how it interacts with other electrons and/or particles, provides the foundation of many subjects in physics including quantum mechanics, and especially the field of *magnetism*. Thus, we'll also devote some attention to the beginning of our recorded knowledge of this wondrous and somewhat mysterious entity known as the electron. Our awareness of this fundamental particle was only made possible by light itself.

The origin of the word electron, or elektron, is Greek and translates to the word *amber*. This is no coincidence because *amber* is the name of an old resin found on evergreen trees, known even by the ancient Greeks. An example of a natural yellow resin is shown in Fig. 1.9. An interesting observation was made with amber. The ancient Greeks noticed that if you rubbed the amber against something like a dry cloth, it would cause some interesting forces to appear that would cause the amber to attract relatively light objects close by, such as papyrus. In today's context, these peculiar forces are known as *electrostatic forces*, caused by the accumulated *free electrons* on the surface of the amber. The electrons are liberated from their atoms because of the thermo-mechanical energy generated from the act of rubbing amber with another material. At that time, there was no knowledge of concepts like atoms and their constituents of electrons, protons, and neutrons. Therefore, humanity's insight and understanding of these peculiar observations was limited. All that can be said, with certainty, was that there was something special going on with the amber material and materials like it.

Fast forwarding several hundred years to the 19th century, significant advancements had been made moving through the Industrial Age. A

1.6 Electron Revealed By Electromagnetic Waves

Figure 1.9: Examples of aged amber resin. The properties of amber provided early clues towards the existence of the electron.

process known as *glass blowing* had been developed to a relatively high level of technical capability. This development included perfecting methods that could *empty* out the interior gas volume of blown glass tubes (by lowering the pressure), and inserting electrodes in its interior. A diagram of an early Crooke's tube is illustrated in Fig. 1.10 Fig. 1.11

Figure 1.10: Schematic of an early Crooke's tube like the one used by J.J. Thomson. Image from https://history.aip.org/exhibits/electron/jjcrooke.htm (American Institute of Physics).

shows these tubes in action, illustrating the manipulation of a beam of electrons, or cathode ray. These special kinds of low-pressure tubes are

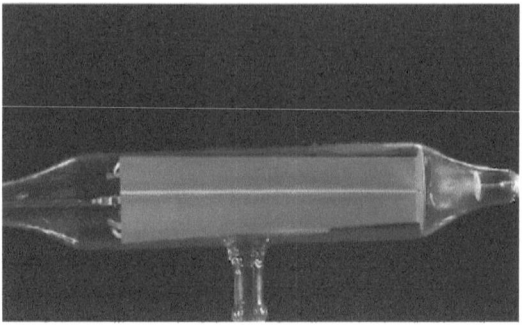

(a) An image of an experimentally generated cathode ray beam (the blue horizontal beam) inside a Cathode ray tube. The beam enters from the right, and travels to the left. The top figure is the direct beam without external influence

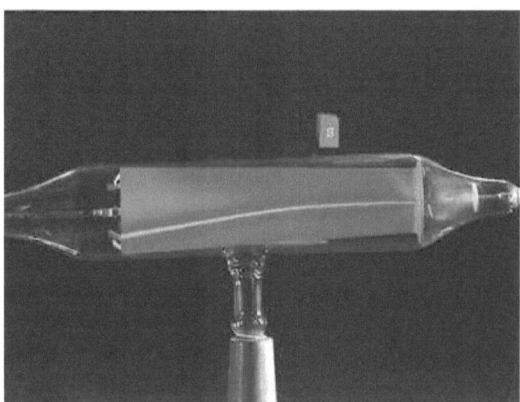

(b) A deflected cathode ray from a Lorenz force actuated by a horizontal magnet (poles oriented from back to front). The deflection evidences the charged nature of the luminous beam of particles.

Figure 1.11: Illustration of electron beams or cathode rays generated and manipulated with a magnetic field in a Crooke's tube. From work at University of New South Wales: url formerly: http://lrrpublic.cli.det.nsw.edu.au

known as **Crooke's tubes**, named after the British chemist and physicist, William Crooke (1832-1919), who pioneered a way to greatly reduce the pressure inside these tubes. A number of experiments took place using these vacuum tubes during the 19th century. This lead to the observation of an unusual bright visible beam of light-energy being emitted from the *negatively charged cathode* inside of the tube, like that shown in Fig.

1.6 Electron Revealed By Electromagnetic Waves

1.11. Since it apparently emanated from the cathode, this type of beam became known as a *cathode ray*. At that time, it was postulated that cathode rays may be due to a distinct form of matter, not being liquid, solid, nor gas. It was not until the close of the 19th century, in 1897, that British experimental physicist Joseph John Thomson (1856-1940) would be one of the first to quantify properties of the cathode ray.

J.J. Thomson demonstrated that this beam of matter consisted of *negatively charged entities*, meaning that the observed entities are discrete and charged. He designed experiments using cathode ray tubes, balancing their motion with electric and magnetic fields. This allowed him to estimate both the charge-to-mass ratio of the electron, as well as the charge on the electron. Fig. 1.12 is a photograph of one of Thomson's cathode ray tubes from his laboratory. The horizontal plates near the

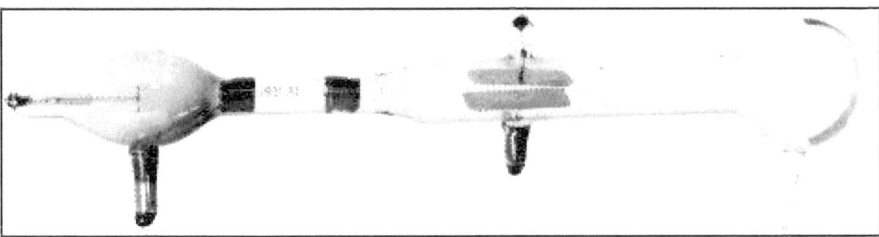

Figure 1.12: Photograph of a cathode ray tube used by Thomson in discovering electrons.

center of the cathode ray tube were connected to an electric battery to generate an electric field across the plates. Notice the glass finger hanging from the bottom of the right-most glass sphere of the ray tube. This is where the vacuum was drawn down in the ray tube. Fig. 1.13 is a diagram of Thomson's cathode ray tube used in his famous 1897 paper, also depicting the cathode ray bending due to the electric and/or magnetic field used in their experiments. From his experiments, J. J. Thomson calculated the charge-to-mass ratio of the electron to be around 1.7588×10^{11} C/kg, which is close to today's accepted value of the ratio which is 1.7588196×10^{11} C/kg. However, the estimate of the charge on the electron e erred by 50%, compared to today's accepted value. Although there was significant error in the charge estimation, it was a significant accomplishment that pointed to the existence of something new, and ultimately directed others to improve upon these estimates.

Eleven years later in 1908, American physicists Robert Millikan (1868-1953) and his student Harvey Fletcher (1884-1981) would improve upon

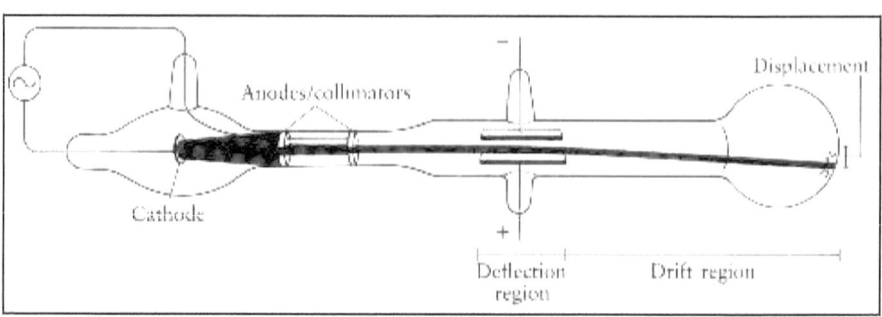

Figure 1.13: A diagram of Thomson's cathode ray tube used in Thomson's 1897 paper.

the estimates made by J.J. Thomson. For Millikan's experiments, they made use of the fact that charged particles behave as a nucleus around which water vapor droplets will condense. Their experiments involved dropping oil, in the form of a mist, between electrically charged plates, or electrodes, through a hole in the top charged plate. Their droplet experimental apparatus is shown in Fig. 1.14. The force on the droplets passing through the electric field could be measured, as well as the electric field between the electrodes. By repeating the experiment for many droplets, they were able to confirm that the charges were all small integer multiples of a certain value $-e$, now known as the **elementary charge**. By varying the charge on different drops, they noticed that the charge was always a multiple of -1.6×10^{-19} C. This meant that the *particles* discovered by Thomson were carrying a *unit* charge of this amount. Their refined determination of the electron charge using their now-famous oil drop experiments lead to a more accurate determination of e, as well as the mass of the electron m_e, since the ratio was known. This marked a pivotal point for the development of quantum mechanics because now we had gained an awareness of tiny charged particles. Many more inquiries would soon follow that would ultimately lead to an even better understanding of this remarkable discovery of tiny fundamental particles of matter.

Later, another property of the electron known as the *spin* would also be revealed. Somewhat like charge, this second property of the electron is a *quantized* amount of intrinsic angular momentum. It was discovered, unwittingly, not long after the electron charge. In the case of spin, before its observation, it was postulated by more than one person. We will discuss this topic in more detail in Chap. 9. However, with the known existence of the *electron* and its charge, there was waning doubt of the

1.6 Electron Revealed By Electromagnetic Waves

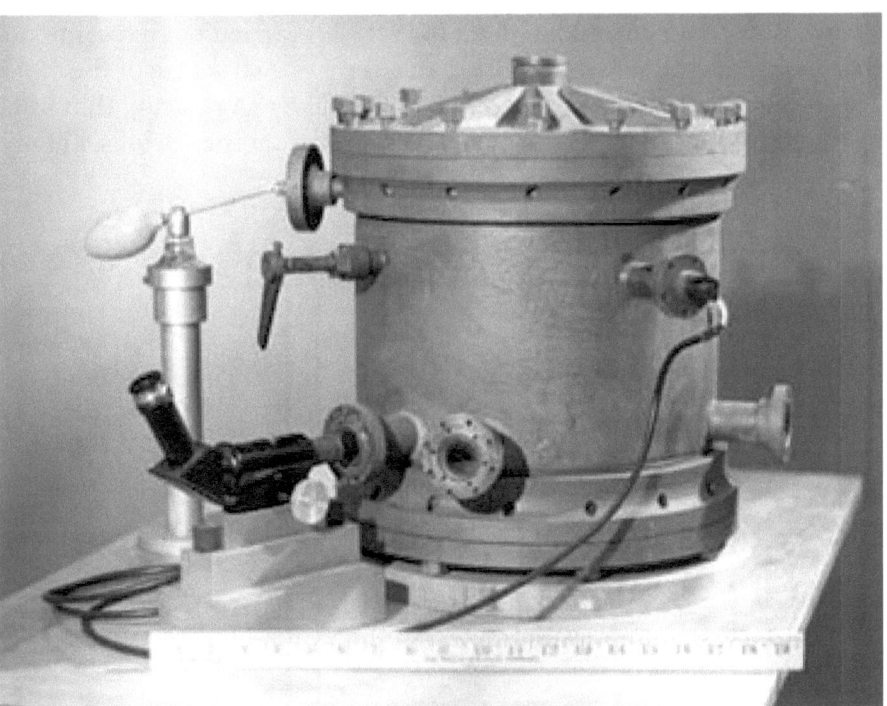

Figure 1.14: Photograph of Millikan and Fletcher's oil droplet experimental setup. By varying the charge on different droplets, they observed the charge to always a multiple of $-e = -1.6 \times 10 - 19$ C, which is the charge on a single electron.

existence of discrete matter, such as atoms. And if the atom possesses some number of electrons with charge $-Ze$, where Z is some integer, it stood to reasoning that there might also be some positive charges within it with charge $+Z|e|$, as well, to balance the charge, since most stable matter is neutral in charge. However, there were still debates on the structure of the atom. How was it arranged with the electron as part of it, and how would these so-called counter charges be arranged in tandem with electrons?

J.J. Thompson, who discovered the electron, also had proposed an early hypothetical structure of the atom. He suggested that the positive charges form a cloud of distributed charge around a centered cluster of negatively charged electrons. This model became known as the **plum-pudding model**. However, in 1910, under the guidance of New Zealand-British physicist Ernest Rutherford (1871-1937), two researchers, German physicist Johannes Wilhelm Geiger (1882-1945) and

British physicist Ernest Marsden (1889-1970) carried out experiments that more conclusively determined an alternative structure of the atom. The atoms they studied were gold (Au) and silver (Ag), using thin foils, as illustrated in Fig. 1.15. Their success relied on the use of what are

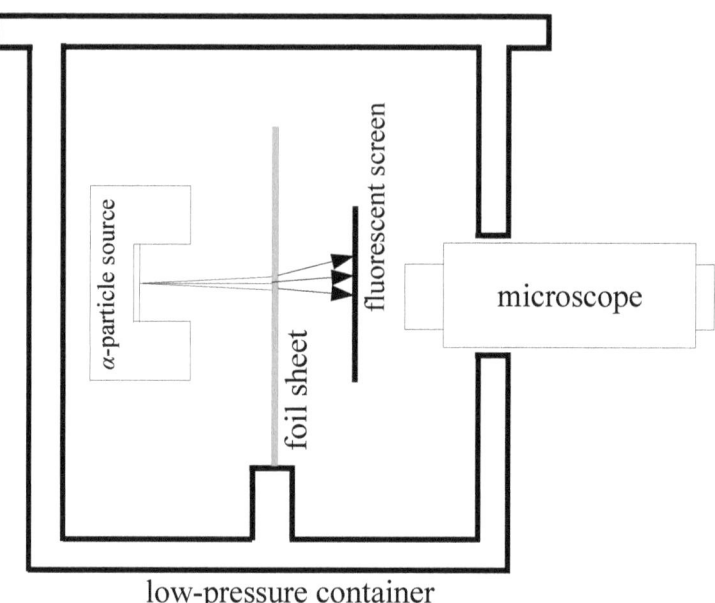

Figure 1.15: Illustration of the experimental setup used by Ernest Rutherford, to probe atomic nuclei using α-particles. After the particles passed through the thin foils of gold (Au) or silver (Ag), they were detected on a zinc-sulfide screen (fluorescent), upon which there was a slight scintillation, observable with a microscope.

known as α-particles, which are unstable particles that consist of two protons and two neutrons bound together into a particle, like a helium nucleus. They can be created from a process known as *radioactive decay*, discovered in 1896, with uranium salts. The process involves the breaking down of radioactive heavy elements like uranium, thorium and radium, and gives rise to three kinds of emitted rays, α, β, and γ. Though this was not known at the time, these elements have nuclei that are neutron rich, meaning that they possess an excess of neutrons compared to protons, and neutron rich elements tend to decay and emit α-particles. Their experiments fired these tiny energetic particles at thin foils of Au and Ag. The α-particles end up colliding with the atoms of the foil, and undergo deflections and mostly back-scattering at a variety of angles. By using conservation of the kinetic energy and momentum of the α-particles, the characteristics of the measured angles provided evidence of the structure of the atom. They found that the suggested structure of the atom was a tiny ball of positive charges concentrated centrally, with electrons broadly distributed around the central positively charged core.

Atomic structure: By 1910, the existence of electrons, protons, and neutrons was established, and the atomic structure was known to be a centrally compact positive nucleus with electrons whizzing about it.

This brings us to the point of understanding near the turn of the century, of electromagnetics and of the dawning of a knowledge of new forms of fundamental particles with charge and without, as well as their structure in the *atom*.

1.7 Chapter Summary

In this chapter, the historical backdrop to the development of quantum mechanics has been discussed. The developments in the understanding of light as an electromagnetic wave turned out to be crucial. The generation of electromagnetic waves by Hertz triggered new possibilities and confirmed the existence of a diverse electromagnetic waves. They all could be described by the complete Maxwell or electromagnetic equations. Concepts such as wave-particle duality began with the historical developments of light and electromagnetic waves, subtly introducing these notions to humanity for the first time. The study and growing understanding of light also correlates with the first observation of electrons in a luminous cathode ray. By the turn of the century, electromagnetic waves were known, and the first known subatomic particles has begun to be characterized via laboratory experiments. This lead to the first measurement of charge and mass of the electron, and its relationship to the atomic structure. Something new and unfamiliar was unfolding in the world of physics. But, we'll need to connect thermodynamics to this history to better understand the beginnings of quantum mechanics. This topic is discussed in the next chapter. Table 1.1 lists some of the key results from this chapter.

Table 1.1: Chapter 1 Summary Equations. Note that H = Henry, N = Newton, C = Coulomb, J = Joule, m = meter.

Name	Equation
Snell's law	$\frac{\sin\theta_1}{\sin\theta_2} = \frac{v_1}{v_2} = \text{constant}$
Faraday's Law	$\nabla \times \mathbf{E} = -\frac{\partial \mathbf{B}}{\partial t}$
Ampére-Maxwell Equation	$\nabla \times \mathbf{H} = \mathbf{J} + \partial \mathbf{D}/\partial t$
Wave Equation	$\frac{\partial^2 w}{\partial t^2} = v^2 \nabla^2 w$
vacuum electric permittivity	$\epsilon_0 = 8.854 \times 10^{-12} \text{C}^2 \cdot \text{J}^{-1} \cdot \text{m}^{-1}$
vacuum magnetic permeability	$\mu_0 = 4 \times \pi \times 10^{-7} \text{N} \cdot \text{A}^{-2}$
speed of light in vacuum c	$c = \frac{1}{\sqrt{\epsilon_0 \mu_0}} = 299,792,457 \text{m/s}$
electron charge e	$e = 1.60217662 \times 10^{-19}$ C

1.8 Chapter Problems

Problem 1.1 From the discussion in this chapter, make a list of supporting observations of why light behaves like a particle; Then, make a list for how light behaves as a wave. Can they both be right?

Problem 1.2 If Thomas Young would have used a single slit to observe light, could he have found evidence of light behaving as a wave, and not a particle? What would he have found different from a stream of particles going through a slit?

Problem 1.3 A current I is flowing through a wire of circular cross-section. Integrate Ampére's equation over the area of the wire to show

$$H = \frac{I}{2\pi R}$$

R is the radius of the wire and H is the value of the magnetic field on the surface of the wire.

Problem 1.4 The index of refraction n for air and water are $n_w = 1.33$ and $n_a = 1.00$. Using Eq. (1.1), determine what is the constant Ptolemy should have found?

Problem 1.5 Based on the previous problem, how much does light bend entering water from air at an angle of $\theta_1 = 15°$ and $\theta_1 = 0°$ (exactly normal)?

Problem 1.6 Define the vector quantity $\mathbf{S} = \mathbf{E} \times \mathbf{H}$, where $\mathbf{E}$ is the electric field and $\mathbf{H}$ is the magnetic field described by the Maxwell equations. Show that

$$\nabla \cdot \mathbf{S} = -\frac{1}{2}\frac{\partial}{\partial t}(\mathbf{E} \cdot \mathbf{D} + \mathbf{B} \cdot \mathbf{H}) - \mathbf{J} \cdot \mathbf{E}$$

$\mathbf{D}$ is the electric displacement vector and $\mathbf{J}$ is the current density. hint: Use the vector identity

$$\nabla \cdot \mathbf{F} \times \mathbf{G} = \nabla \times \mathbf{F} \cdot \mathbf{G} - \mathbf{F} \cdot \nabla \times \mathbf{G}$$

The vector $\mathbf{S}$ is known as the **Poynting vector**, and it describes the direction of motion of the wave, or the direction of wave energy flow.

Problem 1.7 A displacement wave vector is given by $y = A\cos(kx - \omega t)$. The displacement wave also has a potential energy of $V = kx^2/2$. With a velocity $v = dx/dt$ and energy $E = K + V$, show the following:

$$\frac{dE}{dx} = m\omega^2 A^2 \sin^2(kx - \omega t) \quad \text{and} \quad P = \frac{dE}{dt} = vm\omega^2 A^2 \sin^2(kx - \omega t)$$

1.9 Suggested Readings & References

[1] M. Born, *Über Quantenmechanik*, Z. Phys. **26**, pp. 379–395, (1924)
[2] William A. Fedaka and Jeffrey J. Prentis, Am. J. Phys. **77**, No. 2, (2009)
[3] A. Ghatak and S. Lokanathan, *Quantum Mechanics: Theory and Applications*, Springer-Science (2004)
[4] *Understanding: Magnetism, Understanding* Series, TLC/Discovery Channel, narrated by Candice Bergen (1994-2004)

CHAPTER 2

Blackbody Radiation & The Discovery Of $\hbar$

In this chapter, we turn our attention to a somewhat different, but equally important historical development for quantum mechanics. This chapter goes hand-in-hand with the development of our understanding of light, which gives us a clue about certain kinds of *waves*. It turns out that we need another branch of physics to teach us more about *discreteness*. Actually, it was observations of confined light in a heated box that also perplexed many physicists. Between light and heat, these two histories came together to boldly suggest a path for a revolution in physics, involving the connection and reconciliation between wave and particle. Since we've talked about light-waves, but in *the absence of heat*, already, now we'll focus on developments involving both the fields of *thermodynamics* and light. One of the goals is to underscore the important role thermodynamics played in the development of quantum mechanics. Moreover, thermodynamics was not only necessary for the birth of quantum mechanics, but *is* necessary to the application of QM to most real-world problems. This is because physics is *always* experienced at finite temperatures. As far as experience goes, we only know $T > 0$, in temperature units of K=Kelvin. Later, useful examples of how thermodynamics and quantum mechanics *come* together will be demonstrated to some extent in Chap. 12, when we discuss temperature dependent spin-based potential energy, and to a greater extent in Chap. 15, when we discuss the topic of *superconductivity*. For this, we have to understand the relation between temperature (in thermodynamic equilibrium) and energy. This was one of the many gifts of thermodynamics. Of interest here, is *a famous quantum hypothesis* that must be dealt with. It lead to the discovery of one of the most important fundamental constants in quantum mechanics. Let us discuss the pieces that came together revealing how this discovery was made.

2.1 Blackbody Radiation Dilemma

Around 1900, a German physicist named Max Karl Ernst Ludwig Planck was investigating a problem concerning something known as *blackbody radiation*. A *blackbody* is an ideal material or body capable of absorbing all the thermal radiation falling onto it. When such radiation is contained within a cavity, poking a hole in it permits the *heated* radiation to escape containing signature information about the blackbody radiation *inside* the cavity. The intensity of the exiting light or *irradiance* from within the cavity can be measured, and this is known as **blackbody radiation**. It equals the radiant exitance of a blackbody which is in thermodynamic equilibrium or constant temperature T. Planck was an expert in thermodynamics. At the time, like many others, he was puzzling over a better way to explain *blackbody radiation* over the entire frequency spectrum, from theory. The timing was off the heels of German physicist Wilhelm Wien (1864-1928), whose model was close to correct, but contained phenomenological constants. By 1900, measurement data existed revealing the behavior of radiation emitted from a blackbody. An example is shown in Fig. 2.1.

An experiment is a question which science poses to Nature, and a measurement is the recording of Nature's answer.
M. Planck

Nature had answered the question pertaining to blackbody radiation, however, Planck was not satisfied with the attempts of science to explain Nature's response, which demonstrated significant disagreement (with Nature). Planck set about gaining better agreement between the two. In treating the *blackbody radiation problem*, we will maintain a significant degree of fidelity with Planck's original derivation because it has the advantage of illustrating the clear connection between thermodynamics and the quantum hypothesis. This is an important aim we will empha-

2.1 Blackbody Radiation Dilemma

size throughout this chapter. At the time, scientists had figured out experimentally that the light coming from a tiny hole in the side of a fire-burning stove, for example, had distinctive properties. Scientists came to call this special behavior of a heated cavity *Hohlraum radiation* (Hohlraum is German for cavity), and it came to be a fundamental problem in the physics of heat and light. Several people of notoriety analyzed this difficult problem, while others became notable because they worked on this problem. Planck was among the latter.

The heat is described using thermodynamics, while the light by electromagnetic equations discussed in Chap. 1. Belief in the existence

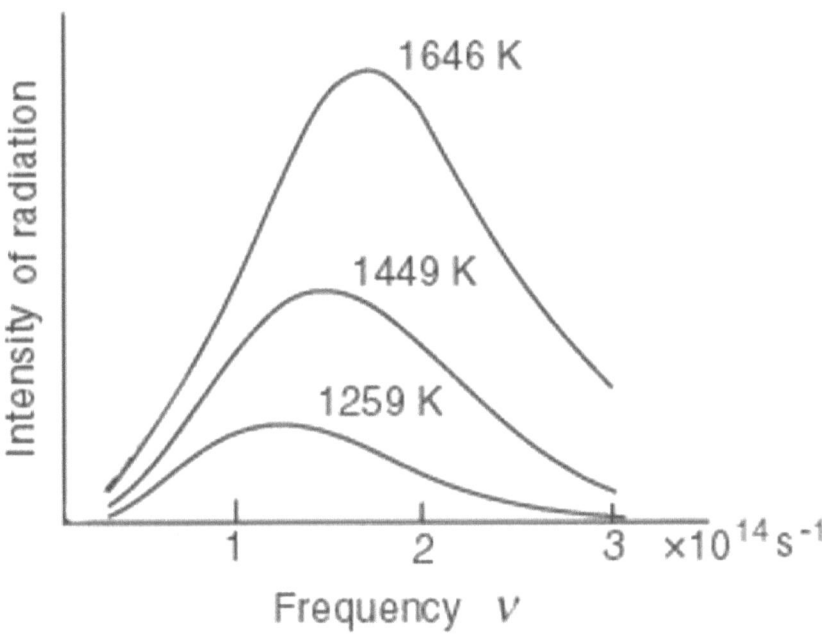

Figure 2.1: Example of measured data from a blackbody or cavity. Such data began being observed in the late nineteenth century and illustrated a unique behavior of blackbody radiation. Contriving a predictive model for this observation proved to be a challenge.

of *atoms* played its role, as not all physicists at the time accepted the notion. For the *atomists*, however, many theories had been developed based on the idea of a large countably many atoms making up matter. A former student of German physicist, Josef Stefan (who also investigated blackbody radiation), named Ludwid Boltzmann (1844-1906), ardently believed in the existence of atoms. Using the fundamental concept of the atom, along with mathematical principles of random motion, Boltzmann applied statistical methods to thermodynamics. His work was motivated by the idea that gases consisted of lots of tiny particles. It involved bringing fundamental thermodynamics concepts into the discussion, including concepts known as *entropy S* and *internal energy U*. Planck, who knew Boltzmann, used Boltzmann's work to treat the problem of blackbody radiation. The choice to use a thermodynamic theory based on statistical mechanics is ultimately what lead to Planck's required quanta assumption, now known the world over. It arose from Planck's decision to use Boltzmann's statistical thermodynamics principles to tackle the fundamental problem of blackbody radiation. Let's now discuss some important concepts in thermodynamics that played a role in not only solving the blackbody radiation problem, but provide part of the platform for the development of quantum mechanics.

2.2 Classical Thermodynamics of Ideal Gas

The preceding development of thermodynamics was fortuitous for quantum mechanics. It turned out that drawing insight from the thermodynamics of ideal gases was essential. A box of hot gas particles was likened to a heat cavity of photons, however, this approach was only valid up to a point. Planck figured this out. For an ideal gas in a box of fixed volume V, where there is work done on the system by the cavity walls, the **first law of thermodynamics**, which balances energy, gives

$$\delta Q = \partial U - \delta W = \partial U - P dV \tag{2.1}$$

Eq. (2.1) says that the energy of the system is balanced between the heat transfer δQ, the variation of internal energy U, and work done, which is PdV. P is the pressure of the system. For a rigid container of fixed volume V, the net work PdV done to the system is zero, so the first law gives

$$\delta Q = \partial U$$

If we also assume that our system is in thermodynamic equilibrium at a fixed temperature T, we can also write

$$\partial S = \frac{\delta Q}{T} = \frac{\partial U}{T}$$

Note that this equation is defined in terms of an absolute temperature T, given in K=Kelvins. We can define a variation of a quantity called **entropy** $\partial S = \delta Q/T$. Then, we find that for an ideal gas in a box of fixed size V, in thermodynamic equilibrium

$$\frac{\partial S}{\partial U} = \frac{1}{T} \qquad (2.2)$$

So, where does this choice of definition come from and why do we care to define such a quantity? The term *entropy* (Greek word for *transformation*) was coined in 1865 by German physicist Rudolf Julius Emanuel Clausius (1822-1888), during the heyday of very inefficient thermodynamic heat engines or heat cycles. Clausius noticed that a particular ratio was constant in reversible, or ideal, heat cycles. That ratio was, in fact, heat exchanged to absolute temperature. Clausius thought that this conserved ratio must correspond to a real, physically significant quantity, and he thus, named it "entropy". It has become an integral concept in thermodynamics, however, we will only require limited understanding for our purposes. We will have no occasion to require *more* than what we discuss in this chapter. Moreover, as you will see, we will not need to determine the entropy explicitly, but it will useful as a bridge to connecting some relations involving entropy. Continuing on with the second derivative of entropy, we have

$$\frac{\partial^2 S}{\partial U^2} = \frac{\partial}{\partial U}\left(\frac{1}{T}\right) \qquad (2.3)$$

This is the extent of relationships from classical thermodynamics, used by Planck to obtain his famous solution to the blackbody radiation problem. The remainder was from statistical thermodynamics. Let us next consider how *statistical thermodynamics* describes the entropy S. Used with the thermodynamics of an ideal gas, it lead to a result that indicated that, in fact, *a box of photons does not behave identically to a box of gas particles*, but there are some similarities.

2.3 STATISTICAL INTERPRETATION OF ENTROPY

The concept of entropy S is formally introduced into thermodynamics through the *second law of thermodynamics*, which states that

$$\Delta S \geq 0 \quad \text{Second Law Of Thermodynamics} \qquad (2.4)$$

The *equality* in Eq. (2.4) only holds for **reversible processes**. In thermodynamics, a *reversible process* is a process whose direction can be reversed by applying infinitesimal changes to a property of the system through its surroundings. This leads to no change in entropy. Throughout such a process, the system is also said to be in **thermodynamic equilibrium** with its surroundings. This law was first put into words for reversible systems by Rudolph Clausius and William Thomson (1824-1907), after nineteenth century scientists and engineers had noticed that heat does not pass from a colder body to a warmer body by itself. As mentioned, in 1865, Clausius noticed that a particular ratio of quantities was always constant in reversible, or ideal, heat cycles, namely, heat exchanged δQ to absolute temperature T. Over time, the concept of entropy evolved in meaning and even interpretation after developments in the field of *statistical mechanics*. A relevant interpretation for quantum mechanics is that, based on a statistical mechanical approach (based on countably many atoms), *entropy can be thought of as a measure of disorder*. Since a colder body only sees heat transfer into the system in nature, it only increases its disorder. Of course, this can be avoided, but only at the expense of *work*.

The disorder is more probable in systems with larger numbers of degrees of freedom. You've probably noticed that metal will rust, rocks can crumble, etc. and all of these are examples of how larger systems tend to transition towards greater disorder. Processes like this are only enhanced at higher temperatures. Since the disorder transition tendency turns out to be higher for systems with a larger number of degrees of freedom, this suggests that entropy can be measured by quantities such as volume which correlates to the number of possible **microstates** of a 3D physical arrangement of atoms. This kind of correlation is a consequence of the large number of *discrete* atoms. A large combination of microstates constitutes a given *macrostate* for a body, say with N constituent particles. The entropy of a system, like energy, is therefore *extensive*, meaning that it scales with volume. Such quantities are added for combined systems. For example, if a system A has entropy S_A, while a separate system B has entropy S_B, the total entropy of the combined system is

$$S = S_A + S_B \tag{2.5}$$

The number of ways to achieve a *macrostate* is also known as the **thermodynamic probability**, and **number of complexions**, denoted by Ω. The former name is an unfortunate misnomer because Ω *counts*, and this is why it satisfies $\Omega \geq 1$. However, like a statistical probability, for two systems A and B with respective thermodynamic probabilities, Ω_A and Ω_B, the combined system is given by the product of the probabilities of A and B where

$$\Omega = \Omega_A \cdot \Omega_B \tag{2.6}$$

If we assume S to be a function of Ω (all thermodynamic variables are), it follows that

$$f(\Omega) = f(\Omega_A \cdot \Omega_B) = f(\Omega_A) + f(\Omega_B) \tag{2.7}$$

Similar log relations had already been obtained using classical systems, but they involved volumes V. The only function to behave as f does is the log function. This prompted Boltzmann, around 1875, to propose that the entropy S must be proportional to the *log* of the thermodynamic probability, where the entropy S is given by

---

Boltzmann Statistical Entropy Relation

$$S = k_B \ln \Omega \tag{2.8}$$

Boltzmann's relation turns out to be more general, since a body at constant volume can still undergo changes in its thermodynamic probability. Eq. (2.8) introduced the Boltzmann constant k_B to the world. It is an indispensable parameter in thermodynamics and heat transfer. In the next section, let us review a fundamental statistical thermodynamic concept relating energy and temperature. It was used in the *blackbody radiation* problem, first applied to gas particles. The concept is known as *equipartition*, and refers to the equal distribution of energy between the three degrees of freedom x, y, and z, in space.

2.4 IDEAL GAS LAW AND KINETIC ENERGY

For an ideal gas, consider the problem of a large, countably many number N of almost identical parts moving around inside of a container of fixed volume V, at equilibrium temperature T. The coordinate system we'll use is spherical with coordinates (r, φ, θ), as illustrated in Fig. 2.2. A fundamental assumption is that *the motion of any particle in the container is random, and thus, has no preferred direction*. Alternatively, it is equally likely to move in any direction. This assumption leads to each angle or orientation being equally probable, and this detail enables the *proportions of particles* to be completely described by the proportions of surface areas. Since the velocity vector **v** does not have a preferred direction, we may imagine this vector directed from a point O to the elementary area dA', as shown in Fig. 2.2. This areal patch contains vectors in the *neighborhood* of **v**.

This is sufficient to determine the amount of particles having velocity vector **v** within the patch. It involves the concept of the **solid angle**, illustrated in Fig. 2.2. The solid angle is defined as the ratio of the

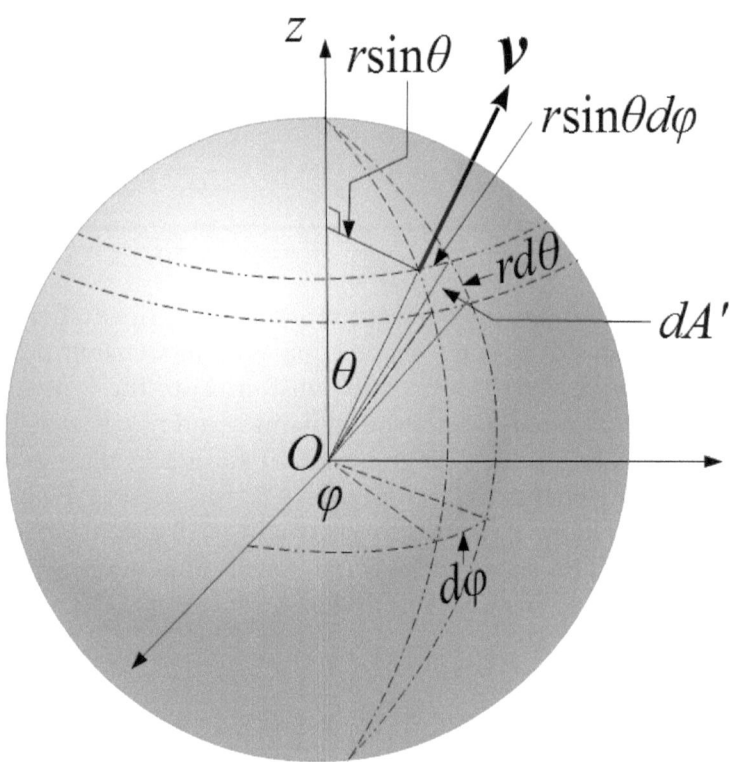

Figure 2.2: Illustration of solid angle and relationship between velocity of a particle and its differential surface.

subtended area on a spherical surface to the square of the radius of the

2.4 Ideal Gas Law and Kinetic Energy

sphere, or

$$d\Omega' = \frac{dA'}{r^2} \tag{2.9}$$

The definition for the solid angle, in Eq. (2.9), is a generalization of an ordinary (planar) angle θ which may be defined as

$$d\theta = \frac{rd\theta}{r} = \frac{ds}{r}$$

s is the arc length. With solid angles, the numerator in the ratio extends to the maximum possible surface area subtended by dA' relative to r^2. Using this definition, the subtended surface area on a sphere of radius r, using polar coordinates r, θ, and φ, is given by the integral of the differential solid angle $d\Omega'$

$$\Omega' = \frac{\int_0^\theta \int_0^\varphi r^2 \sin\theta \, d\varphi d\theta}{r^2} = \int_0^\theta \int_0^\varphi \sin\theta \, d\varphi d\theta \tag{2.10}$$

If we integrate Eq. (2.10) from 0 to the limits of $\theta = \pi$ and $\varphi = 2\pi$, we have

$$\Omega' = \int_0^\pi \int_0^{2\pi} \sin\theta \, d\varphi d\theta = 2\pi \int_0^\pi \sin\theta \, d\theta = 2\pi(-\cos\theta)\Big|_0^\pi = 4\pi$$

The full range for any solid angle Ω' is then given by

$$0 \leq \Omega' \leq 4\pi$$

The physical units suggested by the definition are dimensionless, however, the conventional name for this dimensionless unit is the **steradian** (sr), and it is useful in *suggesting* the information being described. This also means that for 1 steradian (sr) the corresponding area subtended on a sphere is exactly equal to r^2. Applying this idea, with O as the origin, we imagine **v** pointing from the origin where the tip sweeps along the surface of a corresponding sphere. Note that any randomly oriented vector may be described in this frame. The elementary area dA' containing neighboring velocities is formed by two circles of latitude differing by an amount $d\theta$, as well as two other circles of longitude differing by an amount $d\varphi$ also illustrated in Fig. 2.2. The physical areal element dA' is given by the product of the two defining differential *arc-lengths*, $rd\theta$ and $r\sin\theta d\varphi$. Thus, dA' is given by

$$dA' = (rd\theta)(r\sin\theta d\varphi) = r^2 \sin\theta d\theta d\phi \tag{2.11}$$

The solid angle $d\Omega'$ corresponding to dA' is given by the ratio of the elemental surface area dA' to r^2, in units of *steradians*. Thus, $d\Omega'$ is given by

$$d\Omega' = \frac{dA'}{r^2} = \sin\theta d\theta d\varphi$$

We may use these concepts to draw two important correlations. First, the element of area dA' that we have described correlates with the number of particles $dN_{v,\theta,\varphi}$ with velocity **v** having angles between $\theta + d\theta$ and $\phi + d\varphi$. Second, the maximum solid angle Ω' corresponds to the total number of particles dN_v with velocity **v**, so we may write

$$\frac{d\Omega'}{4\pi} = \frac{dN_{v,\theta,\varphi}}{dN_v} \tag{2.12}$$

or

$$dN_{v,\theta,\varphi} = dN_v \frac{d\Omega'}{4\pi} = \frac{dN_v}{4\pi}\sin\theta d\theta d\varphi \tag{2.13}$$

dN_v is the total of number of particles over all combinations of angles, with velocity v. Since the particles have no preferred direction, the ratios of the surface areas also provide the ratios of the relative portions, etc. Let us apply Eq. (2.13) to the surface of the container in which the particles are moving. Specifically, how many of these particles will collide with the element area da (of the container, not to be confused with dA'), because we ultimately want to count the number of collisions per unit area. Here is where we have to get more creative. Of the particles counted in Eq. (2.13), the amount that will collide with the container surface is determined by the fraction of particles that are within a distance $v\tau$ of the wall. τ can be regarded as a mean time between collisions. Thus, they will strike da, provided they are close enough that no collisions with other particles take place within that distance/time. Those particles will only collide with da. The volume element dV containing these colliding particles from $dN_{v,\theta,\phi}$ that will collide with element area da is, then, given by

$$dV = \tau\mathbf{v}\cdot da = \tau v\cos\theta da \tag{2.14a}$$

Therefore, the proportion of particles that will strike differential area da of the container of the particles is

$$\frac{dV}{V} \tag{2.14b}$$

2.4 Ideal Gas Law and Kinetic Energy

Combining results to obtain *numbers of particles*, we can multiply this proportion by the number of particles $dN_{v,\theta,\varphi}$. Then, we have

$$dN'_{v,\theta,\varphi} = dN_{v,\theta,\varphi}\frac{dV}{V} \tag{2.15}$$

$dN'_{v,\theta,\varphi}$ denotes the number within the differential element dA' that actually strike the container. Now that we have the amount of particles colliding with the container differential area da, the number of *collisions* can be determined. The last piece we need to the puzzle is to determine the change in momentum *per collision*. For an *elastic* collision involving a particle striking the surface of a container, only the component of momentum normal to the surface will change, and it will only reverse itself, as illustrated in Fig. 2.3. Thus, the change in momentum **p** for a particle traveling towards the colliding wall oriented perpendicular to a polar-axis (z), is given by

$$\Delta p_{\text{per collision}} = p_f - p_i = 2m\mathbf{v} \cdot \mathbf{n} = 2mv\cos\theta \tag{2.16}$$

In Eq. (2.16), note that the range of θ is 0 to $\pi/2$, describing the angle between **p** and the unit *surface normal* vector **n**. Using Eq. (2.16) along with Eq. (2.13), the number of collisions striking da with particles of velocity v within the differential area dA', is given by

$$\text{number of collisions}_{v,\theta,\varphi} = \left(\frac{dN_v}{4\pi}\sin\theta d\theta d\varphi\right)\left(\frac{\tau v\cos\theta}{V}da\right) \tag{2.17}$$

To obtain the total change in momentum for particles with velocity v, we multiply Eq. (2.16) by Eq. (2.17) and integrate over θ and φ, to obtain

$$\text{change in momentum}_v = \frac{dN_v}{2\pi V}\tau mv^2\left[\int_0^{2\pi}d\varphi\int_0^{\pi/2}\sin\theta\cos^2\theta d\theta\right]da \tag{2.18}$$

The integral in Eq. (2.18) with respect to φ gives 2π, since the integrand is only a function of θ. For the remaining integral, using substitution, let $u = \cos\theta$, then the integral becomes

$$\int_0^{\pi/2}\sin\theta\cos^2\theta d\theta = -\int_1^0 u^2 du = -\frac{u^3}{3}\bigg|_1^0 = \frac{1}{3}$$

With this result, Eq. (2.18) becomes

$$\text{total change in momentum}_v = \frac{dN_v}{3V}\tau mv^2 da \tag{2.19}$$

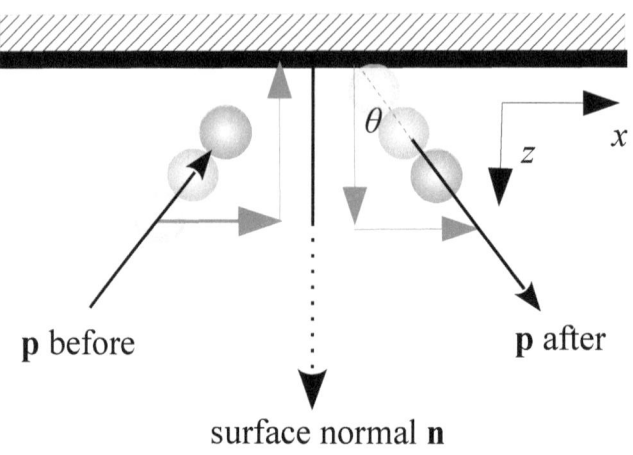

$$\Delta \mathbf{p} = (p_x + p_z) - (p_x - p_z) = 2p_z = 2\mathbf{p} \cdot \mathbf{n}$$

Figure 2.3: Illustration of the momentum change in an elastic collision of particle with a surface. The change is always in the component perpendicular to the surface.

The rate of change of the momentum is the differential force dF_v. It is obtained by dividing Eq. (2.19) by the corresponding amount of time taken to achieve the change in momentum, which is just τ. Thus, using the definition for pressure $dP = dF/da$ leads us to

$$dP_v = \frac{dN_v}{3} mv^2$$

Integrating over all possible velocities v, we have for the pressure P (force per unit area of the container)

$$P = \int dP_v = \frac{m}{3V} \int v^2 dN_v \qquad (2.20)$$

2.4 Ideal Gas Law and Kinetic Energy

Since dN_v is the distribution of the number of particles with speed v, the integral in Eq. (2.20) is proportional to the definition of the mean velocity $<v^2>$, where

$$\langle v^2 \rangle = \frac{1}{N} \int v^2 dN_v \qquad (2.21)$$

Thus, Eq. (2.20) becomes

$$PV = \frac{N}{3} m \langle v^2 \rangle = \frac{2}{3} N \left(\frac{1}{2} m \langle v^2 \rangle \right) \qquad (2.22)$$

The quantity in parentheses is the average kinetic energy per particle. We may compare Eq. (2.22) to the empirical ideal gas law given by

$$PV = nRT = nN_a \frac{R}{N_a} T = Nk_B T \qquad (2.23)$$

n is the number of moles and $N_a = 6.022 \times 10^{23}$ (number of atoms in a mole) is the **Avogadros number**. Eq. (2.23) relates the Boltzmann constant k_B to the **universal gas constant** $R = 8.3144 \text{JK}^{-1}\text{mol}^{-1}$ where

$$k_B = \frac{R}{N_a} \qquad (2.24)$$

Eq. (2.23) also tells us that an association can be made between the RHS of Eqs. (2.22) and (2.23) such that

$$Nk_B T = \frac{2}{3} N \left(\frac{1}{2} m \langle v^2 \rangle \right)$$

or

$$\frac{3}{2} k_B T = \frac{1}{2} m \langle v^2 \rangle = \frac{1}{2} m (\langle v_x^2 \rangle + \langle v_y^2 \rangle + \langle v_z^2 \rangle) \qquad (2.25)$$

Because we have three components of velocity, each with equal probability, Eq. (2.25) means that each degree of freedom contributes an equal amount to the kinetic energy. The amount is given by

Equipartition of Kinetic Energy For Ideal Gas

$$\text{each degree of freedom contributes } \frac{1}{2} k_B T \qquad (2.26)$$

While this result was one of the first applied to the blackbody radiation problem, it leads to an incorrect description of photons, but works well for gas particles. Before we get to the application of Eq. (2.26), lets take a moment to discuss more details around how one obtains the thermodynamic probability for a system of countably many particles. It turns out to be an important part to dealing with the blackbody radiation problem.

2.5 STATISTICAL THERMODYNAMICS AND STATE OCCUPATIONS

In a system with a *finite* number of particles all seeking to occupy a number of *available* energy states, it is possible to determine how the particles will fill up these states. This is one of the key objectives of *statistical* thermodynamics. A set of, say, N particles having to play out all the possible arrangements leads to a number of microstates that are possible. There is an important number in this problem that can only be determined from *counting principles*. For the total number of microstates Ω, there are *three important well-known formulas* obtained from counting, compulsory for thermodynamics. They play a critical role in both statistical thermodynamics and quantum mechanics. We will see later that Planck relied on one of them obtaining the missing link in the blackbody radiation problem. This enabled him to derive what is known as the **Bose-Einstein distribution** (BE), named after Indian physicist Satyendra Nath Bose (1894-1974). Einstein played a role in helping publish Bose's paper entitled "Planck's Law and the hypothesis of light quanta" in the German journal *Zeitschrift für Physik*. Einstein added a contribution to Bose's paper in the submission and the combined result was published in 1924. The *Bose-Einstein distribution* applies to *indistinguishable* 'particles' possessing whole-integer amounts of spin angular momentum. It was Bose that first recognized the significance of being indistinguishable. Note that the *spin* topic will be discussed in more detail beginning in Chap. 9. These kind of 'particles' are called **bosons**. *Photons are examples of bosons*, and this is a relevant point to the blackbody radiation problem.

Bosons: Bosons are fundamental particles possessing whole-integer amounts of *spin* angular momentum

2.5 Statistical Thermodynamics and State Occupations

When a system has more than one distinguishable state that possess the same energy, these same-energy distinguishable states are known as **degenerate states**. Moreover, for any degenerate state, the number of states that all have the same energy E_i is known as the **degeneracy level**. In a B.E. distribution, each so-called degenerate state having a total of N_i bosons (or particles) distributed in those states, with degeneracy levels d_i, the total number of possible microstates (or thermodynamic probability) $\Omega = \Omega_{BE}$ for the ith degenerate state is

Bose-Einstein Thermodynamic Probability

$$\Omega_{BE} = \frac{(N_i + d_i)!}{N_i! d_i!} \tag{2.27}$$

Eq. (2.27) is the **Bose-Einstein thermodynamic probability**, equaling the number of microstates.

In addition to *bosons*, another important type of particle we will need to know is the **fermion**. Fermions are fundamental particles that possess half-integer amounts of spin angular momentum. *Electrons are examples of fermions.* As long as the electrons are *indistinguishable*, the corresponding number of possible microstates for the fermions is given by

Fermi-Dirac Thermodynamic Probability

$$\Omega_{FD} = \frac{d_i!}{N_i!(d_i - N_i)!} \tag{2.28}$$

Eq. (2.28) is known as the **Fermi-Dirac thermodynamic probability**. Each of the number of microstates or thermodynamic probabilities above apply to *indistinguishable* fundamental particles, such as in a gas of fermions or bosons (when they are free or not bound to an atom). Along with the Boltzmann entropy formula given by Eq. (2.8), one can determine the way in which the particles will occupy the available degenerate energy states. As an example of this, for the case of a gas of fermions, let's obtain such a distribution for the occupation of energy states $f(E_\lambda)$.

2.5.1 Fermi-Dirac Distribution

Let's determine how the thermodynamic number of microstates Ω can be used to obtain the energy states occupation distribution. The Boltzmann entropy formula serves as the starting point using Eq. (2.28) to express the number of microstates Ω. In this case, the entropy for a gas of indistinguishable fermions becomes

$$S_i = k_B \ln \Omega_{FD} = k_B \ln \frac{d_i!}{N_i!(d_i - N_i)!} \tag{2.29}$$

For a degenerate state with total energy $E_i = N_i \epsilon$ and degeneracy level d_i (number of distinguishable states with same energies), the number of particles in that degenerate state can be obtained by

$$N_i = \frac{E_i}{\epsilon} = \frac{d_i \langle E_m \rangle}{\epsilon} \tag{2.30}$$

ϵ is the energy of each particle or fermion. $\langle E_m \rangle$ is the average energy per degenerate state or mode. Using the identity for log that states $\ln(A \div B) = \ln A - \ln B$, Eq. (2.29) becomes

$$S_i = k_B [\ln d_i! - \ln N_i! - \ln(d_i - N_i)!] \tag{2.31}$$

For large N_i, we can make use of an approximation called *Stirling's approximation*, which states that for any large integer $\ln L! \approx L \ln L - L$. Substitution of this relation into Eq. (2.31), we get

$$S = k_B [d_i \ln d_i - N_i \ln N_i - (d_i - N_i) \ln(d_i - N_i)]$$

Substituting the relations between energy and N_i, given by Eq. (2.30), we can write

$$S = k_B \left[d_i \ln d_i - \frac{d_i \langle E_m \rangle}{\epsilon} \ln \frac{d_i \langle E_m \rangle}{\epsilon} - \left(d_i - \frac{d_i \langle E_m \rangle}{\epsilon} \right) \ln \left(d_i - \frac{d_i \langle E_m \rangle}{\epsilon} \right) \right]$$

Using the identity $\ln AB = \ln A + \ln B$, some terms cancel and the above equation for S becomes

$$S = k_B \left[\left(\frac{d_i \langle E_m \rangle}{\epsilon} - d_i \right) \ln \left(1 - \frac{\langle E_m \rangle}{\epsilon} \right) - \left(\frac{d_i \langle E_m \rangle}{\epsilon} \right) \ln \frac{\langle E_m \rangle}{\epsilon} \right] \tag{2.32}$$

In this form, all logarithm arguments now have a factor of d_i, thus, we may write an equation for $S/d_i = \langle S_d \rangle$, where $\langle S_d \rangle$ is the average entropy per degenerate state, thus, Eq. (2.32) becomes

$$\langle S_d \rangle = k_B \left[\left(\frac{\langle E_m \rangle}{\epsilon} - 1 \right) \ln \left(1 - \frac{\langle E_m \rangle}{\epsilon} \right) - \left(\frac{\langle E_m \rangle}{\epsilon} \right) \ln \frac{\langle E_m \rangle}{\epsilon} \right] \tag{2.33}$$

2.5 Statistical Thermodynamics and State Occupations

We now have the entropy of a degenerate state as a function of its mean energy. This enables us to use Eq. (2.2), which says differentiate Eq. (2.33) and the result equals $1/T$. Note that $\langle E_m \rangle$ plays the role of the energy U. After differentiation, we obtain the following relation

$$\frac{1}{T} = \frac{k_B}{\epsilon} \ln\left[\frac{\epsilon - \langle E_m \rangle}{\langle E_m \rangle}\right]$$

Solving for $\langle E_m \rangle$, we get the result

$$\langle E_m \rangle = \frac{\epsilon}{e^{\frac{\epsilon}{k_B T}} + 1} \quad (2.34)$$

Using Eq. (2.30), for N_i, we can write a dimensionless probability distribution function $f(\epsilon; T)$ as

$$f_{FD}(\epsilon; T) = \frac{\langle E_m \rangle}{\epsilon} = \frac{N_i}{d_i}$$

This leads to

Fermi-Dirac Probability Distribution

$$f_{FD}(\epsilon; T) = \frac{1}{e^{\frac{\epsilon}{k_B T}} + 1} \quad (2.35)$$

Since there are d_i available states, and N_i particles, there can be *at most* $N_i = d_i$ particles per mode. We'll see that $f_{FD} \leq 1$, which means that $d_i \geq N_i$, or that the degeneracy level is always larger than the number of fermions. Eq. (2.35) is called the **Fermi-Dirac distribution**, and it describes how many of the d_i states will get occupied as a function of ϵ. It is the probability that a fermion of energy ϵ will occupy a state. Lower energy states are more probable up to a certain energy level *represented by zero*. In this form, the result is general because N_i/d_i is defined for any degenerate state. Fig. 2.4 plots the *Fermi-Dirac distribution* for three different temperatures, $T = 1K$, $T = 150K$, and $T = 300K$. From Fig. 2.4, we can see that the higher the temperature, the more shallow are the probability variations with $\epsilon/k_B T$. This indicates that the additional thermal energy afforded by higher temperature is sufficient to change the populations by taking lower energy states and raising their energies up to higher energy states. This is because in this region, the relative

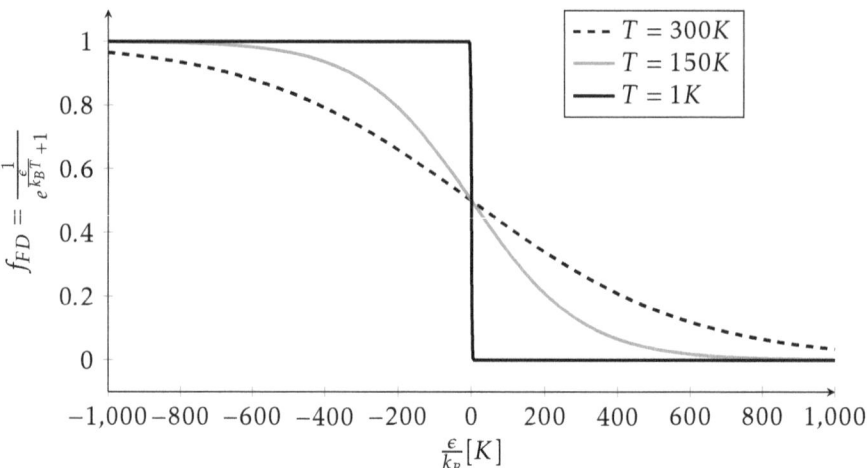

Figure 2.4: Fermi-Dirac distribution for three temperatures 1K, 150K, and 300K.

energy spacings $\Delta \epsilon$ are the smallest. This fact, however, is revealed by quantum mechanics. Thus, these states demand the least amount of energy to excite them into higher energy states.

All the curves cross at the same point $(0, 0.5)$. The energy level corresponds to what is called the **Fermi level** ϵ_F. From the statistical thermodynamics point of view, this unique energy corresponds to the energy having a probability of 1/2 for being occupied/unoccupied. The *Fermi level at zero* at zero temperature represents the energy region at the highest occupied energy state, just before no further fermionic states are occupied.

 Fermi Level: Most generally, the Fermi level is the energy having a probability of 1/2 for being occupied/unoccupied. The Fermi level at $T = 0K$ coincides identically with the highest filled energy level.

The ratio N_i/d_i is different for different degenerate states. We can, therefore, expect that this value depends on these quantities in some way. This connection will be shown more clearly in the next section where we treat the *Boltzmann distribution*. With this definition for f_{FD}, an alternative form can be written with the energy defined relative to the Fermi level ϵ_F, so that the zero point in Fig. 2.4 is redefined explicitly at

2.5 Statistical Thermodynamics and State Occupations

the *Fermi level* where we have

> **Fermi-Dirac Distribution (w/Fermi Level)**
>
> $$f_{FD}(\epsilon) = \frac{1}{e^{(\epsilon-\epsilon_F)/k_B T} + 1} \qquad (2.36)$$

This distribution is named after Italian-American physicist Enrico Fermi (1901-1952) who published it in 1926, and Paul Dirac, who independently published on it the same year. This form becomes particularly useful when we consider electrical conduction because the energy of electrons in motion, or the probability of motion, is determined by the energy relative to a defined energy point. This will be discussed in more detail later in Sec. 12.6. We have, thus, determined the energy occupation for *a gas of indistinguishable fermions*. There is an analogous distribution that can be derived for indistinguishable *bosons*, however, we will save this until Sec. 2.9, where we discuss how Planck derived his famous blackbody radiation distribution. If you have followed how we have arrived to this point, then you will follow the mind of Planck as he unraveled the solution to the blackbody radiation problem.

Next, we discuss another important distribution, also found using the number of microstates Ω. It is useful for systems with *distinguishable* particles, such as those in a lattice of a solid state material. They have distinct locations in space and are therefore, distinguishable. It was first introduced by Boltzmann, who applied it to atoms of solid state matter.

2.5.2 Boltzmann Distribution

If we know the counted thermodynamic probability Ω, the **Boltzmann probability distribution** for fundamental particles can be found. The result is given by

> **Boltzmann Probability Distribution**
>
> $$f_B(E_i) = \frac{e^{-E_i/k_B T}}{\sum_{i=-J}^{+J} e^{-E_i/k_B T}} \qquad (2.37)$$

We shall derive this result here, however, we are going to use a slightly different approach, which is more common now. Note, that it is not how Boltzmann originally carried out his derivation. Notice the different form of the distribution, involving a *discrete* sum over the possible states of the fundamental particle, contrary to the Fermi-Dirac distribution. It provides the probability of finding a particle in an energy state E_i. We'll need two constraints to enforce in this approach using Lagrange multipliers. First, assume we have a system of atoms capable of maintaining any one of N possible states. In equilibrium, the total energy E_T may be expressed as

$$E_T = \sum_{i=-J}^{+J} N_i E_i \quad \text{(constraint 1)} \tag{2.38}$$

The summation is over a finite number of states J, which must be an integer. We must also enforce that the system maintains the same number of total particles N, so

$$N = \sum_{i=-J}^{+J} N_i \quad \text{(constraint 2)} \tag{2.39}$$

The counting principle, as we observed with the Fermi-Dirac distribution, is an important part of the derivation of the Boltzmann or Maxwell-Boltzmann distribution. It provides the number of ways that N atoms can be arrange in all the available energy levels each being distinct, such that the ith level has N_i atoms. The number of microstates for this case is given by

Boltzmann Thermodynamic Probability

$$\Omega_B = N! \left(\prod_i \frac{d_i^{N_i}}{N_i!} \right) \tag{2.40}$$

The symbol $\prod$ denotes multiplication or a product of terms. The distribution is obtained by determining the N_i values that maximize/minimize Ω. To do this, we may use the fact that the same values of N_i that minimize Ω, also minimize $\ln\Omega$, which is proportional to the entropy S. This observation is useful because maximizing $\ln\Omega$ provides an easier path to obtain the corresponding N_i. $\ln\Omega$ is minimized along with the

2.5 Statistical Thermodynamics and State Occupations

constraints that N and E_T are both constant. Let us define a Lagrangian that maximizes Eq. (15.53) and satisfies these two constraints. The Lagrangian becomes

$$L(N_i) = \ln\left[N!\left(\prod_i \frac{d_i^{N_i}}{N_i!}\right)\right] + \lambda_N\left(N - \sum_{i=-J}^{+J} N_i\right) + \lambda_E\left(E_T - \sum_{i=-J}^{+J} N_i E_i\right) \quad (2.41)$$

Expanding Eq. (2.41) by exploiting the logarithm properties leads to

$$L(N_i) = \ln N! + \sum_{i=-J}^{+J}(N_i \ln d_i - \ln N_i!) + \lambda_N\left(N - \sum_{i=-J}^{+J} N_i\right) + \lambda_E\left(E_T - \sum_{i=-J}^{+J} N_i E_i\right) \quad (2.42)$$

Since this problem *assumes a large number of atoms* in the ensemble, it is useful to further simplify Eq. (2.42) by, again, making use of *Stirling's approximation* for the logarithm of a factorial where $\ln N! \approx N \ln N - N$. Using this approximation in the Lagrangian given by Eq. (2.42) gives

$$L(N_i) = N \ln N - N +$$

$$\sum_{i=-J}^{+J}(N_i \ln d_i - (N_i \ln N_i - N_i)) + \lambda_N\left(N - \sum_{i=-J}^{+J} N_i\right) + \lambda_E\left(E_T - \sum_{i=-J}^{+J} N_i E_i\right) \quad (2.43)$$

We may take the derivatives of the Lagrangian with respect to N_i. Taking the partial derivatives gives

$$\frac{\partial L}{\partial N_i} = \ln d_i - (\ln N_i + 1) + 1 - \lambda_N - \lambda_E E_i = \ln d_i - \ln N_i - \lambda_N - \lambda_E E_i = 0 \quad (2.44)$$

Eq. (2.44) can be rearranged to obtain an equation for the $\ln N_i$, where we have

$$\ln N_i = \ln d_i - \lambda_N - \lambda_E E_i \Rightarrow N_i = e^{\ln d_i - \lambda_N - \lambda_E E_i} = d_i e^{-(\lambda_E E_i + \lambda_N)} \quad (2.45)$$

Defining a dimensionless quantity analogous to the Fermi-Dirac distribution, $f = N_i/d_i$ becomes

$$f = \frac{N_i}{d_i} = e^{-(\lambda_E E_i + \lambda_N)} \quad (2.46)$$

If we compare the exponential term with that of Eq. (2.36), we see that $\lambda_E = 1/k_B T$ and $\lambda_N = -E_F/k_B T$, where E_F denotes the Fermi level. This

is the only drawback of this approach is that it does not explicitly reveal λ_E. It is known from the original approach used with the Fermi-Dirac distribution. However, the relation for λ_N evidences a relation between the Fermi level E_F and the numbers N_i in the respective states. Thus, the **Boltzmann probability** becomes

Boltzmann Probability

$$P_B(E_i) = e^{-\frac{E_i - E_F}{k_B T}} \tag{2.47}$$

Three examples of the Boltzmann probability are shown in Fig. 2.5, for $T = 1$ K, 150 K, and 300 K. As a probability which must satisfy $P_B \leq 1$,

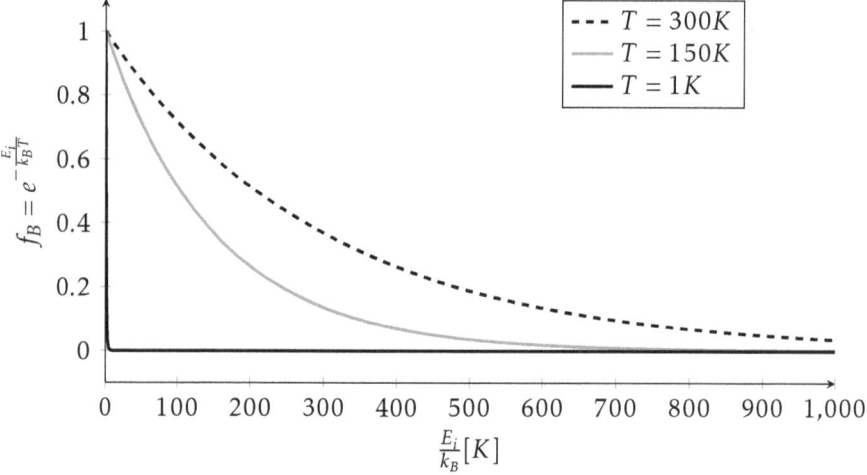

Figure 2.5: The Boltzmann distribution for three temperatures $T = 1$K, 150K, and 300K. The higher the temperature, the more probable is occupation of the energy E_i.

the Boltzmann probability is only defined for positive energies since negative energies lead to greater than unity. The occupation probability for states above E_F becomes more probable at higher temperatures, however, P_B always progressively diminishes with increasing energy E_i.

Since the values of N_i are different for each energy, we may define a probability distribution f_B for E_i given by N_i/N. This gives

$$f_B(E_i) = \frac{N_i}{N} = \frac{d_i e^{-(E_i - E_F)/k_B T}}{\sum_i d_i e^{-(E_i - E_F)/k_B T}} \tag{2.48}$$

2.6 Maxwell Equations Give Blackbody Modal Density

As a probability distribution, f_B satisfies the condition

$$\sum f_B(E_i) dN_i = 1$$

If we assume that the degeneracy level is the same for all the states E_i, then $d_i = d$ and Eq. (2.48) becomes

$$f_B(E_i) = \frac{N_i}{N} = \frac{e^{-(E_i - E_F)/k_B T}}{\sum_i e^{-(E_i - E_F)/k_B T}} \tag{2.49}$$

Eq. (2.49) is identical to Eq. (2.37), which is the *Boltzmann probability distribution*. Now that we have illustrated how to make use of statistical thermodynamics relevant to the blackbody radiation problem, let's change gears and jump to the physical description of the photons in the cavity of the *blackbox*. This is another important component for this problem. They are described by the Maxwell equations.

2.6 MAXWELL EQUATIONS GIVE BLACKBODY MODAL DENSITY

Maxwell equations can be applied to the light bouncing around the cavity walls, prior to exiting the cavity's hole. This analysis was done originally by Lord Raleigh and Sir James Jeans in 1897. The walls of the cavity were assumed to be perfect conductors, and this meant that in order to achieve thermodynamic equilibrium, the electric field E must vanish at the walls (otherwise there would be forces present to drive non-equilibrium processes). The optical problem definition for the cavity is, therefore, described by

$$\frac{\partial^2 E}{\partial t^2} = c^2 \nabla^2 E \quad \text{(volume)} \quad E = 0 \quad \text{(bounding surface)} \tag{2.50}$$

This equation, along with the boundary condition $E = 0$ at the walls of the cavity, has a solution given by

$$E = E_0 \sin\left(\frac{n_1 \pi}{L_x} x\right) \sin\left(\frac{n_2 \pi}{L_y} y\right) \sin\left(\frac{n_3 \pi}{L_z} z\right) \sin\left(\frac{2\pi c}{\lambda} t\right) \tag{2.51}$$

Plugging equation (2.51) into (2.50) confirms that it is a solution that also enforces the boundary condition $E = 0$ at the walls. Since our solution for E is a wave which vanishes at the boundaries, it means that the waves in the cavity must always be integer multiples of half-wavelengths ($n(\lambda/2) = L$), because, otherwise, the field would be nonzero.

After substituting the solution back into the wave equation in Eq. (2.50), and dividing out common factors, we get the following result:

$$\frac{n_1^2 \pi^2}{L^2} + \frac{n_2^2 \pi^2}{L^2} + \frac{n_3^2 \pi^2}{L^2} = \frac{\omega^2}{c^2} \tag{2.52}$$

It is useful if we express Eq. (2.52) as

$$n_1^2 + n_2^2 + n_3^2 = \frac{L^2 \omega^2}{\pi^2 c^2} \tag{2.53}$$

Using the relation $\omega = 2\pi f = (2\pi c)/\lambda$, Eq. (2.53) becomes

$$n_1^2 + n_2^2 + n_3^2 = \frac{L^2 \omega^2}{\pi^2 c^2} = \frac{4 L^2 \pi^2 c^2}{\pi^2 c^2 \lambda^2} = \frac{4 L^2}{\lambda^2} = R_n^2 \tag{2.54}$$

Eq. (2.54) is a suggestive result because its mathematical form resembles the form of an equation for a sphere with radius $R_n = 2L/\lambda$ (recall $x^2 + y^2 + z^2 = R^2$, for a sphere). Based on this concept, we may come up with a way to count the number of modes as a function of the wavelength λ, by computing relevant volumes under our sphere in the first octant, which uses only positive integers. The axes count in integers. To visualize the concept, consider a simplified 2D version of the idea illustrated in Fig. 2.10.

The total number of modes N_m is a function of the radius R_n. The importance of having this relationship is that it allows us to determine the number of modes as a function of the wavelength λ lying between λ and $\lambda + d\lambda$, where $d\lambda$ corresponds to dR_n. The differential is really what we are after. For example, in the 2D case illustrated in Fig. 2.10, we can obtain the number of modes N_m from the first quandrant's area A_{q1} (only positive integers). We are exploiting the fact that the integers, which determine the modes, serve as units along the respective axes. Instead of the x-axis measuring, say, distance, it measures a count n_i along an ith dimension. Using this analogy, for the total number of modes in the first quandrant $N_m = A_{q1} = (\pi R_n^2)/4$, and from this relation, if follows that $dN_m = dA_q = (\pi R_n d R_n)/2$, etc. Extending this idea to the 3D case, we equate the total number of modes N_m in the first octant to it's volume V_o, which is an eighth of that of a sphere. For the 3D case, we have

$$N_m = V_o = \frac{1}{8} \frac{4\pi R_n^3}{3} = \frac{4\pi R_n^3}{24} = \frac{32 \pi L^3}{24 \lambda^3} \tag{2.55}$$

In Eq. (2.55), we have used the relation $R_n = 2L/\lambda$. For each optical mode counted (combination of integers n_1, n_2, n_3), there are two existing

2.6 Maxwell Equations Give Blackbody Modal Density

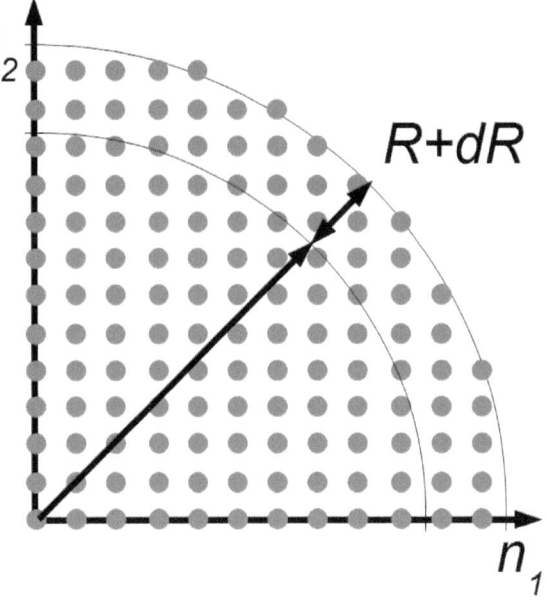

Figure 2.6: 2D illustration of the geometrical relationship between a circle and the modal integers, n_1 and n_2. For large n, this allows defining R and dR, which describe the number of modes with energy up to E_λ and with energies between E_λ and $E_\lambda + dE_\lambda$, respectively.

radiation modes, namely a *transverse electric* mode (*TE*) and a *transverse magnetic* (*TM*) mode. All optical modes can be written as a linear combination of these two independent modes. They both may exist inside the cavity. It means that we must multiply our result by a factor of 2, thus giving

$$N_m = \frac{64\pi L^3}{24\lambda^3} = \frac{8\pi L^3}{3\lambda^3} \tag{2.56}$$

Since our result is proportional to L^3, which is the physical volume of the cavity, we can write the number of modes per unit volume, or *density of modes* $\rho_m = N_m/L^3$ as

$$\rho_m(\lambda) = \frac{8\pi}{3\lambda^3} \tag{2.57}$$

The result in Eq. (2.57) can be used to obtain the differential modal density, which is given by

$$\frac{d\rho_m}{d\lambda} = -\frac{8\pi}{\lambda^4} \qquad (2.58)$$

or

> **Wavelength Differential Modal Density**
>
> $$d\rho_m(\lambda) = -\frac{8\pi}{\lambda^4} d\lambda \qquad (2.59)$$

Thus, the modal density decreases with increasing wavelength. Since measurements at the time of Planck where often taken in terms of frequency f, we can also express (2.59) in terms of f. Using $\lambda = c/f$ and $d\lambda = -(c/f^2)df$, Eq. (2.59) becomes

$$d\rho_m(f) = -\frac{8\pi f^4}{c^4}(-(c/f^2)df) \qquad (2.60)$$

So, we have for the frequency differential modal density, in terms of frequency

> **Frequency Differential Modal Density**
>
> $$d\rho_m(f) = \frac{8\pi f^2}{c^3} df \qquad (2.61)$$

The number of modes per unit volume turns out to be useful in solving the blackbody radiation problem. It's utility lies in the fact that we then only need to determine the mean energy per mode $\langle E_m \rangle$. The product of the two $\rho_m \times \langle E_m \rangle$ leads to the final energy density distribution as a function of wavelength λ or frequency f. This corresponds to the existing experimental data driving this problem of blackbody radiation. The difference between all the distributions we consider below lies in the treatment of the mean energy per mode $\langle E_m \rangle$. Among the early problem-solvers to offer a solution in this regard was Rayleigh and Jeans.

2.7 Rayleigh-Jeans Distribution

Knowing the number of modes per unit volume leaves only the task of having to find the average energy per mode $\langle E_m \rangle$. Lord Rayleigh (1842-1919) and Sir James Hopwood Jeans (1877-1946) had the idea to look to thermodynamics for the average energy per mode utilizing the result we obtained in Eq. (2.26). Doing this, however, implicitly assumes that its reasonable to regard the modes of light as particles of a gas in a box. We found in Sec. 2.4 that each particle will have a contribution of $k_B T/2$ to the mean kinetic energy, where k_B is Boltzmann's constant. Rayleigh and Jeans went with $\langle E_m \rangle = k_B T/2$. And since each mode, in our case, has two independent optical modes of TE (transverse electric) and TM (transverse magnetic), the energy density distribution is then $du_m(f) = d\rho_m(f) \cdot \langle E_m \rangle df$. From this, we obtain the result from Rayleigh and Jeans, given by

Rayleigh-Jean's Distribution

$$\frac{du_m}{df}(f) = \frac{8\pi f^2}{c^3} k_B T \tag{2.62}$$

Eq. (2.62) was obtained by Rayleigh and Jeans c.1900, although it was not the first result, chronologically, of a blackbody radiation distribution. Examples of Rayleigh-Jeans distribution are shown in Fig. 2.7. The R-J distribution varies quadratically with f. Increasing temperature causes the energy density to rise more sharply. However, this form ultimately diverges as frequency continues increasing, contrary to experiments, which approaches zero as $f \to \infty$. Therefore, the distribution was only in agreement with experiments at low frequencies (or long wave-lengths), but diverged at higher frequencies, sharply contradicting experimental observations. This divergence, approaching higher frequencies towards ultraviolet, became known as the *ultraviolet catastrophe*. This suggested that the assumption of treating the modes within the cavity exactly the same as heated particles in a box was not completely correct. There was more to it. As we mentioned early, this was not the first distribution proposed. The first one was done around 1896, and is the subject of the next section.

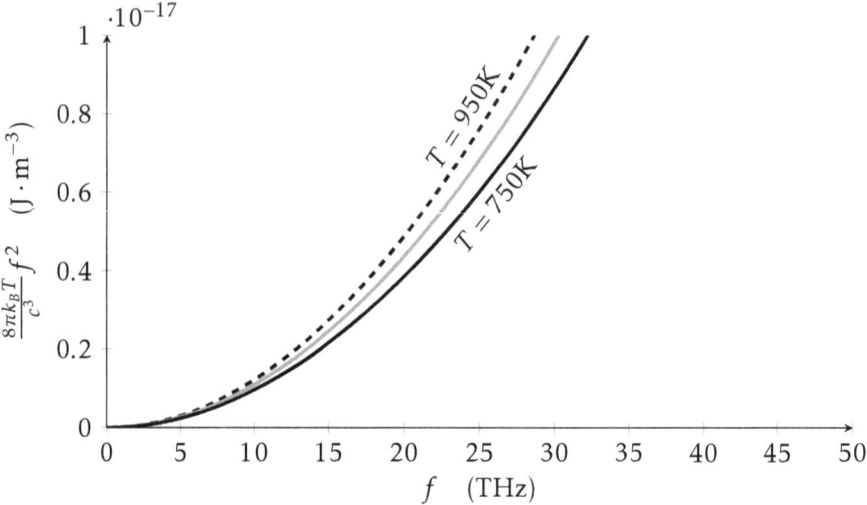

Figure 2.7: Rayleigh-Jeans distribution at temperatures 750 (dashed), 850 (middle-gray), and 950 (solid black) K. The R-J distribution is a quadratic function in f, and rises sharply with increasing temperatures. The distribution ultimately diverges as frequency continues increasing, contrary to experiments, which approaches zero as $f \to \infty$.

2.8 Wiens Distribution

German physicist Wilhelm Carl Werner Otto Fritz Franz Wien (1864-1928) also proposed a solution to the blackbody radiation problem, prior to the work of Rayleigh and Jeans. Thus, the modal density approach was not established. Wien's approach combined both modal density and mean energy per mode $\langle E_m \rangle$, however, Wien used the statistical-mechanical results of Boltzmann, particularly, from the Boltzmann distribution given by Eq. (2.47), casting, originally, in terms of wavelength λ. The energy distribution, utilizing the Boltzmann distribution was

$$\epsilon(\lambda) = N_m(\lambda) e^{-\epsilon/k_B T} \tag{2.63}$$

Wien assumed a coefficient of the form

$$N_m = \frac{C_1}{\lambda^n} \quad \text{and} \quad \epsilon = \frac{C_2}{\lambda} \tag{2.64}$$

Evidently, Wien was onto something by taking the energy to be inversely proportional to the wavelength of the mode. With these relations, Wien worked out a distribution of the form

$$\frac{du_m}{d\lambda} = \frac{C_1}{\lambda^n} e^{-C_2/\lambda k_B T}$$

2.8 Wien's Distribution

C_1 and C_2 were treated as phenomenological constants relying on experimental data to fit. Using measurement data to fit n, Wien found that the best agreement was found when $n = 5$, giving Wien's distribution as

Wien's Blackbody Distribution

$$\frac{du_m}{d\lambda}(\lambda) = \frac{C_1}{\lambda^5} e^{-C_2/\lambda k_B T} \tag{2.65}$$

In terms of frequency f, we have

Wien's Blackbody Distribution (in frequency)

$$\frac{du_m}{df}(f) = \frac{C_{f1}}{f^3} e^{-C_{2f} f/k_B T} \tag{2.66}$$

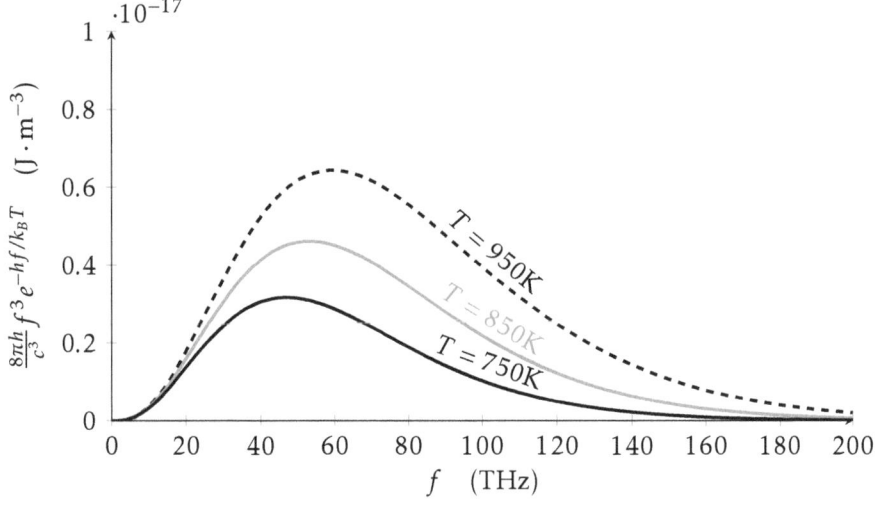

Figure 2.8: Wien's distribution at three different temperatures $T = 750$, 850, and 950K.

Note that $C_1 \neq C_{1f}$ and $C_2 \neq C_{1f}$. Because of the reliance on phenomenological constants, Wien's distribution was still less than ideal. Fig. 2.8 illustrates Wien's distribution for the temperatures 950 K, 850 K, and 750 K, correcting for the values of the constants, which were not known at the time. Though Wien's result had the potential to describe experimental data fairly well, in comparison to that of Rayleigh and Jeans, it lacked rigor in its derivation. With Wien and R-J, more was needed to fully predict the blackbody radiation phenomenon over the entire spectrum. This is where Planck stepped into the picture.

2.9 Planck's Distribution Law

Planck was aware of both the ultraviolet catastrophe as well as Wien's results, and knew an alternative approach was needed. In addition to classical thermodynamics, Planck also chose to use the advanced developments in statistical thermodynamics, much of which came from Boltzmann. A key result from Boltzmann, that was used by Planck, involved a statistical interpretation of entropy S for a system of countably many particles. The application of statistical thermodynamics gave a new path (as opposed to statistical mechanics) to solving this problem.

Unlike any of the previous efforts, Planck started with Boltzmann's formula for entropy S, given by

$$S = k_B \ln \Omega \tag{2.67}$$

S is the entropy and ω is the number of microstates. This is where Boltzmann's famous constant k_B was originally introduced. However, using Boltzmann's entropy formula requires counting the number of microstates Ω that the system can have. To count the number of microstates, we must first determine how the number of microstates Ω can be determined for a system of N_i photons. This is the topic of the next section.

2.9.1 Bose-Einstein Statistics For Photons

This particular problem of light quanta in a cavity requires that we determine how many ways can a given number of photons be distributed across energy levels. It is a counting problem at its core. Recall that we have the number of modes per unit volume ρ_m obtained from the electromagnetic equations. We require, additionally, the mean energy per mode, $\langle E_m \rangle$. Each frequency constitutes a mode and each mode has its own number of photons N_f in that mode. Each mode, which has

2.9 Planck's Distribution Law

a fixed amount of energy, therefore has a number of degenerate states d_f, where they have different combinations of integers that result in the same energy (recall our modal density). Thus, for each mode, we must count the number of ways that N_f photons can be distributed across a degenerate state with degeneracy d_f. To determine the number of possible arrangements, it is useful to recast this counting problem into a slightly different, but more definitive counting problem that provides a known solution.

Imagine that we spread out the total number of photons N_f per mode along a very long *fictitious line*. Then, we ask how many ways can we place $d_f - 1$ separators between photons along the line (to create d_f groups, we need $d_f - 1$ separators). It is important to recognize that the particles or photons and the d_f separators are both *indistinguishable*, so their order does not matter.

Let's consider placing the first separator along the line. We can place it at any of the inter-spaces between the photons, which accounts for $N_f - 1$ choices. But, since we also have d_f degenerate states, we have to be mindful of the fact that placing the separator on the far end of all the photons is equivalent to putting all the photons into a single degenerate state. With this caveat, there are an additional d_f choices for where we can place the separator on the extreme edge. A tally reveals that we have a total of $N_f - 1 + d_f$ choices to pick from. This is analogous to having $N_f - 1 + d_f$ objects that are nonrepeatable (meaning after it's used, the set becomes one-less), which has a total of $N_f - 1 + d_f!$ possibilities for the separator.

But we have $d_f - 1$ separators, and, like our photons, they are indistinguishable. When the order does not matter, we must reduce to total possibilities to account for this indistinguishability. In this frame, it is now a conventional combinatorial problem which is described by the combination's formula $_{N_f-1+d_f}C_{d_f-1} = (N_f - 1 + d_f)!/d_f - 1)!N_f!$. So, we finally have the number of possible microstates Ω, for distributing N_f photons among d_f degenerate states as

$$\Omega = \frac{(N_f - 1 + d_f)!}{(d_f - 1)!N_f!} \tag{2.68}$$

From here, two approximations of Eq. (2.68) are useful. We can neglect the −1 since for a classical system, N_f and d_f are very large compared to

1, giving

$$\Omega \approx \frac{(N_f + d_f)!}{N_f! d_f!} \quad (2.69)$$

Taking the log of (2.69) gives

$$\ln\Omega = \ln(N_f + d_f)! - \ln(d_f!) - \ln(N_f!) \quad (2.70)$$

We can use what is known as **Stirling's approximation**, which states that for a large integer L, $\ln L!$ can be approximated as $\ln L! \approx L\ln L - L$. We'll have you prove this in the Chapter problem 2.2. Applying Stirling's approximation to (2.70) gives

$$\ln\Omega = (d_f + N_f)\ln(d_f + N_f) - (d_f + N_f) - d_f \ln d_f + d_f - N_f \ln N_f + N_f \quad (2.71)$$

The non-log terms in Eq. (2.71) cancel each other, and the Boltzmann entropy S for the cavity is given by

$$S = k\ln\Omega = k[(d_f + N_f)\ln(d_f + N_f) - d_f \ln d_g - N_f \ln N_f] \quad (2.72)$$

The total energy E of a mode or frequency state with degeneracy d_f is given by

$$E = d_f \cdot \langle E_m \rangle \quad (2.73)$$

$\langle E_m \rangle$ is the average energy for a degenerate state or mode, which is what we are after, however, we will omit the brackets $\langle \rangle$ moving forward, for convenience. The number of photons N_f for a given mode is the total energy/photon energy $= d_f E_m/\epsilon_p$ where ϵ_p is the photon energy. This is where a distinction can be made from the approach of the Rayleigh-Jeans law. Substituting of this result into the Boltzmann entropy equation in Eq. (2.72) leads to

$$S = k\left[\left(d_f + \frac{d_f E_m}{\epsilon_p}\right)\ln\left(d_f + \frac{d_f E_m}{\epsilon_p}\right) - d_f \ln d_f - \frac{d_f E_m}{\epsilon_p}\ln\frac{d_f E_m}{\epsilon_p}\right] \quad (2.74)$$

Factoring d_f out, leads to the mean entropy $\langle S_m \rangle$ for a degenerate state, given by

$$S_m = \frac{S}{d_f} = k\left[\left(1 + \frac{E_m}{\epsilon_p}\right)\ln\left(d_f + \frac{d_f E_m}{\epsilon_p}\right) - \ln d_f - \frac{E_m}{\epsilon_p}\ln\frac{d_f E_m}{\epsilon_p}\right] \quad (2.75)$$

2.9 Planck's Distribution Law

Treating the first and third log argument terms on the right side as pure d_f products, $d_f(1 + E_m/\epsilon_p)$ and $d_f(U_d/\epsilon_p)$, and using the property of logarithms, $\log(AB) = \log A + \log B$, Eq. (2.75) becomes

$$S_m = k\left[(1 + \frac{E_m}{\epsilon_p})\ln(1 + \frac{E_m}{\epsilon_p}) - \frac{E_m}{\epsilon_p}\ln\frac{E_m}{\epsilon_p}\right] \tag{2.76}$$

Using our earlier thermodynamics relations (see Sec. 2.2), setting $1/T = \partial S_m/\partial E_m$, and solving for E_m, we obtain

$$E_m = \frac{\epsilon_p}{e^{\epsilon_p/k_B T} - 1} \tag{2.77}$$

Using the relation $N_f = d_f E_m/\epsilon_p$, we find that the number of photons per modal frequency f becomes

$$N_f = \frac{d_f}{e^{\epsilon_p/k_B T} - 1} \Rightarrow f_{BE} = \frac{N_f}{d_f} = \frac{1}{e^{\epsilon_p/k_B T} - 1} \tag{2.78}$$

The probability of finding a photon in a given frequency mode is N_f/d_f (ratio of number of particles to available states). Therefore, the mean energy for a mode becomes

$$\text{avg energy per mode} = \langle E_m \rangle = \frac{N_f \cdot \epsilon_p}{d_f} = \frac{\epsilon_p}{e^{\epsilon_p/k_B T} - 1} \tag{2.79}$$

We must also use Planck's assumption that the energy ϵ_p is proportional to f (i.e. $\epsilon_p = hf$), inline with Wien's parameter. This, along with our earlier expression for the number of modes per unit volume given by Eq. (2.61) leads us home. Putting it all together, we obtain Planck's final result for the overall blackbody energy distribution $du_f = d\rho_m E_m df$, or

Planck's Blackbody Distribution

$$u_f(f,T)df = \frac{8\pi h}{c^3}\frac{f^3}{e^{hf/k_B T} - 1}df \tag{2.80}$$

Eq. (2.80) is **Planck's distribution**, and the form of the distribution Planck found is generally known as the **Bose-Einstein distribution**. Planck's distribution represents a remarkable achievement given the level of rigor and level of resulting correctness. Planck found that agreement with all experimental results came if he took the arbitrary constant h to have the single value $h = 6.55 \times 10^{-27} erg \cdot sec = 6.626 \times 10^{-34} J \cdot s$. This value is known today as **Planck's constant** and is arguably the single most important constant in quantum mechanics. Its value today, in SI units, is known to be

Planck's constant h

$$h = 6.62607004 \times 10^{-34} \text{m}^2\text{kgs}^{-1} \tag{2.81}$$

In his famous 1901 paper, Planck also devoted a section to providing numerical values for other constants appearing in his theoretical derivations. In addition to the value for h, Planck also gave us one of the first estimates for Boltzmann's constant $k_B = 1.346 \times 10^{-16} erg/\text{grad} = 1.381 \times 10^{-23} J/K$. This value we know today to be

Boltzmann's constant k_B

$$k_B = 1.38064852 \times 10^{-23} \text{m}^2\text{kgs}^{-2}\text{K}^{-1} \tag{2.82}$$

Fig. 2.9 illustrates Planck's distribution for the temperatures 950 K, 850 K, and 750 K, bearing close resemblance to Wiens. Planck's distribution, however, has the advantage of no phenomenological parameters and it fit all the data over the whole range of f. For contrast, all three distributions, Rayleigh-Jeans, Wien, and Planck, are all shown in Fig. 2.10. Planck's distribution law can alternatively be expressed in terms of the wavelength λ, using the relations for light $c = \lambda f$, from whence it follows $df = -(c/\lambda^2)d\lambda$. With these substitutions, Planck's distribution law in terms of wavelength λ becomes

$$\rho(f,T)df = \rho_\lambda(\lambda,T)d\lambda = \frac{8\pi hc}{\lambda^5} \frac{1}{e^{hc/(k_B \lambda T)} - 1} d\lambda \tag{2.83}$$

The form given in Eq. (2.83) allowed Planck to reproduce another empirical relationship known as the **Wien's displacement law**. This law states that

2.9 Planck's Distribution Law

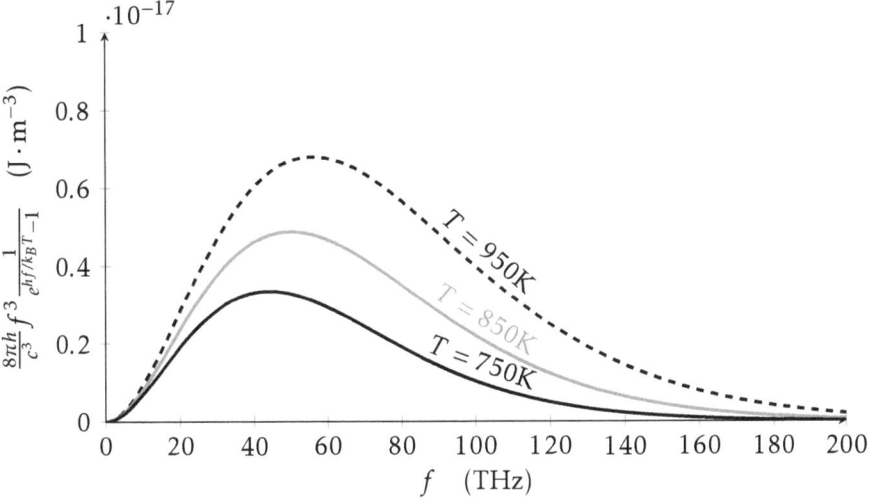

Figure 2.9: Planck's blackbody distribution at three different temperatures 750, 850, and 950 K.

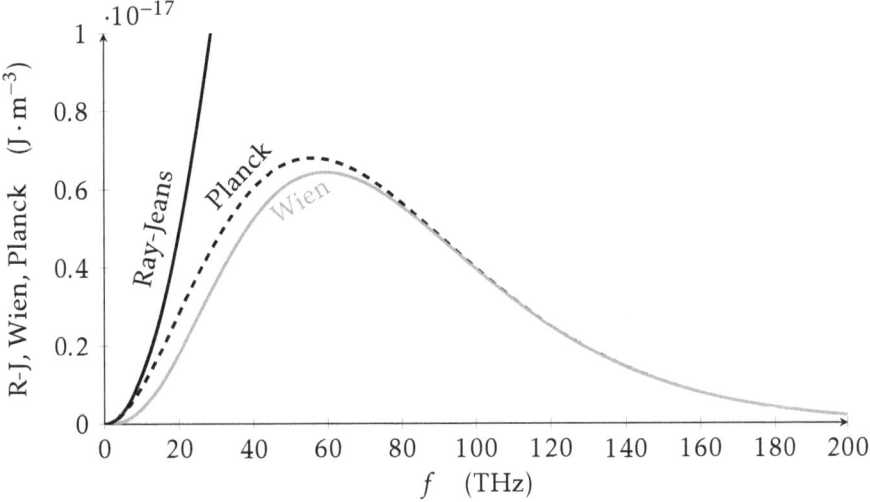

Figure 2.10: Comparison of three blackbody distribution formulas, including Planck, Wien, and the Rayleigh-Jeans distribution at $T = 950$ K. Same parameters used for all three curves.

Wien's Displacement Law

$$\lambda_{max} T = 2.90 \times 10^{-3} \, \text{m} \cdot \text{K} \qquad (2.84)$$

It was named so after Wien noticed that with a temperature change, there was always a corresponding shift or displacement in the λ at peak energy. With Planck's law in the form of Eq. (2.83), the max wavelength λ_{max} at which the maximum average energy takes place can be obtained by differentiating and setting the result equal to zero, then solving for λT. Carrying this out with Eq. (2.83), inserting values for the constants Planck obtained, leads to

$$\lambda_{max} T = \frac{hc}{4.965 k_B} \tag{2.85}$$

Substitution of the constants h, c, and k_B give a value of $2.898 \times 10^{-3} m \cdot K$, which is very close to the empirical value in Eq. (2.84). It's amazing that Planck's resulting distribution came from utilizing such a simple equation as the entropy given by Eq. (2.67). This was indeed, a sensational result. Moreover, there is the origin Planck's constant h, being discovered in the course of solving the blackbody radiation problem. Clearly, he and others were onto something with the quantum hypothesis.

2.10 Distribution Deducible From Pure Thermodynamics

Although, a few different tools were used to obtain Planck's final result, ultimately, the form of Planck's distribution, which was part of the novelty, is purely a thermodynamic consequence. We will show this fact in an alternative fashion using the known empirical entropy trends available at the time of Planck. In Planck's 1901 paper, he utilizes the following observed empirical relationship:

$$\frac{\partial^2 S}{\partial U^2} = -\frac{C_0}{U} \tag{2.86}$$

C_0 is a constant. This relationship is based purely on experiments. Additionally, others had found a different emperical relation from Eq. (2.86), particularly when λT was large. In these conditions, it was observed that

$$\frac{\partial^2 S}{\partial U^2} = -\frac{C_1}{U^2} \tag{2.87}$$

Based on these two experimental observations with the entropy S, we can express a generalized mathematical relation containing both behaviors as

$$\frac{\partial^2 S}{\partial U^2} = -\frac{C_2}{U(C_3 U + C_4)} \tag{2.88}$$

2.10 Distribution From Pure Thermodynamics

C_2, C_3, and C_4 are also constants. Combining Eq. (2.88) with our earlier result from Eq. (2.3) leads to

$$\frac{\partial}{\partial U}\left(\frac{1}{T}\right) = -\frac{C_2}{U(C_3U + C_4)} \tag{2.89}$$

The argument on the left-hand side of Eq (2.89) is strictly a function of temperature, and U is also strictly a function of temperature which allows us to write (2.89) as

$$\frac{d}{dU}\left(\frac{1}{T}\right) = -\frac{C_2}{U(C_3U + C_4)} \tag{2.90}$$

This modification to the derivative in (2.90) enables it to be integrated in a straight-forward fashion where we obtain

$$\frac{1}{T} = -C_2 \int \frac{1}{U(C_3U + C_4)} dU \tag{2.91}$$

To carry out the integral in Eq. (2.91), we first express the ratio in the integrand in a more convenient form, as a difference of two simpler fractions, as follows:

$$\frac{1}{U(C_3U + C_4)} = \frac{p(C_3U + C_4)}{U(C_3U + C_4)} - \frac{qU}{U(C_3U + C_4)} \tag{2.92}$$

p and q are two parameters for which we will solve for, that guarantee our desired result. Note that for Eq. (2.92) to be correct, the following must be true.

$$p(C_3U + C_4) - qU = 1 \tag{2.93}$$

Since the left-hand side is a polynomial in U, and it must be true for all values of U, the coefficients on both sides of Eq. (2.93) have to be equal. So we have

$$pC_3 - q = 0 \tag{2.94a}$$

$$pC_4 = 1 \tag{2.94b}$$

From the conditions in Eqs. (2.94a) and (2.94b), we get $p = 1/C_4$ and $q = pC_3 = C_3/C_4$. We can now write our integrand as

$$\frac{1}{T} = -\frac{C_2}{C_4} \int \left[\frac{C_3U + C_4}{U(C_3U + C_4)} - \frac{C_3U}{U(C_3U + C_4)}\right] dU$$

Or

$$\frac{1}{T} = -\frac{C_2}{C_4} \int \left[\frac{1}{U} - \frac{1}{U + C_4/C_3} \right] dU$$

After integration, we find that

$$\frac{C_4}{C_2 T} = -[\ln U - \ln(U + C_4/C_3)] = \ln \frac{U + C_4/C_3}{U}$$

Solving for internal energy U, we get

$$U = \frac{C_3/C_4}{e^{\frac{C_4}{C_2 T}} - 1} \tag{2.95}$$

The result in Eq. (2.95) is the same form as Planck's energy spectrum per mode or the Bose-Einstein distribution. Thus, the form above is *purely* thermodynamic in origin. Since Planck was an expert in thermodynamics, it was, perhaps, his destiny, to come across this problem. Next, let's consider some experimental observations that required Planck's newly discovered constant h.

2.11 PHOTOELECTRIC EFFECT NEEDS PLANCK'S CONSTANT

After Planck's groundbreaking work with blackbody radiation, h would, scientifically speaking, start becoming a household number. Around the same time, at the turn of the 19th century, physicists had observed a phenomenon known as the **photoelectric effect**, which was observed using tubes which were evolved from Crooke's tubes. Now, they could be incorporated into an electrical circuit. By directing light onto the electrodes inside the tube, usually made of sodium (Na) or cesium (Cs), one could observe a ray of glowing particles ejected from one electrode and directed to the other electrode. This was demonstrated convincingly by the German physicist Philipp Eduard Anton von Lenard (1862-1947).[1]

Leonard was one of the first to find that when shining light onto these ray tubes, that an electrical response was generated and it was found to be independent of the light intensity, and rather depended only on the

---
[1] Philipp Lenard was a known supporter of Nazi Germany, and was known to heavily criticize the works of Albert Einstein, who later explained his results theoretically.

2.11 Photoelectric Effect Needs Planck's Constant

wavelength λ. This finding was puzzling at the time, as it was counter to the understanding of electromagnetic waves, which dictated that higher power absorption was due to higher wave intensity. This photoelectric effect presented a new puzzle to the world of physics. It was Albert Einstein that would go on to explain the photoelectric effect successfully. Summed up nicely in his famous 1905 paper, Einstein wrote:

...when a light ray starting from a point is propagated, the energy is not continuously distributed over an ever increasing volume, but it consists of a finite number of energy quanta, localized in space, which move without being divided and which can be absorbed or emitted as a whole.
A. Einstein

Einstein was describing the fundamental particle we now call the *photon*, a term which was later coined by American chemist Gilbert Lewis in 1926. Based on this understanding, Einstein postulated that each quanta of light has an energy given by

Einstein Relation

$$E = hf \qquad (2.96)$$

Using this postulate, Einstein went on to write the following equation to correctly describe the photoelectric effect.

$$KE_{max} = hf - \phi_{thresh} \qquad (2.97)$$

h is Planck's constant which Einstein found necessary, and the parameter ϕ_{thresh} represents a threshold energy that depends on the metal of the electrode. It represents the binding energy of the electron to the metal atoms. Einstein's relation is also expressed as $E = h/(2\pi) \times 2\pi f = \hbar\omega$, where $\hbar$ is known as Planck's reduced constant. As pointed out by American physicist Robert Andrews Millikan (1968-1953), Einstein's equation turned out to be an equation of exact validity and was thoroughly demonstrated by Millikan himself to be of general applicability. It was shown through very careful experiments done by Millikan, to be correct for different metals including cesium, sodium, as well as copper. Initially, even Max Planck did not believe Einstein's theory describing this effect. Eventually, this changed and Einstein later received the Nobel prize in 1921, for his work on the photoelectric effect.

This discrete nature of electromagnet waves, as presumed by Einstein, was later buttressed by the work of American physicist Arthur Holly Compton (1892-1962). In 1922, through a series of experiments using x-ray tubes known as Molybdenum K-tubes, in combination with carbon targets for scattering the rays, calcite crystals for diffracting, and ionization chambers, for detection, Compton devised a method to study scattering properties of x-rays. He found something remarkable. Using Planck's quanta concept, he observed that there were shifts in wavelength between the original x-rays and the scattered rays. These shifts were a function of the angle of scattering θ. Based on what he observed, he found that he could describe his findings well by assuming conservation of momentum and relativistic energy of the collisions, exactly as is done with a nonrelativistic particle, to obtain the following result.

Compton Scattering Wavelength Shift

$$\lambda^0 - \lambda = \frac{h}{m_e c} \cdot (1 - \cos\theta) \tag{2.98}$$

λ^0 is the initial wavelength, before scattering, λ is after scattering. Compton's result demonstrated that light quanta apparently possessed particle-like properties and it was somehow tied to this constant h, and this was significant in light of the pervasive belief that light was *strictly* a wave, rather, than a particle, but certainly not both.

2.12 RYDBERG FORMULA AND THE DE BROGLIE RELATION

In parallel with many of these activities, observing the emission spectrum of hydrogen atoms in the late 1800s, Swedish physicist Johannes Robert Rydberg (1854-1919) noticed that a certain empirical relationship describes the discrete wavelengths observed in *hydrogen spectral lines*. He found that the empirical observations for hydrogen were well described by

$$\frac{1}{\lambda} = R_d \cdot \left(\frac{1}{n_1^2} - \frac{1}{n_2^2}\right) \tag{2.99}$$

The parameter R_d is known as the *Rydberg constant* and he found it to have a value of approximately $10,965,609 \text{m}^{-1}$. n_1 and n_2 are integers

2.12 Rydberg Formula And The De Broglie Relation

greater then zero. The first series with $n_1 = 1$ (the second integer $n_2 = n_1 + 1$ or larger) is called the *Lyman series*. When $n_1 = 2$, the series is called the *Balmer series*. For $n_1 = 3$, it is called the *Paschen series*. When $n_1 = 4$, it is called the *Brackett series*, $n_1 = 5$, it is called the *Pfund series*, and when $n_1 = 6$, it is called the *Humphreys series*. The first three series are illustrated in Fig. 2.11. Table 2.1 summarizes the list of the hydrogen

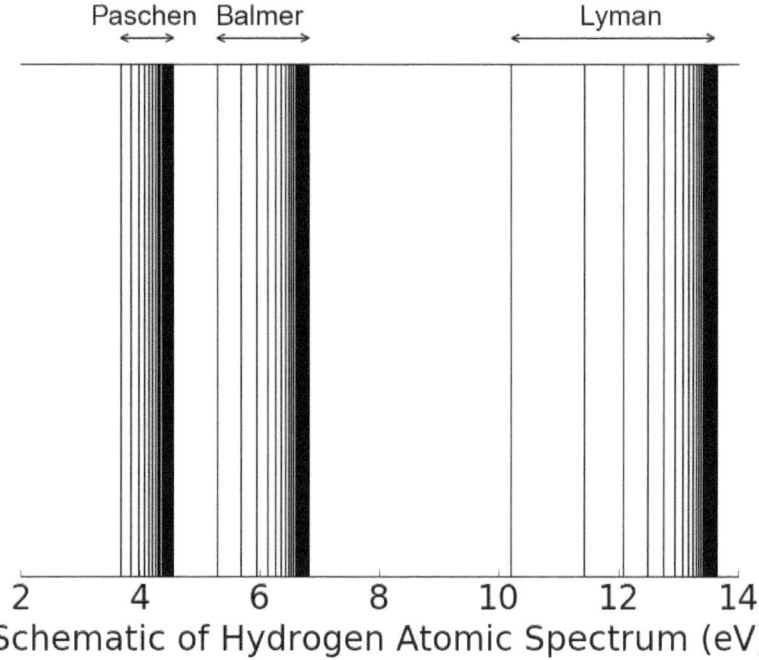

Figure 2.11: Atomic spectrum for the Hydrogen atom. The first three series are illustrated, the Lyman, Balmer, and Paschen series. They go from higher energy differences towards lower energy differences. As n_i increases, the energy spacing or separation becomes smaller.

series. The empirical formula by Rydberg was eventually predicted with remarkable accuracy by some of the earlier theories in quantum mechanics, including the most elementary model by Neils Bohr, as well as the quantum theory introduced by Erwin Schrödinger.

To crown the events taking place up until the early 1920s, ultimately building the momentum for the birth of quantum mechanics, French physicist Louis-Victor-Pierre-Raymond de Broglie (1892-1987) wrote in 1922:

Series Name	n_1	n_2
Lyman series	1	2,3,4,...
Balmer series	2	3,4,5,...
Paschen series	3	4,5,6,...
Brackett series	4	5,6,7,...
Pfund series	5	6,7,8,...
Humphreys series	6	7,8,9,...

Table 2.1: Summary of the hydrogen spectrum series. They are defined by the value of n_1, while n_2 takes integer values larger than n_1.

I was convinced that the wave-particle duality discovered by Einstein in this theory of light quanta was absolutely general and extended to all of the physical world, and it seemed certain to me, therefore, that the propagation of a wave is associated with the motion of a particle of any sort...photon, electron, proton, or any other.
L. de Broglie

de Broglie postulated that all matter has associated with it, both a wavelength λ, as surely as it will have a mass. An important relationship he put forward was

de Broglie Relation

$$p = \frac{h}{\lambda} \tag{2.100}$$

Eq. (2.100) is known as the **de Broglie relation**. What is remarkable about de Broglie's postulation was that he proposed the relation around 1922, but it was not until c.1927 that the first experiments were done by American physicists Clinton Joseph Davisson (1881-1958) and Lester Halbert Germer (1896-1971), with electrons, that umambiguously demonstrated this to be true. Davisson and Germer both designed and built a vacuum apparatus for measuring the energies of electrons that would be scattered from a metal surface. In their experiment, electrons from a heated filament were accelerated by a voltage and allowed to strike the surface of a nickel target.

The electron beam was directed at the target, which could also be rotated

2.12 Rydberg Formula And The De Broglie Relation

to observe angular dependencies. The electron detector they used is called a Faraday box and it was placed on an arc to allow observations of electrons at different angles. The electrons were reflected into the detector and they were surprised to find that at certain angles there was a maximum in the intensity. They had observed electron wave-like behavior, and their results could be described using what is known as *Braggs law*, which had been applied to x-rays successfully. In their application, they were also able to obtain values for the lattice spacing in the nickel crystal target. de Broglie later won the Nobel Prize for this work in 1929, after the convincing results of Germer, Davisson, and others. Experiments continued demonstrating the wave-like behavior of particles, like electrons. Fig. 2.12 shows more recent experimental data from electrons published by Tonomura, *et al* in the American Journal of Physics in 1988. With astonishing clarity, their results unambiguously demonstrate the validity of de Broglie's relation.

In Fig. 2.12, each frame is organized as follows: (*a*) is after 10 electrons were fired; (*b*) is after 100; (*c*) 3000, (*d*) 20,000, and (*e*) 70,000 electrons. Coming to understand the wave-nature of electrons was an important achievement for quantum mechanics. This understanding has guided humans to numerous indispensable technologies. One of them is known as **transmission electron microscopy** or TEM. This marvelous technology, widely used today, uses a beam of electrons from a source, directed at a thin film sample, inline with the electron beam. The electron waves, in passing through the sample, experience changes in their energies as they interact with the thin film sample atoms. The transmitted electron beam, thus, carries information about the sample, and it is then captured on a special film, such as a fluorescent screen, to image the information carried from the sample. Its challenging to argue against the fact that technologies like TEM would likely not exist as it does, without the contributions of de Broglie.

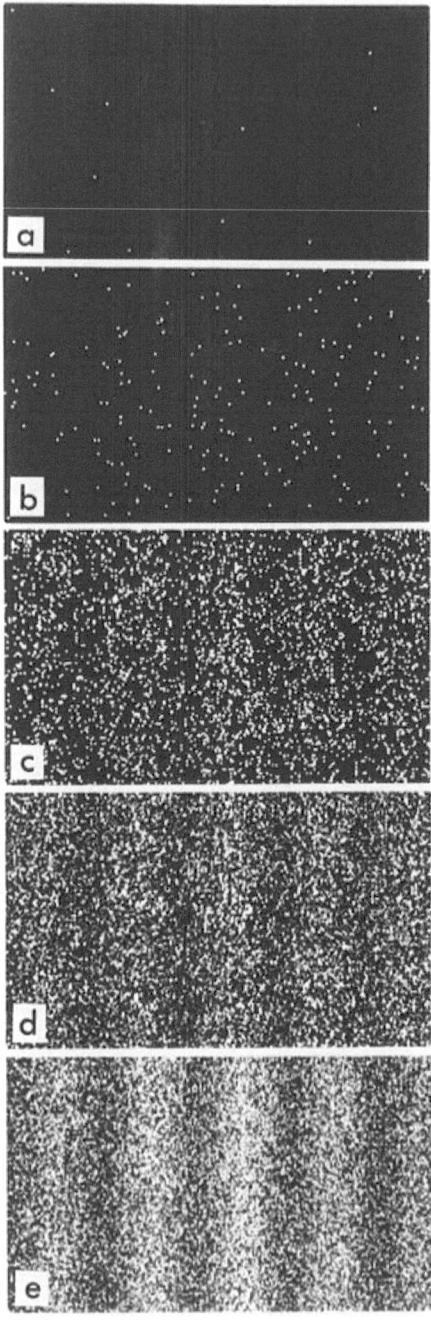

Figure 2.12: Experimental demonstration of electron interference. Frame (*a*) is after 10 electrons were fired; (*b*) is after 100; (*c*) 3000, (*d*) 20,000, and (*e*) 70,000 electrons. Reproduced from A. Tonomura, J. Endo, T. Matsuda, and T. Kawasaki, Demonstration of single-electron buildup of an interference pattern, American J. of Phys , 57, 117 (1989), with the permission of the American Association of Physics Teachers.

2.13 CHAPTER SUMMARY

This chapter has focused on another important historical development for quantum mechanics, specifically, that of thermodynamics and statistical thermodynamics. Boltzmann's work laid a foundation for a novel path for Planck to take with the blackbody radiation problem. He not only found the correct distribution for the mean energy per mode, but in so doing, he also discovered one of the most important parameters in quantum mechanics, now known as Planck's constant h. The validity of Planck's results were further solidified when Einstein discovered that h was also required to correctly describe the photoelectric effect, as well as other experimental works. Table 2.2 lists some of the key results from this chapter.

Table 2.2: Chapter 2 Summary Equations.

Name	Equation
Boltzmann entropy formula	$S = k_B \ln \Omega$
mean energy of gas particle per DOF	$\langle E_m \rangle = \frac{1}{2} k_B T$
Bose-Einstein number of microstates	$\Omega_{BE} = \frac{(N_i + d_i)!}{N_i! d_i!}$
Fermi-Dirac number of microstates	$\Omega_{FD} = \frac{d_i!}{N_i!(d_i - N_i)!}$
Bose-Einstein distribution	$f_{BE} = \frac{1}{e^{\epsilon_p / k_B T} - 1}$
Fermi-Dirac distribution	$f_{FD}(\epsilon) = \frac{N_i}{d_i} = \frac{1}{e^{(\epsilon - \epsilon_F)/k_B T} + 1}$
Planck's blackbody distribution	$u_f(f, T) = \frac{8\pi h}{c^3} \frac{f^3}{e^{hf/k_B T} - 1}$

2.14 CHAPTER PROBLEMS

Problem 2.1 What is the entropy of a system that *has only a single state*? Use the Boltzmann entropy equation. What does the answer suggest about minimum entropy changes while increasing the number of states?

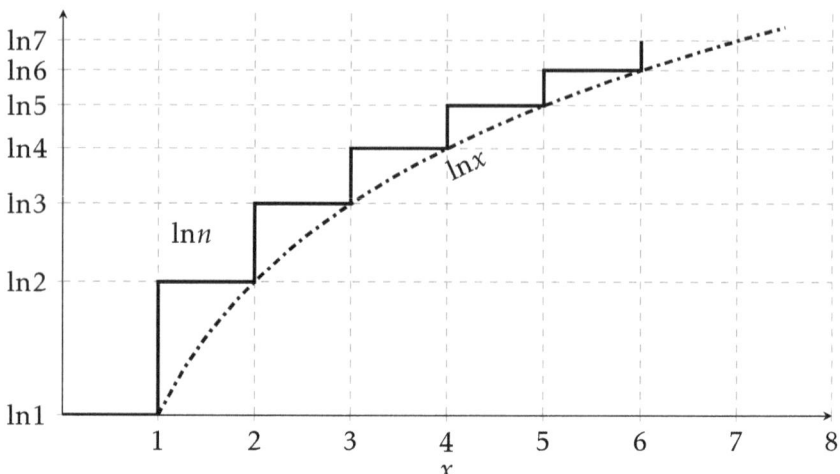

Problem 2.2 According to the figure, considering integers along the x axis, since $\Delta n = 1$, the area under each step becomes $\ln n$. Based on this, the summed areas become

$$\sum_{1}^{n} \ln n \, dn = \ln n! \approx \int_{1}^{n} \ln n \, dn$$

Use integration by parts to show that

$$\ln x! \approx x \ln x - x + 1$$

For large $x \gg 1$, the result is known as *Stirling's approximation*.

Problem 2.3 How many ways can $n = 1$ particles fill a total of m available states? Start with 1, 2, 3, and so one to obtain the more general answer.

Problem 2.4 Show that $n = 3$ *distinguishable* particles fill a total of $m = 3$ available states in 6 ways? How does the answer change if the particles are *indistinguishable*?

Problem 2.5 How many ways can $n = 2$ *distinguishable* particles fill a total of $m = 3$ available states? How about for $n = 2$ *indistinguishable* particles?

Problem 2.6 How does whether *distinguishable* or *indistinguishable* particles affect how the available states are filled? Make the case for why one always leads to more possibilities?

Problem 2.7 Heat transfer is accompanied by a change in energy E of a system. The heat capacity of a system is a measure of the change in energy of a system per unit change in temperature T. That is

$$C = \frac{dE}{dT}$$

The heat capacity per unit mass, or *specific heat*, is $c = C/m$. The heat transfer ΔQ is given by

$$\Delta Q = mc\Delta T$$

A system of $m = 1$kg is *stopped* by a frictional system, where its initial speed is $v_i = 15$m/s. Assuming all the work done by the friction stops the system, calculate the change in temperature ΔT for the mass, assuming a specific heat of iron $c = 450$J/kg-K. Hint: use the *First Law of Thermodynamics* and assume all the energy converts to heat.

2.15 SUGGESTED READINGS & REFERENCES

[1] M. Planck, *On the Law of the Energy Distribution in the Normal Spectrum*, Ann. Phys., **4**, 553 (1901)(english translation)
[2] J. Crepeau , *A Brief History of the T4 Radiation Law* Proceedings HT, DOI: 10.1115/HT2009-88060 (2009)
[3] M. Zemansky and R. Dittman, *Heat and Thermodynamics*, McGraw-Hill (1981)
[4] E. Fermi, *Sulla quantizzazione del gas perfetto monoatomico*. Rendiconti Lincei (in Italian) (1926). 3: pg. 145–9., translation Zannoni, Alberto (1999-12-14). *On the Quantization of the Monoatomic Ideal Gas*
[5] P.A.M. Dirac, *On the Theory of Quantum Mechanics*. Proceedings of the Royal Society A. **112** (762), pg. 661–77 (1926)
[6] A. Tonomura, J. Endo, T. Matsuda, and T. Kawasaki, *Demonstration of single-electron buildup of an interference pattern*, American J. of Phys. **57**, 117 (1989)

CHAPTER 3

A Masterful Equation With Wave-Particle Duality

Leading up to the mid-1920s, there was a *certain uncertainty* about the state of matter, as well as in the confidence of some domains of physics. In the case of light, for example, through events that transpired, we have seen that the momentum of belief swung from the corpuscular model of light in the days of René Descartes and Isaac Newton, to a wave theory of light nucleated by Huygens and ultimately solidified by Thomas Young's experiment, and the results of Maxwell. However, history also lead us to the discovery of the photoelectric effect, for which Einstein showed that the phenomenon could only be described if we ascribe discrete properties to what he referred to as a *quanta of light*.

We also learned about the discovery of subatomic particles, like electrons, first detected by J.J. Thomson. These tiny wave-particles were seen to behave as a stream of particles being pushed around by electric and magnetic fields, for example, in Crooke's tubes. All of this was crowned by de Broglie postulating that *all matter* would have an associated wave λ. This was confirmed for the first time experimentally in 1927. De Broglie's elegant postulation would go on to spur new directions for developing theories for quantum mechanics. Physics was beginning to suggest that we may not be able to consider the laws of physics, themselves, as describing *strictly* waves, nor strictly particles. Something was missing. No theory yet existed that could describe particles that possessed wave-like properties, or conversely, waves that possessed particle-like properties and whose scope could explain the breadth of experimental data that had been observed. With all of these events leading up to the mid-1920s, that was about to change.

Matter and energy seem granular in structure, and so does life, but not so mind.
E. Schrödinger

In this chapter, an objective lies in merging the granularity and the wave-nature of matter into a single physical description or equation. But, what kind of theory could simultaneously contain both aspects of matter? More than one person came up with powerful answers to this question. One of them was Erwin Schrödinger. We'll focus on his answer to this question, starting in the next section.

3.1 AN EQUATION DESCRIBING PARTICLES AND WAVES

What might an equation look like that can describe a wave with particle-like properties simultaneously? This question can only be answered heuristically, at least partially, by taking what we've learned so far and putting some things together. It cannot be derived *from* a more fundamental theory, for example. To begin, consider a wave function $\psi(x,t)$, somewhat like the kind described by Maxwell's equations, having the form

$$\psi(x,t) = A e^{i(kx-\omega t)} \tag{3.1}$$

Based on the de Broglie's relation for the momentum $p = h/\lambda$, we can also write the momentum as

$$p = \frac{h}{\lambda} = \frac{h}{2\pi} \frac{2\pi}{\lambda} = \hbar k \tag{3.2}$$

$k = 2\pi/\lambda$ is the *wavenumber*. Analogously, Einstein's relation can be used to write the energy E as

$$E = hf = \frac{h}{2\pi}(2\pi f) = \hbar \omega \tag{3.3}$$

3.1 An Equation Describing Particles And Waves

Substitution of Eqs. (3.2) and (3.3) into Eq. (3.1) leads to a wave function given by

$$\psi(x,t) = Ae^{i(\frac{p}{\hbar}x - \frac{E}{\hbar}t)} = Ae^{\frac{i}{\hbar}(p \cdot x - E \cdot t)} \qquad (3.4)$$

The advantage of this form is that p and E are more familiar to particles experienced in laboratories, but also translate to wave properties, as well. With a wave function that is a function of momentum p and energy E, let's consider what we can do to ψ, that in turn provides information about p and E. In quantum mechanics, this is the idea behind a **continuous linear operation** or **operator**. A continuous linear operator is a continuous linear transformation between vector spaces, and conserves both multiplication and addition operations. For example, if $\widehat{O}$ is a linear operator, operating on $\psi = A\psi_1 + B\psi_2$, it follows that

$$\widehat{O}\psi = A\widehat{O}\psi_1 + B\widehat{O}\psi_2 \qquad (3.5)$$

All *operators* that we shall encounter are *continuous linear operators*. As our first example, one operator can be defined to expose the momentum p and, likewise, another for the total energy E from the wave function ψ. More specifically, we seek a momentum operator $\widehat{\mathbf{p}}$, such that

$$\widehat{\mathbf{p}}\psi = \mathbf{p}\psi$$

The LHS is an *operator* operating on ψ and, under certain conditions, leads to the real-world measurable momentum $\mathbf{p}$ multiplied by the wave function ψ. In finding $\widehat{\mathbf{p}}$, we only need a first order derivative, as we have

$$\frac{\partial \psi}{\partial x} = \frac{ip}{\hbar} Ae^{\frac{i}{\hbar}(px - Et)} = \frac{ip}{\hbar}\psi \qquad (3.6)$$

This leads to the following result for the wave-particle **momentum operator** along the x-direction:

$$\widehat{p}_x \psi = -i\hbar \frac{\partial \psi}{\partial x} \qquad (3.7)$$

To denote an **operator**, we shall use the **notation of the hat above** the operator symbol, as in the LHS of Eq. (3.7). In general, this *symbolic notation shall be used throughout this book to denote an operator*. This result can be interpreted as prescribing an operation that can be applied to the wave function ψ, in order to expose the physical momentum p_x of the system. In other words, the momentum information is completely contained in the wave function. It is convention to alternatively express this relationship as

x-component Momentum Operator

$$\widehat{p}_x \Rightarrow -i\hbar \frac{\partial}{\partial x} \tag{3.8}$$

This notation informs us of an operation that can be performed on the wave function ψ in order to obtain the momentum p. Since there is a way to obtain p from the wave function, it should also be possible to obtain p^2. This provides us with more information, such as the kinetic energy K, which is proportional to p^2. Differentiating $\partial \psi / \partial x$ again leads to

$$\frac{\partial^2 \psi}{\partial x^2} = \frac{ip}{\hbar} \frac{\partial \psi}{\partial x} = \frac{ip}{\hbar} \cdot \frac{ip}{\hbar} \psi = -\frac{p^2}{\hbar^2} \psi \tag{3.9}$$

This leads to the following result for an operator for $\widehat{p}_x^2$

$$\widehat{p}_x^2 \psi = -\hbar^2 \frac{\partial^2 \psi}{\partial x^2} \tag{3.10}$$

Using the shorthand notation for the operator, as introduced in Eq. (3.8), we have

x-component Momentum Operator Squared

$$\widehat{p}_x^2 \Rightarrow -\hbar^2 \frac{\partial^2}{\partial x^2} \tag{3.11}$$

Thus, $\widehat{p}_x^2$ is obtained by second order differentiation of the wave function ψ, with respect to x, then multiplying by $-\hbar^2$. Since the kinetic energy K is given by $p^2/2m$, the operator $\widehat{K}$ is given by

$$\widehat{K}\psi = -\frac{\hbar^2}{2m} \frac{\partial^2 \psi}{\partial x^2} \tag{3.12}$$

or

1D Kinetic Energy Operator

3.1 An Equation Describing Particles And Waves

$$\widehat{K} \Rightarrow -\frac{\hbar^2}{2m}\frac{\partial^2}{\partial x^2} \tag{3.13}$$

Since the momentum can be obtained from the wave function psi, as well as the kinetic energy K, so too, is the total energy E of the system. Following a similar process to obtain the energy E, we can apply the following operation:

$$\frac{\partial \psi}{\partial t} = -\frac{iE}{\hbar} A e^{\frac{i}{\hbar}(p \cdot x - E \cdot t)} = -\frac{iE}{\hbar}\psi \tag{3.14}$$

E is then obtained by the following operation on the wave function $\psi(x,t)$.

$$\widehat{E}\psi = i\hbar \frac{\partial \psi}{\partial t} \tag{3.15}$$

or

Total Energy Operator

$$\widehat{E} \Rightarrow i\hbar \frac{\partial}{\partial t} \tag{3.16}$$

These operations are interesting because they show how the wave function ψ contains all the information for the system. By applying certain operations, we can access this information.

So far, we have established a way to extract information based on the spatial dependence x (momentum), and time dependence t (total energy). However, we can also construct a total energy equation using only spatial dependence. The total energy E is given by

$$E = K + V \tag{3.17}$$

In Eq. (3.17), V is the potential energy, which when added to the kinetic energy K, gives the total energy E. The combination $T + V$, is known as the Hamiltonian, being independent of time.

From here, we can introduce the wave function ψ by multiplying Eq. (3.17) by ψ, leading to

$$E\psi = K\psi + V\psi \tag{3.18}$$

Substituting our operators from Eqs. (3.13) and (3.16) into Eq. (3.18) leads to the following energy operator equation, operating on the wave function ψ:

1D Hamiltonian Operator Equation

$$i\hbar \frac{\partial \psi}{\partial t} = -\frac{\hbar^2}{2m} \frac{\partial^2 \psi}{\partial x^2} + V\psi \qquad (3.19)$$

Both sides of Eq. (3.19) equate to total energy operators. The LHS is the total energy obtained through a first order operation with respect to time. The RHS provides an operator only involving space. The total energy spatial operator on the RHS is composed of both the kinetic energy and the potential energy. We, therefore, have established a possible equation that operates on our wave function ψ, and in so doing, provides a wealth of information about both the *particle-like* properties, as well as a *wave-like* properties of the system. Moreover, we have an equation that describes the dual nature of matter. It is contains, by design, both the particle and wave characteristics.

The equation above is written for a one dimensional (1D) system, or for the case in which the wave motion is along a straight line, in the *x*-direction. It is straight forward to show that for the full 3D case, where the wave is described by

$$\psi(\mathbf{r}, t) = A e^{\frac{i}{\hbar}(\mathbf{p} \cdot \mathbf{r} - Et)} \qquad (3.20)$$

The vector $\mathbf{r}$ is the 3D position vector, $\mathbf{r} = [x, y, z]^T$. From this, the 3D version of the Hamiltonian operator equation in (3.19) becomes

3D Hamiltonian Operator Equation

$$i\hbar \frac{\partial \psi}{\partial t} = -\frac{\hbar^2}{2m} \nabla^2 \psi + V\psi \qquad (3.21)$$

The first operator on the right-hand side of equation (3.21) is called the *Laplacian* and is defined, in Cartesian coordinates as

$$\nabla^2 \psi = \frac{\partial^2 \psi}{\partial x^2} + \frac{\partial^2 \psi}{\partial y^2} + \frac{\partial^2 \psi}{\partial z^2} \qquad (3.22)$$

3.1 An Equation Describing Particles And Waves

Because ψ contains all the information of the system, we need only operate on our wave function $\psi(\mathbf{r}, t)$ to get whatever information we desire. The following operators also correspond to the 3D case:

3D Momentum Operator

$$\hat{\mathbf{p}}\psi(\mathbf{r}, t) = -i\hbar \nabla \psi \qquad (3.23)$$

For $\hat{p}^2$, we have

3D Momentum-Squared Operator

$$\hat{\mathbf{p}}^2 \psi = -\hbar^2 \nabla^2 \psi \qquad (3.24)$$

An important point that should be highlighted about the wave functions ψ that we have introduced is that *they are not dimensionless functions*. As we continue, this fact will become more apparent. Specifically, for a 1D wave function, in SI units, the wave function ψ has units of $1/\sqrt{m}$, where m denotes units of length in meters. In 2D, the units become $(1/\sqrt{m})^2 = 1/m$, and for 3D, the units become $(1/\sqrt{m})^3 = 1/m\sqrt{m}$. Then $\psi^*\psi$ will have units of $(1/m\sqrt{m})^2 = 1/m^3$.

To illustrate the units of ψ with a specific example, let's consider the units involved for the 1D momentum operator, given by Eq. (3.8). For the LHS of this 1D equation, we have

$$\hat{p}_x \psi \Rightarrow \frac{kg \cdot m}{s} \frac{1}{\sqrt{m}} = \frac{kg \cdot \sqrt{m}}{s}$$

For the RHS, substituting the unit for $\hbar$ as Joule-seconds (J·s) and that for $\partial \psi/\partial x$ to be $\partial/\partial x(m^{(-1/2)}) = m^{(-3/2)}$, we may write

$$-i\hbar \frac{\partial}{\partial x} \Rightarrow J \cdot s \frac{1}{\sqrt{m^3}} = N \cdot m \cdot s \frac{1}{\sqrt{m^3}}$$

$$= \frac{kg \cdot m^2 \cdot s}{s^2} \frac{1}{\sqrt{m^3}}$$

$$= \frac{kg \cdot \sqrt{m}}{s}$$

We can see that the resulting units for both the LHS and RHS, both agree. Most of the time, this unit is buried in what is known as the *normalization constant* or coefficient for the function, which we shall discuss in more detail in Chap. 3. Next, we move on to discuss another special operator relation which is so special, in fact, that it bears the name of the physicist who introduced it.

3.2 THE SCHRÖDINGER EQUATION

In 1926, Austrian physicist Erwin Rudolf Josef Alexander Schrödinger published a series of three papers not only suggesting a wave-equation of the form discussed above, but he also showed that such an equation lead to exact solutions for hydrogen's atomic heat-induced spectra, reproducing the results of Rydberg empirical and Bohr theoretical models. His equation is identical to Eq. (3.21), equating the operators given by the total energy E_λ and the Hamiltonian. This equation is better known as the **Schrödinger equation** and is given by

> **Schrödinger's Equation**
>
> $$i\hbar \frac{\partial \psi}{\partial t} = \widehat{H} \psi \qquad (3.25)$$

$\widehat{H}$ is the Hamiltonian operator, discussed above. Eq. (3.25) is the foundation of *non-relativistic quantum mechanics* (relativistic quantum mechanics is introduced in Chap. 11). In this context, *non-relativistic* means that the velocities involved are sufficiently less than the speed of light in vacuum, which Maxwell showed to be $c \approx 3 \times 10^8 m/s$. As long as this condition is met, the Schrödinger equation is appropriate. The Schrödinger equation is one of the *master equations* in quantum mechanics. It's solutions, characteristics, and qualities turns out to yield a wealth of accurate information about fundamental particles.

Solving the Schrödinger equation often demands useful mathematics tools. We, therefore, will take time to introduce some of them for analyzing and solving the Schrödinger equation, and other equations, as well. The next section begins to develop tools for working with the wave functions described by the Schrödinger equation.

3.3 Introduction to Dirac delta function

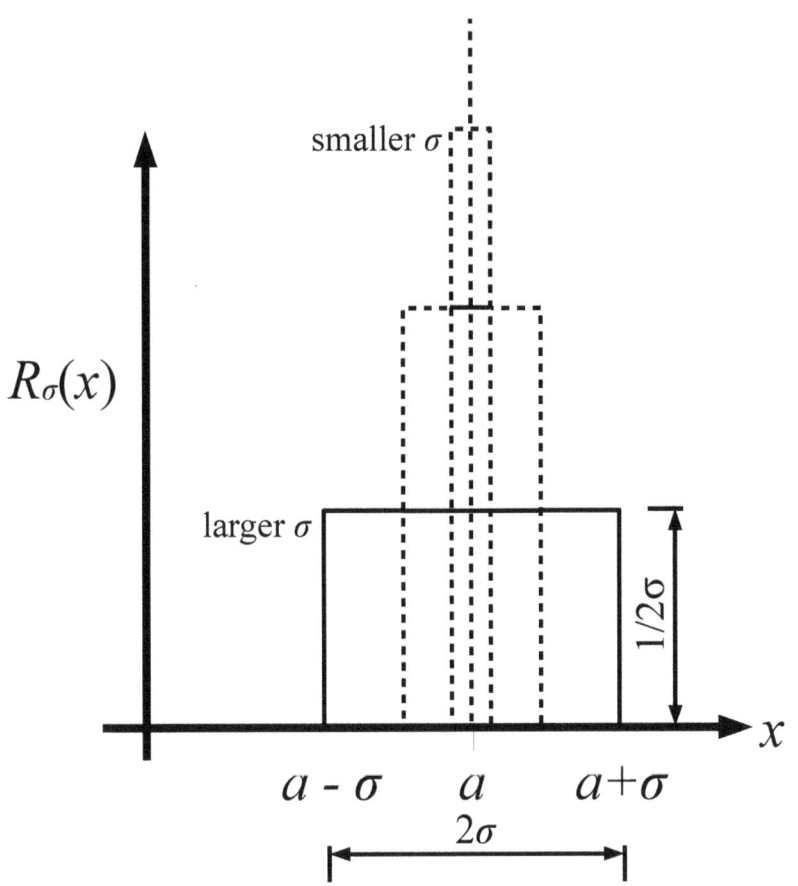

Figure 3.1: Illustration of rectangle function $R_\sigma(x)$ at different widths σ.

It is fortuitous that several developments in mathematics had taken place well before the advent of quantum mechanics. One area in mathematics that was utilized relates to periodic functions, and it provides a foundation for one of the first functions we will consider, known as the **Dirac delta function** (δ-function), named after the British physicist Paul Adrien Maurice Dirac (1902-1984), who first referred to it as a δ-function. The original form of this function was enabled by the work of French polymath Jean-Baptiste Joseph Fourier (1768-1830), with the Fourier integral theorem (which we shall get to). This lead to the first form of the Dirac δ-function. We will develop the connection between the Dirac δ-function and Fourier's work. Let's begin by considering a function known as the **rectangle function** R_σ, defined as

$$R_\sigma(x) = \begin{cases} \frac{1}{2\sigma} & \text{if } x \in [a-\sigma, a+\sigma] \\ 0 & \text{otherwise} \end{cases} \qquad (3.26)$$

The *rectangle function* $R_\sigma(x)$ has a height $h = 1/(2\sigma)$. The width w of the nonzero part of the function is $w = 2\sigma$, illustrated in Fig. 3.1.

The definition of $R_\sigma(x)$ ensures that the area under its curve is always equal to unity because

$$\int_{-\infty}^{\infty} R_\sigma(x)dx = \int_{a-\sigma}^{a+\sigma} \frac{1}{2\sigma} dx = \frac{1}{2\sigma} \int_{a-\sigma}^{a+\sigma} dx = \frac{1}{2\sigma}[a+\sigma-(a-\sigma)] = 1 \qquad (3.27)$$

In taking the limit letting $\sigma \to 0$, the result is the sharp vertical line shown at $x = a$, as shown in Fig. 3.1. By definition, the height $h \to \infty$ as $\sigma \to 0$. Still, for any value of σ, we may define as one requirement for the **Dirac delta function** $\delta(x-a)$, that, like the rectangle function, it possesses the important property

Dirac Delta Function Property 1

$$\int_{-\infty}^{\infty} \delta(x-a)dx = 1 \qquad (3.28)$$

For the integral to be defined, the integrand, which is the Dirac δ-function, must be continuous. The δ-function is zero everywhere except at $x = a$. Let's now consider the integral of the product of R_σ with a well-behaved function $f(x)$, as illustrated in Fig. 3.2. The integral of this product is given by

$$\int_{-\infty}^{\infty} R_\sigma(x-a) \cdot f(x)dx \qquad (3.29)$$

If we, again, consider the asymptotic limit $\sigma \to 0$, the effect will be to reduce and focus the width w towards the value $f(a)$, as illustrated in

3.3 Introduction to Dirac delta function

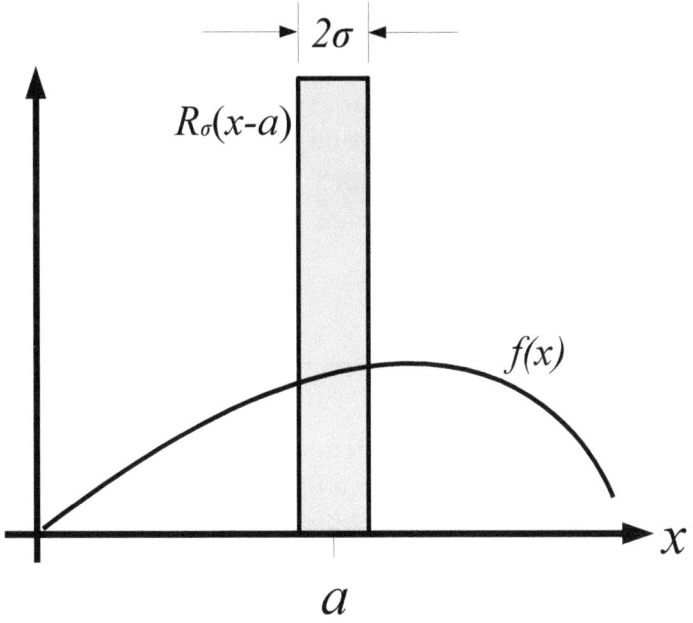

Figure 3.2: Rectangle function and well-behaved function f(x), over the same range. An important property follows from the integral of the product.

Fig. 3.2. Taking the limit of the integral of this product as $\sigma \to 0$, we have

$$\lim_{\sigma \to 0} \int_{-\infty}^{\infty} R_\sigma(x-a) \cdot f(x) dx = \lim_{\sigma \to 0} \int_{a-\sigma}^{a+\sigma} R_\sigma(x-a) \cdot f(x) dx \quad (3.30)$$

$$= f(a) \int_{a-\sigma}^{a+\sigma} R_\sigma(x-a) \cdot dx \quad (3.31)$$

Using the result from Eq. (3.27), we then have

$$\lim_{\sigma \to 0} \int_{-\infty}^{\infty} R_\sigma(x-a) \cdot f(x) dx = f(a) \quad (3.32)$$

Therefore, we have found an important property of the asymptotic limit of $R_\sigma(x)$ given by

$$\lim_{\sigma \to 0} \int_{-\infty}^{\infty} R_\sigma(x-a) \cdot f(x) dx = f(a)$$

Since the first property holds for any σ, based on this property of the asymptotic limit, we may further define the Dirac δ-function $\delta()$ as

$$\delta(x-a) = \lim_{\sigma \to 0} R_\sigma(x-a)$$

With the definition in terms of an asymptotic limit of R_σ, a second important property of the Dirac δ-function becomes

> **Dirac Delta Function Property 2**
>
> $$\int_{-\infty}^{\infty} \delta(x-a) \cdot f(x) dx = f(a) \qquad (3.33)$$

Because the Dirac δ-function is defined through integral properties, and is defined in an asymptotic limit of a function, the function whose asymptotic limit leads to $\delta(x-a)$, as observed with $R_\sigma(x-a)$, is *not unique*. Many other functions satisfy these properties where asymptotic limits lead to $\delta(x-a)$.

Another function that we may consider is the **Ramp** or **Slope function** $S_\sigma(x)$ defined as

$$S_\sigma(x) = \begin{cases} 0 & \text{if } x < a-\sigma \\ \frac{1}{2\sigma} \cdot [x-(a-\sigma)] & \text{if } x \in [a-\sigma, a+\sigma] \\ 1 & \text{if } x > a+\sigma \end{cases} \qquad (3.34)$$

The slope function S_σ is illustrated in Fig. 3.3, for varying values of σ. The derivatives of S_σ in the low and high values of x are zero, since the values are constant. The derivative in the region $x \in [a-\sigma, a+\sigma]$ is $1/2\sigma$. Thus, the derivative of S_σ is R_σ, or

$$\frac{dS_\sigma}{dx} = R_\sigma(x-a) \qquad (3.35)$$

Additionally, if we consider the limiting case of $\sigma \to 0$, the slope function becomes a step function known as the **Heaviside function** $H(x-a)$. So we have

$$\lim_{\sigma \to 0} S_\sigma(x-a) = H(x-a) \qquad (3.36)$$

It follows that

$$\lim_{\sigma \to 0} \frac{dS_\sigma}{dx} = \lim_{\sigma \to 0} R_\sigma(x-a) = \frac{dH}{dx} = \delta(x-a) \qquad (3.37)$$

3.3 Introduction to Dirac delta function

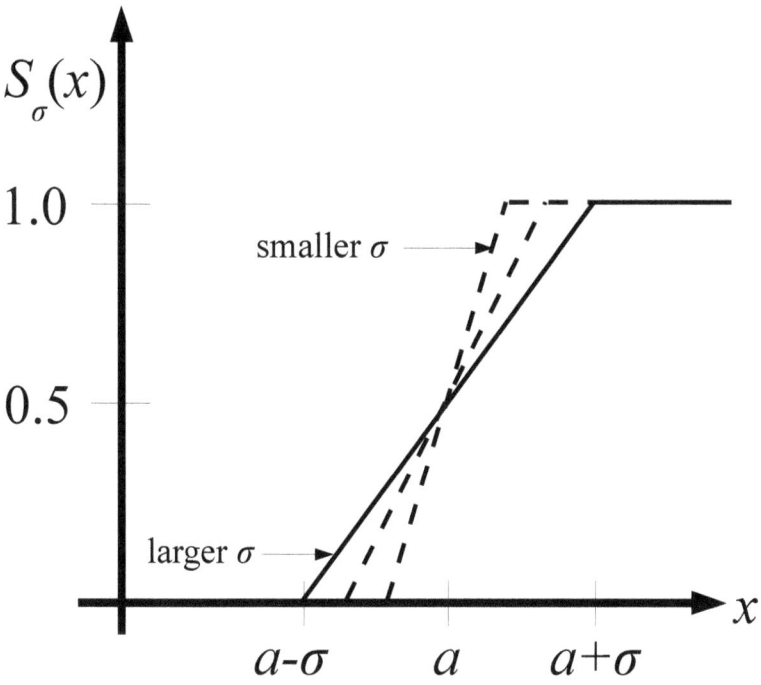

Figure 3.3: Slope function $S_\sigma(x-a)$ at small, medium, and large σ

The δ-function is, then, the derivative of the *Heaviside function*, named after British polymath Oliver Heaviside (1850-1925), and it can also be defined in terms of the asymptotic limit of the slope function S_σ. This asymptotic limit known as the Heaviside function is illustrated in Fig. 3.4.

Another important function that can be used to define $\delta(\cdot)$ is a function called the **normalized Gaussian Distribution** or **Normal Distribution** $G_\sigma(x)$. It turns out that Gaussian functions are common in several of the wave functions described by the Schrödinger equation. Therefore, it is useful to develop some tools that allow us to handle these functions. Before we discuss the properties of the Gaussian distribution, let us work out a useful mathematical result that allows the *Gaussian distribution* to

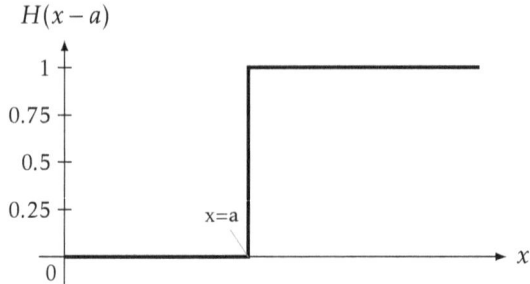

Figure 3.4: The Heaviside function $H(x-a)$. The step or discontinuity occurs at $x = a$.

be integrated in such a way that the result is always unity, obtaining the *normal distribution*. Toward this end, consider the following integral:

$$I = \int_{-\infty}^{\infty} e^{-x^2} dx \qquad (3.38)$$

Since the integrand is *positive-definite*, the integral I is also positive-definite. Then, we can obtain I from I^2, by taking the square root. I^2 is given by

$$I^2 = \left[\int_{-\infty}^{\infty} e^{-x^2} dx\right]^2 \qquad (3.39)$$

I^2 can also be written as

$$I^2 = \int_{-\infty}^{\infty} e^{-x^2} dx \int_{-\infty}^{\infty} e^{-y^2} dy = \int_{-\infty}^{\infty} e^{-(x^2+y^2)} dxdy \qquad (3.40)$$

The next step is to transform the integral from Cartesian coordinates to polar coordinates (r, θ). x and y are related to the polar coordinates as

$$x = r\cos\theta$$
$$y = r\sin\theta \qquad (3.41)$$
$$r^2 = x^2 + y^2$$

Substituting polar coordinates using (r, θ) leads to

$$I^2 = \int_{-\infty}^{\infty} e^{-(x^2+y^2)} dxdy = \int_{-\infty}^{\infty} e^{-r^2} dxdy$$

The next step is subtle and takes advantage of the fact that we are integrating from $-\infty$ to $+\infty$ in the 2D space. This restricts our result

3.3 Introduction to Dirac delta function

to the case of integration over the infinite plane. In this case, instead of direct substitution of dx and dy, where $da = dxdy$ represents the 2D differential elemental area, multiplying by the function $\exp(-r^2)$, where we are summing up these differentials over the entire 2D plane, any suitable differential in the (r, θ) plane will work. This allows us to represent the integral such that $da = dsdr = rd\theta dr$, thus

$$I^2 = \int_{-\infty}^{\infty} e^{-r^2} dr ds = \int_{-\infty}^{\infty} e^{-r^2} r dr d\theta \qquad (3.42)$$

Note that this is not the same as direct substitution and is only justified when we consider integrating over the infinite plane.

$$I^2 = \int_{-\infty}^{\infty} e^{-r^2} dr ds = \int_{0}^{2\pi} d\theta \int_{0}^{\infty} e^{-r^2} r dr \qquad (3.43)$$

Letting $u = -r^2$, then $du = -2rdr$, and the integral in (3.43) becomes

$$I^2 = \int_{0}^{2\pi} d\theta \int_{0}^{\infty} e^{-r^2} r dr = -2\pi \frac{\int_{0}^{-\infty} e^u du}{2} = -\pi(e^{-\infty} - e^0) = \pi \quad (3.44)$$

From the result of I^2, we have that

$$I = \sqrt{\pi} \qquad (3.45)$$

However, if $u = -ar^2$, then $du = -2ardr$, and the integral in (3.43) becomes

$$I^2 = \int_{0}^{2\pi} d\theta \int_{0}^{\infty} e^{-ar^2} r dr = -2\pi \frac{\int_{0}^{-\infty} e^u du}{2a} = -\pi(e^{-\infty} - e^0) = \frac{\pi}{a} \quad (3.46)$$

Then,

$$I = \sqrt{\frac{\pi}{a}} \qquad (3.47)$$

Then, if $G_\sigma = e^{-\frac{x^2}{2\sigma^2}}$, then $a = 1/(2\sigma^2)$, and the *normal distribution* centered at a, $G_\sigma(x-a)$ is then given by

---

Normal Distribution

$$G_\sigma(x-a) = \frac{1}{\sigma\sqrt{2\pi}} e^{-\frac{(x-a)^2}{2\sigma^2}} \qquad (3.48)$$

$$\int_{-\infty}^{+\infty} G_\sigma(x-a)dx = \int_{-\infty}^{+\infty} G_\sigma(x)dx = 1 \qquad (3.49)$$

Fig. 3.5 illustrates the normalized Gaussian distribution centered at $x = a$. Because $\int G_\sigma dx = 1$, increasing σ, for example, widens the distribution, leading to a reduction in the maximum value, since the area remains unchanged. This effect from σ is illustrated in Fig. 3.6. In each of the examples of G_σ shown, only the value of σ has changed. For all values of σ, the area under G_σ remains unity. Because of this condition for the *normal distribution*, as σ is reduced and the width decreases, the height h of the distribution simultaneously increases. In fact, as the value of σ approaches and even goes below unity, the values of G_σ exceeds unity $(G_\sigma(x) > 1)$ in the vicinity of $x = a$, but $G_\sigma \cdot dx$ remains sufficiently small such that $\int G_\sigma dx = 1$. In the limiting case for $G_\sigma(x-a)$, as $\sigma \to 0$, as we had with the rectangle function R_σ, we again have a distribution of zero width, which leads to

$$\lim_{\sigma \to 0} \int G_\sigma(x-a) f(x) dx = \lim_{\sigma \to 0} \int G_\sigma(x-a) f(a) dx$$

$$= f(a) \times \lim_{\sigma \to 0} \int G_\sigma(x-a) dx = f(a)$$

Therefore, the δ-function can be defined by

$$\lim_{\sigma \to 0} G_\sigma(x-a) = \delta(x-a) \qquad (3.50)$$

Therefore, we have established that the δ-function can be defined using a variety of functions, such as the rectangle function R_σ, the slope function S_σ, as well as the normal Gaussian distribution G_σ because they all can satisfy the defining integral properties of the δ-function. In the next section, this concept will be further applied to connect the δ-function with the Fourier transforms.

3.3.1 Extension Of Integral With Gaussian Functions

Another useful integral can be obtained using the result in (3.47). Let us extend the integral to be

$$I' = \int_{-\infty}^{\infty} e^{(-ax^2+bx)} dx \qquad (3.51)$$

3.3 Introduction to Dirac delta function

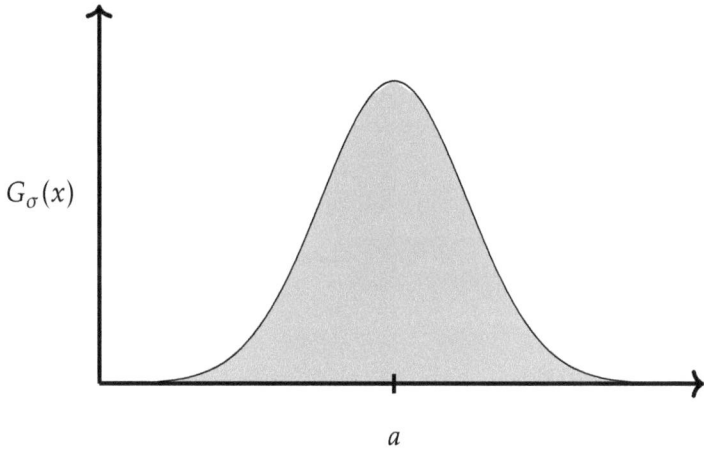

Figure 3.5: Normal Gaussian distribution $G_\sigma(x-a)$, centered at a.

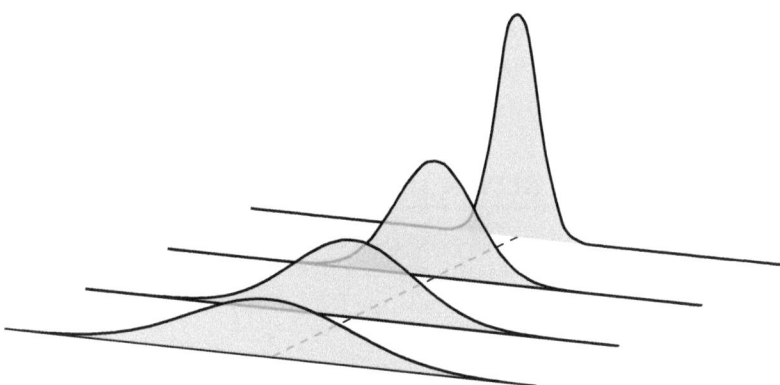

Figure 3.6: $G_\sigma(x)$ for different values of σ. Because the area (shaded region under each curve) is unity for all values of σ, reducing σ tends to increase the peak value (trend from front curve towards rear-most curve), with respect to $x = a$ (the dashed line).

For the integral in (3.51), we can rewrite the integrand by adding and subtracting to the exponent term. That allows us to write I' in the following form:

$$I' = \int_{-\infty}^{\infty} e^{(-ax^2+bx)} dx = \int_{-\infty}^{\infty} e^{-a(x^2-(b/a)x+(b^2/4a^2)-(b^2/4a^2))} dx \quad (3.52)$$

We can write this as

$$I' = e^{(b^2/4a)} \int_{-\infty}^{\infty} e^{-a(x-b/2a)^2} dx \quad (3.53)$$

Substituting $\beta = \sqrt{a}\xi \rightarrow \beta^2 = a\xi^2$, which also means that $d\xi = (1/\sqrt{a})d\beta$, Eq. (3.53) becomes

$$I' = e^{-(b^2/4a^2)} \int_{-\infty}^{\infty} e^{-a\xi^2} d\xi = \frac{e^{-(b^2/4a^2)}}{\sqrt{a}} \int_{-\infty}^{\infty} e^{-\beta^2} d\beta = \sqrt{\frac{\pi}{a}} \times e^{b^2/4a^2} \quad (3.54)$$

Thus, the extended integral is given by

Generalized Gaussian Integral

$$\int_{-\infty}^{\infty} e^{(-ax^2+bx)} dx = \sqrt{\frac{\pi}{a}} \cdot e^{b^2/4a} \quad (3.55)$$

Because of its non-strict definition, the Dirac delta function turns out to be defined by many functions. In the next section, we introduce yet another function fit for the definition of the Dirac delta function, and is very compatible with wave functions, or periodic functions.

3.4 Dirac Delta Function and Fourier Transforms

To introduce the Fourier transform, let's first consider a function known as the **Sine Cardinal** or **Sinc** function. It is defined as

$$\text{Sinc} = \frac{\sin(x)}{x} \quad (3.56)$$

It is well-known that $\lim_{x \to 0} \text{Sinc} = 1$, because if we express the Maclaurin series for $\sin(x)$, we obtain

$$\sin(x) = x + \frac{1}{3!}x^3 + \frac{1}{5!}x^5 + \cdots \Rightarrow \frac{\sin(x)}{x} = 1 + \frac{1}{3!}x^2 + \frac{1}{5!}x^4 + \cdots$$

Then, as $x \to 0$, Sinc $\to 1$. Thus, Sinc is bounded at $x = 0$, and since $|\sin x| \le 1$ for all x, as $x \to \infty$, Sinc $\to 0$. These qualities remain true even if we scale the sine argument x by a real number g, and the denominator x by a real number b, where we have

$$S_g(x) = \frac{\sin(gx)}{bx} \quad (3.57)$$

Examples of the function $S_g(x - a)$, with $b = \pi$ are illustrated in Fig. 3.7, for $g = 1$ (Sinc), $g = 2$, and $g = 3$. Fig. 3.7 illustrates how changing

3.4 Dirac Delta Function and Fourier Transforms

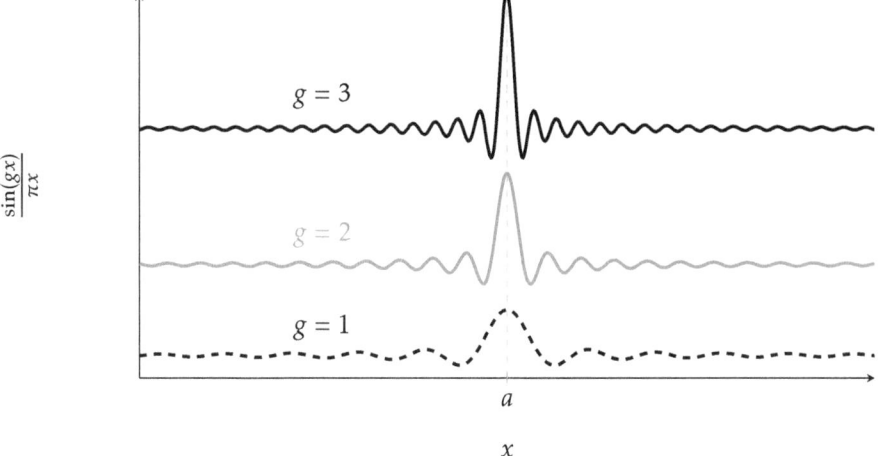

Figure 3.7: Examples of the shifted Sinc function sin$g(x-a)/\pi(x-a)$ for different values of g.

g causes distinct changes in S_g, where increasing g leads to higher oscillation rates, which also causes more narrow sharply peaked center regions. The choice of $b = \pi$, here, was not a coincidence. It is chosen based on an evaluation of the integral of the function given by

Dirichlet Integral

$$\int_{-\infty}^{+\infty} \frac{\sin(gx)}{x} dx = \pi \quad (3.58)$$

This integral is known as the **Dirichlet Integral**. We shall obtain the result of this integral by first getting the result for Sinc and then obtaining our final result for S_g by scaling. For the first step, we use the following integral result for e^{-ax}:

$$\int_0^{+\infty} e^{-ax} dx = -\frac{1}{\alpha} e^{-ax} \Big|_0^{+\infty} = -\frac{1}{\alpha}[0-1] = \frac{1}{\alpha} \quad (3.59)$$

Thus, by letting $1/x$ be represented by Eq. (3.59), we may rewrite the Dirichlet integral as a double integral given by

$$\int_{-\infty}^{+\infty} \frac{\sin x}{x} dx = \int_0^{+\infty} e^{-xy} dy \int_{-\infty}^{+\infty} \sin x\, dx = \int_{-\infty}^{+\infty} \int_0^{+\infty} e^{-xy} \sin x\, dx\, dy$$

(3.60)

Let's determine the inner-most integral I_1 given by

$$I_1 = \int_0^{+\infty} e^{-xy} \sin x \, dx \tag{3.61}$$

Rewriting $\sin x$ using *Euler's formulas* $e^{i\theta} = \cos\theta + i\sin\theta$ and $e^{-i\theta} = \cos\theta - i\sin\theta$, we have

$$I_1 = \int_0^{+\infty} e^{-xy} \left[\frac{e^{ix} - e^{-ix}}{2i} \right] dx = \frac{1}{2i} \int_0^{+\infty} e^{-x(y-i)} - e^{-x(y+i)} dx \tag{3.62}$$

Using Eq. (3.59), we can evaluate this readily since our integral is only a function of x, and, therefore, the coefficients depending on y and i are constants. Then we have

$$I_1 = \frac{1}{2i} \left[\frac{-1}{y-i} e^{-x(y-i)} \Big|_0^\infty - \frac{-1}{y+i} e^{-x(y+i)} \Big|_0^\infty \right] = \frac{1}{2i} \left[\frac{1}{y-i} - \frac{1}{y+i} \right] \tag{3.63}$$

Multiplying by $y+i/y+i$ and $y-i/y-i$, respectively, we obtain

$$I_1 = \frac{1}{2i} \left[\frac{y+i}{y^2+1} - \frac{y-i}{y^2+1} \right] = \frac{1}{y^2+1} \tag{3.64}$$

With this result for I_1, Eq. (3.60) becomes

$$\int_{-\infty}^{+\infty} \frac{\sin x}{x} dx = \int_{-\infty}^{+\infty} \frac{1}{y^2+1} dy \tag{3.65}$$

Let $y = \tan\theta$, leading to $dy/d\theta = \sec^2\theta$. With this substitution, we must recognize that although y goes from $-\infty$ to $+\infty$, because we have chosen $y = \tan\theta$, it follows that θ only goes from $-\pi/2$ to $\pi/2$ (or we could have chosen any interval where $\tan\theta$ is defined). Using this substitution, our integral becomes

$$\int_{-\infty}^{+\infty} \frac{\sin(x)}{x} dx = \int_{-\pi/2}^{+\pi/2} \frac{\sec^2\theta}{\tan^2\theta + 1} d\theta \tag{3.66}$$

Using the identity, $1 + \tan^2\theta = \sec^2\theta$, we have our final result

$$\int_{-\infty}^{+\infty} \frac{\sin(x)}{x} dx = \int_{-\pi/2}^{+\pi/2} \frac{\tan^2\theta + 1}{\tan^2\theta + 1} d\theta = \int_{-\pi/2}^{+\pi/2} d\theta = \pi \tag{3.67}$$

3.4 Dirac Delta Function and Fourier Transforms

Now that we have the integral of the Sinc function, we will show that scaling x also leads to the same result. Let $x = gy$ and $dx = gdy$. Substituting this into the Sinc integral leads to

$$\int_{-\infty}^{+\infty} \frac{\sin(x)}{x} dx = \int_{-\infty}^{+\infty} \frac{\sin(gy)}{gy} gdy = \int_{-\infty}^{+\infty} \frac{\sin(gy)}{y} dy = \pi$$

So, we have that the Dirichlet integral is a constant π for any value of g, hence our choice for $b = \pi$ in Fig. 3.7, because dividing the integral by π gives another integral whose area is equal to unity, or

$$\int_{-\infty}^{+\infty} \frac{\sin(gx)}{\pi x} dx = \int S_g(x) dx = 1 \qquad (3.68)$$

As shown in Fig. 3.7, as g increases, the function becomes more narrow, with rising sharpness at the center. A measure of the width of $\sin(gx)/\pi x$ is given by the distance between the first $\pm$ roots, given by the condition

$$\frac{\sin(gx)}{\pi x} = 0 \Rightarrow gx = \pm\pi$$

From this, the width Δx becomes

$$\Delta x = \frac{2\pi}{g} \qquad (3.69)$$

Therefore, in the limit of $g \to \infty$, the width is zero, and we have

$$\lim_{g \to \infty} \frac{\sin(gx)}{\pi x} = \delta(x) \qquad (3.70)$$

This particular function, involving a harmonic component, has some interesting properties which can be exploited. Consider the following integral I_g given by

$$I_g = \frac{1}{2\pi} \int_{-g}^{+g} e^{ikx} dk \qquad (3.71)$$

From what we have already seen, this integral can be evaluated to obtain

$$\frac{1}{2\pi} \int_{-g}^{+g} e^{ikx} dk = \frac{1}{2\pi} \left[\frac{e^{igx} - e^{-igx}}{ix} \right] = \frac{1}{\pi x} \left[\frac{e^{igx} - e^{-igx}}{2i} \right] = \frac{\sin(gx)}{\pi x} \qquad (3.72)$$

We have used Euler's formula again. Considering the integral properties of S_g obtained above, it follows that

$$\lim_{g \to \infty} \frac{\sin(gx)}{\pi x} = \frac{1}{2\pi} \lim_{g \to \infty} \int_{-g}^{+g} e^{ikx} dk = \frac{1}{2\pi} \int_{-\infty}^{+\infty} e^{ikx} dk = \delta(x) \quad (3.73)$$

We can also write

$$\frac{1}{2\pi}\int_{-g}^{+g} e^{-ikx}dk = \frac{1}{2\pi}\left[\frac{e^{-igx}-e^{igx}}{-ix}\right] = \frac{1}{\pi x}\left[\frac{e^{igx}-e^{-igx}}{2i}\right] = \frac{\sin(gx)}{\pi x} \quad (3.74)$$

Therefore, we have obtained another way to express the δ-function as

$$\delta(x) = \frac{1}{2\pi}\int_{-\infty}^{+\infty} e^{(\pm)ikx}dk \quad (3.75)$$

Moving forward, it will be useful to denote the argument x as follows

---

Integral Representation of Dirac Delta Function

$$\delta(x-x') = \frac{1}{2\pi}\int_{-\infty}^{+\infty} e^{(\pm)ik(x-x')}dk \quad (3.76)$$

Eq. (3.76) is known as the **Integral Representation of the Dirac Delta function**. Recall the *Sifting property* of $\delta(x)$ where

$$f(x) = \int_{-\infty}^{+\infty} \delta(x-x')f(x')dx' \quad (3.77)$$

Using the integral representation of the Dirac delta function, we can also write

$$f(x) = \frac{1}{2\pi}\int_{-\infty}^{+\infty}\int_{-\infty}^{\infty} e^{ik(x-x')}f(x')dkdx' \quad (3.78)$$

Let us group the x' terms together and the x term as follows

$$f(x) = \frac{1}{2\pi}\int_{-\infty}^{+\infty}\int_{-\infty}^{\infty} e^{-ikx'}f(x')dx'e^{ikx}dk \quad (3.79)$$

With this form of the function $f(x)$, we may define a new function $F(k)$ given by

---

Fourier Transform $(x-k)$

$$F(k) = \frac{1}{\sqrt{2\pi}}\int_{-\infty}^{+\infty} f(x')e^{-ikx'}dx' \quad (3.80)$$

3.4 Dirac Delta Function and Fourier Transforms

Eq. (3.80) is known as the **Fourier transform** of the function $f(x')$. Given the Fourier transform $F(k)$, Eq. (3.79) becomes

Inverse Fourier Transform $(x-k)$

$$f(x) = \frac{1}{\sqrt{2\pi}} \int_{-\infty}^{+\infty} F(k) e^{ikx} dk \qquad (3.81)$$

Eq. (3.81) is the **Inverse Fourier Transform**, and it allows us to obtain the function $f(x)$ from having its Fourier transform $F(k)$. Both of these relations together constitute what is known as the **Fourier Integral Theorem**. Next, let us do an example illustrating an application of the Fourier transform.

Exercise 3.1 Compute the Fourier transform $F(k)$ for the function $f(x) = G_\sigma = \frac{1}{\sigma\sqrt{2\pi}} e^{-\frac{x^2}{2\sigma^2}}$, using Eq. (3.80).

Solution

Use Eq. (3.80) along with the integral from Eq. (3.55), the Fourier transform is

$$F(k) = \frac{1}{\sqrt{2\pi}} \int_{-\infty}^{+\infty} f(x) e^{-ikx} dx$$

Substituting of $f(x)$, we have

$$F(k) = \frac{1}{\sqrt{2\pi}} \int_{-\infty}^{+\infty} \frac{1}{\sigma\sqrt{2\pi}} e^{-\frac{x^2}{2\sigma^2}} e^{-ikx} dx = \frac{1}{2\sigma\pi} \int_{-\infty}^{+\infty} e^{-\frac{x^2}{2\sigma^2} - ikx} dx$$

This integral has the form of Eq. (3.55) where

$$I' = \int_{-\infty}^{\infty} e^{(-ax^2 + bx)} dx = \sqrt{\frac{\pi}{a}} e^{b^2/4a}$$

Comparing the above integral to the integral of the Fourier transform, we have that $a = 1/2\sigma^2$ and $b = -ik$, and plugging this in leads to

$$F(k) = \frac{1}{2\sigma\pi}\sqrt{\frac{\pi}{a}}e^{b^2/4a} = \frac{1}{2\sigma\pi}\sqrt{2\sigma^2\pi}e^{-\sigma^2k^2/2} = \frac{1}{\sqrt{2\pi}}e^{-\sigma^2k^2/2}$$

Thus, we find that the Fourier transform of a normal Gaussian in x-space is also a Gaussian, but in k-space.

Also, note that the width or spread Δx in the Gaussian is of the order of σ. And, with the Fourier transform, we find that the spread Δk is on the order of $1/\sigma$, which means that $\Delta x \cdot \Delta k \approx 1$.

■

3.5 From Fourier Transform To Parseval's Theorem

Now that we have introduced the Fourier transform $F(k)$, we can look at some of the consequences of this transform. To do this, let us first note that the complex conjugate $f^*(x)$ of the function $f(x)$, in terms of its Fourier transform, is given by

$$f^*(x) = \frac{1}{\sqrt{2\pi}} \int F^*(k)e^{-ik'x}dk \qquad (3.82)$$

Let us compute the integral of the square of the amplitude of $f(x)$, i.e. $|f|^2 = ff^*$. Using the definition of the Inverse Fourier transform, $\int |f|^2 dx$ is given by

$$\int |f|^2 dx = \frac{1}{\sqrt{2\pi}}\frac{1}{\sqrt{2\pi}} \int \int F(k)e^{ikx}dk \int F^*(k')e^{-ik'x}dk'dx \qquad (3.83)$$

All integrals are from $-\infty$ to $+\infty$. Because we are combining two independent integrals for $F(k)$ and $F^*(k')$, respectively, different dummy variables must be used, hence k and k'. Regrouping our integral terms, we have

$$\int |f|^2 dx = \frac{1}{2\pi} \int \int \int F(k)F^*(k')e^{i(k-k')x}dk'dkdx \qquad (3.84)$$

Recall the *Integral representation of the Dirac Delta function* given by Eq. (3.76). The Dirac Delta function is contained in our result above, with the differential dx. This means that we can write our integral to be

$$\int |f|^2 dx = \int \int F(k)F^*(k')\delta(k-k')dk'dk \qquad (3.85)$$

3.5 From Fourier Transform To Parseval's Theorem

We have reduced three integrals down to two. Now, we'll reduce two down to one by using the property of the Dirac Delta function given by Eq. (3.77). With this, we have

$$\int |f|^2(x)dx = \int F(k)F^*(k)dk \tag{3.86}$$

This result is known as **Parseval's Theorem** and states that

Parseval's Theorem

$$\int |f(x)|^2 dx = \int |F(k)|^2 dk \tag{3.87}$$

Therefore, if $f(x)$ is normalized so $\int |f|^2 dx = 1$, the $F(k)$ is also normalized. Up to now, we have only used the notation x and k as variables associated with the Fourier transform pair. Since $k \cdot x$ is in radians, then if x is in spatial units such as m (for meters), then k has units of m^{-1} (or 2πm^{-1}). However, the mathematics doesn't care what we call these variables. If we decided to call the argument t for time (instead of x), with units of time in s (for seconds), we have $f(t)$ and ω with units of 2πs^{-1}, instead of k. Since time-dependent functions are also common, this generality is useful. With time-dependent functions, the *Fourier transform* of the function $f(t)$ is given by

Fourier Transform $(t - \omega)$

$$F(\omega) = \frac{1}{\sqrt{2\pi}} \int_{-\infty}^{+\infty} f(t)e^{-i\omega t}dt \tag{3.88}$$

The corresponding *Inverse Fourier transform* becomes

Inverse Fourier Transform $(t - \omega)$

$$f(t) = \frac{1}{\sqrt{2\pi}} \int_{-\infty}^{+\infty} F(\omega)e^{i\omega t}d\omega \tag{3.89}$$

Chapter 3. A Masterful Equation With Duality

To illustrate how this is applied in the time domain, let us consider another example.

Exercise 3.2 Using Eq. (3.88), compute the Fourier transform $F(\omega)$, for the function illustrated, given by

$$f(t) = e^{-\frac{t^2}{2\tau^2}} e^{i\omega_0 t}$$

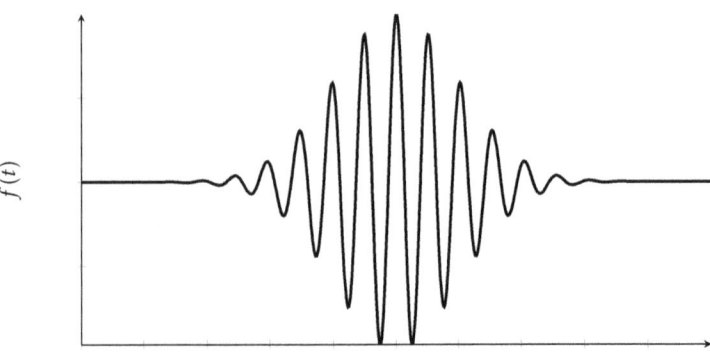

Solution

Using Eq. (3.88) and substituting $f(t)$, we have

$$F(\omega) = \frac{1}{\sqrt{2\pi}} \int_{-\infty}^{+\infty} e^{-\frac{t^2}{2\tau^2}} e^{i\omega_0 t} e^{-i\omega t} dt$$

Combining the exponents of the exponential leads to

$$F(\omega) = \frac{1}{\sqrt{2\pi}} \int_{-\infty}^{+\infty} e^{-\frac{t^2}{2\tau^2} - i(\omega - \omega_0)t} dt$$

Again, our Fourier transform has the form of the integral in Eq. (3.55), only now, the variable x is called t, so we have

$$I' = \int_{-\infty}^{\infty} e^{(-at^2 + bt)} dt = \sqrt{\frac{\pi}{a}} e^{b^2/4a}$$

Comparing the above integral to the integral of our Fourier transform, we have that $a = 1/2\tau^2$ and $b = -i(\omega - \omega_0)$. Plugging this into our solution leads to

$$F(\omega) = \frac{1}{\sqrt{2\pi}}\sqrt{\frac{\pi}{a}}e^{b^2/4a} = \tau e^{-\frac{(\omega-\omega_0)^2\tau^2}{2}}$$

Thus, we again find that the Fourier transform of a Gaussian in $t-space$ is also a Gaussian, but in $\omega-space$. $F(\omega)$ is illustrated below. Looking at the spread Δt which is given by τ, we find that the spread $\Delta\omega$ is on the order of $1/\tau$, which means that $\Delta t \cdot \Delta\omega \approx 1$, as we saw before, in the $x-k$ space. If we also use Einstein's relation for the energy, $E = \hbar\omega$, then $\Delta\omega = \Delta E/\hbar$. Substitution leads to $\Delta t \cdot \Delta E \approx \hbar$. This is also not a coincidence, and we will be coming back to this point. ∎

Let's turn our attention back to the notion of operators in quantum mechanics. Then, we'll apply these lessons to our first problem in the domain of quantum mechanics, namely, the free particle problem.

3.6 GENERALIZATION OF OPERATORS AND OPERATIONS

The idea of an *operator* plays a central role in quantum mechanics. We have introduced examples such as the energy operator known as the Hamiltonian, and the momentum operator **p**. In Sec. 3.2, these operators operated on the wave function ψ. A more general name for these special functions is **eigenstate vectors** (they can be scalar or vectors) or **eigen-functions**, and when operated on they convey state information about the system. Thus, operating on these eigenstates leads to parallel state vectors. This relation is known to as an **eigenstate problem**, and thus, these operators are also known as *eigenoperators*. Solving the eigenstate problem leads to both the *eigenstate* solutions as well as their *eigenvalues*. In quantum mechanics, when the operator has real eigenvalues, these eigenvalues always correspond to some quantity that exists in the physical world. It is, therefore, called a **measurable** or **observable**. For example, with the Hamiltonian, the *measurable* corresponds to the available states of energy E_λ. Likewise, with the momentum operator $\mathbf{p} = -i\hbar\nabla\cdot$, one obtains the available states of momentum, etc. The simplest *operator* is that of multiplication. If we have an eigenstate ψ, then the multiplication operator is defined as

$$y\psi \Rightarrow y \times \psi \tag{3.90}$$

In this case, the observable is the location along an axis y. Posed as an eigenoperator equation, it becomes

$$y\psi = \lambda\psi$$

In this simple case, since y is a measurable quantity, then

$$\lambda = y$$

y is the *observable*. This goes for any of the real coordinates, where we have

$$x\psi \Rightarrow x \times \psi \quad \text{and} \quad \lambda_x = x$$

$$z\psi \Rightarrow z \times \psi \quad \text{and} \quad \lambda_z = z$$

The concept of the real operator is that the operator $\widehat{O}$ operates on *something*, and the result of this operation gives a measurable quantity times the same *something*. This concept is expressed as an eigenoperator equation, given by

Eigenoperator/Eigenvalue Equation

$$\widehat{O}\psi = \lambda\psi \tag{3.91}$$

The operator $\widehat{O}$ may operate on a scalar or vector state. Examples of this that we have seen involving the Hamiltonian and momentum operators $\widehat{O} = \widehat{H}$, $\widehat{O} = \widehat{p}^2 = -\hbar^2 \nabla^2$, so we have

$$\widehat{O}\psi = \lambda\psi \Rightarrow \widehat{H}\psi = E_\lambda \psi$$

$$\widehat{O}\psi = \lambda\psi \Rightarrow \widehat{\mathbf{p}} = -i\hbar\nabla\psi = \mathbf{p}\psi$$

$$\widehat{O}\psi = \lambda\psi \Rightarrow \widehat{\mathbf{p}}^2 = -\hbar^2\nabla^2\psi = p^2\psi$$

The relations above are examples of **eigenoperator equations**. In quantum mechanics, it is taken as *axiom*, or without proof, that Next, these ideas are applied to the problem of *the free particle*. However, now, we understand this is not simply a corpuscular body, but has properties of a wave associated with it. Let's see what the Schrödinger equation tells us about such a novel particle.

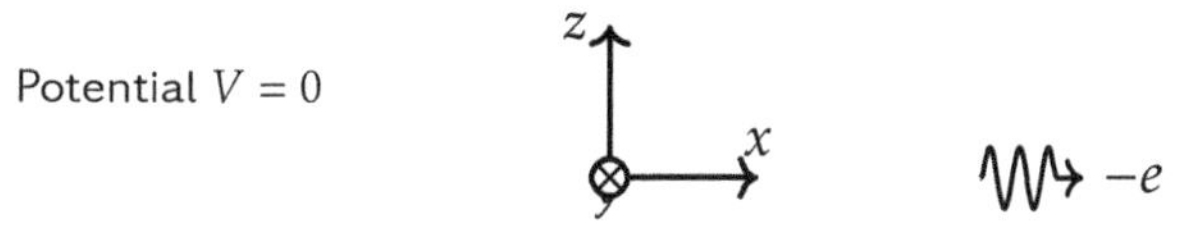

3.7 The Free Particle Problem

Figure 3.8: The free particle problem, which moves in zero potential, or $V = 0$.

3.7 THE FREE PARTICLE PROBLEM

Consider a quantum particle, such as an electron, traveling along the x-direction, in a space with a potential energy $V = 0$, as illustrated in Fig. 3.8. Confining the motion of the quantum particle to the x-direction, the problem is one-dimensional. The 1D time-dependent Schrödinger equation for the case of $V = 0$ is given by

$$i\hbar \frac{\partial \Psi}{\partial t} = -\frac{\hbar^2}{2m} \frac{\partial^2 \Psi}{\partial x^2} \qquad (3.92)$$

The form of Eq. (3.92) has an x-dependence on the RHS, and a time dependence on the LHS. This is a well-known form in differential equations because it means that the equation may be expressed in the form

$$f(t) \frac{\partial F}{\partial t} = g(x) \frac{\partial G}{\partial x} \qquad (3.93)$$

The wave function $\psi = \psi(x, t)$ depends on both variables. We will show that an equation of the form in Eq. (3.93) follows from the condition that $\psi(x, t) = F(t) \cdot G(x)$. Taking this form of ψ, we have that

$$\frac{\partial \psi}{\partial t} = \frac{\partial F}{\partial t} G(x)$$

$$\frac{\partial \psi}{\partial x} = F(t) \frac{\partial G}{\partial x}$$

Therefore, we have

$$i\hbar \frac{\partial \psi}{\partial t} = -\frac{\hbar^2}{2m} \frac{\partial^2 \Psi}{\partial x^2} \Rightarrow i\hbar G(x) \frac{\partial F}{\partial t} = -\frac{\hbar^2}{2m} F(t) \frac{\partial G}{\partial x}$$

This can be re-arranged to give

$$\frac{i\hbar}{F(t)} \frac{\partial F}{\partial t} = -\frac{\hbar^2}{2m} \frac{1}{G(x)} \frac{\partial G}{\partial x} \Rightarrow f(t) \frac{\partial F}{\partial t} = g(x) \frac{\partial G}{\partial x}$$

This process is known as **separation of variables**, because the wave function can be expressed as the product of two independent functions in x and t, respectively. As we will see, this also leads to two independent equations in x and t, respectively. We will see this idea often when solving the Schrödinger equation analytically. The *separation of variables* allows us to use a trial function of the form

$$\Psi = X_f(x) T_f(t) \tag{3.94}$$

Substitution into the Schrödinger equation and dividing by $X_f T_f$ leads to

$$i\hbar \frac{1}{T_f} \frac{dT_f}{dt} = -\frac{\hbar^2}{2m} \frac{1}{X_f} \frac{d^2 X_f}{dx^2} \tag{3.95}$$

The LHS of Eq. (3.95) is a function in x, while the RHS is a function in t. Since both variables are independent variables, this equality can only be true if both sides are equal to a constant. That constant is the eigenvalue $\lambda = E_\lambda$ since both sides equate to the Hamiltonian operator, whose eigenvalue is the energy E_λ. So we have

$$i\hbar \frac{1}{T_f} \frac{dT_f}{dt} = E_\lambda \tag{3.96a}$$

$$-\frac{\hbar^2}{2m} \frac{1}{X_f} \frac{d^2 X_f}{dx^2} = E_\lambda \tag{3.96b}$$

Eq. (3.96a) can be rearranged as

$$\frac{dT_f}{dt} + \frac{iE_\lambda}{\hbar} T_f$$

The solution of this equation is given by

$$T_f(t) = T_0 e^{-i \frac{E_\lambda}{\hbar}(t - t_0)}$$

For Eq. (3.96b), we can write

$$\frac{d^2 X_f}{dx^2} + k^2 X_f = 0 \tag{3.97}$$

We have introduced the parameter k^2, defined by

$$k^2 = \frac{2mE_\lambda}{\hbar^2} \tag{3.98}$$

3.7 The Free Particle Problem

The general solution to Eq. (3.98) is given by

$$X_f(x) = Ae^{-ikx} + Be^{ikx} = Ae^{-\frac{i}{\hbar}\sqrt{2mE_\lambda}x} + Be^{\frac{i}{\hbar}\sqrt{2mE_\lambda}x} \tag{3.99}$$

Therefore, we have $\Psi = T_f X_f$. Let us apply the momentum operator $\widehat{p}_x$ to $\Psi(x, t = 0) = X_f$ to determine the momentum of the system. Recall the momentum operator is given by

$$\widehat{p}_x \rightarrow -i\hbar \frac{\partial}{\partial x}$$

Applying this to the first independent solution Ae^{-ikx}, we find that

$$-i\hbar \frac{\partial X_f}{\partial x} = -i\hbar(-\frac{i}{\hbar}\sqrt{2mE_\lambda})Ae^{-ikx} = -i\hbar(-\frac{i}{\hbar}\sqrt{\frac{2mp^2}{2m}})Ae^{-ikx} = -|p|Ae^{-ikx}$$

We have use the relations

$$p = -\sqrt{2mE_\lambda} < 0 \rightarrow E_\lambda = \frac{p^2}{2m} \tag{3.100}$$

For a free particle, E_λ is the kinetic energy K of the free particle, however, since $p < 0$, this part of the solution corresponds to the particle moving to the 'left' along x, in a coordinate system with $+x$ pointing towards the right. Then, it is straightforward to show that the other solution leads to

$$p = +\sqrt{2mE_\lambda} > 0 \tag{3.101}$$

Based on these relations, we can also write X_f as a function of the real momentum p, where we have

$$X_f = Ce^{\frac{i}{\hbar}px} \tag{3.102}$$

If E_λ is allowed to take negative values, then $p = \sqrt{2mE_\lambda} = i\sqrt{2m|E_\lambda|}$, for which e^{-ipx} diverges. Thus, for a well-behaved solution, the kinetic energy values must only take positive values. That is

$$E_\lambda = \frac{p^2}{2m} > 0 \quad \text{(for a free single-body particle)} \tag{3.103}$$

The total solution $\psi = X_f T_t$ is given by

$$\psi = X_f T_f = Ce^{\frac{i}{\hbar}px} T_0 e^{-i\frac{E_\lambda}{\hbar}t}$$

or

> **Free Particle Eigenfunction**
>
> $$\psi = \psi_0 e^{\frac{i}{\hbar}(px - E_\lambda t)} \qquad (3.104)$$

Eq. (3.104) is not the most general form of solution for the free particle. Before we determine the most general from, let's express the integral representation of the δ-function, by using de Broglie's relation $p = \hbar k$, which gives

$$\delta(x - x') = \frac{1}{2\pi} \int_{-\infty}^{+\infty} e^{(\pm)ik(x-x')} dk = \frac{1}{2\pi\hbar} \int_{-\infty}^{+\infty} e^{-\frac{ip}{\hbar}(x-x')} dp \qquad (3.105)$$

Using the second property of the δ-function, the wave function $\psi(x)$ can be written as

$$\psi(x) = \int_{-\infty}^{+\infty} \psi(x')\delta(x - x') dx$$

Substitution of Eq. (3.105) (with + sign) then gives

$$\psi(x) = \frac{1}{2\pi\hbar} \int_{-\infty}^{+\infty} \psi(x') \int_{-\infty}^{+\infty} e^{\frac{ip}{\hbar}(x-x')} dp\, dx'$$

We may define the Fourier transform in $x - p$ space, rather than $x - k$. The $x - p$ Fourier transform $F(p)$ becomes

> **Fourier Transform $(x - p)$**
>
> $$F(p) = \frac{1}{\sqrt{2\pi\hbar}} \int_{-\infty}^{+\infty} \psi(x') e^{-\frac{ip}{\hbar}x'} dx' \qquad (3.106)$$

Combining this result with the time-dependent solution $e^{-iE_\lambda t/\hbar}$, the most general form for the free particle solution becomes

$$\Psi(x, t) = \frac{1}{\sqrt{2\pi\hbar}} \int_{-\infty}^{+\infty} F(p) e^{-\frac{i}{\hbar}(px - E_\lambda t)} dp$$

or

3.7 The Free Particle Problem

General Free Particle Solution

$$\Psi(x,t) = \frac{1}{\sqrt{2\pi\hbar}} \int_{-\infty}^{+\infty} F(p) e^{-\frac{i}{\hbar}\left(px - \frac{p^2}{2m}t\right)} dp \qquad (3.107)$$

In practice, $F(p)$ is found from the initial condition $\psi(x,0)$ using Eq. (3.106), since it does not depend on time t. The wave function for the free particle turns out to possess an unusual property in quantum mechanics, and it has to do with our ability to localize the wave function of the free particle. Let's discuss this point in the next section.

3.7.1 Non-Normalizable Wave Functions

Given the free particle solution obtained from the Schrödinger equation, let's consider the following property

$$\int_{-\infty}^{+\infty} \psi^*\psi \, dx = 1 \quad \text{Normalization Condition} \qquad (3.108)$$

A wave function that satisfies Eq. (3.108) is said to be **normalized** (this topic will be treated in more detail in Chap. 4). Substitution of the free particle wave function ψ, we get

$$(\psi_0)^2 \int_{-\infty}^{+\infty} e^{\frac{i}{\hbar}(px - E_\lambda t)} e^{\frac{i}{\hbar}(E_\lambda t - px)} dx = 1 \qquad (3.109)$$

Since the exponential terms are of the form $e^{-i\phi}e^{+i\phi}$, we have

$$\int_{-\infty}^{+\infty} \psi^*\psi \, dx = (\Psi_0)^2 \int_{-\infty}^{+\infty} dx \qquad (3.110)$$

Eq. (3.110) tells us that

$$\frac{1}{\psi_0^2} = \infty \qquad (3.111)$$

Or that $\psi_0 = 0$. This result says that only a trivial solution can have the form of the free particle solution, and be normalizable. In other words, *the free particle wave function is not normalizable*. This result may seem puzzling at first sight because it departs from most problem in quantum mechanics. But, what does this mean? The underlying reason for this is due to the lack of variation in the potential V.

Even if $V = V_0 > 0$, but still uniform, everything in our analysis above remains the same, except the parameter E_λ needs to be replaced by $E_\lambda - V_0 = K - V_0$ (K is the kinetic energy). In our Hamiltonian, we would instead define a value k' as

$$k'^2 = \frac{2m(K - V_0)}{\hbar^2} \qquad (3.112)$$

This also means that for the free particle, $K \geq V_0$, and the spatially-dependent part of our wave function X_f becomes

$$X_f = Ae^{-ik'x} + Be^{ik'x} = Ae^{-\frac{i}{\hbar}\sqrt{2m(E_\lambda - V_0)}x} + Be^{\frac{i}{\hbar}\sqrt{2m(E_\lambda - V_0)}x} \qquad (3.113)$$

Applying the momentum operator to Eq. (3.113) leads to

$$p' = \pm\sqrt{2m(K - V_0)} \qquad (3.114)$$

There is a corresponding reduction in the momentum from p to p'. However, since the shift is real-valued, our result for the normalization condition remains the same since we have

$$1 = \int_{-\infty}^{+\infty} \psi^*\psi \, dx =$$

$$|\psi_0|^2 \int_{-\infty}^{+\infty} e^{\frac{i}{\hbar}(p'x - (E_\lambda - V_0)t)} e^{\frac{i}{\hbar}((E_\lambda - V_0)t - p'x)} dx = |\psi_0|^2 \int_{-\infty}^{+\infty} dx \quad (3.115)$$

The exponent in the exponential term of the integrand sums to 0 and we still find that the wave function ψ is *not* normalizable. This motivates clarification of two important assumptions that must be kept in mind in quantum mechanics. (1) The first interpretation from this result is that *only a nonuniform potential is synonymous with a normalizable wave function solution to the Schrödinger equation*. The potential $V \equiv V(\mathbf{x})$ varies with position. In practice, it is the Coulomb potential that satisfies this condition. However, note that nonuniform generally means that the potential is, at least, not the same everywhere in the domain described by the Schrödinger equation, as well as at domain boundaries. We will see an example of this in the problem of the *infinitely deep potential well* (IDPW), also known as the *particle in a box problem*. When the potential has this property, a *normalizable* wave function may be obtained. In this case, approximations are always possible, however, strictly speaking, the Schrödinger equation essentially requires nonuniform potential V to yield a normalizable wave function. And this is, in no way, a shortcoming of the Schrödinger equation, but rather an intriguing quality that can only be revealed with it. And (2), each time we encounter a wave function of this form in practice, the total wave function *is always multiplied by another wave function which is normalizable*. It is often left out, strictly for convenience. We'll have occasions to observe this shorthand throughout the text. Thus, this must be kept in mind.

3.8 EXPECTATION VALUES AND UNCERTAINTIES

For any normalizable wave function Ψ, we have already hinted that such a wave function has a statistical interpretation. Recall the idea of a *probability distribution* $f(p)$, as a distribution of probabilities p of states. For example, a coin has two states, *heads* and *tails*. Let's say that each state has a probability $p_i = 0.5$, so the probability distribution is uniform, and summarized below:

Heads: $p_1 = 0.5$, $\quad$ Tails: $p_2 = 0.5$, $\quad\quad f(p_i) = \{p_1, p_2\} = \{0.5, 0.5\}$

Probability distributions in quantum mechanics can be distributed over *discrete* or *continuous* variables. In the coin example, notice that

$$\sum_{i=1}^{2} p_i = .5 + .5 = 1$$

This property can be generalized for any probability distribution $f_D(p(v))$, where v is *any* continuous variable on which p depends. Then f_D satisfies the more general condition given by

Probability Distribution Condition

$$\int_v f_D(p(v)) dv = 1 \quad\quad (3.116)$$

Eq. (3.116) says that when adding up all the possible probabilities, it always sums to unity. In other words, there is a 100 % chance of finding the coin, or the system, in one of the states of existence described by f_D.

Therefore, for any normalizable wave function depending on space x where $\int \psi^*\psi dx = 1$, we can define a *probability density function* as

$$f_D(p(x)) = f_D(x) = \psi^* \cdot \psi = |\psi|^2$$

Additionally, when applying Parseval's theorem to a normalized wave function, it also follows that

$$1 = \int_{-\infty}^{+\infty} \psi^* \cdot \psi dx = 1 \Rightarrow \int_{-\infty}^{+\infty} F^* \cdot F dp = 1$$

$F(p)$ is the Fourier transform of the wave function $\psi(x)$. This just means that the probability of finding the particle somewhere in physical space is 100%, where the limits of integration are often finite, though this is not necessary. More finely, however, the probability of finding the particle between x and $x + dx$ is given by $\psi(x)^* \cdot \psi(x)dx$. This is in the spatial domain. Congruently, in momentum space, the probability of finding the particle with a momentum between p and $p + dp$ becomes $F(p)^* \cdot F(p)dp$. These ideas can be extended to define statistical concepts like *average*. The definition of an **average** or **expectation value** $\langle \cdot \rangle$ for a variable q with a probability distribution described by f_p is given by

$$\langle q \rangle = \frac{\int_v q f_p(v)dv}{\int_v f_p(v)dv} = \int_v q f_p(v)dv$$

Applying this to the most general wave function, the *expectation value* of a quantity q becomes

Expectation Value Using A Distribution

$$\langle q \rangle = \frac{\int_V \psi^* q \psi dV}{\int_V \psi^* \psi dV} \tag{3.117}$$

If the wave function is normalized, then the denominator is unity, and the expectation value simplifies to

$$\langle q \rangle = \int_V \psi^* q \psi dV = \iiint_V \psi^* q \psi dx dy dz \tag{3.118}$$

Note that on many occasions, we shall use a shorthand for the volume integral and write it as

$$\langle q \rangle = \int \psi^* q \psi dx$$

3.8 Expectation Values and Uncertainties

	Average and Uncertainties				
Devices	D1	D2	D3	D4	D5
measurement 1	0.5	0.68	0.698	0.6998	0.69998
measurement 2	0.6	0.69	0.699	0.6999	0.69999
measurement 3	0.7	0.70	0.700	0.7000	0.70000
measurement 4	0.8	0.71	0.701	0.7001	0.70001
measurement 5	0.9	0.72	0.702	0.7002	0.70002
Average	0.7	0.7	0.7	0.7	0.7
Uncertainty	0.1	0.01	0.001	0.0001	0.00001

Table 3.1: Examples of averages and uncertainties from device measurements. All devices have the same expectation value or average, with different uncertainties. All measurement devices possess some uncertainty, necessitating multiple measurements to quantify these statistical parameters.

Eq. (3.118) also means that any function of the *expectation value* can be defined, as well. In addition to the expectation value, we also want to consider the **uncertainty** in a variable q which is known from *measurements*. The *uncertainty*, denoted by Δq, in q is defined as

Uncertainty Definition

$$\Delta q = \sqrt{\langle q^2 \rangle - \langle q \rangle^2} \tag{3.119}$$

Therefore, the **uncertainty** Δq is the square root of the average of the square *minus* the square of the average. For example, consider a set of measurements taken with five different devices D1, D2, D3, D4, and D5. Each device will measure *something* five times. The measured numbers from all devices are summarized in Table 3.1 Table 3.1 contains examples of averages and uncertainties from device measurements. All devices have the same *expectation value* or *average*, however, they have different *uncertainties*. In the physical world, these concepts are inescapable. All measurement devices possess some uncertainty meaning that repeat measurements don't yield the same number every time (assuming infinite precision). So what does this mean? It means that in a given set of measurements, there is variation. The larger the uncertainty, the larger the variation or spread in values, and therefore, there will be significant numbers of measurements far from the *expectation value*. This fact usually necessitates multiple measurements to quantify these statistical parameters for the measurement device.

Similarly, in processes where very large numbers of things (like products to be sold) are being made, *statistical parameters* become very important. For example, if a company has processes resulting in measurements like D1, they will have much more uncertainty, and this would lead to large quantities of products being disposed or re-worked, *etc*, potentially leading to a more costly process if the deviants cannot be sold.

Of course, the largest maker of things in the world is nature! For example, there are very large numbers of atoms in most things we encounter on a day-to-day basis. It turns out that nature inherently understands these statistical concepts we have discussed. Its informative to understand how uncertainty is managed by nature as we will see that there is, in fact, a *certain* way that nature manages her *expectations*, as well as *uncertainty*. So, let us get to that.

3.9 THE UNCERTAINTY PRINCIPLE

The **uncertainty principle** was first demonstrated by German physicist Werner Karl Heisenberg (1901-1976) in a famous paper he published in 1927, while Heisenberg was at an institution now known as Neils Bohr's Institute (known then as The Institute for Theoretical Physics), at the University of Copenhagen in Denmark. Owing to the wave nature of matter, this principle exposes nonintuitive behaviors of quantum mechanical systems, which are known as experimental facts today. It's original form involved the product of uncertainties of the location or position Δx and momentum Δp, and gave additional meaning to Planck's constant. Before we prove this principle, its formal statement is given by

---
Uncertainty Principle

$$\Delta x \Delta p \geq \frac{\hbar}{2} \qquad (3.120)$$

3.9 The Uncertainty Principle

The proof of Ineq. (3.120) can be carried out using a mathematical theorem known as **Schwartz' Inequality**. This inequality states that for any two single valued integrable complex functions f and g, with complex conjugates f^* and g^*, the following is true:

Schwartz' Inequality

$$\int_{-\infty}^{+\infty} f^* \cdot f \, d\tau \int_{-\infty}^{+\infty} g^* \cdot g \, d\tau \geq \frac{1}{4}\left[\int_{-\infty}^{+\infty} (f^* \cdot g + f \cdot g^*) d\tau\right]^2 \quad (3.121)$$

First, we will prove *Schwartz' Inequality*, then use it to obtain the final result for Heisenberg's *Uncertainty Principle*. Let λ be a nonzero real number, and for convenience, let us also define the terms a, b, and c as

$$a = \int_{-\infty}^{+\infty} f^* \cdot f \, d\tau \quad (3.122a)$$

$$b = \int_{-\infty}^{+\infty} f^* \cdot g \, d\tau \quad (3.122b)$$

$$c = \int_{-\infty}^{+\infty} g^* \cdot g \, d\tau \quad (3.122c)$$

Since we are utilizing nonzero functions, we can write

$$\int_{-\infty}^{+\infty} |\lambda f + g|^2 d\tau = \int_{-\infty}^{+\infty} (\lambda f^* + g^*)(\lambda f + g) d\tau > 0 \quad (3.123)$$

Multiplying the terms of the integrand in (3.123), we obtain

$$\int_{-\infty}^{+\infty} (\lambda f^* + g^*)(\lambda f + g) d\tau = \int_{-\infty}^{+\infty} (\lambda^2 f^* f + (\lambda(f^* g + f g^*))g^* g) d\tau \quad (3.124)$$

Using the definitions from Eqs. (3.122a)-(3.122c), we obtain the following real quadratic inequality, in λ

$$\lambda^2 a + \lambda(b + b^*) + c \geq 0 \quad (3.125)$$

It is important to realize that all the coefficients in Ineq. (3.125) are real numbers. This means that we can utilize the rules of real quadratic equations. In the case where we have a strict inequality in (3.125), we have

$$\lambda^2 a + \lambda(b + b^*) + c \neq 0$$

Stated another way, it means that there are no real roots to the equation

$$\lambda^2 a + \lambda(b+b^*) + c = 0$$

So the discriminant, which in this case, is given by $\sqrt{(b+b^*)^2 - 4ac}$ is negative, i.e.

$$4ac > (b+b^*)^2$$

Note that in the case of an equality in (3.125), since λ is any nonzero and real number, it follows that $a,b,$ and c must be zero in this case. Combining these results leads us to Schwartz' Inequality, where we have

$$ac \geq \frac{1}{4}(b+b^*)^2 \qquad (3.126)$$

Using the definitions from Eqs. (3.122a)-(3.122c), we obtain, identically, Ineq. (3.121). Next, we will combine Ineq. (3.126) with ideas of the expectation value to discuss how it applies to the domain of quantum mechanics. Let f and g be defined utilizing the operators we defined in Chap. 3 in Eqs. (3.7) and (3.10), as

$$f = p\Psi = -i\hbar \frac{\partial \Psi}{\partial x} \qquad (3.127a)$$

$$g = ix\Psi \qquad (3.127b)$$

Using this, let us now determine the values of a, b, and c, from Eqs. (3.122a) - (3.122c). Determining a, we get

$$a = \int_{-\infty}^{+\infty} f^* f \, d\tau = \int_{-\infty}^{+\infty} i\hbar \frac{\partial \Psi^*}{\partial x} \left(-i\hbar \frac{\partial \Psi}{\partial x}\right) d\tau \qquad (3.128)$$

This becomes

$$a = \hbar^2 \int_{-\infty}^{+\infty} \frac{\partial \Psi^*}{\partial x} \frac{\partial \Psi}{\partial x} d\tau \qquad (3.129)$$

Using integration by parts $\int u \, dv = uv - \int v \, du$, Eq. (3.129) can be written as

$$a = \hbar^2 \left[\Psi^* \frac{\partial \Psi}{\partial x} \Big|_{-\infty}^{+\infty} - \int_{-\infty}^{+\infty} \Psi^* \frac{\partial^2 \Psi}{\partial x^2} d\tau \right] \qquad (3.130)$$

Eq. (3.130) requires that we evaluate the wave function Ψ at ∞. We established ealier, when we discussed the continuity equation, that the

3.9 The Uncertainty Principle

wave functions must always vanish at ∞, in order to be *normalizable*. So, the evaluated part of Eq. (3.130) vanishes and we are left with

$$a = -\hbar^2 \int_{-\infty}^{+\infty} \Psi^* \frac{\partial^2 \Psi}{\partial x^2} d\tau = \int_{-\infty}^{+\infty} \Psi^* \left(-\hbar^2 \frac{\partial^2}{\partial x^2}\right) \Psi d\tau \quad (3.131)$$

This is the expectation value of the operator $\hat{p}^2$, defined in Eq. (3.10). Thus, we have

$$a = \langle \hat{p}^2 \rangle \quad (3.132)$$

The next term we will determine is c, simply given by

$$c = \int_{-\infty}^{+\infty} g^* \cdot g \, d\tau = \int_{-\infty}^{+\infty} -ix\Psi^*(ix\Psi) d\tau = \int_{-\infty}^{+\infty} \Psi^* x^2 \Psi \, d\tau = \langle x^2 \rangle \quad (3.133)$$

Because the uncertainty is a difference in quantities, shifting the abscissa uniformly does not affect the difference that is computed. It is equivalent to translating the functions leftward or rightward, which does not influence the area or integral of the function. This can be used to conveniently choose $\langle x \rangle = 0$ and $\langle p \rangle = 0$.

We have left, the 3rd and final term $I_3 = f^*g + fg*$ that will be integrated. For the integrand I_3, we get

$$f^*g + fg^* = -\hbar \left(\frac{\partial \Psi^*}{\partial x} x\Psi + \frac{\partial \Psi}{\partial x} x\Psi^* \right) \quad (3.134)$$

We can make use of the following fact

$$\frac{\partial}{\partial x}(\Psi^* x \Psi) = \frac{\partial \Psi^*}{\partial x} x\Psi + \Psi^* x \frac{\partial \Psi}{\partial x} + \Psi^* \Psi \quad (3.135)$$

Then, we can write Eq. (3.134) as

$$f^*g + fg^* = -\hbar \left(\frac{\partial}{\partial x}(\Psi^* x \Psi) - \Psi^* \Psi \right) \quad (3.136)$$

Upon integration, we have

$$\int_{-\infty}^{+\infty} f^*g + fg^* d\tau = -\hbar \left((\Psi^* x \Psi) \Big|_{-\infty}^{+\infty} - \int_{-\infty}^{+\infty} \Psi^* \Psi d\tau \right) \quad (3.137)$$

We recognize, again, that the first term on the RHS will require evaluation of Ψ^* and Ψ at ∞. With the same argument as above, this term will also vanish (note: although x also goes to ∞, the product of $\Psi x \Psi^*$ still vanishes). So, we are left with

$$\int_{-\infty}^{+\infty} f^*g + fg^* d\tau = \hbar \int_{-\infty}^{+\infty} \Psi^* \Psi d\tau \tag{3.138}$$

Because of the normalization condition, the integral in Eq. (3.138) is unity, so we get the remarkable result that

$$\int_{-\infty}^{+\infty} f^*g + fg^* d\tau = \hbar \tag{3.139}$$

Putting the result together to express Eq. (3.126), we have

$$\left(\langle \hat{x}^2 \rangle - \langle \hat{x} \rangle^2\right)\left(\langle \hat{p}^2 \rangle - \langle \hat{p} \rangle^2\right) = \langle x^2 \rangle \langle p^2 \rangle \geq \frac{\hbar^2}{4} \tag{3.140}$$

Applying the square root $\sqrt{\cdot}$ to both sides of Ineq. (3.140) leads to our final result

$$\Delta x \Delta p \geq \frac{\hbar}{2} \tag{3.141}$$

The *uncertainty principle* is a fundamental property of quantum mechanical systems. It arises from the wave nature of fundamental particles, which is a quality that is altogether missing from conventional classical descriptions (e.g. Newton's laws). When the system is large enough, this distinction becomes less important because the energy levels are all so close, compared to the energy of the system. However, when the system approaches a small enough number of quantum states, this detail becomes crucial, and quantum mechanics is necessary. In the next section, we'll discuss more consequences from the Schrödinger equation that relate time derivatives of operators to the Hamiltonian $\widehat{H}$.

3.10 THE GENERALIZED-GENERALIZED EHRENFEST THEOREM

It is useful to be able to obtain the time derivative of a quantum mechanical operator. In this section, we will determine the relation that gives us this quantity. We can determine directly, the time-derivative of an operator $\widehat{O}$, based on the Hamiltonian operator in the Schrödinger equation. Using the definition of the expectation value of an operator, total time derivative becomes

$$\frac{\partial \langle \widehat{O} \rangle}{\partial t} = \frac{\partial}{\partial t} \left[\frac{\int_{-\infty}^{+\infty} \psi^\dagger \widehat{O} \psi dx}{\int_{-\infty}^{+\infty} \psi^\dagger \psi dx} \right] \tag{3.142}$$

3.10 The Generalized-Generalized Ehrenfest Theorem

Because $\langle \widehat{O} \rangle$ is a ratio of two integral functions, we can use the quotient rule to evaluate the derivative, given by

$$\frac{\partial}{\partial t}\left(\frac{N(t)}{D(t)}\right) = \frac{\frac{\partial N}{\partial t}D - N\frac{\partial D}{\partial t}}{D^2} = \frac{1}{D}\frac{\partial N}{\partial t} - \frac{N}{D}\frac{\frac{\partial D}{\partial t}}{D} \tag{3.143}$$

This relation will be used to obtain the derivative of an operator $\langle O \rangle$, with respect to time. Let's start with the numerator integral term $N(t)$. Bringing the time derivative into the integral (which is not integrated over time, but space), we have

$$\frac{\partial N}{\partial t} = \frac{\partial}{\partial t}\int_{-\infty}^{+\infty} \psi^\dagger \widehat{O} \psi \, dx$$

$$= \int_{-\infty}^{+\infty} \frac{\partial \psi^\dagger}{\partial t}\widehat{O}\psi \, dx + \int_{-\infty}^{+\infty} \psi^\dagger \frac{\partial \widehat{O}}{\partial t}\psi \, dx + \int_{-\infty}^{+\infty} \psi^\dagger \widehat{O} \frac{\partial \psi}{\partial t} \, dx$$

The second term in the last line is the expectation value of the derivative of the operator. This applies to any operator that has an explicit time dependence. Fortunately, many operators are independent of time, and therefore, this term often vanishes in practice. Nonetheless, $\partial N/\partial t$ becomes

$$\frac{\partial N}{\partial t} = \int_{-\infty}^{+\infty} \frac{\partial \psi^\dagger}{\partial t}\widehat{O}\psi \, dx + \int_{-\infty}^{+\infty} \psi^\dagger \widehat{O} \frac{\partial \psi}{\partial t} \, dx + \left\langle \frac{\partial \widehat{O}}{\partial t} \right\rangle \tag{3.144}$$

In the next step, the time derivatives of the wave functions are replaced by using the Schrödinger equation, which relates the time derivatives to the Hamiltonian $\widehat{H}$. The partial derivatives of ψ and $\psi^\dagger$ can be replaced by using the following relations:

$$i\hbar \frac{\partial \psi}{\partial t} = \widehat{H}\psi \Rightarrow \frac{\partial \psi}{\partial t} = -\frac{i}{\hbar}\widehat{H}\psi$$

For $\psi^\dagger$, we also have

$$\frac{\partial \psi^\dagger}{\partial t} = \frac{i}{\hbar}\psi^\dagger \widehat{H}^\dagger$$

Multiplying by $1/D$, and substituting these relations into Eq. (3.144), we obtain

$$\frac{1}{D}\frac{\partial N}{\partial t} = \frac{i}{\hbar}\left[\frac{\int_{-\infty}^{+\infty} \psi^\dagger \widehat{H}^\dagger \widehat{O}\psi \, dx}{\int_{-\infty}^{+\infty} \psi^\dagger \psi \, dx} - \frac{\int_{-\infty}^{+\infty} \psi^\dagger \widehat{O}\widehat{H}\psi \, dx}{\int_{-\infty}^{+\infty} \psi^\dagger \psi \, dx}\right] + \left\langle \frac{\partial \widehat{O}}{\partial t} \right\rangle \tag{3.145}$$

We can carry out a similar process for the denominator integral $D(t)$, using the relations for ψ and $\psi^\dagger$. The partial derivative of the denominator becomes

$$\frac{\partial D}{\partial t} = \frac{\partial}{\partial t} \int_{-\infty}^{+\infty} \psi^\dagger \psi \, dx$$

$$= \int_{-\infty}^{+\infty} \frac{\partial \psi^\dagger}{\partial t} \psi \, dx + \int_{-\infty}^{+\infty} \psi^\dagger \frac{\partial \psi}{\partial t} \, dx$$

$$= \frac{i}{\hbar} \left[\int_{-\infty}^{+\infty} \psi^\dagger \widehat{H}^\dagger \psi \, dx - \int_{-\infty}^{+\infty} \psi^\dagger \widehat{H}^\dagger \psi \, dx \right]$$

Thus, expressing the second term on the RHS of Eq. (3.143), we have

$$\frac{N}{D} \frac{\frac{\partial D}{\partial t}}{D} = \frac{i}{\hbar} \left[\frac{\int_{-\infty}^{+\infty} \psi^\dagger \widehat{O} \psi \, dx}{\int_{-\infty}^{+\infty} \psi^\dagger \psi \, dx} \right] \left[\frac{\int_{-\infty}^{+\infty} \psi^\dagger \widehat{H}^\dagger \psi \, dx}{\int_{-\infty}^{+\infty} \psi^\dagger \widehat{O} \psi \, dx} - \frac{\int_{-\infty}^{+\infty} \psi^\dagger \widehat{H}^\dagger \psi \, dx}{\int_{-\infty}^{+\infty} \psi^\dagger \widehat{O} \psi \, dx} \right] \quad (3.146)$$

Putting the results for both terms together using the definition of an expectation value, we have the following general result for the time derivative of an operator $\langle O \rangle$:

Generalized-Generalized Ehrenfest Theorem

$$\frac{\partial \langle \widehat{O} \rangle}{\partial t} = \frac{i}{\hbar} \left[\langle \widehat{H}^\dagger \widehat{O} \rangle - \langle \widehat{O} \widehat{H} \rangle \right] - \frac{i}{\hbar} \langle O \rangle \left[\langle \widehat{H}^\dagger \rangle - \langle \widehat{H} \rangle \right] + \left\langle \frac{\partial \widehat{O}}{\partial t} \right\rangle \quad (3.147)$$

Eq. (3.147) is a *generalized* version of another general relation. For this reason, we refer to it as the **Generalized-Generalized Ehrenfest theorem** or G² Ehrenfest theorem. It is the time-derivative of a non-Hermitian operator. In the case that $\widehat{H}^\dagger = \widehat{H}$ (i.e. $\widehat{H}$ is Hermitian, adjoint, or real), the term in brackets then becomes the commutator, and Eq. (3.147) reduces to

Generalized Ehrenfest Theorem

$$\frac{\partial \langle \widehat{O} \rangle}{\partial t} = \frac{i}{\hbar} \langle [\widehat{H}, \widehat{O}] \rangle + \left\langle \frac{\partial \widehat{O}}{\partial t} \right\rangle \quad (3.148)$$

3.11 Fundamental Particle Through A Single Slit

Eq. (3.148) was originally derived by Werner Heisenberg, and is known as the **Generalized Ehrenfest Theorem**. The stricter *Ehrenfest Theorem* is an application of Eq. (3.148) to the position **x** and momentum $\hat{\mathbf{p}}$ operators, named after Austrian physicist Paul Ehrenfest (1880-1933). In this sense, we regard Eq. (3.147) as the *Generalized Generalized Ehrenfest Theorem* or G^2-*Ehrenfest theorem*.

In the next couple sections, we'll examine problems that reveal unambiguously that the Schrödinger equation contains wave-particle behavior, in agreement with experimental evidence we've seen so far.

3.11 FUNDAMENTAL PARTICLE THROUGH A SINGLE SLIT

Now, we examine a somewhat subtle capability of quantum mechanics. We do this by considering a physical problem related to the observations made by Thomas Young, in 1801. We start with an ensemble of fundamental particles like photons, electrons, etc., propagating to the right, say along the +x direction. Each particle passes through a single slit of width w, then lands on a wall-screen located a distance D from the slit, as illustrated in Fig. 3.9.

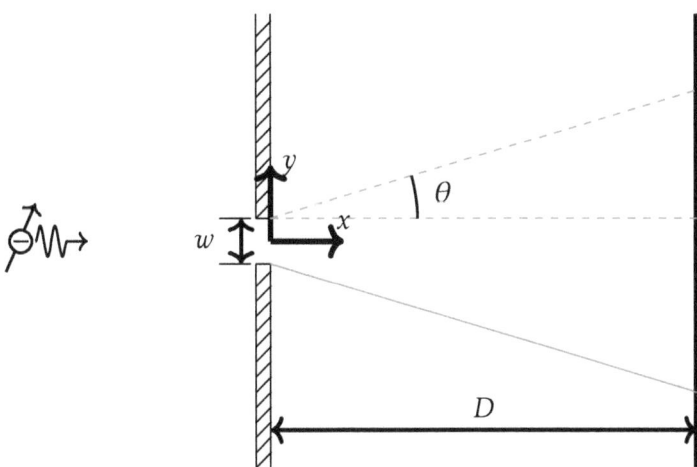

Figure 3.9: Itinerant fundamental particle moving through a single slit, following by hitting a wall-screen that captures the particle, allowing observations of the probability distribution across the y axis.

Before entering the slit, the normalized wave function for the fundamen-

Chapter 3. A Masterful Equation With Duality

tal particle is taken as a localized wave-packet described by

$$\psi(x,t) = \frac{1}{(\pi\sigma_0^2)^{\frac{1}{4}}} e^{-\frac{x^2}{2\sigma_0^2}} \cdot e^{\frac{i}{\hbar}(p_0 x - Et)} \quad (3.149)$$

Due to the $1/\sigma_0^{1/2}$ term, because σ_0 has units of length, the units of $\psi(x,t)$ are inverse units of $\sqrt{\text{length}}$. Moreover, for a valid description of any fundamental particle, the localization provided by the normalized Gaussian function is necessary, although we shall see, that it does not play an important role in the resulting behavior in the transverse or y direction. Nevertheless, after passing through the slit, ψ becomes further localized along y, and the wave function becomes

$$\psi(x,y,t) = \left[\frac{1}{(\pi\sigma_0^2)^{\frac{1}{4}}} e^{-\frac{x^2}{2\sigma_0^2}} \cdot e^{\frac{i}{\hbar}(p_0 x - Et)}\right] \times \psi_w(y) = \psi_x(x) \cdot \psi_y(y) \quad (3.150)$$

ψ_w represents the wave function associated with the slit. Once the particle passes through the slit, the slit wave function ψ_w becomes a factor in the total wave function ψ. ψ_w is given by

$$\psi_w(y) = \begin{cases} \frac{1}{\sqrt{w}} & \text{if } -\frac{w}{2} < y < \frac{w}{2} \\ 0 & \text{otherwise} \end{cases} \quad (3.151)$$

In order to have a normalized total wave function ψ, the value in the slit, of $\psi_w = C = 1/\sqrt{w}$ is obtained from the normalization condition, which gives

$$\int_{-\infty}^{+\infty} \psi_w^* \psi_w dy = \int_{-w/2}^{+w/2} |\psi_w|^2 dy = C^2 \int_{-w/2}^{+w/2} dy = 1 \quad (3.152)$$

From this condition, we obtain $C = \frac{1}{\sqrt{w}}$, used in Eq. (3.151). This amplitude also leads to the wave function having the expected units determined by $1/\sqrt{\sigma_0} \times 1/\sqrt{w} = 1/\text{length}$. We know that the more general solution for a free particle, obtained earlier, is given by Eq. (3.107). In this problem with a fundamental particle going through a slit, it will become evident that we must use the general solution given by Eq. (3.107). Eq. (3.150) alone, is insufficient. For the free particle, the energy is purely kinetic, and thus, the general *Fourier transform* representation of the wave function Ψ becomes

$$\Psi(x,t) = \frac{1}{\sqrt{2\pi\hbar}} \int_{-\infty}^{+\infty} F(p) e^{-\frac{i}{\hbar}\left(\mathbf{p}\cdot\mathbf{r} - \frac{p^2}{2m}t\right)} dp$$

3.11 Fundamental Particle Through A Single Slit

$F(p)$ is the Fourier transform of the wave function in the $x-p$ (or $x-k$) space. We can obtain $F(p)$ using Eq. (3.106), which gives

$$F(p) = \frac{1}{\sqrt{2\pi\hbar}} \int_{-\infty}^{+\infty} \psi(x,y,t) e^{-\frac{i p \cdot r}{\hbar}} dx\, dy \qquad (3.153)$$

The definition of $F(p)$ does not depend on the time variable t. This fact becomes useful because, then, we are free to choose a single value of time for convenience. If we choose the initial condition such that $t = t_0 = 0$, Eq. (3.153) simplifies to

$$F(p) = \frac{1}{\sqrt{2\pi\hbar}} \int_{-\infty}^{+\infty} \psi_x(x) e^{-\frac{i p_x x}{\hbar}} dx \int_{-\infty}^{+\infty} \psi_y(y) e^{-\frac{i p_y y}{\hbar}} dy \qquad (3.154)$$

$$= \frac{1}{\sqrt{2\pi\hbar}} f_x(p_x) f_y(p_y) \qquad (3.155)$$

Substitution of Eq. (3.150) into the above then leads to

$$F(p) = \frac{1}{\sqrt{2\pi\hbar}} \int_{-\infty}^{+\infty} \left[\frac{1}{(\pi\sigma_0^2)^{\frac{1}{4}}} e^{-\frac{x^2}{2\sigma_0^2}} \cdot e^{\frac{i}{\hbar} p_0 x} \right] e^{-\frac{i p_x x}{\hbar}} dx \int_{-\infty}^{+\infty} \frac{1}{\sqrt{w}} e^{-\frac{i p_y y}{\hbar}} dy$$

$$= \frac{1}{\sqrt{2\pi\hbar}} \frac{1}{(\pi\sigma_0^2)^{\frac{1}{4}}} \frac{1}{\sqrt{w}} \int_{-\infty}^{+\infty} e^{-\frac{x^2}{2\sigma_0^2}} \cdot e^{-\frac{i(p_x - p_0)x}{\hbar}} dx \int_{-w/2}^{+w/2} e^{-\frac{i p_y y}{\hbar}} dy$$

The first of the two integrals above has the form of Eq. (3.55), therefore, we may evaluate this integral exactly. Using the notation of Eq. (3.55), $b = -i(p_x - p_0)/\hbar$ and $a = 1/\sigma_0^2$. Then, the first integral works out to be

$$\sqrt{\frac{\pi}{a}} \cdot e^{b^2/4a} = \sigma_0 \sqrt{2\pi} e^{-\frac{(p_x - p_0)^2 \sigma_0^2}{2\hbar^2}} \qquad (3.156)$$

Substitution of this first integral result into Eq. (3.154), we then have

$$F(p) = \frac{1}{\sqrt{2\pi\hbar}} \frac{1}{(\pi\sigma_0^2)^{\frac{1}{4}}} \frac{1}{\sqrt{w}} \sigma_0 \sqrt{2\pi} e^{-\frac{(p_x - p_0)^2 \sigma_0^2}{2\hbar^2}} \times \int_{-w/2}^{+w/2} e^{-\frac{i p_y y}{\hbar}} dy$$

For the last integral, we can evaluate it as

$$\int_{-w/2}^{+w/2} e^{-\frac{i p_y y}{\hbar}} dy = \frac{e^{-\frac{i p_y w}{2\hbar}} - e^{\frac{i p_y w}{2\hbar}}}{-i p_y/\hbar} = \frac{e^{\frac{i k_y w}{2}} - e^{-\frac{i k_y w}{2}}}{i k_y} = \frac{2\sin\left(\frac{w}{2} k_y\right)}{k_y}$$

We have used the relation $p_y = \hbar k_y$. Including the factor $1/\sqrt{w}$, and using the relation $k_y = 2\pi/\lambda$, we have the result for $f_y(p_y)$ given by

$$f_y(p_y) = \frac{2\sin\left(\frac{w}{2}k_y\right)}{\sqrt{w}k_y} = \sqrt{w}\frac{\sin\beta}{\beta} \tag{3.157}$$

β is the dimensionless parameter $\beta = wk_y/2$. In our earlier discussion of $F(p)$, we also know from Parseval's theorem (c. Sec. 3.5), that it can be used to quantify the probability of finding the system with a momentum between p_y and $p_y + dp_y$, where such a quantity is proportional to $|f(p_y)|^2 dp_y$. What is surprising about this result is that although the incoming wave has no momentum along y, the very act of *passing through the slit introduces non-zero momentum along the y direction*. In other words, the wave *tends to spread out* after passing through the slit. This *spreading-out* phenomenon is known as **diffraction**, and we have arrived at it using quantum mechanical results describing fundamental particles.

Diffraction was first predicted and explained for electromagetic waves, by Fresnel and Fraunhofer after modifying Huygen's theory of light (c.Chap. 1). Einstein later helped to figure out that electromagnetic waves are comprised of *photons*, which are examples of fundamental particle known as bosons. There is no need to distinguish between this form of *diffraction* obtained by quantum mechanics, from that explained by Fresnel and Fraunhofer. They are only different way's of looking at the same phenomenon. Quantum mechanics, however, is more general, in that it applies to electrons, as well as photons, and any other fundamental particle. We can use this result to determine what an experiment would turn up, if carried out, pumping these particles through a slit. In the experiment, after the slit, the fundamental particles can be directed towards a surface oriented normal to the propagation direction, as shown in Fig. 3.9. If we denote the angle between the x axis (propagation direction) and spreading direction, containing both x and y components of momentum, the probability density *per unit width* can be written as

$$\frac{1}{w}|f_y(p_y)|^2 = \frac{\sin^2(\pi v \sin\theta)}{\pi^2 v^2 \sin^2\theta} \tag{3.158}$$

The dimensionless parameter v is the ratio of the slit width w to the wavelength λ of the incoming fundamental particles. If, for example, the fundamental particles are photons, then the wavelength λ is that of

3.11 Fundamental Particle Through A Single Slit

the light, etc. The probability density distribution in Eq. (3.158) can be used to determine how the probability density will vary in the transverse or y-direction, at a distance D, measured along the x-direction, which is the initial propagation direction. This can be measured by detecting the particles distributed across y. The angle θ is related to y by

$$\frac{y}{D} = \tan\theta$$

In most experiments, the parameter D is very large compared to the spatial extent of the distribution along the transverse. Therefore in most experimental conditions, to a good approximation, we can write

$$\frac{y}{D} \approx \sin\theta \Rightarrow y \approx D\sin\theta$$

A plot of $|f_y(p_y)|^2/w$ versus the transverse distance $y \approx D\sin\theta$ away from the center, is illustrated in Fig. 3.10. The distribution can be compared to an experimentally observed distribution, illustrated in Fig. 3.11, obtained by passing a helium-neon laser beam through a single narrow slit.

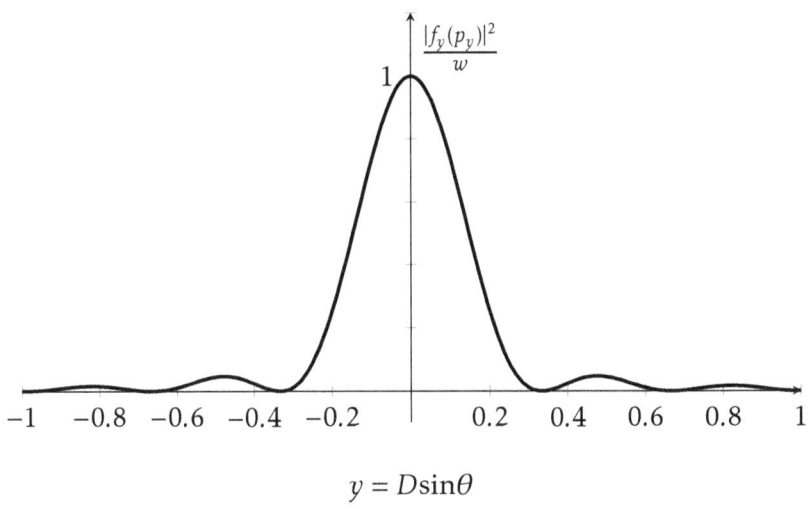

Figure 3.10: Probability distribution along spatial coordinate y for the fundamental particle passing through a single slit. The wavelength $\lambda = 632$nm and the width $w = 3\lambda$.

The spreading effect taking place in diffraction is contained in the $\sin\theta$ term, and follows when we make use of the *uncertainty principle*, discussed in Sec. 3.9. The uncertainty in the position along y is determined

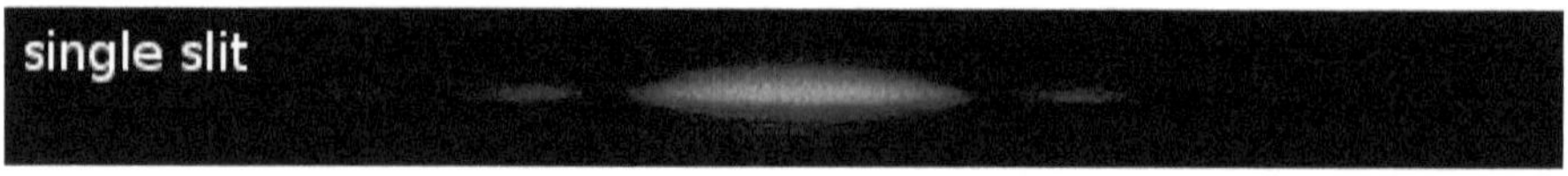

single slit

Figure 3.11: Example of experimental diffraction pattern of a helium-neon laser passing through a narrow slit. Image from http://hyperphysics.phy-astr.gsu.edu/hbase/phyopt/sinslit.html.

by the slit width w. And since $\langle p_y \rangle = 0$ (due to the asymmetric form of $\sin\theta$), and the analogous quantity for the momentum p_y, so we have

$$\Delta p_y \Delta y = \frac{\hbar}{2} \approx |f_y| w$$

The analysis for passing a wave through one slit can be extended to determine the distribution that can be expected from any number of slits. In the next section, we will extend the analysis to *two slits* since its sufficient to illustrate how more than one slit is treated, and it allows us to describe the observations of Thomas Young, with his two slit experiment discussed in Chap. 1, Sec. 1.3.

3.12 Revisiting Young's Double Slit Experiment

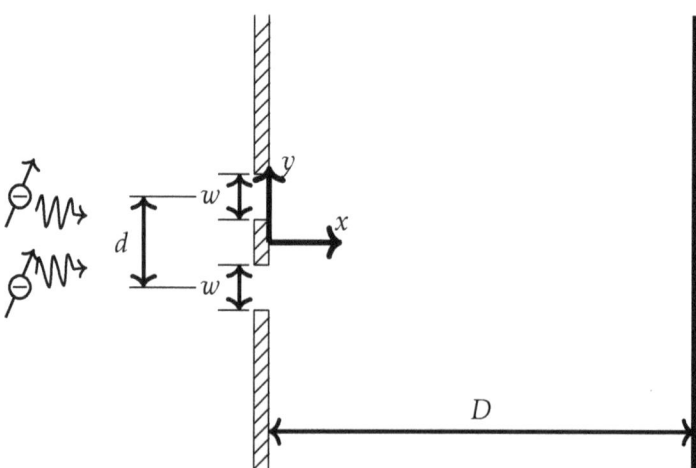

Figure 3.12: Diagram of fundamental particles moving through a *double slit*, then landing on a screen a distance D away from the slits. The screen 'captures' the particles, allowing observations of the probability distribution across the transverse y axis.

3.12 Revisiting Young's Double Slit Experiment

Thomas Young passed light through two tiny openings to convincingly demonstrate light to be a wave. We can now consider Young's experiment using what we learned in the previous section. We just need to consider two slits, instead of only one. For two slits, the wave still proceeds along the $+x$ direction toward both slits. The only difference that is required in setting up the analysis is the introduction of one additional parameter. While w is still the width of both slits, we must also define the distance between the slits. Let the distance between the *centers* of both slits be denoted by d. The origin is now at the point centered between both slits, as illustrated in Fig. 3.12.

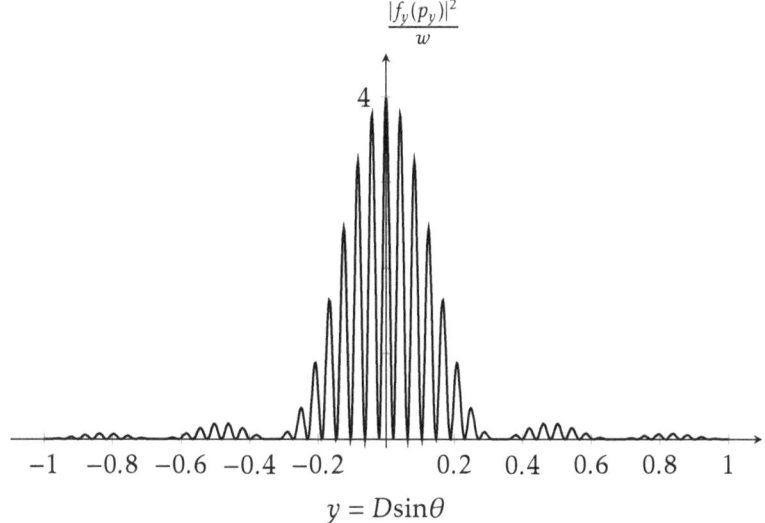

(a) Computed example of the probability density as a function of transverse distance y.

(b) Experimental example corresponding to computed example above.

Figure 3.13: Example of observed diffraction pattern from a helium-neon laser passing through two slits, with a separation distance $d = 0.7$mm.

Fig. 3.13 shows an example of what one would observe after carrying out a double slit experiment. The illustration shows (a) a computed *and* (b) experimental result utilizing a helium-neon laser of wave-length $\lambda = 632$nm. With the tools we've developed so far, we can work out the result observed by Thomas Young, like that shown in Fig. 3.13.

Chapter 3. A Masterful Equation With Duality

For two slits, we may add the wave functions from both slits, just has we had in the single slit. Thus, the initial total wave function $\psi(x, y, t)$ for two slits can be written as

$$\psi(x, y, 0) = \psi_1(x, y, 0) + \psi_2(x, y, 0) \tag{3.159}$$

ψ_1 is the wave function associated with slit 1, and ψ_2 with slit 2. The two wave functions associated with slit 1 and slit 2, respectively, are given by

$$\psi_1(x, y, 0) = \left[\frac{1}{(\pi\sigma_0^2)^{\frac{1}{4}}} e^{-\frac{x^2}{2\sigma_0^2}} \cdot e^{\frac{i}{\hbar}P_0 x}\right] \times \psi_{w1}(y) \tag{3.160a}$$

$$\psi_2(x, y, 0) = \left[\frac{1}{(\pi\sigma_0^2)^{\frac{1}{4}}} e^{-\frac{x^2}{2\sigma_0^2}} \cdot e^{\frac{i}{\hbar}P_0 x}\right] \times \psi_{w2}(y) \tag{3.160b}$$

The Fourier transform $F(p)$ of the total wave function given by Eq. (3.159) becomes

$$F(p) = \frac{1}{\sqrt{2\pi\hbar}} \int_{-\infty}^{+\infty} \psi(x, y) e^{-\frac{ip \cdot x}{\hbar}} dx dy$$

$$= \frac{1}{\sqrt{2\pi\hbar}} \int_{-\infty}^{+\infty} \psi(x) e^{-\frac{ip_x x}{\hbar}} dx \int_{-\infty}^{+\infty} (\psi_1(y) + \psi_2(y)) e^{-\frac{ip_y y}{\hbar}} dy$$

$$= \frac{1}{\sqrt{2\pi\hbar}} f_x(p_x) \cdot \left(\int_{-d/2-w/2}^{-d/2+w/2} \psi_1(y) e^{-\frac{ip_y y}{\hbar}} dy + \int_{+d/2-w/2}^{+d/2+w/2} \psi_2(y) e^{-\frac{ip_y y}{\hbar}} dy \right)$$

$$= \frac{1}{\sqrt{2\pi\hbar}} f_x(p_x) f_y(p_y)$$

The integral $f_x(p_x)$ is identical to that for the single slit, given by Eq. (3.156). However, because the slits are at different locations in space with respect to y, integrals of $\psi_{wi}(y)$ differ. Since we already have the first integral, we need only be concerned with the y-dependent part, given by

$$f_y(p_y) = \int_{-d/2-w/2}^{-d/2+w/2} \psi_1(y) e^{-\frac{ip_y y}{\hbar}} dy + \int_{+d/2-w/2}^{+d/2+w/2} \psi_2(y) e^{-\frac{ip_y y}{\hbar}} dy$$

$$= \frac{e^{-\frac{ip_y(-d+w)}{2\hbar}} - e^{-\frac{ip_y(-d-w)}{2\hbar}}}{-ip_y/\hbar} + \frac{e^{-\frac{ip_y(+d+w)}{2\hbar}} - e^{-\frac{ip_y(+d-w)}{2\hbar}}}{-ip_y/\hbar}$$

3.12 Revisiting Young's Double Slit Experiment

$$= \frac{e^{-\frac{ip_y(-d)}{2\hbar}} e^{-\frac{ip_y w}{2\hbar}} - e^{-\frac{ip_y(-d)}{2\hbar}} e^{-\frac{ip_y(-w)}{2\hbar}}}{-ip_y/\hbar} + \frac{e^{-\frac{ip_y d}{2\hbar}} e^{-\frac{ip_y w}{2\hbar}} - e^{-\frac{ip_y d}{2\hbar}} e^{-\frac{ip_y(-w)}{2\hbar}}}{-ip_y/\hbar}$$

$$= e^{-\frac{ip_y(-d)}{2\hbar}} \left[\frac{e^{-\frac{ip_y w}{2\hbar}} - e^{-\frac{ip_y(-w)}{2\hbar}}}{-ip_y/\hbar} \right] + e^{-\frac{ip_y d}{2\hbar}} \left[\frac{e^{-\frac{ip_y w}{2\hbar}} - e^{-\frac{ip_y(-w)}{2\hbar}}}{-ip_y/\hbar} \right]$$

$$= \left[e^{-\frac{ip_y(-d)}{2\hbar}} + e^{-\frac{ip_y d}{2\hbar}} \right] \left[\frac{e^{-\frac{ip_y w}{2\hbar}} - e^{-\frac{ip_y(-w)}{2\hbar}}}{-ip_y/\hbar} \right]$$

$$= 2\cos\left(\frac{k_y d}{2}\right) \times 2 \frac{\sin\left(\frac{w}{2} k_y\right)}{\sqrt{w} k_y}$$

$$= 2\sqrt{w} \times \cos\left(\frac{k_y d}{2}\right) \times \frac{\sin\beta}{\beta}$$

The observable probability density per unit width that can be expressed as a function of y, is now proportional to

$$\frac{1}{w}|f_y(p_y)|^2 = 4\cos^2\left(\frac{k_y d}{2}\right) \frac{\sin^2\beta}{\beta^2}$$

The result is proportional to the probability density, and like we had done for the single slit in the previous section, we can express it as a function of the transverse distance y, using the approximation $y \approx D\sin\theta$. Fig. 3.14 shows another simulated example for the two slit distribution. In Fig. 3.14, the wavelength used is also $\lambda = 632$nm, and the slit width $w = 50\lambda$. The centers are separated by a length $d = 3w$. Notice that the gray dotted line shows the result for a single slit, which modulates the two-slit function.

The analysis of a fundamental particle through a double slit, as well as the illustrated examples, demonstrate that the phenomenon of *diffraction* is, indeed, relevant to any fundamental particle. It is not restricted to light photons. Recall the example illustrating the direct evidence for electrons in Fig. 2.12. Thus, all fundamental particles exhibit behavior of both waves and particles, and a theory like that devised by Schrödinger proves to be capable of producing these behaviors. Surely, Schrödinger, as well as Heisenberg[1], were on to something.

---
[1] Heisenberg created an equivalent theory of quantum mechanics called *matrix me-*

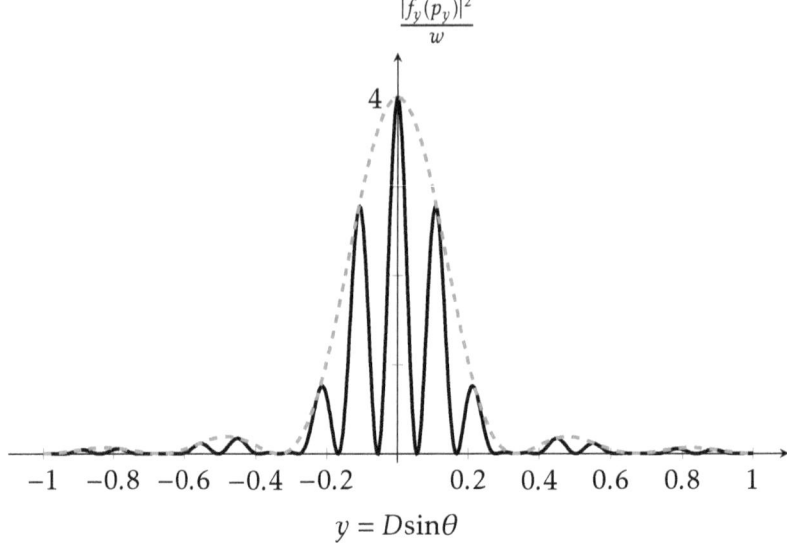

Figure 3.14: Example of a probability distribution along the transverse spatial coordinate y for a fundamental particle passing through two slits.

3.13 Chapter Summary

This chapter has introduced one of the first quantum mechanics equations of motion, known as the *Schrödinger equation*. This equation expresses a relationship between operators for the energies, including the kinetic energy K and the potential energy V. In quantum mechanics, operators are more abstract mathematical concepts that are applied to wave functions to yield eigenvalues multiplied by the same wave function. If the eigenvalues are real-valued, then the eigenvalues are called *observables* for the operator.

We also developed some mathematical tools that lead to the Fourier transform pair of a function, relating this pair to the Dirac δ-function. To get our feet wet with the Schrödinger equation, we then considered the free particle problem, obtaining the expected propagating wave as the solution. However, we also found that this form of wave function is not normalizable, meaning $\int \psi^* \psi dx \neq 1$. The chapter ended by demonstrating that quantum mechanics, indeed, leads to two well-observed phenomena from waves, namely, diffraction and interference, indicating

---

chanics. Schrödinger later proved the equivalence between his and Heisenberg's formulation.

3.13 Chapter Summary

that particles also have these properties.

Hopefully, we have been able to impress upon the reader that quantum mechanics starts to achieve something special for an entity endowed with wave-particle duality. That is, they posses both wave and particle behavior. In the coming chapters, we shall continue to explore what properties and behaviors are predicted. Table 3.2 lists some of the key results from this chapter.

Table 3.2: Chapter 3 Summary Equations.

Name	Equation
Schrödinger Equation	$i\hbar \frac{\partial \psi}{\partial t} = -\frac{\hbar^2}{2m}\nabla^2 \psi + V\psi$
Momentum Operator (x)	$\hat{p}_x \Rightarrow -i\hbar \frac{\partial}{\partial x}$
Momentum Squared Operator	$\hat{p}_x^2 \Rightarrow -\hbar^2 \frac{\partial^2}{\partial x^2}$
δ-Function	$\delta(x-x') = \frac{1}{2\pi}\int_{-\infty}^{+\infty} e^{(\pm)ik(x-x')}dk$
Fourier Transform	$F(k) = \frac{1}{\sqrt{2\pi}}\int_{-\infty}^{+\infty} f(x')e^{-ikx'}dx'$
Inverse Fourier Transform	$f(x) = \frac{1}{\sqrt{2\pi}}\int_{-\infty}^{+\infty} F(k)e^{ikx}dk$
Free Particle wave function	$\psi(x,t) = \psi_0 e^{\frac{i}{\hbar}(px-E_\lambda t)}$
General Free Particle Solution	$\Psi(x,t) = \frac{1}{\sqrt{2\pi\hbar}}\int_{-\infty}^{+\infty} F(p)e^{-\frac{i}{\hbar}\left(px-\frac{p^2}{2m}t\right)}dp$
Expectation value $\langle \cdot \rangle$	$\langle q \rangle = \frac{\int_V \psi^* q \psi \, dV}{\int_V \psi^* \psi \, dV}$
Uncertainty in quantity q	$\Delta q = \sqrt{\langle q^2 \rangle - \langle q \rangle^2}$
Uncertainty principle	$\Delta x \Delta p \geq \frac{\hbar}{2}$

3.14 CHAPTER PROBLEMS

Problem 3.1 Using Eq. (3.117), show that the expectation value of any variable x whose distribution is described by a shifted normal distribution given by $G_\sigma(x) = \frac{1}{\sigma\sqrt{2\pi}} e^{-\frac{(x-a)^2}{2\sigma^2}}$ is given by the parameter a.

Problem 3.2 The general wave equation is given by

$$\frac{\partial^2 w}{\partial t^2} = v^2 \frac{\partial^2 w}{\partial x^2}$$

v is the wave velocity. Let $w = e^{i(px-Et)/\hbar}$. Show that with $p = \hbar k = mv$, that the energy $E = \hbar\omega$ is given by

$$E^2 = \frac{\hbar^4 k^4}{m^2}$$

How is this result different from the ordinary kinetic energy of a fundamental particle?

Problem 3.3 Determine whether the following operators $\widehat{O}$ are linear operators ($\widehat{O}(f_1(x) + f_2(x)) = \widehat{O}f_1(x) + \widehat{O}f_2(x)$) or not:

(a) $\widehat{O}f(x) = 3x$

(b) $\widehat{O}f(x) = 3x + \sqrt{x}$

(c) $\widehat{O}f(x) = \frac{1}{x}$

Problem 3.4 Show that e^{ax} is an eigenfunction of the operator:

$$\widehat{O} = \frac{d^n}{dx^n}$$

Also, determine the eigenvalues.

Problem 3.5 Consider two operators $\widehat{O}$ given by

$$\widehat{O}_1 = \frac{d}{dx} \quad \text{and} \quad \widehat{O}_2 = xx^2 \text{(multiply function by)}$$

Show that

$$\widehat{O}_1^2 f(x) \neq \left(\widehat{O}_1 f(x)\right)^2$$

Is the same statement true for $\widehat{O}_2$?. And does $\widehat{O}_1 \widehat{O}_2 = \widehat{O}_2 \widehat{O}_1$?

Problem 3.6 Let the Hamiltonian be given by the kinetic energy Hamiltonian

$$\widehat{K} = -\frac{\hbar^2}{2m}\frac{\partial^2}{\partial x^2}$$

Since it is Hermitian, use the Generalized Ehrenfest theorem given by Eq. (3.148) to prove that

$$\frac{\partial \widehat{p_x}}{\partial t} = 0$$

Problem 3.7 Use the kinetic energy Hamiltonian

$$\widehat{K} = -\frac{\hbar^2}{2m}\frac{\partial^2}{\partial x^2}$$

and the *Generalized Ehrenfest theorem* given by Eq. (3.148) to prove that

$$\frac{\partial \langle x \rangle}{\partial t} = \frac{-i\hbar}{m}\frac{\partial \psi}{\partial x}$$

This result is known as the *Ehrenfest theorem*.

Problem 3.8 Consider the Schrödinger equation given by

$$i\hbar\frac{\partial \psi}{\partial t} = -\frac{\hbar^2}{2m}\frac{\partial^2 \psi}{\partial x} + V\psi$$

Let $\psi(x,t) = \left(Ae^{ikx} + Be^{-ikx}\right)Ce^{-i\beta t}$. Substitute ψ into the Schrödinger equation and show that

$$\beta = \frac{\hbar k^2}{2m} + V$$

3.15 Suggested Readings & References

[1] W. Heisenberg, "*Uber die quantentheoretische Umdeutung kinematischer und ¨ mechanischer Beziehungen / On the quantum theoretical re-interpretation of kinematic and mechanical relations* Zeitschrift für Physik A Hadrons and Nuclei, **33**, No. 1. (1 December 1925), pp. 879-893, doi:10.1007/bf01328377

Table 3.3

[2] E. Schrödinger, *Quantisierung als Eigenwertproblem. (Erste Mitteilung.) / Quantisation as an eigenvalue problem. (First Communication.)*, Annalen der Physik **384**, (4), 361-376 (1926)

[3] E. Schrödinger, *Quantisierung als Eigenwertproblem. (Zweite Mitteilung.) / Quantisation as an eigenvalue problem. (Second Communication)*, Annalen der Physik **384**, (6), 489-527 (1926)

[4] E. Schrödinger, *Quantisierung als Eigenwertproblem. (Dritte Mitteilung.) / Quantisation as an eigenvalue problem. (Third Communication)*, Annalen der Physik **385**, (13), 437-490 (1926)

[5] E. Schrödinger, *Quantisierung als Eigenwertproblem. (Vierte Mitteilung) / Quantisation as an eigenvalue problem. (Fourth Communication)*, Annalen der Physik **385**, (13), 437-490 (1926)

[6] E. Schrödinger, *Über das Verhaeltnis der Heisenberg-Born-Jordanschen Quantenmechanik zu der meinen / On the relationship of the Heisenberg-Born-Jordan quantum mechanics to mine*, Annalen der Physik **384**, (8), 734-756 (1926)

CHAPTER 4

Infinitely Deep Potential Well

Now that we have discussed the developments leading to the Schrödinger equation, and we have reviewed some useful mathematics that can help us to deal with the solutions known as the *wave function* ψ, we'll begin to analyze problems more relevant to matter. In other words, what does the Schrödinger equation have to say about how matter behaves, based on wave concepts. We will see that with the wave function and Hamiltonian operator available, formally nonintuitive kinds of information can be obtained. In fact, in the next few chapters, we will continue focusing on how to use analytical methods with powerful mathematical tools and techniques to solve some important problems in quantum mechanics. However, the reader should be aware that *most real-world problems treated in quantum mechanics are much too difficult to obtain exact solutions*. However, to study closely those problems that offer exact solutions provides an enormous amount of insight and develops new intuition for more complex problems. Toward this aim, this chapter discusses a problem at the opposite end of the spectrum from the *free particle problem*...namely, we consider a infinitely confined or localized fundamental particle in space, and explore what the Schrödinger equation has to say about localized particles, being relevant to bound electrons existing within atoms.

4.1 INFINITELY DEEP 1D POTENTIAL WELL

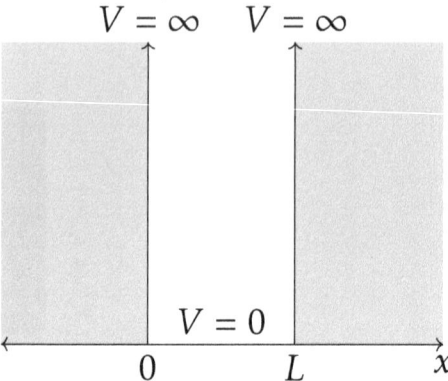

Figure 4.1: Infinitely deep potential well. At $x = 0$ and $x = L$, the potential energy V is infinite, while $V = 0$ in the range $0 \leq x \leq L$. This arrangement dictates that the particle exists within the range $0 < x < L$.

An electron can only be localized by *forces* keeping it localized. In many classical cases concerning charged particles, this force can be described with effective potential energy gradients. This means that for localization to be possible, there must be a nonuniform potential energy $V(x)$ interacting with the electron. This idea of interaction can also be observed in quantum mechanics. This problem provides a simpler way to observed these kind of effects. This problem, specifically, is known as the *infinitely deep potential well (IDPW)* problem. It is also known as the *particle in a (one-dimensional) box* problem, illustrated in Fig. 4.1.

The IDPW problem includes both the potential energy V as well as the kinetic energy K. Fig. 4.1 illustrates the potential energy in the 1D **infinitely deep potential well** as a function $V(x)$ and it is defined as

$$V(x) = \begin{cases} 0 & \text{if } 0 < x < L \\ +\infty & \text{at } x = 0 \text{ and } x = L \end{cases} \quad (4.1)$$

An intention behind the definition for the potential function $V(x)$ in the *IDPW* is so that the wave function $\psi(x)$, which is the solution to the Schrödinger equation, vanishes at the boundaries of the problem, namely at $x = 0$ and $x = L$. The electric potential energy function is consistent with a classical potential energy function, and we know that a *conservative* classical potential energy (e.g. electrostatic potential) gives rise to a force proportional to $-\partial V/\partial x$. This also means that $V(x)$ is

4.1 Infinitely Deep 1D Potential Well

taken to be a real-valued function of space, but not time. Therefore, both infinite potentials generate effectively infinite forces along $+x$ at $x = 0$, and along $-x$ at $x = L$. They, therefore, confine the particle to the box. We will develop this argument more rigorously later in the chapter, however, let us write the following boundary conditions for now:

$$\psi(x) = \{0 \quad \text{at } x = 0 \text{ and } x = L \tag{4.2}$$

Since the IDPW problem only varies in the x direction, the 1D version of the Schrödinger equation can be used, which is given by

$$i\hbar \frac{\partial \psi}{\partial t} = -\frac{\hbar^2}{2m} \frac{\partial^2 \psi}{\partial x^2} + V\psi$$

The assumption that the potential energy V is strictly a function of x, and is *not dependent on time t* will be the taken in all cases considered in this book. However, they will not always be real-valued. We will explore examples of complex valued potentials in Chap. 12.

To solve this problem, we can test the separable form of the Schrödinger equation discussed in the previous chapter. So, let us assume $\psi(x, t)$ can be written as the product of a time dependent function $T(t)$ and a spatially varying function $X(x)$, where we have

$$\psi(x, t) = T(t) \cdot X(x) \tag{4.3}$$

Substitution of Eq. (4.3) into the 1D Schrödinger equation leads to

$$i\hbar X(x) \frac{dT}{dt} = -\frac{\hbar^2}{2m} T(t) \frac{d^2 X}{dx^2} + V T(t) \cdot X(x) \tag{4.4}$$

Because $X(x)$ and $T(t)$, our trial functions, are strictly functions of a single variable, respectively, we have replaced partial derivatives with total derivatives. After dividing both sides by $\psi = T \cdot X$, Eq. (4.4) becomes

$$i\hbar \frac{1}{T(t)} \frac{dT}{dt} = -\frac{\hbar^2}{2m} \frac{1}{X(x)} \frac{d^2 X}{dx^2} + V(x) \tag{4.5}$$

Eq. (4.12) is telling us something interesting. It says that the left-hand side (LHS), which is strictly a function of t, is always equal to the right-hand side (RHS), which is strictly a function of x. Because x and t are independent variables, this can only be true if both sides are equal to a constant (otherwise, they can be changed independently of each other).

This constant is the eigenvalue E_λ. We know that it must be a form of energy because this constant will have the same units as V, which is potential energy. Thus, we can write the following two equations based on Eq. (4.12):

$$i\hbar \frac{1}{T(t)} \frac{dT}{dt} = E_\lambda \tag{4.6a}$$

$$-\frac{\hbar^2}{2m} \frac{1}{X(x)} \frac{d^2 X}{dx^2} + V(x) = E_\lambda \tag{4.6b}$$

Let's solve Eq. (4.13) first, where we write it as

$$\frac{dT}{dt} = -\frac{iE_\lambda}{\hbar} T(t) \tag{4.7}$$

From integrating Eq. (4.7), the solution is given by

$$T(t) = T_0 e^{-i\frac{E_\lambda}{\hbar}(t-t_0)} \tag{4.8}$$

T_0 is the initial condition of the wave function T, and t_0 is the initial time. Next, we solve Eq. (4.13). We can write it as

$$\frac{d^2 X}{dx^2} + \frac{2m(E_\lambda - V)}{\hbar^2} X(x) = 0 \tag{4.9}$$

Introducing a constant $k^2 = 2m(E_\lambda - V)/\hbar^2$, we have

$$\frac{d^2 X}{dx^2} + k^2 X(x) = 0 \tag{4.10}$$

The general solution to Eq. (4.9) is given by

$$X(x) = A\sin(kx) + B\cos(kx) \tag{4.11}$$

A and B are determined from boundary conditions, while $k = \sqrt{2m(E_\lambda - V)/\hbar^2}$. Using the boundary conditions given by Eq. (4.9), we have

$$X(0) = 0 = B$$

$$X(L) = 0 = A\sin(kL)$$

The second boundary condition says that our sine function has to vanish at the boundary. This can only happen if kL is an integer multiply of π, i.e.

$$kL = n\pi \Rightarrow k^2 = \frac{2m(E_\lambda - V)}{\hbar^2} = \frac{n^2 \pi^2}{L^2} \tag{4.12}$$

4.1 Infinitely Deep 1D Potential Well

Solving for $E_\lambda - V$, we have

$$kL = n\pi \Rightarrow k^2 = E_\lambda - V = \frac{n^2\pi^2\hbar^2}{2mL^2} \tag{4.13}$$

Since $V = 0$ for $0 < x < L$, we can write the resulting eigenvalues E_λ to be

1D Infinitely Deep Potential Well Energy Levels

$$E_\lambda = \frac{n^2\pi^2\hbar^2}{2mL^2} \tag{4.14}$$

Eq. (4.17) can also be written in terms of the constant $E_1 = \pi^2\hbar^2/2mL^2$, where we have

$$E_\lambda = n^2 E_1 \tag{4.15}$$

E_1 represents the lowest energy state and the energy E_λ only takes discrete values, rather than continuous ones. This tells us that a change from state 1 to state 2 will correspond to a change in energy ΔE_λ given by

$$\Delta E_\lambda = \Delta E_{12} = \frac{\pi^2\hbar^2}{2mL^2}\left(n_2^2 - n_1^2\right) \tag{4.16}$$

Since k depends on E_λ and E_λ is discrete, then we have discrete solutions $X_n(x)$, given by

$$X_n(x) = A\sin(k_n x) = A\sin\left(\frac{n\pi x}{L}\right) \tag{4.17}$$

n takes any integer value, $n = 1, 2, 3, ...$, etc. Adding the time factor $T(t) = T_n(t)$ to form the complete solution, we then have

1D Infinitely Deep Potential Well Eigenfunctions

$$\psi_n(x,t) = \psi_0 e^{-i\frac{E_n}{\hbar}(t-t_0)}\sin\left(\frac{n\pi x}{L}\right) \tag{4.18}$$

In Eq. (4.18), $\psi_0 = T_0 \cdot A$. Fig. 4.2 illustrates the first five lowest energy states and corresponding wave functions. Note that, relative to the

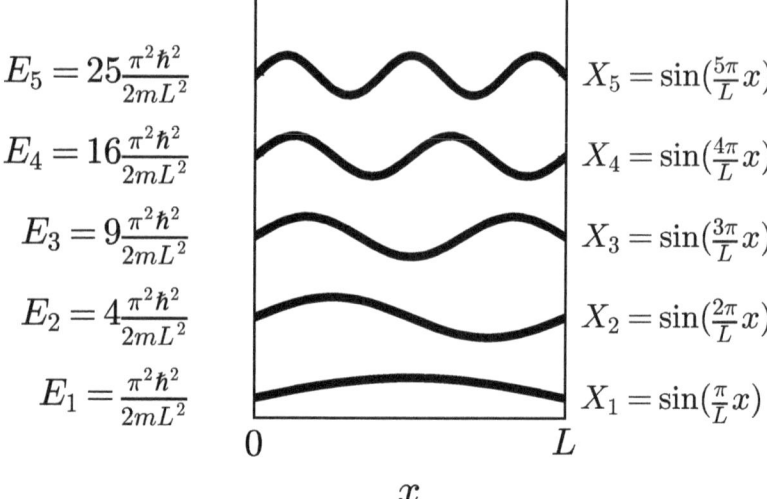

Figure 4.2: First five 1D infinitely deep potential well solutions, $\psi_n(0,x) = X_n(x)$. From the lowest energy $n = 1(E_1)$, to $n = 5(E_5)$, all solutions vanish at $x = 0$ and $x = L$. In this case, the quantum number n also corresponds to the number of inflection points in the wave function.

center of the box $x = L/2$, the wave functions are alternating symmetric and anti-symmetric. The point at $x = L/2$ is a point of symmetry for the problem. We will exploit this fact in a later chapter (Chap. 6), when dealing with the potential well problem.

Based on Eq. (4.18), there is also a variation in $\psi_n(x,t)$ with time, given by $T_n(t)$ (Eq. (4.8)). The frequency f of the time variation is just $\omega = E_\lambda/\hbar$ or $f = E_\lambda/2\pi\hbar$, with period $T = 1/f$. Fig. 4.3 shows an example of the time variation of the real part of $\psi_n(x,t)$, for the case where $n = 2$, $L = 1nm$, at different fractions of a period T. The wave ψ_n undergoes variations in time like a string pinned on both ends, at $x = 0$ and $x = L$, with an electron mass. At $t = .25T$, where T is the period of the wave, the phase has shifted $.25 \times 2\pi = \pi/2$. Since the real part of $e^{\pi/2}$ is zero, the real part of the wave vanishes at this instant of time. Likewise at $3\pi/2$, corresponding to $t = .75T$. At $.5T$, the phase is π, which leads to a factor of -1. Given the set of functions $X_n(x)T_n(t)$, the most general solution of the IDPW problem is given by a linear combination of all of

4.1 Infinitely Deep 1D Potential Well

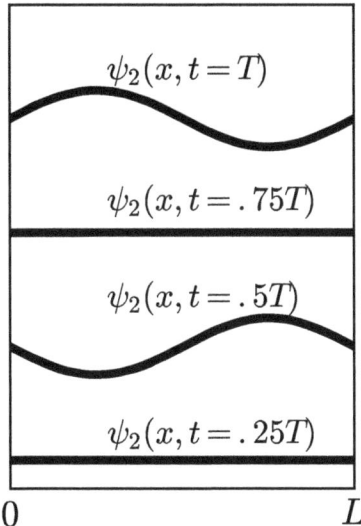

Figure 4.3: Compete cycle of the real part of the state corresponding to $n = 2$, (E_2). The wave undergoes variations in time like a string pinned on both ends, at $x = 0$ and $x = L$, with an electron mass. At $t = .25T$, where T is the period of the wave, the phase has shifted $.25 \times 2\pi = pi/2$. Since the real part of $e^{-\pi/2}$ is zero, the real part of ψ vanishes at this instant of time. Likewise, at $t = .75T$. At $.5T$, the phase is π, where $e^{-\pi} = -1$.

our solutions. This is written as

$$\Psi(x,t) = \sum_{n=0}^{n=\infty} c_n \cdot \psi_n(x,t) \tag{4.19}$$

Using the results obtained above for the separate solutions, we also have

---
1D Infinitely Deep Potential Well General Solution

$$\Psi(x,t) = \sum_{n=0}^{n=\infty} c_n \cdot X_n(x) e^{-\frac{iE_n}{\hbar}(t-t_0)} \tag{4.20}$$

From Eq. (4.20), the question arises as to whether the coefficients c_n have any physical interpretation or not. To answer this question, we must first take Ψ to be normalized, that is

$$\int_{-\infty}^{+\infty} \Psi^*\Psi d\mathbf{x} = 1$$

Let us also simplify the problem by considering by only considering two possible eigen-states superposed to give Ψ (it will be straightforward to generalize the result from this case). The initial state at $t = 0$ is then given by

$$\Psi(x,t) = c_1 X_1 + c_2 X_2$$

Substitution into the normalization condition gives

$$1 = \int_{-\infty}^{+\infty} \Psi^*\Psi d\mathbf{x}$$

$$= \int_{-\infty}^{+\infty} (c_1^* X_1^* + c_2^* X_2^*)(c_1 X_1 + c_2 X_2) d\mathbf{x}$$

$$= \int_{-\infty}^{+\infty} (c_1^* c_1 X_1^* X_1 + c_1^* c_2 X_1^* X_2 + c_2^* c_1 X_2 X_1^* + c_2^* c_2 X_2^* X_2) d\mathbf{x}$$

$$= c_1^* c_1 \int_{-\infty}^{+\infty} X_1^* X_1 d\mathbf{x} + c_1^* c_2 \int_{-\infty}^{+\infty} X_1^* X_2 d\mathbf{x}$$

$$\ldots + c_2^* c_1 \int_{-\infty}^{+\infty} X_2^* X_1 d\mathbf{x} + c_2^* c_2 \int_{-\infty}^{+\infty} X_2^* X_2 d\mathbf{x}$$

If the eigenfunctions X_n are normalized, then we have

$$1 = |c_1|^2 + |c_2|^2 + c_1^* c_2 \int_{-\infty}^{+\infty} X_1^* X_2 d\mathbf{x} + c_2^* c_1 \int_{-\infty}^{+\infty} X_2^* X_1 d\mathbf{x}$$

The two remaining integrals contain terms of the form $X_n \cdot X_m$ where $n \neq m$. It turns out that *if the Hamiltonian contains a real-valued potential function $V(x)$*, then for any pair of distinct eigensolutions where $(n \neq m)$, the pair always satisfies the condition

$$\int_{-\infty}^{+\infty} \psi_i^* \psi_j d\mathbf{x} = 0 \quad \text{if} \quad i \neq j \tag{4.21}$$

This will be proven rigorously in Sec. 4.4, but for now, we simply note that if this statement is true, then we have for our simply case, the result

$$1 = |c_1|^2 + |c_2|^2$$

4.1 Infinitely Deep 1D Potential Well

This provides an interpretation of $|c_n|^2$. They represent the probability of finding the particle in the nth state. This result, indeed, extends to the most general case of Ψ given by Eq. (4.20), to give

$$\sum_{n=1}^{\infty} |c_n|^2 = 1 \tag{4.22}$$

Two important points must be made in us making the leap from 2 eigensolutions to any number. Firstly, the product $\Psi^*\Psi$, for the general case becomes

$$1 = \int_{-\infty}^{+\infty} \Psi^*\Psi dx$$

$$= \int_{-\infty}^{+\infty} (c_1^* X_1^* + c_2^* X_2^* + \cdots)(c_1 X_1 + c_2 X_2 + \cdots) dx$$

$$= c_1^* c_1 \int_{-\infty}^{+\infty} X_1^* X_1 dx + \sum_{m \neq 1} c_1^* c_m \int_{-\infty}^{+\infty} X_1^* X_m dx +$$

$$c_2^* c_2 \int_{-\infty}^{+\infty} X_2^* X_2 dx + \sum_{m \neq 2} c_2^* c_m \int_{-\infty}^{+\infty} X_2^* X_m dx +$$

$$c_3^* c_3 \int_{-\infty}^{+\infty} X_3^* X_3 dx + \sum_{m \neq 3} c_3^* c_m \int_{-\infty}^{+\infty} X_3^* X_m dx +$$

$$\vdots$$

We still have integrals of only two types, one being $X_n^* \cdot X_n$, as before, while the other is also of the form $X_m \cdot X_n$ where $n \neq m$. In this particular case, even though there are infinitely many of them, as long as the condition in Eq. (4.22) holds, the result given by Eq. (4.21) also holds. The second important point to be made is regarding the time-dependent factor $e^{-iE_n t/\hbar}$, which we have conveniently omitted in the above analysis. This term will also have two distinct forms. For the terms $X_n^* X_n$, the time-dependent factor in this product becomes $e^{iE_n t/\hbar} \cdot e^{-iE_n t/\hbar} = 1$. The other time-dependent form appears with terms having $X_n^* X_m$ where $n \neq m$. The time-dependent factor becomes $e^{iE_n t/\hbar} \cdot e^{-iE_m t/\hbar} = e^{-i(E_n - E_m)t/\hbar}$. However, since it is independent of x, it comes out of the integral, and becomes a factor multiplied by $\int X_n^* X_m dx$. Since this integral vanishes, the result is the same.

We thus, find a meaningful interpretation of $|c_n|^2$. The square root of the probability $|c_n|$ (or c_n) is known as the **probability amplitude**, though $|c_n|$ is within a phase factor of c_n. Next, we show how the 1D IDPW problem can readily be extended to a 3D analog, as well.

4.2 Extending The IDPW To 3D

Extending the *Infinitely Deep Potential Well* problem to 3D, conceptually, only requires us to replace the 1D box with a 3D box, as illustrated in Fig. 4.4. Fortunately, as long as we stick with a Cartesian coordinate system, the mathematics does not change much. The 3D potential function $V(x, y, z)$ now becomes

$$V(x,y,z) = \begin{cases} 0 & \text{if } 0 < x < L, 0 < y < L, \text{ and } 0 < z < L \\ +\infty & \text{at } x, y, z = 0 \text{ and } x, y, z = L \end{cases} \quad (4.23)$$

For the 3D case we, again, make use of the *method of separation of variables*. We assume that the wave function $\psi(\mathbf{r}, t)$ takes the form

$$\psi(\mathbf{r}, t) = T(t)X(x)Y(y)Z(z) \quad (4.24)$$

In Eq. (4.24), r is the 3D vector $\mathbf{r} = (x, y, z)$. The 3D Schrödinger equation that we must now solve is given by

$$i\hbar \frac{\partial \psi}{\partial t} = -\frac{\hbar^2}{2m} \nabla^2 \psi + V\psi \quad (4.25)$$

Substitution of Eq. (4.24) into Eq. (4.25) leads to

$$i\hbar (XYZ) \frac{dT}{dt} =$$

$$-\frac{\hbar^2}{2m} \left[T\left(YZ \frac{d^2 X}{dx^2} + XZ \frac{d^2 Y}{dy^2} + XY \frac{d^2 Z}{dz^2} \right) \right] + V(TXYZ) \quad (4.26)$$

All the partial derivatives have been replaced by total derivatives because the test functions T, X, Y, and Z only depend on a single variable. Next, dividing the $\psi = TXYZ$, we have

$$i\hbar \frac{1}{T} \frac{dT}{dt} = -\frac{\hbar^2}{2m} \left(\frac{1}{X} \frac{d^2 X}{dx^2} + \frac{1}{Y} \frac{d^2 Y}{dy^2} + \frac{1}{Z} \frac{d^2 Z}{dz^2} \right) + V(r) \quad (4.27)$$

4.2 Extending The IDPW To 3D

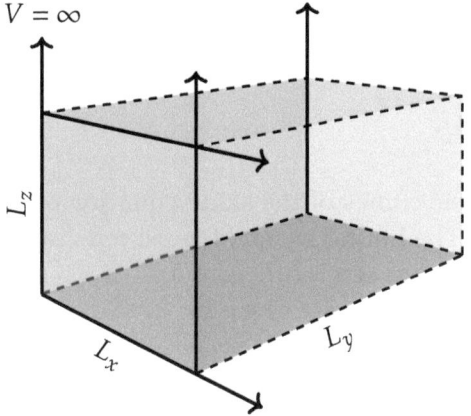

Figure 4.4: Illustration of a 3D particle in a box function. Each dimension is now a 1D infinite potential well.

The LHS of Eq (4.27) is now strictly a function of time t, while the RHS is strictly a function of space $r = (x, y, z)$, and each is an independent variable. Just as we have seen previously, this is only possible if both sides are equal to a constant, given by the eigenvalue E_λ. Therefore, we may write

$$i\hbar \frac{1}{T}\frac{dT}{dt} = E_\lambda \tag{4.28a}$$

$$\left(\frac{1}{X}\frac{d^2X}{dx^2} + \frac{1}{Y}\frac{d^2Y}{dy^2} + \frac{1}{Z}\frac{d^2Z}{dz^2} \right) = -\frac{2m(E_\lambda - V)}{\hbar^2} = -k^2 \tag{4.28b}$$

We encountered Eq. (4.28a) earlier in the 1D case (see Eq. (4.13)), and we found then, that the time-dependent function $T(t)$ is given by

$$T(t) = T_0 e^{-\frac{i}{\hbar}E_\lambda t}$$

Thus, Eq. (4.28b), describing the spatially dependent part, is what remains to be solved. Since each spatial variable is also independent, it follows that each term $f_1(x) = (1/X)d^2X/dx^2$, $f_2(y)$, and $f_3(z)$, each equals their own respective *sub*constants, which all sum to k^2. Let us separate the constant k^2 as

$$k^2 = k_x^2 + k_y^2 + k_z^2 \tag{4.29}$$

Writing three separate equations, we then have

$$\frac{1}{X}\frac{d^2X}{dx^2} = -k_x^2 \tag{4.30a}$$

$$\frac{1}{Y}\frac{d^2Y}{dy^2} = -k_y^2 \tag{4.30b}$$

$$\frac{1}{Z}\frac{d^2Z}{dx^2} = -k_z^2 \tag{4.30c}$$

We now have three copies of the same equation obtained in Eq. (4.10) for the 1D case. The boundary conditions remain the same where the wave function vanishes at $x = 0, L_x$, $y = 0, L_y$, $z = 0, L_z$. Based on what we found in 1D, the solutions are given by

$$X(x) = A\sin(k_x x) \tag{4.31a}$$

$$Y(y) = C\sin(k_y y) \tag{4.31b}$$

$$Z(z) = E\sin(k_z z) \tag{4.31c}$$

Since the wave function must vanish at the box edges in all three directions, it necessitates the following conditions for all three directions:

$$k_x L_x = n_x \pi \Rightarrow k_x^2 = \frac{n_x^2 \pi^2}{L_x^2} \tag{4.32a}$$

$$k_y L_y = n_y \pi \Rightarrow k_y^2 = \frac{n_y^2 \pi^2}{L_y^2} \tag{4.32b}$$

$$k_z L_z = n_z \pi \Rightarrow k_z^2 = \frac{n_z^2 \pi^2}{L_z^2} \tag{4.32c}$$

Thus, for the 3D particle in a box, there are three integers, $n_x = 1, 2, 3, 4, ...$, $n_y = 1, 2, 3, 4, ...$, and $n_z = 1, 2, 3, 4, ...$. Using this result along with Eq. (4.29), we can write the following relation

$$\frac{n_x^2 \pi^2}{L_x^2} + \frac{n_y^2 \pi^2}{L_y^2} + \frac{n_z^2 \pi^2}{L_z^2} = k^2 \tag{4.33}$$

Then, using the definition of k^2 in terms of the energy E_λ, we have the energy states for the 3D IDPW problem, given by

3D Infinitely Deep Potential Well Energy States

4.2 Extending The IDPW To 3D

$$E_\lambda = \frac{\pi^2 \hbar^2}{2m}\left(\frac{n_x^2}{L_x^2} + \frac{n_y^2}{L_y^2} + \frac{n_z^2}{L_z^2}\right) \qquad (4.34)$$

Since E_λ depends on three integers, namely, n_x, n_y, and n_z, rather alternatively denoting the energy by E_n, we now have $E_{nx,ny,nz}$. In the case where $L_x = L_y = L_z = L$, Eq. (4.34) simplifies to the

$$n_x^2 + n_y^2 + n_z^2 = \frac{k^2 L^2}{\pi^2} = \frac{2mL^2 E_\lambda}{\pi^2 \hbar^2} \qquad (4.35)$$

Additionally, since k is the wave number, let us now substitute $k = 2\pi/\lambda$ so that Eq. (4.35) becomes

$$n_x^2 + n_y^2 + n_z^2 = \frac{4\pi^2 L^2}{\lambda^2 \pi^2} = \frac{4L^2}{\lambda^2} \qquad (4.36)$$

Recall in Chap. 2, we had discussed the blackbody radiation problem tackled by Planck and others. Eq. (4.36) to the result in Eq. (2.54), where it lead Raleigh, Jeans, and Planck to a way to count the number of optical modes per unit volume. The earlier result, however, was obtained using Maxwell's equations. Here, quantum mechanics has been used to find the same result for the number of modes for quanta confined to a box (or cavity). However, we also showed that the other part of Planck's work, relating to the mean energy per mode was purely thermodynamic in origin. We cannot obtain that particular result from the Schrödinger equation, as is, because it does not account directly, for temperature. This also means that even if quantum mechanics had existed when Planck was solving the *blackbody radiation problem*, he still would not have been able to do any better!

For the *3D Infinitely Deep Potential Well* problem, the complete solution given by

$$\psi(\mathbf{r},t) = TXYZ = A\sin\left(\frac{n_x \pi}{L_x}x\right) C\sin\left(\frac{n_y \pi}{L_y}y\right) E\sin\left(\frac{n_z \pi}{L_z}y\right) T_0 e^{-\frac{i}{\hbar}E_\lambda t} \qquad (4.37)$$

or

3D Infinitely Deep Potential Well Eigenfunctions

$$\Psi_{n_x,n_y,n_z}(\mathbf{r},t) = C_n \sin\left(\frac{n_x\pi}{L_x}x\right)\sin\left(\frac{n_y\pi}{L_y}y\right)\sin\left(\frac{n_z\pi}{L_z}z\right)e^{-\frac{i}{\hbar}E_\lambda t} \quad (4.38)$$

We have introduced a normalization constant C_n in Eq. (4.38). The normalization constant is obtained by enforcing the *normalization condition* over the box (since wave functions vanish outside the box). This leads to

$$\int_V \psi^* \cdot \psi \, dV = 1 \quad (4.39)$$

Note that the time-dependent factor becomes $T^*T = e^{iE_\lambda t/\hbar}e^{-iE_\lambda t/\hbar} = e^0 = 1$. So, we then have

$$\int_{-\infty}^{\infty} \psi^*\psi \, d\mathbf{r} = C_N^2 \times$$

$$\int_0^{L_x} \sin^2\left(\frac{n_x\pi}{L_x}x\right)dx \int_0^{L_y} \sin^2\left(\frac{n_y\pi}{L_y}y\right)dy \int_0^{L_z} \sin^2\left(\frac{n_z\pi}{L_z}z\right)dz \quad (4.40)$$

To obtain this integral, let us consider each integral separately. We denote the first integral to evaluate as

$$I_x = \int_0^{L_x} \sin^2\left(\frac{n_x\pi}{L_x}x\right)dx \quad (4.41)$$

The evaluation of Eq. (4.41) can be done using the trigonometric identity

$$2\sin\theta\sin\varphi = \cos(\theta-\varphi) - \cos(\theta+\varphi) \quad (4.42)$$

Since both arguments are the same, we get $\sin^2\theta = (1+\cos 2\theta)/2$. Then, I_x becomes

$$I_x = \int_0^{L_x}\sin^2\left(\frac{n_x\pi}{L_x}x\right)dx = \int_0^{L_x}\frac{1+\cos\left(\frac{2n_x\pi}{L_x}x\right)}{2}dx =$$

$$\frac{L_x}{2} + \frac{1}{2}\int_0^{L_x}\cos\left(\frac{2n_x\pi}{L_x}x\right) = \frac{L_x}{2} + \frac{L_x}{4n_x\pi}\sin\left(\frac{2n_x\pi}{L_x}x\right)\Big|_0^{L_x} = \frac{L_x}{2} \quad (4.43)$$

4.2 Extending The IDPW To 3D

Because the last term evaluates sin at 0 and $2n_x\pi$ (integer multiples of π), this term vanishes, and Eq. (4.43) becomes

$$I_x = \frac{L_x}{2} \tag{4.44}$$

The evaluation is identical for I_y and I_z (just replace x with y, then with z and they look the same!). So, our integral in Eq. (4.40) becomes

$$1 = \int_{-\infty}^{\infty} \psi * \psi d\mathbf{x} = C_n^2 \times$$

$$\int_0^{L_x} \sin^2\left(\frac{n_x\pi}{L_x}x\right)dx \int_0^{L_y} \sin^2\left(\frac{n_y\pi}{L_y}y\right)dy \int_0^{L_z} \sin^2\left(\frac{n_z\pi}{L_z}z\right)dz =$$

$$C_n^2 \times \left(\frac{L_x}{2}\frac{L_y}{2}\frac{L_z}{2}\right) \tag{4.45}$$

This gives the following result for the normalization constants C_n in 1D, 2D, and 3D, respectively:

$$C_n = \sqrt{\frac{2}{L_x}} \quad \text{(1D)} \tag{4.46a}$$

$$C_n = \sqrt{\frac{2}{L_x}\frac{2}{L_y}} \quad \text{(2D)} \tag{4.46b}$$

$$C_n = \sqrt{\frac{2}{L_x}\frac{2}{L_y}\frac{2}{L_z}} \quad \text{(3D)} \tag{4.46c}$$

With the normalization constants, and having the solution for the 1D IDPW, as well as the 3D solution, it is straightforward write all forms of the solution from 1D, 2D, to 3D, where they are given by

$$\psi_{1D}(x) = \sqrt{\frac{2}{L_x}}\sin\left(\frac{n_x\pi x}{L_x}\right) \tag{4.47a}$$

$$\psi_{2D}(x,y) = \sqrt{\frac{2^2}{L_xL_y}}\sin\left(\frac{n_x\pi x}{L_x}\right)\sin\left(\frac{n_y\pi y}{L_y}\right) \tag{4.47b}$$

$$\psi_{3D}(x,y,z) = \sqrt{\frac{2^3}{L_xL_yL_z}}\sin\left(\frac{n_x\pi x}{L_x}\right)\sin\left(\frac{n_y\pi y}{L_y}\right)\sin\left(\frac{n_z\pi z}{L_z}\right) \tag{4.47c}$$

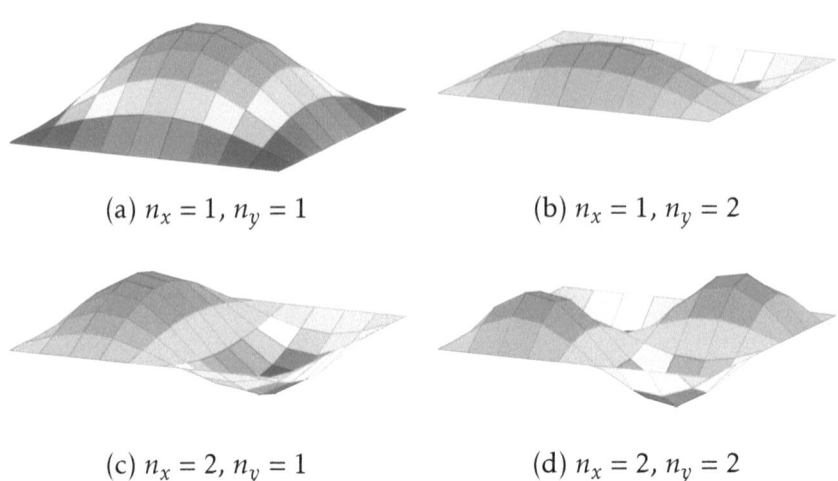

Table 4.1: Illustration of four eigen-solutions from the 2D IDPW. The quantum number n_x and n_y are: (a) $n_x = 1, n_y = 1$, (b) $n_x = 1, n_y = 2$, (c) $n_x = 2, n_y = 1$, and (c) $n_x = 2, n_y = 2$.

For visualization of eigensolutions of the IDPW problem, we show 2D examples below, since they are easier to visualize. Four cases are shown in Table 4.1, corresponding to $(n_x = 1, n_y = 1)$, $(n_x = 1, n_y = 2)$, $(n_x = 2, n_y = 1)$, and $(n_x = 2, n_y = 2)$. For all eigensolutions, because of the boundary conditions, they vanish at the boundaries of the box, just as we had in the 1D case. These solutions, obtained via quantum mechanics, are the same as 2D eigen-modes one can obtain by solving the classical 2D string wave-equation (the application to strings leads to boundary conditions of IDPW problem), given earlier by Eq. (1.33).

This completes the solution to the *infinitely deep potential well* (or *particle in a box*) problem. The problem also lends itself to the introduction of other useful concepts in quantum mechanics. In the next section, we'll introduce one of them using the IDPW problem as a basis.

4.3 Introduction to Density of States

The IDPW problem revealed that the eigen-energy E_λ can be directly related to the quantum numbers n_x, n_y, and n_z. This is evident from Eq.

4.3 Introduction to Density of States

(4.35), which leads to

$$E_\lambda(n_x, n_y, n_z) = \frac{\pi^2 \hbar^2}{2mL^2}\left(n_x^2 + n_y^2 + n_z^2\right) \tag{4.48}$$

Eq. (4.48) also suggests that there are several different choices of (n_x, n_y, n_z) that lead to the same energy state. For example, $(7, 7, 11)$ leads to the same energy as $(11, 7, 7)$ because $7^2 + 7^2 + 11^2 = 11^2 + 7^2 + 7^2$. Distinct energy states that have the same energy are called **degenerate states**. Different quantum states have different quantum coordinate (n_x, n_y, n_z), but they may still have the same energy. When the quantum numbers are large, there are very many degenerate states with the same energy. It therefore, becomes useful to have a way to keep track of the *number of available states per unit energy*. Let's explore how this can be achieved.

As we encountered in Chap. 2, with Eq. (2.54), there is a resemblance, in the mathematical form, to a 3D sphere of radius R (recall $x^2 + y^2 + z^2 = R^2$), such that

$$R(E_\lambda) = \left(\frac{2mL^2 E_\lambda}{\pi^2 \hbar^2}\right)^{\frac{1}{2}} \tag{4.49}$$

The x, y, and z axes, here, are replaced by the quantum coordinate axes n_x, n_y, and n_z. For the case of large quantum numbers, where $\Delta n_i / n_i$ is small, these kinds of classical analogies are permitted. In this case, it provides a means to readily estimate the number of discrete states N with energies $\leq E_\lambda$. N is estimated by the volume V_o of the 3D sphere, but only in the first octant, where n_x, n_y, and n_z are positive. Since an eighth of the volume of a sphere is given by

$$V_o = \frac{1}{8}\left(\frac{4}{3}\pi R^3\right) = \frac{1}{6}\pi R^3$$

We then have

$$N(E_\lambda) = \frac{1}{6}\pi \left(\frac{2mL^2 E_\lambda}{\pi^2 \hbar^2}\right)^{\frac{3}{2}} \tag{4.50}$$

However, because fundamental particles also possess spin states (two per particle), the number in Eq. (4.50) multiplies by a factor of 2. So, the number of states $N(E_\lambda)$ for a particle in a box, is given by

$$N(E_\lambda) = \frac{2}{6}\pi \left(\frac{2mL^2 E_\lambda}{\pi^2 \hbar^2}\right)^{\frac{3}{2}} \tag{4.51}$$

We will introduce the spin formally in Chap. 9. For any finite energy differential dE_λ, or range given by ΔE_λ, the corresponding differential number of possible states dN over the range can be obtained. Integration of this differential gives the total number of states with energies found in this range. The number of states per unit energy is defined as the **density of states** $D(E_\lambda)$, given by

Density Of States

$$D(E_\lambda) = \frac{dN}{dE_\lambda} \tag{4.52}$$

Differentiating Eq. (4.51) to get the *density of states*, we get

$$D(E_\lambda) = \frac{dN}{dE_\lambda} = \frac{\sqrt{2}m^{\frac{3}{2}}L^3}{\pi^2\hbar^3}E_\lambda^{\frac{1}{2}} \tag{4.53}$$

Being proportional to the volume L^3, the *density of states* can also be expressed per unit volume, which we'll denote by $D_v(E)$, as

$$D_v(E_\lambda) = \frac{dn}{dE_\lambda} = \frac{\sqrt{2}m^{\frac{3}{2}}}{\pi^2\hbar^3}E_\lambda^{\frac{1}{2}} \Rightarrow dn = \frac{\sqrt{2}m^{\frac{3}{2}}}{\pi^2\hbar^3}E_\lambda^{\frac{1}{2}}dE_\lambda$$

Fig. 4.5 plots the number of states $N(E)$, given by Eq. (4.51), for a cubic nanometer of volume as a function of the energy E_λ. The number of available states rises relatively sharply with the energy. Even for a sample volume of 1 cubic nanometer, the number of states can be fairly large. Eq. (4.53) is an example of the *density of states* or **DOS** per unit volume $d(E)$, also plotted in Fig. 4.6. The *DOS* provides the distribution of *available* states across the energy spectrum. When solving the Schrödinger equation used in quantum mechanics, one of the principle results from it includes the available energies and their corresponding quantum numbers. So, the relationship between N and E_λ can just about always be determined from quantum mechanics. For many systems of practical interest, because the quantum numbers (e.g. n_x, n_y, n_z) can be very large, it is useful to have a systematic way to express the DOS of a system. Then, integrals may be used with the DOS, and this leads to more utility.

Another form that may be used for the DOS is in the form of a probability distribution for states between *any finite range of energies*. For the 3D

4.3 Introduction to Density of States

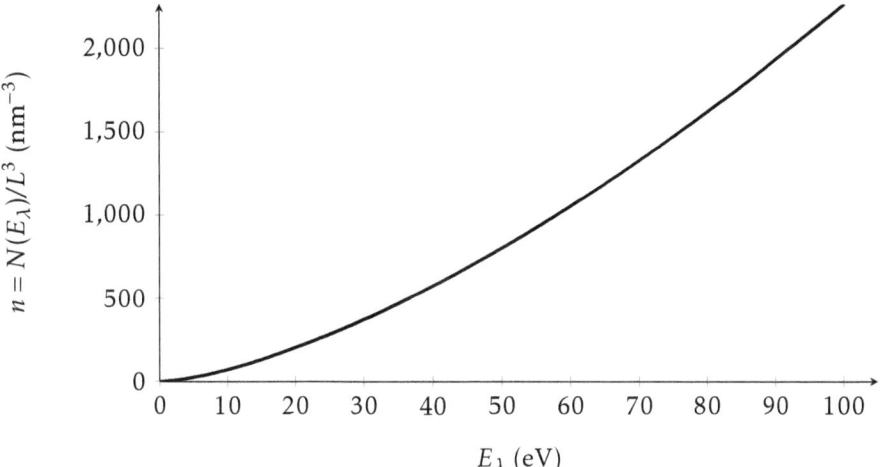

Figure 4.5: Number of states per unit volume (cubic nanometer) n with energy E_λ, plotted as a function of E_λ.

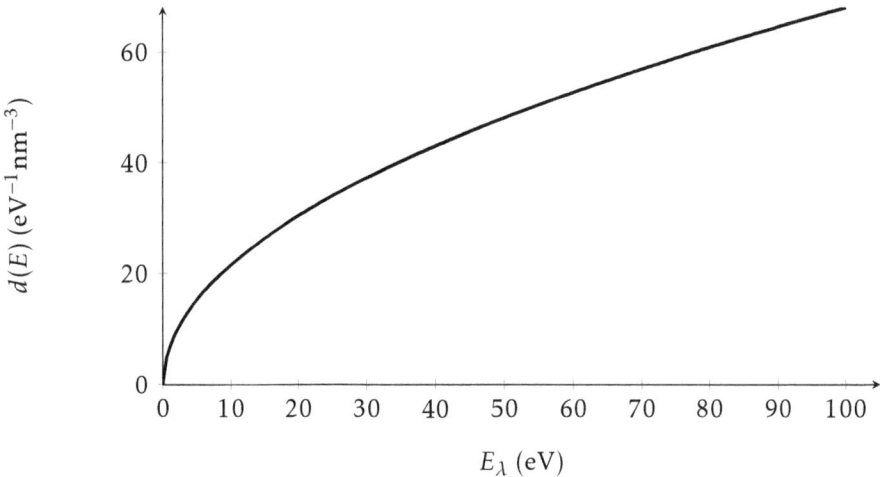

Figure 4.6: Density of states $d(E)$ (per unit volume of a cubic nanometer) as a function of energy E_λ.

particle in a box, the DOS given by Eq. (4.53), allows the total number of available states N is given by

$$N = \int_0^\infty D(E_\lambda)dE_\lambda = \int_0^\infty \frac{dN}{dE_\lambda}dE_\lambda = \int_0^\infty dN \tag{4.54}$$

Then, the number of available states ΔN with energies within the range

of $\Delta E = E_2 - E_1$ becomes

$$N_\Delta = \int_{E_1}^{E_2} D(E_\lambda) dE_\lambda = \int_0^{E_2} D(E_\lambda) dE_\lambda - \int_0^{E_1} D(E_\lambda) dE_\lambda = N_2 - N_1 \quad (4.55)$$

Since N_Δ is the total number of states in the interval between E_1 and E_2, dividing Eq. (4.55) by the total N_Δ gives a **density of states probability distribution** f_D, defined as

Density of States Probability Distribution

$$f_D(E_\lambda) = \frac{D(E_\lambda)}{N_\Delta} \quad (4.56)$$

This definition leads to

$$1 = \int_{E_1}^{E_2} \frac{D(E_\lambda)}{N_\Delta} dE_\lambda = \int_{E_1}^{E_2} f_D(E_\lambda) dE_\lambda \quad (4.57)$$

The *density of states* over a finite interval is, therefore, proportional to the probability distribution of the states across the energy spectrum, particularly, over the range of energies between E_1 and E_2.

 Density of states: The DOS can be used to express the probability distribution of states over any finite energy range ΔE.

This result follows from the fact that the probability of a state with energy E_λ is proportional to the number of states with that energy over the total number of possible states (within that range). Recall this was the case for the probability distributions we obtained in Chap. 2, such as the *Fermi-Dirac* and *Bose-Einstein* distributions. The *DOS* provides additional information about a system.

However, as mentioned, the density of states only accounts for the number of *available* energy states, and *provides no information on which energy states are occupied*. For that, we need to know the energy occupation relative to the number of available states available at an energy level. This relates to the thermodynamic distributions discussed in Chap. 2. Generally, a system will seek to minimize it's energy, therefore filling

4.3 Introduction to Density of States

the lower energy states first, then subsequent higher energy levels are filled until all the fundamental particles have a home energy state. This process is limited because energy is needed to access these states. Eventually, there is a limit to the states reached due to available energy limitations. Let us denote this relative occupancy probability distribution by $f_N(E_\lambda) = d\aleph/dN$. It represents out of how many available states there are, the ratio of how many of them are occupied. Like f_D and $D(E_\lambda)$, it depends on the energy, and it is a dimensionless quantity. The product of the two $f_D \cdot f_N$ provides probability of an occupied state for the system $\aleph(E_\lambda)$ as a function of E_λ. In other words, it is proportional to the system's energy spectrum. We can use this to express the total occupied number of states $\aleph$ for a system as

$$\aleph_\Delta = \int_{E_1}^{E_2} d\aleph = \int_{E_1}^{E_2} \frac{d\aleph}{dN} \frac{dN}{dE_\lambda} dE_\lambda = \int_{E_1}^{E_2} f_O(E_\lambda) D(E_\lambda) dE_\lambda \quad (4.58)$$

$d\aleph$ represents the probability distribution of the occupancy of certain energy levels. $f_N(E_\lambda)$ becomes the temperature dependent statistical probability distribution, such as the Fermi-Dirac distribution. It now becomes possible to express the *probabilities* of finding the system with an energy E_λ for any thermodynamic equilibrium temperature T. The **occupation probability distribution** as a function of energy E_λ is given by

Temperature Dependent Occupation Probability Distribution

$$f_O(E_\lambda) = \frac{f_N(E_\lambda) D(E_\lambda)}{\aleph_\Delta} \quad (4.59)$$

It follows from Eq. (4.59) that

$$\int f_O(E_\lambda) dE_\lambda = \frac{1}{\aleph_\Delta} \int_{E_1}^{E_2} f_N(E_\lambda) D(E_\lambda) dE_\lambda = \frac{\aleph_\Delta}{\aleph_\Delta} = 1 \quad (4.60)$$

Eq. (4.60) defines a more useful probability distribution for a physical system. This is because the product $f_N(E_\lambda) D(E_\lambda)$ forms a distribution function whose area is bounded. At low energies, it is zero, while it approaches zero towards higher energies, as well. We saw examples of functions with this property in Secs. 3.4 and 3.3 with the Gaussian distribution, the Sinc function, etc. Fig. 4.7 illustrates the distribution

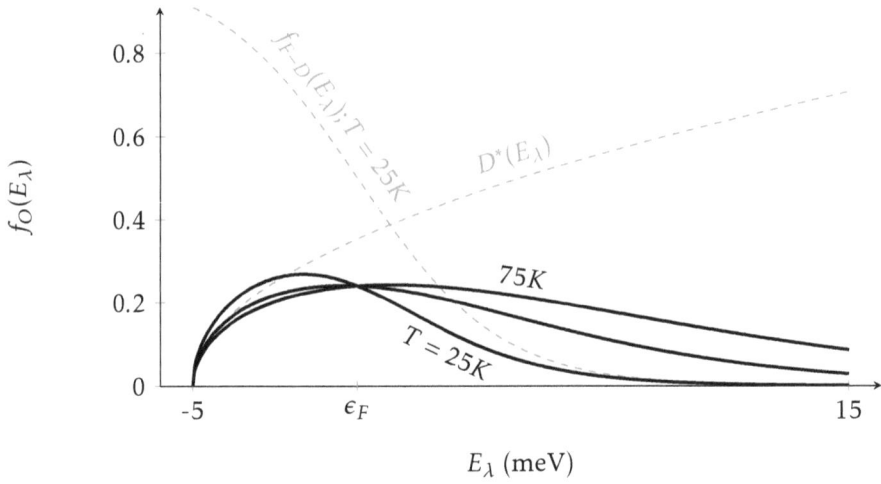

Figure 4.7: Example of $f_O = D^* \cdot f_N$ where $f_N = f_{F-D}$ (Fermi-Dirac distribution) as a function of energy E_λ. D^* is a scaled version of D to aid in the illustration (not normalized). The gray dotted lines are the density of states D and Fermi-Dirac distribution respectively. The product $D \cdot f_N$ is a probability distribution over the energy range in the vicinity of the Fermit energy. Note that $E_\lambda = 0$ meV corresponds to the Fermi level in this illustration, for $T = 25, 50, 75$K.

f_O for the case $f_N(E_\lambda) = f_{F-D} = N_i/d_i$ being the *Fermi-Dirac distribution*. Because of this property, $f_O = D \cdot f_N$ can be normalized to unit area where $\int f_O(E)dE = 1$. In normalized form, the distribution approaches the Dirac delta function in the limit $T \to 0$, since reducing T narrows the distribution while raising the peak, analogous to how reducing σ narrows the *Normal Gaussian distribution*. The product function $f_O = D \cdot f_N$, therefore, enables a definition for the expectation value of a quantity $\langle E_\lambda \rangle$ *over any range of energies* or $\langle E_\lambda - \mu \rangle$. μ is any energy reference point chosen, e.g. the Fermi level. The expectation for the energy becomes

$$\langle E_\lambda \rangle = \frac{\int_{E_1}^{E_2} E_\lambda d\aleph(E_\lambda)}{\aleph_\Delta} \tag{4.61}$$

In terms of the density of states, Eq. (4.58) leads to

Energy Expectation Value Using Density of States

$$\langle E_\lambda \rangle = \frac{\int_{E_1}^{E_2} E_\lambda D(E_\lambda) f_N(E_\lambda) dE_\lambda}{\int_{E_1}^{E_2} D(E_\lambda) f_N(E_\lambda) dE_\lambda} \qquad (4.62)$$

If the system is comprised of indistinguishable electrons, which are *fermions*, then the thermodynamic equilibrium energy probability distribution $f_N(E_\lambda)$ is given by the *Fermi-Dirac distribution*, derived in Chap. 2, Sec. 2.5.1, given by Eq. (2.36). Together, they can describe the expectation value for a system of energies. We will come back to these ideas in later chapters, as well. Let's now look at some additional important properties of the eigensolutions to the Schrödinger equation.

4.4 ORTHOGONALITY OF SCHRÖDINGER'S EIGENSOLUTIONS

We return to another question relating to the probabilities, namely, the integral of the product of any two *distinct* eigen-solutions $\psi_m(x,t)$ and $\psi_n(x,t)$, where $m \neq n$. For the 1D IDPW problem, we can evaluate the integral (in space) of the product of any two different eigen-solutions, as

$$\int_{-\infty}^{+\infty} \psi_m^* \psi_n dx = \int_0^L X_m^* e^{i\frac{E_m}{\hbar}(t-t_0)} X_n e^{-i\frac{E_n}{\hbar}(t-t_0)} dx$$

$$= e^{-i\frac{E_n-E_m}{\hbar}(t-t_0)} \int_0^L X_m^* \cdot X_n dx$$

Substitution of the solutions $X_m(x)$ and $X_n(x)$ given by Eq. (4.47a), we have

$$\int_{-\infty}^{+\infty} \psi_m^* \psi_n dx = e^{-i\frac{E_n-E_m}{\hbar}(t-t_0)} \int_0^L \sin\left(\frac{m\pi x}{L}\right) \cdot \sin\left(\frac{n\pi x}{L}\right) dx \qquad (4.63)$$

To evaluate the integral, we can make use of the following trigonometric identity:

$$\sin(nu) \cdot \sin(mu) = \frac{1}{2}[\cos(n-m)u - \cos(n+m)u]$$

Substitution of this identity into Eq. (4.63), we then have

$$\int_{-\infty}^{+\infty} \psi_m^* \psi_n dx = \frac{e^{-i\frac{E_n-E_m}{\hbar}(t-t_0)}}{2} \int_0^L \left[\cos\left(\frac{(m-n)\pi x}{L}\right) - \cos\left[\frac{(m+n)\pi x}{L}\right]\right] dx$$

(4.64)

Evaluation of both terms gives

$$\int_{-\infty}^{+\infty} \psi_m^* \psi_n dx = \frac{e^{-i\frac{E_n-E_m}{\hbar}(t-t_0)}}{2} \int_0^L \left[\cos(\frac{(m-n)\pi x}{L}) - \cos(\frac{(m+n)\pi x}{L})\right] dx$$

$$= \frac{e^{-i\frac{E_n-E_m}{\hbar}(t-t_0)}}{2} \Bigg(\frac{L}{(m-n)\pi} \sin\left[\frac{(m-n)\pi x}{L}\right]$$

$$\ldots - \frac{L}{(m+n)\pi} \sin\left[\frac{(m+n)\pi x}{L}\right] \Bigg) \Bigg|_0^L$$

Because x takes the value of 0 and L, the arguments of the sine functions become integer multiples of π, and since $\sin(n\pi) = 0$, we have the result

$$\int_{-\infty}^{+\infty} \psi_m^* \psi_n dx = 0 \qquad (4.65)$$

The result above holds for any choice of integer pair where $m \neq n$. Whenever such an integral of the product of two distinct functions is zero, as we have here, the two functions are said to be **orthogonal**. In this case, the entire set of independent solutions ψ_n that we have found, solving both Eq. (4.11) and Eq. (4.9), all satisfy this condition, pairwise. When a set of functions satisfies *orthogonality* for any two integers (or states), the entire set of functions satisfies what is known as the **orthogonality condition**. The orthogonality condition can be expressed as

Orthogonality Condition

$$\int \psi_m(x) \cdot \psi_n(x) dx = 0 \quad \forall m \neq n \qquad (4.66)$$

The *orthogonality condition* turns out to be much more general than to just this example of ψ. We can show that it is a satisfied condition for *any* solution to the Schrödinger equation, provided the potential energy V is real-valued. Let us proceed to show this. We will show that if $E_n \neq E_m$, then it follows that

$$\int_{-\infty}^{+\infty} \psi_m(\mathbf{r}) \cdot \psi_n(\mathbf{r}) d\mathbf{r} = 0$$

4.4 Orthogonality Of Schrödinger's Eigensolutions

Provided that the potential energy is real-valued, then the Schrödinger equation is separable is space and time. Then, the Schrödinger equation for both wave functions ψ_m and ψ_n are described by

$$\nabla^2 \psi_m + \frac{2m}{\hbar^2}(E_m - V)\psi_m(\mathbf{r}) = 0 \tag{4.67a}$$

$$\nabla^2 \psi_n + \frac{2m}{\hbar^2}(E_n - V)\psi_n(\mathbf{r}) = 0 \tag{4.67b}$$

First, let's take the complex conjugate of the second equation. Then multiply the first equation by the complex conjugate ψ_n^*, and the new second equation by ψ_m. This leads to

$$\psi_n^* \nabla^2 \psi_m + \frac{2m}{\hbar^2}(E_m - V)\psi_n^* \psi_m(\mathbf{r}) = 0$$

$$\psi_m \nabla^2 \psi_n^* + \frac{2m}{\hbar^2}(E_n - V)\psi_m \psi_n^*(\mathbf{r}) = 0$$

Taking the difference between the two equations gives

$$\psi_n^* \nabla^2 \psi_m - \psi_m \nabla^2 \psi_n^* + \frac{2m}{\hbar^2}(E_m - E_n)\psi_n^* \psi_m(\mathbf{r}) = 0 \tag{4.68}$$

The first term can be replaced by making use of vector calculus identity

$$\phi \nabla^2 \alpha - \alpha \nabla^2 \phi = \nabla \cdot (\phi \nabla \alpha - \alpha \nabla \phi) \tag{4.69}$$

Substitution into Eq. (4.70) gives

$$\nabla \cdot (\psi_n^* \nabla \psi_m - \psi_m \nabla \psi_n^*) + \frac{2m}{\hbar^2}(E_m - E_n)\psi_n^* \psi_m(\mathbf{r}) = 0 \tag{4.70}$$

The next step will be to integrate Eq. (4.70) over 3D space, or volume. This leads to

$$\int_{-\infty}^{+\infty} \nabla \cdot (\psi_n^* \nabla \psi_m - \psi_m \nabla \psi_n^*) d\mathbf{r} + \frac{2m}{\hbar^2}(E_m - E_n) \int_{-\infty}^{+\infty} \psi_n^* \psi_m d\mathbf{r} = 0 \tag{4.71}$$

The first term can be replaced using the identity known as the *Divergence Theorem*, which states

$$\int_{Vol} \nabla \cdot \mathbf{A} \, dV = \int_{Surf} n \cdot \mathbf{A} \, dS$$

This integral equates a volume integral to a surface integral. The surface integral is on the bounding surface of Vol. Note that $\int_{-\infty}^{+\infty} \psi_n^* \psi_m d\mathbf{r} =$

$\int_{-\infty}^{+\infty}\int_{-\infty}^{+\infty}\int_{-\infty}^{+\infty} \psi_n^* \psi_m dV$. It is the 3D integral over all space. The Divergence Theorem, therefore, allows us to replace the volume integral with an integral on the boundaries. So, Eq. (4.71) becomes

$$\int_{Surf} n \cdot (\psi_n^* \nabla \psi_m - \psi_m \nabla \psi_n^*) dS + \frac{2m}{\hbar^2}(E_m - E_n) \int_{-\infty}^{+\infty} \psi_n^* \psi_m d\mathbf{r} = 0 \quad (4.72)$$

We can determine the exact value of this substitute term, because the integral sums values of the two terms that are proportional to values of the wave functions ψ_n^* and ψ_n^* on the boundary. But these values are precisely zero, at the boundaries! Therefore, the integrand is identically zero, and thus, so is the integral.

Note that all eigen-solutions for any fundamental particle is taken to vanish at the boundary, whether it's localized or itinerant (traveling from atom to atom). So, this entire term vanishes and we are left with

$$\frac{2m}{\hbar^2}(E_m - E_n) \int_{-\infty}^{\infty} \psi_n^* \psi_m d\mathbf{r} = 0 \quad (4.73)$$

Since we assume that $E_m \neq E_n$, the coefficient is nonzero, and our final result follows, such that

$$\int_{-\infty}^{\infty} \psi_n^* \psi_m d\mathbf{r} = 0 \quad (4.74)$$

We have the result we were after, where we find that the *orthogonality condition* is, indeed, satisfied by any solution ψ, satisfying the Schrödinger equation with real spatially dependent potential.

4.5 REAL EIGENVALUES OF SCHRÖDINGER'S EQUATION

You may have noticed that while we carried out our derivation of the result in Eq. (4.74), we quietly used the fact that the energies E_m and E_n are real numbers. We did this when we took the complex conjugate of the Schrödinger equation in Eq. (4.67b). This turns out to be true as long as the potential V is real-valued, and we will demonstrate this point in this section. In fact, it follows a very similar flow to the above proof, except here, we will only consider 1D. Extending to 3D is straightforward based on the analysis in the previous section, so we leave this as an exercise for the reader.

4.5 Real Eigenvalues Of Schrödinger's Equation

 Real Eigenvalues: Because the kinetic energy operator is defined already as a real-valued operator, as long as the potential V is also real, then the eigenvalues will always be real-valued.

We wish to show that under the same conditions as above (real-valued $V(x)$) *the eigenvalues of the Schrödinger equation are necessarily real*, and not complex. We can do this by showing that E_n equals its complex conjugate E_n^*, or $E_n = E_n^*$. This statement is only true for real numbers. We start with the 1D Schrödinger equation for the energy E_n.

$$\frac{d^2\psi_n}{dx^2} + \frac{2m}{\hbar^2}(E_n - V)\psi_n(x) = 0 \tag{4.75}$$

Let's take the complex conjugate of Eq. (4.75), now allowing E_n to be complex, while V is still real. Then, we have

$$\frac{d^2\psi_n^*}{dx^2} + \frac{2m}{\hbar^2}(E_n^* - V)\psi_n^*(x) = 0 \tag{4.76}$$

Multiply the complex equation by ψ_n and the other by ψ_n^*, and take the difference, which leads to

$$\psi_n^*\frac{d^2\psi_n}{dx^2} - \psi_n\frac{d^2\psi_n^*}{dx^2} + \frac{2m}{\hbar^2}(E_n - E_n^*)\psi_n^*\psi_n(x) = 0 \tag{4.77}$$

The first two terms have a form that allows the following substitution:

$$\psi_n^*\frac{\partial^2\psi_n}{\partial x_i^2} - \psi_n\frac{\partial^2\psi_n^*}{\partial x_i^2} = \frac{\partial}{\partial x_i}\left(\psi_n^*\frac{\partial\psi_n}{\partial x_i} - \psi_n\frac{d\psi_n^*}{\partial x_i}\right) \tag{4.78}$$

Using this result in Eq. (4.77), then integrating gives

$$\left(\psi_n^*\frac{d\psi_n}{dx} - \psi_n\frac{d\psi_n^*}{dx}\right)\Big|_{-\infty}^{+\infty} + \frac{2m}{\hbar^2}(E_n - E_n^*)\int_{-\infty}^{\infty}\psi_n^*\psi_m(x) = 0 \tag{4.79}$$

Because the wave functions vanish on the boundaries, the first term vanishes. Then, we are left with

$$\frac{2m}{\hbar^2}(E_n - E_n^*)\int_{-\infty}^{+\infty}\psi_n^*\psi_n(x)dx = 0 \tag{4.80}$$

The integral term in Eq. (4.80) is positive definite because the eigensolutions to the Schrödinger Equation are always non-trivial solutions. Since the integral gives the square of the amplitude, the result is always > 0. The integral is the addition of a bunch of positive numbers, which is still a positive number. It follows that $E_n = E_n^*$, which means that E_n is, in fact, a real number.

We have only used *one side* of the Schrödinger equation, which contains the spatial-dependent Hamiltonian operator. It contains both the kinetic energy operator and potential energy. Note that this is possible because we have assumed that V is independent of time, or $V = V(x)$. Under this assumption, the separability of time and space may be exploited, enabling us to restrict our attention to the Hamiltonian operator. In general, whenever an operator (in this case, the Hamiltonian) leads to real eigenvalues, it is known as a **Hermitian operator**, a **real-operator**, or **self-adjoint operator**. Therefore, the Hamiltonian considered here is an example of a *Hermitian operator*. Though Hermitian operators are very common, there is nothing restricting the possibility of a non-Hermitian operator. That is, an operator that leads to complex eigen-values. We'll see examples of these in later chapters.

4.6 CURRENT DENSITY AND A CONTINUITY EQUATION

In this section, we will show another important result from the Schrödinger equation. Specifically, we'll show that the Schrödinger equation in the Hermitian form leads to a general continuity equation. We begin with the Schrödinger equation and it's complex conjugate, given by

$$i\hbar \frac{\partial \psi}{\partial t} = -\frac{\hbar^2}{2m} \nabla^2 \psi + V\psi \qquad (4.81a)$$

$$-i\hbar \frac{\partial \psi^*}{\partial t} = -\frac{\hbar^2}{2m} \nabla^2 \psi^* + V\psi^* \qquad (4.81b)$$

Next, multiply Eq. (4.81a) by ψ^*, and Eq. (4.81b) by ψ, to obtain

$$i\hbar \psi^* \frac{\partial \psi}{\partial t} = -\frac{\hbar^2}{2m} \psi^* \nabla^2 \psi + V\psi^*\psi$$

$$-i\hbar \psi \frac{\partial \psi^*}{\partial t} = -\frac{\hbar^2}{2m} \psi \nabla^2 \psi^* + V\psi\psi^*$$

Taking the difference between the two equations leads to the following equation:

$$i\hbar \left(\psi^* \frac{\partial \psi}{\partial t} + \psi \frac{\partial \psi^*}{\partial t} \right) = i\hbar \frac{\partial}{\partial t}(\psi^*\psi) = \frac{\hbar^2}{2m}\left(\psi \nabla^2 \psi^* - \psi^* \nabla^2 \psi\right) \qquad (4.83)$$

4.6 Current Density And A Continuity Equation

Note that the last terms involving the potential energy V cancel one another, as long as V is real. Based on Eq. (4.83), let us define a quantity $\rho(x)$ as

$$\rho(\mathbf{r}) \equiv \psi \cdot \psi^* = \psi^* \cdot \psi = |\psi|^2(\mathbf{r}) \tag{4.84}$$

To evaluate the integral of ρ over its occupied volume, it must be that

$$s = \int_{\text{Vol}} \rho \, dV = \int_{\text{Vol}} \psi^* \cdot \psi \, dV = \int_{\text{Vol}} |\psi|^2 dV = \text{finite number} \tag{4.85}$$

This condition turns out to be necessary in order to take construct any meaning behind Eq. (4.83). Specifically, for any wave function that satisfies the Schrödinger equation, the following condition for ψ must be satisfied:

Square Integrable Condition

$$\int_{-\infty}^{+\infty} |\psi(x)|^2 dx < \infty \tag{4.86}$$

The condition given by Ineq. (4.86) means that ψ is **square integrable**. Note that for any square integrable function, it is a *necessary condition* that

$$\lim_{|x| \to \infty} \psi(x) = 0 \tag{4.87}$$

We may then take Eq. (4.87) as the general boundary condition for the wave function ψ satisfying the Schrödinger equation of real-valued non-uniform potential. If Eq. (4.88) holds, then it guarantees that we can always choose a convenient scale ($C_n^2 = 1/s$) such that

Normalization Condition

$$\int_{\text{Vol}} C_n^2 |\psi|^2 dV = 1 \tag{4.88}$$

For any *square-integrable* function we have a finite integral result. The scaled wave function $C_n \psi$ justifies a **normalized wave function**. With this understanding, Eq. (4.83) becomes

$$i\hbar \frac{d\rho}{dt} = \frac{\hbar^2}{2m} \left(\psi \nabla^2 \psi^* - \psi^* \nabla^2 \psi \right) \tag{4.89}$$

A lasting interpretation of Eq. (4.88) which is now standard, is due to German physicist Max Born (1982 - 1970). Born formulated his interpretation as follows: The integrand $\Psi^*\Psi$ is interpreted as **the probability density function**, since integrating it over all of space gives back unity, just as any probability density function does. Born proposed the following conditions as necessary properties of the wave function Ψ. These conditions are known as **Born's conditions**:.

Table 4.2: Born's Conditions.

Born's Conditions for the Wave-Function Ψ
1. The wave function must be *single-valued*, defined once for any x.
2. The wave function must be *square-integrable*.
3. The first derivative (proportional to momentum) of Ψ must be finite everywhere.
4. The second derivative (proportional to energy) must also be finite everywhere.

Born received the Nobel Prize for his work in this area, in 1954. Now, let us consider the right-hand side of Eq. (4.83), where we have

$$RHS = \frac{\hbar^2}{2m}(\psi \nabla^2 \psi^* - \psi^* \nabla^2 \psi) \qquad (4.90)$$

Let's consider the following relation involving the *divergence* operator $\nabla \cdot ()$:

$$\nabla \cdot (\psi \nabla \psi^*) = \psi \nabla \cdot (\nabla \psi^*) + \nabla \psi \cdot \nabla \psi^*$$

Likewise for the term $\psi^* \nabla \psi$, we have

$$\nabla \cdot (\psi^* \nabla \psi) = \psi^* \nabla \cdot (\nabla \psi) + \nabla \psi^* \cdot \nabla \psi$$

Since the last terms of both equations are the same, taking the difference of these divergences, we get

$$\nabla \cdot (\psi \nabla \psi^*) - \nabla \cdot (\psi^* \nabla \psi) = \psi \nabla \cdot (\nabla \psi^*) - \psi^* \nabla \cdot (\nabla \psi) = \psi(\nabla^2 \psi^*) - \psi^*(\nabla^2 \psi)$$

4.6 Current Density And A Continuity Equation

But, this is identical to the RHS of Eq. (4.90), which means that we can also write Eq. (4.83) as

$$i\hbar \frac{d\rho}{dt} = \frac{\hbar^2}{2m}\nabla \cdot (\psi \nabla \psi^*) - \nabla \cdot (\psi^* \nabla \psi) = \frac{\hbar^2}{2m}\nabla \cdot (\psi \nabla \psi^* - \psi^* \nabla \psi) \quad (4.91)$$

Or, we can also write Eq. (4.91) as

$$\frac{d\rho}{dt} + \nabla \cdot \left[\frac{i\hbar}{2m}(\psi \nabla \psi^* - \psi^* \nabla \psi)\right] = 0 \quad (4.92)$$

This equation suggests that we can define a **current density vector operator** per unit charge $\widehat{J}$ by

Current Density (Per Unit Charge) Operator

$$\widehat{J} = \frac{i\hbar}{2m}(\psi \nabla \psi^* - \psi^* \nabla \psi) \quad (4.93)$$

Note that since the current density operator given by Eq. (4.93) is defined *per unit charge carrier*, the current density in conventional units $\widehat{J}_c$ becomes

$$\widehat{J}_c = \langle n \rangle q \widehat{J} \quad (4.94)$$

$\langle n \rangle$ is the average number of charge carriers per unit volume having unit charge q. Because *the term is parentheses is the difference between complex conjugates*, it is always purely imaginary. Then, the current density operator $\widehat{J}$ becomes necessarily real with the factor of i. The Schrödinger equation, therefore, leads to a *real valued* continuity equation given by

Continuity Equation (Hermitian Hamiltonian)

$$\frac{d\rho}{dt} + \nabla \cdot \mathbf{J} = 0 \quad (4.95)$$

Eq. (4.95) is a remarkable result, in that, we have only utilized the Schrödinger equation, along with Born's condition to derive a well-known classical equation relating a change in density with time, to a current divergence. Let's examine more closely, the meaning of the divergence operator which appears in Eq. (4.95).

4.7 MEANING OF DIVERGENCE

To look more closely at the divergence operator, let us consider a small volume element ΔV, as illustrated in Fig. 4.8. We first consider the

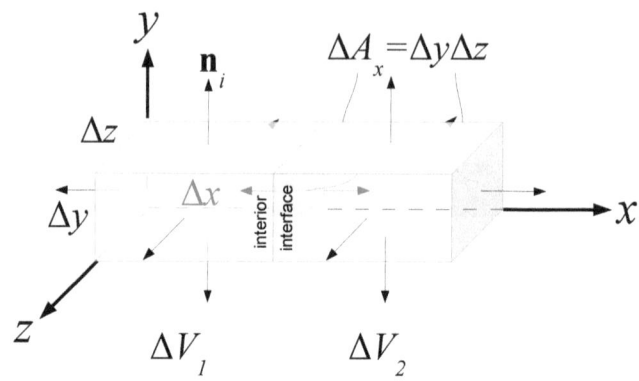

Figure 4.8: Differential element through which **J** flows.

component of **J** flowing along the x-direction, through face ΔA_x. Let **J** represent the rate of flow of *something that can be measured or counted*. The *something* is distributed, per unit area and is expressed per unit time, t. For example, mass m can be expressed per unit area, per unit time. Let's call it s for *something*.

Along the x direction, there are two faces. The first will be the inlet where J_x enters, at say $J_x(x)$. Then, the exit location will be $J_x(x + \Delta x)$. Assuming there are no *sources* or *sinks* of s inside the volume element, the net change (from in-flow) of s within the element dV over a time Δt, due to flow along the x-direction is

$$\text{net change due to } x\text{-flow} = \Delta s^x \tag{4.96}$$

The change in s can be expressed as

$$\Delta s^x = [J_x(x) - J_x(x + \Delta x)] \Delta y \Delta z \Delta t \tag{4.97}$$

Since J_x is, generally, a function of x, we can write $J_x(x + \Delta x)$ using a Taylor Series expansion given by

$$J_x(x + \Delta x) = J_x(x) + \frac{\partial J_x}{\partial x} \Delta x + O(\Delta x^2) \tag{4.98}$$

4.7 Meaning of Divergence

We have lumped all higher order terms into $O(\Delta x^2)$. It is of little concern because we will take the limit, and in doing so, this term will vanish. Using Eq. (4.98) (ignoring the higher order terms), we have

$$\Delta s^x = -\frac{\partial J_x}{\partial x}\Delta x \Delta y \Delta z \Delta t \tag{4.99}$$

So far, we have only accounted for the change in s due to flow in the x-direction. In general, there is flow in all three directions (x, y, and z), so the total change $\Delta s = \Delta s^x + \Delta s^y + \Delta s^z$, which gives

$$\Delta s^{total} = \Delta s = \Delta s^x + \Delta s^y + \Delta s^z \tag{4.100}$$

We can do the same analysis for the other two directions to obtain their contributions to the total change in s. Then, all the contributions from x, y, and z, using Eqs. (4.100) and (4.99), leads to

$$\Delta s = -\left(\frac{\partial J_x}{\partial x} + \frac{\partial J_y}{\partial y} + \frac{\partial J_z}{\partial z}\right)\Delta x \Delta y \Delta z \Delta t \tag{4.101}$$

Note that $\Delta x \Delta y \Delta z = \Delta V$. We may use this relation to express *the change in something per unit volume* $= \Delta s/\Delta V = \Delta \rho_s$. ρ_s is the volume density of s. Then, dividing by Δt, we get

$$\frac{\Delta \rho_s}{\Delta t} = -\frac{\partial J_x}{\partial x} - \frac{\partial J_y}{\partial y} - \frac{\partial J_z}{\partial z} \tag{4.102}$$

Taking the limit of Eq. (4.102) as $\Delta x, \Delta y, \Delta z$, and Δt all approaching 0, we obtain

$$\frac{\partial \rho_s}{\partial t} + \frac{\partial J_x}{\partial x} + \frac{\partial J_y}{\partial y} + \frac{\partial J_z}{\partial z} = 0 \tag{4.103}$$

or

$$\frac{\partial \rho_s}{\partial t} + \nabla \cdot \mathbf{J} = 0 \tag{4.104}$$

Therefore, we see that the divergence operator ($\nabla \cdot \mathbf{J}$) represents the negative rate of change of something, per unit volume, assuming no sinks or sources exist within the differential volume element. A positive divergence, then, corresponds to a loss of s from the unit volume, etc. The divergence of the current density can be related to other useful quantities, as well. In the next section, we'll use the result here to deduce another relation between involving the current density $\mathbf{J}$.

4.8 Continuity Links to Gauss' Theorem

We can also approach the analysis above in an alternative manner, beginning with equating Eq. (4.96) with Eq. (4.99), which both express Δs^x. This gives

$$J_x(x)\Delta A_x \Delta t - J_x(x+\Delta x)\Delta A_x \Delta t = -\frac{\partial J_x}{\partial x}\Delta x \Delta y \Delta z \Delta t \qquad (4.105)$$

or

$$J_x(x)\Delta A_x - J_x(x+\Delta x)\Delta A_x = -\frac{\partial J_x}{\partial x}\Delta x \Delta y \Delta z \qquad (4.106)$$

The LHS of Eq. (4.106) can also be described in the following way: *out minus in* can alternatively be expressed by multiplying the x-component of J_x by it's respective area dA_x using the unit vectors **n** that point *away* from the differential area dA_x. A difference, again, results because although $J_x(x)$ and $J_x(x+\Delta x)$ point in the same direction (+x), the area normal vectors $\mathbf{n}(x)$ are 180° apart. Thus, J_x has the normal vector pointing in the $-x$ direction at x, while the **n** is directed along $+x$ at $x+dx$. Therefore, we can also write Eq. (4.106) as

$$-(J_x(x)\cdot \mathbf{n}(x) + J_x(x+\Delta x)\cdot \mathbf{n}(x+\Delta x))dA = -\frac{\partial J_x}{\partial x}\Delta x \Delta y \Delta z \qquad (4.107)$$

Accounting for all contributions from x, y, and z, Eq. (4.107) becomes

$$-\sum_{i=1}^{6} J_i \cdot \mathbf{n}_i dA = -\left(\frac{\partial J_x}{\partial x} + \frac{\partial J_y}{\partial y} + \frac{\partial J_z}{\partial z}\right)\Delta x \Delta y \Delta z \qquad (4.108)$$

Hence, for a volume element, summing the flux $J_i \cdot \mathbf{n}_i$ over the bounding surface equals the divergence of **J** for that volume element (multiplied by the volume). So what happens if we place two differential volume elements side by side? Two elements side-by-side are illustrated in Fig. 4.9. Here, the area normal vectors of the *interior* interface always point away from each other. And since they are at the same location, they will have the same flow of J_i, which means summing $J_i \cdot n_i$ will always vanish at an interior interface. Thus, the only terms which will contribute to the summation of the fluxes are, again, the terms for the exterior bounding surface of the two-volume elements. And this will be true at any interior interface in x, y, and z directions. So, we may write Eq. (4.108) for an arbitrary number of volume elements filling a volume V, with elements

4.8 Continuity Links to Gauss' Theorem

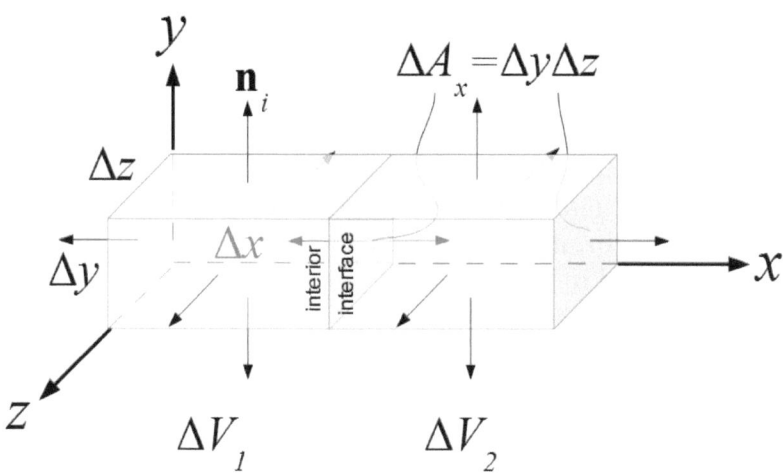

Figure 4.9: Illustration of two differential element through which **J** flows. At the interior interface, the contributions from $\mathbf{J} \cdot \mathbf{n} dA$ because J is the same, however, the normals $\mathbf{n}_i$ always point opposite to one another.

dV_j.

$$\sum_j \sum_{i=1}^{6} \mathbf{J}_i \cdot \mathbf{n}_i dA = \sum_j \left(\frac{\partial J_x}{\partial x} + \frac{\partial J_y}{\partial y} + \frac{\partial J_z}{\partial z} \right) dV_j \qquad (4.109)$$

Though the left hand side looks slightly complicated with a double sum, it reduces to the sum of only those differential areas dA that are on the bounding surface, which we will denote by dS. So, we can write Eq. (4.109) more succinctly as

$$\sum_i \mathbf{J}_i \cdot \mathbf{n}_i dS_i = \sum_j \left(\frac{\partial J_x}{\partial x} + \frac{\partial J_y}{\partial y} + \frac{\partial J_z}{\partial z} \right) dV_j \qquad (4.110)$$

We again take the limit as dV_j vanishes, and we have the result

Gauss' Theorem

$$\int_S \mathbf{J}\cdot\mathbf{n}\,dS = \int_V \nabla\cdot\mathbf{J}\,dV \qquad (4.111)$$

Eq. (4.111) is known is **Gauss' Theorem**, and it should not be confused with Gauss' law, in the Maxwell equations. It says that summing the fluxes of $\mathbf{J}\,dS$ through the bounding surface S is the same as adding up the divergences throughout the entire volume. This follows from the same picture used to obtain the continuity equation. So, we find that there is, indeed, a link between the continuity equation as given by Eqs. (4.95) and (4.104), and *Gauss's theorem*.

4.9 Chapter Summary

In this chapter we have explored an important problem in quantum mechanics which allows an exact solution, known as the *Infinitely Deep Potential Well* problem, or the *Particle in a Box* problem. We have obtained the exact solutions up to 3D, and used this problem to introduce some important properties of the Schrödinger equation and the wave functions described by it. We learned that orthogonality is a property of wave functions, as well as real-valued eigenvalues, when we use real-valued potential energy $V(x)$.

We also showed that the Schrödinger equation leads to a continuity equation, which lead boldly to definitions of a probability density and current density operator. Table 4.3 lists some of the key results from this chapter.

Table 4.3: Chapter 4 Summary Equations.

Name	Equation
1D IDPW eigen-function	$\psi_n(x,t) = C_n \sin\left(\frac{n\pi x}{L}\right) e^{-i\frac{E_\lambda t}{\hbar}}$
1D IDPW energy states	$E_\lambda = \frac{n^2 \pi^2 \hbar^2}{2mL^2}$
Density of states (DOS)	$D(E_\lambda) = \frac{dN}{dE_\lambda}$
3D IDPW eigen-function Ψ/C_N	$\sin\left(\frac{n_x \pi x}{L_x}\right)\sin\left(\frac{n_y \pi y}{L_y}\right)\sin\left(\frac{n_z \pi z}{L_z}\right) e^{-i\frac{E_\lambda t}{\hbar}}$

Table 4.3

3D IDPW energy states	$E_\lambda = (n_x^2 + n_y^2 + n_z^2)\frac{\pi^2 \hbar^2}{2mL^2}$
Expectation with density of states	$\langle E_\lambda \rangle = \frac{\int_0^\infty E_\lambda g(E_\lambda) F(E_\lambda) dE_\lambda}{\int_0^\infty g(E_\lambda) F(E_\lambda) dE_\lambda}$
wave function orthogonality	$\int \psi_m(x) \cdot \psi_n(x) dx = 0 \quad \forall m \neq n$
Normalization condition	$\int_V |\Psi|^2 dV = 1$
Definition of current density J	$J \equiv \frac{i\hbar}{2m}(\psi \nabla \psi^* - \psi^* \nabla \psi)$

4.10 Chapter Problems

Problem 4.1 Using the solution for the 1D particle in a box, show that the average position of a particle confined to the IDPW is given by $L/2$ by evaluating

$$\langle x \rangle = \frac{2}{L} \int_0^L x \sin^2\left(\frac{n\pi x}{L}\right) dx$$

Problem 4.2 Discuss the possibility of degenerate states for the 1D, 2D, and 3D infinitely deep potential well eigenstate. What is the trend with degeneracy as the dimensional order increases?

Problem 4.3 Using the solution for the 1D particle in a box, show that the average momentum of a particle is

$$\langle p \rangle = 0$$

Problem 4.4 Show that the following set of functions given by

$$\psi_n = \frac{1}{\sqrt{2\pi}} e^{in\varphi} \quad 0 \leq \varphi 2\pi$$

are orthogonal, or that

$$\int_0^L \psi_n^* \psi_m d\varphi = \delta_{nm}$$

δ_{nm} is the *Kronecker delta* defined here as

$$\delta_{mn} = \begin{cases} 1 & \text{if } m = n \\ 0 & \text{if } m \neq n \end{cases}$$

Problem 4.5 Calculate the current density $\widehat{J}$ for a system with a wavefunction given by

$$\psi = Ae^{ikx}$$

Use the definition of the *current density* (per unit charge) given by Eq. (4.93) by evaluating

$$\widehat{J} = \frac{i\hbar}{2m}(\psi \nabla \psi^* - \psi^* \nabla \psi)$$

And what is the *current density* for a wavefunction given by

$$\psi = Ae^{-kx}$$

Problem 4.6 Earlier, when we discussed the current density, it was pointed out that any real wave function has zero current density J. Using the definition of current density, show that *any purely imaginary wave function* also has zero probability current.

Problem 4.7 Use the momentum operator on the eigenstate of the 1D infinitely deep potential well given by

$$\psi_n = \sqrt{\frac{2}{L}} \sin\left(\frac{n\pi x}{L}\right)$$

and determine

$$\widehat{p} = -i\hbar \frac{\partial}{\partial x} \psi_n$$

Is ψ_n and eigenfunction of the momentum operator? Use the result for $\widehat{p}$ to determine

$$\langle p^2 \rangle = \int_0^L \widehat{p}^2 \psi_n^* \psi_n dx$$

4.10 Chapter Problems

Problem 4.8 Calculate the current density $\widehat{J}$ for a system with a wavefunction given by

$$\psi = Ae^{i(k_r + ik_k)x}$$

Use the definition of the *current density* (per unit charge) given by Eq. (4.93).

CHAPTER 5

The Harmonic Oscillator

Quantum mechanics is widely known for its predictions of quantum states. We saw an example of this with the *infinitely deep potential well problem* in the previous chapter. There, we constructed a *fictitious* potential function at the boundaries, for our convenience. This was sufficient to reveal this particular characteristic of quantum mechanical systems. In more realistic systems, quantum states result when the wave-particle does not possess more kinetic energy K compared to negative potential V. This will be demonstrated most convincingly by the hydrogen atom problem in Chap. 8. In this chapter, we are going to consider a problem that has a well-known classical counterpart. As such, we utilize a well-known classical potential function, which is smoothly varying in a manner resembling a real physical system. However, this change in the potential function turns out to demand greater mathematical rigor. Therefore, we will have an opportunity to see how much mathematics it can sometimes take to expose the quantum nature of a system. And secondly, we begin to explore the role of time and how it can be treated to provide time-dependent information. This special problem is known as the *harmonic oscillator problem*. Before determining what the Schrödinger equation predicts, let's first review some results from the classical harmonic oscillator.

5.1 THE CLASSICAL HARMONIC OSCILLATOR

Any solution in quantum mechanics, when the system is scaled up to a sufficiently large one, must correspond to a classical one. This sort of compulsory *correspondence* is known as **the Principle of Correspondence (PoC)** or the **Correspondence Principle**. To demonstrate aspects of this principle, it helps to recapitulate classical results for the harmonic oscillator, then explore, in detail, the corresponding quantum mechanical system, and determine the extent to which the PoC is maintained.

Consider a classical harmonic oscillator having a mass m, connected to a spring with stiffness or force constant k_s, as shown in Fig. 5.1. The

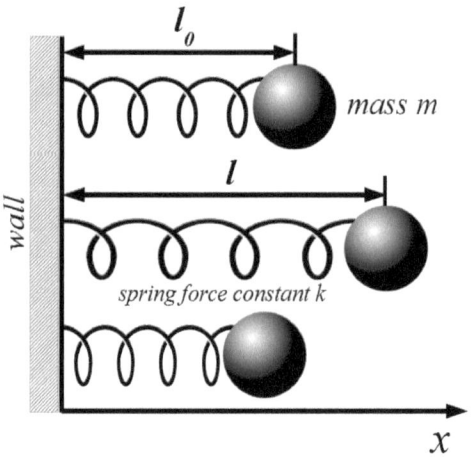

Figure 5.1: A *classical* harmonic oscillator, consisting of a mass m attached to a spring. There is an equilibrium position at which the system desires to be, and displacing the spring produces a force that counters the displacement $l - l_0$.

spring acts to counter any displacement $l - l_0$, from equilibrium. For convenience, from here on out, we will denote such displacement as $x = l - l_0$. The origin of our coordinate system, therefore, corresponds to the equilibrium position $x = 0 = l_0 - l_0$. This also allows us to write the change in potential energy $\Delta V = V - V(0)$ as just V, since $V(0) = 0$. If the force of the spring is linear in the displacement, the spring is said to

5.1 The Classical Harmonic Oscillator

follow *Hooke's law* which is given by

$$F_k = -k_s(l - l_0) = -k_s x \tag{5.1}$$

For a conservative system, the potential energy V of the spring is found by obtaining the work done *by the spring* to bring its mass from l back to its equilibrium position l_0, i.e. the negative of the work done *to* the spring. The change in potential energy ΔV is given by

$$\Delta V = V(x) = \int_l^{l_0} F_k dx = \int_l^{l_0} (-k_s x) dx = \frac{k_s x^2}{2} \tag{5.2}$$

The potential energy of the spring becomes

Harmonic Oscillator Potential Energy

$$V(x) = \frac{1}{2} k_s x^2 \tag{5.3}$$

Having the force F_k taken along x allows us to express the classical ordinary differential equation of motion using $\sum F = F_k = ma$, which leads to

$$-k_s x = m \frac{d^2 x}{dt^2} \Rightarrow \frac{d^2 x}{dt^2} + \frac{k_s}{m} x = 0 \tag{5.4}$$

Eq. (5.4) has a general time-dependent solution given by

$$x(t) = A \sin \omega t + B \cos \omega t \tag{5.5}$$

This solution can be plugged into Eq. (5.5) to verify it is a solution. The parameter ω has units of radians per seconds s^{-1} and is called the *natural frequency*. It is given by

Harmonic Oscillator Natural Frequency

$$\omega = \sqrt{\frac{k_s}{m}} \tag{5.6}$$

Using Eq. (5.6), the potential energy $V(x)$ can also be written as

$$V(x) = \frac{k_s x^2}{2} = \frac{m \omega^2 x^2}{2} \tag{5.7}$$

Since the equation of motion given by Eq. (5.2) is second order in time, it requires specification of both the initial displacement $x(t_0) = x_0$ as well as the initial velocity $dx/dt(t_0) = v(t_0) = v_0$ at time $t = t_0 = 0$. For x_0, in order for any motion to develop and we avoid the trivial solution, it must be that $x_0 > 0$. Then, the initial condition is *some* displacement away from equilibrium. We can, however, take $v_0 = 0$. At $t = 0$, we then have

$$x(0) = A \cdot 0 + B \cdot 1 = x_0 \Rightarrow B = x_0$$

$$\frac{dx}{dt}(0) = A\omega \cdot 1 - B\omega \cdot 0 = 0$$

The result of applying both of these initial conditions to Eq. (5.5) results in $A = 0$, and a solution given by

Classical Harmonic Oscillator Solution

$$x(t) = x_0 \cos \omega t \tag{5.8}$$

Therefore, whatever the initial displacement x_0, it determines the amplitude of the harmonic displacement oscillation. Fig. 5.2 illustrates both the solutions for different initial conditions x_0 (top) and different spring stiffness constants k_s, which changes the *natural frequency*. However, changing the spring stiffness k_s has no effect on the amplitude of the oscillation. Now, let's consider the total energy of the harmonic oscillator, which includes both the kinetic energy $K(t)$ and the potential energy $V(t)$, given by Eq. (5.7). The kinetic energy $K(t)$ is given by

$$K(t) = \frac{1}{2}m\left(\frac{dx}{dt}\right)^2 = \frac{mx_0^2\omega^2}{2}\sin^2\omega t$$

Therefore, the total energy $E = K + V$ becomes

$$E = \frac{mx_0^2\omega^2}{2}\sin^2\omega t + \frac{k_s}{2}x_0^2\cos^2\omega t \tag{5.9}$$

Substitute of the spring stiffness k_s using Eq. (5.6), the total energy $E(t)$ becomes

$$E = \frac{mx_0^2\omega^2}{2}\left(\sin^2\omega t + \cos^2\omega t\right)$$

or

5.1 The Classical Harmonic Oscillator

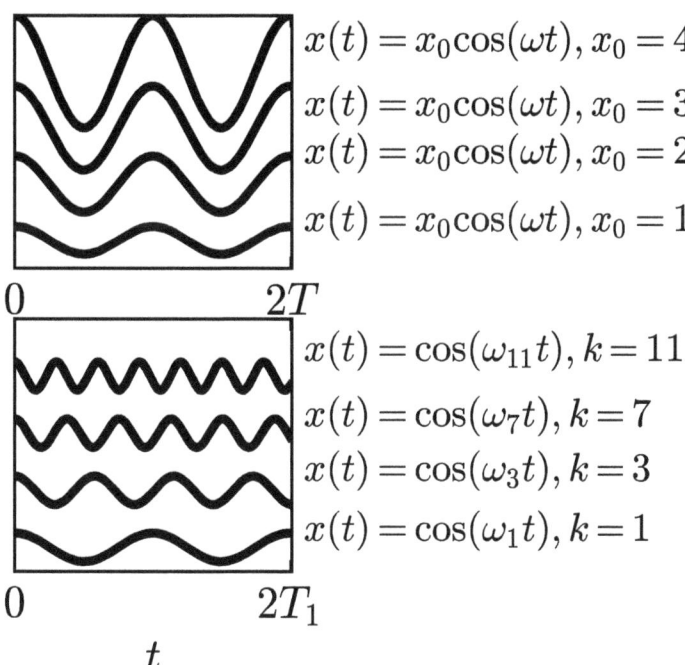

Figure 5.2: Illustrations of the time variation of the displacement $x(t)$, of a classical oscillator, under (top) different initial displacements, x_0 and (bottom) different spring stiffnesses k.

Conserved Harmonic Oscillator Energy

$$E = \frac{kx_0^2}{2} \tag{5.10}$$

We have used the fact that $\sin^2 \omega t + \cos^2 \omega t = 1$. Even though the mass is oscillating in time, the system's total energy is always *conserved*. Using Eqs. (5.6) and (5.10), it follows that the amplitude x_0 of the classical harmonic oscillator can be expressed as

Harmonic Oscillator Amplitude

$$x_0^2 = \frac{2E}{m\omega^2} \qquad (5.11)$$

The results summarized above are well-established classical relations for a harmonic oscillator without any dissipation, or energy loss. For the quantum mechanical treatment of this same problem, we expect a consistent set of relations so we satisfy the *PoC*. So, let's find out what the Schrödinger equation has to say about the harmonic oscillator.

5.2 Quantum Mechanical Harmonic Oscillator

For the quantum mechanical harmonic oscillator problem, we solve the Schrödinger equation for a system with the potential energy V given by Eq. (5.2) or Eq. (5.7), which is second order in x. The Schrödinger equation in this case is spatially 1D, and is given by

$$i\hbar \frac{\partial \psi}{\partial t} = \left[-\frac{\hbar^2}{2m} \frac{\partial^2}{\partial x^2} + \frac{1}{2} m\omega^2 x^2 \right] \psi \qquad (5.12)$$

As we have seen in previous problems, Eq. (5.12) can also be solved using *separation of variables*. We, therefore, assume the wave function ψ can be expressed as $\psi = X(x)T(t)$. Substituting this form into Eq. (5.12), we have

$$i\hbar \frac{1}{T} \frac{dT}{dt} = -\frac{\hbar^2}{2m} \frac{1}{X} \frac{d^2 X}{dx^2} + \frac{1}{2} m\omega^2 x^2 \qquad (5.13)$$

Since the LHS depends solely on time t, while the RHS depends on x, both being independent variables, it follows that both sides must be equal to a constant given by the eigenenergy E_λ. This leads to the following two separated equations:

$$i\hbar \frac{1}{T} \frac{dT}{dt} = E_\lambda \qquad (5.14a)$$

$$-\frac{\hbar^2}{2m} \frac{1}{X} \frac{d^2 X}{dx^2} + \frac{1}{2} m\omega^2 x^2 = E_\lambda \qquad (5.14b)$$

We have seen Eq. (5.14a) before, for example, in the particle in a box (or infinitely deep potential well) problem in Chap. 4 (c. Eq. (4.13)). There,

5.3 Spatial Dependence Solvable By Frobenius Method

we found the solution for $T(t)$ to be given by

$$T(t) = T_0 e^{-i\frac{E_\lambda}{\hbar}(t-t_0)} \tag{5.15}$$

This gives the time-dependent part of the wave function $\psi = X(x)T(t)$. The determination of the spatial part of the wave function turns out to be a little more complicated, but, solvable nonetheless. In the next section, we'll utilize a particular method invented by mathematicians to determine $X(x)$.

5.3 SPATIAL DEPENDENCE SOLVABLE BY FROBENIUS METHOD

We can solve Eq. (5.14b), for $X(x)$, using what is known as the *Method of Frobenius* or *Frobenius method*, named after German mathematician Ferdinand Georg Frobenius (1849-1917). This method solves a second order ordinary differential equation by assuming that the solution, in this case $X(x)$, takes the form of an infinite series. Then, after substitution into the differential equation and setting up appropriate conditions, all the parameters associated with the series are determined. For a second order differential equation having the form of the harmonic oscillator equation, the infinite series solution has a special name which we shall discuss. Using this method, the complete solution can be obtained, with the caveat that it is, generally, in the form of an infinite series. We'll see that this fact has further implications in quantum mechanics.

Eq. (5.14b) can be written as

$$\frac{d^2 X}{dx^2} + \frac{2m}{\hbar^2}\left[E_\lambda - \frac{1}{2}m\omega^2 x^2\right]X(x) = 0 \tag{5.16}$$

Let's introduce useful parameters to both simplify the notation as well as modify its form to a more exploitable ordinary differential equation being susceptible to the *Frobenious Method*. Let the dimensionless parameter ζ, replacing x, be defined as

$$\zeta = \alpha x \Rightarrow d\zeta = \alpha dx \tag{5.17}$$

We will define the parameter α shortly. From this definition, the first derivative of X becomes

$$\frac{dX}{dx} = \frac{dX}{d\zeta}\frac{d\zeta}{dx} = \alpha \frac{dX}{d\zeta}$$

The second derivative then becomes

$$\frac{d^2 X}{dx^2} = \frac{d\zeta}{dx}\frac{d}{d\zeta}\left(\alpha\frac{dX}{d\zeta}\right) = \alpha^2\frac{d^2 X}{d\zeta^2}$$

After substitution of the derivatives into Eq. (5.16), we have the following form of the Schrödinger equation:

$$\alpha^2 \frac{d^2 X}{d\zeta^2} + \frac{2m}{\hbar^2}\left[E_\lambda - \frac{1}{2\alpha^2}m\omega^2\zeta^2\right]X(\zeta) = 0 \tag{5.18}$$

or

$$\frac{d^2 X}{d\zeta^2} + \left[\frac{2mE_\lambda}{\alpha^2 \hbar^2} - \frac{m^2\omega^2}{\alpha^4 \hbar^2}\zeta^2\right]X(\zeta) = 0 \tag{5.19}$$

This last form is the equation that motivates how the parameter α is defined. It is defined such that the coefficient for ζ^2 (the last term on the LHS) is 1. Then, let

$$\alpha^4 = \frac{m^2\omega^2}{\hbar^2}\zeta^2 \Rightarrow \alpha = \sqrt{\frac{m\omega}{\hbar}} \tag{5.20}$$

The units of α should be recognized. It is inverse units of length so $\alpha x = \zeta$ is dimensionless. This can be confirmed as follows:

$$\alpha = \sqrt{\frac{m\omega}{\hbar}} \Rightarrow \text{units}(\alpha) = \sqrt{\frac{\text{kg}\cdot\text{rad}}{\text{s}\cdot\text{J}\cdot\text{s}}}$$

$$= \sqrt{\frac{\text{kg}\cdot\text{rad}\cdot\text{s}^2}{\text{s}^2\cdot\text{kg}\cdot\text{m}^2\cdot\text{s}}}$$

$$= \sqrt{\frac{1}{\text{m}^2}} = \frac{1}{\text{m}} = \text{m}^{-1}$$

Thus, $\zeta = \alpha x$ is a dimensionless quantity. We shall see this quantity used extensively in the following analysis of this problem in this chapter and later chapters, too. Eq. (5.19) then becomes

$$\frac{d^2 X}{d\zeta^2} + \left[\frac{2E_\lambda}{\hbar\omega} - \zeta^2\right]X(\zeta) = 0 \tag{5.21}$$

Let us define an additional dimensionless parameter Ω as

$$\Omega = 2E_\lambda/\hbar\omega \tag{5.22}$$

5.3 Spatial Dependence Solvable By Frobenius Method

In SI units, note that the units of $\hbar\omega$ =J-s×s^{-1} =J (for Joules), as are the units of E_λ. The Schrödinger equation then becomes

$$\frac{d^2 X}{d\zeta^2} + \left[\Omega - \zeta^2\right] X(\zeta) = 0 \tag{5.23}$$

This turns out to be a more useful form of the ordinary differential equation describing $X(x)$. We can transform Eq. (5.23) into the following form:

$$x\frac{d^2 y}{dx^2} + (c-x)\frac{dy}{dx} - ay(x) = 0 \quad \text{Confluent Hypergeometric Equation} \tag{5.24}$$

The reason we want this form is because we know how to obtain the solution from Eq. (5.24). This equation is known as the **Confluent Hypergeometric Equation(CHE)**. If we are able to get Eq. (5.23) into the form of Eq. (5.24), then we can identify parameters a and c, for the harmonic oscillator problem. To do this, we must make a few more transformations. Then, let us define

$$\gamma = \zeta^2 \rightarrow \frac{d\gamma}{d\zeta} = 2\zeta \tag{5.25}$$

With this definition, the first derivative is given by

$$\frac{dX}{d\zeta} = \frac{dX}{d\gamma}\frac{d\gamma}{d\zeta} = \frac{dX}{d\gamma} 2\zeta$$

For the second derivative, we have

$$\frac{d^2 X}{d\zeta^2} = \frac{d}{d\zeta}\left[\frac{dX}{d\gamma} 2\zeta\right] = \frac{d}{d\gamma}\left[\frac{dX}{d\gamma} 2\zeta\right]\frac{d\gamma}{d\zeta} \tag{5.26}$$

This gives

$$\frac{d^2 X}{d\zeta^2} = 4\zeta^2 \frac{d^2 X}{d\gamma^2} + 2\frac{dX}{d\gamma} = 4\gamma\frac{d^2 X}{d\gamma^2} + 2\frac{dX}{d\gamma} \tag{5.27}$$

Let's now consider an asymptotic limit of Eq. (5.27). Substituting the result in Eq. (5.27) and dividing by 4γ, leads to

$$\frac{d^2 X}{d\gamma^2} + \frac{1}{2\gamma}\frac{dX}{d\gamma} + \left[\frac{\Omega}{4\gamma} - \frac{1}{4}\right] X(\gamma) = 0 \tag{5.28}$$

Taking the limit as $\gamma \to \infty$, Eq. (5.28) becomes

$$\frac{d^2 X}{d\gamma^2} - \frac{1}{4} X(\gamma) = 0 \tag{5.29}$$

Eq. (5.29) has the solution X_∞ given by

$$X_\infty = Ce^{-\frac{1}{2}\gamma} \tag{5.30}$$

C is an integration constant, independent of γ. For convenience, let $C = 1$. This result gives us an idea. Since we have already used the idea of separating the variables successfully to express $\psi = X(x)T(t)$, let's use the idea once again by setting $X(\gamma) = X_0 \cdot X_\infty = X_0(\gamma)e^{-\frac{1}{2}\gamma}$. If we are able to obtain a solution from this assumption, then it works! With X taking this form, its first derivative is given by

$$\frac{dX}{d\gamma} = \left[\frac{dX_0}{d\gamma} - \frac{1}{2}X_0\right]e^{-\frac{1}{2}\gamma} \tag{5.31}$$

Using Eq. (5.31) to get the second derivative, we have

$$\frac{d^2X}{d\gamma^2} = \left[\frac{d^2X_0}{d\gamma^2} - \frac{dX_0}{d\gamma} + \frac{1}{4}X_0\right]e^{-\frac{1}{2}\gamma} \tag{5.32}$$

All terms have a factor $e^{-\frac{1}{2}\gamma}$, so this just factors out. Substituting these results leads to

$$4\gamma\left[\frac{d^2X_0}{d\gamma^2} - \frac{dX_0}{d\gamma} + \frac{1}{4}X_0\right] + 2\left[\frac{dX_0}{d\gamma} - \frac{1}{2}X_0\right] + [\Omega - \gamma]\gamma X_0(\gamma) = 0 \tag{5.33}$$

This can be rearranged into the following *Confluent Hypergeometric differential equation*

Oscillator Confluent Hypergeometric Equation

$$\gamma\frac{d^2X_0}{d\gamma^2} + \left(\frac{1}{2} - \gamma\right)\frac{dX_0}{d\gamma} - \left(\frac{1-\Omega}{4}\right)X_0 = 0 \tag{5.34}$$

In Eq. (5.34), $\Omega = 2E_\lambda/\hbar\omega$. By comparing Eq. (5.24) and Eq. (5.34), we find that

$$a = \frac{1-\Omega}{4} \quad \text{and} \quad c = \frac{1}{2} \tag{5.35}$$

Summarizing what we have established so far, we have the complete time dependent solution factor $T(t)$. For the spatially dependent part X, we have found that the exact solution is in the form $X = X_0(\gamma)e^{-\frac{1}{2}\gamma}$

5.3 Spatial Dependence Solvable By Frobenius Method

where X_0 is described by the ordinary differential equation given in Eq. (5.34). Thus, all that remains is to solve for $X_0(\gamma)$.

We are now in a position to find the solution to Eq. (5.24) in terms of parameters a and c by using the *method of Frobenius*. Assume that $X_0(\gamma)$ takes the form of an infinite power series given by

$$X_0(\gamma) = \sum_{r=0}^{\infty} d_r \gamma^{p+r} \tag{5.36a}$$

The parameter p is introduced artificially so that by taking the derivatives, it ends up in the coefficients of the equation. With Eq. (5.36a), the first derivative of $X_0(\gamma)$ is given by

$$\frac{dX_0}{d\gamma} = \sum_{r=0}^{\infty} (p+r) d_r \gamma^{p+r-1} \tag{5.36b}$$

Continuing on, the second derivative is given by

$$\frac{d^2 X_0}{d\gamma^2} = \sum_{r=0}^{\infty} (p+r-1)(p+r) d_r \gamma^{p+r-2} \tag{5.36c}$$

Substituting Eqs. (5.36a) - (5.36c) into Eq. (5.34), we have

$$\gamma \left[\sum_{r=0}^{\infty} (p+r-1)(p+r) d_r \gamma^{p+r-2} \right] + (c-\gamma) \left[\sum_{r=0}^{\infty} (p+r) d_r \gamma^{p+r-1} \right] - a \left[\sum_{r=0}^{\infty} d_r \gamma^{p+r} \right] = 0 \tag{5.37}$$

Multiplying the coefficients depending on γ, we get

$$\left[\sum_{r=0}^{\infty} (p+r-1)(p+r) d_r \gamma^{p+r-1} \right] + c \left[\sum_{r=0}^{\infty} (p+r) d_r \gamma^{p+r-1} \right] - \left[\sum_{r=0}^{\infty} (p+r) d_r \gamma^{p+r} \right] - a \left[\sum_{r=0}^{\infty} d_r \gamma^{p+r} \right] = 0 \tag{5.38}$$

Some of the terms in Eq. 5.38 have the same exponent powers so they can be combined. In fact, there are only two unique terms in the above result. Thus, the combining of terms reduces Eq. (5.38) to

$$\sum_{r=0}^{\infty} d_r(p+r+c-1)(p+r)\gamma^{p+r-1} + \sum_{r=0}^{\infty} d_r(p+r+a)\gamma^{p+r} = 0 \quad (5.39)$$

Since the parameter p is taken as constant for each independent solution to the second order differential equation, we can multiply Eq. (5.39) by γ^{1-p}. This gives

$$\sum_{r=0}^{\infty} d_r(p+r+c-1)(p+r)\gamma^r - \sum_{r=0}^{\infty} d_r(p+r+a)\gamma^{r+1} = 0$$

or

$$\sum_{r=0}^{\infty} d_r(p+r+c-1)(p+r)\gamma^r = \sum_{r=0}^{\infty} d_r(p+r+a)\gamma^{r+1} \quad (5.40)$$

If Eq. (5.40) is to be valid *for any value of* γ, it follows that all the coefficients on the left must be equal to those on the right having the same exponent. This leads to useful results. The power or exponent on the RHS is $r+1$, while on the LHS it is r. This means that the very first term on the left ($r = 0$) has no counterpart on the RHS. In other words, this term must be equal to zero, or

---

Indicial Equation

$$d_0 p(p+c-1) = 0 \quad (5.41)$$

The only way that Eq. (5.41) can equal zero is for the following conditions for p to be true:

$$p = 1-c \quad \text{or} \quad p = 0 \quad (5.42)$$

Eq. (5.41) is known as the **indicial equation**, and it determines the allowed values of p defining the two independent solutions to the second order ordinary differential equation we are solving. We must find *both* solutions corresponding to each value of p, given by Eq. (5.42).

For now, we shall keep p in our analysis as a constant parameter to allow for a general solution in terms of p. Let us write out the first few terms of our series from both sides, where we have

5.3 Spatial Dependence Solvable By Frobenius Method

$$d_0(p+c-1)(p) + d_1(p+c)(p+1)\gamma + d_2(p+c+1)(p+2)\gamma^2 + \ldots =$$

$$d_0(p+a)\gamma + d_1(p+a+1)\gamma^2 + d_2(p+a+2)\gamma^3 + \ldots \quad (5.43)$$

We observe that the d_0 coefficient on the RHS equates with (has the same exponent) as the d_1 coefficient on the LHS, the d_1 term on the RHS equates with the d_2 term, and so on. More generally, by equating the coefficients of equal powers, we obtain the general relation

$$d_{r-1}(p+r-1+a) = d_r(p+r+c-1)(p+r)$$

or

Recurrence Relation

$$\frac{d_r}{d_{r-1}} = \frac{(p+r-1+a)}{(p+r)(p+r+c-1)} \quad (5.44)$$

Eq. (5.44) is known as the **recurrence relation** or **recurrence ratio**, and it defines an iterative sequence that determines all the coefficients for the infinite power series. Thus, we only need to determine d_0. Fortunately, its a simple matter because the solution will need to normalized later (i.e. must satisfy $\int_{-\infty}^{\infty} |X|^2 dx = 1$), so we can assume $d_0 = 1$. Then, for both independent solutions, all coefficients d_r can be computed sequentially, or in a recurring manner by using Eq. (5.44). This provides for two independent solutions for $X_0(\gamma)$. Let us denote the independent solution depending on parameter p by $S_p(\gamma)$.

Let's look more closely at the form of our independent general solutions S_p. Using Eq. (5.44), each of the independent solutions is given by

$$S_p(\gamma; a, c) = 1 + d_0 \frac{(p+a)}{(p+1)(p+c)} \gamma + d_1 \frac{(p+a+1)}{(p+2)(p+c+1)} \gamma^2 +$$

$$d_2 \frac{(p+a+2)}{(p+3)(p+c+2)} \gamma^3 + d_3 \frac{(p+a+3)}{(p+4)(p+c+3)} \gamma^4 +$$

$$d_4 \frac{(p+a+4)}{(p+5)(p+c+4)} \gamma^5 + d_5 \frac{(p+a+5)}{(p+6)(p+c+5)} \gamma^6 + \ldots \quad (5.45)$$

Since we know the values for p, found from the *indicial equation*, we can readily write S_p a more explicit form for the first independent solution S_0, where $p = 0$. This solution becomes

$$S_0(\gamma;a,c) = 1 + d_0 \frac{a}{1(c)}\gamma + d_1 \frac{(a+1)}{2(c+1)}\gamma^2 +$$

$$d_2 \frac{(a+2)}{3(c+2)}\gamma^3 + d_3 \frac{(a+3)}{4(c+3)}\gamma^4 +$$

$$d_4 \frac{(a+4)}{5(c+4)}\gamma^5 + d_5 \frac{(a+5)}{6(c+5)}\gamma^6 + \dots \quad (5.46)$$

Substitution of the coefficients in terms of the preceding term (using the recurrence relation), e.g. $d_0 \equiv 1$, $d_1 = a/c$, etc., we have the following series for the first independent solution S_0:

$$S_0(\gamma;a,c) = 1 + \frac{1}{1!}\frac{a}{c}\gamma + \frac{1}{2!}\frac{a(a+1)}{c(c+1)}\gamma^2 +$$

$$\frac{1}{3!}\frac{a(a+1)(a+2)}{c(c+1)(c+2)}\gamma^3 + \frac{1}{4!}\frac{a(a+1)(a+2)(a+3)}{c(c+1)(c+2)(c+3)}\gamma^4 +$$

$$\frac{1}{5!}\frac{a(a+1)(a+2)(a+3)(a+4)}{c(c+1)(c+2)(c+3)(c+4)}\gamma^5 +$$

$$\frac{1}{6!}\frac{a(a+1)(a+2)(a+3)(a+4)(a+5)}{c(c+1)(c+2)(c+3)(c+4)(c+5)}\gamma^6 + \dots \quad (5.47)$$

The infinite series given by Eq. (5.47) is a type of **Confluent Hypergeometric Series** (CHS). This series form is known as the **Confluent Hypergeometric Function(CHF)** (of the first kind). It is often denoted by $_1F_1(x;a,c)$. The first argument x is the independent variable, and the second and third are only constant input parameters over the range of x. The CHF is sometimes written using a shorthand notation with symbols known as **Pochhammer symbols** or **rising factorials**, defined as

> **Pochhammer Symbol**
>
> $$a(a+1)(a+2)\dots(a+r) \equiv (a)^r \quad (5.48)$$

5.3 Spatial Dependence Solvable By Frobenius Method

In terms of *Pochhammer symbols*, Eq. (5.47) becomes

$$S_0(\gamma; a, c) = {}_1F_1(x; a, c) = \sum_{r=0}^{\infty} \frac{1}{r!} \frac{(a)_{r-1}}{(c)_{r-1}} \gamma^r \qquad (5.49)$$

Any coefficient d_r, of S_0, starting from $r = 1$ can be written as

$$d_r = \frac{1}{r!} \frac{\prod_{k=1}^{r}(a+k-1)}{\prod_{k=1}^{r}(c+k-1)} \qquad (5.50)$$

$\prod$ is the product symbol representing multiplication of terms. From this, we now have our first independent solution of the *Confluent Hypergeometric Equation*, S_0 where

$${}_1F_1(\gamma; a, c) = \sum_{r=0}^{\infty} d_r \gamma^{p+r} = \gamma^p \sum_{r=0}^{\infty} d_r \gamma^r = \sum_{r=0}^{\infty} d_r \gamma^r \qquad (5.51)$$

Then S_0 becomes

First Independent General Solution

$$S_0 = {}_1F_1(\gamma; a, c) = \sum_{r=0}^{\infty} d_r \gamma^r \qquad (5.52)$$

The zero order coefficient is taken to be $d_0 = 1$, so all other subsequent coefficients d_r can be determined using Eq. (5.44). To express the second independent solution S_{1-c} explicitly, we can take advantage of the fact that a and c are unspecified parameters at this point. When $p = 0$, Eq. (5.44) gives

$$\frac{d_r}{d_{r-1}} = \frac{(p+r-1+a)}{(p+r)(p+r+c-1)} = \frac{(r-1+a)}{r(r+c-1)} \qquad (5.53)$$

But, if we substitute $p = 1 - c$, for the second solution, we have

$$\frac{d_r}{d_{r-1}} = \frac{((1-c)+r-1+a)}{((1-c)+r)((1-c)+r+c-1)} = \frac{(r-1+(a-c+1))}{r(r+(2-c)-1)} \qquad (5.54)$$

Eq. (5.54) takes the same form as Eq. (5.53), with modified parameters a' and c'. Then, we have

$$\frac{d_r}{d_{r-1}} = \frac{(r-1+a')}{r(r+c'-1)} \qquad (5.55)$$

The modified parameters a' and c' for the second independent solution, in terms of the parameters for the first independent solution a and c, become

$$a' = a - c + 1 \tag{5.56a}$$

$$c' = 2 - c \tag{5.56b}$$

We can now write the second independent solution S_{1-c} as

$$S_{1-c} = \sum_{r=0}^{\infty} d'_r \gamma^{p+r} = \gamma^p \sum_{r=0}^{\infty} d'_r \gamma^r = \gamma^{1-c} {}_1F_1(\gamma; a', c') \tag{5.57}$$

Note that the coefficient γ^p is a power expressed in terms of c, not c', because we have replaced the value of p for this particular solution. Given the modified parameters, the coefficients for the second independent solution become

$$d_r = \frac{1}{r!} \frac{\prod_{k=1}^{r}(a'+k-1)}{\prod_{k=1}^{r}(c'+k-1)} \tag{5.58}$$

The complete general solution to Eq. (5.34) is given by the linear combination of both independent solutions S_0 and S_{1-c}. Therefore, the complete general solution becomes

---

Spatial Eigensolution Harmonic Oscillator

$$X_0(\gamma; a, c) = C_1 \cdot {}_1F_1(\gamma; a, c) + C_2 \cdot \gamma^{1-c} {}_1F_1(\gamma; a-c+1, 2-c) \tag{5.59}$$

Note that the general relation between solutions S_p and the *Confluent Hypergeometric Function* ${}_1F_1$ is given by

$$S_p(x; a, c) = x^p \cdot {}_1F_1(x; a, c) \tag{5.60}$$

If $p = 0$, as it is with the first independent solution, then they are identical. For this reason, the independent solutions in the form given here *is* the Confluent Hypergeometric Function, where the CHF is the fundamental solution to the Confluent Hypergeometric ordinary differential equation given in Eq. (5.24). Though we have obtained the general solution, we have not yet considered boundary conditions for the wave function X. Therefore, further analysis is needed for the harmonic oscillator to obtain the specific solution also satisfying boundary conditions. This turns out to be an important step before we can claim to have the complete solution to the Schrödinger equation. This point of satisfying the boundary conditions is discussed in the following subsection.

5.3 Spatial Dependence Solvable By Frobenius Method

5.3.1 Conditions For CHF To Satisfy Schrödinger Equation

For the quantum harmonic oscillator, we have identified the parameters $c = 1/2$, and therefore $c' = 2 - c = 3/2$. Simple enough, since these are explicit constants. However, the value for $a = (1 - \Omega)/4$ is unknown because a is a function of the energy E_λ. It is, therefore, not known explicitly because we need to identify the values for $\Omega = 2E_\lambda/\hbar\omega$. The final solution to the Schrödinger equation is tied to the values of a and a'. So, we must take a closer look at the ability of the Confluent Hypergeometric function to satisfy the boundary conditions of the Schrödinger equation.

The CHF is a generalization of the infinite series for the exponential function $\exp(x) = e^x$. This can be seen if we consider the specific case when $a = c$, and the CHF reduces to

$$_1F_1(\gamma; a, c) = 1 + \frac{1}{1!}\frac{a}{a}\gamma + \frac{1}{2!}\frac{a(a+1)}{a(a+1)}\gamma^2 +$$

$$\frac{1}{3!}\frac{a(a+1)(a+2)}{a(a+1)(a+2)}\gamma^3 + \frac{1}{4!}\frac{a(a+1)(a+2)(a+3)}{a(a+1)(a+2)(a+3)}\gamma^4 +$$

$$\frac{1}{5!}\frac{a(a+1)(a+2)(a+3)(a+4)}{a(a+1)(a+2)(a+3)(a+4)}\gamma^5 +$$

$$\frac{1}{6!}\frac{a(a+1)(a+2)(a+3)(a+4)(a+5)}{a(a+1)(a+2)(a+3)(a+4)(a+5)}\gamma^6 + \ldots$$

$$= 1 + \frac{1}{1!}\gamma + \frac{1}{2!}\gamma^2 + \frac{1}{3!}\gamma^3 + \frac{1}{4!}\gamma^4 + \frac{1}{5!}\gamma^5 + \frac{1}{6!}\gamma^6 + \ldots \quad (5.61)$$

With this case, it is clear that

$$_1F_1(\gamma; a, a) = 1 + \frac{1}{1!}\gamma + \frac{1}{2!}\gamma^2 + \frac{1}{3!}\gamma^3 + \frac{1}{4!}\gamma^4 + \frac{1}{5!}\gamma^5 + \frac{1}{6!}\gamma^6 + \ldots = e^\gamma$$

This demonstrates that $_1F_1$ possesses e^γ behavior. But, since our solution also involves a factor of $e^{-\gamma/2}$ (c.Eq. 5.30), if we leave $_1F_1$ in its general infinite series form, our solution behaves like $e^{-\gamma/2} \cdot e^\gamma = e^{\gamma/2}$. But, this function *cannot* satisfy the normalization condition of a wave function where $\lim_{|x|\to\infty}\psi^*\psi = 0$. In fact, the only way that the CHF can satisfy the normalization condition is if it is turned into a polynomial p_n of finite order n because then $\lim_{|x|\to\infty}\psi^*\psi = 0$ *is* satisfied. In other words, the infinite series must be truncated somehow. Any finite polynomial in γ multiplied by $e^{-\gamma/2}$ is normalizable because the exponential terms converges more rapidly, compared to a finite order polynomial p_n. By inspection of the CHF, we can ensure that $_1F_1$ is a finite order polynomial by requiring the numerators d_r to vanish at some finite order n in the series. Hence, the condition to achieve a finite polynomial becomes

$$a = \frac{1-\Omega}{4} = -n \tag{5.62}$$

The integer n takes values $n = 0, 1, 2, 3,...$ Since $a' = a - c + 1$ is a function of a and c, once a is known, a' is also determined. Determining a' using Eq. (5.62) gives

$$a' = a - c + 1 = a + \frac{1}{2} \tag{5.63}$$

Eq. (5.63) tells us that when a is a negative integer, ensuring that $_1F_1$ satisfies the normalization condition, a' is *not*, and vice versa. Since a and a' both can't be negative integers simultaneously, it means that it is not possible for both solutions S_0 and S_{1-c} to satisfy the required normalization condition simultaneously! This, therefore, necessitates that for any complete solution satisfying the boundary conditions, only one of the two coefficients C_1, C_2 of the general solution from Eq. (5.59) can be nonzero. In the case where C_2 is nonzero, the condition will be

$$a' = a - c + 1 = a + \frac{1}{2} = \frac{1-\Omega}{4} + \frac{1}{2} = \frac{3-\Omega}{4} = -n \tag{5.64}$$

For both cases combined, solving for Ω gives

$$1 - \Omega = -4n \Rightarrow \Omega = 4n + 1 = 1, 5, 9, 13,... \tag{5.65a}$$

$$3 - \Omega = -4n \Rightarrow \Omega = 4n + 3 = 3, 7, 11, 15,... \tag{5.65b}$$

Hence, we find that for the complete set of solutions, Ω *only* takes odd integer values, i.e.

$$\Omega = 2n + 1 \tag{5.66}$$

Using the definition of Ω given by Eq. (5.22), we can determine the corresponding eigenenergy values E_λ as

$$\Omega = \frac{2E_\lambda}{\hbar\omega} = 2n + 1 \tag{5.67}$$

Solving for the allowed energy values for the 1D harmonic oscillator, we obtain the result

5.3 Spatial Dependence Solvable By Frobenius Method

1D Harmonic Oscillator Energy

$$E_\lambda = \left(n+\frac{1}{2}\right)\hbar\omega \qquad (5.68)$$

Consequently, a transition between any two consecutive states ΔE_λ becomes

$$\Delta E_\lambda = \left(n+1+\frac{1}{2}\right)\hbar\omega - \left(n+\frac{1}{2}\right)\hbar\omega = \hbar\omega$$

or transitions are given by

Energy Transition Of Harmonic Oscillator

$$\Delta E_\lambda = \hbar\omega \qquad (5.69)$$

Combining this result with the asymptotic limit solution, we can *now* write the complete spatially dependent solution to the Schrödinger equation for the harmonic oscillator $X = X_0 e^{-\frac{1}{2}\gamma}$, or

$$X(\gamma; a, c) = \left[C_1 \cdot {}_1F_1\left(\gamma; \frac{1-\Omega}{4}, \frac{1}{2}\right) + C_2 \cdot \gamma^{\frac{1}{2}} {}_1F_1\left(\gamma; \frac{3-\Omega}{4}, \frac{3}{2}\right) \right] e^{-\frac{1}{2}\gamma}$$

Substituting $\gamma = \zeta^2$, where $\zeta = \alpha x$, in terms of the spatial variable x, we have

1D Harmonic Oscillator Spatial Eigensolution

$$X(\alpha x) = \left[C_{11} F_1\left(\alpha x; \frac{1-\Omega}{4}, \frac{1}{2}\right) + C_2 \alpha \cdot x \cdot {}_1F_1\left(\alpha x; \frac{3-\Omega}{4}, \frac{3}{2}\right) \right] e^{-\frac{1}{2}\alpha^2 x^2}$$

$$(5.70)$$

$\alpha = \sqrt{(m\omega)/\hbar}$, in Eq. (5.70), remember that either $C_1 = 0$ or $C_2 = 0$.

There is a more convenient way to represent the solution given by Eq. (5.70) without having to keep track of whether $C_1 = 0$ or $C_2 = 0$. In the next section, we'll introduce an alternative form of the truncated series functions when a and/or a' are set to $-n$. It turns out that the sequence of finite order polynomials reproduces well-known polynomials introduced by mathematicians in the mid-1800s.

5.3.2 Truncated CHF Leads to Hermite Polynomials H_n

In the previous section, we showed that in order for the wave function ψ to satisfy the normalization condition, it is necessary that a and/or a' be a negative integer $-n$. This results in the truncation of terms in the CHF, vanishing of all subsequent series terms $n, n+1, n+2$, etc. Note that there is no approximation needed in satisfying these conditions. The solution is still exact. The truncation of the *Confluent Hypergeometric Series* results in polynomials of order n, where $n = |a|$ or $n = |a'|$. Let us compute, explicitly, the first five polynomials. The smallest value n can take is 0. In this case, Eq. (5.66) implies $\Omega = 1$. Since truncating $_1F_1$ leads to a polynomial, the second term in the solution with C_2 is at least linear in x, due to the αx factor. For the lowest order polynomial, which is a constant, since one of the coefficients C_1 and C_2 must be zero, it follows that $C_2 = 0$. This consequently ensures the solution satisfies the normalization condition. In this case, the truncated series then gives

$$_1F_1\left(\gamma; a = 0, \frac{1}{2}\right) = 1 + \frac{1}{1!}\frac{a}{c}\gamma + \frac{1}{2!}\frac{a(a+1)}{c(c+1)}\gamma^2 + \ldots = 1 + 0\gamma + 0\gamma^2 + \ldots = 1$$

To express the resulting polynomials in the variable ζ, the convention that we will use is to make the coefficient of the highest power (the leading coefficient) of γ, which is n, equal to $+2^n$. Doing this scales the equation by a constant. This is allowed because the normalization process for the wave function rescales the resulting form, anyway. Scaling the truncated polynomials in this way enables the systematic production of what are known as **Hermite Polynomials** H_n. Thus, by knowing the CHF $_1F_1$, any Hermite polynomial can be generated. For the case of $|a| = n = 0$, the leading coefficient becomes $2^0 = 1$, which is what we already have. Then, the lowest order Hermite polynomial we have just obtained is given by

$$H_0(\zeta) = 1 \tag{5.71}$$

This turns out to be the only case where $_1F_1 = H_n$. Next, in the case of $n = 1$, we must, again, use Eq. (5.67) which gives $\Omega = 2n + 1 = 3$,

5.3 Spatial Dependence Solvable By Frobenius Method

giving $a' = 0$. This also leads to $a = -1/2$, which is not a negative integer. Therefore, C_1 must be zero in order to satisfy the boundary conditions. Then, for the case of $n = 1$, using the above result for $a = 0$, we already know

$$_1F_1\left(\gamma; a' = 0, \frac{3}{2}\right) = 1$$

The second term corresponding to $p = 1 - c$, however, also has a factor $\gamma^{1-c} = \gamma^{\frac{1}{2}} = \zeta$. So, we have the complete truncated polynomial given by

$$S_{1-c} = \gamma^{1-c}{}_1F_1\left(\gamma; a' = 0, \frac{3}{2}\right) = \gamma^{\frac{1}{2}}(1 - 0\gamma + 0...) = \zeta \tag{5.72}$$

Scaling the resulting polynomial to obtain the Hermite polynomial H_1, the highest power coefficient is $2^n = 2^1 = 2$. Then, for $n = 1$, we have

$$H_1(\zeta) = 2^1 \gamma^{\frac{1}{2}} = 2\zeta \tag{5.73}$$

Continuing on to the case of $n = 2$, we have $\Omega = 2n + 1 = 5$. This yields $a = -1$, thus a' cannot be a negative integer. Therefore, C_2 is zero. This gives

$$S_0\left(\gamma; a = -1, \frac{1}{2}\right) = 1 - \frac{1}{1/2}\gamma + 0... = 1 - 2\gamma \tag{5.74}$$

Scaling Eq. (5.74) so that the highest power has a coefficient $2^n = 2^2 = 4$, the Hermite polynomial H_2 becomes

$$H_2(\zeta) = 4\gamma - 2 = 4\zeta^2 - 2 \tag{5.75}$$

Hopefully, the pattern is becoming clear. We'll do two more examples to drive the point home. For $n = 3$, we find that $a' = -1$, while $a = -3/2$. Therefore, since a is not a negative integer, to satisfy normalization conditions, it follows that C_1 must be zero. We then have

$$S_{1-c}(\gamma; a' = -1, \frac{3}{2}) = \gamma^{\frac{1}{2}}\left(1 - \frac{1}{3/2}\gamma + 0...\right) = \gamma^{\frac{1}{2}}\left(1 - \frac{2}{3}\gamma\right) \tag{5.76}$$

We want the leading coefficient to be $+2^3$. For this, we can scale by $-\frac{3}{2} \times +2^3 = -12$. With this, further substituting $\gamma = \zeta^2$, the Hermite polynomial H_3 becomes

$$H_3(\zeta) = 8\zeta^3 - 12\zeta \tag{5.77}$$

As the last example, we'll consider the case of $n = 4$. In this case, $\Omega = 2n + 1 = 9$, leading to $a = -2$, and $a' = 6/4$. Since a' is not a negative integer, $C_2 = 0$. Then, we have

$$S_0\left(\gamma; a = -2, \frac{1}{2}\right) = 1 + \frac{-2}{1/2}\gamma + \frac{1}{2!}\frac{-2(-2+1)}{1/2(1/2+1)}\gamma^2 + 0...) = 1 - 4\gamma + \frac{8}{6}\gamma^2$$

Scaling our result so that the leading coefficient is $+2^4 = 16$, we can scale the above result by $+\frac{6}{8} \times 16 = 12$. With this, substituting $\gamma = \zeta^2$, we obtain the result for H_4 given by

$$H_4(\zeta) = 16\zeta^4 - 48\zeta^2 + 12 \tag{5.78}$$

This process can continue indefinitely to obtain any Hermite polynomial H_n, provided the values for a and c are known. The first ten Hermite polynomials are plotted in Fig. 5.3.

Figure 5.3: First ten Hermite polynomials H_n. The polynomials iterate through symmetric and anti-symmetric states, as n increases. This pattern establishes sequential symmetric and anti-symmetric quantum states for the harmonic oscillator.

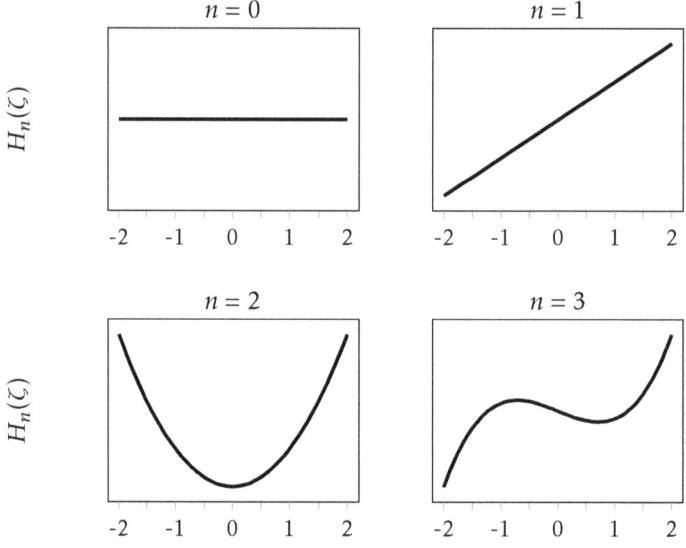

5.3 Spatial Dependence Solvable By Frobenius Method

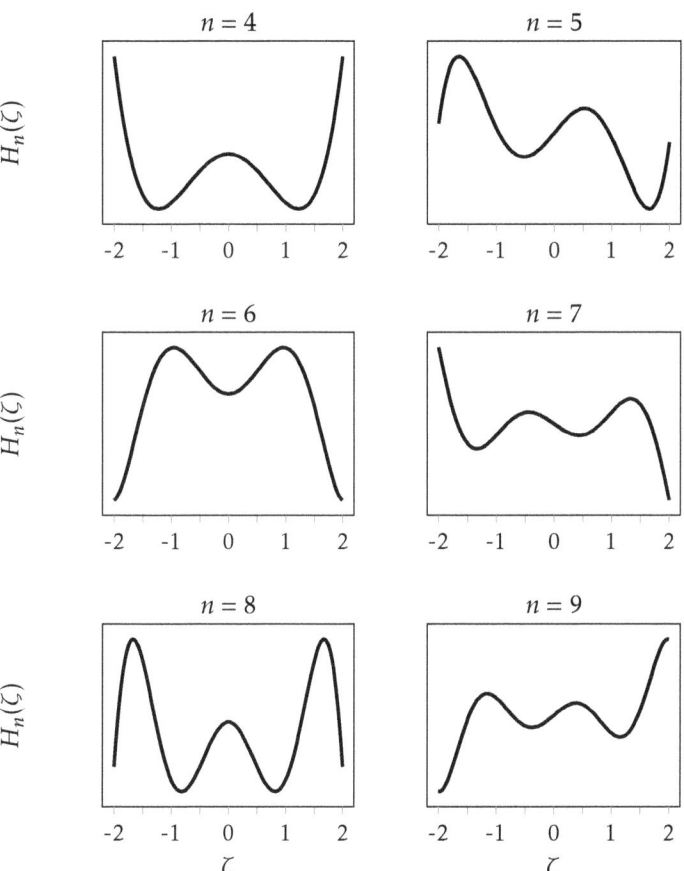

The Hermite polynomials iterate through symmetric and anti-symmetric eigenstates, as n increases. With the Hermite polynomials H_n being proportional to the Confluent Hypergeometric Function $_1F_1$, we can express the eigen-solution to the CHE in Eq. (5.34) as

> **1D Harmonic Oscillator Spatial Solution**
>
> $$X_n(\zeta) = C_n H_n(\zeta) e^{-\frac{1}{2}\zeta^2} \tag{5.79}$$

The first ten normalized eigensolutions $X_n(\zeta)$, given by Eq. (5.79), are plotted in Fig. 5.4.

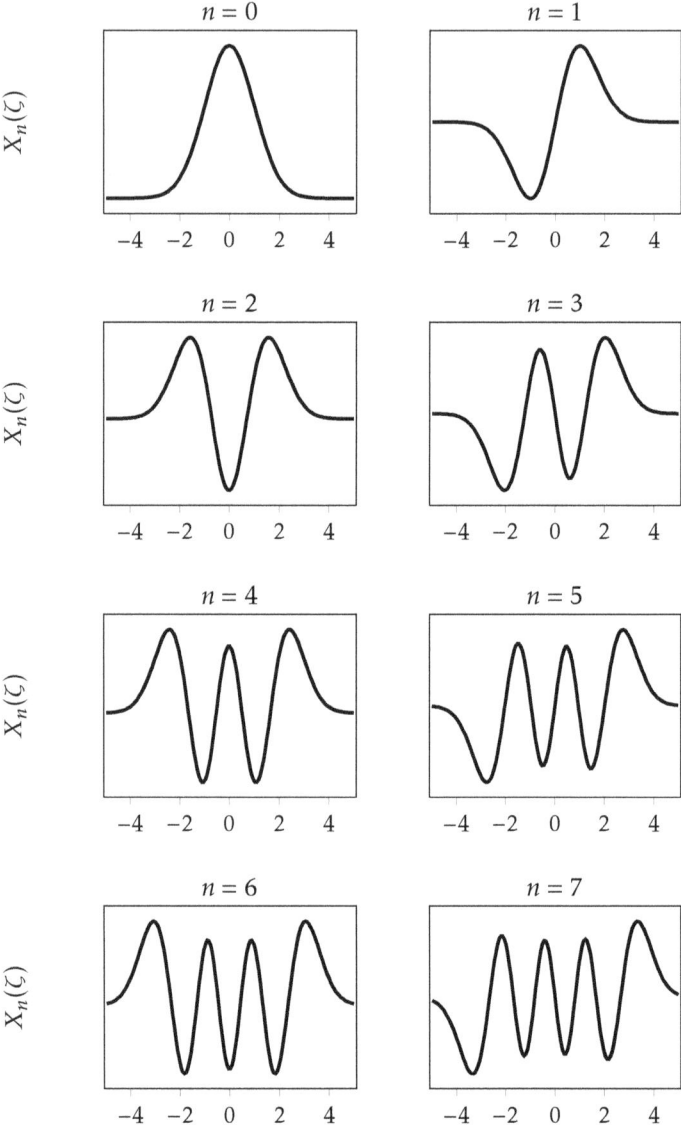

Figure 5.4: First ten eigensolutions $X_n(\zeta)$ of the harmonic oscillator problem. Owing to the Hermite polynomials being multiplied by a symmetric Gaussian function, the eigenstates iterate through symmetric and anti-symmetric spatial states as n increases.

5.3 Spatial Dependence Solvable By Frobenius Method

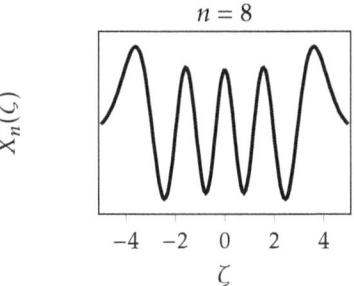
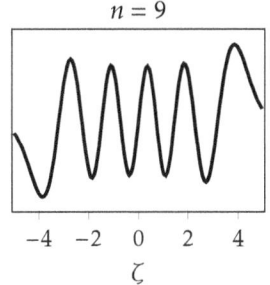

Because the Hermite polynomials are multiplied by a symmetric Gaussian function, we have that the eigenstates iterate through symmetric and anti-symmetric states, as n increases. Being able to generate Hermite polynomials from the CFE is very useful for two primary reasons: firstly, because the CHF is easy to remember, we can readily write the solution in terms of these well-known Hermite polynomials; lastly, we do not have to be concerned with the two independent solutions S_0 and S_{1-c}, tracking whether $C_1 = 0$ or $C_2 = 0$. Combining this result for space with the time-dependent solution, the complete eigen-function $\psi(x,t)$ of the 1D Schrödinger

Space-Time 1D Harmonic Oscillator Eigenfunctions

$$\psi_n(\zeta, t) = C_n H_n(\zeta) e^{-\frac{1}{2}\zeta^2} e^{-i\frac{E_\lambda}{\hbar}(t-t_0)} \tag{5.80}$$

C_n is the normalization factor, which is unique for each eigensolution since it, generally, depends on n. For convenience, the first fourteen ($n = 0, 1, ..., 13$) Hermite polynomials are given in Table 5.4. We can determine C_n from the *normalization condition*. To do this, we can evaluate the normalization integral using a well-known property of the Hermite polynomials, given by

Orthogonality Property Of Hermite Polynomials

$$\int_{-\infty}^{\infty} H_m(x) H_n(x) e^{-x^2} dx = 2^n n! \sqrt{\pi} \delta_{mn} \qquad (5.81)$$

Using the *orthogonality property of the Hermite polynomials*, evaluation of the normalization condition leads to

$$1 = \int_{-\infty}^{+\infty} \psi_n(\zeta)^\dagger \psi_n(\zeta) dx$$

$$= \int_{-\infty}^{+\infty} C_n^\dagger H_n^\dagger(\zeta) e^{-\zeta^2/2} C_n H_n(\zeta) e^{-\zeta^2/2} dx$$

$$= \frac{|C_n|^2}{\alpha} \int_{-\infty}^{+\infty} H_n(\zeta) H_n(\zeta) e^{-\zeta^2} d\zeta$$

$$= \frac{|C_n|^2}{\alpha} \left(2^n n! \sqrt{\pi} \right)$$

We have used the earlier relation $\zeta = \alpha x$. This leads us to our final result for the magnitude of the normalization factor $|C_n|$, given by

1D Oscillator Normalization Factor

$$|C_n| = \sqrt{\frac{\alpha}{2^n n! \sqrt{\pi}}} \qquad (5.82)$$

$\alpha = \sqrt{(m\omega)/\hbar}$ is a parameter having units of m^{-1}, defined earlier in Eq. (5.20). m is the mass, ω is the natural frequency of the harmonic oscillator, and $\hbar = h/2\pi$ is Planck's reduced constant.

The most general solution for the harmonic oscillator is a linear combination of the countably infinite eigenstates $\psi_n(\zeta, t)$, given by Eq. (5.80). We will see that it is this general solution that leads to the classical results we obtained in the first section. We can write the most general solution as

$$\Psi(x, t) = \Psi(\zeta, t) = \sum_{n=0}^{\infty} b_n \psi_n(\zeta, t)$$

5.3 Spatial Dependence Solvable By Frobenius Method

$$= \sum_{n=0}^{\infty} b_n X_n(\zeta) T_n$$

$$= \sum_{n=0}^{\infty} b_n X_n(\zeta) e^{-\frac{i}{\hbar} E_n t}$$

b_n coefficient are scalars, generally complex. In the above form, $X = X_n$ and $T = T_n$ are used to denote that they both depend on the integer n. $T(t)$ is determined by the energy E_n given in Eq. (5.68), while for $X(x)$, is determined the order of the truncated polynomial, which is the Hermite polynomial H_n. Substitution of the solution for E_n, given by Eq. (5.68), we have the general solution given by

1D Harmonic Oscillator General Solution

$$\Psi(x,t) = \sum_n b_n X_n e^{-i\left(n+\frac{1}{2}\right)\omega t} \tag{5.83}$$

Based the general solution given by Eq. (5.83), the initial state $\Psi(x, t = 0)$ is

$$\Psi(x,0) = \sum_n b_n X_n \tag{5.84}$$

In Chap. 4, we proved that any pair of eigensolutions of the Schrödinger equation are always mutually orthogonal, provided we have a real-valued Hermitian Hamiltonian operator. These assumptions are met by the harmonic oscillator, composed of kinetic energy and the real-valued classical potential energy. Therefore, that for any two distinct integers n and m, it must be that

$$\int_{-\infty}^{\infty} X_m^\dagger X_n dx = \delta_{mn}$$

$X_m^\dagger$ is more general notation for the *transposed complex conjugate* of X_m. Since X_m is a scalar function in this case, it is identical to the complex conjugate of X_m, or $X_m^\dagger = X_m^*$. δ_{mn} is the Kronecker delta function. The integral is nonzero only when $m = n$, and in this case, the integral becomes unity since we assume normalized eigensolutions. Otherwise, the result is zero, which means that the eigenfunctions are *orthogonal*.

This property can be used to obtain the coefficients b_n. Multiplying the initial state by $X_m^\dagger$ and integrating over space, we obtain

$$\int_{-\infty}^{\infty} X_m^\dagger \Psi(x,0) dx = \int_{-\infty}^{+\infty} \sum_n b_n X_m^\dagger X_n$$

$$= \sum_n b_n \int_{-\infty}^{+\infty} X_m^\dagger X_n dx$$

$$= \sum_n b_n \delta_{mn}$$

$$= b_m$$

Then, having the initial condition of the general wave function, we can obtain any of the coefficients by the result

State Coefficient Formula

$$b_n = \int_{-\infty}^{\infty} X_n^\dagger \Psi(x,0) dx \qquad (5.85)$$

Therefore, the initial condition for the most general solution provides sufficient information to obtain the total general solution given by Eq. (5.83). It turns out that because of the rich properties of the Hermite polynomials and the form of the Hamiltonian, so much more information can be revealed. We'll demonstrate some of this, beginning in the next section.

5.4 Symmetry & Annihilation/Creation Operators

If we look closely at the Hamiltonian for the harmonic oscillator, we find an interesting and useful symmetry that can be exploited. Recall that the Schrödinger equation for the harmonic oscillator has the form

$$i\hbar \frac{\partial \psi}{\partial t} = \left[-\frac{\hbar^2}{2m} \frac{\partial^2}{\partial x^2} + \frac{1}{2} m\omega^2 x^2 \right] \psi \Rightarrow \widehat{H} = \frac{1}{2} m\omega^2 x^2 - \frac{\hbar^2}{2m} \frac{\partial^2}{\partial x^2} \qquad (5.86)$$

The Hamiltonian $\widehat{H}$ can be written in the following form:

$$\widehat{H} = \frac{1}{2m} \left[m^2 \omega^2 x^2 - \left(\hbar \frac{\partial}{\partial x} \right)^2 \right] \qquad (5.87)$$

5.4 Symmetry & Annihilation/Creation Operators

We can readily identify a symmetry in Eq. (5.87) that allows an expression of lower order operators, as it resembles a scalar equation of the form

$$a^2 + b^2 = c_0(a+ib)(a-ib)$$

c is a scalar corresponding to $1/2m$. The real number a corresponds to $m\omega x$ in Eq. (5.87), and we can show a correspondence between the real number b and the momentum operator $-i\hbar \partial/\partial x$. Therefore, let us define the following operator

$$\widehat{A} = \frac{1}{\sqrt{2m}}[m\omega x + i\widehat{p}_x] \tag{5.88}$$

Since x is real-valued, and so is $\widehat{p}_x$ as a Hermitian operator, the first order operator $\widehat{A}$ necessarily has complex eigenvalues because of the purely imaginary momentum operator term. Moreover, because x and p_x change with time, the eigenvalues of the operator $\widehat{A}$ are generally time-dependent. For the other first order operator, we can define $\widehat{A}^\dagger$ as

$$\widehat{A}^\dagger = \frac{1}{\sqrt{2m}}[m\omega x - i\widehat{p}_x] \tag{5.89}$$

The eigenvalues of $\widehat{A}^\dagger$ are, therefore, complex conjugates of those of $\widehat{A}$. We know that any complex number multiplied by its complex conjugate is always a real number, guaranteeing the Hamiltonian remains Hermitian or having real eigenvalues. This is an important property resulting from the symmetry. Any operator $\widehat{A}^\dagger$ with complex eigenvalues that are complex conjugates of an operator $\widehat{A}$ is known as the **adjoint** of the operator $\widehat{A}$. The topic of *adjoint operators* is discussed in more detail in Chap. 14. Let's express the Hamiltonian for the harmonic oscillator in terms of the operators $\widehat{A}$ and $\widehat{A}^\dagger$. This leads to

$$\widehat{A}\widehat{A}^\dagger = \frac{1}{\sqrt{2m}}[m\omega x + i\widehat{p}_x]\frac{1}{\sqrt{2m}}[m\omega x - i\widehat{p}_x]$$

$$= \frac{1}{2m}\left[m^2\omega^2 x^2 + \widehat{p}_x^2 - im\omega(x\widehat{p}_x - \widehat{p}_x x)\right]$$

Using the definition for the square of the momentum operator, given by Eq. (3.11), we have

$$\widehat{A}\widehat{A}^\dagger = \frac{1}{2m}\left[m^2\omega^2 x^2 - \hbar^2\frac{\partial^2}{\partial x^2} - im\omega(x\widehat{p}_x - \widehat{p}_x x)\right]$$

Comparing the above result with Eq. (5.87), we find that the first two terms by themselves equate to the Hamiltonian of the harmonic oscillator $\widehat{H}$, however, we also have an additional term in parentheses. Let's take a closer look at this third operator, which becomes

$$\left[x\left(-i\hbar\frac{\partial}{\partial x}\right)-\left(-i\hbar\frac{\partial}{\partial x}\right)x\right]\psi_n = x\left(-i\hbar\frac{\partial \psi_n}{\partial x}\right)-\left(-i\hbar\frac{\partial (x\psi_n)}{\partial x}\right)$$

$$= -ix\hbar\frac{\partial \psi_n}{\partial x} + ix\hbar\frac{\partial \psi_n}{\partial x} + i\hbar\psi_n$$

$$= i\hbar\psi_n$$

Hence, the operator $x\widehat{p}_x - x\widehat{p}_x$ turns out to be a constant, or has the eigenvalue given by

---
Position-Momentum Commutation Relation

$$x\widehat{p}_x - \widehat{p}_x x = i\hbar \tag{5.90}$$

The result given in Eq. (5.90) illustrates an important property of quantum mechanical operators. Specifically, their *order* matters because Eq. (5.90) means that $x\widehat{p}_x \neq \widehat{p}_x x$ or $x\widehat{p}_x - \widehat{p}_x x \neq 0$. In quantum mechanics, this general rule must always be remembered when considering a sequence of operators. Because $x\widehat{p}_x - \widehat{p}_x x \neq 0$, x and $\widehat{p}_x$ are said *not* to **commute** with one another. We discuss the topic of *commutation* between operators in more detail in Chap. 7. Using the above result, we have the following relation between $\widetilde{AA^\dagger}$ and $\widehat{H}$:

$$\widetilde{AA^\dagger} = \widehat{H} - \frac{1}{2m}[im\omega(i\hbar)]$$

$$= \widehat{H} + \frac{\hbar\omega}{2}$$

Thus, the Hamiltonian becomes

$$\widehat{H} = \widetilde{AA^\dagger} - \frac{\hbar\omega}{2} \tag{5.91}$$

If we apply the operator $\widetilde{AA^\dagger}$ on the eigensolutions ψ_n, using the relation $\widehat{H}\psi_n = E_\lambda \psi_n$, we then have

$$\widetilde{AA^\dagger}\psi_n = \left[\hbar\omega\left(n+\frac{1}{2}\right) + \frac{\hbar\omega}{2}\right]\psi_n$$

5.4 Symmetry & Annihilation/Creation Operators

$$= (n+1)\hbar\omega\psi_n$$

Adding a constant of $\hbar\omega/2$ to both sides, we can write the following relation:

$$\left[\widehat{AA^\dagger} + \frac{\hbar\omega}{2}\right]\psi_n = \hbar\omega\left[(n+1) + \frac{1}{2}\right]\psi_n \tag{5.92}$$

A similar analysis for $\widehat{A^\dagger A}$, instead, gives

$$\widehat{A^\dagger A} = \frac{1}{2m}\left[m^2\omega^2 x^2 - \hbar^2\frac{\partial^2}{\partial x^2} + im\omega(x\widehat{p_x} - \widehat{p_x}x)\right]$$

$$= \widehat{H} - \frac{\hbar\omega}{2}$$

$$= \hbar\omega\left(n + \frac{1}{2}\right) - \frac{\hbar\omega}{2}$$

$$= \hbar\omega n$$

In this case, subtracting a constant from both sides, we can write

$$\left[\widehat{A^\dagger A} - \frac{\hbar\omega}{2}\right]\psi = \hbar\omega\left[(n-1) + \frac{1}{2}\right]\psi \tag{5.93}$$

The above analysis shows that the operators $\widehat{AA^\dagger}$ and $\widehat{A^\dagger A}$ tend to shift the $n + 1/2$ term of the eigenvalues. Additionally, we find that changing the order of the operators leads to different results. In this case, changing the order affects the direction in which the integer n effectively changes. For the first case, there is a shift of $+\hbar\omega/2$. Adding and subtracting the constants we have chosen, lead to clean integers on the RHS.

It is more common to define these operators in dimensionless form as $\widehat{a} = \widehat{A}/\sqrt{\hbar\omega}$ and $\widehat{a}^\dagger = \widehat{A}^\dagger/\sqrt{\hbar\omega}$. In this dimensionless form, we have the following two first-order symmetric operators

$$\widehat{a} = \frac{1}{\sqrt{2m\hbar\omega}}[m\omega x + i\widehat{p_x}] \tag{5.94a}$$

$$\widehat{a}^\dagger = \frac{1}{\sqrt{2m\hbar\omega}}[m\omega x - i\widehat{p_x}] \tag{5.94b}$$

The dimensionless form leads to the following dimensionless eigenvalue equations, given by

$$\left[\widehat{aa^\dagger} + \frac{1}{2}\right]\psi = \left[(n+1) + \frac{1}{2}\right]\psi \tag{5.95a}$$

$$\left[\hat{a}^\dagger\hat{a} - \frac{1}{2}\right]\psi = \left[(n-1) + \frac{1}{2}\right]\psi \qquad (5.95b)$$

Having defined the first order operators $\hat{a}$ and $\hat{a}^\dagger$, that relate directly to the second order Hamiltonian $\widehat{H}$ for the harmonic oscillator, we are in a position to examine new questions such as what these lower order operators do to an eigenstate $\psi_n(\zeta)$ of the harmonic oscillator. That is, what do we get when we compute $\hat{a}\psi_n$? Substituting the operator $\hat{p}_x = -i\hbar\partial/\partial x$, and using the definition of $\hat{a}$, given by Eq. (5.94a), we have

$$\hat{a}\psi_n = \frac{1}{\sqrt{2m\hbar\omega}}[m\omega x + i\hat{p}_x]\psi_n$$

$$= \frac{1}{\sqrt{2}}\left[\sqrt{\frac{m\omega}{\hbar}}x\psi_n + \sqrt{\frac{\hbar}{m\omega}}\frac{\partial\psi_n}{\partial x}\right]$$

Using the earlier definition of $\alpha = \sqrt{m\omega/\hbar}$ and $\zeta = \alpha x$, the above becomes

$$\hat{a}\psi_n = \frac{1}{\sqrt{2}}\left(\zeta\psi_n + \frac{\partial\psi_n}{\partial\zeta}\right) \qquad (5.96)$$

We can evaluate these terms explicitly using the exact solution for ψ_n given by Eq. (5.79). First, for the term $\partial\psi_n/\partial\zeta$, we have

$$\frac{\partial\psi_n}{\partial\zeta} = C_n\left[-\zeta e^{-\frac{\zeta^2}{2}}H_n(\zeta) + e^{-\frac{\zeta^2}{2}}\frac{\partial H_n}{\partial\zeta}\right] \qquad (5.97)$$

Now, we make use of another special property of Hermite polynomials known as a recurrence relation, namely

Hermite Polynomial Derivative Relation

$$\frac{dH_n}{d\zeta} = 2nH_{n-1}(\zeta) \qquad (5.98)$$

Substituting this relation into Eq. (5.97) gives

$$\frac{\partial\psi_n}{\partial\zeta} = C_n\left[-\zeta e^{-\frac{\zeta^2}{2}}H_n(\zeta) + 2ne^{-\frac{\zeta^2}{2}}H_{n-1}(\zeta)\right] \qquad (5.99)$$

Then, substituting this result into Eq. (5.107), we obtain

$$\hat{a}\psi_n = \frac{C_n}{\sqrt{2}}\left(\zeta e^{-\frac{\zeta^2}{2}}H_n(\zeta) + -\zeta e^{-\frac{\zeta^2}{2}}H_n(\zeta) + 2ne^{-\frac{\zeta^2}{2}}H_{n-1}(\zeta)\right)$$

5.4 Symmetry & Annihilation/Creation Operators

$$= \frac{2nC_n}{\sqrt{2}} e^{-\frac{\zeta^2}{2}} H_{n-1}(\zeta)$$

Using the definition for the normalization constant $C_n = \sqrt{\alpha/2^n n! \sqrt{\pi}}$, we can simplify further as follows:

$$\widehat{a}\psi_n = \sqrt{\frac{\alpha}{2^n n!}} \sqrt{2n} e^{-\frac{\zeta^2}{2}} H_{n-1}(\zeta)$$

$$= \sqrt{n} \sqrt{\frac{2n\alpha}{2^n n!}} e^{-\frac{\zeta^2}{2}} H_{n-1}(\zeta)$$

$$= \sqrt{n} \sqrt{\frac{\alpha}{2^{n-1}(n-1)!}} \sqrt{2n} e^{-\frac{\zeta^2}{2}} H_{n-1}(\zeta)$$

$$= \sqrt{n} C_{n-1} e^{-\frac{\zeta^2}{2}} H_{n-1}(\zeta)$$

Therefore, we obtain the following result for $\widehat{a}\psi_n$:

> **Harmonic Oscillator Annihilation Operator Relation**
>
> $$\widehat{a}\psi_n = \sqrt{n}\psi_{n-1} \tag{5.100}$$

The first order operator $\widehat{a}$ takes any eigensolution ψ_n and returns a state proportional to the eigensolution ψ_{n-1}. Because of this property of effectively reducing the energy from n to $n-1$, $\widehat{a}$ earns the name of being the **annihilation operator**, and is also an example of a **lowering operator**.

A similar analysis can be done for $\widehat{a}^\dagger \psi_n(\zeta)$, which gives

$$\widehat{a}^\dagger \psi_n = \frac{1}{\sqrt{2m\hbar\omega}} [m\omega x - i\widehat{p}_x] \psi_n$$

$$= \frac{1}{\sqrt{2}} \left[\sqrt{\frac{m\omega}{\hbar}} x\psi_n - \sqrt{\frac{\hbar}{m\omega}} \frac{\partial \psi_n}{\partial x} \right]$$

In terms of $\alpha = \sqrt{m\omega/\hbar}$ and the dimensionless parameter $\zeta = \alpha x$, we now have

$$\widehat{a}^\dagger \psi_n = \frac{1}{\sqrt{2}} \left(\zeta \psi_n - \frac{\partial \psi_n}{\partial \zeta} \right) \tag{5.101}$$

Then, after substitution of the result for $\partial \psi_n / \partial x$ given by Eq. (5.97), we get

$$\hat{a}^{\dagger} \psi_n = \frac{C_n}{\sqrt{2}} \left(2\zeta e^{-\frac{\zeta^2}{2}} H_n(\zeta) - 2n e^{-\frac{\zeta^2}{2}} H_{n-1}(\zeta) \right) \tag{5.102}$$

In the case of $\hat{a}^{\dagger}$, we make use of a different *recurrence relation* for Hermite polynomials, namely

Hermite Polynomial Recurrence Relation

$$H_{n+1}(\zeta) = 2\zeta H_n(\zeta) - 2n H_{n-1}(\zeta) \tag{5.103}$$

Substitution of this relation into Eq. (5.102) gives

$$\hat{a}^{\dagger} \psi_n = \frac{C_n}{\sqrt{2}} e^{-\frac{\zeta^2}{2}} H_{n+1}(\zeta) \tag{5.104}$$

Using the definition for the normalization constant C_n, we can simplify further as

$$\hat{a}^{\dagger} \psi_n = \frac{\sqrt{n+1} \cdot C_n}{\sqrt{n+1}\sqrt{2}} e^{-\frac{\zeta^2}{2}} H_{n+1}(\zeta)$$

$$= \sqrt{n+1} \sqrt{\frac{\alpha}{2^{n+1}(n+1)!}} e^{-\frac{\zeta^2}{2}} H_{n-1}(\zeta)$$

$$= \sqrt{n+1}\, C_{n+1} e^{-\frac{\zeta^2}{2}} H_{n+1}(\zeta)$$

Thus, we obtain the following result for $\hat{a}^{\dagger} \psi_n$:

Harmonic Oscillator Creation Operator Relation

$$\hat{a}^{\dagger} \psi_n = \sqrt{n+1}\, \psi_{n+1} \tag{5.105}$$

In this case, $\hat{a}^{\dagger}$ takes any eigensolution ψ_n and returns a state proportional to the eigensolution ψ_{n+1}, effectively increasing the energy from n to $n+1$. Hence, $\hat{a}^{\dagger}$ is known as the **creation operator**, and is an example of a **raising operator**. We can see that the unique properties that we have found for the eigensolutions of the harmonic oscillator are due, entirely, to the unique properties of the Hermite polynomials. It is straight forward to show, with Eqs. (5.100) and (5.105), that the following eigenvalues for $\widehat{a}\widehat{a}^\dagger$ and $\widehat{a}^\dagger\widehat{a}$ are given by

$$\widehat{a}\widehat{a}^\dagger \psi_{n-1} = n\psi_{n-1} \tag{5.106a}$$

$$\widehat{a}^\dagger\widehat{a}\psi_{n+1} = (n+1)\psi_{n+1} \tag{5.106b}$$

The first of these, given by Eq. (5.106a) also means that the eigenvalue of $\widehat{a}\widehat{a}^\dagger$ is the quantum number n.

5.5 INTRODUCTION TO THE COHERENT STATE

One question that we have not yet answered is what are the general eigenvalues λ_a and eigenfunctions ψ_a of $\widehat{a}$ itself. Due to the form of $\widehat{a}$ and $\widehat{a}^\dagger$, we expect that the eigenvalues of $\widehat{a}^\dagger$ are complex conjugates of λ_a. This answer to this question defines the so-called **coherent state** of the harmonic oscillator. Thus, the *coherent state* is defined by the following eigenvalue problem:

Coherent State Definition

$$\widehat{a}\psi_a = \lambda_a \psi_a \tag{5.107}$$

Using the operator definition for $\widehat{a}$, Eq. (5.107) becomes

$$\frac{1}{\sqrt{2}}\left(\zeta\psi + \frac{\partial \psi}{\partial \zeta}\right) = \lambda_a \psi \Rightarrow \frac{\partial \psi}{\partial \zeta} = \left(\sqrt{2}\lambda_a - \zeta\right)\psi$$

The above result leads to the following equation relating differentials:

$$\frac{d\psi}{\psi} = \left(\sqrt{2}\lambda_a - \zeta\right)d\zeta \tag{5.108}$$

Generally, $\lambda_a(\zeta(t), p_x(t))$ is a function of ζ and the expected value for the momentum p_x. To get past this difficulty, let us only consider $\lambda_a(t = 0) = \lambda$, or *the eigenvalue associated with the initial condition*. We can further assume that the initial condition for the momentum is $\widehat{p}_x = 0$. This simply means that we are starting from rest. Under these conditions for λ_a, we have

$$\lambda_a(t=0) = \frac{1}{\sqrt{2m\hbar\omega}}[m\omega x + i\widehat{p}_x]$$

$$= \frac{1}{\sqrt{2m\hbar\omega}} m\omega x$$

$$= \frac{\zeta_0}{\sqrt{2}}$$

We find that the initial condition of the eigenvalue λ_a for the coherent state is given by

Coherent State Eigenvalue Initial Condition

$$\lambda_a(0) = \frac{\zeta_0}{\sqrt{2}} \tag{5.109}$$

Using this initial eigenvalue, and integrating Eq. (5.108) leads to a solution corresponding to the initial condition, given by

$$\psi_a(\zeta, 0) = C_n e^{-\frac{\zeta^2}{2} + \sqrt{2}\lambda_a \zeta}$$

$$= C_n e^{-\frac{\zeta^2}{2} + \zeta_0 \zeta}$$

To facilitate the normalization of $\psi_a(\zeta, 0)$, let us write the result for $\psi_a(\zeta)$ as

$$\psi_a(\zeta, 0) = C_n e^{-\frac{1}{2}(\zeta^2 - 2\zeta_0\zeta + \zeta_0^2)} e^{\frac{\zeta_0^2}{2}}$$

$$= C_n e^{\frac{\zeta_0^2}{2}} e^{-\frac{1}{2}(\zeta - \zeta_0)^2}$$

We determine the normalization constant C_n or $|C_n|$ using the normalization condition, which gives

$$1 = \int_{-\infty}^{+\infty} \psi_a^\dagger(\zeta) \psi_a(\zeta) dx$$

$$= \frac{C_n^2}{\alpha} e^{\zeta_0^2} \int_{-\infty}^{+\infty} e^{-(\zeta-\zeta_0)^2} d\zeta$$

5.5 Introduction To The Coherent State

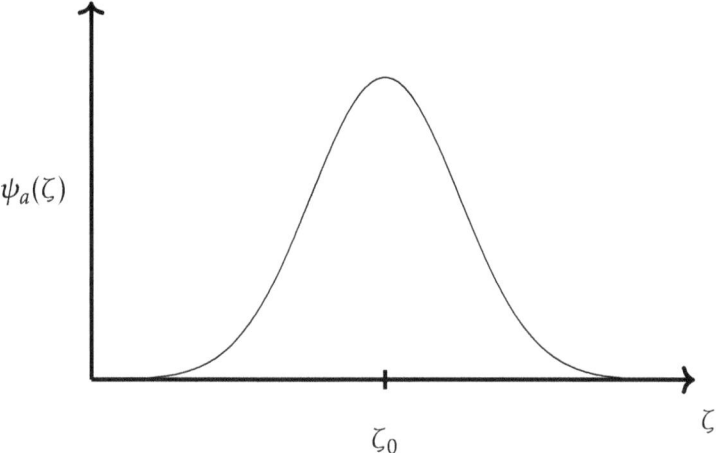

Figure 5.5: The harmonic oscillator coherent state, $\psi_a(\zeta)$ is a Gaussian distribution centered at the initial dimensionless position ζ_0.

$$= C_n^2 \frac{\sqrt{\pi}}{\alpha} e^{\zeta_0^2}$$

Solving for C_n, we obtain

$$C_n = \sqrt{\frac{\alpha}{\sqrt{\pi}}} e^{-\frac{\zeta_0^2}{2}} \tag{5.110}$$

Substitution of this result into the above expression for $\psi_a(\zeta, 0)$ leads to our final form for the initial condition of the eigenstate of $\widehat{a}$, given by

Harmonic Oscillator Coherent State

$$\psi_a(\zeta, 0) = \sqrt{\frac{\alpha}{\sqrt{\pi}}} e^{-\frac{1}{2}(\zeta - \zeta_0)^2} \tag{5.111}$$

The coherent state given in Eq. (5.111) is illustrated in Fig. 5.5.

$\psi_a(\zeta)$ is a Gaussian distribution centered at the initial dimensionless position $\zeta_0 = \alpha x_0$.

The coherent state of the harmonic oscillator is an important quantum state, and most of the direct evidence of this statement is given in Chap.

14, using a more concise notation known as *Bra-Ket* notation. There, the coherent state will be shown to be the most general form of the initial state of the oscillator demonstrating the following:

1. The *coherent state* has an identical eigenvalue (given by Eq. (5.109)) to the most general eigenstate for $\widehat{H}$, at time $t = 0$.

2. The expectation value of $x(t)$ using $\psi_a(\zeta)$ leads to the classical solution for the harmonic oscillator, therefore, establishing agreement with the *principle of correspondence*.

3. The first two conditions prove that $\psi_a(\zeta,0) = \Psi(\zeta,0)$, or is the initial state of the harmonic oscillator.

The second point is somewhat similar to our finding that *the most general solution* for a free fundamental particle through a slit leads to the physical phenomenon of diffraction, as well as interference. We found this in the double slit problem. The most general form of the solution, in both cases, turns out to be necessary in order to be in agreement with the *principle of correspondence*. These points will be discussed further in Chap. 14. Let's extend our results for the 1D harmonic oscillator to the domain of 3D.

5.6 Extending To 3D Harmonic Oscillator

Extending the harmonic oscillator to higher dimensions assumes the potential energy is dependent on all three spatial coordinates x, y, and z. Therefore, the potential energy $V(\mathbf{r})$ now has the form

$$V(\mathbf{r}) = \frac{1}{2}m\omega^2 r^2 = \frac{1}{2}m\omega^2(x^2 + y^2 + z^2) \tag{5.112}$$

The spatial part of the 3D Schrödinger equation is then given by

$$\widehat{H} = \nabla^2 \psi + \frac{2m}{\hbar^2}[E_\lambda - V(\mathbf{r})]\psi = 0 \tag{5.113}$$

Substituting Eq. (5.112) into Eq. (5.113) and expanding, we have

$$\left(\frac{\partial^2 \psi}{\partial x^2} + \frac{\partial^2 \psi}{\partial y^2} + \frac{\partial^2 \psi}{\partial z^2}\right) + \frac{2m}{\hbar^2}\left[E_\lambda - \frac{1}{2}m\omega^2(x^2 + y^2 + z^2)\right]\psi = 0 \tag{5.114}$$

5.6 Extending To 3D Harmonic Oscillator

Using the *method of separation of variables* again, we assume the wave function has the form

$$\psi(\mathbf{r}) = X(x)Y(y)Z(z) \tag{5.115}$$

Substitution of Eq. (5.115) into Eq. (5.114), then dividing by ψ leads to the following:

$$\left(\frac{1}{X}\frac{d^2X}{dx^2} - \frac{m^2\omega^2}{\hbar^2}x^2\right) + \left(\frac{1}{Y}\frac{d^2Y}{dy^2} - \frac{m^2\omega^2}{\hbar^2}y^2\right) +$$

$$\left(\frac{1}{Z}\frac{d^2Z}{dz^2} - \frac{m^2\omega^2}{\hbar^2}z^2\right) = -\frac{2mE_\lambda}{\hbar^2} = -k^2 \tag{5.116}$$

We have an equation that sums a function of x, a function of y, and a function of z, and they sum to equal a constant. Since they are all functions of independent variables x, y, and z, a necessary condition is that each function, separately, must equal to a constant. Thus, we have

$$\frac{1}{X}\frac{d^2X}{dx^2} - \frac{m^2\omega^2}{\hbar^2}x^2 = -k_x^2 \tag{5.117a}$$

$$\frac{1}{Y}\frac{d^2Y}{dy^2} - \frac{m^2\omega^2}{\hbar^2}y^2 = -k_y^2 \tag{5.117b}$$

$$\frac{1}{Z}\frac{d^2Z}{dz^2} - \frac{m^2\omega^2}{\hbar^2}z^2 = -k_z^2 \tag{5.117c}$$

We can define k^2 in terms of three x, y, and z components where we have

$$k_x^2 + k_y^2 + k_z^2 = k^2 \tag{5.117d}$$

Eqs. (5.117a), (5.117b), and (5.117c) have the same form as what we have seen in the 1D harmonic oscillator problem, where the corresponding 1D equation was given by Eq. (5.14b). Since we have worked out the solution to that equation (using the method of Frobenius), given in Eq. (5.80), X, Y, and Z each take a corresponding form, based on the 1D solution. The 3D solution $\psi = X(x)Y(y)Z(z)T(t)$ becomes

$$\psi_{n_x,n_y,n_z} = C_{n_x}C_{n_y}C_{n_z}H_{n_x}(\zeta_1)H_{n_y}(\zeta_2)H_{n_z}(\zeta_3)e^{-\frac{1}{2}(\zeta_1^2+\zeta_2^2+\zeta_3^2)}e^{-i\frac{E_\lambda}{\hbar}(t-t_0)}$$

$$\tag{5.118}$$

Eq. (5.118) can also be written as

3D Harmonic Oscillator Eigensolutions

$$\psi_{n_x,n_y,n_z} = C_{n_x,n_y,n_z} H_{n_x}(\alpha x) H_{n_y}(\alpha y) H_{n_z}(\alpha z) e^{-\frac{\alpha^2}{2}\mathbf{r}^2} e^{-i\frac{E_\lambda}{\hbar}(t-t_0)} \quad (5.119)$$

The parameter α is identical to the value in the 1D problem where $\alpha = \sqrt{m\omega/\hbar}$. To determine the normalization constant for the 3D harmonic oscillator, we use the same property of the Hermite polynomials, given by Eq. (5.81). In the case of the 3D oscillator, we end up with the product of three integrals, one in x, one in y, and one in z. Then, we can use Eq. (5.81) for each integral separately since they have identical form. This leads to

$$1 = \frac{C_{nx}^2}{\alpha}\left(2^{n_x} n_x! \sqrt{\pi}\right) \cdot \frac{C_{ny}^2}{\alpha}\left(2^{n_y} n_y! \sqrt{\pi}\right) \cdot \frac{C_{nz}^2}{\alpha}\left(2^{n_z} n_z! \sqrt{\pi}\right) \quad (5.120)$$

We can define the 3D normalization constant as

$$C_{n_x,n_y,n_z}^2 = C_{nx}^2 \cdot C_{ny}^2 \cdot C_{nz}^2 \quad (5.121)$$

Then, solving for C_{n_x,n_y,n_z}, we get

3D Harmonic Oscillator Normalization Constant

$$C_{n_x,n_y,n_z} = \sqrt{\frac{\alpha^3}{\pi^{3/2} \cdot 2^{n_x+n_y+n_z} \cdot n_x! n_y! n_z!}} \quad (5.122)$$

Also, using Eq. (5.117d), it follows that

$$k_x^2 + k_y^2 + k_z^2 = \frac{2mE_{\lambda x}}{\hbar^2} + \frac{2mE_{\lambda y}}{\hbar^2} + \frac{2mE_{\lambda z}}{\hbar^2} = \frac{2mE_\lambda}{\hbar^2} \quad (5.123)$$

Using the solution for the energy for the 1D problem, given in Eq. (5.68), the 3D energy solution becomes

$$E_\lambda = \left(n_x + \frac{1}{2}\right)\hbar\omega + \left(n_y + \frac{1}{2}\right)\hbar\omega + \left(n_z + \frac{1}{2}\right)\hbar\omega \quad (5.124)$$

Summing the terms of Eq. (5.124) and rearranging, the 3D eigenenergy $E_\lambda = E_{n_x,n_y,n_z}$ becomes

> **3D Quantum Mechanical Harmonic Oscillator Energy**
>
> $$E_{n_x,n_y,n_z} = \left(n_x + n_y + n_z + \frac{3}{2}\right)\hbar\omega \qquad (5.125)$$

As in the case of the 1D harmonic oscillator problem, the lowest energy state or *ground state* is also nonzero, given by $(3/2)\hbar\omega$. We see that the wave function for the 3D harmonic oscillator is essentially the product of the 1D wave functions, while the energy is the sum. Having this pattern from 1D to 3D, facilitates expressing the solutions for the 2D case, as well. Note that unlike in the 1D case, there is *degeneracy* in any oscillator in a dimension greater than 1D, because $n_x + n_y$ and $n_x + n_y + n_z$ can be satisfied by several unique combinations, depending on the quantum numbers. For example, the energy E_λ for the quantum state $(n_x, n_y, n_z) = (2, 5, 7)$ is the same as that for the quantum state $(10, 2, 2)$, since both combinations sum to 14. Any other combination summing to 14 also has the same eigenenergy. The number of degenerate states grows rapidly as these integers grow. Although their energies are the same, their wave functions can differ with respect to space.

5.7 CHAPTER SUMMARY

In this chapter, we have discussed the problem of the harmonic oscillator. The classical oscillator was first reviewed to provide a reference frame, to which the quantum mechanical results can be compared. Using the *Method of Frobenius*, we found the exact eigensolutions for the harmonic oscillator $\psi_n(\zeta, t)$, which involves the Hermite Polynomials $H_n(\zeta)$. The special properties of the Hermite polynomials were found to lead to a number of interesting characteristics for the eigensolutions ψ_n.

Then, exploiting symmetry in the Hamiltonian, we introduced the first order *annihilation* and *creation operators* for the oscillator and used them to define and solve for the *coherent state*. The coherent state is the most general initial condition for the harmonic oscillator, but we had to defer some additional details involving the coherent state, when this work can be done more concisely. Lastly, we extended the 1D result for the eigensolutions to 3D. Table 5.3 lists some of the key results from this chapter.

Table 5.3: Chapter 5 Summary Equations. $\zeta = \alpha \cdot x = \sqrt{m\omega/\hbar} \cdot x$; $\zeta_0 = \alpha \cdot x_0 = \sqrt{m\omega/\hbar} \cdot x_0$; m is mass, ω is the natural frequency, x_0 is the initial position.

Name	Equation
1D Oscillator Energy States	$E_\lambda = \left(n + \frac{1}{2}\right)\hbar\omega$
Eigenstates 1D harmonic oscillator	$\psi_n(\zeta, t) = C_n H_n(\zeta) e^{-\frac{1}{2}\zeta^2} e^{-i\frac{E_\lambda}{\hbar}(t-t_0)}$
1D Normalization Constant	$C_n = \sqrt{\frac{\alpha}{2^n n! \sqrt{\pi}}}$
Position-Momentum Commutation	$x\widehat{p}_x - \widehat{p}_x x = i\hbar$
Annihilation Operator	$\widehat{a} = \frac{1}{\sqrt{2m\hbar\omega}}[m\omega x + i\widehat{p}_x]$
Creation Operator	$\widehat{a}^\dagger = \frac{1}{\sqrt{2m\hbar\omega}}[m\omega x - i\widehat{p}_x]$
Annihilation Operator Relation	$\widehat{a}\psi_n = \sqrt{n}\psi_{n-1}$
Creation Operator Relation	$\widehat{a}^\dagger \psi_n = \sqrt{n+1}\psi_{n+1}$
Oscillator Coherent State	$\psi_a(\zeta, 0) = \sqrt{\frac{\alpha}{\sqrt{\pi}}} e^{-\frac{1}{2}(\zeta-\zeta_0)^2}$
3D Oscillator Energy States	$E_\lambda = \left(n_x + n_y + n_z + \frac{3}{2}\right)\hbar\omega$
3D Normalization Constant	$C_{n_x, n_y, n_z} = \sqrt{\frac{\alpha^3}{(\pi^{3/2}) 2^{(n_x+n_y+n_z)} n_x! n_y! n_z!}}$

5.7 Chapter Summary

n	Hermite Polynomial H_n
0	$H_0(\zeta) = 1$
1	$H_1(\zeta) = 2\zeta$
2	$H_2(\zeta) = 4\zeta^2 - 2$
3	$H_3(\zeta) = 8\zeta^3 - 12\zeta$
4	$H_4(\zeta) = 16\zeta^4 - 48\zeta^2 + 12$
5	$H_5(\zeta) = 32\zeta^5 - 160\zeta^3 + 120\zeta$
6	$H_6(\zeta) = 64\zeta^6 - 480\zeta^4 + 720\zeta^2 - 120$
7	$H_7(\zeta) = 128\zeta^7 - 1344\zeta^5 + 3360\zeta^3 - 1680\zeta$
8	$H_8(\zeta) = 256\zeta^8 - 3584\zeta^6 + 13440\zeta^4 - 13440\zeta^2 + 1680$
9	$H_9(\zeta) = 512\zeta^9 - 9216\zeta^7 + 48384\zeta^5 - 80640\zeta^3 + 30240\zeta$
10	$H_{10}(\zeta) = 1024\zeta^{10} - 23040\zeta^8 + 161280\zeta^6 - 403200\zeta^4 + 302400\zeta^2 - 30240$
11	$H_{11}(\zeta) = 2048\zeta^{11} - 56320\zeta^9 + 506880\zeta^7 - 1774080\zeta^5 + 2217600\zeta^3 - 665280\zeta$
12	$H_{12}(\zeta) = 4096\zeta^{12} - 135168\zeta^{10} + 1520640\zeta^8 - 7096320\zeta^6 + 13305600\zeta^4 - 7983360\zeta^2 + 665280$
13	$H_{13}(\zeta) = 8192\zeta^{13} - 319488\zeta^{11} + 4392960\zeta^9 - 26357760\zeta^7 + 69189120\zeta^5 - 7983360\zeta^2 + 17297280\zeta$

Table 5.4: List of First 14 Hermite Polynomials.

5.8 Chapter Problems

Problem 5.1 The first two eigenstates for the harmonic oscillator can be written as

$$\psi_1 = \left(\frac{\alpha}{\pi}\right)^{1/4} e^{-\alpha x^2/2} \quad \text{and} \quad \psi_2 = \left(\frac{4\alpha^3}{\pi}\right)^{1/4} x e^{-\alpha x^2/2}$$

Show that these two states are orthogonal, or that

$$\int_{-\infty}^{\infty} \psi_1^* \psi_2 \, dx = 0$$

Problem 5.2 Show that the harmonic oscillator solution $x(t) = A\sin\omega t + A\cos\omega t$, where $\omega = \sqrt{k/m}$, is also a solution to Newton's equation $F = m\ddot{x}$ for the harmonic oscillator.

Problem 5.3 Show that for the harmonic oscillator, the expectation of x is zero, and that the expectation $\langle x^2 \rangle$ for $n = 2$ is given by

$$\langle x^2 \rangle = \frac{5}{2} \frac{\hbar}{(mk)^{1/2}}$$

Problem 5.4 Show that the expectation of p is zero, and the squared momentum $\langle p^2 \rangle$ for $n = 2$ is given by

$$\langle p^2 \rangle = \frac{5}{2} \hbar (mk)^{1/2}$$

Problem 5.5 We know the Hamiltonian operator $\widehat{H}$ is given by

$$\widehat{H} = -\frac{\hbar^2}{2m} \nabla^2 + V(x, y, z)$$

Prove that for any eigensolution ψ of $\widehat{H}$, and for any linear operator $\widehat{O}$ that

$$\int_V \psi^* \left(\widehat{H}\widehat{O} - \widehat{O}\widehat{H} \right) \psi \, dV = 0$$

Problem 5.6 Using the above result from the previous problem, define the linear operator $\widehat{O}$ to be

$$\widehat{O} = -i\hbar \left(x \frac{\partial}{\partial x} + y \frac{\partial}{\partial y} + z \frac{\partial}{\partial z} \right)$$

Show that

$$\widehat{HO} - \widehat{OH} = i\hbar\left(x\frac{\partial}{\partial x} + y\frac{\partial}{\partial y} + z\frac{\partial}{\partial z}\right) - 2i\hbar\widehat{K}$$

$\widehat{K}$ is the kinetic energy operator.

Problem 5.7 The harmonic oscillator solution can be written as

$$x(t) = A\sin(\omega t + \phi)$$

Solving for t, one obtains

$$t = \frac{1}{\omega}\sin^{-1}\left(\frac{x}{A}\right) - \phi$$

Use both of these equations, differentiating the equation of t to prove that

$$\frac{dx}{dt} = \omega\sqrt{A^2 - x^2} = A\cos(\omega t + \phi)$$

Now, use the relation

$$p(x)dx = \frac{dx}{\pi\sqrt{A^2 - x^2}}$$

with the above to show that $p(x)$ satisfies the probability distribution condition given by

$$1 = \frac{\omega}{\pi}\int_0^{\pi/\omega} dt = \int p(x)dx$$

5.9 Suggested Readings & References

[1] A. Ghatak and S. Lokanathan, *Quantum Mechanics: Theory and Applications*, Springer-Science (2004)

CHAPTER 6

Evanescence, Scattering, & Tunneling

When light travels along an optical wave-guide, the light continually bounces off of the bounding interface of the waveguide. The interface is formed between the central core and the surrounding material known as cladding, which has a lower index of refraction. This confinement to the core, where the light continually reflects, is concomitant with a *special condition at the core-clad interface*. It turns out that the electric field of the electromagnetic wave does not vanish abruptly at the interface. Instead, there is a finite distance from the interface extending out into the cladding, over which the electric field decays to zero. This unique decaying behavior occurs when waves reflect at an interface. Building on these concepts from electromagnetic waves, we are going to show in this chapter, that in similar conditions, the eigenfunctions of the Schrödinger equation behave in an analogous manner to the electromagnetic waves described by Maxwell's equations. Consequently, this leads to some interesting behavior in quantum systems. We are going to analyze a few problems carefully that reveal just how wave functions can behave at a particular interface, namely, where there is an abrupt change in potential V. We will begin by solving the Schrödinger equation for the case of a single potential step, followed by two more complex examples combining two sequential potential steps. This will allow us to uncover some rather remarkable properties

Chapter 6. Evanescence, Scattering, & Tunneling

of quantum systems, which are indeed exploited in several well-known devices today. We'll also examine an application of the current density J, derived from the continuity equation derived in Chap. 4, assuming a Hermitian Hamiltonian.

6.1 Creation Of Evanescent wave functions

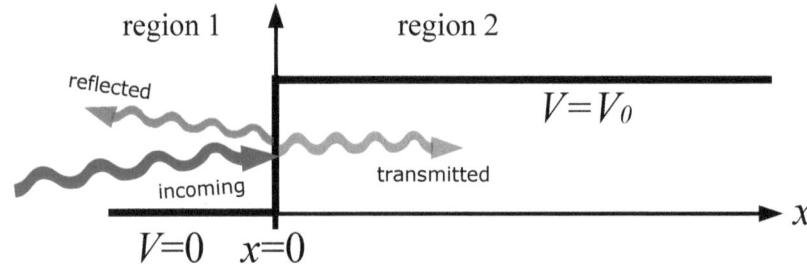

Figure 6.1: Illustration of a fundamental particle approaching a step change in the potential energy V. At $x = 0$, there is a discontinuous jump from $V = 0$ to $V = V_0$. For $x < 0$, $V = 0$ and for $x > 0$, $V = V_0$.

Consider a potential energy function like that shown in Fig. 6.1. Let us solve the Schrödinger equation for a free particle whose motion to the right leads to an encounter with this potential energy step function $V(x)$. We can imagine this to represent how a fundamental particle in motion encounters an interface where, on one side of the interface a distinct environment from the first side exists. Because the potential only depends on x, the problem is spatially 1D. The Schrödinger equation for the spatially 1D problem is given by

$$i\hbar \frac{\partial \psi}{\partial t} = \widehat{H}\psi = \left(-\frac{\hbar^2}{2m}\frac{\partial^2}{\partial x^2} + V(x)\right)\psi \tag{6.1}$$

For this problem, we may again use the *method of separation of variables* between the spatial variable x and the time variable t. Using this approach, the wave function ψ assumes the form

$$\psi(t,x) = T(t) \cdot X(x) \tag{6.2}$$

Substituting Eq. (6.2) into the 1D Schrödinger equation given by Eq.

6.1 Creation Of Evanescent wave functions

(6.1), we have

$$i\hbar X \frac{dT}{dt} = T\left(-\frac{\hbar^2}{2m}\frac{d^2X}{dx^2} + V \cdot X\right) \tag{6.3}$$

Dividing both sides of Eq. (6.3) by the wave function $\psi = T \cdot X$ gives

$$i\hbar \frac{1}{T}\frac{dT}{dt} = -\frac{\hbar^2}{2m}\frac{1}{X}\frac{d^2X}{dx^2} + V(x) \tag{6.4}$$

The LHS of Eq. (6.4) is purely a function of t while the RHS is purely a function of x. Since both are independent variables, this can only be true if both sides are equal to a constant. That constant is the eigenenergy E_λ. Then, we have the following two equations:

$$i\hbar \frac{1}{T}\frac{dT}{dt} = E_\lambda \tag{6.5a}$$

$$-\frac{\hbar^2}{2m}\frac{1}{X}\frac{d^2X}{dx^2} + V(x) = E_\lambda \tag{6.5b}$$

After integration of Eq. (6.5a) with respect to time t to obtain T, the first equation yields the solution

$$T(t) = T_0 e^{-\frac{i}{\hbar}E_\lambda(t-t_0)} \tag{6.6}$$

T_0 is the initial condition $T(0)$. For the equation describing space, Eq. (6.5b) can be written as

$$\frac{d^2X}{dx^2} + \frac{2m(E_\lambda - V)}{\hbar^2} X = 0 \tag{6.7}$$

Just as we had done in the free particle problem where $V = 0$, and we defined $k^2 = 2mE/\hbar^2$ (c. Eq. (3.98)), here, we can define the parameter k by

$$k^2 = \frac{2m(E_\lambda - V)}{\hbar^2} \tag{6.8}$$

k is the wave-number of the fundamental particle, analogous to the wave-number of an electromagnetic wave, where $k = 2\pi/\lambda$. λ is the associated *wavelength*. Then, we can write Eq. (6.7) as

$$\frac{d^2X}{dx^2} + k^2 X = 0 \tag{6.9}$$

Eq. (6.9) has the general solution given by

$$X(x) = Ae^{ikx} + Be^{-ikx} \tag{6.10}$$

Eq. (6.10) is the sum of the two independent solutions of a second order differential equation. Consistent with the general solution of the classical wave equation (c.Eq.(1.34)), we have two propagating waves, with wave numbers $+k$ and $-k$. $+k$ denotes wave motion to the right, while $-k$ is to the left. However, a wave traveling to the left in the region $x > 0$ is only possible if there exists an opportunity for the wave to be reflected backwards. This is because we only input one of the two waves, namely, a wave traveling to the right. This possibility *can* exists at the position $x = 0$ and to the left of the potential jump. Thus, the general solution will have both terms in the region $x < 0$. However, whatever part of the solution propagates past $x = 0$ will not be reflected as it is assumed that the right side extends indefinitely. Therefore, only a right propagating wave exists in the region $x > 0$. Thus, the solutions will be different in the two respective regions. Specifically, we have

$$X_1(x) = Ae^{ik_1 x} + Be^{-ik_1 x} \quad (x < 0) \tag{6.11a}$$

$$X_2(x) = Ce^{ik_2 x} \quad (x > 0) \tag{6.11b}$$

Since we have two distinct regions, we also two respective wavenumbers k, denoted by k_1 and k_2. k_1 corresponds to the region $x < 0$ and k_2 corresponds to $x > 0$. Based on Eq. (6.8), k_1^2 and k_2^2 become

$$k_1^2 = \frac{2mE_\lambda}{\hbar^2} \quad (x < 0) \tag{6.12a}$$

$$k_2^2 = \frac{2m(E_\lambda - V_0)}{\hbar^2} \quad (x > 0) \tag{6.12b}$$

Now we have three unknowns $A, B,$ and C, all independent of x. However, we can only write two equations using the continuity of both the wave function X_i and its first derivative at $x = 0$. Thus, we can only solve for two of them, simultaneously. Fortunately, the form of Eqs. (6.11a) and (6.11b) allow us to, instead, solve for the ratios $A' = A/C$ and $B' = B/C$, which are still useful. This is because A/C is inversely proportional to the transmission probability T. And $B/A = B'/A'$ turns out to be proportional to the reflection probability R. We will discuss these quantities in more detail shortly. At $x = 0$, equating $X_1(x)$ and $X_2(x)$ and their derivatives $X_1'(x)$ and $X_2'(x)$ gives

$$A + B = C \quad (x = 0) \tag{6.13a}$$

6.1 Creation Of Evanescent wave functions

$$ik_1 A - ik_1 B = ik_2 C \quad (x = 0) \tag{6.13b}$$

Dividing both equations by C to give the desired form in terms of A' and B', we have

$$\frac{A}{C} + \frac{B}{C} = 1 \tag{6.14a}$$

$$ik_1 \left(\frac{A}{C} - \frac{B}{C} \right) = ik_2 \tag{6.14b}$$

This gives two equations in two unknowns, which can be solved. The solutions are

$$\frac{A}{C} = \frac{k_1 + k_2}{2k_1} \tag{6.15a}$$

$$\frac{B}{C} = 1 - \frac{A}{C} = \frac{k_1 - k_2}{2k_1} \tag{6.15b}$$

Using Eqs. (6.15a) and (6.15b), B/A becomes

$$\frac{B}{A} = \frac{B}{C} \cdot \frac{C}{A} = \frac{k_1 - k_2}{2k_1} \cdot \frac{2k_1}{k_1 + k_2} = \frac{k_1 - k_2}{k_1 + k_2} \tag{6.15c}$$

Using the definitions for k_1 and k_2, given by Eqs. (6.12a) and (6.12b), the three ratios in Eqs. (6.15a), (6.15b), and (6.15c) can be expressed in terms of the parameters in the Schödinger equation, which gives

$$\frac{A}{C} = \frac{1}{2} + \frac{1}{2}\sqrt{1 - \frac{V_0}{E_\lambda}} \tag{6.16a}$$

$$\frac{B}{C} = \frac{1}{2} - \frac{1}{2}\sqrt{1 - \frac{V_0}{E_\lambda}} \tag{6.16b}$$

$$\frac{B}{A} = \frac{1 - \sqrt{1 - \frac{V_0}{E_\lambda}}}{1 + \sqrt{1 - \frac{V_0}{E_\lambda}}} \tag{6.16c}$$

B/A is the ratio of the *amplitude* of the reflected wave to that of the incoming wave. By inspection, we see that when $V_0 = 0$, then $B/A = 0$, and when $V_0 = E_\lambda$, then $B/A = 1$. This indicates the range of B/A, so $0 \leq B/A \leq 1$. Fig. 6.2 illustrates the dependence of these three parameters on the potential energy V_0 in the region $0 < V_0 \leq E_\lambda$.

In the scenario illustrated in Fig. 6.2, the potential energy is positive, but less than the incoming energy. For a free particle, encountering a

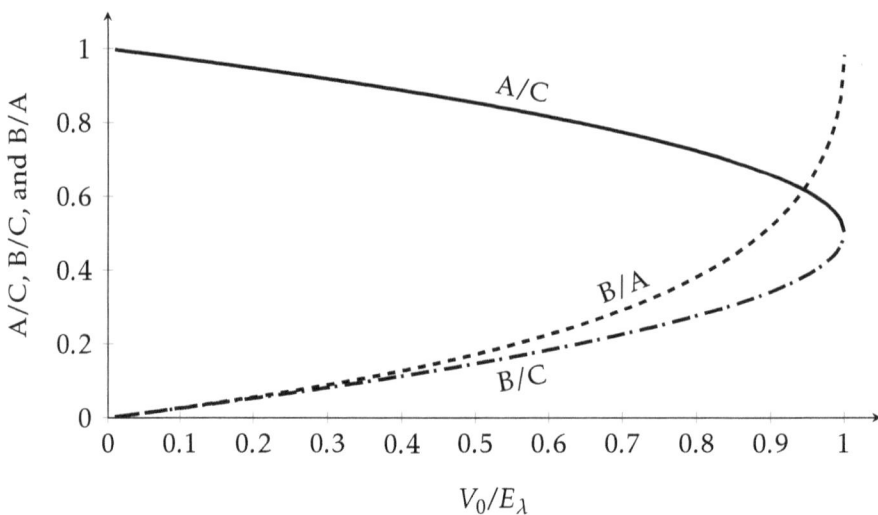

Figure 6.2: Ratios A/C, B/C, and A/B as functions of the dimensionless ratio V_0/E_λ. The case of $B/A = 1$ means that the amplitude of the reflecting wave matches that of the incoming wave.

positive potential energy, the kinetic energy is reduced. To the charge carrier, this is a **potential hill**.

 Positive Potentials Are Hills To Free Particles: At it's most basic level, for a free fundamental particle, a momentum reduction becomes probable whenever there is a positive or uphill potential encountered. Note that the positive potential does not *have* to be spatially varying. It is also accompanied by a scattering event.

Since the problem involves a free particle initially propagating to the right, recall that the eigenvalues for E_λ are continuous. This is the result we found in the free particle problem in Chap. 3. The chosen range of $E_\lambda > V_0 > 0$ in Fig. 6.2 ensures that the terms underneath the radical in A/C, B/C, and B/A are non-negative. However, there are two more important cases that we will distinguish that extend beyond this range. Another case is when $V_0 < 0$. In this case, the term underneath the radical is *real and positive definite*. Therefore, the fundamental particle encounters an effectively *downhill* potential variation, or a **potential hole**. As suggested by Eq. (6.12b), this leads to an increase in wave number k_2, hence, the particle *gains* momentum. A good example of this is in superconductors, which, under the proper conditions, are able to

6.1 Creation Of Evanescent wave functions

conduct charge carriers practically without loss. We'll discuss this topic in detail in Chap. 15. And note that these are unique conditions because *they apply strictly to propagating charge carriers*, because we know that for a bound electron, a negative potential tends to stabilize the bound state further, where as for free electrons, their propagation as free particles is made more probable.

 Negative Potentials Are Holes For Free Particles: When a propagating wave encounters a negative potential step, downward potential or hole, it gains momentum. This can also be interpreted as no scattering.

The other important case to be considered is when $V_0 \geq E_\lambda$. We know that if $V_0 > 0$, the incoming wave with amplitude A *always* gives rise to *some* reflected wave component. Connecting this to a more general phenomenon of matter, in a 3D system, where there is an angular distribution to the incoming waves of fundamental particles, this finite reflection is what corresponds to general **scattering**, or **scattering events**. Additionally, in a lattice of periodic potential hills, it leads to the phenomenon of *band structures* because there is constructive interference in these scattered waves, which can completely cancel the incoming wave, corresponding to an ability to conduct in such conditions. The extent of scattering is determined by the extent of the positive potential variation V_0. However, now, we discuss more an extreme case of a positive potential V_0, namely, when $V_0 > E_\lambda$. When $V_0 > E_\lambda$ in region two, the wave-number k_2 is then given by

$$k_2^2 = -\frac{2m(V_0 - E_\lambda)}{\hbar^2} \quad (x > 0)$$

Since k_2^2 is absolutely negative, the wave-number k_2, in this region, becomes

$$k_2 = \sqrt{-\frac{2m(V_0 - E_\lambda)}{\hbar^2}}$$

$$= i\sqrt{\frac{2m(V_0 - E_\lambda)}{\hbar^2}}$$

$$= ik_2'$$

Note that k_2' is a strictly positive real quantity. The solution in region two then becomes

$$X_2(x) = Ce^{ik_2 x} = Ce^{i(ik_2')x} = Ce^{-k_2' x} \quad (x > 0) \tag{6.17}$$

Therefore, when $V_0 > E_\lambda$, we end up with a wave function having a decreasing amplitude as a function of $+x$, moving away from $x = 0$. The wave function also decays more rapidly for larger $V_0 - E_\lambda$. Examples of a decaying $X_2(x)$ are illustrated in Fig. 6.3. This type of decaying

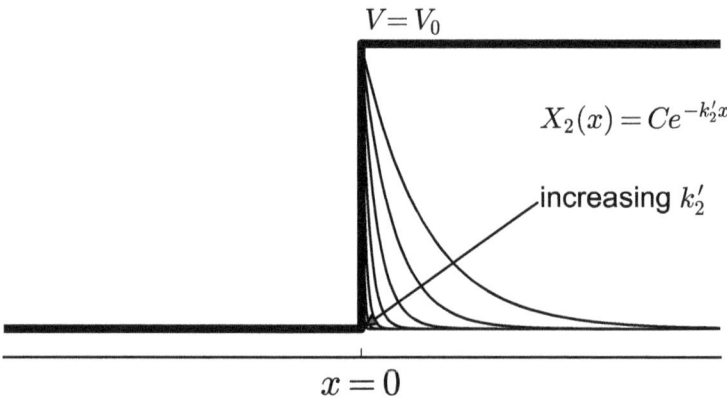

Figure 6.3: Examples of evanescent wave functions with decaying amplitude as a function of position x, resulting from the condition $V_0 > E_\lambda$. The wave function is real-valued with a decay length $\delta \sim 1/k_2'$. As k_2' increases, the decay length decreases. This also means that the probability of finding the particle at a point beyond $x = 0$ diminishes.

wave function is known as an **evanescent wave function**, analogous to an **evanescent wave**, and it is a result of a potential rise, where the height of the potential energy change is larger than the incoming energy possessed by the incoming fundamental particle. The resulting *evanescent* wave function is purely real-valued, and exponentially decays over a length scale on the order of $\delta \sim 1/k_2'$. As k_2' increases, the decay length decreases. This means that the probability of finding the particle at a point beyond $x = 0$, in region two, diminishes as k_2' increases. And in the limit of $k_2' \to \infty$ or $V_0 \to \infty$, we recover the particle in a box condition, where the wave function has no coverage in region two, and is therefore confined to region one. This trend is also illustrated in Fig. 6.3.

In region one, the real and imaginary components of B/A are also func-

6.2 Transmission and Reflection Probabilities

tions of the dimensionless ratio V_0/E_λ. The real component starts at $Re(B/A) = 1$, then decreases monotonically, while approaching $Re(B/A) = -1$ asymptotically. The imaginary component starts at 0, when $V_0/E = 1$, then sharply rises (negatively), then proceeds to decay back towards 0, asymptotically. Meanwhile, the amplitude is fixed at 1, corresponding to full reflection of the wave. Therefore, in the same conditions that lead to

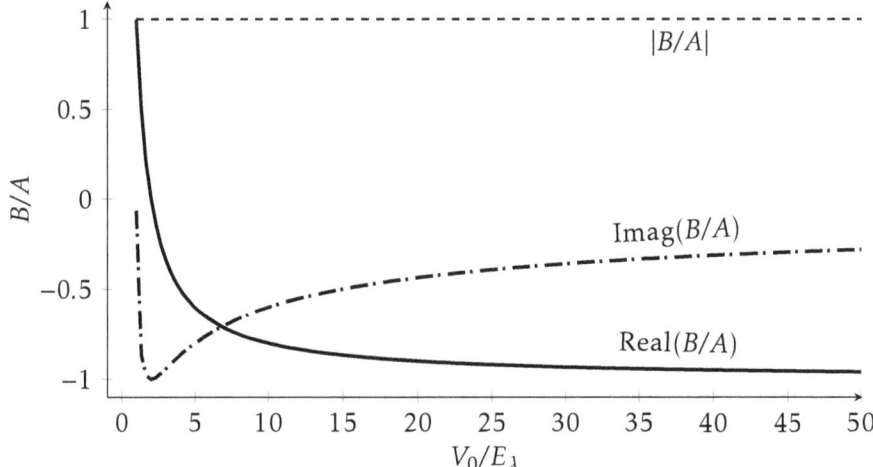

Figure 6.4: Real and imaginary components of B/A as functions of the dimensionless ratio V_0/E_λ. The real component starts at 1, then decreases while approaching −1 asymtotically. The imaginary component starts at 0, when $V_0/E = 1$, then sharply rises (negatively), then proceeds to decay back towards 0, asymptotically. Meanwhile, the amplitude is fixed at 1, corresponding to full refelection of the wave.

an evanescent wave, the amplitude of B/A is constant at unity, however the complex quantity B/A changes phase, based on the value of V_0. The real part transitions from +1 to −1, while the imaginary part starts at zero, peaks at −1, then transitions back to zero. While we can observe the behavior of the ratios, we still need to tie this information to physical quantities. This is done by using the current density. In the next section, we'll look at how this information relates to current densities, and other useful parameters defined in terms of the current density.

6.2 TRANSMISSION AND REFLECTION PROBABILITIES

With some of the incoming wave function traveling to the right (while some also gets reflected), we may view the problem in another way.

Instead, we may ask whether a current density $\widehat{J}$ is flowing in region two, particularly, in the conditions where an evanescent wave ($x > 0$) forms. The answer is *no*, and this answer can readily be obtained from looking closely at the equation we derived in Chap. 4, Eq. (4.93), defining the current density **J**, given by

$$\widehat{J} \equiv \frac{i\hbar}{2m}(\psi \nabla \psi^* - \psi^* \nabla \psi) \tag{6.18}$$

Since, with an *evanescent wave*, the wave function is real-valued, it equals its complex conjugate, so $\psi^* = \psi$. Therefore, $\nabla \psi = \nabla \psi^*$. So, we obtain

$$\psi \nabla \psi^* - \psi^* \nabla \psi = \psi \nabla \psi - \psi \nabla \psi = 0$$

Thus, any strictly real-valued wave function *always* leads to the term in parentheses vanishing!

 Real-Value Wave Function Current Density: Real-valued wave functions always have zero current density.

Hence, the current density will always be zero in a region with only an evanescent wave function. This lack of current in region two is because an evanescent wave function is a consequence of the *total reflection* process of the incoming wave given by $Ae^{ik_1 x}$. However, in the reflection process, we see that the reflecting wave has a finite *penetration depth* $\delta \sim 1/|k_2| = 1/k_2'$ beyond the interface of reflection ($x = 0$). This is where the evanescent wave lives.

As we have alluded in the chapter introduction, the phenomenon of evanescence is not exclusive to quantum mechanics. It is also a well-known behavior of electromagnetic waves when undergoing what is called **total internal reflection**, which occurs when an electromagnetic wave continually reflects off the walls of an optical wave-guide, while propagating through the wave-guide. These reflections keep the wave confined to the interior of the wave-guide known as the *core*, while the evanescent wave forms just beyond the surface, where the exterior overclad material meets the core of the wave-guide. Another example of when an evanescent wave forms with electromagnetic waves is in the generation of what is known as a *surface-plasmon-polariton*, which is a collective interfacial oscillation involving certain metallic-dielectric

6.2 Transmission and Reflection Probabilities

interfaces. Though this process is more complicated because the surface-plasmon-polariton is a result of tapping into the evanescent wave region of the wave-guide, leading to an interface with evanescent waves on both sides of the metallic-dielectric interface. Both of these are examples of the application of *evanescent waves*. Thus, these unique decaying waves are also found in other physical systems.

We can take the ideas we have been discussing further, but we will need to utilize the current density $\widehat{\mathbf{J}}$. In quantum mechanics, it can be used to define useful parameters known as reflection R and transmission T probabilities. The **reflection probability** is a dimensionless ratio defined in terms of current densities, given by

Reflection Probability

$$R = \frac{|J_R|}{|J_I|} \tag{6.19}$$

Along with the reflection probability, we can define the **transmission probability** T as

Transmission Probability

$$T = \frac{|J_T|}{|J_I|} \tag{6.20}$$

The current density given by Eq. (6.18) is an appropriate choice because they obey conservation. This follows from the definition being defined to satisfy the continuity equation. Thus, whenever there are no sources or sinks interacting with the propagating wave function within the domain described by the Schrödinger equation, it follows that, along the direction of flow taken normal to the interface, we have the following relation between the incoming flow J_I, the transmitted flow J_T, and the reflected flow J_R:

$$J_I = J_R + J_T \tag{6.21}$$

This just states that from the incoming wave function, some is reflected while the rest is transmitted. If none is reflected, then all is transmitted, etc. From this balance of the currents, it follows that

$$R + T = \frac{|J_R|}{|J_I|} + \frac{|J_T|}{|J_I|} = 1 \qquad (6.22)$$

In the potential step problem we are discussing, since the incoming wave and reflected wave are generally complex functions, they will, in general, have nonzero current densities. For the incoming wave, the associated current density J_I can be found from

$$J_I = \frac{i\hbar}{2m}\left[(-ik_1)Ae^{ik_1x}A^*e^{-ik_1x} - ik_1A^*e^{-ik_1x}Ae^{ik_1x}\right]$$

$$= \frac{i\hbar k_1}{2m}\left(2|A|^2\right)$$

$$= \frac{\hbar k_1}{m}|A|^2$$

A similar calculation for the reflected and transmitted wave's component current density J_R and J_T leads to

$$J_R = \frac{\hbar k_1}{m}|B|^2$$

$$J_T = \frac{\hbar k_2}{m}|C|^2$$

Then, using the current densities J_I, J_R, and J_T, the reflection and transmission probabilities R and T become

$$R = \frac{J_R}{J_I} = \frac{|B|^2}{|A|^2} \qquad (6.23a)$$

$$T = \frac{J_T}{J_I} = \frac{k_2}{k_1}\frac{|C|^2}{|A|^2} \qquad (6.23b)$$

From Eq. (6.23b), we see that non-penetrating reflection takes place when $k_2 = 0$, which is when $V_0 = E_\lambda$. This condition represents the transition point leading into evanescent wave function behavior. Beyond this transition point, the transmission T is not given by Eq. (6.23b) since the wave function becomes real-value, thus $T = 0$. The reflection and transition probabilities are plotted in Fig. 6.5, for the range $0 < V_0 \leq E_\lambda$. Beyond this range, $R = 1$ and $T = 0$. Using Eqs. (6.15a) and (6.15c), it is straightforward to show that given the solutions for A/C and B/A, that $R + T = 1$, also illustrated in Fig. 6.5.

6.3 Electrons Across A Rectangular Potential Barrier

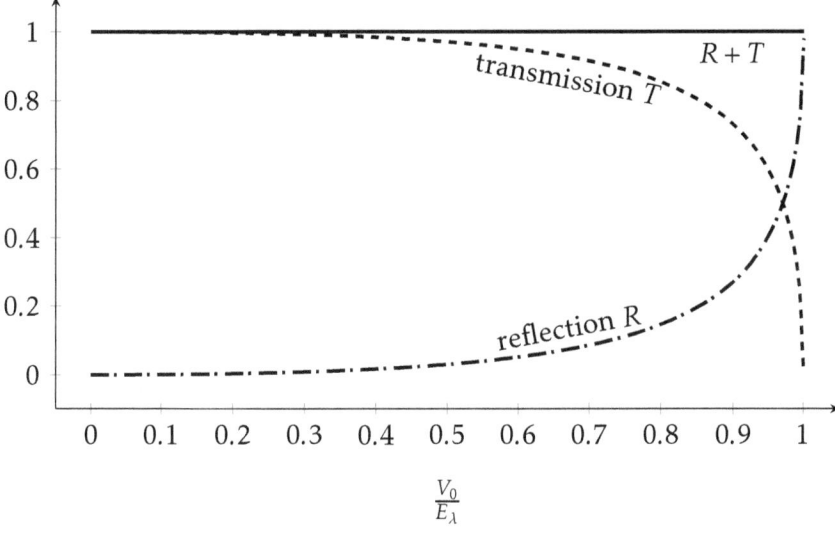

Figure 6.5: Transmission and reflection probabilities T and R. Because $R+T = 1$, they have correlated behavior, where as R increases, T necessarily decreases.

In this section, we found that when a free particle encounters variations in potential energy $V(x)$, a multitude of rich phenomena becomes possible. Specifically, with a single potential step, we have found that transmission, reflection, or more generally scattering, and evanescent wave functions are possible, depending on the conditions. Let's extend the complexity of the potential energy variation from a single step, to two sequential steps, one up, followed by a step down to the original starting potential. This type of potential variation is known as a **rectangular potential barrier** and is the subject of the next section.

6.3 ELECTRONS ACROSS A RECTANGULAR POTENTIAL BARRIER

A *rectangular potential barrier* is illustrated in Fig. 6.6. The analysis for the rectangular potential barrier is as it was for the single potential step. But, now we have an additional potential step down, located a distance L away from the first step at $x = 0$. It turns out that this simple change to the problem leads to significantly more mathematics, though it is straightforward. The second potential step, which is an example of a *potential hole*, goes from V_0 down to 0. The Schrödinger equation is the

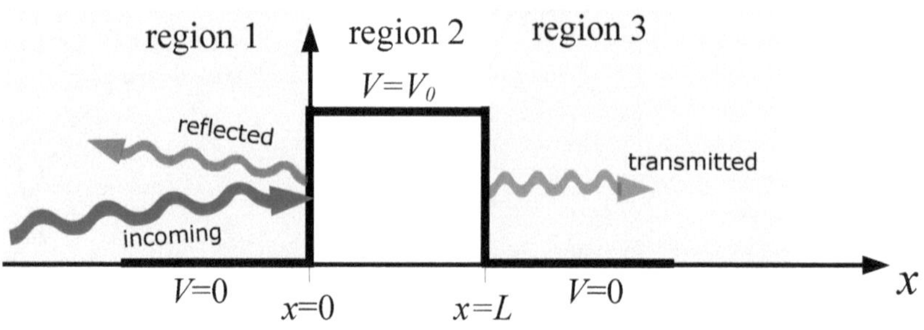

Figure 6.6: Illustration of a rectangular potential barrier. It is two consecutive potential steps (first up, then second one down) separated by a distance L. The incoming wave is traveling to the right in region 1, while the first transmitted wave function exists in region 2, and a second transmitted wave function exists in region 3.

same as in Sec. 6.1 (one-dimensional in space), where we have

$$i\hbar \frac{\partial \psi}{\partial t} = \left(-\frac{\hbar^2}{2m}\frac{\partial^2}{\partial x^2} + V(x)\right)\psi$$

We again use the method of separation of variables by setting the wave function $\psi(x,t) = X(x)T(t)$. After separation into two independent equations for space and time, $T(t)$ is found to be identical to our earlier result given by Eq. (6.6). The Hamiltonian for each layer leads to a spatial differential equation given by

$$\frac{\partial^2 X_i}{\partial x^2} + \frac{2m}{\hbar^2}(E_\lambda - V_i)X_i = 0 \Rightarrow \frac{\partial^2 X_i}{\partial x^2} + k^2 X_i = 0$$

The Schrödinger equation is applied in all the regions, and V generally takes a different in each material region. V is also assumed to be independent of space, or is *uniform* in each region. For the spatial part of the tunneling problem, or $X(x)$, we have the same result given by Eq. (6.9). Since we have three regions, we have the following three solutions corresponding to regions 1, 2, and 3:

$$X_1 = Ae^{ik_1 x} + Be^{-ik_1 x} \quad \text{(region 1)} \qquad (6.24a)$$

$$X_2 = Ce^{ik_2 x} + De^{-ik_2 x} \quad \text{(region 2)} \qquad (6.24b)$$

$$X_3 = Ee^{ik_3 x} \quad \text{(region 3)} \qquad (6.24c)$$

6.3 Electrons Across A Rectangular Potential Barrier

$A, B, C, D,$ and E are all determined from boundary conditions. Toward this objective, we set the wave functions and their derivatives equal at $x = 0$ and $x = L$. This results in four equations in five unknowns. To reduce the number of unknowns down to four, we divide all equations by amplitude A, or assume $A = 1$. This leads to the following four equations:

$$1 + \frac{B}{A} = \frac{C}{A} + \frac{D}{A} \quad (x = 0) \tag{6.25a}$$

$$ik_1\left(1 - \frac{B}{A}\right) = ik_2\left(\frac{C}{A} - \frac{D}{A}\right) \quad (x = 0) \tag{6.25b}$$

$$\frac{C}{A}e^{ik_2 L} + \frac{D}{A}e^{-ik_2 L} = \frac{E}{A}e^{ik_3 L} \quad (x = L) \tag{6.25c}$$

$$ik_2\left(\frac{C}{A}e^{ik_2 L} - \frac{D}{A}e^{-ik_2 L}\right) = ik_3\frac{E}{A}e^{ik_3 L} \quad (x = L) \tag{6.25d}$$

Let us rename the four unknowns $B' = B/A$, $C' = C/A$, $D' = D/A$ and $E' = E/A$. Then, we have

$$1 + B' = C' + D' \quad (x = 0) \tag{6.26a}$$

$$1 - B' = \frac{k_2}{k_1}(C' - D') \quad (x = 0) \tag{6.26b}$$

$$C'e^{ik_2 L} + D'e^{-ik_2 L} = E'e^{ik_3 L} \quad (x = L) \tag{6.26c}$$

$$C'e^{ik_2 L} - D'e^{-ik_2 L} = \frac{k_3}{k_2}E'e^{ik_3 L} \quad (x = L) \tag{6.26d}$$

The above equations can be combined to give the following three equations, where the combinations are shown on the left and resulting equation on the right:

(6.26a) + (6.26b) $\Rightarrow \quad 2 = (1 + k_2/k_1)C' + (1 - k_2/k_1)D'$

(6.26c) + (6.26d) $\Rightarrow \quad 2C'e^{ik_2 L} = (1 + k_3/k_2)E'e^{ik_3 L}$

(6.26c) − (6.26d) $\Rightarrow \quad 2D'e^{-ik_2 L} = (1 - k_3/k_2)E'e^{ik_3 L}$

Amplitude ratios C' and D' become

$$C' = \frac{E'}{2}\left(1 + \frac{k_3}{k_2}\right)e^{ik_3 L}e^{-ik_2 L} \qquad (6.27a)$$

$$D' = \frac{E'}{2}\left(1 - \frac{k_3}{k_2}\right)e^{ik_3 L}e^{ik_2 L} \qquad (6.27b)$$

C' and D' can be substituted into the first combination of equations above to yield the solution for E' given by

$$E' = \frac{4k_1 k_2 e^{-ik_3 L}}{(k_1 + k_2)(k_2 + k_3)e^{-ik_2 L} + (k_1 - k_2)(k_2 - k_3)e^{ik_2 L}} \qquad (6.27c)$$

From this result, we can see that the probability amplitude is generally a complex number and it depends on the properties of the material region, as well as adjacent regions. The quantity $E' = E/A$ gives us the transmission probability T, which is proportional to $(E')^* E' = |E'|^2$. With this solution available, we can obtain the last amplitude ratio B' from Eq. (6.26a), which gives

$$B' = C' + D' - 1 = \frac{(k_2 + k_3)e^{-ik_2 L} + (k_2 - k_3)e^{ik_2 L} - 2k_2 e^{-ik_3 L}}{2k_2 e^{-ik_3 L}} \qquad (6.27d)$$

B' provides the reflection probability R. While we now have all four amplitude ratios for the most general forms of wave numbers, it is common to treat the two possible cases for $k_2^2 = 2m(E_{\lambda,2} - V_0)/\hbar^2$. By doing so, we can obtain two distinct expressions for transmission T and reflection R probabilities. We'll do this next.

6.3.1 Evanescent Barrier Solutions $V_0 > E_\lambda$

The first case we'll consider is the case when $V_0 > E_{\lambda,2}$ in material two. The transmission and reflection probabilities T and R are defined as

$$T = \frac{J_T}{J_I} \quad \text{and} \quad R = \frac{J_R}{J_I}$$

J_I is the component we refer to as the incoming current given by $\psi_I = Ae^{ik_1 x}$, while the transmitted current is determined from $\psi_T = Ee^{ik_3 x}$. The reflected current is determined from $\psi_R = Be^{ik_1 x}$. Using the definition for the current density given by Eq. (6.18), we have

$$J_I = \frac{\hbar |A|^2 k_1}{m}, \quad J_R = \frac{\hbar |B|^2 k_1}{m} \quad \text{and} \quad J_T = \frac{\hbar |E|^2 k_3}{m}$$

6.3 Electrons Across A Rectangular Potential Barrier

From this, the probabilities become

$$R = \frac{J_R}{J_I} = \frac{|B|^2}{|A|^2} = |B'|^2 \tag{6.28}$$

$$T = \frac{J_T}{J_I} = \frac{k_3|E|^2}{k_1|A|^2} = \frac{k_3}{k_1}|E'|^2 \tag{6.29}$$

Therefore, the above results lead to R and T. However, to calculate them analytically, we must know whether k_2 is real or complex. In the condition $V_0 > E_{\lambda,2}$, it follows that $k_2^2 = 2m(E_\lambda - V_0)/\hbar^2 < 0$, and thus $\sqrt{2m(E_\lambda - V_0)}/\hbar = k_2 = ik_2'$. While k_2 is purely imaginary, k_2' is real. This can be substituted into Eqs. (6.27a) - (6.27d) to determine a more compact expression for T and R. E' and its complex conjugate become

$$E' = \frac{4ik_1 k_2' e^{-ik_3 L}}{(k_1 + ik_2')(ik_2' + k_3)e^{k_2 L} + (k_1 - ik_2')(ik_2' - k_3)e^{-k_2 L}} \tag{6.30a}$$

and

$$(E')^* = \frac{-4e^{ik_3 L} ik_1 k_2'}{e^{k_2 L}(-ik_2' + k_3)(k_1 - ik_2') + e^{-k_2 L}(-ik_2' - k_3)(k_1 + ik_2')} \tag{6.30b}$$

The calculation of $|E'|^2$ is straightfoward, however, takes a little algebra. Along the way, one may also make use of the hyperbolic cosine function relations $\cosh x = (e^x + e^{-x})/2$ and $\cosh 2x = 2\sinh^2 x + 1$. After some algebra, T works out to be

Rectangular Barrier Transmission Probability ($k_2^2 < 0$)

$$T = \frac{4k_1 k_2'^2 k_3}{(k_1^2 + k_2'^2)(k_2'^2 + k_3^2)\sinh^2(k_2' L) + (k_1 k_2' + k_2' k_3)^2} \tag{6.31}$$

For the reflection probability R, we obtain

Rectangular Barrier Reflection Probability ($k_2^2 < 0$)

$$R = \frac{(k_1^2 + k_2'^2)(k_2'^2 + k_3^2)\sinh^2(k_2' L) + (k_1 k_2' - k_2 k_3)^2}{(k_1^2 + k_2'^2)(k_2'^2 + k_3^2)\sinh^2(k_2' L) + (k_1 k_2' + k_2' k_3)^2} \tag{6.32}$$

We can check by adding Eqs. (6.31) and (6.32) to confirm that $R + T = 1$. In the above results, k_1, k_2, and k_3 are allowed to be different, but k_1, k_3 are real, while $k_2 = ik_2'$ is purely imaginary is assumed. If we let $k_1 = k_3$ (i.e. same material on both sides of the barrier), Eq. (6.31) reduces to

$$T = \frac{4k_1^2 k_2'^2}{(k_1^2 + k_2'^2)^2 \sinh^2(k_2' L) + 4k_1^2 k_2'^2} \qquad (6.33a)$$

Likewise, the reflection probability R for the case $k_1 = k_3$ reduces to

$$R = \frac{(k_1^2 + k_2'^2)^2 \sinh^2(k_2' L)}{(k_1^2 + k_2'^2)^2 \sinh^2(k_2' L) + 4k_1^2 k_2'^2} \qquad (6.33b)$$

Figs. 6.7, 6.8, and 6.9 illustrate the transmission and reflection probabilities, T and R, varying the height of the potential barrier V_0. In Figs.

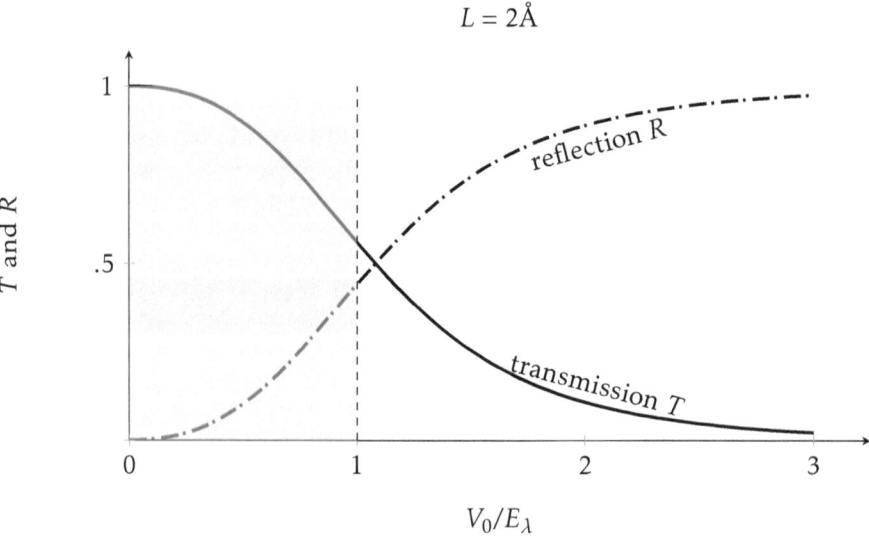

Figure 6.7: Illustration of the transmission and reflection probabilities R and T as functions of V_0/E_λ. In the case shown, $L = 2\text{Å}$ and $E_\lambda = 3eV$. Note that the region to the right of the shaded rectangle is described by Eqs. (6.33a) and (6.33b).

6.8 and 6.9, the vertical dashed line at $V_0/E_\lambda = 1$ represents the classical limit. However, all examples here illustrate nonzero transmission at the classical limit. It is this property leads to the potential for *tunneling* into

6.3 Electrons Across A Rectangular Potential Barrier

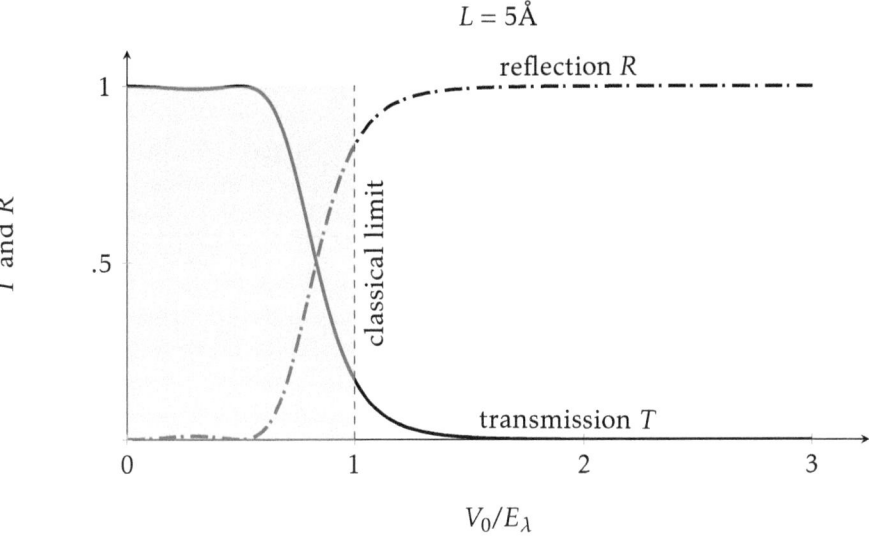

Figure 6.8: Transmission probabilities T versus V_0/E, for the case $V > E_0$. $L = 5\text{Å}$, and $E_\lambda = 3eV$.

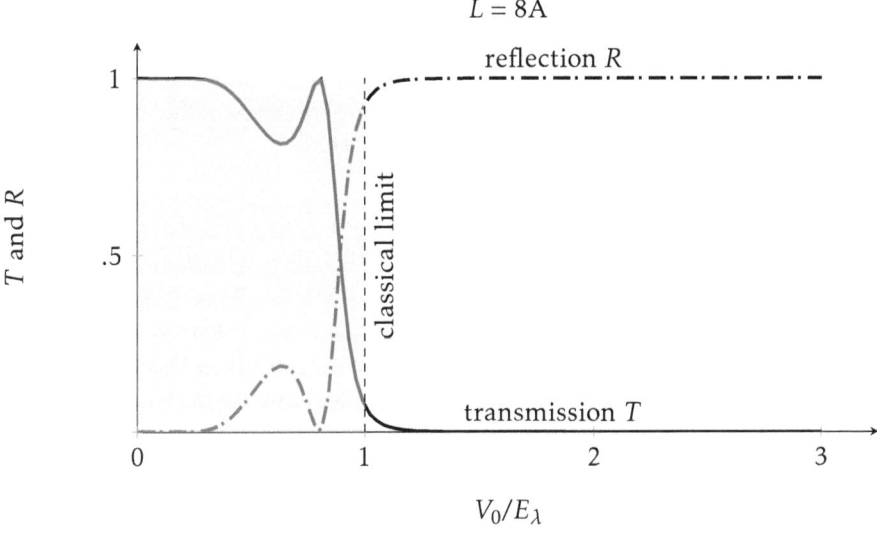

Figure 6.9: Reflection and transmission probabilities R and T as a function of V_0/E_λ. In this illustration $L = 8\text{Å}$ and $E_\lambda = 3eV$. As the thickness increases, so does the sharpness of the transitions. Thinner barriers transmit more than thicker barriers.

any subsequent material adjacent to the barrier. As the barrier thickness increases, so does the sharpness of the transitions, which means transmission drops more efficiently and/or reflection increases more efficiently for thicker barriers. This is illustrated more plainly in Fig. 6.10, plotting the dependence on the dimensionless parameter $k_2' L$. The

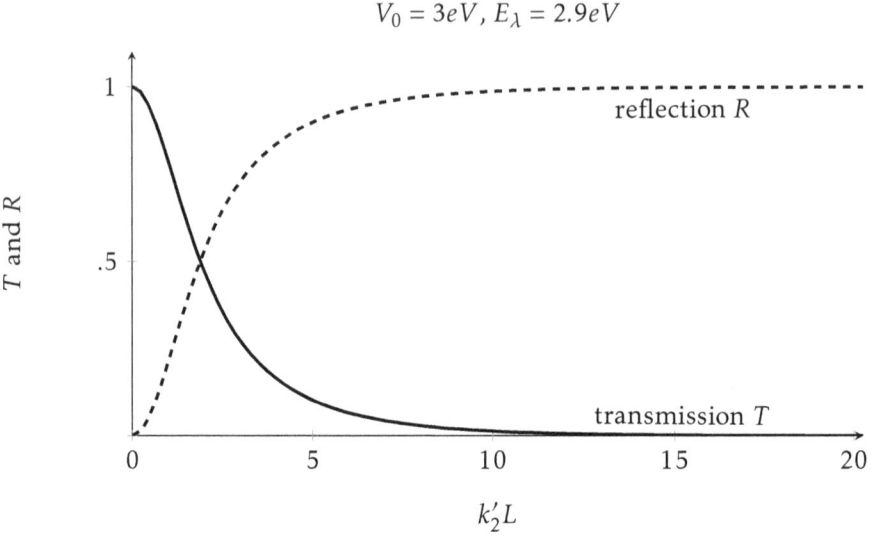

Figure 6.10: Reflection and transmission probabilities R and T as functions of barrier thickness L. Here, $V_0 = 3eV$, and $E_\lambda = 2.9 eV$, using the electron mass. For a given energy and barrier height, increasing thickness reduces the amount of transmission, and therefore, increases reflection.

abscissa $k_2' L$ indicates that the transmission decays over a length scale of the order of $1/k_2' \sim \lambda_2$, where λ_2 is the wave-length in the rectangular barrier region. This phenomenon that we have obtained from the Schrödinger equation, indicating a finite transmission $T > 0$ in the conditions where the system energy E_λ is strictly less than the potential barrier, is known as **tunneling** or **quantum tunneling**, and it is absolutely enabled by the existence of an evanescent wave function in the region of the rectangular potential. Tunneling is, thus, manifestly, most applicable for small or thin devices. Today, this phenomenon is exploited in several technological devices, such as in more recent types of transistors, used in computers, and *magnetic tunnel junctions* utilized as the magnetic sensors many applications like hard disk drives and global-positioning-satellite (GPS) systems.

You may have noticed the nonmonotonic bahavior of T and R in the shaded region corresponding to $V_0/E_\lambda \leq 1$. This unique behavior arises due to a phenomenon known as **interference**, which takes place in any

6.3 Electrons Across A Rectangular Potential Barrier

region where the wave function has the general form $\psi(x) = Ae^{ikx} + Be^{-ikx}$, and a boundary which can reflect the incoming wave described by Ae^{ikx}, so $B > 0$. Let us consider the other important case, namely, when $V_0 < E_\lambda$.

6.3.2 Propagating Wave Barrier Solutions $V_0 < E_\lambda$

For when $V_0 < E_\lambda$, it follows that $k_2 = \sqrt{2m(E_\lambda - V_0)}/\hbar > 0$, because $k_2^2 > 0$. The processes of determining the solutions for the amplitude ratios B', C', D', and E' are the same as in the previous case. Since k_2 is real, the explicit form of B', C', D', and E' is given by Eqs. (6.27a), (6.27b), (6.27c), and (6.27d). Instead of evanescent waves described by $e^{-k_2'x}$ in region 2 (corresponding to 100% reflection at the first interface), the wave function contains purely propagating waves $e^{ik_2 x}$. For the transmission and reflection probability, the conjugates of E' and B' given by Eqs. (6.27c) and (6.27d), for this case, become

$$(E')^* = \frac{4k_1 k_2 e^{ik_3 L}}{(k_1 + k_2)(k_2 + k_3) e^{ik_2 L} + (k_1 - k_2)(k_2 - k_3) e^{-ik_2 L}} \quad (6.34a)$$

$$(B')^* = \frac{(k_2 + k_3) e^{ik_2 L} + (k_2 - k_3) e^{-ik_2 L} - 2k_2 e^{-ik_3 L}}{2k_2 e^{ik_3 L}} \quad (6.34b)$$

$(E')^* E' = |E'|^2$ and $(B')^* B' = |B'|^2$ can be substituted into Eqs (6.28) and (6.29) to obtain R and T. Note that in obtaining $|E'|^2$ and $|B'|^2$, we also use the relations $\cos x = (e^{ix} + e^{-ix})/2$ and $\cos^2 k_2 L = 1 - \sin^2 k_2 L$. Then, we obtain the following result for the transmission probability given by $T = k_3 |E'|^2 / k_1$:

Barrier Transmission Probability ($k_2^2 > 0$)

$$T = \frac{4k_1 k_2^2 k_3}{(k_2^2 - k_1^2)(k_2^2 - k_3^2)\sin^2(k_2 L) + (k_1 k_2 + k_2 k_3)^2} \quad (6.35)$$

And the reflection probability $R = |B'|^2$ is found to be

Barrier Reflection Probability ($k_2^2 > 0$)

$$R = \frac{(k_2^2 - k_1^2)(k_2^2 - k_3^2)\sin^2(k_2 L) + (k_1 k_2 - k_2 k_3)^2}{(k_2^2 - k_1^2)(k_2^2 - k_3^2)\sin^2(k_2 L) + (k_1 k_2 + k_2 k_3)^2} \tag{6.36}$$

In the simplified case of $k_1 = k_3$, Eqs. (6.35) and (6.36) become

$$T = \frac{4k_1^2 k_2^2}{(k_2^2 - k_1^2)^2 \sin^2(k_2 L) + 4k_1^2 k_2^2} \tag{6.37}$$

$$R = \frac{(k_2^2 - k_1^2)^2 \sin^2(k_2 L)}{(k_2^2 - k_1^2)^2 \sin^2(k_2 L) + 4k_1^2 k_2^2} \tag{6.38}$$

When $k_2^2 > 0$, we find that the solutions for T and R contain a $\sin^2(k_2 L)$ term (as opposed to a $\sinh^2(k_2' L)$ term when $k_2^2 < 0$). This term, in particular, can lead to oscillations and/or undulations in T and R, as functions of L and V_0. Examples of R and T from Eqs. (6.37) and (6.37) are illustrated in Figs. 6.14, 6.12, and 6.13. Both the transmission T

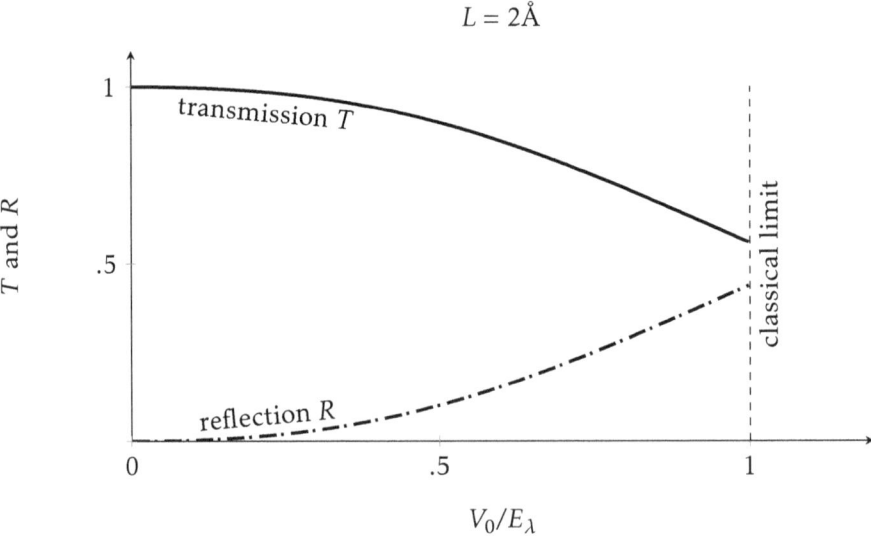

Figure 6.11: Illustration of the transmission and reflection probabilities R and T as functions of V_0/E_λ, with $L = 2\text{Å}$ and $E_\lambda = 3eV$.

and reflection R probabilities as a function of barrier thickness L are illustrated in Fig. 6.15. The potential barrier problem has different consequences depending on whether the system has more or less energy

6.3 Electrons Across A Rectangular Potential Barrier

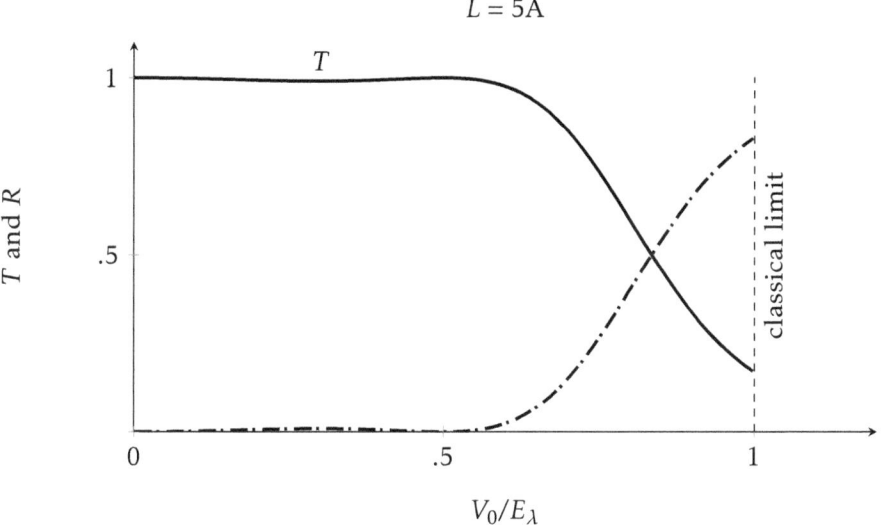

Figure 6.12: Illustration of the transmission and reflection probabilities R and T as functions of V_0/E_λ, with $L = 5\text{Å}$ and $E_\lambda = 3eV$.

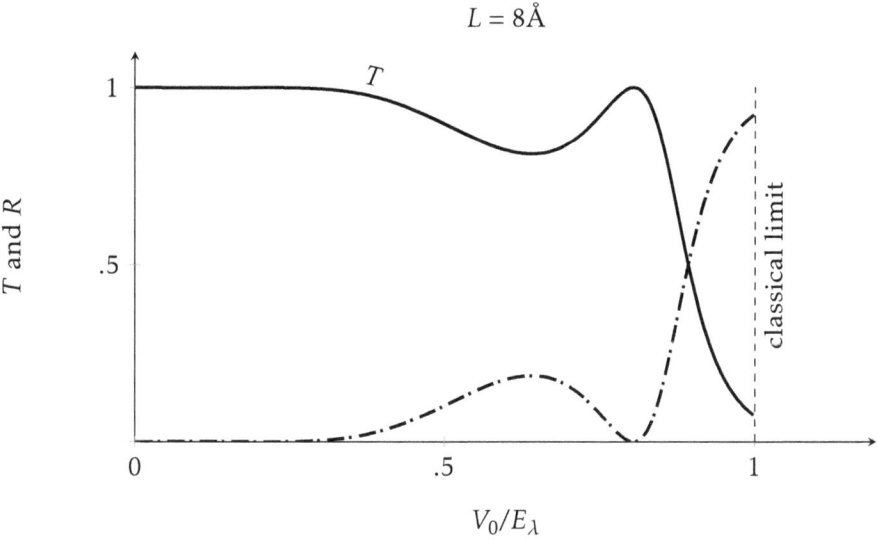

Figure 6.13: Illustration of the transmission and reflection probabilities R and T as functions of V_0/E_λ, with $L = 8\text{Å}$ and $E_\lambda = 3eV$.

than the rectangular potential barrier V_0. One case leads to tunneling across the barrier, even though the system energy is less than that of the barrier. In the other case where the system energy is larger than

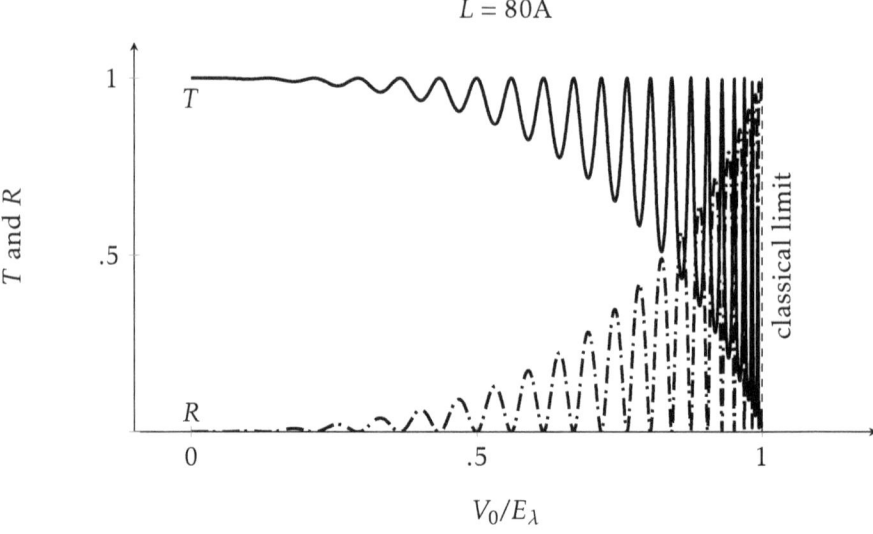

Figure 6.14: Illustration of the transmission and reflection probabilities R and T as functions of V_0/E_λ, with $L = 80\text{Å}$ and $E_\lambda = 3eV$.

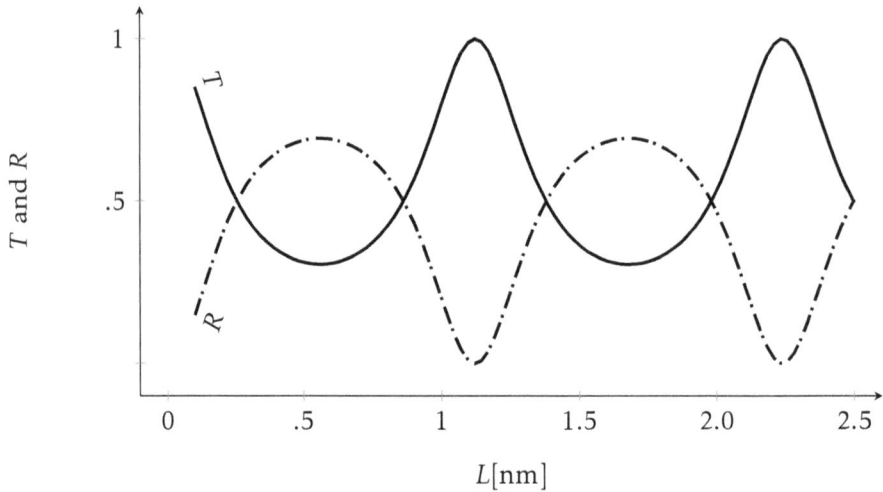

Figure 6.15: Illustration for the case $k_2^2 > 0$, R and T as a function of a barrier thickness L. The potential barrier height $V_0 = 3eV$, $E_\lambda = 1.1 V_0$, and electron mass m_e is also assumed. This behavior corresponds to interference of the electron wave function with itself.

that of the potential barrier, an interference phenomenon takes place for a wave propagating through a material with an interface. Fig. 6.16 illustrates the transmission for both cases $k_2^2 > 0$ and $k_2^2 < 0$, for direct

6.3 Electrons Across A Rectangular Potential Barrier

comparison. The oscillatory nature of T and R are *subtle* consequences

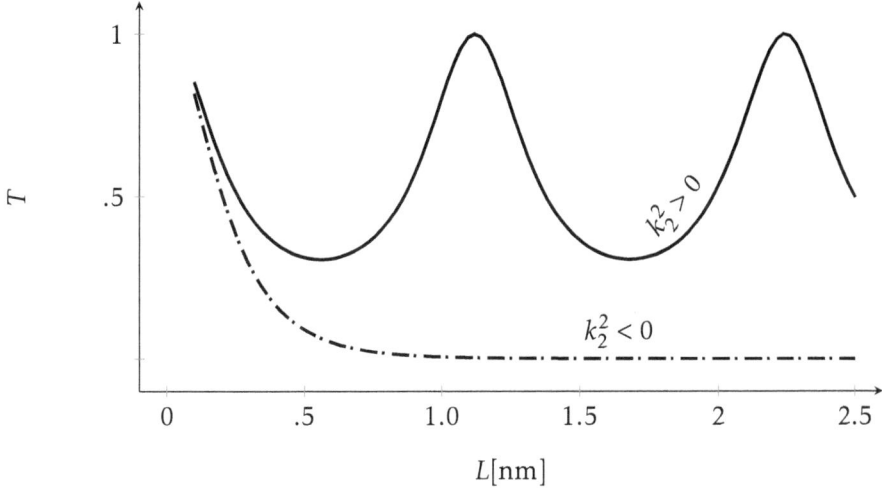

Figure 6.16: Illustration of the transmission probabilities T for both cases $k_2^2 > 0$ (solid black) and $k_2^2 < 0$, varying the rectangular potential barrier thickness L. For both cases, $V_0 = 3eV$, and in the first case $k_2^2 > 0$, $E_\lambda = 1.1 V_0$, while $E_\lambda = 0.9 V_0$ for the case $k_2^2 < 0$. The electron mass m_e is assumed. In one case, we have an oscillating transmission, while in the other case, we have a monotonically decaying transmission.

of **wave function interference**. However, *the probability density does not reveal interference*. Let the wave function be given by

$$X(x) = Ce^{ikx} + De^{-ikx}$$

We'll have you show in Chapter problem 6.6 that

$$X^*X = |C|^2 + |D|^2 + (C^*D)_R \left[\cos^2 kx - \sin^2 kx\right] + (C^*D)_I \cos kx \sin kx \quad (6.39)$$

In the above, $k > 0$. It is the second order trigonometric terms that give rise to an interference phenomenon.

Both conditions we have considered $V_0 < E_\lambda$ and $V_0 > E_\lambda$ turn out to be very relevant to the electrical conduction properties of materials. We will have an occasion to explore this more in Chap. 11, when we discuss asymmetric spin-dependent transport. Additionally, this characteristic of an electron, which back-scatters part of it's wave, superposes with all the other occurrences in the lattice (periodic or regular arrangement of atoms). Conditions are reached where the combined electron back-scattered wave cancels the incoming wave. When there is cancellation of this kind, standing waves form and this gives rise to *forbidden energy bands* where electrons no longer can propagate forward or backwards in the material. But, these bands essentially arise from the partial scattering of charge carriers having wave properties. Indeed, this is de Broglie's pioneering work in action.

This completes our analysis of the *rectangular potential barrier* problem. Next, we will consider a problem with a potential energy function that is an inversion of the potential barrier just discussed, known as a **finite potential well**.

6.4 THE FINITE POTENTIAL WELL PROBLEM

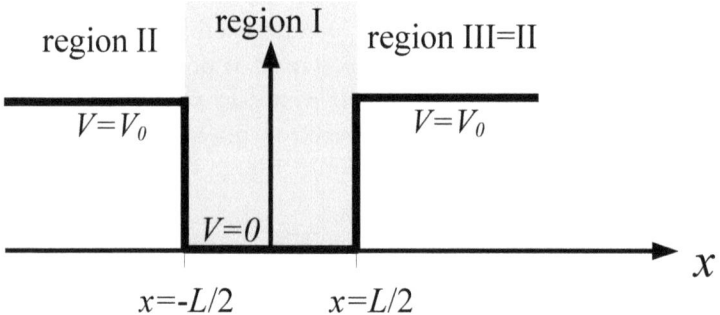

Figure 6.17: Illustration of the potential well problem, where the origin is in the symmetric center of the well. In this problem, the fundamental particle begins in the well region where $V = 0$. The potential energy in region two is also taken to be the same as that of region three, but they are larger than in the well.

In the *potential well problem*, the potential well is taken to be zero potential energy within the symmetrically centered lower potential region. However, in the exterior regions one and three, there is a finite potential energy $V_0 > 0$. In appearance, the potential dependence on x is an inversion of the rectangular potential barrier. Inverting the potential, however, also dictates that we start with the fundamental particle in region one, where $V = 0$. With this picture, it is more useful to view the potential well problem as a combination of the infinitely deep potential well problem (IDPW) discussed in Chap. 4, and the free particle

6.4 The Finite Potential Well Problem

problem discussed in Chap. 5. Recall that in the free particle problem, there is no potential energy, or $V_0 = 0$. However, we saw in the IDPW problem, that with an infinite potential, a fundamental particle will be confined to the box, with eigen-solutions proportional to $\sin(n\pi x/L)$, with an origin located at the first step transition. In the IDPW coordinate system $\psi(0) = \psi(L) = 0$, however, what happens when the potential is *finite* instead? This frame helps to develop some initial intuition for the problem. Fortunately, we can find out exactly what Schrödinger's equation reveals about the potential well by exploiting the symmetry of the potential well.

The application of the Schrödinger equation is similar to what we have seen previously, where we apply the spatially 1D equation to both regions one and two. This leads to

$$i\hbar \frac{\partial \psi}{\partial t} = \left(-\frac{\hbar^2}{2m}\frac{\partial^2}{\partial x^2} + V(x)\right)\psi \tag{6.40}$$

Again, using the method of separation of variables for space-x and time-t, the wave function ψ for the potential well problem assumes the form

$$\psi = T(t)X(x) \tag{6.41}$$

Substituting Eq. (6.41) into Eq. (6.40) gives

$$i\hbar X \frac{dT}{dt} = T\left(-\frac{\hbar^2}{2m}\frac{d^2 X}{dx^2} + VX\right) \tag{6.42}$$

Then, dividing Eq. (6.42) by the wave function $\psi = T \cdot X$ leads to

$$i\hbar \frac{1}{T}\frac{dT}{dt} = -\frac{\hbar^2}{2m}\frac{1}{X}\frac{d^2 X}{dx^2} + V \tag{6.43}$$

The LHS of Eq. (6.43) is purely a function of t while the RHS is purely a function of x. Since both are independent variables, this can only be true when both sides are equal to a constant E_λ, so we have

$$i\hbar \frac{1}{T}\frac{dT}{dt} = E_\lambda \tag{6.44a}$$

$$-\frac{\hbar^2}{2m}\frac{1}{X}\frac{d^2 X}{dx^2} + V = E_\lambda \tag{6.44b}$$

Integration of (6.44a) leads to a solution for $T(t)$ given by

$$T(t) = T_0 e^{-\frac{i}{\hbar}E_\lambda(t-t_0)} \tag{6.45}$$

This is the first of the two functions we need for the method of separation of variables. To obtain the second function X, we can write Eq. (6.44b) as

$$\frac{d^2X}{dx^2} + \frac{2m(E_\lambda - V)}{\hbar^2}X = 0 \tag{6.46}$$

In the two distinct regions of the potential well, Eq. (6.46) becomes

$$\frac{d^2X}{dx^2} + \frac{2mE_\lambda}{\hbar^2}X = 0 \quad \text{(region 1)} \tag{6.47a}$$

$$\frac{d^2X}{dx^2} + \frac{2m(E_\lambda - V_0)}{\hbar^2}X = 0 \quad \text{(region 2)} \tag{6.47b}$$

Thus, the corresponding wave-numbers k_1 and k_2 are defined by

$$k_1^2 = \frac{2m|E_\lambda|}{\hbar^2} \tag{6.48a}$$

$$k_2^2 = -\frac{2m(V_0 - E_\lambda)}{\hbar^2} = -k_2'^2 \tag{6.48b}$$

If $E_\lambda < V_0$, our previous experience tells us that we will have evanescence at the exterior of the potential well, and this also indicates continual reflection processes within the well. So, this corresponds to bound states. These conditions and corresponding bound states within the well are of primary interest, here.

Since the potential $V_0 < \infty$ at $x = -L/2, L/2$, we cannot specify that $X(-L/2) = X(L/2) = 0$, as we did in the 1D *particle in a box* problem. In fact, because we have a generally finite potential, we cannot directly specify any value of the wave function at $x = -L/2, L/2$ as it will depend on V_0 and E_λ. In order to find the energy states E_n corresponding to $E_\lambda < V_0$, we can take advantage of the fact that if we choose the origin as a symmetric point $x = 0$, the eigen-solutions form a sequence of symmetric and anti-symmetric states (c. Sec. 4.1). Then, each available energy state can be associated with either a symmetric or anti-symmetric wave function. Let's see how we can use this information to solve the problem. In region one, Eq. (6.47a) becomes

$$\frac{d^2X}{dx^2} - k_1^2 X = 0 \tag{6.49}$$

Eq. (6.49) has the general solution

$$X_1 = A\cos k_1 x + B\sin k_1 x \tag{6.50}$$

6.4 The Finite Potential Well Problem

As we had found in the Sec. 6.3, in region two, the general solution is a combination of *evanescent* wave functions given by

$$X_2 = Ce^{-k_2'x} + De^{k_2'x}$$

The normalization condition for X requires that the wave function vanishes at ∞. It follows that $D = 0$, otherwise, this term diverges at $+\infty$, and we have

$$X_2 = Ce^{-k_2'x} \quad (6.51)$$

Eq. (6.50) is the sum of a symmetric and anti-symmetric function. Using the fact that any energy state E_n corresponds to *either* a symmetric or anti-symmetric solution, let us consider both separately. For the odd or anti-symmetric solutions, the wave function X_1 is

$$X_1 = B\sin(k_1 x) \quad (6.52)$$

Using the continuity boundary conditions of X and its first derivative at $x = L/2$, we have

$$B\sin\left(k_1 \frac{L}{2}\right) = Ce^{-k_2' \frac{L}{2}} \quad (6.53a)$$

$$k_1 B\cos\left(k_1 \frac{L}{2}\right) = -k_2' Ce^{-k_2' \frac{L}{2}} \quad (6.53b)$$

Forming a ratio of Eqs. (6.53a) and (6.53b), for the anti-symmetric states, we obtain the relation

$$k_1 \cot\left(k_1 \frac{L}{2}\right) = -k_2'$$

We can also express the above in dimensionless form as follows:

$$k_1 \frac{L}{2} \cot\left(k_1 \frac{L}{2}\right) = -k_2' \frac{L}{2} \quad (6.54)$$

Using Eq. (6.54), let us define the following relation:

$$\frac{(k_1 L)^2}{4} + \frac{(k_2' L)^2}{4} = \frac{2mV_0 L^2}{4\hbar^2} \Rightarrow \alpha^2 + \beta^2 = \eta^2 \quad (6.55)$$

We are going to use Eq. (6.55) to define three parameters α, β, and η, by writing Eq. (6.55) as follows:

$$\alpha^2 + \beta^2 = \eta^2 \quad (6.56)$$

This allows us to write Eq. (6.54) as

$$-\alpha\cot(\alpha) = \sqrt{\eta^2 - \alpha^2} = \beta = \beta(\alpha) \quad \text{(anti-symmetric)} \quad (6.57a)$$

For the symmetric states, carrying out a similar procedure using $X_1 = A\cos(k_1 x)$, we obtain the relation

$$\alpha\tan(\alpha) = \sqrt{\eta^2 - \alpha^2} = \beta(\alpha) \quad \text{(symmetric)} \quad (6.57b)$$

For a chosen value of V_0 and L, Eq. (6.57a) is a function of the single variable E_λ. Therefore, in principle, it can be solved for E_λ. However, since both equations are nonlinear in α, it can only be solved using numerical or computational methods, which will not be done here. Instead, we can visualize the solutions first, by plotting both sides of Eqs. (6.57a) and (6.57b) and identifying where they overlap with . Fig. 6.18 shows both sets of symmetric (tangent) and anti-symmetric (cotangent) curves as functions of α. Fig. 6.18 illustrates that both the anti-symmetric negative cotangent curves and the symmetric tangent curves run mostly parallel to one another, separated by a distance of $\pi/2$. The dashed lines are $\alpha\tan\alpha$ curves corresponding to the symmetric states, while the blue solid lines are $-\alpha\cot\alpha$ curves, corresponding to the anti-symmetric states. These curves are given by the LHS of Eqs. (6.57a) and (6.57b). They intersect the black quarter circle lines, which are $\beta(\alpha)$ or (α, β) points, a finite number of times N_η, for any choice of η. For a given η, the solutions to the finite potential well problem correspond only to those discrete values (α_n, β_n) given by the intersections. The discrete solutions (α_n, β_n) can be seen in Fig. 6.18, illustrated by the open circles.

The potential V_0 and potential well length L influence η, which is the radius of the circles. Increasing η, or increasing the size of the intersecting circle increases the number of quantum states N_η, which is the number of intersections with the anti-symmetric cotangent and symmetric tangent curves. Determining these discrete values of α_n provides the bound quantum energy state E_n, given by

Bound-state Energy Levels of 1D Potential Well

$$E_n = -\frac{2\hbar^2 \alpha_n^2}{mL^2} \quad (6.58)$$

6.4 The Finite Potential Well Problem

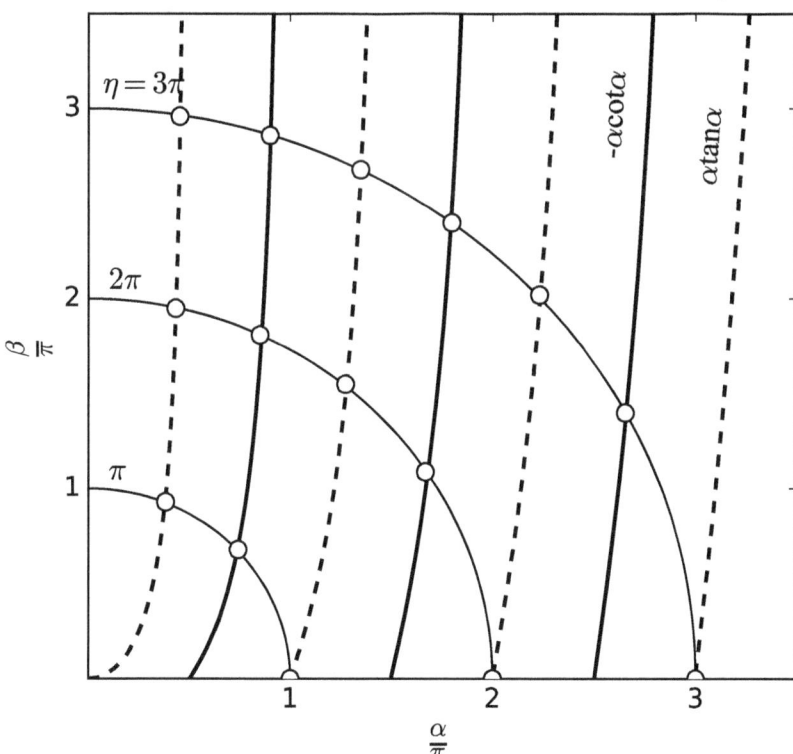

Figure 6.18: The potential well symmetric and anti-symmetric curves, along with the intersecting (α,β) curve, and the discrete solutions corresponding to the solutions to the finite potential well problem.

We can determine the number of quantum states N_η corresponding to a radius of η using the fact that each positive branch of *tan* and *cot* functions spans horizontally a width of $\pi/2$, and there is no overlap in their respective regions. This is so because $\pi/2$ is the distance between the x-axis crossing and its asymptote to the immediate right of it. Since the first branch, which corresponds to a symmetric state, starts at the origin, for all circles of radius $\eta < \pi/2$, there will be one symmetric state. From here, each addition to η of $\pi/2$ corresponds to an additional state. Thus, the number of discrete states N_η, corresponding to η, is given by

$$N_\eta = \frac{\eta}{\pi/2} + 1$$

Substitution of the expression for η leads to

> **Number of Quantum States On A η Radius**
>
> $$N_\eta = \frac{L}{\pi\hbar}\sqrt{2mV_0} + 1 \qquad (6.59)$$

For the values of η ($\pi, 2\pi$, and 3π) illustrated in Fig. 6.18, Eq. (6.59) gives 3, 5, and 7, respectively, in agreement with the number of quantum states corresponding to each open circles in that radius. There will be a wave function for each value of α_n. For example, in the case of $\eta = 2\pi$ (the middle quarter circle), there are five discrete solutions, thus, there are five corresponding wave functions. They are illustrated in Fig. 6.19. The potential well width L used in Fig. 6.19 is $L = 5$Å, and

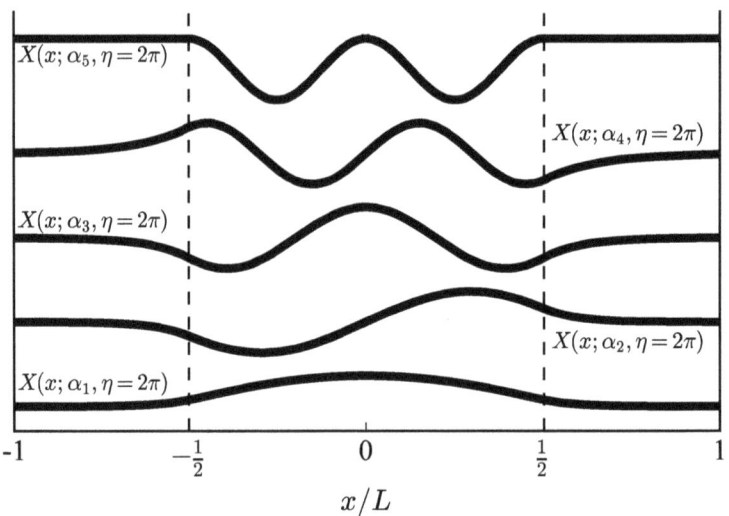

Figure 6.19: The five potential well symmetric and anti-symmetric wave functions for the case of $\eta = 2\pi$, corresponding to the second quarter circle in Fig. 6.18. The potential width used in this example is $L = 5$Å, and the corresponding energies $E_\lambda = E_n = 1.1, 4.3, 9.8, 16.8, 24\,eV$. There are five α_n solutions, from left to right (ordered by increasing energy).

the corresponding energies for the eigenstates shown from bottom to top are $E_\lambda = E_n = 1.1, 4.3, 9.8, 16.8, 24\,eV$. From the bottom to the top wave function corresponds to going clockwise on the $\eta/\pi = 2$. The

6.4 The Finite Potential Well Problem

wave functions partly *leak* outside of the potential well, described by an evanescent wave function at the well exterior. This is in contrast to the infinite potential well we discussed in Chap. 3, where all wave functions vanish at the edges of the well. For comparison, the first five eigensolutions are shown in Fig. 6.20. Fig. 6.19 also provides visual

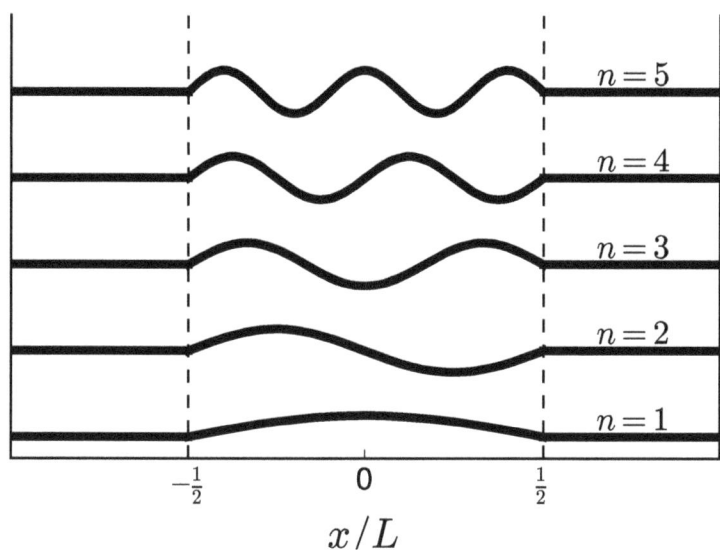

Figure 6.20: The first five ($n = 1, 2, 3, 4, 5$) infinitely deep potential well wave functions shown for comparison to those of the finite potential well, shown in Fig. 6.19. In the finite potential well, the wave functions do not vanish at the boundaries, as with infinite potential.

insight into the possibility of degenerate states because each branch of the cot() and tan() functions does *not* overlap. Since each branch is one-to-one, meaning that there is only a single value for any α, all the discrete solutions are guaranteed to be unique. There are no two eigenstates alike, or the eigenstates of the problem are non-degenerate, as in the case of the infinitely deep potential well.

This chapter has revealed rich phenomena for both free particles and bound particles, interacting with the presence of sharp potential changes. Quantum mechanics predicts that such potential variations can lead to anything from bound states, which is accompanied by complete reflection, to various combinations of partially free states (reflected or scattered) accompanied by partial reflection and transmission. In next chapter, we introduce an additional property of electrons known as angular momentum. It turns out to be important in understanding the subject of *magnetism*, which is a central subject for this text.

6.5 Chapter Summary

In this chapter, we have discussed the consequences of propagating particles in the presence of potential steps. First, a single sharp rising potential step was considered, where we found that reflection and/or scattering, to some extent, always occurs in a finite potential rise. We also found that when $V_0 > E_lambda$, an special type of wave function forms known as an evanescent wave function forms at the boundary location of the rising potential. We then considered the *rectangular potential barrier*, and obtained the probabilities of transmission T and reflection R. Lastly, we considered the finite potential well, though we analyzed it in a different manner, it turns out to be very similar to the infinitely deep potential well problem, however, with evanescent wave function properties at the boundaries. Some key results are summarized in Table 6.1.

Table 6.1: Chapter 6 Summary Equations.

Name	Equation
Reflection definition	$R = \frac{|J_R|}{|J_I|}$
Transmission definition	$T = \frac{|J_T|}{|J_I|}$
Reflection ($k_2^2 < 0$)	$R = \frac{(k_1^2 + k_2'^2)(k_2'^2 + k_3^2)\sinh^2(k_2'L) + (k_1 k_2' - k_2 k_3)^2}{(k_1^2 + k_2'^2)(k_2'^2 + k_3^2)\sinh^2(k_2'L) + (k_1 k_2' + k_2' k_3)^2}$
Transmission ($k_2^2 < 0$)	$T = \frac{4 k_1 k_2'^2 k_3}{(k_1^2 + k_2'^2)(k_2'^2 + k_3^2)\sinh^2(k_2'L) + (k_1 k_2' + k_2' k_3)^2}$
Reflection ($k_2^2 > 0$)	$R = \frac{(k_2^2 - k_1^2)(k_2^2 - k_3^2)\sin^2(k_2 L) + (k_1 k_2 - k_2 k_3)^2}{(k_2^2 - k_1^2)(k_2^2 - k_3^2)\sin^2(k_2 L) + (k_1 k_2 + k_2 k_3)^2}$
Transmission ($k_2^2 > 0$)	$T = \frac{4 k_1 k_2^2 k_3}{(k_2^2 - k_1^2)(k_2^2 - k_3^2)\sin^2(k_2 L) + (k_1 k_2 + k_2 k_3)^2}$
Potential Well Eigenenergy	$E_n = -\frac{2\hbar^2 \alpha_n^2}{mL^2}$

6.6 Chapter Problems

Problem 6.1 The wave number for an electron with kinetic energy and potential energy can be described by

$$k = \sqrt{\frac{2m(E_\lambda - V)}{\hbar^2}}$$

Calculate the wave number k for an electron with $E = 1$ eV in a potential of $V = +.5$ eV? (a hill). Will the wave slow down or speed up? Same question for $V = -.5$ eV (a hole).

Problem 6.2 For two cases above, calculate the reflection and transmission probabilities using the appropriate (simplified)equations (e.g. Eqs. (6.33a) and (6.33b)) for R and T.

Problem 6.3 Earlier, in the problem of the rectangular barrier, the following transmission and reflection probabilities were obtained:

$$T = \frac{4k_1^2 k_2'^2}{(k_1^2 + k_2'^2)^2 \sinh^2(k_2' L) + 4k_1^2 k_2'^2}$$

and

$$R = \frac{(k_1^2 + k_2'^2)^2 \sinh^2(k_2' L)}{(k_1^2 + k_2'^2)^2 \sinh^2(k_2' L) + 4k_1^2 k_2'^2}$$

k_1 is the wave number in the first region and k_2' is the wave number in the barrier region of width L. The dependence on the barrier width L only appears in the sinh terms. Look at the asymptotic case of $L \to 0$, and show that $T = 1$ and $R = 0$. Then, take $L \to \infty$ and show that $T = 0$ and $R = 1$.

Problem 6.4 Use the above relations for reflection and transmission given by Eqs. (6.33a) and (6.33b) and show that

$$R + T = 1$$

Now show that the condition for $R = T = .5$ is given by

$$k_2' L = \sinh^{-1}\left[\frac{4k_1^2 k_2'^2}{k_1^2 + k_2'^2}\right]$$

Problem 6.5 Show the same for the more general reflection and transmission probabilities given by Eqs. (6.31) and (6.32).

Problem 6.6 Show that for any wave function of the form $X = Ae^{ikx} + Be^{-ikx}$, where A and B are the probability amplitudes, that $X = \overline{A}\cos kx + \overline{B}\sin kx$ where $\overline{A} = A + B$ and $\overline{B} = i(A - B)$. *Hint*: Use the Euler identity $e^{i\theta} = \cos\theta + i\sin\theta$. Also show that

$$X^*X = |A|^2 + |B|^2 + (A^*B)_R\left[\cos^2 kx - \sin^2 kx\right] + (A^*B)_I \cos kx \sin kx$$

Discuss the conditions $B = 0$ and $A = B$.

Problem 6.7 Use Fig. 6.18 to estimate the energy of an electron in a potential well 1 angstrom wide, where $\alpha_n = 2\pi$.

For this case, also compute the ratio of wave numbers k_2/k_1 for a 2 angstrom wide potential well. Use one of the cot/tan trigonometric relations.

Problem 6.8 Show that for any wave function of the form $\psi = Ae^{ikx} + Be^{-ikx}$, that the current density is given by

$$J(x) = \frac{\hbar k}{m}\left(|A|^2 - |B|^2\right)$$

Comment on the meaning and conditions that lead to $J(x) = 0$?

Problem 6.9 For a propagating wave function $\psi_0 = Ae^{ikx}$, we have seen that when A is independent of space, ψ_0 is *non-normalizable*. To extricate the wave function from this dilemma, let

$$\psi(x) = f_c(x)\psi_0$$

$f_c(x)$ is any real valued compact (or square integrable) function of x, which satisfies the condition

$$\int_{-\infty}^{+\infty} |f_c(x')|^2 dx' = 1$$

Show that

$$\psi\frac{\partial \psi^*}{\partial x} - \psi^*\frac{\partial \psi}{\partial x} = -2ik\psi^*\psi$$

This result is identical in form to the result when $\psi = \psi_0$, however, the wave function we start with *is normalizable*. This represents the

6.6 Chapter Problems

more correct form of an electron in motion, particularly, coming from a bound electronic state. The bound state energy information is, therefore, contained in the real-valued function $f_c(x)$.

CHAPTER 7

Orbital Angular Momentum

Magnetism based on a quantum mechanical framework is one of the central objectives of this text. In this chapter, we'll be getting warm with this topic discussing one of the most basic concepts for understanding magnetism. You may have noticed that up to now, the problems that we have considered have only involved kinetic energy associated with translational straight-line 1D motion. However, all atoms, which consist of various fundamental particles including electrons, *also* have rotational degrees of freedom, for example, with electrons orbiting around the nucleus. This motion, in particular, is a type of *orbital angular momentum*, and we'll consider here, in some detail, this fundamental concept. Angular momentum entered quantum mechanics in one of the very first papers on the so-called "new" quantum mechanics in a paper by Born, Heisenberg and Jordan in 1926. Though the eigenfunctions were already known, orbital angular momentum and its eigenstates were fully covered. To paint a more complete picture of the entire solutions, we'll cover what was known prior to 1926, as well as the eigenvalues, etc. First, we discuss the classical conceptual origins of a *magnetically-induced* angular momentum. Then, we'll look at how the associated classical ideas can be extended to the domain of quantum mechanics, which is our ultimate objective. We'll also consider a few other related topics.

Understanding the angular momentum of a single electron goes along way to understanding that of a large number of electrons. In fact, the physics of a single electron must be consistent with that of a system of many electrons, much like a group of singers depends, to a large extent, on the characteristics of each individual in the group. This idea represents a fundamental one in quantum mechanics known as the *Principle of Correspondence*, which simply dictates consistency between a quantum system and a classical system. This was discussed some with the harmonic oscillator problem. This idea shall also be in the backdrop of what comes in this and subsequent chapters, as well. So, let us begin by reviewing some classical concepts relating to orbital angular momentum. Then, we will build from there.

7.1 Orbital Angular Momentum Of A Rotating System

The non-relativistic *classical* kinetic energy K of a mass m with motion of speed v, is given by

$$K = \frac{1}{2}mv^2 = \frac{p^2}{2m} \tag{7.1}$$

For a rotating mass on a circular path, the instantaneous linear velocity is given by $v = r\omega$, where the angular velocity is $\omega = d\theta/dt =$, as illustrated in Fig. 7.1. Using these relations, we may express the kinetic energy K as

$$K = \frac{1}{2}mv^2 = \frac{1}{2}mr^2\omega^2 = \frac{1}{2}I\omega^2 \tag{7.2}$$

If we define the parameter I as

$$I = mr^2$$

Then, the kinetic energy becomes

$$K = \frac{1}{2}I\omega^2 \tag{7.3}$$

From Eqs. (7.2) and (7.3), we can draw an analogous relationship between the **rotational inertia** I and the **translational inertia** m. Likewise, there is a correspondence between the linear velocity v and the angular velocity ω. Based on this, we can formally define a *classical* quantity

7.1 Orbital Angular Momentum Of A Rotating System

known as the **angular momentum** denoted by L, analogous to the linear momentum p, given by

$$L = I\omega \Leftrightarrow p = mv \tag{7.4}$$

Then, the kinetic energy K can be written analogous to Eq. (7.31), as

$$K = \frac{1}{2}I\omega^2 = \frac{L^2}{2I} \tag{7.5}$$

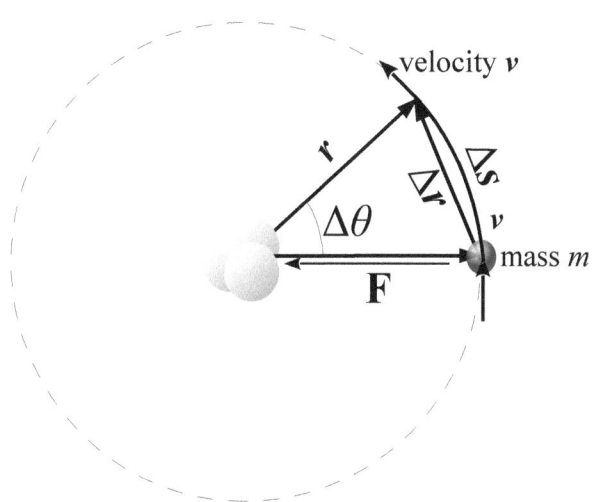

Figure 7.1: Illustration of angular momentum L in circular motion. There is a correspondence between the translational degrees of freedom and the rotational, summarized in Table 7.1.

When a mass m moves in circular motion, it can be shown that there is a radial force **F** acting on the mass keeping it from going in a straight (tangential) trajectory. This can be shown by considering the mass m rotating in a circular orbit, noting that in taking the limits toward δr, δs, and $\delta\theta$ approaching zero, the change in the radial vector r approaches the change in the arc length s. Therefore, in the limit, $|dr| = |ds| = |rd\theta|$.

It follows that

$$\mathbf{v} = r\frac{d\theta}{dt} = \mathbf{r}\omega \tag{7.6}$$

Now, we will find the derivative of the velocity **v**. For some, this step can be a little non-intuitive because one might look at the equation above and conclude that since $|\mathbf{r}| = r$ and ω are constants, there will be no acceleration. However, *there will be a nonzero acceleration* due to a changing orientation of the velocity vector **v**. Since we are dealing with vector quantities, defined by both magnitude and direction, for a vector *not* to change, it must be that neither the direction nor magnitude changes. In circular motion, though the magnitude of the position vector does *not* change, its direction does. Therefore, there is always a nonzero variation in **r**, and therefore, in **v**. The magnitude of this variation leads to nonzero acceleration. Thus, we have

$$\mathbf{a} = \frac{d\mathbf{v}}{dt} = \frac{d}{dt}(\mathbf{r}\omega) = \frac{d\mathbf{r}}{dt}\omega = \frac{d\mathbf{s}}{dt}\omega = \mathbf{v}\omega \tag{7.7}$$

The above can be used to obtain the resulting force responsible for the circular motion of the mass m. We know from Newton's second law that

$$|\mathbf{F}| = \left|\frac{d\mathbf{p}}{dt}\right| = m|\mathbf{a}| = mv\omega = mr\omega^2 = \frac{mv^2}{r} \tag{7.8}$$

This type of force is known as a **centripetal force**, and it is present in any system with orbital angular momentum. The force acts centrally, keeping the particle from ordinarily traveling along a straight path, tangentially.

With this, we may obtain an analog of $\mathbf{F} = md\mathbf{v}/dt$, for the orbital angular momentum **L**. Since $\mathbf{L} = \mathbf{r} \times \mathbf{p}$, taking the derivative of **L**, we have

$$\frac{d\mathbf{L}}{dt} = \frac{d\mathbf{r}}{dt} \times \mathbf{p} + \mathbf{r} \times \frac{d\mathbf{p}}{dt}$$

But, since we have already established that $\mathbf{v} = d\mathbf{s}/dt = d\mathbf{r}/dt$, and since $\mathbf{p} = m\mathbf{v}$, the first term is zero because it involves the cross-product of two parallel vectors ($\mathbf{v} \times \mathbf{v}$). Therefore, we have the result

$$\frac{d\mathbf{L}}{dt} = \mathbf{r} \times \frac{d\mathbf{p}}{dt} = \mathbf{r} \times \mathbf{F} = \boldsymbol{\tau} \tag{7.9}$$

From this result, we have that

$$\mathbf{F} = \frac{d\mathbf{p}}{dt} \Leftrightarrow \boldsymbol{\tau} = \frac{d\mathbf{L}}{dt} \tag{7.10}$$

Translational	Rotational
inertia m	$I = mr^2$
velocity v	ω
momentum p	L
kinetic energy $\frac{1}{2}mv^2$	$\frac{1}{2}I\omega^2$
force $\mathbf{F} = \frac{d\mathbf{p}}{dt}$	$\boldsymbol{\tau} = \frac{d\mathbf{L}}{dt}$

Table 7.1: Correspondence between parameters for translational and rotational motion.

That is, the time-derivative of linear momentum **p** is the force **F**, and analogously, the time-derivative of angular momentum **L** is the rotational force or torque $\boldsymbol{\tau}$, causing rotational motion. The correspondences between the linear and rotational motion that we have discussed so far, are summarized in Table 7.1. Let's continue the discussion of angular momentum with a specific application to the electron in circular motion in the presence of a magnetic field **B**.

7.2 CANONICAL PROBLEM OF THE MAGNETIC MOMENT

This section introduces the **canonical** (or basic, standard) **problem** of the magnetic moment. It provides the proper framework for some important concepts in the subject of *magnetism*. The first concept that is defined in this problem is the *magnetic moment*. It has its conceptual origins in classical *magnetostatics*. However, later, in Chap. 12, we'll extend these ideas to include electrodynamics, accounting for time-dependent changes. This ultimately leads to a more complete description, more in-line with the observed dissipated behavior of magnetic moments of electrons. In physical systems, the magnetic moment vectors associated with electrons align themselves with net magnetic fields, particularly, when they are not initially aligned with the magnetic moment.

Magnetic moment: the magnetic moment of an electron, when coupled to a magnetic field, tends to align with the magnetic field, opposite to the angular momentum vector, which anti-aligns with a magnetic field.

This is why the canonical problem is important, as it defines a *nonequilibrium initial condition*. This has to be treated by a dynamical description that will ultimately lead us to a Hamiltonian with such capability. Here, we'll restrict the discussion to an introduction of a magnetic moment and how it couples to a magnetic field. We shall show how this coupling arises, ignoring any dissipation or time-dependent phenomena, for simplicity (until Chap. 12). Using the *canonical problem of the magnetic moment*, we can lay the classical foundation for the behavior of a magnetic moment, denoted by $\boldsymbol{\mu}$, when introduced to a magnetic field $\mathbf{B}$.

When a particle of charge q moves with velocity v, in a uniform magnetic field $\mathbf{B}$, the field exerts a force on the moving charge. This empirical force is known as the *Lorentz force*, which can be derived more than one way. For example, it can be obtained from Faraday's law of electromagnetics, given by Eq. (1.23), or from a relativistic correction to the electric force $\mathbf{F}_e = q\mathbf{E}$. Both lead to the **Lorentz force** given by

$$\mathbf{F} = q\mathbf{v} \times \mathbf{B} \tag{7.11}$$

With Eq. (7.11), consider if, instead of a point charge q, the charge is distributed uniformly throughout a circular loop, with the distributed charge moving uniformly in the wire loop (e.g. an electrical current I_0), as illustrated in Fig. 7.2. The figure illustrates the *initial condition* of the magnetic moment $\boldsymbol{\mu}$ in a magnetic field. It is this condition that defines the **canonical magnetic moment problem**. With a distributed charge, we may speak of a differential force $\delta\mathbf{F}$ acting on an elemental length of the wire Δs, possessing a *differential* charge δq. Taking the proper limit $\Delta\mathbf{s} \to d\mathbf{s}$, $\delta\mathbf{F}$ is given by

$$\delta\mathbf{F} = \delta q \frac{d\mathbf{s}}{dt} \times \mathbf{B} = \frac{\delta q}{dt} d\mathbf{s} \times \mathbf{B} = I_0 d\mathbf{s} \times \mathbf{B} \tag{7.12}$$

In the top and bottom parts of the loop shown in Fig. 7.2, the currents move in opposite directions while the motion is perpendicular to the magnetic field $\mathbf{B}$, which has the same direction at the top and bottom of the loop. Meanwhile, the components of current parallel/anti-parallel to $\mathbf{B}$ (on the left and right)) do not contribute to the force. From this,

7.2 Canonical Problem Of The Magnetic Moment

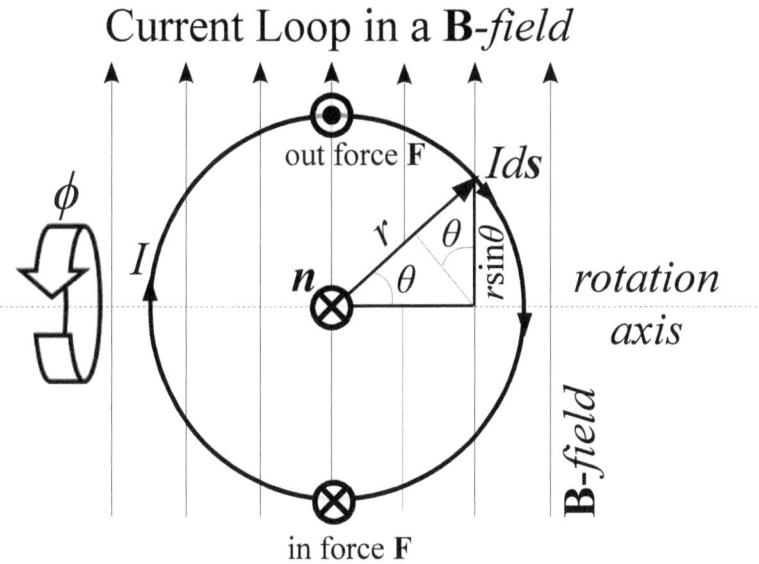

Figure 7.2: Illustration of current I_0 circulation in a closed loop, in a uniform magnetic field **B**. Because the loop contains charged particles (in motion), the *Lorentz force* will begin to act on the loop, imparting a torque τ or moment on the loop.

we can deduce that there will be *two* Lorentz forces acting on the loop, being focused at the top and bottom. One of these forces acts 'into the diagram', at the top of the loop, while the other acts 'out of the diagram' at the bottom of the loop. This constitutes a *magnetically-induced moment* or torque τ on the loop caused by the two equal and opposite forces at the top and bottom of the loop. Consequently, this will tend to cause the loop to want to rotate with an angle ϕ, about the rotation axis. We can write the differential torque $d\tau$ given by the sum of both force-moments acting on top and bottom $d\mathbf{s}$ portions as

$$d\tau = 2 \cdot r\sin\theta \times \delta F = 2r\sin\theta \times I_0 d\mathbf{s} \times \mathbf{B} \tag{7.13}$$

$r\sin\theta$ is the moment arm perpendicular to the axis of rotation, parallel to the magnetic field axis. Eq. (7.13) is a double cross product involving two distinct angles, namely θ (from the right-most product $d\mathbf{s} \times B$) and the angle of the loop rotation ϕ about the rotation axis (from the other cross product). Using the relation $ds = rd\theta$, $d\tau$ becomes

$$d\tau = 2r\sin\theta \times I_0 rd\theta \times \mathbf{B} = 2I_0 Br^2 \sin^2\theta \sin\phi d\theta \tag{7.14}$$

To obtain the total torque τ, we must integrate Eq. (7.14) with respect to θ from 0 to π (we only need to compute over one half the loop since we include the factor of 2 in the torque). This gives

$$\tau = \int_{\theta=0}^{\theta=\pi} d\tau = 2I_0 Br^2 \sin\phi \int_0^{\pi} \sin^2\theta d\theta \tag{7.15}$$

The integral of $\sin^2\theta$ is well-known and is given by $\theta/2 - \sin2\theta/4$, which, upon evaluating at the limits of 0 to π, gives the result $\int \sin^2\theta d\theta = \pi/2$. Therefore, Eq. (7.15) becomes

$$\tau = (\pi r^2) I_0 B \sin\phi \tag{7.16}$$

Since πr^2 is simply the orbital area A_o of the loop, we may also write Eq. (7.16) as

$$\tau = A_o I_0 B \sin\phi = I_0 (A_o \mathbf{n}) \times \mathbf{B} = I_0 \mathbf{A}_o \times \mathbf{B} \tag{7.17}$$

n is the unit vector normal to the orbital area A_o of the loop. Based on Eq. (7.17), we may define an **orbital magnetic moment** μ_L, having units of Amps × m² as

---
Classical Orbital Magnetic Moment

$$\mu_L = I_0 \mathbf{A}_o \tag{7.18}$$

A magnetic moment is also known as a *orbital magnetic dipole*. Therefore, using Eq. (7.18), the total torque τ becomes

---
Magnetostatic Torque On Orbital Magnetic Moment

$$\tau = \mu_L \times \mathbf{B} \tag{7.19}$$

$\mathbf{A}_o = A_o \mathbf{n}$ is a vector quantity with magnitude given by the orbital area $A_o = \pi r^2$. It points along the unit normal to the orbital area. The

7.2 Canonical Problem Of The Magnetic Moment

convention for the direction is by using the right-hand rule with the loop current I. Though we have obtained Eq. (7.18) for a circular loop, the final result is also applicable to other closed loop shapes.

Since the current loop has a torque acting on it from an external magnetic field **B**, and begins to rotate, there is an associated *magnetic potential energy* V_M of the loop, due to the magnetic field. This additional or variation in the energy ΔE_0 of the loop of the moving charge is given by $\int \tau d\phi$, which leads to

$$\Delta E = V_M = \int_{\phi_i}^{\phi_f} \tau d\phi = \int_{\phi_i}^{\phi_f} A_o I_0 B \sin\phi \, d\phi = -I_0 A_o B \cos\phi \Big|_{\pi/2}^{\phi} \quad (7.20)$$

Thus, we have the following result for the magnetic potential energy V_M:

Magnetic Potential Energy Of Magnetic Moment

$$V_M = -\boldsymbol{\mu}_L \cdot \mathbf{B} \quad (7.21)$$

The above result given by Eq. (7.20) uniquely defines the *initial condition* for the *canonical magnetic moment problem* to be

Defined Initial Condition Of Magnetic Moment

$$\phi_i = \frac{\pi}{2} \quad (7.22)$$

The initial condition turns out to be very important when considering dynamical processes, which we will discuss in Chap. 12. $\Delta E = V_M$ is known as the **magnetic potential energy** and it arises because charged particles in circular motion in a magnetic field **B** experience a torque. This particular concept of magnetic potential energy turns out to be widely used in the description of any magnetic moment μ. We shall not only have occasion to see these relations many times in later chapters, but we will also build on these ideas, later, in order to extend the contributions to the magnetic potential energy. Since the concept involves moving charged bodies orbiting in a loop, there must also be some angular momentum **L** associated with the loop. With both orbital angular momentum **L** and magnetic moment **μ**, we want to relate the two. This is done of the next section.

7.3 CLASSICAL ORBITAL MAGNETIC MOMENT

Since the current I is circulating around the loop of orbital area A_o, and $\boldsymbol{\mu}_L$ is pointing along the direction of the area normal unit vector **n**, we can expect that the magnetic moment is also proportional to the orbital angular momentum **L**, at any orientation relative to **B**. We may determine the constant of proportionality between the *orbital magnetic moment* $\boldsymbol{\mu}_L$ and the *orbital angular momentum* $\mathbf{L} = \mathbf{r}_o \times \mathbf{p}$, where $\mathbf{r} = \mathbf{r}_o$ is the orbiting radius. For a single electron orbiting a circular loop in a time period T, the linear velocity v is given by

$$\mathbf{v} = v\mathbf{e}_\varphi = \frac{2\pi r}{T}\mathbf{e}_\varphi \Rightarrow \frac{1}{T} = \frac{v}{2\pi r} \tag{7.23}$$

$\mathbf{e}_\varphi$ is the unit vector along the loop tangential direction. Using Eq. (7.23), the current I for a single electron charge is

$$\mathbf{I} = -\frac{e}{T}\mathbf{e}_\varphi \Rightarrow I = -\frac{ev}{2\pi r_o} \tag{7.24}$$

$e > 0$ is the *elementary charge* for the proton and electron. Eq. (7.24) tells us that the conventional current I is opposite in direction to the motion of the electron, described by **v**. Expressing the magnetic moment $\boldsymbol{\mu}_L = IA_o\mathbf{n}$, we then have

$$\mu_L = IA_o = -\frac{ev}{2\pi r_o}\pi r_o^2 = -\frac{evr_o}{2}$$

For the more general case that **v** is *not perpendicular* to $\mathbf{r}_o$, we may write

$$\boldsymbol{\mu}_L = I\mathbf{A}_o = -\frac{e}{2}\mathbf{r}_o \times \mathbf{v} \tag{7.25}$$

Only the component perpendicular to $\mathbf{r}_o$ contributes to the magnetic moment. Therefore, the magnetic moment becomes

$$\boldsymbol{\mu}_L = -\frac{e}{2}\mathbf{r}_o \times \mathbf{v} = -\frac{e}{2m_e}\mathbf{r}_o \times m_e\mathbf{v} \tag{7.26}$$

Using the definition of $\mathbf{L} = \mathbf{r}_o \times \mathbf{p}$, we have the following relation between $\boldsymbol{\mu}_L$ and **L**:

7.3 Classical Orbital Magnetic Moment

Orbital Magnetic Moment-Momentum Relation

$$\boldsymbol{\mu}_L = -\frac{e}{2m_e}\mathbf{L} \tag{7.27}$$

The constant of proportionality is typically denoted by γ, defined as

Single Electron Gyromagnetic Ratio

$$\gamma = -\frac{e}{2m_e} \tag{7.28}$$

It is known as the *single* **electron gyromagnetic ratio**. It is often defined as a positive quantity $\gamma = \frac{e}{2m_e}$, as well.

For an electron, in particular, we will show later in this chapter that the *quantum orbital angular momentum* is given by

$$L = |\mathbf{L}| = \sqrt{\ell(\ell+1)}\hbar$$

ℓ takes any *integer* value $\ell = 0, 1, 2, \ldots$ Since we know the magnetic moment $\boldsymbol{\mu}_L$ is proportional to $\mathbf{L}$, we should not be required to solve an additional eigenvalue equation for $\boldsymbol{\mu}_L$. In other words, because of this proportional relationship, knowing L is sufficient to know μ_L. Using Eq. (7.27), we may write the orbital magnetic moment as

$$\mu_L = |\boldsymbol{\mu}_L| = \sqrt{\ell(\ell+1)}\gamma\hbar \tag{7.29}$$

Just as $\sqrt{\ell(\ell+1)}$ can be regarded as counting how many $\hbar$ units are in the orbital angular momentum, we may identify the quantity $\gamma\hbar = e\hbar/2m_e$ as an analogous quantity for counting magnetic moment units. The constant $\gamma\hbar$ is known as the single electron **Bohr magneton** denoted by μ_B.

Single Electron Bohr Magneton

$$\mu_B = |\gamma|\hbar \tag{7.30}$$

From Eq. (7.29), μ_B can be regarded as a natural unit for counting magnetic moments, analogous to $\hbar$ for the orbital angular momentum L. Let's now examine how angular momentum and magnetic moment relate to an electron's potential energy.

These quantities relating to orbital angular momentum were originally introduced in the classical domain based on the electromagnetic equations. The classical frame provides a conceptual basis of their origins and some level of physical meaning. However, the classical discussion ends here. Next, we are going to take these concepts and apply them onto the domain of quantum mechanics. Thus, let us begin the quantum mechanics of orbital angular momentum, which is an *operator*.

7.4 Quantum Mechanical Angular Momentum $\widehat{L}$

We see that vector of orbital angular momentum **L** is a property of a rotating body. If there are multiple bodies, then the orbital angular momentum superposes for the total system, meaning that their vector quantities L_1, L_2, etc. add together. In the quantum mechanical framework, the rotation or orbit still has a radial vector of orbit **r** and an instantaneous linear momentum $\widehat{\mathbf{p}}$, however, now, they are quantum mechanical operators. Classically, we have seen that **L** is given by the vector product

$$\mathbf{L} = \mathbf{r} \times \mathbf{p} \tag{7.31}$$

Eq. (7.31) can be expressed as three scalar equations given by

$$L_x = yp_z - zp_y \tag{7.32}$$

$$L_y = zp_x - xp_z \tag{7.33}$$

$$L_z = xp_y - yp_x \tag{7.34}$$

In the quantum mechanical framework, the classical orbital angular momentum becomes a quantum mechanical operator when we replace classical p_i with the corresponding quantum mechanical momentum operators. We have already discussed **r** and $\widehat{\mathbf{p}}$ as quantum mechanical operators in Chap. 3. **r** is the same as in the classical variable of a

7.4 Quantum Mechanical Angular Momentum $\widehat{L}$

coordinate system. In Chap. 3, we found that the linear momentum vector operator $\widehat{\mathbf{p}} = \left[\widehat{p}_x, \widehat{p}_y, \widehat{p}_z\right]^T$ is given by the three operators

$$\widehat{p}_x = -i\hbar \frac{\partial}{\partial x} \tag{7.35a}$$

$$\widehat{p}_y = -i\hbar \frac{\partial}{\partial y} \tag{7.35b}$$

$$\widehat{p}_z = -i\hbar \frac{\partial}{\partial z} \tag{7.35c}$$

Using the quantum mechanical operators for the linear momentum, given by Eqs. (7.35a)-(7.35c), the orbital angular momentum operators become

$$\widehat{L}_x = -i\hbar \left(y \frac{\partial}{\partial z} - z \frac{\partial}{\partial y} \right) \tag{7.36a}$$

$$\widehat{L}_y = -i\hbar \left(z \frac{\partial}{\partial x} - x \frac{\partial}{\partial z} \right) \tag{7.36b}$$

$$\widehat{L}_z = -i\hbar \left(x \frac{\partial}{\partial y} - y \frac{\partial}{\partial x} \right) \tag{7.36c}$$

Just as with the energy operator known as the Hamiltonian $\widehat{H}\psi$, the angular momentum operators operate on their corresponding *wave functions* ψ, that satisfy their respective eigenvalue equation. In this case, having Hermitian operators, the eigenvalues correspond to the observable angular momentum L. Let us take the angular momentum component $\widehat{L}_z$, for which the eigenvalue equation is defined as

$$\widehat{L}_z \psi = \lambda_z \hbar \psi \tag{7.37}$$

Substitution of Eq. (7.36c) gives

$$-i\hbar \left(x \frac{\partial \psi}{\partial y} - y \frac{\partial \psi}{\partial x} \right) = \lambda_z \hbar \psi \tag{7.38}$$

Because orbital angular momentum involves circular motion, it is more convenient to transform the coordinate system from Cartesian (x,y,z) into polar coordinates (r,θ,φ), before solving the equations involving orbital angular momentum. Fig. 7.3 illustrates the relationship between the two coordinate systems that we will utilize. From Fig. 7.3, the rela-

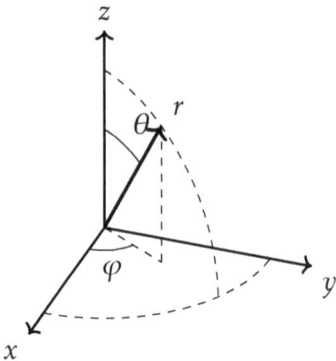

Figure 7.3: Relation between the Cartesian coordinates and polar coordinates, used in the orbital angular momentum problem.

tionships between the two coordinate systems is given by the following set of relations:

$$x = r\sin\theta\cos\varphi, \quad y = r\sin\theta\sin\varphi$$

$$z = r\cos\theta, \quad x^2 + y^2 + z^2 = r^2$$

The above equations also lead to the following relations that will be used:

$$\frac{y}{x} = \tan\varphi, \quad \text{and} \quad \frac{x^2+y^2}{z^2} = \tan^2\theta$$

These relations can be used to make substitutions into Eq. (7.38) to render it as an explicit equation of (r,θ,φ). We make substitutions as follows:

$$x\frac{\partial\psi}{\partial y} = r\sin\theta\cos\varphi\left(\frac{\partial\psi}{\partial r}\frac{\partial r}{\partial y} + \frac{\partial\psi}{\partial\theta}\frac{\partial\theta}{\partial y} + \frac{\partial\psi}{\partial\varphi}\frac{\partial\varphi}{\partial y}\right) \tag{7.39a}$$

and

$$y\frac{\partial\psi}{\partial x} = r\sin\theta\sin\varphi\left(\frac{\partial\psi}{\partial r}\frac{\partial r}{\partial x} + \frac{\partial\psi}{\partial\theta}\frac{\partial\theta}{\partial x} + \frac{\partial\psi}{\partial\varphi}\frac{\partial\varphi}{\partial x}\right) \tag{7.39b}$$

From Eqs. (7.39), we can see that all terms involving x, y, and z must be replaced with a function of (r,θ,φ). This means the six terms $\partial r/\partial y$, $\partial\theta/\partial y$, $\partial\varphi/\partial y$, $\partial r/\partial x$, $\partial\theta/\partial x$, and $\partial\varphi/\partial x$ must be replaced. Using our relations above, the following relations can be used in substitution:

$$\frac{\partial r}{\partial x} = \frac{x}{r} = \sin\theta\cos\varphi, \quad \frac{\partial r}{\partial y} = \frac{y}{r} = \sin\theta\sin\varphi, \quad \frac{\partial\theta}{\partial x} = \frac{\cos\theta\cos\varphi}{r}$$

$$\tag{7.40a}$$

7.4 Quantum Mechanical Angular Momentum $\widehat{L}$

$$\frac{\partial \theta}{\partial y} = \frac{\cos\theta \sin\varphi}{r}, \quad \frac{\partial \varphi}{\partial x} = -\frac{1}{r}\frac{\sin\varphi}{\sin\theta}, \quad \frac{\partial \varphi}{\partial y} = \frac{1}{r}\frac{\cos\varphi}{\sin\theta} \qquad (7.40\text{b})$$

After substitution of Eqs. (7.39) and Eq. (7.40) into Eq. (7.36), we obtain the result

$\widehat{L}_z$ Operator

$$\widehat{L}_z \psi = -i\hbar \frac{\partial \psi}{\partial \varphi} \qquad (7.41)$$

Therefore, the *eigenvalue equation* $\widehat{L}_z = \lambda_z \hbar \psi$ for $\widehat{L}_z$ becomes

$\widehat{L}_z$ Eigenvalue Equation

$$-i\hbar \frac{\partial \psi}{\partial \varphi} = \lambda_z \hbar \psi \qquad (7.42)$$

It is straightforward to check that the solution to the eigenvalue equation for $\widehat{L}_z$ is given by

$$\psi = C_n e^{i\lambda_z \varphi} \qquad (7.43)$$

C_n is the normalization constant, independent of φ. The limits of φ range from 0 to 2π, and therefore, in order for the wave function ψ to be continuous at these limits, any number of complete revolutions must repeat itself at the limits. For this condition to be satisfied, it must be that

$$C_n e^{i\lambda_z \varphi} = C_n e^{i\lambda_z(\varphi+2\pi)} \qquad (7.44)$$

This means that

$$e^{i2\pi\lambda_z} = 1 \qquad (7.45)$$

Considering Euler's theorem $e^x = \cos x + i\sin x$, Eq. (7.45) tells us that $e^{i2\pi\lambda_z} = \cos(2\pi\lambda_z) + i\sin(2\pi\lambda_z) = 1$. But 1 being a real number, means it can only be true if $\sin(2\pi\lambda_z) = 0$. For this to be satisfied, λ_z can only be an integer $m = -2, -1, 0, 1, 2$, etc. The eigenfunction for L_z, denoted by $\psi = \Phi(\varphi)$, is given by

$$\Phi(\varphi) = C_n e^{im\varphi} \quad \text{where} \quad m = ..., -2, -1, 0, 1, 2, ...$$

The normalization constant C_n is found from the *normalization condition* $\int \Phi^*\Phi d\varphi = 1$, which leads to

$$1 = \int_0^{2\pi} C_n e^{im\varphi} C_n^* e^{-im\varphi} d\varphi = |C_n|^2 \int_0^{2\pi} d\varphi \Rightarrow |C_n| = \frac{1}{\sqrt{2\pi}}$$

Using the normalization condition prohibits an exact knowledge of C_n within a phase factor $e^{i\theta}$. This is because we solve for $|C_n|^2$. This sort of result is said to be *within a phase-factor*. Any complex number having the same amplitude, with arbitrary phase, still satisfies our normalization condition. The normalized eigenfunction Φ for the z component of the angular momentum is then given by

Normalized Eigenfunction Of $\widehat{L}_z$ Operator

$$\Phi(\varphi) = \frac{1}{\sqrt{2\pi}} e^{im\varphi} \qquad m = \cdots, -3, -2, -1, 0, 1, 2, 3, \cdots \qquad (7.46)$$

It turns out that the eigenfunction for $\widehat{L}_z$ is not a function of r nor θ. This made the analysis simpler. However, the other two components of angular momentum are not strict functions of φ. In a similar manner as we've done for $\widehat{L}_z$, the other components of the orbital angular momentum can also be determined. Substitution of Eqs. (7.39) and Eq. (7.40) into Eq. (7.36), we obtain an operator $\widehat{L}_x$ given by

$\widehat{L}_x$ Operator

$$\widehat{L}_x = i\hbar \left(\sin\varphi \frac{\partial}{\partial \theta} + \cot\theta \cos\varphi \frac{\partial}{\partial \varphi} \right) \qquad (7.47)$$

And for the $\widehat{L}_y$ component, we have

$\widehat{L}_y$ Operator

$$\widehat{L}_y = i\hbar \left(-\cos\varphi \frac{\partial}{\partial \theta} + \cot\theta \sin\varphi \frac{\partial}{\partial \varphi} \right) \qquad (7.48)$$

7.4 Quantum Mechanical Angular Momentum $\widehat{L}$

Having $\widehat{L}_x$, $\widehat{L}_y$, and $\widehat{L}_z$ enables us to express any equation that depends on $\widehat{L}$. Since kinetic energy involves the square of the momentum (i.e. $\widehat{p}^2$ or $\widehat{L}^2$), we can also find the operator for $\widehat{L}^2$. As we saw in Chap. 3 with the operator $\widehat{p}^2$, the square of an operator in quantum mechanics means that the operator is applied twice or consecutively to the object on the right of it. It does not mean multiplying something by itself, as it does with scalars. Therefore, $\widehat{L}^2$ is given by

$$\widehat{L}^2 = \widehat{L}_x^2 + \widehat{L}_y^2 + \widehat{L}_z^2 \tag{7.49}$$

From Eq. (7.42), the operator for L_z^2 is given by

$$\widehat{L}_z^2 \psi = \widehat{L}_z(\widehat{L}_z \psi) = -i\hbar \frac{\partial}{\partial \varphi}\left(-i\hbar \frac{\partial \psi}{\partial \varphi}\right)$$

Therefore, we have the following eigenvalue equation $\widehat{L}_z^2 = \lambda \psi$, which becomes

L_z^2 Operator Eigenvalue Equation

$$-\hbar^2 \frac{\partial^2 \psi}{\partial \varphi^2} = \lambda \hbar^2 \psi \tag{7.50}$$

It is a matter of convenience to include the factor of $\hbar^2$, or not on the RHS of Eq. (7.50). Including it leads to dimensionless eigenvalues that are pure integer functions. Eqs. (7.47) and (7.47) allow us to write $\widehat{L}_x^2$ and $\widehat{L}_y^2$ as

$$\widehat{L}_x^2 \psi = -\hbar^2 \left(\sin\varphi \frac{\partial}{\partial \theta} + \cot\theta \cos\varphi \frac{\partial}{\partial \varphi}\right)\left(\sin\varphi \frac{\partial}{\partial \theta} + \cot\theta \cos\varphi \frac{\partial}{\partial \varphi}\right)\psi \tag{7.51a}$$

$$\widehat{L}_y^2 \psi = -\hbar^2 \left(-\cos\varphi \frac{\partial}{\partial \theta} + \cot\theta \sin\varphi \frac{\partial}{\partial \varphi}\right)\left(-\cos\varphi \frac{\partial}{\partial \theta} + \cot\theta \sin\varphi \frac{\partial}{\partial \varphi}\right)\psi \tag{7.51b}$$

Eqs. (7.49), (7.51), (7.51b), and (7.50), can be used to write the operator $\widehat{L}^2$, which works out to be

$$\widehat{L}^2 \psi = -\hbar^2 \left[\frac{1}{\sin\theta}\frac{\partial}{\partial \theta}\left(\sin\theta \frac{\partial \psi}{\partial \theta}\right) + \frac{1}{\sin^2\theta}\frac{\partial^2 \psi}{\partial \varphi^2}\right] \tag{7.52}$$

Eq. (7.52) defines the operator for $\widehat{L}^2$. Thus, we can solve the corresponding eigenvalue equation for $\widehat{L}^2$ by solving the eigenvalue equation given by

$$\widehat{L}^2 \psi = -\hbar^2 \left[\frac{1}{\sin\theta} \frac{\partial}{\partial\theta} \left(\sin\theta \frac{\partial\psi}{\partial\theta} \right) + \frac{1}{\sin^2\theta} \frac{\partial^2\psi}{\partial\varphi^2} \right] = \lambda \hbar^2 \psi$$

or, after factoring $\hbar^2$, we have

$\widehat{L}^2$ Operator Eigenvalue Equation

$$L^2\psi = -\left[\frac{1}{\sin\theta} \frac{\partial}{\partial\theta} \left(\sin\theta \frac{\partial\psi}{\partial\theta} \right) + \frac{1}{\sin^2\theta} \frac{\partial^2\psi}{\partial\varphi^2} \right] = \lambda \psi \quad (7.53)$$

We must remember, with this form, the resulting eigenvalues must be scaled by $\hbar^2$. Before we get to the solution of Eq. (7.53), which entails finding both the eigenvalues λ and the associated eigenfunctions or wave functions ψ, let us first discuss a relevant historical topic that provides some insight into the origin of the solutions for the $\widehat{L}^2$ operator, namely, the *Laplacian equation* and *Legendre polynomials*. This is the subject of the next section.

7.5 THE LAPLACIAN AND LEGENDRE EQUATIONS

It turns out that Eq. (7.53) has been seen before its appearance in the quantum mechanical orbital angular momentum operator $\widehat{L}^2$. The part if brackets is a part of the 3D Laplacian (∇^2) in spherical coordinates, given by

$$\nabla^2 \phi = \frac{1}{r^2} \left[\frac{\partial}{\partial r} \left(r^2 \frac{\partial}{\partial r} \right) + \frac{1}{\sin\theta} \frac{\partial}{\partial\theta} \left(\sin\theta \frac{\partial}{\partial\theta} \right) + \frac{1}{\sin^2\theta} \frac{\partial^2}{\partial\varphi^2} \right] \phi \quad (7.54)$$

This relation generally describes a 3D potential ϕ (e.g. gravitational, electrical) of the form

$$\phi(r) = \frac{a}{|\mathbf{r} - \mathbf{r}_0|}$$

a is a constant independent of position. Whenever there is electrical *charge balance*, given by equal numbers of positive and negative charges

7.5 The Laplacian and Legendre Equations

so total charge $Q = 0$ (e.g. a dipole, quadrupole, octupole, etc.), anywhere in the distant space *outside* the body of charges, the Laplacian describes the potential by the equation $\nabla^2 \phi = 0$. Since ϕ only depends on the distance between the position of interest (**r**) and the source (**r**$_0$), by choosing the origin of the coordinate system conveniently, the potential function can be transformed to the following form

$$\phi(\mathbf{r}) = \frac{a}{|\mathbf{r} - \mathbf{r}_0|} = \frac{1}{|r - r\cos\theta|} = \frac{1}{r} \frac{a}{|1 - \cos\theta|} = f(r)g(\cos\theta) \quad (7.55)$$

θ is the inclusive angle between the vectors **r** and **r**$_0$. In mathematics, there is an infinite series known as the **geometric series** given by

$$g(\cos\theta) = g(\rho) = \sum_{n=0}^{\infty} a\rho^n = \frac{a}{1-\rho}$$

The name of this series originates from the early mathematical idea of using known geometrical shapes to obtain integrals or areas of mathematical functions such as parabolas. The idea was expressed as early as 250 B.C. by Greek mathematician Archimedes of Syracuse (287 B.C.-212 B.C.), who studied in Egypt. Each term in the series can be represented as a factor times the previous term. $g(\cos\theta)$ in Eq. (7.55) is in the form of a *geometric series*, with an interval of convergence $\rho = \cos\theta \in (-1, 1)$. This implies that the potential function can be represented by an infinite series.

While investigating gravitational potential, French mathematician Adrien-Marie Legendre (1752-1833) found that such a potential can, indeed, be represented by a series of the form

$$\phi(r, \cos\theta) = 1 + rP_1(\cos\theta) + r^2 P_2(\cos\theta) + r^3 P_3(\cos\theta) + \ldots = \sum_{n=0}^{\infty} r^n P_n(\cos\theta) \quad (7.56)$$

P_n is a polynomial of order n, where $n = 0, 1, 2, 3, \ldots$, whose argument is $\cos\theta$. These polynomials are known as **Legendre polynomials**. It turns out that an equation describing only the polynomials P_n can be obtained by plugging the series form proposed by Legendre into Eq. (7.54), and carrying out some calculus. To do this, we'll consider the three parts of the Laplacian separately, then sum the results from these three parts together.

The radial part (the first term) of the Laplacian gives

$$\frac{\partial}{\partial r}\left(r^2 \frac{\partial}{\partial r}\right) \sum_{n=0}^{\infty} r^n P_n(\cos\theta) = \sum_{n=0}^{\infty} n(n+1) r^n P_n(\cos\theta) \quad \text{(radial part)}$$

Note that since the Laplacian equation is equal to zero on the RHS, we may ignore the factor $1/r^2$, for now. The second term of the Laplacian then leads to

$$\frac{1}{\sin\theta} \frac{\partial}{\partial \theta}\left(\sin\theta \frac{\partial}{\partial \theta}\right) \sum_{n=0}^{\infty} r^n P_n(\cos\theta) =$$

$$-2\cos\theta \sum_{n=0}^{\infty} r^n \frac{dP_n}{d(\cos\theta)} + \sin^2\theta \sum_{n=0}^{\infty} r^n \frac{d^2 P_n}{d(\cos\theta)^2}$$

Because the series is a function of r and θ only, and does not depend on φ, all derivatives w.r.t. φ are 0. Thus, the last term in the Laplacian is 0. Putting the results together, we have that the Laplacian $\nabla^2 \phi$ is given by

$$\nabla^2 \phi = \sum_{n=0}^{\infty} n(n+1) r^n P_n(\cos\theta) - 2\cos\theta \sum_{n=0}^{\infty} r^n \frac{dP_n}{d(\cos\theta)} +$$

$$\sin^2\theta \sum_{n=0}^{\infty} r^n \frac{d^2 P_n}{d(\cos\theta)^2} = 0$$

Since the RHS is zero, and our equation must hold for any value of r, each coefficient of r^n in the series must also be zero. So we have that for all $n = 0, 1, 2, \ldots$,

$$n(n+1) P_n(\cos\theta) - 2\cos\theta \frac{dP_n}{d(\cos\theta)} + \sin^2\theta \frac{d^2 P_n}{d(\cos\theta)^2} = 0 \quad (7.57)$$

Letting $x = \cos\theta$, Eq. (7.57) becomes

Legendre Differential Equation

$$(1-x^2) \frac{d^2 P_n}{dx^2} - 2x \frac{dP_n}{dx} + n(n+1) P_n(x) = 0 \quad (7.58)$$

Eq. (7.58) is known as the **Legendre differential equation**, and it is the defining equation for the Legendre polynomials P_n. Therefore, the

7.5 The Laplacian and Legendre Equations

infinite series proposed by Legendre satisfies the Laplace equation if and only if P_n satisfies Eq. (7.58). The question remains as to what are the solutions for the Legendre polynomials P_n? One clever way to answer this question is to use a mathematical trick. Consider the function

$$f_n = (x^2 - 1)^n \tag{7.59}$$

We, first, want to generate a second order differential equation for f_n. Starting with differentiating Eq. (7.59) once, we obtain

$$\frac{df_n}{dx} = n(x^2 - 1)^{n-1}(2x)$$

Multiplying both sides by $(x^2 - 1)$, and rearranging, we then have

$$(1 - x^2)\frac{df_n}{dx} + 2xnf_n = 0$$

Differentiating once again for a second order differential equation, we obtain

$$(1 - x^2)\frac{d^2 f_n}{dx^2} + 2x(n-1)\frac{df_n}{dx} + 2nf_n = 0 \tag{7.60}$$

Though we have obtained a second order differential equation with f_n, we are not done yet. As a final step, let us continue to differentiate Eq. (7.60) n times. The result will be a differential equation for the $(n+2)^{nd}$ derivative of f_n. This is done using the generalized product rule for differentiation, also known as the **Leibniz rule**, which states that for the product of two n-times differentiable functions $y_1(x)$ and $y_2(x)$

Leibniz' Rule of Differentiation

$$\frac{d^n}{dx^n}(y_1 \cdot y_2) = \sum_{k=0}^{n} \frac{n!}{k!(n-k)!} \frac{d^k y_1}{dx^k} \frac{d^{n-k} y_2}{dx^{n-k}} \tag{7.61}$$

The formula may look complicated, but fortunately, it turns out to be fairly simple and straightforward. Let's apply the Liebniz' rule to Eq. (7.60), term by term. Applying Leibniz' rule to the first term in Eq. (7.60), we obtain

$$\text{first term} \rightarrow f_n^{n+2} - 2xnf_n^{n+1} - n(n-1)f_n^n$$

The superscripts $n+2$, $n+1$, and n denote the $n+2$, $n+1$, and n derivatives of f_n, respectively. Applying to the second term, we get

$$\text{second term} \rightarrow 2x(n-1)f_n^{n+1} + 2n(n-1)f_n^n$$

And for the last term, we simply have

$$\text{third term} \rightarrow 2xnf_n^n$$

Adding all three results together, we have

$$(1-x^2)f_n^{n+2} - 2xf_n^{n+1} + n(n+1)f_n^n = 0$$

We are getting close! Let us define another function $D_n(x)$ as the nth derivative of f_n or

$$D_n(x) \equiv \frac{d^n}{dx^n}(x^2-1)^n = f_n^n \tag{7.62}$$

Then, the differential equation for $D_n(x)$ becomes

$$(1-x^2)\frac{d^2 D_n}{dx^2} - 2x\frac{dD_n}{dx} + n(n+1)D_n = 0 \tag{7.63}$$

Eqs. (7.63) and Eq. (7.58) are identical, which means that D_n satisfies the differential equation for P_n, and therefore, any function proportional to D_n is a solution to Eq. (7.58), or

$$P_n = C_0 \frac{d^n}{dx^n}(x^2-1)^n \tag{7.64}$$

The convention for expressing Legendre polynomials P_n is to write P_n such that $P_n(1) = 1$. In evaluating $d^n P_n/dx$, it will be a sum of terms where all terms but one have a factor $x^2 - 1$. So, when $x = 1$, they all vanish and our condition becomes

$$C_0 n! 2^n x^n + 0 = 1 \rightarrow C_0 = \frac{1}{2^n n!} \tag{7.65}$$

Therefore, the resulting Legendre polynomials that satisfy the condition $P_n(1) = 1$ are defined by

Rodrigues' Formula For Legendre Polynomials

$$P_n = \frac{1}{2^n n!} \frac{d^n}{dx^n}(x^2-1)^n \tag{7.66}$$

7.5 The Laplacian and Legendre Equations

Eq. (7.66) is known as the **Rodrigues' formula**, first given in 1816 by the French mathematician Benjamin Olinde Rodrigues (1795-1851). It is the exact solution for the Legendre polynomials, and has been known for over 200 years! Table 7.2 lists the first few Legendre polynomials.

Table 7.2: First ten Legendre polynomials P_n. The Legendre polynomials arise as factors in the solution to the orbital angular momentum problem, and are part of the infinite series solution to the Laplace equation. They can be obtained directly and iteratively from the Rodriques formula.

Name	Equation
$n = 0$	$P_0(x) = 1$
$n = 1$	$P_1(x) = x$
$n = 2$	$P_2(x) = \frac{1}{2}(3x^2 - 1)$
$n = 3$	$P_3(x) = \frac{1}{2}(5x^3 - 3x)$
$n = 4$	$P_4(x) = \frac{1}{8}(35x^4 - 30x^2 + 3)$
$n = 5$	$P_5(x) = \frac{1}{8}(63x^5 - 70x^3 + 15x)$
$n = 6$	$P_6(x) = \frac{1}{16}(231x^6 - 315x^4 + 105x^2 - 5)$
$n = 7$	$P_7(x) = \frac{1}{16}(429x^7 - 693x^5 + 315x^3 - 35x)$
$n = 8$	$P_8(x) = \frac{1}{128}(6435x^8 - 12012x^6 + 6930x^4 - 1260x^2 + 35)$
$n = 9$	$P_9(x) = \frac{1}{128}(12155x^9 - 25740x^7 + 18018x^5 - 4620x^3 + 315x)$

A corresponding plot of the first ten Legendre polynomials is shown in Fig. 7.4.

Figure 7.4: First ten Legendre polynomials P_n, plotted over the range of interest $(-1, 1)$.

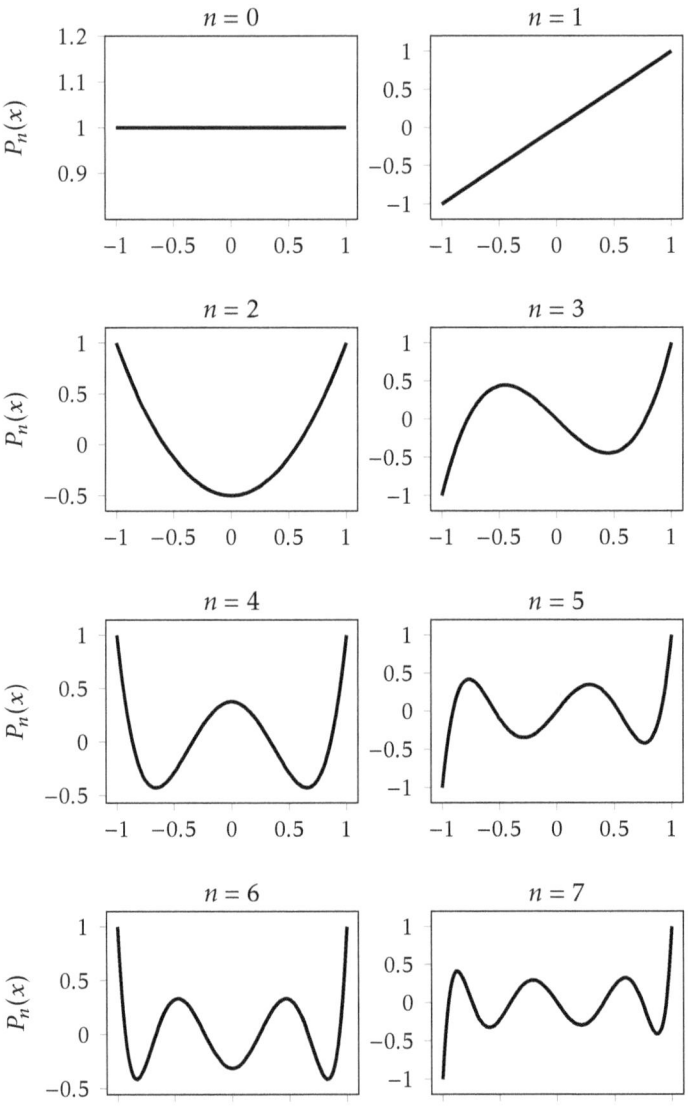

7.6 Extending To Associated Legendre Functions

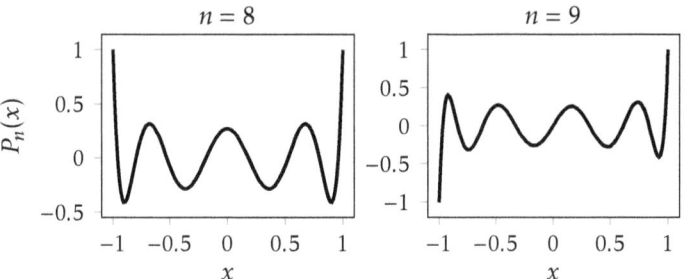

Because of the symmetry of the potential function, consecutive values of n alternate between anti-symmetric and symmetric states. Let's extend these ideas a bit further to obtain a result that will be directly applicable to the orbital angular momentum problem.

7.6 EXTENDING TO ASSOCIATED LEGENDRE FUNCTIONS

So we have discussed from whence the Legendre polynomials originate, and we have also seen that, through a clever choice of functions, we were able to use Leibniz' rule to obtain the exact solutions for P_n, the *Legendre polynomials*. Taking this idea further, lets see what happens when we apply Leibniz' rule to Eq. (7.53), differentiating m additional times. After carrying this out for all three terms as we did for the Legendre polynomials, then adding the results, we obtain

$$(1-x^2)\frac{d^2 P_n^m}{dx^2} - 2x(m+1)\frac{dP_n^m}{dx} + (n(n+1) - m(m+1))P_n^m = 0 \quad (7.67)$$

P_n^m now represents the mth derivative of P_n. We now have a second order differential equation describing P_n^m. Using yet another mathematical trick, let us define a function Q by the following:

$$P_n^m = (1-x^2)^{-\frac{m}{2}} Q \quad \text{or} \quad Q = (1-x^2)^{\frac{m}{2}} P_n^m \quad (7.68)$$

After substitution of Eq. (7.68) into Eq. (7.67), multiplying the result by $(1-x^2)^{m/2}$, and after some algebra, we obtain the following second order differential equation for $Q(x)$.

$$(1-x^2)\frac{d^2 Q}{dx^2} - 2x\frac{dQ}{dx} + \left[n(n+1) - \frac{m^2}{1-x^2}\right]Q = 0 \quad (7.69)$$

Using the definition for Q, given by Eq. (7.68), and using Rodrigues' formula for P_n, the function for Q, is given by

> **Associated Legendre Functions**
>
> $$A_n^m = \frac{C_n}{2^n n!}(1-x^2)^{\frac{m}{2}}\frac{d^{n+m}}{dx^{n+m}}(x^2-1)^n \qquad (7.70)$$

Eq. (7.70) defines the **Associated Legendre Functions**, and they are closely *associated* with the Legendre polynomials P_n, however, their are significant differences. Just as with P_n, since $x = \cos\theta$, n, and m are all real numbers, A_n^m is a real-valued function. Among the differences between P_n and A_n^m, the *Associated Legendre functions* are only polynomials when m is an even integer. This condition leads to $(1-x^2)^{m/2}$ becoming a polynomial, multiplied by another polynomial. For our interests, the normalization condition is used to find the normalization constant C_n. To evaluate $\int (A_n^m)^2 dx = 1$, we can use a well-known mathematical property of associated Legendre functions, namely,

$$\int_{-1}^{1} A_n^m \cdot A_k^m \, dx = \frac{2(n+m)!}{(2n+1)(n-m)!}\delta_{n,k} \qquad (7.71)$$

$\delta_{n,k}$ is the Kronecker delta function which is defined to be equal to 1 only if $n = k$, and 0 otherwise. Using this property, the normalization factor C_n for the Associated Legendre function is found to be

$$C_n = \sqrt{\frac{(2n+1)[(n-m)!]}{2(n+m)!}} \qquad (7.72)$$

Then, we can write the explicit normalized associated Legendre function as

> **Normalized Associated Legendre Functions**
>
> $$A_n^m = \frac{1}{2^n n!}\sqrt{\frac{(2n+1)[(n-m)!]}{2(n+m)!}}(1-x^2)^{\frac{m}{2}}\frac{d^{n+m}}{dx^{n+m}}(x^2-1)^n \qquad (7.73)$$

We should point out that we have departed from the usual convention in the notation for the associated Legendre functions. The common convention is P_n^m (rather than A_n^m, as we have used), but we intentionally avoid this notation here because we prefer to use P_n^m to denote the mth

7.6 Extending To Associated Legendre Functions

derivative of the Legendre polynomial P_n, which is also a polynomial. A_n^m, generally, is not a polynomial.

Eq. (7.70) implicitly dictates the allowed values of m, that ensures we always have a non-zero function. If these are to be useful as a basis set like Legendre polynomials, they cannot be zero-functions. They must be a constant (zeroth order) and/or higher order. A_n^m is the product of two terms, a explicit polynomial-*like* term and a pure polynomial derivative term. For A_n^m to result in a non-zero function, it is dictated by the derivative term on the RHS that the highest value m can take is that in which the $(n+m)^{th}$ derivative yields a non-zero constant, otherwise, the product vanishes. We must consider all possible values of m, being $m > 0$, $m < 0$, and $m = 0$. If $m > 0$, since $(x^2 - 1)^n$ is a polynomial of degree $2n$, we therefore have the condition that $n + m \leq 2n$ or $m \leq n$. If $m < 0$, $n - |m| \geq 0$, which gives $m \geq -n$ in order to ensure a zeroth derivative or higher. Thus, the general relation between the *degree* n and the *order* m, becomes

Relation Between Degree n and Order m Integers

$$|m| \leq n \qquad (7.74)$$

So, $m = -n, -n+1, ..., -1, 0, 1, ..., n-1, n$, and any other values beyond this range lead to a trivial solution, or equals zero. When $m = 0$, we regain the Legendre polynomials P_n. As examples for $m = 1$, Table 7.4 lists some Associated Legendre functions corresponding to $1 \leq n \leq 9$.

Table 7.4: Associated Legendre functions A_n^m for $m = 1$. The Associated Legendre functions are factors in the solution to the orbital angular momentum eigen-value problem.

n	Some Associated Legendre Functions $A_n^1(x)$ ($m=1, n \leq 9$)
1	$A_1^1 = (1 - x^2)^{1/2}$
2	$A_2^1 = 3x(1 - x^2)^{1/2}$
3	$A_3^1 = \frac{1}{2}(15x^2 - 3)(1 - x^2)^{1/2}$

Table 7.4

4 $A_4^1 = \frac{1}{2^3}(140x^3 - 60x)(1-x^2)^{1/2}$

5 $A_5^1 = \frac{1}{2^3}(315x^4 - 210x^2 + 15)(1-x^2)^{1/2}$

6 $A_6^1 = \frac{1}{2^4}(1386x^5 - 1260x^3 + 210x)(1-x^2)^{1/2}$

7 $A_7^1 = \frac{1}{2^4}(3003x^6 - 3465x^4 + 945x^2 - 35)(1-x^2)^{1/2}$

8 $A_8^1 = \frac{1}{2^7}(51480x^7 - 72072x^5 + 27720x^3 - 2520x)(1-x^2)^{1/2}$

9 $A_9^1 = \frac{1}{2^7}(109395x^8 - 180180x^6 + 90090x^4 - 13860x^2 + 315)(1-x^2)^{1/2}$

Fig. 7.5 plots the same associated Legendre functions. The reader should compare these curves to those of the Legendre polynomials in Fig. 7.4

Figure 7.5: First few normalized Associated Legendre functions $A_n^m(x)$ for $m = 1$ ($= Q_n^1$), plotted over the range of interest $(-1, 1)$. The functions sequence through symmetric and anti-symmetric functions, corresponding to symmetric and anti-symmetric states. Associated Legendre functions tend to vanish at the limits of $x = \cos\theta$.

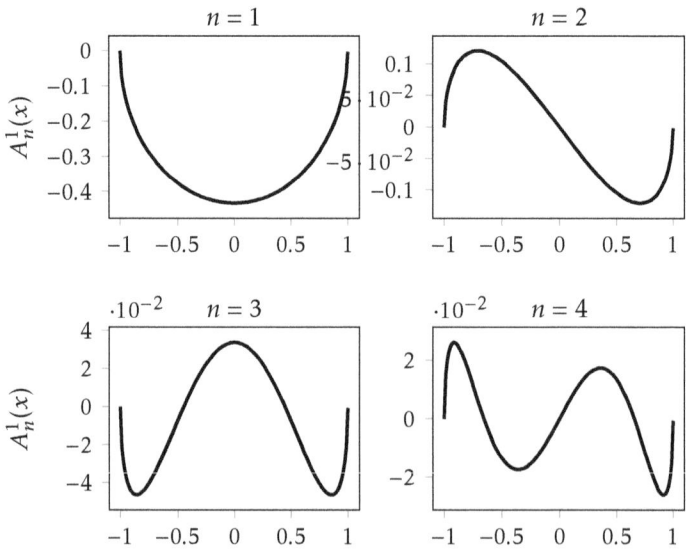

7.6 Extending To Associated Legendre Functions

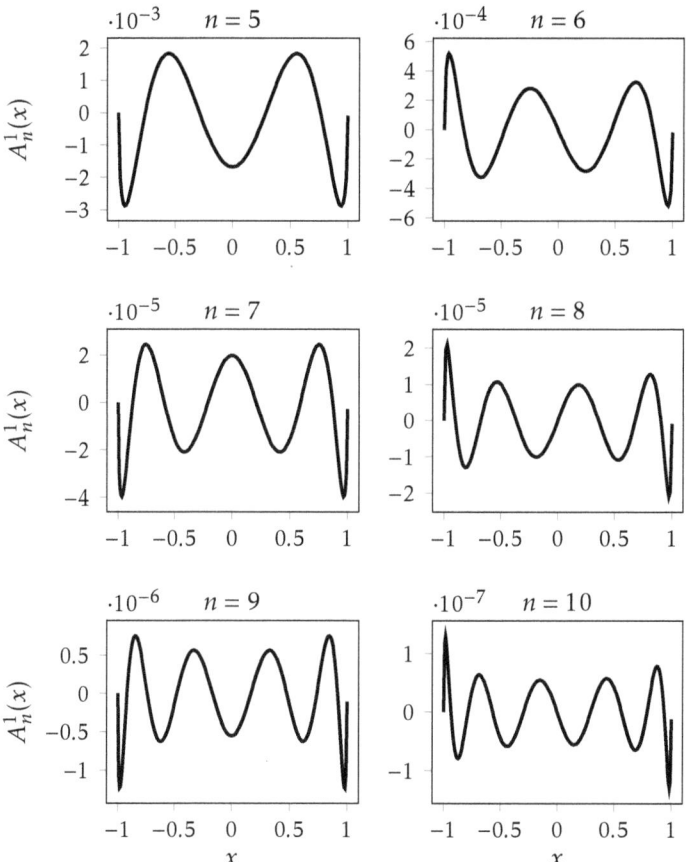

The functions sequence through symmetric and anti-symmetric functions with respect to the origin $x = 0$. We will see later that this characteristic will be inherited by the eigenfunctions for orbital angular momentum, corresponding to symmetric and anti-symmetric states. Unlike the Legendre polynomials, the Associated Legendre functions vanish at the limits of $x = \cos\theta$. The equations and solutions we have been dealing with have relevance to the orbital angular momentum problem we are considering in this chapter. We will see that the Associated Legendre functions emerge, once again, in solving Eq. (7.53).

7.7 Assoc. Legendre Functions & Orb. Angular Momentum

Now that we have discussed some background of the Legendre and Associated Legendre functions, let us return to the angular momentum problem we defined by Eq. (7.53). This equation is separable. We can always try this and if, after substitution of a separable function, we successfully obtain separated equations of independent variables that we can solve, then the method works. Thus, let us assume $\psi = Y_{\theta,\varphi} = F(\theta)\Phi(\varphi)$. Substitution into Eq. (7.53), multiplying by $\sin^2\theta$, then dividing by $Y_{\theta,\varphi}$, we obtain

$$\frac{\sin\theta}{F}\frac{d}{d\theta}\left(\sin\theta\frac{dF}{d\theta}\right) + \lambda\sin^2\theta = -\frac{1}{\Phi}\frac{d^2\Phi}{d\varphi^2} \tag{7.75}$$

The LHS is purely a function of θ while the RHS is purely a function of φ. As we have seen previously, when using the method of separation of variables, since both variables θ and φ are independent variables, it follows that both sides of Eq. (7.75) must be equal to a constant. Remember that we have dropped the $\hbar^2$. So, we set this constant equal to a perfect square m^2. When using the method of separation of variables to solve differential equations, we choose, for first order equations, equating to a constant to the first power, and for second order differential equations, equating to a constant to the second power, etc. So, we have

$$\frac{\sin\theta}{F}\frac{d}{d\theta}\left(\sin\theta\frac{dF}{d\theta}\right) + \lambda\sin^2\theta = m^2 \tag{7.76a}$$

$$-\frac{1}{\Phi}\frac{d^2\Phi}{d\varphi^2} = m^2 \tag{7.76b}$$

Eq. (7.76b) is the second order version of Eq. (7.42), describing L_z. The solution is given by Eq. (7.46). It can be shown that any function satisfying Eq. (7.42) must also satisfy Eq. (7.76b). This part of the equation for L^2 corresponds to contributions from L_z^2. We also know that m^2 is the square of an integer $-2, -1, 0, 1, 2, 3$, etc. We therefore only need to determine λ, as well as the function $F(\theta)$. Before we obtain the solution to Eq. (7.76a), let us transform the equation using

$$\cos\theta = x \rightarrow dx = -\sin\theta d\theta$$

Multiplying Eq. (7.76a) by $F/\sin^2\theta$, Eq. (7.76a) becomes

$$\frac{1}{\sin\theta}\frac{d}{d\theta}\left(\frac{\sin^2\theta}{\sin\theta}\frac{dF}{d\theta}\right) + \lambda F - \frac{m^2}{\sin^2\theta}F = 0 \tag{7.77}$$

7.7 Assoc. Legendre Funcs & Orb. Ang. Momentum

Making use of the transformation relations for x and dx, given above, we then have

$$\frac{d}{dx}\left[(1-x^2)\frac{dF}{dx}\right] + \lambda F - \frac{m^2}{1-x^2} = 0 \qquad (7.78a)$$

or

$$(1-x^2)\frac{d^2F}{dx^2} - 2x\frac{dF}{dx} + \lambda F - \frac{m^2}{1-x^2}F = 0 \qquad (7.78b)$$

With this transformation, we need to solve for $F(x) = F(\cos\theta)$. Eqs. (7.78a) and (7.78b) must hold for all values of m. Let us consider, then, the simplest case of $m = 0$. In this case, Eq. (7.78b) becomes

$$(1-x^2)\frac{d^2F}{dx^2} - 2x\frac{dF}{dx} + \lambda F = 0 \qquad (7.79)$$

Eq. (7.79) has the form of the *Legendre differential equation* given in Eq. (7.58). The only difference is that in Eq. (7.79), λ is in place of $n(n+1)$. If we can show that $\lambda = n(n+1)$, then we will have the complete solution, based on our analysis in the previous section, because then, F must be equal to the Associated Legendre function, given by Eq. (7.68).

To determine λ, Eq. (7.79) is sufficient. Our objective is to solve Eq. (7.79) for λ. The *Method of Frobenious* can be used, where we assume the solution takes the form of an infinite series. Thus

$$F = x^s \sum_{n=0}^{\infty} a_n x^n = \sum_{n=0}^{\infty} a_n x^{n+s} \qquad (7.80)$$

The parameter s must be ≥ 0 in order to ensure a normalizable wave function at $x = 0$. Recall that our interval of convergence for x is $(-1, 1)$. Substituting our infinite series solution into Eq. (7.79), we have

$$(1-x^2)\frac{d^2}{dx^2}\sum_{n=0}^{\infty} a_n x^{n+s} - 2x\frac{d}{dx}\sum_{n=0}^{\infty} a_n x^{n+s} + \lambda \sum_{n=0}^{\infty} a_n x^{n+s} = 0$$

After carrying out the derivatives of the series, and multiplying the factors, we obtain

$$\sum_{n=0}^{\infty} (n+s-1)(n+s)a_n x^{n+s-2} =$$

$$\sum_{n=0}^{\infty}[(n+s+1)(n+s)+2(n+s)-\lambda]a_n x^{n+s}$$

Multiplying by x^{2-s}, and rewriting the term on the RHS so that $(n+s)$ is factored, we may write

$$\sum_{n=0}^{\infty}(n+s-1)(n+s)a_n x^n = \sum_{n=0}^{\infty}[(n+s)(n+s+1)-\lambda]a_n x^{n+2} \quad (7.81)$$

We can make two observations from Eq. (7.81). First, because both sides are polynomials in x, and the equality must hold for any value of x, it necessitates that corresponding polynomial coefficients must be equal. Thus, x^0, x^1, x^2, etc. must have the same coefficients on both sides of Eq. (7.81). But, there is a difference in the corresponding exponents. The left side has an x^0 and x^1 term, while the right side's first exponent begins at x^2. This means that the coefficients for x^0 and x^1, corresponding to $n=0$ and $n=1$, respectively, must be equal to zero. This leads to the following *indicial equations* for this application of the method of Frobenius.

$$(s-1)s = 0 \quad (7.82a)$$

$$(s+1)s = 0 \quad (7.82b)$$

We have two second order equations that suggest three unique solutions $s=0$ (twice), $s=1$, and $s=-1$. However, the only allowable values for a bounded function are $s=0$ and $s=1$. $s=-1$ is not allowed since it leads to an infinite value at $x=0$ for the lowest order term in the series. This is a divergent series in the center of our interval $(-1,1)$. We therefore have the two solutions

$$s = 0 \quad (7.83a)$$

$$s = 1 \quad (7.83b)$$

When we evaluate the first few terms of Eq. (7.81), we get

$$s(s-1)a_0 + s(s+1)a_1 x + (s+1)(s+2)a_2 x^2 + (s+2)(s+3)a_3 x^3 + ... =$$

$$[s(s+1)-\lambda]a_0 x^2 + [(s+1)(s+2)-\lambda]a_1 x^3 + ... \quad (7.84)$$

Since the LHS coefficient index n is shifted by $+2$ relative to its corresponding power on the RHS, equating the coefficients leads to the following relation

$$(n+s+2)(n+s+1)a_{n+2} = [(n+s)(n+s+1)-\lambda]a_n$$

From this, we obtain a recurrence relation given by

7.7 Assoc. Legendre Funcs & Orb. Ang. Momentum

> **Recurrence Relation For Orbital Angular Momentum**
>
> $$\frac{a_{n+2}}{a_n} = \frac{(n+s)(n+s+1) - \lambda}{(n+s+1)(n+s+2)} \qquad (7.85)$$

Given our two solutions for s, we have the following

$$\frac{a_{n+2}}{a_n} = \frac{n(n+1) - \lambda}{(n+1)(n+2)} \quad (s=0) \qquad (7.86)$$

$$\frac{a_{n+2}}{a_n} = \frac{(n+1)(n+2) - \lambda}{(n+2)(n+3)} \quad (s=1) \qquad (7.87)$$

Evaluating the coefficients a_n in the asymptotic limit of $n \to \infty$, for a given λ, Eq. (7.85) leads to

$$\lim_{n \to \infty} \frac{a_{n+2}}{a_n} = \frac{n^2}{n^2} = 1$$

This tells us that for large n, the coefficients a_n approach a uniform single value. Let us denote this sufficiently large n by n^*. Then, the infinite series solution becomes

$$x^s \sum_{n=0}^{\infty} a_n x^n \to x^s (a_0 + a_1 x + \ldots + a_{n^*-1} x^{n^*-1}) + ax^s \sum_{n=n^*}^{\infty} x^n \qquad (7.88)$$

In Eq. (7.88), we have uncovered a problem. For the solution to be valid for a well-behaved wave function, the limit of the function towards the boundaries of the problem, which are ± 1, must be zero, otherwise it is not normalizable. The series on the RHS is given by

$$\sum_{n=n^*}^{\infty} x^n = a_{n^*} x^{n^*} (1 + a_1^* x + a_2^* x^2 + \ldots) \qquad (7.89)$$

This series diverges at ± 1 because convergence for any series of this form is only possible when $x < 1$. Therefore, the infinite series cannot be allowed as a solution for a wave function of the orbital angular momentum problem. The series, then, must be truncated to result in finite polynomials of order n, instead. Since the coefficients a_n are proportional to the recurrence relation, this is possible if and only if the recurrence relation vanishes at some finite value of n. For this, it must be that

$$\lambda = n(n+1) \quad \text{and} \quad \lambda = (n+1)(n+2) \tag{7.90}$$

Eqs. (7.90) has redundancy because according to Eq. (7.86) for $s = 0$, the possible values are $0 \cdot 1, 1 \cdot 2, 2 \cdot 3,...$, but using Eq. (7.87) for $s = 1$, the possible values are $1 \cdot 2, 2 \cdot 3, 3 \cdot 4,...$ The case of $s = 0$, thus, covers all possible values of λ, being simply $n(n+1)$. The convention that we will use is to denote the integer n, which is associated with the eigenvalues of the orbital angular momentum operator $\widehat{L}$, as $n \to \ell$. With these results, we have achieved our objective in determining that the allowed values of λ, which are the eigenvalues of the total angular momentum L^2, are given by

Total Orbital Angular Momentum L^2 Eigenvalues

$$\lambda = \ell(\ell+1) \quad \ell = 0, 1, 2,... \tag{7.91}$$

Recall that our objective was to determine the values of λ, and to show that $\lambda = n(n+1) = \ell(\ell+1)$. We have therefore, also proven that Eq. (7.78b) becomes

$$(1-x^2)\frac{d^2 A_\ell^m}{dx^2} - 2x\frac{dA_\ell^m}{dx} + \ell(\ell+1)A_\ell^m - \frac{m^2}{1-x^2}A_\ell^m = 0 \tag{7.92}$$

In Eq. (7.92), we have replaced F with the associated Legendre function A_ℓ^m, which is the solution. Thus, we have obtained both $\Phi(\varphi)$ and $F(x = \cos\theta) = A_\ell^m(x)$, so we now have the complete solution for the total orbital angular momentum wave functions. The complete eigensolution for the L^2 operator becomes

Spherical Harmonics - Eigenfunctions of L^2 Operator

$$Y_\ell^m = C_n A_\ell^m e^{im\varphi} \tag{7.93}$$

The convention for these eigensolutions is to denote them by Y_ℓ^m. The normalization factor C_n is just the product of those for A_ℓ^m, given by Eq.

7.7 Assoc. Legendre Funcs & Orb. Ang. Momentum

(7.72), and $\Phi(\varphi)$ determined to be $1/\sqrt{2\pi}$. So C_n becomes

$$C_n = \sqrt{\frac{(2n+1)(n-m)!}{4\pi(n+m)!}} \qquad (7.94)$$

The eigenfunctions of the $\widehat{L}^2$ operator Y_ℓ^m, are known as **spherical harmonics** because they depend only on φ and θ. Given the effort involved in finding the eigenfunctions and corresponding eigenvalues for the orbital angular momentum operators, it should not be too surprising that the *spherical harmonics* are not trivial to visualize. It is helpful to see plots of the orbitals for various values of ℓ and m. But, before we get to that, let us discuss another convention used in expressing the atomic orbitals involving Y_ℓ^m.

We have seen that when we have a set of functions that are solutions to the Schrödinger equation, or eigensolutions, then it follows that any linear combination of them also satisfies the same equation. Thus, we can choose combinations of the spherical harmonics Y_ℓ^m that also satisfy the orbital angular momentum operators. For orbitals with $\ell = 2$, which are called d orbitals, some of the conventional linear combinations are

$$d_{z^2} = Y_2^0 = \left(\frac{5}{16\pi}\right)^{\frac{1}{2}}(3\cos^2\theta - 1) \qquad (7.95a)$$

$$d_{xy} = \frac{1}{\sqrt{2}i}(Y_2^{+2} - Y_2^{-2}) = \left(\frac{5}{16\pi}\right)^{\frac{1}{2}}\sin^2\theta\sin 2\varphi \qquad (7.95b)$$

$$d_{xz} = \frac{1}{\sqrt{2}}(Y_2^{-1} - Y_2^{+1}) = \left(\frac{5}{16\pi}\right)^{\frac{1}{2}}\sin\theta\cos\theta\sin\varphi \qquad (7.95c)$$

$$d_{yz} = \frac{i}{\sqrt{2}}(Y_2^{-1} + Y_2^{+1}) = \left(\frac{5}{16\pi}\right)^{\frac{1}{2}}\sin\theta\cos\theta\cos\varphi \qquad (7.95d)$$

$$d_{x^2-y^2} = \frac{1}{\sqrt{2}}(Y_2^{+2} + Y_2^{-2}) = \left(\frac{5}{16\pi}\right)^{\frac{1}{2}}\sin^2\theta\cos 2\varphi \qquad (7.95e)$$

The d orbitals turn out to play an important role in magnetic materials. The logic behind the above convention will be clear upon visualizing these orbitals. With these linear combinations, the subscripts now tell us the orientation of the d orbitals in a Cartesian coordinate system. For example, the d_{xy} orbitals lie parallel to the xy plane. The d_{xz} is parallel to the xz plane, etc.

Below d ($\ell = 2$) orbitals are the orbitals corresponding to $\ell = 1$, called p orbitals and $\ell = 0$, called s orbitals. This transformation mimics the natural behavior of p orbits, as they align with the x, y, and z axes, respectively. Correspondingly, these are denoted by p_x, p_y, and p_z. All s orbitals are spherically symmetric. The lobular shapes of the orbitals

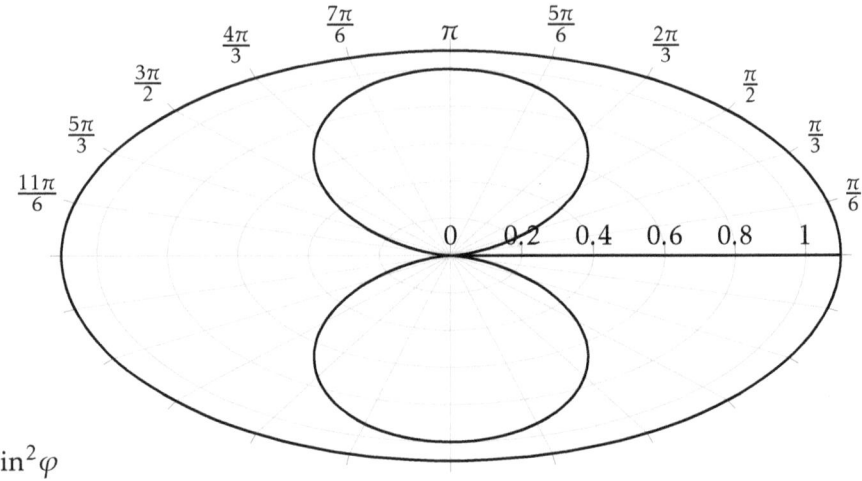

$\sin^2 \varphi$

Figure 7.6: 2D polar plots of $\sin^2 \varphi$.

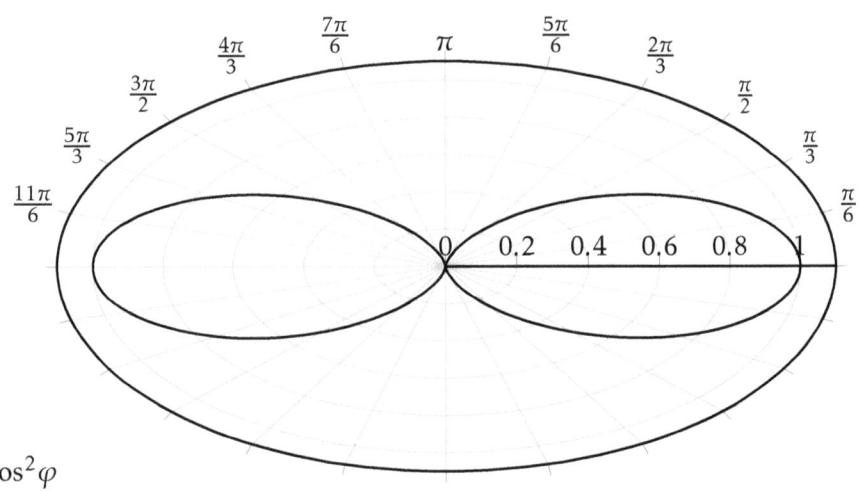

$\cos^2 \varphi$

Figure 7.7: 2D polar plot of $\cos^2 \varphi$.

arise because of the trigonometric functions being plotted in a polar coordinate system. Our intuition tends to lie in Cartesian coordinates, in dealing with transformations between polar and Cartesian. However, in

7.7 Assoc. Legendre Funcs & Orb. Ang. Momentum

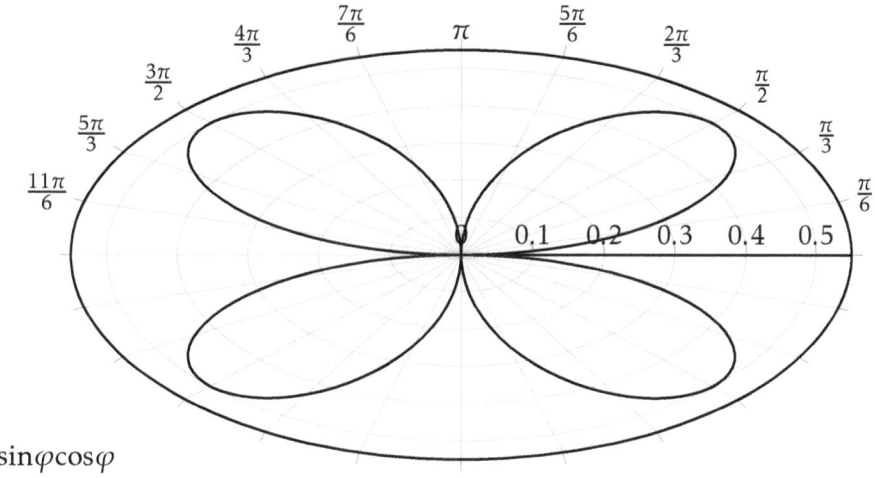

sinφcosφ

Figure 7.8: 2D polar plots of sinφcosφ. Plotting the functions in polar coordinates is the origin of orbital shapes, having an increasing number of lobes as the order increases.

polar coordinates, there is periodicity in the angle argument for sin and cos, and any higher order function of these terms. Hence, the resulting curves produce **polar lobes**.

Figs. 7.6, 7.7, and 7.6 illustrate lobes in 2D polar plots corresponding to some of the components in the d orbitals. They illustrate the origin of orbital lobes, as well as the increasing number of lobes. Two lobes result from symmetric functions such as $\sin^2\varphi$ and $\cos^2\varphi$. The function $\cos^2\varphi$ is just a 90° rotation of $\sin^2\varphi$. And, when there is a $\sin\varphi \cdot \cos\varphi$ term, symmetry is broken, and two lobes become four since the product now has four zeros instead of two. As ℓ increases, more lobes generally can result. Fig. 7.9 illustrates some 3D conventional orbitals, from s to d.

This completes the solution of the orbital angular momentum problem, known as *spherical harmonics* Y_ℓ^m. We started by introducing the quantum mechanical operators $\widehat{L}_x, \widehat{L}_y$, and $\widehat{L}_z$. This allowed us to define eigenvalue equations for both $\widehat{L}_z$ and $\widehat{L}^2$, and solve them exactly. Now, we'll turn our attention to an important topic in quantum mechanics that applies to any operator, but we'll introduce it using momentum operators. The concept is known as *commutation* and it's the subject of the next section.

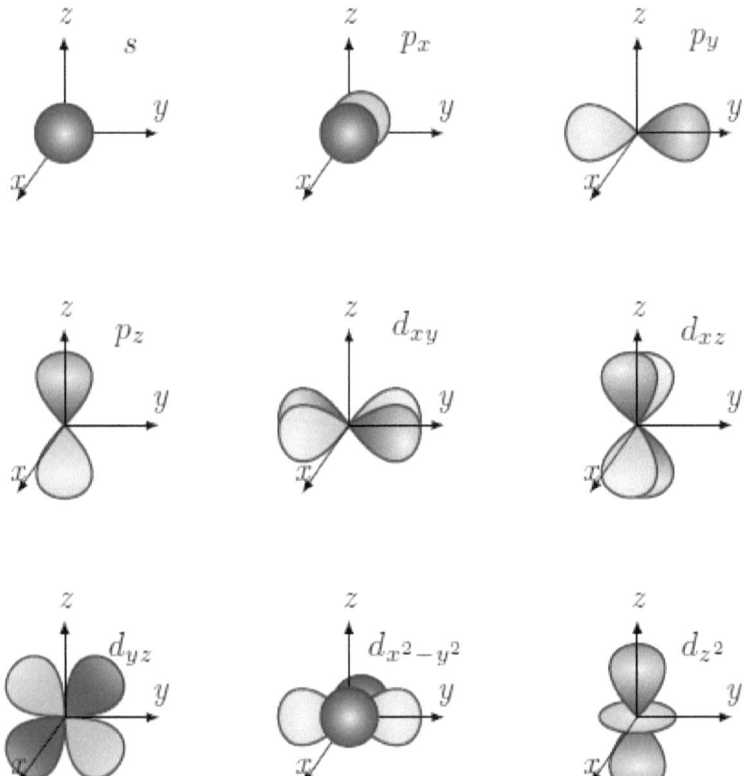

Figure 7.9: Conventional atomic orbitals which are linear combinations of Y_ℓ^m.

7.8 SPECIAL COMMUTATION RELATIONS

In general, the order of operations on a function matters. Let's say I needed to give you directions to my house. Imagine if I told you to turn right, take ten steps, then turn left. This would certainly end up at a different location than if you turned left, took ten steps, then turned right, or reversed the order of operations. Similarly, with operators, there is no reason why changing their order should yield the same result. One linear operator that illustrates this well is matrices. Consider the following two matrices $\widehat{A}$ and $\widehat{B}$ operating on a vector v.

$$\widehat{A} = \begin{bmatrix} 1 & -1 \\ 3 & 0 \end{bmatrix} \quad , \quad \widehat{B} = \begin{bmatrix} 1 & 3 \\ 1 & -2 \end{bmatrix} \quad \text{and} \quad v = \begin{bmatrix} 1 \\ 1 \end{bmatrix}$$

7.8 Special Commutation Relations

If we let $\widehat{B}$ operate on v first, we see that $\widehat{AB}v$ gives

$$\begin{bmatrix} 1 & -1 \\ 3 & 0 \end{bmatrix} \begin{bmatrix} 1 & 3 \\ 1 & -2 \end{bmatrix} \begin{bmatrix} 1 \\ 1 \end{bmatrix} = \begin{bmatrix} 1 & -1 \\ 3 & 0 \end{bmatrix} \begin{bmatrix} 4 \\ -1 \end{bmatrix} = \begin{bmatrix} 5 \\ 12 \end{bmatrix}$$

However, reversing the order of operations, letting $\widehat{A}$ operate on v first, we see that $\widehat{BA}v$ gives

$$\begin{bmatrix} 1 & 3 \\ 1 & -2 \end{bmatrix} \begin{bmatrix} 1 & -1 \\ 3 & 0 \end{bmatrix} \begin{bmatrix} 1 \\ 1 \end{bmatrix} = \begin{bmatrix} 1 & 3 \\ 1 & -2 \end{bmatrix} \begin{bmatrix} 0 \\ 3 \end{bmatrix} = \begin{bmatrix} 9 \\ -6 \end{bmatrix} \neq \begin{bmatrix} 5 \\ 12 \end{bmatrix}$$

Two operators $\widehat{A}$ and $\widehat{B}$ are said to **commute**, if and only if the order of the operations does not matter. Thus, they *commute* when

$$\widehat{AB} = \widehat{BA} \quad (7.96a)$$

Alternatively, we can write

$$\widehat{AB} - \widehat{BA} = 0 \quad (7.96b)$$

In the notation above, the convention is to omit the object on which $\widehat{A}$ and $\widehat{B}$ operates (e.g. v in our example). If two operators commute, then operating with B first, then A, will yield the same result as operating with A first, then B. This leads to the definition of the **commutation operator**, which is defined as

Commutation Operator

$$[\widehat{A}, \widehat{B}] = \widehat{A}(\widehat{B}) - \widehat{B}(\widehat{A}) \quad (7.97)$$

Since quantum mechanics utilizes the notion of operators, the *commutation operator* turns out to be a useful concept that we can apply to linear and angular momentum. In the case of linear momentum, in the x direction, consider

$$[x, \widehat{p}_x] = x\widehat{p}_x - \widehat{p}_x x = x \left(-i\hbar \frac{\partial}{\partial x} \right) - \left(-i\hbar \frac{\partial}{\partial x} \right) x \quad (7.98)$$

Remember that an operator is always applied to a wave function sitting on the right of the operator. For example, applying Eq. (7.98) to a wave function ψ, we have

$$\left[x \left(-i\hbar \frac{\partial}{\partial x} \right) - \left(-i\hbar \frac{\partial}{\partial x} \right) x \right] \psi = x \left(-i\hbar \frac{\partial \psi}{\partial x} \right) - \left(-i\hbar \frac{\partial (x\psi)}{\partial x} \right) =$$

$$-i\hbar\frac{\partial\psi}{\partial x} + i\hbar\frac{\partial\psi}{\partial x} + i\hbar\psi = i\hbar\psi \neq 0$$

The commutation operator $[x,\widehat{p}_x]$ becomes

$$[x,\widehat{p}_x] = [x\widehat{p}_x - \widehat{p}_x x]\psi = i\hbar\psi \tag{7.99}$$

Since the result is not zero, it also means that x and $\widehat{p}_x$ *do not commute*. However, Eq. (7.99) is an eigenoperator equation for $[x,\widehat{p}_x]$. Though they do not commute, we are able to obtain its eigenvalues $\lambda = i\hbar$. Therefore, in many instances when we have the commutation operator $xp_x - p_x x$ involved, it may be replaced by the RHS of Eq. (7.99). In shorthand, we write the operator relation as

$$[x,\widehat{p}_x] = x\widehat{p}_x - \widehat{p}_x x = i\hbar$$

A similar result is found for y and z, where, for all components, we get

$$x\widehat{p}_x - \widehat{p}_x x = i\hbar \tag{7.100a}$$

$$y\widehat{p}_y - \widehat{p}_y y = i\hbar \tag{7.100b}$$

$$z\widehat{p}_z - \widehat{p}_z z = i\hbar \tag{7.100c}$$

These relations can be expressed more compactly, in terms of the commutation operator, as

> **$x,\widehat{p}$ Commutation Relation**
>
> $$[x_i,\widehat{p}_i] = i\hbar \tag{7.101}$$

In contrast, if we determine the commutation relation for different coordinates, such as

$$xp_y - p_y x = x\left(-i\hbar\frac{\partial}{\partial y}\right) - \left(-i\hbar\frac{\partial}{\partial y}\right)x$$

Operating on a wave function ψ, we have

$$[xp_y - p_y x]\psi = \left[x\left(-i\hbar\frac{\partial}{\partial y}\right) - \left(-i\hbar\frac{\partial}{\partial y}\right)x\right]\psi$$

$$= x\left(-i\hbar\frac{\partial\psi}{\partial y}\right) + i\hbar\left(\frac{\partial(x\psi)}{\partial y}\right)$$

7.8 Special Commutation Relations

$$= -ix\hbar\frac{\partial\psi}{\partial y} + ix\hbar\frac{\partial\psi}{\partial y} = 0$$

Since the result is zero, x and p_y commute. It is straightforward to show this result extends to all components where the coordinate of $\mathbf{r} = [x, y, z]^T$ is different from that of $\widehat{\mathbf{p}} = [\widehat{p}_x, \widehat{p}_y, \widehat{p}_z]^T$. So, we have that

---

$x, \widehat{p}$ Commutation Relation

$$[x_j, \widehat{p}_i] = 0 \quad (j \neq i) \tag{7.102}$$

Let us extend these commutation relation ideas to angular momentum $\widehat{\mathbf{L}} = [\widehat{L}_x, \widehat{L}_y, \widehat{L}_z]^T$. Since we know the angular momentum operators for L_x, L_y, and L_z, we can determine the commutation relations for the angular momentum operators *with each other*, using Eq. (7.103). First, let's evaluate

$$\widehat{L}_y\widehat{L}_z - \widehat{L}_z\widehat{L}_y \tag{7.103}$$

The components of the orbital angular momentum are given by

$$\widehat{L}_x = -i\hbar\left(y\frac{\partial}{\partial z} - z\frac{\partial}{\partial y}\right) \tag{7.104a}$$

$$\widehat{L}_y = -i\hbar\left(z\frac{\partial}{\partial x} - x\frac{\partial}{\partial z}\right) \tag{7.104b}$$

$$\widehat{L}_z = -i\hbar\left(x\frac{\partial}{\partial y} - y\frac{\partial}{\partial x}\right) \tag{7.104c}$$

Therefore, Eq. (7.103) becomes

$$\widehat{L}_y\widehat{L}_z - \widehat{L}_z\widehat{L}_y = -\hbar^2\left(z\frac{\partial}{\partial x} - x\frac{\partial}{\partial z}\right)\left(x\frac{\partial}{\partial y} - y\frac{\partial}{\partial x}\right)$$

$$+ \hbar^2\left(x\frac{\partial}{\partial y} - y\frac{\partial}{\partial x}\right)\left(z\frac{\partial}{\partial x} - x\frac{\partial}{\partial z}\right) \tag{7.105}$$

This expands to the following:

$$\widehat{L}_y\widehat{L}_z - \widehat{L}_z\widehat{L}_y = -\hbar^2\left[z\frac{\partial}{\partial x}\left(x\frac{\partial}{\partial y}\right) - z\frac{\partial}{\partial x}\left(y\frac{\partial}{\partial x}\right) - x\frac{\partial}{\partial z}\left(x\frac{\partial}{\partial y}\right) + x\frac{\partial}{\partial z}\left(y\frac{\partial}{\partial x}\right)\right]$$

$$+ \hbar^2\left[x\frac{\partial}{\partial y}\left(z\frac{\partial}{\partial x}\right) - x\frac{\partial}{\partial y}\left(x\frac{\partial}{\partial z}\right) - y\frac{\partial}{\partial x}\left(z\frac{\partial}{\partial x}\right) + y\frac{\partial}{\partial x}\left(x\frac{\partial}{\partial z}\right)\right] \tag{7.106}$$

Though it may appear to be complicated with so many terms, most of them cancel, and we have the result

$$\widehat{L}_y \widehat{L}_z - \widehat{L}_z \widehat{L}_y = \hbar^2 \left[y \frac{\partial}{\partial z} - z \frac{\partial}{\partial y} \right] \quad (7.107)$$

We can also use the operator $\widehat{L}_x$ on the RHS, given in Eq. (7.104a), to write Eq. (7.107) as

$$\widehat{L}_y \widehat{L}_z - \widehat{L}_z \widehat{L}_y = i\hbar \widehat{L}_x \quad (7.108)$$

Evaluating the commutation rules based on Eq. (7.103) for the other two pairs $\widehat{L}_x \widehat{L}_z$ and $\widehat{L}_x \widehat{L}_y$, we find the following commutation rules for the orbital angular momentum.

$$\widehat{L}_y L_z - \widehat{L}_z \widehat{L}_y = i\hbar \widehat{L}_x \quad (7.109a)$$
$$\widehat{L}_z \widehat{L}_x - \widehat{L}_x \widehat{L}_z = i\hbar \widehat{L}_y \quad (7.109b)$$
$$\widehat{L}_x \widehat{L}_y - \widehat{L}_y \widehat{L}_x = i\hbar \widehat{L}_z \quad (7.109c)$$

This is written more compactly as

Orbital Angular Momentum Commutation Relation

$$\left[\widehat{L}_i, \widehat{L}_j \right] = i\hbar \widehat{L}_k \quad (7.110)$$

Because the LHS is also the vector cross-product, it is alternatively written as $\widehat{\mathbf{L}} \times \widehat{\mathbf{L}} = i\hbar \widehat{\mathbf{L}}$. Another set of commutation relations we will consider involves the magnetic vector potential **A**. From the Maxwell equations, the magnetic field **B** can alternatively be expressed in terms of **A**, where $\mathbf{B} = \nabla \times \mathbf{A}$. Let us then define an **extended momentum operator** $\widehat{\mathbf{P}}$ by

Extended Momentum Operator

$$\widehat{\mathbf{P}} \equiv \widehat{\mathbf{p}} + e\mathbf{A} \quad (7.111)$$

e is the elementary charge. We are going to consider two commutation relations involving $\widehat{\mathbf{P}}$. First, lets consider

$$\left[\widehat{P}_i, \widehat{P}_j \right] = \widehat{P}_i \widehat{P}_j - \widehat{P}_j \widehat{P}_i \quad (7.112)$$

7.8 Special Commutation Relations

i and j denote the dimension index $1 \to x$, $2 \to y$, and $3 \to z$. In evaluating this operator, let us start with x and y where we have

$$[\widehat{P}_x, \widehat{P}_y] = (\widehat{p}_x + eA_x)(\widehat{p}_y + eA_y) - (\widehat{p}_y + eA_y)(\widehat{p}_x + eA_x) \tag{7.113}$$

Using the definition for the momentum operator $\widehat{p}_i = -i\hbar \partial/\partial x_i$, the commutation operator in Eq. (7.113) becomes

$$[\widehat{P}_x, \widehat{P}_y]\psi = \left[(\widehat{p}_x + eA_x)(\widehat{p}_y + eA_y) - (\widehat{p}_y + eA_y)(\widehat{p}_x + eA_x)\right]\psi =$$

$$\left[\left(-i\hbar\frac{\partial}{\partial x} + eA_x\right)\left(-i\hbar\frac{\partial}{\partial y} + eA_y\right) - \left(-i\hbar\frac{\partial}{\partial y} + eA_y\right)\left(-i\hbar\frac{\partial}{\partial x} + eA_x\right)\right]\psi \tag{7.114}$$

The products work out to four terms, and the commutation operation becomes

$$[\widehat{P}_x, \widehat{P}_y]\psi = \left[-\hbar^2 \frac{\partial}{\partial x}\frac{\partial}{\partial y} - i\hbar e\frac{\partial}{\partial x}(A_y) - i\hbar e A_x \frac{\partial}{\partial y} + e^2 A_x A_y\right]\psi -$$

$$\left[-\hbar^2 \frac{\partial}{\partial y}\frac{\partial}{\partial x} - i\hbar e\frac{\partial}{\partial y}(A_x) - i\hbar e A_y \frac{\partial}{\partial x} + e^2 A_y A_x\right]\psi \tag{7.115}$$

Each terms operate on the wave function ψ. The last term involves all scalars, thus, it is only multiplication. Since multiplication of scalars is commutative, we can use the following relation:

$$e^2 A_x A_y \psi = e^2 A_y A_x \psi$$

Therefore, this term vanishes. Additionally, the first term involves mixed partials, for which the order also does not matter. Thus, we can use the following relation

$$-\hbar^2 \frac{\partial}{\partial x}\frac{\partial}{\partial y}\psi = -\hbar^2 \frac{\partial^2 \psi}{\partial x \partial y} = -\hbar^2 \frac{\partial^2 \psi}{\partial y \partial x}$$

From this, the first terms vanishes, as well. We now have left

$$[\widehat{P}_x, \widehat{P}_y]\psi = i\hbar e\left[-\frac{\partial}{\partial x}(A_y \psi) - A_x \frac{\partial \psi}{\partial y}\right] -$$

$$i\hbar e\left[-\frac{\partial}{\partial y}(A_x \psi) - A_y \frac{\partial \psi}{\partial x}\right] \tag{7.116}$$

The derivatives of the products must be expanded *using the product rule for differentiation*, which leads to

$$[\widehat{P}_x, \widehat{P}_y]\psi = i\hbar e\left[-\frac{\partial A_y}{\partial x}\psi - A_y\frac{\partial \psi}{\partial x} - A_x\frac{\partial \psi}{\partial y}\right] -$$

$$i\hbar e\left[-\frac{\partial A_x}{\partial y}\psi - i\hbar A_x\frac{\partial \psi}{\partial y} - A_y\frac{\partial \psi}{\partial x}\right] \quad (7.117)$$

The terms involving derivatives of the wave function ψ all cancel, leaving the result

$$[\widehat{P}_x, \widehat{P}_y]\psi = -i\hbar e\frac{\partial A_y}{\partial x}\psi + i\hbar e\frac{\partial A_x}{\partial y}\psi = -i\hbar e\left(\frac{\partial A_y}{\partial x} - \frac{\partial A_x}{\partial y}\right)\psi \quad (7.118)$$

The term in parentheses is just the z component of the magnetic field $\mathbf{B} = \nabla \times \mathbf{A}$, so we have

$$[\widehat{P}_x, \widehat{P}_y]\psi = -i\hbar e B_z \psi \quad (7.119a)$$

The result is similar for the other two pairs of momentum operators, where we get

$$\left[\widehat{P}_y, \widehat{P}_z\right]\psi = -i\hbar e B_x \psi \quad (7.119b)$$

$$\left[\widehat{P}_z, \widehat{P}_x\right]\psi = -i\hbar e B_y \psi \quad (7.119c)$$

The reader should verify that if we take $[\widehat{P}_i, \widehat{P}_i]\psi$, the result would vanish meaning that they commute. Thus, the indexes must be distinct for the result to be nonzero. More compactly, we may write this above results as

Extended Momentum Commutation Relation

$$\left[\widehat{P}_i, \widehat{P}_j\right] = -i\hbar e \varepsilon_{ijk} B_k \quad (7.120)$$

Therefore, we have that the commutation of two *extended momentum* components is a scalar operator, given by Eq. (7.120). ε_{ijk} is known as the *Levi-Civita symbol* defined as

$$\varepsilon_{ijk} = \begin{cases} +1 & \text{if } ijk = (1,2,3), (2,3,1), \text{ or } (3,1,2) \\ -1 & \text{if } ijk = (3,2,1), (1,3,2), \text{ or } (2,1,3) \\ 0 & \text{if } i = j, j = k, \text{ or } i = k \end{cases}$$

7.8 Special Commutation Relations

This notation maintains use of a right-handed coordinate system. We can use these results to determine more commutation relations of higher order in $\widehat{P}_i$. For example, let us determine

$$[\widehat{P}_i^2, \widehat{P}_j] = \widehat{P}_i^2 \widehat{P}_j - \widehat{P}_j \widehat{P}_i^2 \tag{7.121}$$

This will be used to obtain our end goal, which is to determine the following:

$$[\widehat{P}^2, \widehat{P}_j] = [\widehat{P}_x^2 + \widehat{P}_y^2 + \widehat{P}_z^2, \widehat{P}_j]$$

Let's start by considering the component $\widehat{P}_y$ with $\widehat{P}_x^2$, where we first have

$$[\widehat{P}_x^2, \widehat{P}_y] = \widehat{P}_x^2 \widehat{P}_y - \widehat{P}_y \widehat{P}_x^2$$

$$= \widehat{P}_x \widehat{P}_x \widehat{P}_y - \widehat{P}_y \widehat{P}_x \widehat{P}_x$$

$$= \widehat{P}_x \left(-ie\hbar B_z + \widehat{P}_y \widehat{P}_x\right) - \widehat{P}_y \widehat{P}_x \widehat{P}_x \quad \text{(using Eq. (7.120))}$$

$$= \widehat{P}_x (-ie\hbar B_z) + \left(\widehat{P}_x \widehat{P}_y - \widehat{P}_y \widehat{P}_x\right) \widehat{P}_x$$

$$= \widehat{P}_x (-ie\hbar B_z) + (-ie\hbar B_z) \widehat{P}_x \quad \text{(again using Eq. (7.120))}$$

$$= \widehat{P}_x (-2ie\hbar B_z) = (-2ie\hbar B_z) \widehat{P}_x$$

The last step of commuting follows since the term in parentheses is scalar. Carrying out a similar analysis for $[\widehat{P}_z^2, \widehat{P}_y]$, we have

$$[\widehat{P}_z^2, \widehat{P}_y] = \widehat{P}_z^2 \widehat{P}_y - \widehat{P}_y \widehat{P}_z^2$$

$$= \widehat{P}_z \widehat{P}_z \widehat{P}_y - \widehat{P}_y \widehat{P}_z \widehat{P}_z$$

$$= \widehat{P}_z \left(ie\hbar B_x + \widehat{P}_y \widehat{P}_z\right) - \widehat{P}_y \widehat{P}_z \widehat{P}_z \quad \text{(using Eq. (7.120))}$$

$$= \widehat{P}_z (ie\hbar B_x) + \left(\widehat{P}_z \widehat{P}_y - \widehat{P}_y \widehat{P}_z\right) \widehat{P}_z$$

$$= \widehat{P}_z (ie\hbar B_x) + (ie\hbar B_x) \widehat{P}_z \quad \text{(again using Eq. (7.120))}$$

$$= \widehat{P}_z (2ie\hbar B_x) = (2ie\hbar B_x) \widehat{P}_z$$

Again, the last step follows because the term in parentheses is a scalar. For the third case of $[\widehat{P}_y^2, \widehat{P}_y]$, which turns out to be simple because it is the only case that, exploiting Eq. (7.120), we may write

$$[\widehat{P}_y^2, \widehat{P}_y] = \widehat{P}_y [\widehat{P}_y, \widehat{P}_y] = 0$$

Therefore, summing the results for $\widehat{P}_y$, it follows that

$$[\widehat{P}_x^2 + \widehat{P}_y^2 + \widehat{P}_z^2, \widehat{P}_y] = 2ie\hbar\left(B_x\widehat{P}_z - B_z\widehat{P}_x\right) \qquad (7.122a)$$

The reader should confirm that carrying out the same analysis for $\widehat{P}_x$ and $\widehat{P}_z$, the result becomes

$$[\widehat{P}_x^2 + \widehat{P}_y^2 + \widehat{P}_z^2, \widehat{P}_x] = 2ie\hbar\left(B_z\widehat{P}_y - B_y\widehat{P}_z\right) \qquad (7.122b)$$

$$[\widehat{P}_x^2 + \widehat{P}_y^2 + \widehat{P}_z^2, \widehat{P}_z] = 2ie\hbar\left(B_y\widehat{P}_x - B_x\widehat{P}_y\right) \qquad (7.122c)$$

These results can be expressed more compactly as

Commutation Relation of Square Extended Momentum

$$[\widehat{\mathbf{P}}^2, \widehat{\mathbf{P}}] = 2ie\hbar\widehat{\mathbf{P}} \times \mathbf{B} \qquad (7.123)$$

Defining the extended momentum $\widehat{\mathbf{P}}$ in this way turns out to be *necessary* in quantum mechanics because it renders the Schrödinger and Dirac equations to be capable of describing charged particles in both electric and magnetic fields *that are independent of time*. This will shown fully in the next section, where we will use the results above to show that by defining the extended momentum $\widehat{\mathbf{P}} = \widehat{\mathbf{p}} + e\mathbf{A}$, one obtains an equation of motion consistent with Newton's laws, including the response to the *Lorentz Force*, arising from an electric and magnetic field.

7.9 Extended Momentum Leads to Lorentz Force

In Chap. 7, we introduced the extended momentum operator $\widehat{\mathbf{P}}$ given by

$$\widehat{\mathbf{P}} = \widehat{\mathbf{p}} + e\mathbf{A}$$

A is commonly known as the *magnetic vector potential*, whose vector curl is the magnetic field **B**. Mathematically, we say that $\mathbf{B} = \nabla \times \mathbf{A}$. From the above, **A** can also be interpreted as a *magnetic momentum per unit charge*. We had to postpone the full justification for this definition until Eq. (3.148) could be derived. Now, we can provide the justification for the extended definition of the momentum. For the linear momentum in

7.9 Extended Momentum Leads to Lorentz Force

the absence of a magnetic field, the Hamiltonian $\widehat{H}$ of a system has been given by

$$\widehat{H}\psi = \left[\frac{\widehat{\mathbf{p}}^2}{2m} + V\right]\psi \Rightarrow \widehat{H} = \frac{\widehat{\mathbf{p}}^2}{2m} + V$$

Let us now replace $\widehat{\mathbf{p}}$ with the extended momentum $\widehat{\mathbf{P}}$, then apply Eq. (3.148). Thus, we have

$$\widehat{H}' = \frac{\widehat{\mathbf{P}}^2}{2m} + V = \frac{(\widehat{\mathbf{p}} + e\mathbf{A})^2}{2m} + V \tag{7.124}$$

If we assume that the magnetic vector potential $\mathbf{A}$ is independent of time, then the extended momentum operator has no dependence on time. It follows that

$$\left\langle \frac{\partial \widehat{\mathbf{P}}}{\partial t} \right\rangle = 0 \tag{7.125}$$

Using Eq. (3.148) to compute the time derivative of the expectation value of $\widehat{\mathbf{P}}$, we have

$$\frac{d\langle \widehat{\mathbf{P}} \rangle}{dt} = \frac{i}{\hbar} \langle [\widehat{H}', \widehat{\mathbf{P}}] \rangle \tag{7.126}$$

In what follows, we will omit the expectation value notation for convenience, so that we may write

$$\frac{d\widehat{\mathbf{P}}}{dt} = \frac{i}{\hbar} [\widehat{H}', \widehat{\mathbf{P}}]$$

$$= \frac{i}{\hbar}\left[\left(\frac{\widehat{\mathbf{P}}^2}{2m} + V\right)\widehat{\mathbf{P}} - \widehat{\mathbf{P}}\left(\frac{\widehat{\mathbf{P}}^2}{2m} + V\right)\right]$$

$$= \frac{i}{\hbar}\left[\frac{1}{2m}\left(\widehat{\mathbf{P}}^2\widehat{\mathbf{P}} - \widehat{\mathbf{P}}\widehat{\mathbf{P}}^2\right) + \left(V\widehat{\mathbf{P}} - \widehat{\mathbf{P}}V\right)\right]$$

The first term in the last line has the form of the operator given by Eq. (7.123). Using that result, we have

$$\frac{d\widehat{\mathbf{P}}}{dt} = \frac{i}{\hbar}\left[\frac{1}{2m}\left(\widehat{\mathbf{P}}^2\widehat{\mathbf{P}} - \widehat{\mathbf{P}}\widehat{\mathbf{P}}^2\right) + \left(V\widehat{\mathbf{P}} - \widehat{\mathbf{P}}V\right)\right] \tag{7.127a}$$

$$= \frac{i}{\hbar}\left[\frac{1}{2m}\left(2ie\hbar\widehat{\mathbf{P}} \times \mathbf{B}\right) + \left(V\widehat{\mathbf{P}} - \widehat{\mathbf{P}}V\right)\right] \tag{7.127b}$$

Let's now consider the potential term more closely. It can be written as follows:

$$V\widehat{\mathbf{P}} - \widehat{\mathbf{P}}V = V(\mathbf{p} + e\mathbf{A}) - (\mathbf{p} + e\mathbf{A})V \tag{7.128a}$$

$$= -i\hbar V\nabla + eV\mathbf{A} + (i\hbar\nabla)V - e\mathbf{A}V \tag{7.128b}$$

$$= -i\hbar V\nabla + i\hbar V\nabla + i\hbar\nabla V \tag{7.128c}$$

$$= i\hbar\nabla V \tag{7.128d}$$

Using this result in Eq. (7.127b), we have

$$\frac{d\widehat{\mathbf{P}}}{dt} = -\frac{e}{m}\widehat{\mathbf{P}} \times \mathbf{B} - \nabla V = -e\widehat{\mathbf{v}} \times \mathbf{B} - \nabla V \tag{7.129}$$

From Newton's second law, we know that $\mathbf{F} = d\langle\mathbf{P}\rangle/dt$. Thus, we have the general result

General Electromagnetostatic Force

$$\mathbf{F} = -e\langle\widehat{\mathbf{v}} \times \mathbf{B}\rangle - \nabla V \tag{7.130}$$

$\widehat{\mathbf{v}}$ is the velocity operator given by $\widehat{\mathbf{v}} = \widehat{\mathbf{P}}/m$. We recognize the first term as the *Lorentz force* $\mathbf{F}_L$ due to a magnetic field acting on a moving charge, while the second term is the electrostatic force $\mathbf{F}_e = e(\mathbf{F}_e/e) = e\mathbf{E}$ due to the electric field **E**. This is the motivation behind the definition of the *extended momentum* operator $\widehat{\mathbf{P}}$. It provides a more complete description of a charged particle in an electric and magnetic field.

This definition of $\widehat{\mathbf{P}}$ extends quantum mechanics to a level of consistency with magnetostatics. It is straightforward to show that this definition of $\widehat{\mathbf{P}}$ also follows from the Maxwell equations. First, recall that in the full electromagnetics of the Maxwell equations, the most general electric field **E** is given by

$$\mathbf{E} = -\nabla V_e - \frac{\partial \mathbf{A}}{\partial t}$$

V_e is the electric potential. But, **E** is just the force per unit charge $\mathbf{F}_e/q$. Thus, for an electron, the force becomes

$$\mathbf{F}_e = q\mathbf{E} = -q\nabla V_e - q\frac{\partial \mathbf{A}}{\partial t}$$

7.10 Angular Momentum Ladder Operators

If we take $\mathbf{F} = md\mathbf{v}/dt$, we have

$$m\frac{d\mathbf{v}}{dt} = -q\nabla V_e - q\frac{\partial \mathbf{A}}{\partial t} \tag{7.131a}$$

$$= -\nabla V - q\frac{\partial \mathbf{A}}{\partial t} \tag{7.131b}$$

V is the *conservative* electric potential energy. The first term corresponds to motion along the direction of the electrical potential energy variation, while the second term corresponds to motion along the direction of the change in $\mathbf{A}$. The ordinary momentum operator $\widehat{\mathbf{p}}$, used in quantum mechanics, corresponds to the first term. The second term is the new term. Eq. (7.131b), then, equating the most general force $\mathbf{F}_e$ leads to

$$\frac{d\mathbf{P}}{dt} = \frac{d\mathbf{p}}{dt} - \frac{\partial(q\mathbf{A})}{\partial t} \Rightarrow \mathbf{P} = \mathbf{p} + e\mathbf{A}$$

e is the elementary charge relating to the electron charge as $q = -e$. Although the above analysis is general, note that in quantum mechanics, it is always assumed that there is no time dependence for $\mathbf{A}$ (which means no time dependence for $\widehat{\mathbf{P}}$). However, time integration of the above leads to our result $\mathbf{P} = \mathbf{p} + e\mathbf{A}$, which must be satisfied everywhere in space, since $\mathbf{A}$ varies with spatial coordinates. Thus, we see that quantum mechanics and the Maxwell equations are placed on the same footing, in this regard, by defining the extended momentum $\widehat{\mathbf{P}}$.

Next, we discuss a pair of special operators belonging to a class of operators that can be obtained for angular momentum. They are the *lowering* and *raising operators*, and are the subject of the next section.

7.10 ANGULAR MOMENTUM LADDER OPERATORS

The orbital angular momentum operators and eigenfunctions can be determined in one fell swoop, by working out the solution to the eigenvalue equations, which also dictate all the possible values of the quantum numbers. We found earlier that the solution to the eigenequation for $\widehat{L}^2$ is given by

$$Y_{\ell,m}(\theta, \varphi) = A_\ell^m(\theta)e^{im\varphi} \tag{7.132}$$

$A_\ell^m(\theta)$ is the *associated Legendre function* we derived earlier, given by Eq. (7.70). ℓ is the quantum number for the orbital angular momentum, while m is the magnetic quantum number associated with $\widehat{L}_z$, which manifests in the splitting by a magnetic field. Using the operators for $\widehat{L}_x$ given by Eq. (7.47) and $\widehat{L}_y$, given by Eq. (7.48), let us define a new operator $\widehat{L}_-$ as

Orbital Angular Momentum Lowering Operator Definition

$$\widehat{L}_- \equiv \widehat{L}_x - i\widehat{L}_y \qquad (7.133)$$

Eq. (7.133) is known as the **orbital angular momentum lowering operator**. We'll find out shortly why it bears this name. After substituting $\widehat{L}_x$ and $\widehat{L}_y$ from Eqs. (7.47) and (7.47), into the definition for $\widehat{L}_-$, we have

$$\widehat{L}_x - i\widehat{L}_y = i\hbar\left(\sin\varphi\frac{\partial}{\partial\theta} + \cot\theta\cos\varphi\frac{\partial}{\partial\varphi}\right) + \hbar\left(-\cos\varphi\frac{\partial}{\partial\theta} + \cot\theta\sin\varphi\frac{\partial}{\partial\varphi}\right)$$

$$= \hbar(i\sin\varphi - \cos\varphi)\frac{\partial}{\partial\theta} + \hbar(i\cos\varphi + \sin\varphi)\cot\theta\frac{\partial}{\partial\varphi}$$

$$= \hbar\left(-e^{-i\varphi}\frac{\partial}{\partial\theta} + ie^{-i\varphi}\cot\theta\frac{\partial}{\partial\varphi}\right)$$

We have made use of the relations $\cos\varphi - i\sin\varphi = e^{-i\varphi}$ and $\sin\varphi + i\cos\varphi = ie^{-i\varphi}$. Thus, the *orbital angular momentum lowering operator* defined above, is given by

Lowering Operator $\widehat{L}_-$

$$\widehat{L}_- = \hbar e^{-i\varphi}\left[-\frac{\partial}{\partial\theta} + i\cot\theta\frac{\partial}{\partial\varphi}\right] \qquad (7.134)$$

We know the eigenfunction $Y_{\ell,m}$ we obtained for the $\widehat{L}^2$ operator, so if we take the lowest possible energy state of orbital angular momentum, for a given value of ℓ, or

$$Y_{\ell,m=-\ell}(\theta,\varphi) = A_\ell^{-\ell} e^{-i\ell\varphi}$$

Let us explore what happens when we apply the lowering operator $\widehat{L}_-$ to $Y_{\ell,m=-\ell}$. Evaluating this leads to

$$\widehat{L}_- Y_{\ell,-\ell} = \hbar e^{-i\varphi}\left[-\frac{\partial}{\partial\theta} + i\cot\theta\frac{\partial}{\partial\varphi}\right] Y_{\ell,-\ell}$$

7.10 Angular Momentum Ladder Operators

$$= \hbar e^{-i\varphi}\left[-\frac{\partial}{\partial\theta} + i\cot\theta\frac{\partial}{\partial\varphi}\right]A_\ell^{-\ell}e^{-i\ell\varphi}$$

$$= \hbar e^{-i\varphi}\left[-\frac{\partial A_\ell^{-\ell}}{\partial\theta}e^{-i\ell\varphi} + i(-i\ell)\cot\theta e^{-i\ell\varphi}A_\ell^{-\ell}\right]$$

$$= \hbar e^{-i(\ell+1)\varphi}\left[-\frac{\partial A_\ell^{-\ell}}{\partial\theta} + \ell\cot\theta A_\ell^{-\ell}\right]$$

In the above result, we have two terms involved in $\widehat{L}_- Y_{\ell,-\ell}$, with the exponential coefficient and the term in brackets, which is purely a function of θ. For this special case, our concern is with the term in brackets. We will show that this term is identically zero, which means that when the lowering operator acts on a minimum energy state, it returns null. To see this, let us first express the *associated Legendre function* $A_\ell^{-\ell}$ as follows:

$$A_\ell^{-\ell} = C_\ell(1-x^2)^{\frac{-\ell}{2}}\frac{d^{\ell+(-\ell)}}{dx}(x^2-1)^\ell$$

$$= C_\ell(1-x^2)^{\frac{-\ell}{2}}(x^2-1)^\ell$$

$$= (-1)^\ell C_\ell(1-x^2)^{\frac{\ell}{2}}$$

Also, substituting $x = \cos\theta$, evaluation of $dA_\ell^{-\ell}/d\theta$ gives

$$\frac{dA_\ell^{-\ell}}{d\theta} = \frac{dA_\ell^{-\ell}}{dx}\frac{dx}{d\theta}$$

$$= (-1)^\ell \frac{\ell}{2}C_\ell(-\sin\theta)(-2x)\left[(1-x^2)^{\frac{\ell}{2}-1}\right]$$

$$= (-1)^\ell \frac{\ell}{2}C_\ell(-\sin\theta)(-2x)\left[(1-x^2)^{\frac{\ell}{2}}(1-x^2)^{-1}\right]$$

Then, substituting $1 - x^2 = \sin^2\theta$ leads to

$$\frac{dA_\ell^{-\ell}}{d\theta} = (-1)^\ell \ell C_\ell(-\sin\theta)(-\cos\theta)\left[(1-x^2)^{\frac{\ell}{2}}\frac{1}{\sin^2\theta}\right]$$

$$= \ell(-1)^\ell C_\ell \frac{\cos\theta}{\sin\theta}(1-x^2)^{\frac{\ell}{2}}$$

$$= \ell\cot\theta A_\ell^{-\ell}$$

From this result, we can see that the following statement is true:

$$\widehat{L}_- Y_{\ell,-\ell} = \hbar e^{-i(\ell+1)\varphi}\left[-\frac{\partial A_\ell^{-\ell}}{\partial \theta} + \ell\cot\theta A_\ell^{-\ell}\right] = 0$$

This result, being only dependent on $A_\ell^{-\ell}$ gives an important condition for the lowest energy state $A_\ell^{-\ell}$, given by

Lowest-Level Associated Legendre Function Condition

$$-\frac{\partial A_\ell^{-\ell}}{\partial \theta} + \ell\cot\theta A_\ell^{-\ell} = 0 \tag{7.135}$$

This is a property of the associated Legendre function. Moreover, this relation can be used to derive an alternative form for the lowest energy function $A_\ell^{-\ell}$. It we take the relation obtained above, which can be written as

$$\frac{\partial A_\ell^{-\ell}}{\partial \theta} = \ell \cdot \cot\theta A_\ell^{-\ell} = \ell \cdot \frac{\cos\theta}{\sin\theta} \cdot A_\ell^{-\ell}$$

If we let $u = \sin\theta \rightarrow du = \cos\theta d\theta$. Substitution of these relations into the above gives

$$\frac{dA_\ell^{-\ell}}{A_\ell^{-\ell}} = \ell\frac{du}{u} \Rightarrow \ln A_\ell^{-\ell} = \ell\ln(\sin\theta) + \text{constant} = \ln\left(\sin^\ell\theta\right) + \text{constant}$$

From this, we have a result for $A_\ell^{-\ell}$ given by

Lowest-Level Associated Legendre Function $A_\ell^{-\ell}$

$$A_\ell^{-\ell} = C_n \sin^\ell\theta \tag{7.136}$$

As usual, C_n must be determined by the normalization condition. The evaluation of this condition in spherical coordinates gives

$$1 = \int_0^\pi |Y_\ell^{-\ell}|^2 \sin\theta d\theta d\varphi$$

7.10 Angular Momentum Ladder Operators

$$= C_n^2 \int (\sin^\ell\theta)e^{+i\ell\varphi}(\sin^\ell\theta)e^{-i\ell\varphi}\sin\theta d\theta d\varphi$$

$$= 2\pi C_n^2 \int (\sin^{2\ell+1}\theta)d\theta$$

The factor of 2π follows from evaluating the part of the double integral w.r.t. φ. To evaluate the resulting single variable integral in θ, we can use a well-known identity involving integrals of powers of sin functions, known as a *trigonometric reduction formula*. It states that

Trigonometric Reduction Formula

$$\int \sin^n(x)dx = -\frac{1}{n}\sin^{n-1}(x)\cos(x) + \frac{n-1}{n}\int \sin^{n-2}(x)dx \quad (7.137)$$

The first term on the RHS becomes zero because we are evaluating with limits from $x = \theta \in (0, \pi)$, so the sine term vanishes. So, we can continuously evaluate the simplified integral down to the last evaluation, with a couple of little tricks, to obtain C_n. Please follow the patterns closely. This process goes as follows:

$$1 = 2\pi C_n^2 \int_0^\pi \sin^{2\ell+1}\theta d\theta$$

$$= 2\pi C_n^2 \frac{2\ell}{2\ell+1} \int_0^\pi \sin^{2\ell-1}\theta d\theta$$

$$= 2\pi C_n^2 \frac{2\ell(2\ell-2)}{(2\ell+1)(2\ell-1)} \int_0^\pi \sin^{2\ell-3}\theta d\theta$$

$$\vdots$$

$$= 2\pi C_n^2 \frac{2\ell(2\ell-2)(2\ell-4)\ldots(2\ell-2\ell+2)}{(2\ell+1)(2\ell-1)\ldots(2\ell-2\ell+1)} \int_0^\pi \sin\theta d\theta$$

$$= 2\pi C_n^2 \frac{2^\ell \cdot \ell(\ell-1)\ldots 1}{(2\ell+1)(2\ell-1)\ldots(2\ell-2\ell+1)} \times 2$$

$$= 2\pi C_n^2 2^\ell \ell! \frac{1}{(2\ell+1)(2\ell-1)\ldots(2\ell-2\ell+1)} \times 2$$

$$= 2\pi C_n^2 2^\ell \ell! \frac{2\ell(2\ell-2)...(2\ell-2\ell+2)}{(2\ell+1)2\ell(2\ell-1)...1} \times 2$$

$$= 2\pi C_n^2 \left(2^\ell \ell!\right)^2 \frac{2}{(2\ell+1)!}$$

From this, we obtain the result

$$1 = 2\pi C_n^2 \left(2^\ell \ell!\right)^2 \frac{2}{(2\ell+1)!}$$

Solving for C_n, we have

> **Normalization Factor For $Y_\ell^{-\ell}(\theta,\varphi)$**
>
> $$C_n = \frac{1}{2^\ell \ell!} \sqrt{\frac{(2\ell+1)!}{4\pi}} \qquad (7.138)$$

We can define another operator as the *adjoint* operator of $\widehat{L}_-$, denoted by $\widehat{L}_+$ and it is defined as

> **Orbital Angular Momentum Raising Operator Definition $\widehat{L}_+$**
>
> $$\widehat{L}_+ = \widehat{L}_x + i\widehat{L}_y \qquad (7.139)$$

By this definition, the eigenvalues of $\widehat{L}_+$ are complex conjugates to those of $\widehat{L}_-$, and we say the operators are *adjoint operators* of one another. This operator is known as the **orbital angular momentum raising operator** $\widehat{L}_+$. We can find the explicit operator by substituting $\widehat{L}_x$ and $\widehat{L}_y$ into the definition for the $\widehat{L}_+$, where we obtain

$$\widehat{L}_x + i\widehat{L}_y = i\hbar\left(\sin\varphi \frac{\partial}{\partial \theta} + \cot\theta\cos\varphi \frac{\partial}{\partial \varphi}\right) - \hbar\left(-\cos\varphi \frac{\partial}{\partial \theta} + \cot\theta\sin\varphi \frac{\partial}{\partial \varphi}\right)$$

$$= \hbar(i\sin\varphi + \cos\varphi)\frac{\partial}{\partial \theta} + \hbar(i\cos\varphi - \sin\varphi)\cot\theta \frac{\partial}{\partial \varphi}$$

$$= \hbar\left(e^{i\varphi}\frac{\partial}{\partial \theta} + ie^{i\varphi}\cot\theta \frac{\partial}{\partial \varphi}\right)$$

Thus, for $\widehat{L}_+$, we have the following operator:

7.10 Angular Momentum Ladder Operators

Orbital Angular Momentum Raising Operator $\widehat{L}_+$

$$\widehat{L}_+ = \hbar e^{i\varphi}\left[\frac{\partial}{\partial \theta} + i\cot\theta \frac{\partial}{\partial \varphi}\right] \tag{7.140}$$

The *raising operator* $\widehat{L}_+$ is analogous to the lowering operator in its utility, but with the highest energy state. With the lowering operator $\widehat{L}_-$, we found that if we consider the lowest energy state of the spherical harmonics, given by $Y_\ell^{-\ell}$, that $\widehat{L}_- Y_\ell^{-\ell} = 0$. Well, if we now consider the highest energy state Y_ℓ^ℓ for a given ℓ, let us see what happens when $\widehat{L}_+$ operates on this state. We then have

$$\widehat{L}_+ Y_{\ell,\ell} = \hbar e^{i\varphi}\left[\frac{\partial}{\partial \theta} + i\cot\theta \frac{\partial}{\partial \varphi}\right] Y_{\ell,\ell}$$

$$= \hbar e^{i\varphi}\left[\frac{\partial}{\partial \theta} + i\cot\theta \frac{\partial}{\partial \varphi}\right] A_\ell^\ell e^{i\ell\varphi}$$

$$= \hbar e^{i\varphi}\left[\frac{\partial A_\ell^\ell}{\partial \theta} e^{i\ell\varphi} + i(i\ell)\cot\theta e^{i\ell\varphi} A_\ell^\ell\right]$$

$$= \hbar e^{i\varphi} e^{i\ell\varphi}\left[\frac{\partial A_\ell^\ell}{\partial \theta} - \ell\cot\theta A_\ell^\ell\right]$$

Our concern is with the term in brackets. Similar to the case of $\widehat{L}_- Y_\ell^{-\ell}$, because of this term, $\widehat{L}_+ Y_\ell^\ell$ turns out to be identically zero. Let us consider the *associated Legendre function* A_ℓ^ℓ given by

$$A_\ell^\ell = C_\ell (1-x^2)^{\frac{\ell}{2}} \frac{d^{2\ell}}{dx^{2\ell}} (x^2 - 1)^\ell$$

In the derivative term, we are now considering the highest order derivative possible before this term vanishes. Each derivative taken knocks the exponent down to a coefficient, while the exponent lessens by one. From this pattern, one finds this term becomes the constant given by

$$\frac{d^{2\ell}}{dx^{2\ell}}\left[(x^2-1)^\ell\right] = (2\ell)! \tag{7.141}$$

Then, we can write A_ℓ^ℓ as

$$A_\ell^\ell = C_\ell (2\ell)! (1-x^2)^{\frac{\ell}{2}} = D_\ell (1-x^2)^{\frac{\ell}{2}} \tag{7.142}$$

Evaluating the terms involved in $L_+ Y_\ell^\ell$ using Eq. (7.142), we have

$$\frac{dA_\ell^\ell}{d\theta} = \frac{dx}{d\theta} \frac{dA_\ell^\ell}{dx}$$

$$= D_\ell(-\sin\theta)(-2x)\left[\frac{\ell}{2}(1-x^2)^{\frac{\ell}{2}-1}\right]$$

$$= D_\ell(-\sin\theta)(-2x)\left[\frac{\ell}{2}(1-x^2)^{\frac{\ell}{2}}(1-x^2)^{-1}\right]$$

Again, substituting $1 - x^2 = \sin^2\theta$ leads to

$$\frac{dA_\ell^\ell}{d\theta} = D_\ell(-2\cos\theta)(-\sin\theta)\left[\frac{\ell}{2}(1-x^2)^{\frac{\ell}{2}} \frac{1}{\sin^2\theta}\right]$$

$$= \ell D_\ell \frac{\cos\theta}{\sin\theta}(1-x^2)^{\frac{\ell}{2}}$$

$$= \ell \cot\theta A_\ell^\ell$$

It readily follows from this result, that

$$\widehat{L}_+ Y_{\ell,\ell} = \hbar C_n e^{i(\ell+1)\varphi}\left[\frac{\partial A_\ell^\ell}{\partial \theta} - \ell\cot\theta A_\ell^\ell\right] = 0$$

This gives a relation that must be satisfied by A_ℓ^ℓ, namely

Highest-Level Associated Legendre Function Condition

$$\frac{\partial A_\ell^\ell}{\partial \theta} - \ell\cot\theta A_\ell^\ell = 0 \qquad (7.143)$$

Let's now find the eigenvalues of $\widehat{L}_-$ and $\widehat{L}_+$ in relation to those of $\widehat{L}$ and $\widehat{L}_z$. The definitions of $\widehat{L}_\pm$ lead to the fact that they both have complex eigenvalues λ_+, λ_-, respectively, and this leads to the following relations:

$$\widehat{L}_- Y_\ell^m = \lambda_- Y_\ell^m$$

$$\widehat{L}_+ Y_\ell^m = \lambda_+ Y_\ell^m$$

$$\widehat{L}_- \widehat{L}_+ Y_\ell^m = \lambda_- \lambda_+ Y_\ell^m = |\lambda_+|^2 Y_\ell^m = |\lambda_-|^2 Y_\ell^m$$

7.10 Angular Momentum Ladder Operators

Next, let us consider the commutation relation, which we will need, between $\widehat{L}_z$ and $\widehat{L}_+$ given by

$$\left[\widehat{L}_z, \widehat{L}_+\right] = \widehat{L}_z(\widehat{L}_x + i\widehat{L}_y) - (\widehat{L}_x + i\widehat{L}_y)\widehat{L}_z$$

$$= \left(\widehat{L}_z\widehat{L}_x - \widehat{L}_x\widehat{L}_z\right) + i\left(\widehat{L}_z\widehat{L}_y - \widehat{L}_y\widehat{L}_z\right)$$

$$= i\hbar\widehat{L}_y + i(-i\hbar\widehat{L}_x)$$

$$= \hbar\widehat{L}_+$$

In the above, we have used the commutation relations $\widehat{\mathbf{L}} \times \widehat{\mathbf{L}} = i\hbar\widehat{\mathbf{L}}$. This leads to the following equation for $\widehat{L}_+$:

$$\widehat{L}_z\widehat{L}_+Y_\ell^m = \widehat{L}_+\widehat{L}_zY_\ell^m + \left[\widehat{L}_z, \widehat{L}_+\right]Y_\ell^m$$

$$\widehat{L}_z\widehat{L}_+Y_\ell^m = \widehat{L}_+\widehat{L}_zY_\ell^m + \hbar\widehat{L}_+Y_\ell^m$$

$$= m\hbar\widehat{L}_+Y_\ell^m + \hbar\widehat{L}_+Y_\ell^m$$

$$= \hbar(m+1)\widehat{L}_+Y_\ell^m$$

Based on the eigensolutions we obtained for $\widehat{L}_z\Phi = \hbar m\Phi$, which also leads to

$$\widehat{L}_zY_\ell^{m+1} = \hbar(m+1)Y_\ell^{m+1}$$

Since all eigenstates of $\widehat{L}_z$ are distinct orthogonal states, these two states must be proportional to one another, such that

$$\widehat{L}_+Y_\ell^m = \alpha Y_\ell^{m+1} \tag{7.144}$$

A similar conclusion follows for $\widehat{L}_-$, because in this case, the commutation relation gives

$$\left[\widehat{L}_z, \widehat{L}_-\right] = \widehat{L}_z(\widehat{L}_x - i\widehat{L}_y) - (\widehat{L}_x - i\widehat{L}_y)\widehat{L}_z$$

$$= \widehat{L}_z\widehat{L}_x - i\widehat{L}_z\widehat{L}_y - \widehat{L}_x\widehat{L}_z + i\widehat{L}_y\widehat{L}_z$$

$$= i\hbar\widehat{L}_y + i(i\widehat{L}_x)$$

$$= -\hbar\widehat{L}_-$$

This leads to the following equation for $\widehat{L}_-Y_\ell^m$:

$$\widehat{L}_z\widehat{L}_-Y_\ell^m = \widehat{L}_-\widehat{L}_zY_\ell^m + \left[\widehat{L}_z, \widehat{L}_-\right]Y_\ell^m$$

$$= \widehat{L}_-\widehat{L}_z Y_\ell^m - \hbar \widehat{L}_- Y_\ell^m$$
$$= m\hbar \widehat{L}_- Y_\ell^m - \hbar \widehat{L}_- Y_\ell^m$$
$$= \hbar(m-1)\widehat{L}_- Y_\ell^m$$

Therefore, analogous to $\widehat{L}_+$, we find the eigenvalue equation for $\widehat{L}_-$ to be given by

$$\widehat{L}_- Y_\ell^m = \beta Y_\ell^{m-1} \qquad (7.145)$$

The above relations allow us to determine eigenvalues α and β in terms of ℓ and m, as follows:

$$Y_\ell^m \widehat{L}_+^\dagger \widehat{L}_+ Y_\ell^m = Y_\ell^{m+1} |\alpha|^2 \hbar^2 Y_\ell^{m+1} \qquad (7.146)$$

$$Y_\ell^m \widehat{L}_-^\dagger \widehat{L}_- Y_\ell^m = Y_\ell^{m-1} |\beta|^2 \hbar^2 Y_\ell^{m-1} \qquad (7.147)$$

But $L_+^\dagger = \widehat{L}_-$, and $L_-^\dagger = \widehat{L}_+$, since both operators are *adjoint operators* of one another. Then, we have

$$\widehat{L}_+ \widehat{L}_- Y_\ell^m = \left(\widehat{L}_x + i\widehat{L}_y\right)\left(\widehat{L}_x - i\widehat{L}_y\right) Y_\ell^m$$
$$= \left[\widehat{L}_x \widehat{L}_x + i(\widehat{L}_y \widehat{L}_x - \widehat{L}_x \widehat{L}_y) + \widehat{L}_y \widehat{L}_y\right] Y_\ell^m$$
$$= \left[\widehat{L}^2 - \widehat{L}_z^2 + i\left(-i\hbar \widehat{L}_z\right)\right] Y_\ell^m$$
$$= \left(\ell(\ell+1)\hbar^2 - m^2\hbar^2 + m\hbar^2\right) Y_\ell^m$$
$$= (\ell + m)(\ell - m + 1)\hbar^2 Y_\ell^m$$

This result allows us to determine $|\beta|^2$. By substituting the result into Eqs. (7.146) and (7.148), we have

$$(\ell + m)(\ell - m + 1)\left(Y_\ell^m\right)^2 = |\beta|^2 \left(Y_\ell^{m+1}\right)^2 \qquad (7.148)$$

Since the wave functions are normalized, integration of Eq. (7.148) gives the result

$$|\beta|^2 = (\ell + m)(\ell - m + 1) \qquad (7.149)$$

Similarly, for $|\alpha|^2$, we obtain

$$\widehat{L}_- \widehat{L}_+ A_\ell^m = \left(\widehat{L}_x - i\widehat{L}_y\right)\left(\widehat{L}_x + i\widehat{L}_y\right) A_\ell^m$$

7.10 Angular Momentum Ladder Operators

$$= \left[\widehat{L}_x\widehat{L}_x + i(\widehat{L}_x\widehat{L}_y - \widehat{L}_y\widehat{L}_x) + \widehat{L}_y\widehat{L}_y\right]A_\ell^m$$

$$= \left[\widehat{L}^2 - \widehat{L}_z^2 + i\left(i\hbar\widehat{L}_z\right)\right]A_\ell^m$$

$$= \left(\ell(\ell+1)\hbar^2 - m^2\hbar^2 - m\hbar^2\right)A_\ell^m$$

$$= (\ell - m)(\ell + m + 1)A_\ell^m$$

Substitution of this result into Eq. (7.146), integrating it, we have the following results for $|\beta|$ and $|\alpha|$ given by

$$|\alpha| = \sqrt{(\ell - m)(\ell + m + 1)} \qquad (7.150a)$$

$$|\beta| = \sqrt{(\ell + m)(\ell - m + 1)} \qquad (7.150b)$$

Finally, we can state two important relations for the lowering and raising operators $\widehat{L}_-$, $\widehat{L}_+$, defining the two operator eigenvalue equations given by

Angular Momentum Raising Eigenoperator Equations

$$L_+ Y_\ell^m = \sqrt{(\ell - m)(\ell + m + 1)}\, Y_\ell^{m+1} \qquad (7.151)$$

and

Angular Momentum Lowering Eigenoperator Equation

$$L_- Y_\ell^m = \sqrt{(\ell + m)(\ell - m + 1)}\, Y_\ell^{m-1} \qquad (7.152)$$

To demonstrate the utility of the above relations using the lowering and raising operators, let's do a few examples. The idea is that given any state $Y_\ell^{-\ell}$, one can use the *raising operator* $\widehat{L}_+$ as many of 2ℓ times to obtain higher level states Y_ℓ^{m+1}. And likewise, one can use the *lowering operator* $\widehat{L}_-$ as many of 2ℓ times, on Y_ℓ^ℓ, to obtain lower level states Y_ℓ^{m-1}. This is a utility of these kinds of operators. Next, we'll consider an example to illustrate how these operators can be used.

Let's consider the lowest energy state Y_1^{-1} (for $\ell = 1$). Using Eq. (7.137), we have

$$Y_1^{-1} = C_n \sin^1\theta e^{-i\varphi} = \sqrt{\frac{3}{8\pi}} \sin\theta e^{-i\varphi}$$

Using the operator equation for $\widehat{L}_+$ given in Eq. (7.151), we have

$$L_+ Y_1^{-1} = \sqrt{(\ell - m)(\ell + m + 1)} Y_\ell^0 = \sqrt{2 \cdot 1} Y_\ell^0$$

Substituting the operator for $\widehat{L}_+$, the LHS then becomes

$$L_+ Y_1^{-1} = \hbar e^{i\varphi}\left(\frac{\partial}{\partial \theta} + i\cot\theta \frac{\partial}{\partial \varphi}\right) Y_1^{-1}$$

$$= \sqrt{\frac{3}{8\pi}} \hbar e^{i\varphi}\left(\frac{\partial}{\partial \theta} + i\cot\theta \frac{\partial}{\partial \varphi}\right) \sin\theta e^{-i\varphi}$$

$$= \sqrt{\frac{3}{8\pi}} \hbar [\cos\theta + i\cot\theta(-i\sin\theta)]$$

$$= \sqrt{\frac{3}{8\pi}} \hbar (2\cos\theta) = \sqrt{2} Y_1^0$$

From this, we have the following result for Y_1^0:

$$Y_1^0(\cos\theta) = \frac{2}{\sqrt{2}} \sqrt{\frac{3}{8\pi}} \hbar\cos\theta = \sqrt{\frac{3}{4\pi}} \hbar\cos\theta \qquad (7.153)$$

This process could continue, providing a way to obtain all the higher energy states for the orbital $\ell = 1$. These orbitals are examples of p orbitals, defined by the quantum number $\ell = 1$. This demonstrates a powerful utility of the raising (or lowering) operator, particularly, if only the ground state is known.

7.11 Chapter Summary

To examine orbital angular momentum in quantum mechanics, we first reviewed the concept in the classical domain. From there, we took the definitions and applied the quantum mechanical operators for $\widehat{p}$, to obtain the orbital angular momentum operators. We then solved for

7.11 Chapter Summary

the eigenvalues for $\widehat{L}_z^2$, given by m^2 and for $\widehat{L}^2$, given by $\ell(\ell+1)$. We also obtained the corresponding wave functions given by the spherical harmonics $Y_{\ell,m}$. In obtaining the wave functions, we also learned that the solutions to the orbital angular momentum problem were enabled by the earlier work of Legendre and others, in analyzing gravitational potential expansions. It all came together ultimately to provide a path to help solve the angular momentum problem exactly.

We then introduced the commutation operator and relevant relations involving $\widehat{\mathbf{p}}$ and $\widehat{\mathbf{L}}$. This allowed us to introduce the lowering and raising operators for the orbital angular momentum, and demonstrate their utility with the spherical harmonics. This completed our objectives with the orbital angular momentum problem. Table 7.6 lists some of the key results for the total orbital angular momentum.

Table 7.6: Chapter 7 Summary Equations.

Name	Equation
L_z eigen-function	$\Phi(\varphi) = \frac{1}{\sqrt{2\pi}} e^{im\varphi}$
L_z eigenvalues	$m = ..., -2, -1, 0, 1, 2, ...$
L^2 eigenvalues	$\lambda = \ell(\ell+1)$
L^2 quantum numbers	$\ell = 0, 1, 2, 3, ...$
Relation between m and ℓ	$|m| \leq \ell$
Legendre polynomials P_ℓ	$P_\ell = \frac{1}{2^\ell \ell!} \frac{d^\ell}{dx^\ell}(x^2-1)^\ell$
Associated Legendre functions	$A_\ell^m = \frac{1}{2^\ell \ell!}(1-x^2)^{\frac{m}{2}} \frac{d^{\ell+m}}{dx^{\ell+m}}(x^2-1)^\ell$
L^2 eigen-function	$Y_{\ell,m}(\theta,\varphi) = A_\ell^m(x=\cos\theta)e^{im\varphi}$
Lowering Operator	$\widehat{L}_- = \widehat{L}_x - i\widehat{L}_y$
Raising Operator	$\widehat{L}_+ = \widehat{L}_x + i\widehat{L}_y$

Table 7.7 summarizes and illustrates more examples of spherical harmonic functions $Y_{\ell,m}$ in 3D. Note that only positive m values are shown, however, keep in mind that the corresponding $-m$ cases are only rotations of their $+m$ counterparts.

Table 7.7: Real part of the spherical harmonics Y_ℓ^m (projected into cartesian coordinates), indicating both quantum numbers ℓ and m, the Associated Legendre Polynomial factor, and the azimuthal factor from L_z.

ℓ	m	$A_\ell^m(\cos\theta)$	$\Phi(\varphi)$	Real Part$[Y_{\ell,m}]$
0	0	$\frac{1}{(4\pi)^{1/2}}$	1	
1	0	$\left(\frac{3}{4\pi}\right)^{1/2}\cos\theta$	1	
1	1	$-\left(\frac{3}{8\pi}\right)^{1/2}\sin\theta$	$e^{i\varphi}$	
2	0	$\left(\frac{5}{16\pi}\right)^{1/2}(3\cos^2\theta - 1)$	1	
2	1	$-\left(\frac{15}{8\pi}\right)^{1/2}\cos\theta\sin\theta$	$e^{i\varphi}$	
2	2	$\left(\frac{15}{32\pi}\right)^{1/2}\sin^2\theta$	$e^{i2\varphi}$	
3	0	$\left(\frac{7}{4\pi}\right)^{1/2}(5\cos^2\theta - 3)\cos\theta$	1	
3	1	$-\left(\frac{7}{48\pi}\right)^{1/2}(1 - 5\cos^2\theta)\sin\theta$	$e^{i\varphi}$	

Table 7.7

3	2	$\left(\frac{7}{480\pi}\right)^{1/2}\sin^2\theta\cos\theta$	$e^{i2\varphi}$	
3	3	$-\left(\frac{7}{2880\pi}\right)^{1/2}\sin^3\theta$	$e^{i3\varphi}$	
4	0	$\left(\frac{9}{4\pi}\right)^{1/2}(35\cos^4\theta - 30\cos^2\theta + 3)$	1	

7.12 CHAPTER PROBLEMS

Problem 7.1 Show that the two associated Legendre functions A_1^1 and A_1^0 are normal where

$$\int_0^\pi \left(A_1^0(\theta)\right)^* A_1^1(\theta) d\theta$$

Use Table 7.7, which lists these functions.

Problem 7.2 Show that the function

$$u(r) = \frac{1}{r}$$

is a solution to the equation given by

$$\frac{1}{r^2}\frac{\partial}{\partial r}\left(r^2 \frac{\partial u}{\partial r}\right) = 0$$

Substitute $u(r)$ into this equation.

Problem 7.3 Earlier, we used what was called the Trigonometric Recurrence Relation to obtain the normalization constant. Use integration by parts, or

$$\int u \, dv = uv - \int v \, du$$

And prove that

$$\int \sin^n(\theta)d\theta = -\sin^{n-1}\theta\cos\theta + n-1 \int \sin^{n-2}\theta\cos\theta d\theta$$

Hint: Start with

$$\int \sin^n\theta d\theta = \int \sin^{n-1}\theta \sin\theta d\theta = \int u dv$$

Problem 7.4 Let $u(\varphi,\theta) = Y_1^1$. Show that u satisfies the following equation

$$\frac{1}{r^2\sin\theta}\frac{\partial}{\partial\theta}\left(\sin\theta\frac{\partial u}{\partial\theta}\right) + \frac{1}{r^2\sin^2\theta}\frac{\partial^2 u}{\partial\varphi^2} = -\frac{2}{r^2}u$$

We are using only the Laplacian terms dependent on φ and θ, since u is independent of r.

Problem 7.5 In Eq. (7.153), we used the raising operator $\widehat{L}_+$ to obtain $Y_1^0(\cos\theta) = \sqrt{\frac{3}{4\pi}}\hbar\cos\theta$. Use this result along with the raising operator to show that $Y_1^1 = -\sqrt{\frac{3}{8\pi}}\sin\theta e^{i\phi}$.

Problem 7.6 Prove that

$$\left[\widehat{L}^2, \widehat{L}_z\right] = 0$$

Discuss the properties of the Hamiltonian that lead to this result.

Problem 7.7 The lowest spherical harmonic state for $\ell = 2$ is

$$Y_2^{-2}(\varphi,\theta) = \sqrt{\frac{15}{32\pi}}\sin^2\theta e^{-2i\varphi}$$

Use the raising ladder operator $\widehat{L}_+$ to show that

$$Y_2^{-1}(\varphi,\theta) = \sqrt{\frac{15}{8\pi}}\sin\theta\cos\theta e^{-i\varphi}$$

CHAPTER 8

The Hydrogen Atom

Quantum mechanics sheds a great deal of light on the behavior of all the elements of the chemical table. The journey to such a point of understanding began with experiments on the simplest known gas, namely hydrogen, whose chemical symbol is H. In this chapter, we will treat, in sufficient detail, an important application of the Schrödinger equation to the hydrogen atom. Because it is the simplest of all the chemical elements, with a single electron and proton (but no neutron), it can be solved exactly. By 1910, the understanding of the structure of atoms had been established with J.J. Thomson's discovery of the electron and E. Rutherford's discovery of the interior nucleus, *etc*. Because of its simplicity and susceptibility to analytical models, hydrogen has served as an important element for developing and validating quantum mechanics. One of the reasons for this is because of the progress in experimental capability around the early part of the 20th century which allowed for novel observations with gases. So, let's begin by discussing the kind of experimental evidence that had been encountered with hydrogen that ultimately served as an impetus to further extend the realm of quantum physics, to not only predict the observed measurements, but explain them deeply.

8.1 Atomic Spectra of Hydrogen

In addition to having an understanding of the structure of atoms, the development of evacuated *cathode ray tubes* with inserted electrodes turned out to be integral in many historic experiments in physics. With this, physicists learned how to measure the atomic spectra of atoms, using **discharge tubes**. These tubes would contain a trapped gas to be tested, as the tubes were also inline with an electrical circuit. The first methods used were emission-based and involved exciting atoms via high-voltages on the order of several thousand volts. Using an appropriate optical canvas, preceded by a prism to separate the light, and capture the emitted separated light after being excited, a unique set of spectral lines could be observed. This idea is illustrated in Fig. 8.1. A

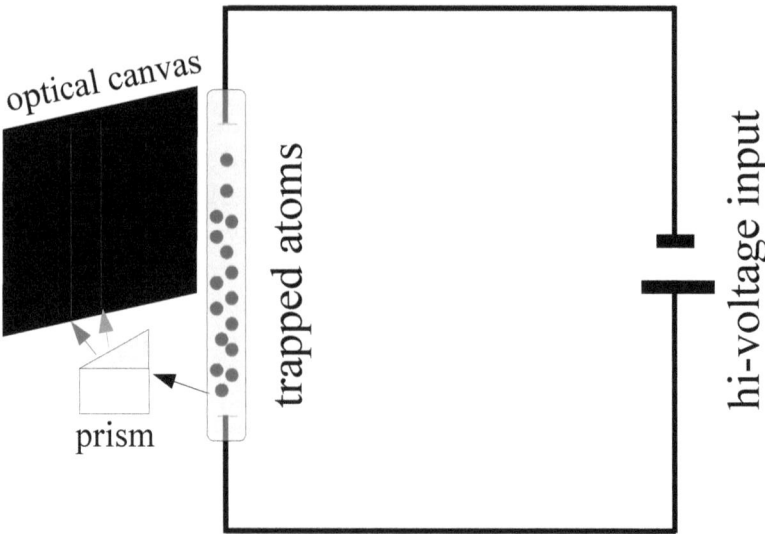

Figure 8.1: Simplified setup of measurements of atomic and/or molecular spectra. Atoms are trapped inside an evacuated tube and placed in a high-voltage circuit. Atoms are excited ohmically, and upon decaying back down to lower energy states, emit characteristic light, serving as a signature of the investigated atoms or molecules.

measured spectrum from hydrogen is shown in Fig. 8.2. In this setup,

8.1 Atomic Spectra of Hydrogen

Figure 8.2: Measured atomic spectra of hydrogen atoms in the visible region. There are typically 3-4 spectral lines, as shown with violet (left), blue-green (middle), and red (right).

atoms are trapped inside an evacuated tube, and this tube is placed in a high-voltage circuit represented on the right. They are then excited electrically with a high voltage, and upon decaying back down to lower energy states, the atoms emit characteristic light, serving as a signature of the investigated gas. Spectral lines were originally observed from only thermally exciting atoms through the electrical heating. This was the first kind of method known to generate **atomic spectra**. An example of a measured hydrogen spectra is shown in Fig. 8.2.

With the human eye, only the lines falling in the visible spectrum can be observed this way. The others will not be visible to the naked eye. Table 8.1 lists some known visible spectral data containing measured wavelengths ($\lambda = c/f$) from hydrogen, indicating the relative intensity, corresponding transition, and colors. The higher energy excited states (top-most) having lower intensities, reflecting their probabilities, while lower energy states have increasing intensity. At the time, there was still missing, a deeper insight or understanding as to the origins of these observations with heated gas. The next section discusses the beginning of understanding *atomic spectra* from physics.

Table 8.1: Measured spectral lines of the Balmer series of hydrogen atoms in the visible light region.

Wavelength (nm)	Relative Intensity	Transition	Color
383.54	5	$9 \to 2$	violet
388.91	6	$8 \to 2$	violet
397.01	8	$7 \to 2$	violet
410.17	15	$6 \to 2$	violet
434.05	30	$5 \to 2$	violet
486.13	80	$4 \to 2$	blue-green

Table 8.1

656.27	120	3 → 2	red
656.29	180	3 → 2	red

Upon observing this kind of spectral data from hydrogen, it became the objective of many to predict such spectral data. Bohr's model was one the first successful ones to do so. It's simple, yet highly accurate. We'll discuss it next.

8.2 Bohr's Atomic Hydrogen Model

Recall that in Chap. 2, we discussed Rydberg's proposed *empirical* model that well-described the behavior of the hydrogen spectral lines, particularly, if only thermal excitation was used to excite the atoms. However, there still lacked any concrete insight as to the physical origin of these observations. Additionally, Rydberg's model turned out to be very limited if one wished to describe other atomic and molecular systems. Around 1913, this changed with the work of Danish physicist Neils Bohr (1885-1962). Bohr imagined the hydrogen atom as illustrated in Fig. 8.3, assuming perfect circular discrete orbitals. At that time, the atomic model of the hydrogen atom was already believed to be a heavy nucleus with a light-weight electron whizzing around it. With this structure in mind, Bohr utilized a quanta assumption in line with what Planck and Einstein had already done. He assumed that in the hydrogen atom, the angular momentum L takes only discrete values. That is to say that the angular momentum L is given by

$$|\mathbf{L}| = L_n = rp = rmv = n\hbar \tag{8.1}$$

In (8.1), n is an integer value given by $1, 2, 3$, and so on which will end up as the principal quantum number appearing in the Rydberg formula. Considering hydrogen, which has a single electron, the Bohr model assumes the single electron is orbiting in planar loops around the nucleus, which contains one proton, but no neutron.

It was shown in Sec. 7.1 that there exists a centripetal force for a mass in a steady circular orbit. For the electron, the Coulomb electric force

8.2 Bohr's Atomic Hydrogen Model

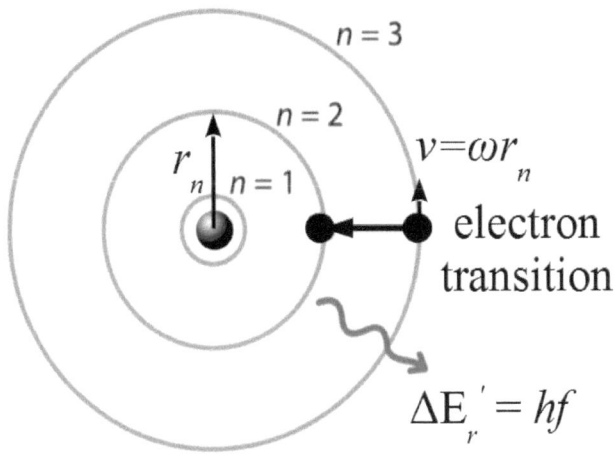

Figure 8.3: Simplified diagram of a hydrogen atom, based on Bohr's model. Perfect circular discrete orbitals are assumed. Light is emitted only when transitioning from one orbital to another.

balances this force. Thus, for a circular orbit, we can sum these forces to obtain the following (in SI units):

$$\sum \mathbf{F} = m\mathbf{a} \Rightarrow \frac{e^2}{4\pi\epsilon_0 r^2} = \frac{mv^2}{r} \tag{8.2}$$

After substituting the Bohr assumption for the angular momentum $L = L_n$ from Eq. (8.1), we have a way to determine the orbital radius r using

$$\frac{e^2}{4\pi\epsilon_0 r^2} = \frac{n^2 \hbar^2}{mr^3} \tag{8.3}$$

Solving for the radius r, we obtain

$$r = \frac{n^2 \hbar^2 4\pi\epsilon_0}{me^2} = \frac{n^2 h^2 \epsilon_0}{m\pi e^2} \tag{8.4}$$

Eq. (8.4) reveals that a consequence of Bohr's assumption of the quanta of angular momentum L, is that the orbital radius r is also quantized because it is a function of the discrete integer n. We can find the total energy E for the hydrogen electron by summing the kinetic and potential energies as

$$E = K + V = \frac{L^2}{2mr^2} - \frac{e^2}{4\pi\epsilon_0 r} \tag{8.5}$$

This can also be written as

$$E = \frac{r^2 v^2 m^2}{2mr^2} - \frac{e^2}{4\pi\epsilon_0 r} = \frac{v^2 m}{2} - \frac{e^2}{4\pi\epsilon_0 r} \tag{8.6}$$

By using the force balance equation in Eq. (8.2), the first term on the far right hand side can be written as

$$E = \frac{e^2}{8\pi\epsilon_0 r} - \frac{e^2}{4\pi\epsilon_0 r} = -\frac{e^2}{8\pi\epsilon_0 r}$$

The energy E is just half the Coulomb potential energy V, for any value of r. It follows that $|V|/K = 2$. This result turns out to be general, and is an example of more fundamental theorem known as the *Virial theorem*. After substituting the result for the orbital radius r, we find that the energy E is given by

Hydrogen Atom Discrete Energy Levels (Bohr Model)

$$E = -\frac{me^4}{8\epsilon_0^2 h^2} \frac{1}{n^2} \tag{8.7}$$

Eq. (8.7) reveals that the energy of the electron in the hydrogen atom is negative. This is the result of the fact that the attractive negative Coulomb's potential energy is the dominating energy which, consequently, keeps the electron *bound* to the hydrogen atom's nucleus. For this reason, negative energies always correspond to **bound energy states**. From his assumption utilizing the little known constant from Plank $\hbar$, Bohr found that the *bound energy states are quantized*, meaning that the available energies the electron can take are *not* arbitrary, but one of infinitely many discrete or quantum states. In the Bohr model, this is inherited from the quantized angular momentum $L = L_n$.

From Bohr's result, the smallest possible orbital radius r_0 for the electron in the hydrogen atom is not zero, but corresponds to setting the integer

8.2 Bohr's Atomic Hydrogen Model

n to its lowest possible value, $n = 1$, which leads to

$$a_0 = \frac{h^2 \epsilon_0}{m\pi e^2} = \frac{4\pi \hbar^2 \epsilon_0}{me^2} = 5.2919 \times 10^{-11} \text{m} = 0.52919 \text{Å} \quad (8.8)$$

Eq. (8.8) is known as the **Bohr radius** and is often used as a unit of measure for atomic distances. Bohr found that the spectral lines that were routinely measured for hydrogen were described well by *transitions* from one energy state to another. This lead to success in predicting the spectral lines of hydrogen, corresponding to changes in the quantum states of the electron, expressed by

Hydrogen Atom Discrete Energy Transitions (Bohr Model)

$$\Delta E = \frac{me^4}{8\epsilon_0^2 h^2} \left(\frac{1}{n_1^2} - \frac{1}{n_2^2} \right) \quad (8.9)$$

In Eq. (8.9), $n_2 > n_1$, which yields a positive change in energy during the transition, so it corresponds to when the electron descends from a higher energy state back to a lower one. Thus, the smaller n value also corresponds to a state closer to the nucleus, or to a more bound state. This state has a lower energy, or is more negative. This provides an interpretation of the observations that could be formulated by Bohr, that as the electron undergoes its transition back down to a lower state, it gives off photons, or quanta of light containing information about the energy levels of the electron. It also suggests that we may use Einstein and Planck's relationship for the light quanta emitted, which was known by 1911, to write the transition equation in terms of the wavelength λ

$$\Delta E(\lambda) = \hbar\omega = h\frac{c}{\lambda} = \frac{me^4}{8\epsilon_0^2 h^2}\left(\frac{1}{n_1^2} - \frac{1}{n_2^2}\right) \Rightarrow \frac{1}{\lambda} = \frac{me^4}{8c\epsilon_0^2 h^3}\left(\frac{1}{n_1^2} - \frac{1}{n_2^2}\right) \quad (8.10)$$

In terms of the frequency f, the energy $E(f)$ is then given by

$$f = \frac{me^4}{8\epsilon_0^2 h^3}\left(\frac{1}{n_1^2} - \frac{1}{n_2^2}\right) \quad (8.11)$$

Another impressive result from Bohr's model was that when this transition equation is expressed in the form of the empirical relation proposed by Rydberg (c. Eq. (2.99)), evaluation of the coefficient obtained by Bohr worked out to be

> **Rydberg Constant**
>
> $$R_d = \frac{me^4}{8c\epsilon_0^2 h^3} \approx 10965609 \quad \text{m}^{-1} \qquad (8.12)$$

The value from Bohr's model turned out to be the same as the empirical parameter determined by Rydberg, applied to the hydrogen atom. This is the *Rydberg constant* R_d. Thus, Bohr had obtained a result that not only predicted the energy transitions, but he predicted the value of the Rydberg constant in terms of other known fundamental constants. After Planck's success with blackbody radiation, leading to the discovery of h, Einstein's success with the photoelectric effect requiring h, and now Bohr's success with the hydrogen model deploying h, the idea of quanta was becoming an indisputable truth in this new kind of physics. But, now that we have also considered orbital angular momentum in sufficient detail in the previous chapter, and we know about the Schrödinger equation, we can combine this knowledge to solve the hydrogen atom problem using the Schrödinger equation.

8.3 Schrödinger's Hydrogen Atom Model

When Bohr treated the hydrogen atom, planar loops where assumed to express the angular momentum, simply as $L = mrv$. This is valid when p and r are orthogonal. This is not a necessary assumption for an atom, which lives in 3D. Additionally, Bohr treated only the electron energy to solve the problem. The nucleus was not considered at all, though this assumption turns out to be fine given the relatively large mass of the nucleus. But this approximation should be justified somehow. It turns out that with the Schrödinger equation, it is not necessary to ignore the nucleus. It is remarkable that although solving the hydrogen atom problem using quantum mechanics involves significantly more mathematics, we will see that it leads to results that are in complete agreement with Rydberg's empirical model, Bohr's early quantum model, and ultimately with even more general experimental data given by hydrogen's atomic spectra. So, let's work out the results from the Schrödinger equation.

Recall that the potential energy $V(\mathbf{r} = |\mathbf{r}_1 - \mathbf{r}_2|)$ of a system involving a

8.3 Schrödinger's Hydrogen Atom Model

single electron with position $\mathbf{r}_1$ and charge $q_1 = -e$, interacting with a nucleus having position $\mathbf{r}_2$ of charge $q_2 = Ze$, is given (in SI units) by

$$V(r) = -\frac{Ze^2}{4\pi\varepsilon_0|\mathbf{r}_1 - \mathbf{r}_2|} \tag{8.13}$$

This electrostatic potential describes any nuclear system of charge Ze having a single electron, like hydrogen, but also, singly-ionized helium (He^+), doubly-ionized lithium (Li^{++}), etc. Including the nucleus as well as the outer electron, the Hamiltonian eiqenvalue equation becomes

$$\widehat{H}\psi = -\frac{\hbar^2}{2m_1}\nabla^2\psi - \frac{\hbar^2}{2m_2}\nabla^2\psi + V\psi = E_\lambda\psi \tag{8.14}$$

Expanding Eq. (8.14), which contains two Laplacian operators, into its full 3D form gives

$$-\frac{\hbar^2}{2m_1}\left[\frac{\partial^2}{\partial x_1^2} + \frac{\partial^2}{\partial y_1^2} + \frac{\partial^2}{\partial z_1^2}\right]\psi - \frac{\hbar^2}{2m_2}\left[\frac{\partial^2}{\partial x_2^2} + \frac{\partial^2}{\partial y_2^2} + \frac{\partial^2}{\partial z_2^2}\right]\psi + V\psi = E_\lambda\psi \tag{8.15}$$

One difficulty with Eq. (8.15) is that it is *not separable*. Thus, we cannot use the method of separation of variables directly on this equation. This is because of the form of $V(|\mathbf{r}_1 - \mathbf{r}_2|)$. For separability, V should be either purely a function of $\mathbf{r}_1$, purely a function of $\mathbf{r}_2$, or a linear combination of any power of $\mathbf{r}_1$ and $\mathbf{r}_2$. However, since V is a function of $|\mathbf{r}_1 - \mathbf{r}_2|$ (a nonlinear function), we have this roadblock. Fortunately, it turns out that we can transform the terms of the Hamiltonian other than the potential to be an explicit function of $\mathbf{r}_1 - \mathbf{r}_2$, and another variable. So, we can trade the two vectors variables $[x_1, y_1, z_1]^T$ and $[x_2, y_2, z_2]^T$ we have for two transformed variables. This does the trick that will allow us to use the method of separation of variables. Let's begin the transformation of the Hamiltonian using the first variable suggested by the potential function, namely

$$\mathbf{r} = \mathbf{r}_1 - \mathbf{r}_2 = [x_1 - x_2, y_1 - y_2, z_1 - z_2]^T = [x, y, z]^T \tag{8.16}$$

For the second variable, we will use the **center of mass** of the hydrogen atom, denoted by $\mathbf{R}$, defined as

$$\mathbf{R} = [X, Y, Z]^T \tag{8.17a}$$

The center of mass coordinates are defined as

$$X = \frac{m_1 x_1 + m_2 x_2}{m_1 + m_2} \tag{8.17b}$$

$$Y = \frac{m_1 y_1 + m_2 y_2}{m_1 + m_2} \tag{8.17c}$$

$$Z = \frac{m_1 z_1 + m_2 z_2}{m_1 + m_2} \tag{8.17d}$$

Since Eq. (8.15) involves second order derivative terms due to the Laplacian operator $\nabla^2\cdot$, we must replace all the derivative terms with their equivalent in terms of the relative position **r** and the center of mass **R**. We will build up this process to get $\partial^2 \psi / \partial x_1^2$. The process goes the same for $\partial^2 \psi / \partial x_2^2$ and $\partial^2 \psi / \partial x_3^2$, etc. Because x_1 generally depends on both x and X, starting with the first derivative, we have

$$\frac{\partial \psi}{\partial x_1} = \frac{\partial \psi}{\partial x} \frac{\partial x}{\partial x_1} + \frac{\partial \psi}{\partial X} \frac{\partial X}{\partial x_1}$$

Using the definitions for r and R given in Eqs. (8.16)-(8.17d), the first derivative becomes

$$\frac{\partial \psi}{\partial x_1} = \frac{\partial \psi}{\partial x} + \frac{m_1}{m_1 + m_2} \frac{\partial \psi}{\partial X}$$

Differentiation for $\partial \psi / \partial x_1$ to obtain the second order derivative gives

$$\frac{\partial^2 \psi}{\partial x_1^2} = \frac{\partial}{\partial x_1}\left(\frac{\partial \psi}{\partial x}\right) + \frac{m_1}{m_1 + m_2} \frac{\partial}{\partial x_1}\left(\frac{\partial \psi}{\partial X}\right)$$

$$= \frac{\partial^2 \psi}{\partial x^2} + \frac{\partial^2 \psi}{\partial x \partial X} \frac{\partial X}{\partial x_1} + \frac{m_1}{m_1 + m_2}\left(\frac{\partial^2 \psi}{\partial x \partial X} \frac{\partial x}{\partial x_1} + \frac{\partial^2 \psi}{\partial X^2} \frac{\partial X}{\partial x_1}\right)$$

So, we have the following result for $\partial^2 \psi / \partial x_1^2$:

$$\frac{\partial^2 \psi}{\partial x_1^2} = \frac{\partial^2 \psi}{\partial x^2} + \frac{2 m_1}{m_1 + m_2} \frac{\partial^2 \psi}{\partial x \partial X} + \left(\frac{m_1}{m_1 + m_2}\right)^2 \frac{\partial^2 \psi}{\partial X^2} \tag{8.18}$$

The final step is just multiplying Eq. (8.18) by $-\hbar^2/2m_1$, leading to the result

$$-\frac{\hbar^2}{2m_1} \frac{\partial^2 \psi}{\partial x_1^2} = -\frac{\hbar^2}{2m_1} \frac{\partial^2 \psi}{\partial x^2} - \frac{\hbar^2}{m_1 + m_2} \frac{\partial^2 \psi}{\partial x \partial X} - \frac{\hbar^2 m_1}{2(m_1 + m_2)^2} \frac{\partial^2 \psi}{\partial X^2} \tag{8.19}$$

Let us evaluate the corresponding term for the proton with mass m_2 and position x_2. After carrying out a similar process as above, the first derivative for x_2 becomes

$$\frac{\partial \psi}{\partial x_2} = -\frac{\partial \psi}{\partial x} + \frac{m_2}{m_1 + m_2} \frac{\partial \psi}{\partial X}$$

8.3 Schrödinger's Hydrogen Atom Model

Differentiation again w.r.t. x_2, and multiplying by $-\hbar^2/2m_2$ gives

$$-\frac{\hbar^2}{2m_2}\frac{\partial^2\psi}{\partial x_2^2} = -\frac{\hbar^2}{2m_2}\frac{\partial^2\psi}{\partial x^2} + \frac{\hbar^2}{m_1+m_2}\frac{\partial^2\psi}{\partial x \partial X} - \frac{\hbar^2 m_2}{2(m_1+m_2)^2}\frac{\partial^2\psi}{\partial X^2} \quad (8.20)$$

In the Hamiltonian given by Eq. (8.15), both these terms add together. The mixed partial derivative terms end up summing to zero since they have opposite signs. So, for the x- component of the summed Laplacian terms in the Hamiltonian, we have

$$-\frac{\hbar^2}{2m_1}\frac{\partial^2\psi}{\partial x_1^2} - \frac{\hbar^2}{2m_2}\frac{\partial^2\psi}{\partial x_2^2} = -\frac{\hbar^2}{2}\left(\frac{1}{m_1}+\frac{1}{m_2}\right)\frac{\partial^2\psi}{\partial x^2} - \frac{\hbar^2}{2}\left(\frac{1}{m_1+m_2}\right)\frac{\partial^2\psi}{\partial X^2} \quad (8.21)$$

Based on Eq. (8.21), we can define the following mass m^* for the combined system:

Reduced Mass Definition of a Two-Body System

$$\frac{1}{m^*} = \frac{1}{m_1} + \frac{1}{m_2} \quad (8.22)$$

Thus, the masses of the proton nucleus and electron combine in the same way as electrical capacitors in series, since in this case

$$\frac{1}{C_{eq}} = \frac{1}{C_1} + \frac{1}{C_2}$$

Eq. (8.22) defines what is known as the **reduced mass** of a two-body system, and it is *always* smaller than the minimum of (m_1, m_2). From this definition, m^* is given by

$$m^* = \frac{m_1 m_2}{m_1 + m_2}$$

It is straightforward to show that the *reduced mass* $m^* < \min(m_1, m_2)$, hence the name. To see this, let m_1 be the smaller of the two. Then, it follows that

$$m^* = \frac{m_1 m_2}{m_1 + m_2} < \frac{m_1 m_2 + m_1^2}{m_1 + m_2} = m_1\left(\frac{m_2 + m_1}{m_1 + m_2}\right) = m_1$$

This *reduced mass* is associated with the motion of the system, described by the kinetic energy. In the last term of Eq. (8.21), we see the mass

$m_1 + m_2 = M$, which is the total mass of the system. Because all the terms look identical for x, y, and z, the process is the same for the other two components y_1, z_1 and y_2, z_2, respectively. So, our final result transforming the Hamiltonian becomes

$$-\frac{\hbar^2}{2m^*}\left(\frac{\partial^2 \psi}{\partial x^2} + \frac{\partial^2 \psi}{\partial y^2} + \frac{\partial^2 \psi}{\partial z^2}\right) - \frac{\hbar^2}{2M}\left(\frac{\partial^2 \psi}{\partial X^2} + \frac{\partial^2 \psi}{\partial Y^2} + \frac{\partial^2 \psi}{\partial Z^2}\right) + V\psi = E_\lambda \psi \quad (8.23)$$

Since $V = V(r = |\mathbf{r}_1 - \mathbf{r}_2|) = V(x, y, z)$, Eq. (8.23) is separable and therefore, we may use the *method of separation of variables* to solve it.

As we have done in the method of separation of variables, let $\psi = B(r)F(R)$. $B(r)$ describes the relative localized system behavior, corresponding to the bound states of the electron, while $F(\mathbf{R})$ describes the free motion of the center of mass $\mathbf{R}$ of the hydrogen atom. Substitution into Eq. (8.23), then dividing by $B \cdot F$ gives

$$-\frac{\hbar^2}{2m^*}\frac{1}{B(r)}\left(\frac{\partial^2 B}{\partial x^2} + \frac{\partial^2 B}{\partial y^2} + \frac{\partial^2 B}{\partial z^2}\right) + V(\mathbf{r}) -$$

$$\frac{\hbar^2}{2M}\frac{1}{F(R)}\left(\frac{\partial^2 F}{\partial X^2} + \frac{\partial^2 F}{\partial Y^2} + \frac{\partial^2 F}{\partial Z^2}\right) = E_\lambda \quad (8.24)$$

The first part of Eq. (8.24) is purely a function of $\mathbf{r} = \mathbf{r}_1 - \mathbf{r}_2$, while the second part is purely a function of $\mathbf{R}$. Both of them add to equal the eigenvalue constant E_λ. Based on previous problems we have solved using the *method of separation of variables*, this is true if and only if both the function of r and R, respectively, equal to constants. So, we can write the following pair of equations:

$$-\frac{\hbar^2}{2m^*}\frac{1}{B(\mathbf{r})}\left(\frac{\partial^2 B}{\partial x^2} + \frac{\partial^2 B}{\partial y^2} + \frac{\partial^2 B}{\partial z^2}\right) + V(r) = E_r \quad (8.25a)$$

$$\frac{\hbar^2}{2M}\frac{1}{F(\mathbf{R})}\left(\frac{\partial^2 F}{\partial X^2} + \frac{\partial^2 F}{\partial Y^2} + \frac{\partial^2 F}{\partial Z^2}\right) = E_R \quad (8.25b)$$

The relationship between E_r, E_R, and E_λ becomes

$$E_r + E_R = E_\lambda \quad (8.26)$$

Let's focus on the first of these equations, describing $B(r)$ for the bound states. Eq. (8.25a) can be written as

$$\frac{\partial^2 B}{\partial x^2} + \frac{\partial^2 B}{\partial y^2} + \frac{\partial^2 B}{\partial z^2} + \frac{2m^*(E_r - V)}{\hbar^2} B = 0 \quad (8.27)$$

8.3 Schrödinger's Hydrogen Atom Model

Since $\mathbf{r} = \mathbf{r}_1 - \mathbf{r}_2$, the *origin* of the relative electron's position r is the location of the proton-nucleus $\mathbf{r}_2$. Because of this, it is useful to express Eq. (8.27) in spherical coordinates (r, θ, φ). Rewriting the Laplacian in Cartesian coordinates in spherical coordinates gives

$$\frac{1}{r^2}\left[\frac{\partial}{\partial r}\left(r^2\frac{\partial}{\partial r}\right) + \frac{1}{\sin\theta}\frac{\partial}{\partial \theta}\left(\sin\theta\frac{\partial}{\partial \theta}\right) + \frac{1}{\sin^2\theta}\frac{\partial^2}{\partial \varphi^2}\right]B + \frac{2m^*(E_r - V)}{\hbar^2}B = 0 \tag{8.28}$$

The second and third terms in the brackets, on the LHS, is the $\widehat{L}^2$ operator for the angular momentum, given by Eq. (7.52). Since Eq. (8.28) is obviously separable in r and the other two variables θ and φ, we may assume $B(r,\theta,\varphi)$ is given by

$$B(r,\theta,\varphi) = R(r)Y_{\ell,m}(\theta,\varphi) \tag{8.29}$$

$Y_{\ell,m}(\theta,\varphi)$ are the spherical harmonics we obtained in Chap. 7, which are eigenfunctions of the $\widehat{L}^2$ and $\widehat{L}_z$ operators. If we assume this form and get all the way to a final solution successfully, then, it will mean that our assumption works! This is how we will approach the solution here. So, let's roll with it for now.

The spherical harmonics $Y_{\ell,m}$ are given by Eq. (7.93), with some examples in Table 7.6. Substitution into Eq. (8.28), then dividing by $B = R \cdot Y_{\ell,m}$, we obtain the following *separated* differential equation:

$$\frac{1}{R}\frac{\partial}{\partial r}\left(r^2\frac{\partial}{\partial r}\right)R - \frac{2m^*Vr^2}{\hbar^2} + \frac{2m^*E_r r^2}{\hbar^2} +$$

$$\frac{1}{Y_{\ell,m}}\left[\frac{1}{\sin\theta}\frac{\partial}{\partial \theta}\left(\sin\theta\frac{\partial}{\partial \theta}\right) + \frac{1}{\sin^2\theta}\frac{\partial^2}{\partial \varphi^2}\right]Y_{\ell,m} = 0 \tag{8.30}$$

Eq. (8.30) reveals to us that the method of separation of variables has worked, as we have transformed Eq. (8.28) to a sum of a function of r and a function of θ and φ, equal to a constant. This means that both functions separately also equal to constants, and therefore, the θ, φ part becomes equal to a constant. This then becomes identical to the $\widehat{L}^2$ operator. Then we may replace this part of the equation with the eigenvalue of the $\widehat{L}^2$ operator. After doing this, Eq. (8.30) becomes

$$\frac{1}{R}\frac{\partial}{\partial r}\left(r^2\frac{\partial}{\partial r}\right)R - \frac{2m^*Vr^2}{\hbar^2} + \frac{2m^*E_r r^2}{\hbar^2} - \frac{1}{Y_\ell^m}[\ell(\ell+1)]Y_\ell^m = 0 \tag{8.31}$$

Left-multiplying by $R(r)/r^2$, we have the following equation in terms of r:

Radial Part of Schrödinger Equation

$$\frac{1}{r^2}\frac{d}{dr}\left(r^2\frac{dR}{dr}\right) + \frac{2m^*}{\hbar^2}\left[E_r - V(r) - \frac{\ell(\ell+1)\hbar^2}{2m^*r^2}\right]R(r) = 0 \quad (8.32)$$

We have replaced the partial derivatives with total derivatives, since we now have a function of the single variable r. Eq.(8.32) is known as the *radial part of the Schrödinger equation*. In the next section, we shall determine the exact solution to this equation.

8.4 Schrödinger Equation Radial Solution

If we obtain an exact solution to the radial equation, we will have also obtained the solution $B(r,\theta,\varphi) = R(r)Y_{\ell,m}$, described by Eq. (8.28). To solve it, we'll further transform Eq. (8.32). Let's use the following transformation:

$$R(r) = \frac{U(r)}{r}$$

To complete the substitution, we will also need the derivatives of R in terms of U. The first derivative is given by

$$\frac{dR}{dr} = \frac{1}{r}\frac{dU}{dr} - \frac{U(r)}{r^2}$$

Then, multiplying by r^2, we have

$$r^2\frac{dR}{dr} = r\frac{dU}{dr} - U(r)$$

Differentiating the above result, the first order derivative terms cancel, and we have

$$\frac{1}{r^2}\frac{d}{dr}\left(r^2\frac{dR}{dr}\right) = \frac{1}{r}\frac{d^2U}{dr^2}$$

Substituting these results into Eq. (8.32), we have

$$\frac{1}{r}\frac{d^2U}{dr^2} + \frac{2m^*}{\hbar^2}\left[E_r - V(r) - \frac{\hbar^2\ell(\ell+1)}{2m^*r^2}\right]\frac{U(r)}{r} = 0$$

8.4 Schrödinger Equation Radial Solution

Multiplying by r and replacing $V(r)$ with its definition using Eq. (8.13), we get the following corresponding differential equation for $U(r)$:

$$\frac{d^2U}{dr^2} + \left[\frac{2m^*E_r}{\hbar^2} + \frac{Zm^*e^2}{2\pi\hbar^2\varepsilon_0}\frac{1}{r} - \frac{\ell(\ell+1)}{r^2}\right]U(r) = 0 \tag{8.33}$$

In addition to transforming the function R, we will also transform independent variable r to obtain a more convenient *normalized* form. This is done by introducing the dimensionless variable ζ defined by

$$\zeta = \alpha r \Rightarrow d\zeta = \alpha dr$$

We'll define α shortly. Then, the second derivative term becomes

$$\frac{d^2U}{dr^2} = \alpha^2 \frac{d^2U}{d\zeta^2}$$

Eq. (8.33) then becomes

$$\alpha^2 \frac{d^2U}{d\zeta^2} + \left[\frac{2m^*E_r}{\hbar^2} + \frac{Zm^*\alpha e^2}{2\pi\hbar^2\varepsilon_0}\frac{1}{\zeta} - \frac{\alpha^2\ell(\ell+1)}{\zeta^2}\right]U(\zeta) = 0$$

Dividing by α^2, we have

$$\frac{d^2U}{d\zeta^2} + \left[\frac{2m^*E_r}{\hbar^2\alpha^2} + \frac{Zm^*e^2}{2\pi\hbar^2\varepsilon_0\alpha}\frac{1}{\zeta} - \frac{\alpha^2\ell(\ell+1)}{\zeta^2}\right]U(\zeta) = 0 \tag{8.34}$$

We may define α^2 such that the first term in brackets becomes

$$\frac{2m^*E_r}{\hbar^2\alpha^2} = -\frac{1}{4} \Rightarrow \alpha^2 = -\frac{8m^*E_r}{\hbar^2} \tag{8.35}$$

This choice recognizes *a priori*, that $E_r < 0$, corresponding to bound states, and it ensures that $\alpha^2 > 0$. So, Eq. (8.34) may now be written as

$$\frac{d^2U}{d\zeta^2} + \left[-\frac{1}{4} + \frac{\beta}{\zeta} - \frac{\ell(\ell+1)}{\zeta^2}\right]U(\zeta) = 0 \tag{8.36}$$

The parameter β is a constant given by

$$\beta \equiv \frac{Zm^*e^2}{2\pi\hbar^2\varepsilon_0\alpha} \tag{8.37}$$

With this form, let's determine the solution behavior in two asymptotic limits. First, for the asymptotic limit of $\zeta \to \infty$, in which case Eq. (8.36) becomes

$$\frac{d^2U}{d\zeta^2} - \frac{1}{4}U(\zeta) = 0 \tag{8.38}$$

The solution of Eq. (8.38) is given by

$$U_{\zeta \to \infty} = U_0 e^{-\frac{\zeta}{2}} = 0 \tag{8.39}$$

For the other asymptotic limit, when $\zeta \to 0$, Eq. (8.36) becomes

$$\frac{d^2 U}{d\zeta^2} - \frac{\ell(\ell+1)}{\zeta^2} U(\zeta) = 0 \tag{8.40}$$

Eq. (8.40) can be solved assuming a solution of the form

$$U = \zeta^p \tag{8.41}$$

The second derivative is then given by

$$\frac{d^2 U}{d\zeta^2} = p(p-1)\zeta^{a-2}$$

Plugging this into Eq. (8.40), and multiplying by ζ^2, we get

$$p(p-1)\zeta^p - \ell(\ell+1)\zeta^p = 0 \tag{8.42}$$

Since ζ is arbitrary and generally nonzero, it follows that

$$p^2 - p - \ell(\ell+1) = 0$$

The two roots of this quadratic equation are given by

$$p = \frac{1 \pm \sqrt{1 + 4\ell(\ell+1)}}{2} = \frac{1 \pm (2\ell+1)}{2} \to p = \ell+1, -\ell \tag{8.43}$$

For the problem we are considering, the negative exponent is not allowed for describing the wave function because it leads to a $1/\zeta^\ell$ factor, which does not approach zero in the limit of $\zeta \to 0$. This is an important point since this is the dominating term as ζ approaches zero. In this limit, this function must 'carry' the wave function to zero in order to be normalizable in spherical coordinates, just as the other asymptotic limit did for us when $\zeta \to \infty$. So, we have

$$p = \ell+1 \quad \text{and} \quad U_{\zeta \to \infty}(\zeta) = U_0 \zeta^{\ell+1}$$

Incorporating both of asymptotic limits into the solution, let $U(\zeta) = U_{\zeta \to 0} \cdot U_{\zeta \to \infty} \cdot W(\zeta) = G \cdot W(\zeta)$. Since we know $G(\zeta)$ already, let us work to obtain an equation describing only $Y(\zeta)$. Then, if we solve for

8.4 Schrödinger Equation Radial Solution

$W(\zeta)$, we will have the solution for $U = rR$, etc. Using this definition for U, the second derivative becomes

$$\frac{d^2 U}{d\zeta^2} = W\frac{d^2 G}{d\zeta^2} + 2\frac{dG}{d\zeta}\frac{dW}{d\zeta} + G\frac{d^2 W}{d\zeta^2} \qquad (8.44)$$

Since $G = \zeta^{\ell+1} \cdot e^{-\zeta/2}$, the first derivative of G is given by

$$\frac{dG}{d\zeta} = (\ell+1)\zeta^\ell e^{-\frac{\zeta}{2}} - \frac{1}{2}\zeta^{\ell+1}e^{-\frac{\zeta}{2}} = \left[\frac{\ell+1}{\zeta} - \frac{1}{2}\right]G(\zeta) \qquad (8.45)$$

The second derivative then becomes

$$\frac{d^2 G}{d\zeta^2} = \frac{d}{d\zeta}\left[\left(\frac{\ell+1}{\zeta} - \frac{1}{2}\right)G(\zeta)\right]$$

$$= \left(\frac{\ell+1}{\zeta} - \frac{1}{2}\right)\frac{dG}{d\zeta} - \left(\frac{\ell+1}{\zeta^2}\right)G(\zeta)$$

Using the result for $dG/d\zeta$, we have

$$\frac{d^2 G}{d\zeta^2} = -\left(\frac{\ell(\ell+1)}{\zeta^2} - \frac{\ell+1}{\zeta} + \frac{1}{4}\right)G(\zeta)$$

Substitution of these results into Eq. (8.44), then substituting that result into (8.36), and multiplying by ζ, we obtain the following differential equation describing $W(\zeta)$

Hydrogen's Confluent Hypergeometric Equation

$$\zeta\frac{d^2 W}{d\zeta^2} + [2(\ell+1) - \zeta]\frac{dW}{d\zeta} - (\ell+1-\beta)W = 0 \qquad (8.46)$$

Eq. (8.46) is another example of the *Confluent Hypergeometric Equation*, discussed in detail in Chap. 5, obtained in the solution of the Schrödinger equation applied to the harmonic oscillator. The CHE is given by

$$x\frac{d^2 y}{dx^2} + (c-x)\frac{dy}{dx} - ay(x) = 0$$

By inspection, we see that

$$c = 2(\ell+1) \quad \text{and} \quad a = \ell+1-\beta \qquad (8.47)$$

In solving the CHE in Chap. 5, we obtained the solution by using the *method of Frobenius*, which assumes the solution has the form of an infinite series. However, we also found that in order for the solution to satisfy the normalization conditions of a localized wave function, the infinite series must be truncated to a finite polynomial. The infinite series we found is the *Confluent Hypergeometric function* $_1F_1(\zeta)$, and the truncated polynomials were scaled to obtain Hermite polynomials H_n. With this, we can deduce the solution for $W(\zeta)$. Moreover, recall that the truncation of the confluent hypergeometric function was achieved by imposing the important condition on the input parameter a, namely, that

$$a = \ell + 1 - \beta = -n_r \quad n_r = 0, 1, 2, \ldots$$

The parameter β must take values given by

$$\beta = n_r + \ell + 1 \qquad (8.48)$$

Since n_r and ℓ both take values $0, 1, 2, \ldots$, we can simply define β as a **total quantum number** n, whose integer value is

$$n = n_r + \ell + 1 \quad \text{(total quantum number)} \qquad (8.49)$$

In quantum mechanics, the *total quantum number* is also known as the **principal quantum number**, and takes the discrete values $1, 2, 3, \ldots$ (notice it cannot equal zero). Using the definition for β, we get

$$\beta = n = \frac{Zm^*e^2}{2\pi\hbar^2\varepsilon_0\alpha}$$

Squaring the result and substituting α^2 given by Eq. (8.35), then solving for $E_r = E_n$, we obtain the following for the allowed energy values:

Hydrogen Electron Bound Energy States (no magnetic field)

$$E_n = -\frac{Z^2 m^* e^4}{32\pi^2 \hbar^2 \varepsilon_0^2} \frac{1}{n^2} \qquad (8.50)$$

In terms of the Planck constant h and the principal quantum number n, Eq. (8.50) can be written as

$$E_n = -\frac{Z^2 m^* e^4}{8h^2 \varepsilon_0^2} \frac{1}{n^2} = -\frac{E_1}{n^2} \qquad (8.51)$$

8.4 Schrödinger Equation Radial Solution

If we set $Z = 1$, the result is identical to Bohr's result given by Eq. (8.7). It is remarkable that after this relatively complicated analysis, the results have lead exactly to Bohr's result, for the case of the hydrogen atom $Z = 1$. E_1 represents the lowest possible energy where $E_1 \sim 13.6$ eV. Note that Bohr's model can also be generalized to obtain Eq. (8.51), as well. This difference is not arising from a quantum mechanical effect, but from assuming the nucleus in Bohr's model to have a charge of $Z = 1$. Schrödinger demonstrated this result in one of three papers he published in 1926, marking the introduction of his gift of the Schrödinger equation to the world, and the rest is history.

The energy difference ΔE_n, which is observed in the measurement of atomic and/or molecular spectra, is given by

Hydrogen Bound-State Energy Differences

$$\Delta E_n = \frac{Z^2 m^* e^4}{8 h^2 \varepsilon_0^2} \left(\frac{1}{n_1^2} - \frac{1}{n_2^2} \right) \quad (8.52)$$

In Eq. (8.52), $n_2 > n_1$, and this equation leads to the familiar hydrogen spectral series, and when $Z = 1$, gives rise to the series of Lyman, Balmer, Paschen, etc. We are now in a position to write down a complete wave function for the hydrogen atom. The complete solution for $U(\zeta)$ is given by

$$U(\zeta) = C_n \zeta^{\ell+1} e^{-\frac{\zeta}{2}} {}_1F_1(\zeta; a, c)$$

Parameters a and c are input parameters to the *Confluent Hypergeometric Function* described by Eq. (8.46). Because a takes the value of a negative integer, the infinite series is truncated. These truncated polynomials are proportional to Hermite polynomials H_n, however, since c also depends on the integer ℓ, we cannot substitute H_n for the CHF like we did in the harmonic oscillator problem where $c = 1/2$ was independent of any quantum number. Thus, we must keep this form to correctly account for all values of c. Then, with $\zeta = \alpha r$, $U(\zeta)$ becomes

$$U(r) = C_n \alpha_n^{\ell+1} r^{\ell+1} e^{-\frac{\alpha r}{2}} {}_1F_1(\zeta; a, c)$$

Since $R(r) = U(r)/r$, the **radial wave functions** for the hydrogen atom become

Hydrogen Bound-State Radial Wave Functions

$$R_{n,\ell}(r) = C_{n,\ell}\, (\alpha_n r)^{\ell}\, e^{-\frac{\alpha_n r}{2}}\, {}_1F_1(\alpha_n r; \ell + 1 - n, 2\ell + 2) \qquad (8.53)$$

$C_{n,\ell}$ is the normalization factor that must be determined from the normalization condition. This integral, in spherical coordinates, is given by

$$\int_0^\infty R^*_{n,\ell}(r) R_{n,\ell}(r) r^2 \, dr = 1 \qquad (8.54)$$

The normalization factor $C_{n,\ell}$ corresponding to $R_{n,\ell}$ works out to be

Radial Function Normalization Factor

$$C_{n,\ell} = \frac{\alpha_n^{3/2}}{(2\ell + 1)!} \left[\frac{(n+1)!}{2n \cdot (n - \ell - 1)!} \right]^{1/2} \qquad (8.55)$$

$\alpha = \alpha_n$ is given by Eq. (8.35). Fig. 8.4 illustrates some examples of the radial wave functions for $\ell = 0, \ell = 1$ and $\ell = 2$, with $n = 1 - 7$.

Figure 8.4: Hydrogen radial wave functions $R_{n,\ell}$, for cases $\ell = 0$ (left column), $\ell = 1$ (middle column), and $\ell = 2$ (right column). r is normalized by the Bohr radius $a_0 \sim 0.52919\text{Å}$, given by Eq. (8.8).

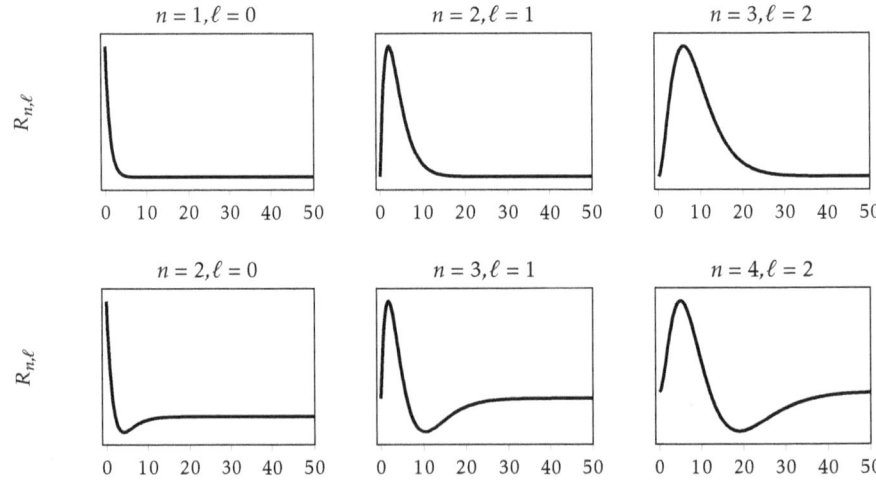

8.4 Schrödinger Equation Radial Solution

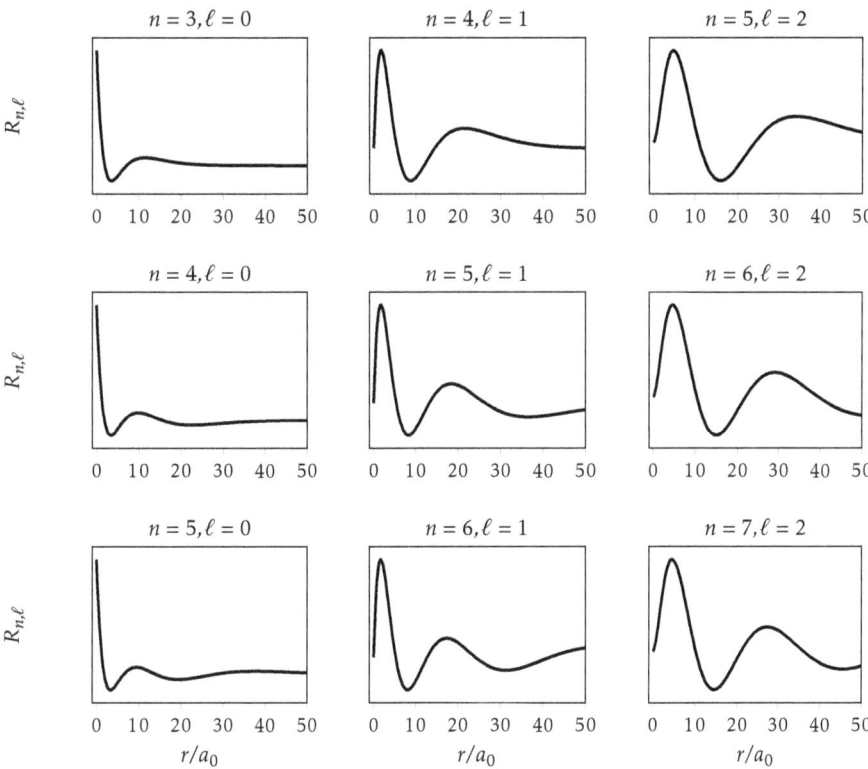

While Fig. 8.4 shows the radial wave functions, Fig. 8.5 illustrates the corresponding probability densities given by $r^2 R_{n,\ell}^2$ since the 3D integration is over spherical coordinates. The left, middle and right columns again correspond to $\ell = 0, 1, 2$, respectively. These functions also convolve with the spherical harmonics $Y_{\ell,m}$ to form the various $s, p,$ and d orbitals, etc. These orbitals will be formally defined in Sec. 8.5. As n and ℓ increase, the probability distributions, depending on $R_{n,\ell}(r)$, increasingly widen or broaden their spatial extent. Therefore, broadening is a characteristic of higher and higher energy states of the hydrogen atom, and it contributes to enlarging the effective atom or molecule size.

Figure 8.5: Examples of hydrogen radial probability densities $r^2 R_{n,\ell}^2$.

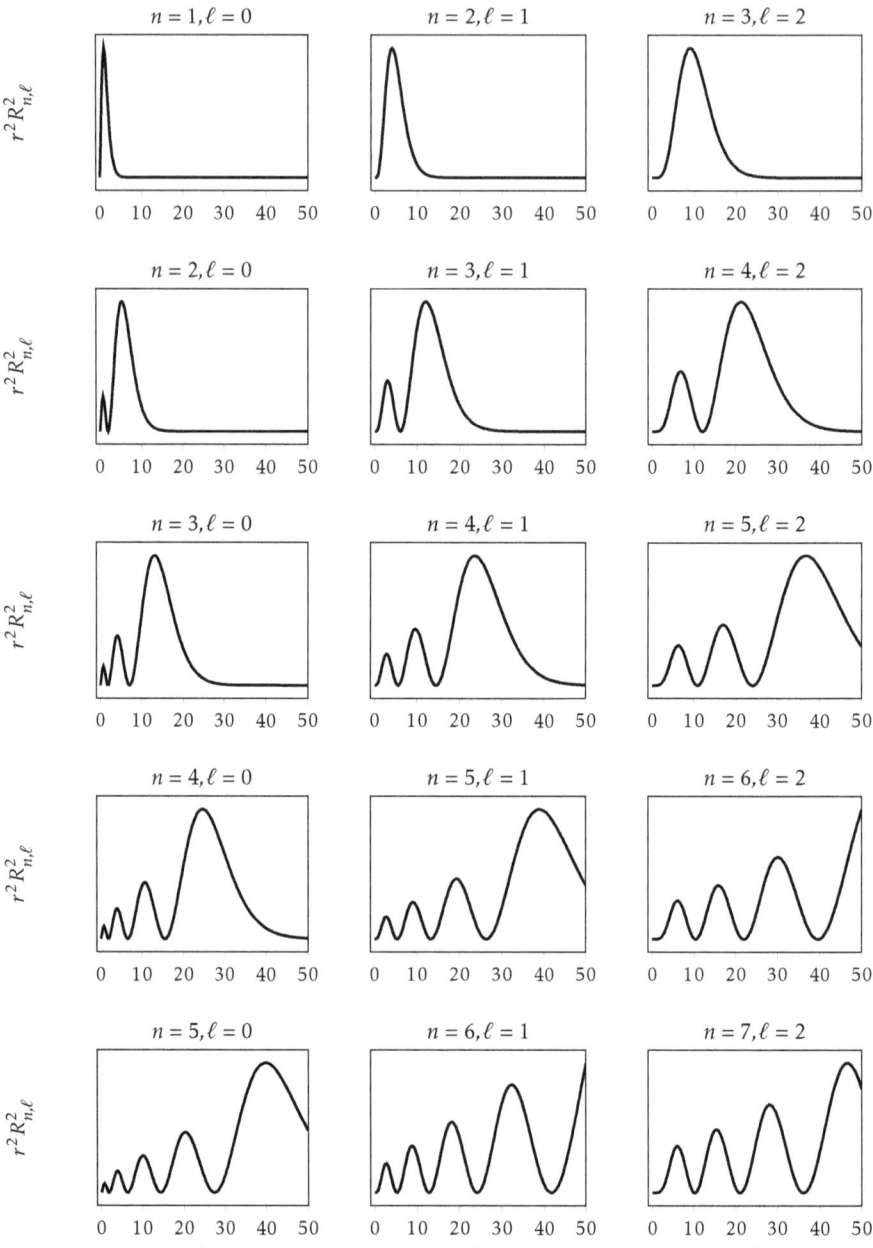

With $R_{n,\ell}(r)$, we can write the complete hydrogen atom wave functions

8.4 Schrödinger Equation Radial Solution

for the bound states which are associated with bound-state energy E_r. The bound-state wave function, $B(r,\theta,\varphi) = R_{n,\ell}Y_{\ell,m}$ are given by

---

Hydrogen Bound-State Wave Functions

$$B_{\ell,m,n}(r,\theta,\varphi) = C_n \alpha_n^\ell r^\ell e^{-\frac{\alpha_n r}{2}} {}_1F_1(\alpha_n r; \ell+1-n, 2\ell+2) Y_{\ell,m}(\theta,\varphi) \quad (8.56)$$

The normalization factor for B is the product of the normalization factors for the radial wave function, given by Eq. (8.55), and that of the spherical harmonics, given by Eq. (7.94). Therefore, $C_n = C_{n,\ell,m}$ becomes

$$C_{n,\ell,m} = \frac{\alpha^{3/2}}{(2\ell+1)!}\left[\frac{(2n+1)(n-m)!}{4\pi(n+m)!}\right]^{1/2}\left[\frac{(n+1)!}{2n\cdot(n-\ell-1)!}\right]^{1/2} \quad (8.57)$$

We now have the complete solution B for the bound states, and they are described by Eq. (8.28). The final part to the hydrogen atom solution from the Schrödinger equation is to obtain the solution for F, described by Eq. (8.25b). Since there is no potential energy term, it is identical to the free particle problem discussed in Sect. 3.7, where $V = 0$. Eq. (8.25b) can be written as

$$\frac{\partial^2 F}{\partial X^2} + \frac{\partial^2 F}{\partial Y^2} + \frac{\partial^2 F}{\partial Z^2} + k^2 F = 0 \quad (8.58)$$

k^2 is the square of the wavenumber defined as

$$k^2 = \frac{2ME_R}{\hbar^2} \quad (8.59)$$

Eq. (8.58) is separable in X, Y, and Z, so F can also be written as $F = F_X F_Y F_Z$. If we substitute this into Eq. (8.58), then divide by F, we obtain

$$\frac{1}{F_X}\frac{\partial^2 F_X}{\partial X^2} + \frac{1}{F_Y}\frac{\partial^2 F_Y}{\partial Y^2} + \frac{1}{F_Z}\frac{\partial^2 F_Z}{\partial Z^2} + k^2 = 0 \quad (8.60)$$

Eq. (8.60) sums three independent functions of X, Y, and Z, equal to a constant. This is true if and only if the respective functions also equal to a constant equal to zero. We may also separate k^2 as

$$k^2 = k_x^2 + k_y^2 + k_z^2 \quad (8.61)$$

Then, Eq. (8.60) leads to the following three equations:

$$\frac{\partial^2 F_X}{\partial X^2} + k_x^2 F_X = 0, \quad \frac{\partial^2 F_Y}{\partial Y^2} + k_y^2 F_Y = 0, \quad \frac{\partial^2 F_Z}{\partial Z^2} + k_x^2 F_Z = 0 \quad (8.62)$$

Each of these are identical to Eq. (3.97), where we found the solution already. Using those results, the hydrogen center of mass solutions are the free-particle wave functions given by

$$F_X = F_{x0} e^{ik_x X}, \quad F_Y = F_{y0} e^{ik_y Y}, \quad F_Z = F_{z0} e^{ik_z Y} \quad (8.63)$$

Therefore, the total solution is given by

$$F = F_X F_Y F_Z = F_0 e^{i(k_x X + k_y Y + k_z Y)} = F_0 e^{i \mathbf{k} \cdot \mathbf{R}} \quad (8.64)$$

Using the de Broglie relation $p = \hbar k$, the solution becomes

---

Hydrogen Center-Of-Mass wave function

$$F(\mathbf{R}) = F_0 e^{\frac{i}{\hbar} \mathbf{p} \cdot \mathbf{R}} \quad (8.65)$$

Eq. (8.65) is the wave function for the hydrogen atom's center of mass, $\mathbf{R}$. It is in the form of the solution to the free particle problem discussed in Sec. 3.7. We found there, however, that such a wave function is *not normalizable*, owing to a uniform potential. Since Eq. (8.65) is only a factor of the total wave function solution, this is not a problem. The complete hydrogen atom wave functions are given by the product of $F(\mathbf{R})$ and B, from Eq. (8.56).

8.5 Conventional Shells and Orbitals

The orbital angular momentum and the hydrogen atom solutions have revealed the existence of three important quantum numbers. These quantum numbers are the *orbital angular momentum quantum number* ℓ, the *magnetic quantum number* $m = m_\ell$, and the *principal quantum number* n. Their values and relationships are described by

$$m = m_\ell = -\ell, \ldots, -2, -1, 0, 1, 2, \ldots, \ell \quad (8.66)$$

$$\ell = 0, 1, 2, \ldots \quad (8.67)$$

8.5 Conventional Shells and Orbitals

$$n = 1, 2, 3, \ldots \quad \text{where} \quad n \geq \ell + 1 \tag{8.68}$$

States defined by these quantum numbers have adopted a general naming convention based on the values of n and ℓ. n determines what is known as the **shell**, while ℓ determines the family of **orbitals**, suggestive of the orbital angular momentum. The convention is that if $n = 1$, then it is defined as the K shell, $n = 2$ is the L shell, $n = 3$ is the M shell, and so on. Therefore, capital letters, *starting with K*, designate the *shell*. The other important designations depend on ℓ. If $\ell = 0$, the *orbital* is said to belong to the *family of orbitals* known as **sharp**, denoted by the first letter s. If $\ell = 1$, the orbital belongs to the **principal** family, denoted by p. $\ell = 2$ defines the **diffuse** family, denoted by d, and $\ell = 3$ is the **fundamental** family, denoted by f. Increasing ℓ further just continues with more letters in alphabetical order, *in lower case*. So, the next family is g, and so on. Fig. 8.6 illustrates and summarizes the relationship between shells and orbitals. The convention for combining both the orbital and shell quantum numbers is to denote the orbital by the first letter of the family name, and indicate the shell by preceding the orbital letter with the value of the principal quantum number n. For example, $\ell = 0$ and $n = 1$ gives the $1s = s$ state, $\ell = 0$ and $n = 2$ gives the $2s$ state. If $\ell = 1$ then $n > 1$, and can take $n = 2$, which is the $2p$ state. Increasing n to $n = 3$ indicates the $3p$ state, and so on. Table 8.4 provides most of the relevant orbital and shell definitions based on their corresponding values of n and ℓ. The first row corresponds to the lowest energy (among the rows), since it is most negative, while the last row is the highest energy (or least negative).

Table 8.4: Conventional s, p, d, f, and g orbitals and shells. They are defined by values of the principal quantum number n and orbital angular momentum quantum number ℓ. The cumulative number of states accounts for how many electrons can be accommodated. The expected number $2\ell + 1$ is shown, but the observed number (in parentheses) is $2 \times 2\ell + 1$.

n	ℓ	Shell	Orbital	$2\ell + 1$	Cumulative $2\ell + 1$
1	0	K	$1s$	1	1 (2)
2	0	L	$2s$	1	2 (4)
2	1	L	$2p$	3	5 (10)
3	0	M	$3s$	1	9 (18)
3	1	M	$3p$	3	12 (24)
3	2	M	$3d$	5	17 (34)

Table 8.4

4	0	N	4s	1	18 (36)
4	1	N	4p	3	21 (42)
4	2	N	4d	5	26 (52)
4	3	N	4f	7	33 (66)
5	0	O	5s	1	34 (68)
5	1	O	5p	3	37 (74)
5	2	O	5d	5	42 (84)
5	3	O	5f	7	49 (98)
5	4	O	5g	9	58 (116)
6	0	P	6s	1	59 (118)

The states are therefore partially defined by values of the principal quantum number n and orbital angular momentum quantum number ℓ. The cumulative number of states accounts for how many electrons can be accommodated. Since for a given combination of n and ℓ, there are $2\ell + 1$ possible states (because $m_\ell = -\ell, ..., -1, 0, 1, ..., \ell$) due to these different orbital angular momentum states, it was expected that the total possible states would be described by the cumulative sum of this number, given in the second to last column to the right. However, it turned out that each orbital (or row) holds $2 \times 2\ell + 1$, shown in the last column. We'll explain the origins of this discrepancy, in detail, in the next chapter.

The element known as Oganesson (Og) has the largest atomic element with $Z = 118$. It is remarkable that a 16 row table, like Table 8.4, generally accounts for all the known stable elements on earth! The way in which the chemical elements in the periodic table are arranged is as follows: starting with hydrogen, with $Z = 1$, the single electron occupies an unfilled 1s orbital (denoted as $1s^1$). Then, helium (He), with two electrons *fills* the 1s ($1s^2$). This *fills* the first shell, which is the K shell. After that, lithium (Li), with three electrons then has its third electron in the 2s orbital ($1s^2 s^1$), and the orbitals are filled in this way from lowest energy to the next highest energy state, and so on. This generates the table of chemical elements. There are a total of 18 columns. The rows increment the atomic number. The columnar arrangement in the table of elements are not coincidental. They contain the elements whose last few electrons lie in similar orbital configurations. For example, as can

8.5 Conventional Shells and Orbitals

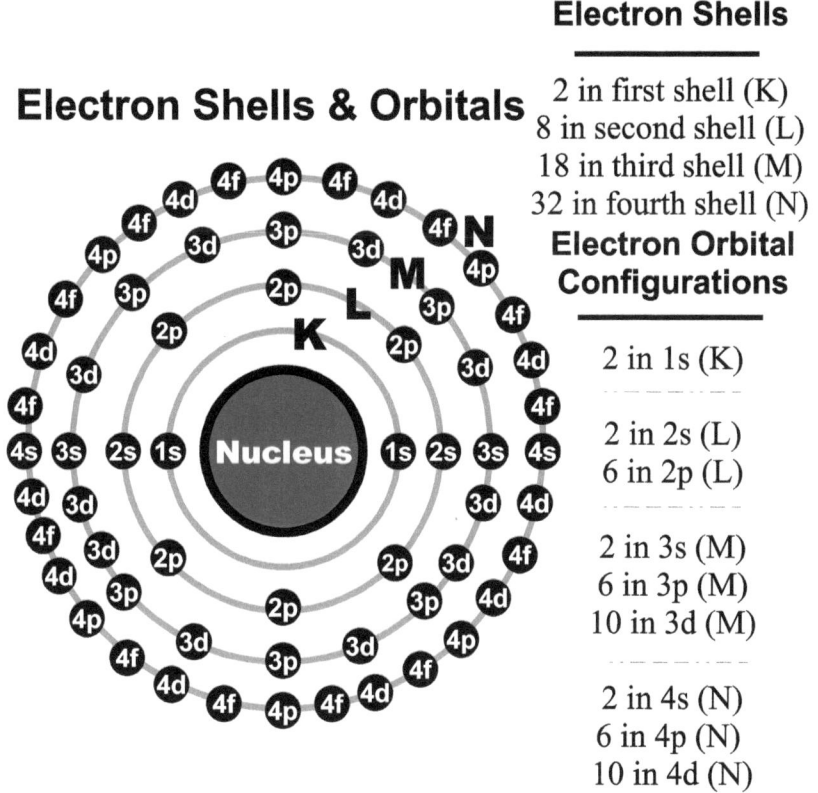

Figure 8.6: Illustration of electron orbitals and shells and their relation.

be seen in the periodic table, chemical elements such as H, Li, Na, etc. are in the same column because they have their last electron contained in an *s* orbital, which are the first orbitals encountered after filling a shell. The filled shells correspond to the rightmost column, containing He, Ne, Ar, etc. Thus, these elements are relatively stable. To account for the columns, the arrangement is as follows: in the first row, there are only two columns corresponding to the two states of the *s*-orbital. Then, there are 8 columns corresponding to the two states in *s* orbitals along with the six states of p-orbitals. Hence, 8=6+2 states. Then, there are 18 columns, which follows from the $2 + 6 + 10$ states, where the ten additional states correspond to the available states of the *d*-orbital, and so on. Beyond this, there are also states (shown via the dashed lines downwards) which are the additional 14 states from the *f*-orbital, etc.

This sort of periodic behavior of the elements leads to its name as the *periodic table*. A periodic table is shown in the *Chapter Summary* section at the end of this chapter.

We have found that the solutions for hydrogen obtained from quantum mechanics includes orbital angular momentum in addition to radial quantized states. Quantum mechanical treatment of the hydrogen atom has not only lead to a deeper insight into the behavior of hydrogen spectra, but has also lead to the an improved understanding of the structure of *all* known chemical elements! However, for us to explain the origins of the observed total number of possible states is often more than the accumulation of $2\ell + 1$, we must discuss how introducing a magnetic field B into the equation affects the system. This is the subject of the next section.

8.6 Observing the Zeeman Effect

So far, we have found that atomic *orbital angular momentum states* are determined by the two quantum numbers, m_ℓ (the magnetic quantum number) and ℓ (orbital quantum number). We found that $\widehat{L^2}$ has eigenvalues $\ell(\ell+1)\hbar^2$, so the orbital angular momentum $\widehat{L}$ takes values of $\sqrt{\ell(\ell+1)}\hbar$. For any value of ℓ, there are $2\ell + 1$ states available because $m_\ell = -\ell, -\ell+1, ..., 0, ..., \ell-1, \ell$.

As we have already pointed out, hydrogen and its atomic spectra played an important role in the development of quantum mechanics. However, there is more to the story of what hydrogen revealed to us. There was another interesting and important observation made with the hydrogen atomic spectra. Recall (*c*.Eq. (8.51)) that the energy levels of a hydrogen atom are described by

$$\Delta E_n = \hbar\omega = 13.6\text{eV} \times \left(\frac{1}{n_1^2} - \frac{1}{n_2^2}\right) \quad n_{1,2} = 1, 2, ..., \infty \quad (8.69)$$

Because $\Delta E_n = \hbar\omega_n = hf_n$, a set of energy transitions between any two states gives rise to a series of spectral lines at a set of discrete transition frequencies when the electron in the hydrogen atom is excited. Fig. 8.7 illustrates hydrogen spectral lines that are observed in experiments where *only heating* of atoms is involved, and therefore, there is no magnetic field experienced by the atoms. The discrete lines correspond to

8.6 Observing the Zeeman Effect

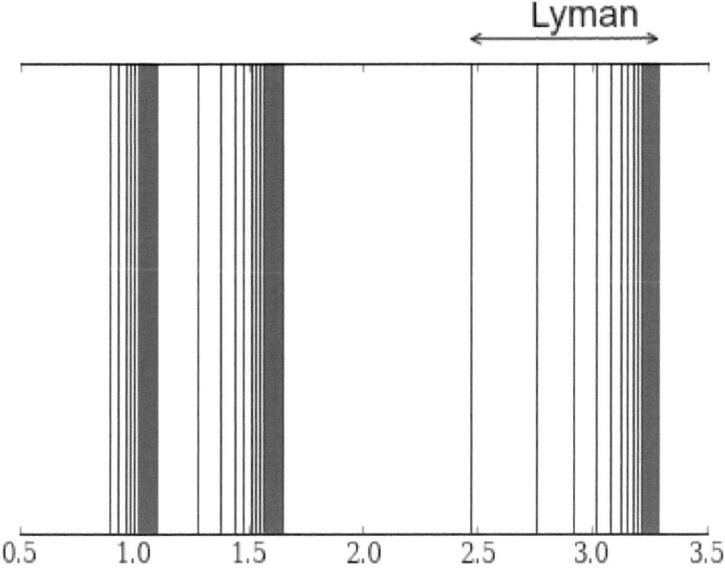

Figure 8.7: Atomic spectrum for heated hydrogen atoms. The first series is label, the Lyman series, for reference.

discrete frequencies whose values are dictated by Eq. (8.69). Although as n increases corresponding to higher and higher energy levels, the *energy difference* between levels gets smaller. This property is visible in the spectral lines. We can also see this by evaluating ΔE_n in the limit $n \to \infty$, which gives

$$\lim_{n \to \infty} (E_{n+1} - E_n) = 13.6 \times \lim_{n \to \infty} \left(\frac{1}{(n+1)^2} - \frac{1}{n^2} \right) \quad (8.70)$$

$$= 13.6 \times \lim_{n \to \infty} \left(\frac{n^2 - (n+1)^2}{n^2(n+1)^2} \right) \quad (8.71)$$

$$= 13.6 \times \lim_{n \to \infty} \left(\frac{-2n-1}{n^2(n+1)^2} \right) \quad (8.72)$$

$$= -13.6 \times \lim_{n \to \infty} \frac{1}{n^3} = 0 \quad (8.73)$$

This means that the energies lump closer and closer together as the system finds higher energy states corresponding to larger quantum numbers n.

Hydrogen bound state energy sequence: The bound state energies of hydrogen form a Cauchy sequence, which is convergent, and therefore, suggests electrons can occupy states indefinitely, and always be in a bound state with negative energy.

This type of sequence is known as a **Cauchy sequence**. A Cauchy sequence is also a convergent sequence. Then, a series S_n composed of the sequence of bound state energies E_n is also *always convergent*, or bounded so that

$$S_n = \sum_n E_n = \text{convergent series} \qquad (8.74)$$

Based on this property, nature could continue adding electrons to occupy electronic states indefinitely, and *always* yield at atom with finite energy.

Returning to the spectral lines, in the beginning, spectral lines like these were produced by resistive heating of hydrogen atoms. The heat excites electrons into higher energy states by raising the expected value of the kinetic energy for each atom, followed by the electron settling down into a lower energy localized state. These relaxing transitions give rise to electromagnetic radiation emission, detectable using a spectrometer. For example, in the Lyman series of the hydrogen atom (see Sec. 2.12)), corresponding to $n_1 = 1, n_2 > n_1$, the generated spectral lines are in the ultraviolet part of the electromagnetic spectrum. If we are only heating the atoms, the spectrum described by Eq. (8.69) is exactly what we would see. But, what if other methods were used to further excite hydrogen atoms? Would Eq. (8.69) be enough? The answer is no, and in the next section, we consider one method used involving a magnetic field, which pointed this out.

8.6.1 Spectral Lines In a Magnetic Field

It was a matter of time before atomic spectra would be able to allow additional means to excite atoms. Dutch scientists Hendrik Antoon Lorentz (1853-1928) and Pieter Zeeman (1865-1943) both took an interest, independently, in the observations from such alternative methods of excitation. They thought to explore whether or not a magnetic field can alter the properties of atomic spectra. It was not at all obvious what the answer to this question should be. The idea came to Zeeman in the wake of studying the Kerr effect. He came up with the idea to

8.6 Observing the Zeeman Effect

Figure 8.8: Illustration of the Zeeman effect observed with sodium (Na) vapor. Zeeman won the Nobel Prize in physics for his discovery of this effect, now bearing his name (shared with Lorentz). The two lines on the top correspond to no magnetic field. They are an example of a *doublet*, also called *D*-lines. When the field is turned on, the widening or splitting is observed.

investigate what would happen if oxyhydrogen (a mixture of hydrogen and oxygen gas) was not only deployed for heating, but also placing the heated gas between the poles of an electromagnet. When observing the resulting spectra from such an experiment, he observed an unusual splitting when applying a magnetic field. The individual lines that formed in the absence of the magnetic field now further split to several equally spaced lines. Zeeman found that when a magnetic field was applied, one line transitioned to a multiplicity of lines that are more closely spaced compared the relatively larger spacing resulting solely from heating. Fig. 8.8 shows a photograph of an experimental result taken by Pieter Zeeman c.1897. As a simpler example, let us consider a $2p(n = 2, \ell = 1)$ orbital, and the result of adding an additional magnetic field **B**. This is illustrated in Fig. 8.9. Where there is no magnetic field, the result is a single spectrum line originally produced from say, heating the gas. However, when a magnetic field is turned on, the single spectral line transitions to a set of three closely spaced spectral lines, known as a *triplet*. This kind of splitting or multiplicity of lines, as observed by Zeeman, provided some of the first experimental evidence of the additional states associated with m_ℓ. Because these states are observed by means of a magnetic field, this is the rationale for calling m_ℓ the magnetic quantum number. Counting these additional states in a magnetic field, one

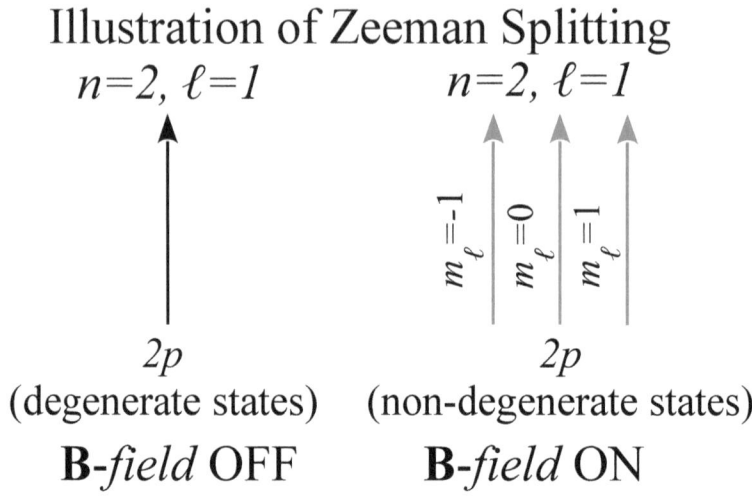

Figure 8.9: Zeeman splitting of 2p orbital in the presence of a magnetic field.

sometimes found that there are up to $2\ell + 1$ of these fine lines associated with a single frequency or value of n. This kind of spectral splitting, due to interactions of the orbital angular momentum L, with a magnetic field, is known as the **fine structure** lines. This particular interaction with the orbital angular momentum L, giving rise to a splitting into $2\ell + 1$ lines, is known as the **ordinary Zeeman effect**. The discrepancy with the maximum number of observable lines being twice this become known as the anomalous Zeeman effect, and this is covered in the next chapter. Let us now find out how the *ordinary Zeeman effect* is predicted by quantum mechanics.

8.7 Quantum Mechanics Of Ordinary Zeeman Effect

To treat the ordinary Zeeman effect, we begin by using our classical result from the discussion of orbital angular momentum and orbital magnetic moment μ in a magnetic field. This was discussed in Secs. 7.2 of Chap. 7. There, we found that the magnetic potential energy of an orbital magnetic moment $\boldsymbol{\mu}_L$, ignoring any dissipation, is given by

$$V_M = -\boldsymbol{\mu}_L \cdot \mathbf{B}$$

8.7 Quantum Mechanics Of Ordinary Zeeman Effect

In Chap. 7, we also found that the orbital magnetic moment μ_L is proportional to the orbital angular momentum L. Thus, for an electron's orbital magnetic moment, we assume a linear relation between $\widehat{\mu}_L$ and $\widehat{L}$ given by $\widehat{\mu}_L = \gamma \widehat{L}$. So, we have

$$V_M = -\gamma \mathbf{L} \cdot \mathbf{B} \tag{8.75}$$

This is sufficient to describe the ordinary Zeeman effect from atomic spectra of gases like hydrogen, placed in a magnetic field. Eq. (8.75) provides a cue for how to express the quantum mechanical Hamiltonian for the *ordinary Zeeman effect* in a way that fulfills the *Principle of Correspondence*. From this, the quantum mechanical Hamiltonian is given by

$$\widehat{H}\psi = \left[\widehat{H}_0 + V_M\right]\psi = \left[\widehat{H}_0 - \gamma \widehat{L} \cdot \mathbf{B}\right]\psi = E_\lambda \psi \tag{8.76}$$

$\widehat{H}_0$ denotes the Hamiltonian that we considered for the hydrogen atom *without the Zeeman interaction*, or the magnetic field. It includes the kinetic energy and the electron-proton potential interaction. This part of the Hamiltonian leads to a spatially dependent state. For this, we have already solved with corresponding eigenvalues $E_{\lambda 0}$ (c.Eq. 8.50) and eigenfunctions ψ_0. Because the magnetic potential V_M is not dependent on spatial coordinates, this Hamiltonian is separable. Hence, we can solve the last term of Eq. (8.76) independently, to determine the additional energy that must be added to $E_{\lambda 0}$. To solve it, let us assume that B is along the z direction, which is the polar axis about which φ rotates. Then, Eq. (8.76) becomes

$$\widehat{H}\psi = \left[\widehat{H}_0 + \frac{|e|B}{2m^*}\widehat{L}_z\right]\psi \tag{8.77}$$

Using our earlier results $\widehat{L}_z\psi = m_\ell \hbar \psi$, Eq. (8.77) leads to an additional energy eigenvalue $\lambda = E_Z$ that must be added to $E_{\lambda 0}$, when a magnetic field is present. Thus, the energy becomes

Hydrogen Energy With Ordinary Zeeman Interaction

$$E_Z = -\frac{Z^2 m^* e^4}{32\pi^2 \hbar^2 \varepsilon_0^2} \frac{1}{n^2} + \frac{|e|B}{2m^*}\hbar m_\ell \tag{8.78}$$

Z is the number of protons within the nucleus about which a single electron orbits. m^* is the reduced mass discussed earlier. The result above can be used to determine the splitting due the electron spin coupling to the magnetic field. This minimum splitting due to the magnetic interaction is given by

$$\Delta E_Z = \frac{|e|\hbar}{2m_e} B \Delta m_\ell = \frac{|e|\hbar}{2m_e} B(1-0) = |\gamma|\hbar B$$

or

Minimum Orbital Energy Splitting In A Magnetic Field

$$\Delta E_Z = \mu_B B \tag{8.79}$$

μ_B the Bohr magneton. So, there is indeed an additional energy contribution when a magnetic field is present. According to this result, for any given n, there are up to $2\ell + 1$ potential splittings given by all the possible values of m_ℓ. This result provides some explanation of the observed splitting, however, there is still another contribution missing from the ordinary Zeeman effect, arising from spin. We'll cover this contribution in the next chapter.

8.8 Chapter Summary

In this chapter, we have used the Schrödinger equation to solve for the eigenenergy states and their corresponding eigensolutions for the hydrogen atom. In obtaining these states, we obtained consistent results with those of Bohr, in predicting atomic spectra. However, we also found that we could treat the additional splitting effects observed when a magnetic field is turned on, by introducing the quantum mechanical Hamiltonian of the Zeeman energy. For each principal quantum number n there are up to additionally $2\ell + 1$ states available from the coupling of the orbital angular momentum to the magnetic field B. Thus, quantum mechanics lead to a more general result predicting that splittings can occur that are proportional to the strength of B. Though there were still some discrepancies observed which will be reconciled in the next chapter, these particular features were in agreement with some observations from experiments. Table 8.5 lists some of the key formulas we obtained in this chapter.

8.8 Chapter Summary

Table 8.5: Chapter 8 Summary Equations. $\zeta = \alpha_n r$, $a = \ell + 1 - n$, $c = 2\ell + 2$

Name	Equation
principal quantum number	$n = 1, 2, 3, \ldots$
Reduced mass	$\frac{1}{m^*} = \frac{1}{m_1} + \frac{1}{m_2}$
Zeeman energy	$E_Z = -\frac{Z^2 m^* e^4}{32\pi^2 \hbar^2 \varepsilon_0^2} \frac{1}{n^2} + \frac{|e|\hbar}{2m^*} B m_\ell$
Radial eigensolutions	$R_{n,\ell}(r) = C_{n,\ell} (\zeta)^\ell e^{-\frac{\zeta}{2}} {}_1F_1(\zeta; a, c)$
Bound-state eigensolutions	$B_{\ell, m_\ell, n}(r, \theta, \varphi) = C_{n, \ell, m_\ell} R_{n, \ell} Y_{\ell, m_\ell}(\theta, \varphi)$

Periodic Table of Elements

Group	1 IA	2 IIA	3 IIIA	4 IVB	5 VB	6 VIB	7 VIIB	8 VIIIB	9 VIIIB	10 VIIIB	11 IB	12 IIB	13 IIIA	14 IVA	15 VA	16 VIA	17 VIIA	18 VIIIA
1	1 H 1.0079 Hydrogen																	2 He 4.0025 Helium
2	3 Li 6.941 Lithium	4 Be 9.0122 Beryllium											5 B 10.811 Boron	6 C 12.011 Carbon	7 N 14.007 Nitrogen	8 O 15.999 Oxygen	9 F 18.998 Flourine	10 Ne 20.180 Neon
3	11 Na 22.990 Sodium	12 Mg 24.305 Magnesium											13 Al 26.982 Aluminium	14 Si 28.086 Silicon	15 P 30.974 Phosphorus	16 S 32.065 Sulphur	17 Cl 35.453 Chlorine	18 Ar 39.948 Argon
4	19 K 39.098 Potassium	20 Ca 40.078 Calcium	21 Sc 44.956 Scandium	22 Ti 47.867 Titanium	23 V 50.942 Vanadium	24 Cr 51.996 Chromium	25 Mn 54.938 Manganese	26 Fe 55.845 Iron	27 Co 58.933 Cobalt	28 Ni 58.693 Nickel	29 Cu 63.546 Copper	30 Zn 65.39 Zinc	31 Ga 69.723 Gallium	32 Ge 72.64 Germanium	33 As 74.922 Arsenic	34 Se 78.96 Selenium	35 Br 79.904 Bromine	36 Kr 83.8 Krypton
5	37 Rb 85.468 Rubidium	38 Sr 87.62 Strontium	39 Y 88.906 Yttrium	40 Zr 91.224 Zirconium	41 Nb 92.906 Niobium	42 Mo 95.94 Molybdenum	43 Tc 96 Technetium	44 Ru 101.07 Ruthenium	45 Rh 102.91 Rhodium	46 Pd 106.42 Palladium	47 Ag 107.87 Silver	48 Cd 112.41 Cadmium	49 In 114.82 Indium	50 Sn 118.71 Tin	51 Sb 121.76 Antimony	52 Te 127.6 Tellurium	53 I 126.9 Iodine	54 Xe 131.29 Xenon
6	55 Cs 132.91 Caesium	56 Ba 137.33 Barium	57-71 La-Lu Lanthanide	72 Hf 178.49 Hafnium	73 Ta 180.95 Tantalum	74 W 183.84 Tungsten	75 Re 186.21 Rhenium	76 Os 190.23 Osmium	77 Ir 192.22 Iridium	78 Pt 195.08 Platinum	79 Au 196.97 Gold	80 Hg 200.59 Mercury	81 Tl 204.38 Thallium	82 Pb 207.2 Lead	83 Bi 208.98 Bismuth	84 Po 209 Polonium	85 At 210 Astatine	86 Rn 222 Radon
7	87 Fr 223 Francium	88 Ra 226 Radium	89-103 Ac-Lr Actinide	104 Rf 261 Rutherfordium	105 Db 262 Dubnium	106 Sg 266 Seaborgium	107 Bh 264 Bohrium	108 Hs 277 Hassium	109 Mt 268 Meitnerium	110 Ds 281 Darmstadtium	111 Rg 280 Roentgenium	112 Uub 285 Ununbium	113 Nh 284 Nihonium	114 Fl 289 Flerovium	115 Mc 288 Moscovium	116 Lv 293 Livermorium	117 Ts 292 Tennessine	118 Og 294 Oganesson

Lanthanide	57 La 138.91 Lanthanum	58 Ce 140.12 Cerium	59 Pr 140.91 Praseodymium	60 Nd 144.24 Neodymium	61 Pm 145 Promethium	62 Sm 150.36 Samarium	63 Eu 151.96 Europium	64 Gd 157.25 Gadolinium	65 Tb 158.93 Terbium	66 Dy 162.50 Dysprosium	67 Ho 164.93 Holmium	68 Er 167.26 Erbium	69 Tm 168.93 Thulium	70 Yb 173.04 Ytterbium	71 Lu 174.97 Lutetium
Actinide	89 Ac 227 Actinium	90 Th 232.04 Thorium	91 Pa 231.04 Protactinium	92 U 238.03 Uranium	93 Np 237 Neptunium	94 Pu 244 Plutonium	95 Am 243 Americium	96 Cm 247 Curium	97 Bk 247 Berkelium	98 Cf 251 Californium	99 Es 252 Einsteinium	100 Fm 257 Fermium	101 Md 258 Mendelevium	102 No 259 Nobelium	103 Lr 262 Lawrencium

Legend:
- ☐ Alkali Metal
- ☐ Alkaline Earth Metal
- ☐ Metal
- ☐ Metalloid
- ☐ Non-metal
- ☐ Halogen
- ☐ Noble Gas
- ☐ Lanthanide/Actinide
- ☐ man-made

Z / Symb / mass / Name

8.9 Chapter Problems

Problem 8.1 The 1s hydrogen atomic orbital is given by

$$\psi_{1s}(r) = \left(\frac{1}{\pi a_0^3}\right)^{1/2} e^{-r/a_0}$$

The probability of finding an electron between r and $r + dr$ can be determined by evaluating the normalization condition only with respect to φ and θ. Evaluate the integral given by

$$f(r) = r^2 dr \int_0^\pi \sin\theta \, d\theta \int_0^{2\pi} d\varphi |\psi_{1s}|^2$$

Show that it gives

$$f(r)dr = \frac{4}{a_0^3} r^2 e^{-2r/a_0} dr$$

Problem 8.2 Show that $f(r)$, given by

$$f(r) = \left(\frac{1}{\pi a_0^3}\right)^{1/2} e^{-r/a_0}$$

satisfies

$$\int f(r)dr = 1$$

Problem 8.3 Using the spherical harmonics $Y_{\ell,m_\ell} = Y_\ell^{m_\ell}$, show that

$$\int_0^\pi d\theta \sin\theta \int_0^{2\pi} d\varphi Y_1^1(\varphi,\theta) Y_1^{-1}(\varphi,\theta) = 0$$

Use Table 7.7, and note that Y_1^{-1} can be written using Y_1^1 by replacing $m = 1$ with $m = -1$, or $e^{i\varphi} \to -e^{-i\varphi}$.

Problem 8.4 According to the illustration of the square of the 1s orbital, the maximum occurs at $r/a_0 = 1$. Since $|1s|^2 dr$ is a probability density, what does this mean?

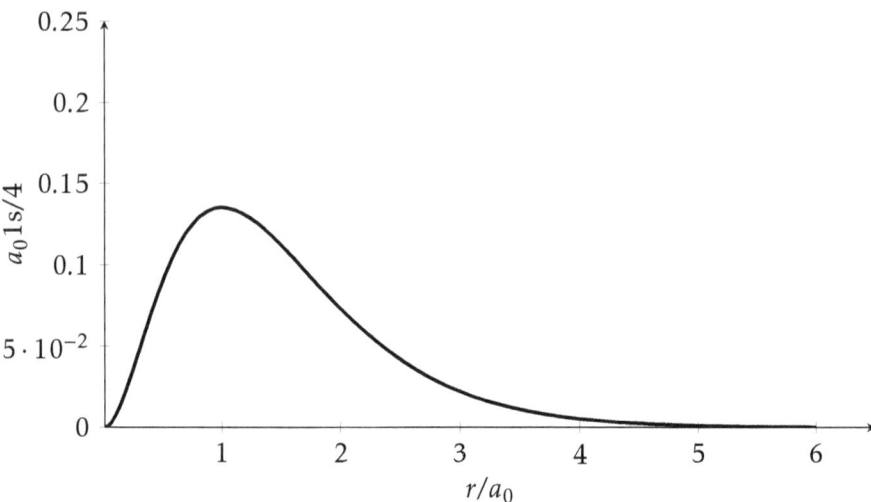

Prove $r/a_0 = 1$ at the maximum by determining where the maximum occurs for $|1s|^2$, given by

$$|1s|^2 = \frac{4}{a_0^3} r^2 e^{-2r/a_0}$$

Problem 8.5 Show that the expectation value of the potential energy $\langle V \rangle$ for an electron in the 2s orbital is given by

$$\langle V \rangle_{2s} = \int_0^{2\pi} d\varphi \int_0^{\pi} d\theta \sin\theta \int_0^{\infty} r^2 \psi_{2s}^* \left(-\frac{e^2}{4\pi\epsilon_0 r}\right) \psi_{2s}$$

$$= -\frac{e^2}{4\pi\epsilon_0 a_0^3} \int_0^{\infty} dr \left(2 - \frac{r}{a_0}\right)^2 r e^{-r/a_0}$$

$$= -\frac{e^2}{16\pi\epsilon_0 a_0}$$

Then, show that this leads to

$$\langle V \rangle_{2s}/\langle K \rangle_{2s} = -2$$

$\langle K \rangle_{2s}$ is the expectation value of the kinetic energy.

Problem 8.6 Use Eq. (9.12) to determine the Zeeman splitting in a magnetic field of .5 Tesla for a hydrogen electron in a 2p orbital. How does this compare to energy of a transition from 1s to the 2p orbital?

Problem 8.7 Use Eq. (9.12) to determine the Zeeman splitting in a magnetic field of .5 Tesla for a hydrogen electron in a 2p orbital. How does this compare to energy of a transition from 1s to the 2p orbital?

8.10 SUGGESTED READINGS & REFERENCES

[1] N. Bohr, *On the Constitution of Atoms and Molecules*, Philos. Mag. 26, 1 (1913)
[2] N. Bohr, "Part II. – Systems containing only a Single Nucleus", Mag. 26, 1 (1913)

CHAPTER 9

The Unveiling of Electron Spin

Angular momentum has been an integral concept in the previous two chapters, where we have, so far, only discussed orbital angular momentum. The operator is denoted by $\widehat{L}$. We also hinted in the previous chapter that this is not the whole story for the angular momentum of any fundamental particle, let alone the electron. Should we find evidence of additional angular momentum, the physics for $\widehat{L}$ should also apply to any other contributions of angular momentum. In the orbital angular momentum problem, we have solved its eigenvalue problem exactly. Additionally, orbital angular momentum re-appeared in the hydrogen problem, when solving the Schrödinger equation. Because the Hamiltonian was separable in the orbital angular momentum variables and the radial coordinate, we could use wholesale, our results. In the hydrogen problem, $\widehat{L}$ describes the localized orbital motion of the electron about the protonic nucleus. It was a matter of time before it would be discovered that this is *not* enough to describe all the contributions to the *total* angular momentum of electrons. We now know that these particles also possess an *intrinsic angular momentum*, analogous to the angular momentum concept of the earth spinning on its own axis, while it is simultaneously orbiting about the sun.

However, there is a notable difference between the angular momentum of the earth and that of the electron. For earth, it's orbital angular momentum is $\approx 3 \times 10^{40}$ kg·m²/s, while its spinning angular momentum is $\approx 6 \times 10^{33}$ kg·m²/s. Therefore, the orbital angular momentum is $\approx 10^7$ times larger than that of the spinning angular momentum. To the contrary, with the electron, these two quantities are amazingly the same order of magnitude. The electron's intrinsic angular momentum is known as *spin*, and in the early 1920s, the evidence of electron *spin* was mounting. In this chapter, we begin the discussion of this additional degree of freedom, which turns out to be crucial in enabling a more complete understanding of electromagnetism, and other more recently emerging technical fields such as the field of *spintronics*. We will get to this topic in Chap. 12, as well. For now, let us begin with the story of how the scientific world discovered spin.

9.1 Evidence of Spin Existence

With all the data coming from the atomic spectra of gases like hydrogen, there was still yet another observation not fully understood after the success of describing the ordinary Zeeman effect. Since ℓ is an integer, the number of splittings or states relative to the *principal quantum number n*, due to the coupling to the orbital angular momentum to the magnetic field is always odd at $2\ell + 1$ states. However, in atoms with odd atomic number Z, like hydrogen, it was often observed that the number of sub-level splittings in a magnetic field was *even, rather than odd*. This observation cannot be explained by the quantum states of orbital angular momentum alone. It suggested the possible existence of an additional angular momentum contribution interacting more directly with the magnetic field. It suggested further that this additional angular momentum can take on half-integer values, while energy splittings for transitions were still whole integer amounts of $\hbar$. If this was true, then for each value of m_ℓ, there would be an additional two possible states where this contribution was either aligned or anti-aligned with the field B. This phenomenon of spectra splitting into even numbers of spectral lines became known as the **anomalous Zeeman effect**, and it arises from the further coupling to the magnetic field of this then little-known degree of freedom known as *spin*.

Around 1925, these observations led German-American physicist, Ralph de Laer Kronig (1904-1995) to postulate that the total angular momentum of hydrogen must also include a contribution from spinning about

9.1 Evidence of Spin Existence

its own axis. The atomic spectra, as well as other experiments pointed out that there are *two and only two states* arising from this intrinsic spin. But, how can one deduce that there must be two states? We can use our intuition based on what we know about orbital angular momentum. Considering that with orbital angular momentum, the number of quantum states is given by $2\ell + 1$, then a corresponding quantum number s for the *spin* angular momentum, should also give exactly two, since two states were observed. In other words, $2s + 1 = 2 \Rightarrow s = 1/2$. Moreover, since we also know that given the value of ℓ, the magnetic quantum number $-\ell \leq m_\ell \leq \ell$, one can postulate an analogous **spin magnetic quantum number** m_s, such that given the single allowed value of s, the only possibility is that $m_s = \pm 1/2$, yielding the two states. Observing a singe valued quantum number s for a single electron turns out to have profound implications. For example, unlike m_ℓ, it cannot take a value of zero. Thus, the spinning contribution to the angular momentum we shall denote by $\widehat{S}$ can be described by

---

Spin Angular Momentum Quantum Numbers

$$S^2 = s(s+1)\hbar^2 \quad \text{and} \quad S_z = m_s \hbar \quad s = \frac{1}{2}, \; m_s = \pm\frac{1}{2} \quad (9.1)$$

---

m_s is analogous to m_ℓ as a magnetic quantum number, which becomes evident in the spectra when a magnetic field is present. Both contribute to the observed *multiplicity* in atomic spectra. These properties of spin follow directly from the established quantum theory of orbital angular momentum where we have only assumed consistency between ℓ and s. This further establishes a physical correspondence between $\widehat{L}$ and $\widehat{S}$. This correspondence can be used to relate their sums to a corresponding total magnetic moment. For the orbital angular momentum, we found the relation between the orbital magnetic moment $\widehat{\mu}_L$ and $\widehat{L}$ to be

$$\mu_L = -\frac{e}{2m_e}L = \gamma L$$

To obtain an analogous relation between the spin magnetic moment $\widehat{\mu}_S$ and spin angular momentum $\widehat{S}$, we cannot use exactly the same approach we used for the orbital angular momentum because the physical picture is different. We do not have an electron orbiting about the nucleus. In fact, we don't know exactly what we have, because this spin is arising from an intrinsic behavior of the electron. However, it is still the electron which has mass m_e and charge e. Therefore, let us assume the following relation for the spin magnetic moment $\widehat{\boldsymbol{\mu}}_S$:

$$\widehat{\boldsymbol{\mu}}_S = g\gamma \mathbf{S} \tag{9.2}$$

Substitution of γ gives

Magnetic Moment-To-Spin Relation

$$\widehat{\boldsymbol{\mu}}_S = -\frac{ge}{2m_e}\widehat{\mathbf{S}} \tag{9.3}$$

The parameter g is a *constant* of proportionality called the **g-factor**, or more generally, the Landé-**factor**, and its value for the electron is a precisely measured experimental constant measured at

Experimentally Measured g-factor

$$g = 2.00231930436146 \tag{9.4}$$

The experimental value of g given by Eq. (9.4) has been repeatably measured with a precision to within 13 digits, making it better than 10^{-12}, or one part per trillion (ppt). Therefore, the ratio of the electron spin magnetic moment to the electron spin angular momentum is twice that for electron orbital angular momentum. We'll show in Chap. 11, that in the approximations of *relativistic quantum mechanics*, g is predicted to take the value $g = 2$, in good agreement with the precisely measured value. Combining both contributions of orbital and spin angular momentum, we may define a total angular momentum operator $\widehat{\mathbf{J}}$ as

Total Angular Momentum Operator

$$\widehat{\mathbf{J}} = \widehat{\mathbf{L}} + \widehat{\mathbf{S}} \tag{9.5}$$

This leads us to a corresponding relation for the total magnetic moment $\widehat{\boldsymbol{\mu}}_T$ given by

$$\widehat{\boldsymbol{\mu}}_T = \widehat{\boldsymbol{\mu}}_L + \widehat{\boldsymbol{\mu}}_S \tag{9.6}$$

9.1 Evidence of Spin Existence

Substitution of the corresponding relations for $\widehat{\boldsymbol{\mu}}_L$ and $\widehat{\boldsymbol{\mu}}_S$, gives the following relation for the electron's total magnetic moment $\widehat{\boldsymbol{\mu}}_T$:

> **Total Electron Magnetic Moment Operator**
>
> $$\widehat{\boldsymbol{\mu}}_T = \gamma\widehat{\mathbf{L}} + g\gamma\widehat{\mathbf{S}} \tag{9.7}$$

Using the additional contribution of spin to the angular momentum, Kronig was able to explain most of the anomalous Zeeman effect observations in the atomic spectra of hydrogen. Unfortunately, Kronig was dissuaded at some point by his advisor at the time, Wolfgang Pauli (who later became a proponent of spin). He consequently discarded his postulations regarding the existence of spin.

One of the arguments fueling the consternation with the early introduction of spin was the absurd consequences when viewed in the framework of *classical* rotations. Estimates of the radius of an electron had been attempted. One particular estimate, from Lorentz, was obtained essentially by setting the electric coulomb potential energy to Einstein's relativistic kinetic energy $E = mc^2$ (this relation will be derived in Chap. 11), with an angular momentum of $\hbar/2 = m_e r_e v_e$. Upon carrying this out and solving for r_e, one obtains an estimated radius of the electron given by $r_e = e^2/4\pi\epsilon_0 mc^2 \sim 10^{-15}$ m. This radius, being much smaller than the Bohr radius, for example, is known as the *Lorentz radius*. Unfortunately, there is no basis for this reasoning used by Pauli because this calculation ignores spin altogether. This result consequently leads to a linear speed of the electron on it's surface given by $v_e \sim 67$ times the speed of light c. This was a problem because it was in lucid contradiction with the postulate of the special theory of relativity, which postulates an upper bound on the speed of any mass at c.

As fate would have it, six months after Kronig discarded his ideas, two Dutch-American physics students by the names of George Eugene Uhlenbeck (1900-1988) and Samuel Abraham Goudsmit (1902-1978) published these same ideas in a couple of articles. By the second article, they pointed out that their postulates of spin did not explain *all* experimental observations of atomic spectra, accurately, but much of it. For example, quantitatively, there was a known discrepancy of a factor of 2 comparing to experimental data. Fortunately, in 1926, British physicist Llewellyn Hilleth Thomas (1903-1992) stepped in for the rectification pointing out that the origin of this discrepancy arises from not considering the acceleration of an electron orbiting a nucleus (due to changing direction), rather than treating it as though it has constant velocity. This additional consideration lead to the needed correction for the spectral frequencies, correcting the missing factor of 2. After this, Kronig, Uhlenbeck, and Goudsmit's ideas of spin now agreed with more experimental observations of atomic spectra, making a believer out of Kronig, Ulenbeck, Goudsmit, and even Wolfgang Pauli.

Although we have somewhat casually made reference to the image of an electron spinning about it's own axis, which is useful in comprehending the concept of spin and even some of it's consequences, the *origin of electron spin remains unknown*. It does not appear that it can be explained successfully quantitatively by any classical argument. Perhaps, the reader shall take up this challenge and play a part in improving upon these uncertainties associated with the intrinsic angular momentum known as *spin*. Fortunately, it does not prevent us from correctly describing its contributions to the total angular momentum J and energy.

Before observations of the even splittings of hydrogen gas, a set of famous experiments had been carried out that incidentally, unwittingly revealed the existence of electron spin. A clear understanding of this fact did not emerge until after the experiments were performed. These experiments became known as the Stern-Gerlach experiment and it turned out to be the first unveiling of the electron spin. Let's discuss these experiments.

9.2   Understanding the Stern-Gerlach Experiment

Although it was not until around 1926 that clear evidence for the existence of electron spin had been observed, it turns out that there was existing evidence of electron spin unknowingly generated by German physicists Otto Stern (1888-1969) and Walther Gerlach (1889-1979). After the discovery of the Zeeman effect in 1896, some scientist developed models to try and explain it. Bohr developed his model and another model by German physicist Arnold Sommerfeld (1868-1951) was introduced in 1916 that generalized Bohr's model allowing for elliptical orbits of electrons (perhaps inspired by the planetary motion). A consequence of these models was that orbital angular momentum

9.2 Understanding the Stern-Gerlach Experiment

was quantized, and it was perceived at the time that this had startling consequences. For example, we know that not only do uniform magnetic fields rotate magnetic moments, but *nonuniform magnetic fields* further translate magnetic moments by a net force. Taking **B** along z, the force in the z-direction, due to the gradient of the magnetic potential energy $V_M = -\boldsymbol{\mu} \cdot \mathbf{B}$ is

$$F_z = -\frac{\partial}{\partial z}(-\boldsymbol{\mu} \cdot \mathbf{B}) = \frac{\partial V_M}{\partial B_z}\frac{\partial B_z}{\partial z} = \mu_z \frac{\partial B_z}{\partial z} \tag{9.8}$$

This is analogous to how an electric dipole behaves in a nonuniform electric field. It was understood classically, that this force could vary from $-\left|\mu_z \frac{\partial B_z}{\partial z}\right|$ to $+\left|\mu_z \frac{\partial B_z}{\partial z}\right|$. However, the models of Bohr and Sommerfeld appeared to contradict this classical understanding because the values of μ were now quantized. It was Stern who came up with the idea of how to get to the bottom of this discrepancy with an experiment. By passing neutral atoms through a nonuniform field, one could distinguish the form of the distribution of angles of μ. The classical expectation: if a beam of atoms is initially randomly oriented upon passing through an inhomogeneous field to be deflected in the +x and −x directions, the distribution of deflection angles will have a maximum value at zero deflection and decrease monotonically in either direction. But this was *not* observed.

Apparently, Stern was not a skilled experimentalist, so he recruited Gerlach for collaboration. Stern never touched the experimental setup. In their experiment, the generation of neutral silver atoms was done in a vacuum so that the silver atoms moved without scattering. They were derived from the effusion of metal in an oven heated to a temperature of ~1273 Kelvin. The atoms were then collimated by slits and directed through the region with a large non-uniform magnetic field **B**, then towards a glass plate for collection. Reasoning based on Bohr and Sommerfelds model lead them to believe that silver atoms possessed nonzero orbital angular momentum where $m_\ell = \pm 1$. They went on to estimate splittings based on this. They expected that the coupling with the magnetic field, which in their experiment, was ~ 0.1 Tesla, would be enough to yield sufficient separation of the particles relative to the beam width. An illustration of the experimental setup of Stern and Gerlach is shown in Fig. 9.1.

In their experiments, the calculated splitting was ~ .2 mm, which necessitated relatively good alignment. This was nontrivial for this time.

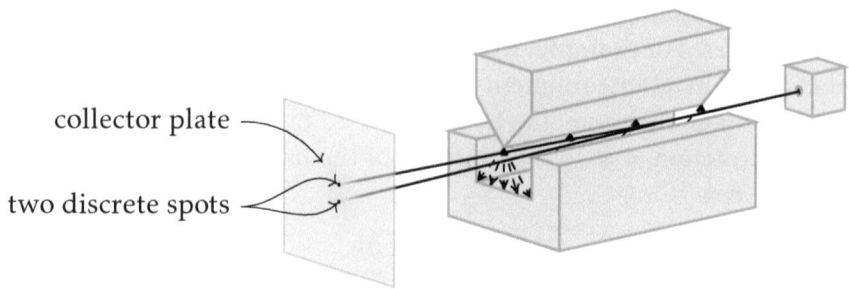

Figure 9.1: Setup for the Stern-Gerlach experiment (not to scale). To the far right, a beam of silver atoms from an oven. They are directed between north and south magnetic poles; the magnets produce a nonuniform fringing magnetic field imposing a force on spins; the result is indicated on a collector plate, upon which the atoms separate a splitting distance, expected to be ~ 0.2 mm.

Based on what we now know from quantum mechanics, the use of silver was not a good choice for their objectives because *neutral silver atoms do not possess orbital angular momentum*. Silver possesses a full $3d$ orbital, and a single valence electron in the $5s$-orbital, which has no orbital angular momentum (s-orbitals are defined by $\ell = 0$). In the first try, there was little to see on the collector plate they used. As the story has it, Stern used to be an avid smoker of cigars and it was his breath that made history as the sulfur interacted with the silver to make silver-sulfide, making their results more visible providing some motivation for their efforts. This is because silver-sulfide is black in color. Historians now tell us that it was more likely that his breadth would not have been enough, but likely some smoke emitted directly from a sulfur-rich cigar in the lab would have been necessary. Still, in light of these observations, the experiment was repeated several more times. It was not until one fateful day in 1922 when Gerlach was delayed in Göttingham, Germany due to a railroad strike, preventing him from getting back to Frankfurt. He decided to repeat the experiment with better alignment, where finally, a clear signature was observed, evidencing two discrete states. Gerlach telegraphed the news to Stern and the rest is history. Gerlach's observations are illustrated in Fig. 9.2. It shows the famous postcard that was sent from Gerlach to Neils Bohr upon successfully completing the experiment in 1922, congratulating him on his theoretical confirmation.

Although Kronig, Uhlenbeck, and Goudsmit's work had postulated that there was more to the atoms' angular momentum than the orbital contribution, it was not pointed out until 1927 by British physicist Ronald

9.2 Understanding the Stern-Gerlach Experiment

Fraser (1899-1985), after re-interpreting the Stern-Gerlach experiment and realizing that their observed splitting was due to the intrinsic angular momentum, and cannot be due to orbital angular momentum. This is because, as we've mentioned, the orbital angular momentum of silver atoms is zero. However it has a net intrinsic spin from the lone electron in the 5s-orbital! The significance of the Stern-Gerlach experiment is that it turned out to be, *unbeknownst to them*, an inaugural unveiling of the existence of an intrinsic angular momentum, made visible for the world to see. It coupled to magnetic fields and experienced forces just as they expected of the orbital angular momentum.

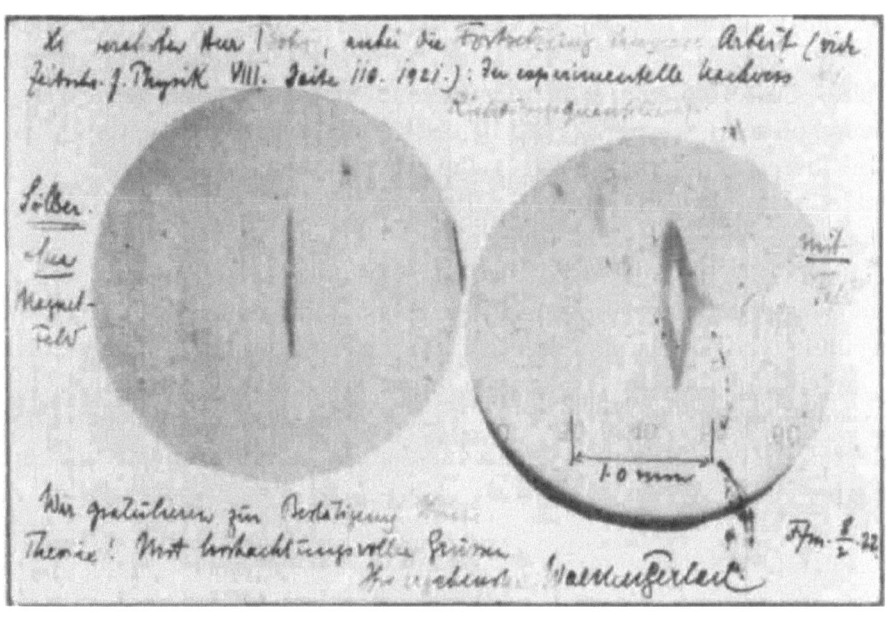

Figure 9.2: Copy of the postcard sent from Walther Gerlach to Neils Bohr, dated Feb 08, 1922, upon successfully completing the experiment in Göttingen. The message on the postcard, in translation, reads "Attached is the experimental proof of directional quantization. We congratulate you on the confirmation of your theory."

We must keep in mind that these experiments were done around three years prior to Kronig, Uhlenbeck, and Goudsmit's enunciation of their ideas of spin, so the concept of spin was not yet formally conceived. Thus, the origin of Stern and Gerlach's misunderstanding with the orbital angular momentum should be taken into context. Nonetheless, all of this amounted to inescapable evidence that there was more to an electron's angular momentum than the orbital angular momentum. It

9.3 Anamolous Zeeman Effect And A 4th Quantum Number

With a background of the experimental developments pointing to the existence of spin, let's turn our attention to how quantum mechanics came to reconcile this discovery. As far as atomic spectra is concerned, the observations had lead to resulting spectral splittings consisting of not just odd numbers, as expected from orbital angular momentum states, but often even splittings were observed, as well. Because the even splittings were observed before the discovery of spin, they were considered anomalous. Hence, when the splittings in a magnetic field were even, it became known as the *anomalous Zeeman effect*. With the postulates above, we can now treat this effect using the Schrödinger equation. To obtain a result incorporating spin, an important step is that we need to determine a suitable Hamiltonian operator. For this, we again look to our classical results from the discussion of orbital angular momentum and orbital magnetic moment $\boldsymbol{\mu}$ in a magnetic field. This was discussed in Chap. 7. There, we found that the magnetic potential energy V_M of an orbital magnetic moment $\boldsymbol{\mu}_L$, ignoring any dissipation, is given by

$$V_M = -\boldsymbol{\mu}_L \cdot \mathbf{B}$$

There is no reason why the form of this result should not extend to include the electron spin magnetic moment. Extending this form of the total potential energy to include a spin magnetic moment $\boldsymbol{\mu}_S$, we can write

$$V_M = -\boldsymbol{\mu}_T \cdot \mathbf{B} = -\boldsymbol{\mu}_L \cdot \mathbf{B} - \boldsymbol{\mu}_S \cdot \mathbf{B}$$

Substituting Eq. (9.7), we have

$$V_M = -\gamma \mathbf{L} \cdot \mathbf{B} - g\gamma \mathbf{S} \cdot \mathbf{B} \qquad (9.9)$$

g is a proportionality factor which can be measured. A good approximation can also be deduced from relativistic quantum mechanics, and we'll do this in Chap. 11. This is sufficient to determine the atomic spectra of hydrogen in a magnetic field. Eq. (9.9) now includes the additional

9.3 Anamolous Zeeman Effect

spin-dependent energy contribution to the quantum mechanical Hamiltonian. With this, the quantum mechanical Hamiltonian $\widehat{H}$ becomes

$$\widehat{H}\psi = \left[\widehat{H}_0 + V_M\right]\psi = \left[\widehat{H}_0 - \gamma \widehat{\mathbf{L}} \cdot \mathbf{B} - g\gamma \widehat{\mathbf{S}} \cdot \mathbf{B}\right]\psi = E_\lambda \psi \qquad (9.10)$$

As we had when we considered the *ordinary* Zeeman effect, $\widehat{H}_0$ denotes the Hamiltonian that we considered for the hydrogen atom *without the Zeeman interaction*, or a magnetic field. It includes the kinetic energy and the electron-nucleus potential interaction. The second terms is the contribution from the orbital angular momentum. For these two contributions, we have already solved. Thus, we only need to solve for the contribution from spin, which goes much like it went for $\widehat{\mathbf{L}}$. Let us assume that **B** is along the z direction. Then, the eigenvalue equation given by Eq. (9.10) becomes

$$\widehat{H}\psi = \left[\widehat{H}_0 + \frac{|e|B}{2m^*}\widehat{L}_z + \frac{g|e|B}{2m^*}\widehat{S}_z\right]\psi \qquad (9.11)$$

Substituting $\widehat{L}_z\psi = m_\ell \hbar \psi$ and $\widehat{S}_z\psi = m_s \hbar \psi$, leads to the more general Zeeman energy eigenvalue $E_\lambda = E_Z$, where we have

Eigenenergy With Anomalous Zeeman Interaction

$$E_Z = -\frac{Z^2 m^* e^4}{32\pi^2 \hbar^2 \varepsilon_0^2} \frac{1}{n^2} + \frac{|e|\hbar}{2m^*} B m_\ell + \frac{g|e|\hbar}{2m^*} B m_s \qquad (9.12)$$

Z is the atomic number, or number of protons in the nucleus. m^* is the *reduced mass* discussed in Chap. 8, and e is the elementary charge. Though $e > 0$, we often include absolute values to enforce a positive value in the equation. We now have the origins of additional splittings in a magnetic field. Also, note that with this result, spin-down electrons are lower in energy compared to spin-up, which turns out to be an important detail in spin-transport.

 Spin-down state: The spin-down state ($m_s = -1/2$) is the lower energy state between both spin states. This is relative to a magnetic field, and leads to electrons occupying spin-down states before spin-up.

When the field is along +z, a spin-down state ($m_s = -1/2$) has lower energy than a spin-up state ($m_s = 1/2$). Thus, the *electron angular momentum improves the stability by oppositely aligning to an external magnetic field*. This also means that the *electron* magnetic moment $\boldsymbol{\mu}$, which is anti-parallel to **S** and **L**, respectively, will point along the B-field direction to lower the energy. The minimum splitting due to the electron spin coupling to a magnetic field is given by

$$\Delta E_Z = \frac{g|e|\hbar}{2m_e} B \Delta m_s = \frac{g|e|\hbar}{2m_e} B \left(\frac{1}{2} - \left[-\frac{1}{2} \right] \right)$$

So, the spin splitting becomes

Spin Splitting In A Magnetic Field

$$\Delta E_Z = g|\gamma|\hbar B \tag{9.13}$$

In terms of the Bohr magneton $\mu_B = |\gamma|\hbar$, Eq. (9.13) may also be written as

$$\Delta E_Z = g\mu_B B \tag{9.14}$$

Thus, for an electron orbiting a nucleus in bounded fashion, given a designated value of the *principal quantum number n*, introducing a magnetic field gives rise to up to $2 \times (2\ell + 1)$ additional available states, which is even when there is spin in the system. This is because for each value of m_ℓ, there are now two more states due to the spin splitting.

This result gives a complete explanation of the pair splittings observed by Gerlach. In the Stern-Gerlach experiment, they used silver atoms which have $\ell = 0$ because the valence electron is in the 5s orbital. Then, the number of splittings is simply given by $2 \times (2\ell + 1) = 2 \times (0 + 1) = 2$. As the postcard in Fig. 9.2 evidences, this was the number of splittings observed by Stern and Gerlach.

Since the spin-dependent Hamiltonian is often independent of space, the spatial variables like **r** and the spin degree of freedom can be treated as separable, and the total wave function ψ can be written as a product given by

$$\psi(x, y, z, S_x, S_y, S_z) = \psi_\mathbf{r}(x, y, z) \cdot \psi_s(\mathbf{S}) \tag{9.15}$$

9.3 Anamolous Zeeman Effect

The addition of the spin angular momentum to the Hamiltonian leads to a total of **four essential quantum numbers** needed to describe the state of an electron. The four numbers include the principal quantum number n, the orbital angular momentum quantum number ℓ (because it determines m_ℓ), the orbital magnetic quantum number $m = m_\ell$, and now we have the fourth quantum number, the spin magnetic quantum number m_s. Although s is a quantum number, since s has only a single value, m_s is sufficient to describe energy splitting due to spin.

This fourth quantum number m_s turned out to provide the missing link to correctly predict the even number of atomic spectra lines by hydrogen (and silver!) atoms in a magnetic field. As with silver, there is a net spin from the single electron in the 1s orbital. Table 9.1 summarizes the *fundamental quantum numbers*, associated eigenvalues, and operators that we have established so far, for the electron.

Table 9.1: Summary of fundamental quantum numbers and eigenvalues for the electron. For the real-valued Hamiltonian here, the eigenvalue λ provide expectation values of the operator and they are generally functions of associated quantum numbers ℓ, m_ℓ, n, and m_s.

Associated Operator	Quantum Number
L_z eigenvalue/quantum number	$\lambda = m_\ell = -\ell, ..., -2, -1, 0, 1, 2, ..., \ell$
L^2 eigenvalues	$\lambda = \ell(\ell+1)$
L^2 quantum numbers	$\ell = 0, 1, 2, 3, ...$
n principal quantum number	$n = 1, 2, 3, ...$
S^2 eigenvalue	$\lambda = s(s+1)$
s spin quantum number	$s = 1/2$
m_s spin quantum number	$m_s = -1/2, 1/2$

9.3.1 Interpretation of Angular Momentum Numbers

There is an interesting interpretation to the quantum numbers of angular momentum. Consider the quantum number ℓ, which determines the orbital angular momentum L. For $\ell = 0$, it's trivial since there is no orbital angular momentum. However, when $\ell = 1$, it means that the electron must rotate a full $2\pi = 2\pi/\ell$ radians, or 1 revolution around the nucleus, according to its periodicity. Thus, everything will look the same to the electron after *one* revolution. Similarly, for $\ell = 2$, the electron needs to rotate only $2\pi/\ell = \pi$ radians before everything repeats again. Continuing on, for $\ell = 3$, $2\pi/\ell = 2\pi/3$ revolutions, etc. Generally, we see there is a $1/\ell$ rotational symmetry, and the electron needs only rotate this amount before repetition. When this concept is extended to the spin degree of freedom, one finds that the electron must rotate $1/s = 2$ parts of a revolution. In other words, the electron spin has to rotate two times before it looks the same again. This is not a very intuitive result, however, this interpretation is inline with the symmetries of angular momentum. Spin, once again, is in a class of its own.

When we want to include the spin degree of freedom in the Hamiltonian, there is a useful way to do this utilizing matrices. Since there are only two states that need to be *carried* around with the Hamiltonian, it does not increase the complexity. In the next section, we will discuss a very common way this is done using *spin matrices*.

9.4 PRINCIPLE SPIN MATRICES AND SPINORS

Because the electron spin magnetic quantum number m_s has only two possible values (as opposed to having infinitely many values like m_ℓ), it provides an opportunity to express the Schrödinger equation as a set of two simultaneous equations, which allows the easy use of matrices.

Since there are only two values of m_s, instead of writing a single equation with a spin angular momentum operator S, we can alternatively write two equations, one corresponding to $\hbar m_s = +\hbar/2$, called the **spin-up** state and the second, corresponding to $\hbar m_s = -\hbar/2$, called the **spin-down** state. Writing Eq. (9.11) as two equations corresponding to both values of m_s gives

$$\widehat{H}\psi = \widehat{H}_+\psi_+ = \left[\widehat{H}_0 + |\gamma|B\widehat{L}_z + \frac{g|\gamma|\hbar B}{2}\right]\psi_+ \qquad (9.16a)$$

$$\widehat{H}\psi = \widehat{H}_-\psi_- = \left[\widehat{H}_0 + |\gamma|B\widehat{L}_z - \frac{g|\gamma|\hbar B}{2}\right]\psi_- \qquad (9.16b)$$

Since each equation is for a corresponding spin-state, it has an associated wave function or eigenfunction, ψ_+ for the *spin-up* state and ψ_- for the

9.4 Principle Spin Matrices and Spinors

spin-down state. So, we can alternatively write Eqs. (9.16) as a vector equation for the pair of spin wave functions ψ_+ and ψ_-. This 2×1 vector of eigenfunctions is known as a **spinor**. Mathematically speaking, since we have a 2×1 vector of two wave functions, the operators in Eqs. (9.16) become 2×2 *matrix operators* (or a matrix of scalar operators) $\widehat{H}$ operating on the 2×1 spinor. Since the only terms that differ are the spin angular momentum term, the identity matrix becomes the matrix coefficient in the first two terms, but not in the last term. Thus, we can write the following matrix-vector form of the Hamiltonian as

$$\widehat{H}\begin{bmatrix}\psi_+\\\psi_-\end{bmatrix}=\begin{bmatrix}\widehat{H}_0 & 0\\0 & \widehat{H}_0\end{bmatrix}\begin{bmatrix}\psi_+\\\psi_-\end{bmatrix}+$$

$$|\gamma|B\begin{bmatrix}\widehat{L}_z & 0\\0 & \widehat{L}_z\end{bmatrix}\begin{bmatrix}\psi_+\\\psi_-\end{bmatrix}+\frac{g|\gamma|\hbar B}{2}\begin{bmatrix}1 & 0\\0 & -1\end{bmatrix}\begin{bmatrix}\psi_+\\\psi_-\end{bmatrix} \quad (9.17)$$

Eq. (9.17) can be written more concisely, in matrix notion as

$$\widehat{H}\begin{bmatrix}\psi_+\\\psi_-\end{bmatrix}=\widehat{H}_0 I(2)\begin{bmatrix}\psi_+\\\psi_-\end{bmatrix}+|\gamma|B\widehat{L}_z I(2)\begin{bmatrix}\psi_+\\\psi_-\end{bmatrix}+\frac{g|\gamma|\hbar B}{2}\widehat{\sigma}_z\begin{bmatrix}\psi_+\\\psi_-\end{bmatrix} \quad (9.18)$$

I(2) denotes the 2×2 identity matrix. For the last term, we have introduced a spin matrix operator $\widehat{\sigma}_z$ defined as

$$\widehat{\sigma}_z=\begin{bmatrix}1 & 0\\0 & -1\end{bmatrix}$$

For a diagonal matrix, we know that its diagonal elements are its eigenvalues. Thus, the matrix $\widehat{\sigma}_z$ has eigenvalues $\lambda=\pm 1$. For a field along the $+z$ direction, Eq. (9.18) provides an alternative form using a matrix notation for the Hamiltonian, *which includes the spin degree of freedom*.

But, what if the field is along a different principle axis, say x or y? For this, we would need different 2×2 matrices, one for each direction. So, how can we obtain valid 2×2 matrices for all three directions? One way to approach this question is to enforce the rules for orbital angular momentum component operators L, since S is also a contribution to the angular momentum. Hence, the same laws governing $\widehat{L}$ must hold for the intrinsic angular momentum, S. This has the advantage that it will involve all three directions x, y, and z. So, we have

$$\widehat{S}_y\widehat{S}_z-\widehat{S}_z\widehat{S}_y=i\hbar\widehat{S}_x \quad (9.19a)$$

$$\widehat{S}_z\widehat{S}_x - \widehat{S}_x\widehat{S}_z = i\hbar\widehat{S}_y \tag{9.19b}$$

$$\widehat{S}_x\widehat{S}_y - \widehat{S}_y\widehat{S}_x = i\hbar\widehat{S}_z \tag{9.19c}$$

Using the condition that the spin angular momentum quantum number $s = 1/2$, which yields $m_s = \pm 1/2$ *along any axis*, we can define a 2×2 matrix which must possess the desired two anti-symmetric eigenvalues as $\widehat{\sigma}_z$, such that along any direction

$$\widehat{S}_i = \frac{\hbar}{2}\widehat{\sigma}_i \tag{9.20}$$

In Eq. (9.20), $\widehat{S}_i$ is now a 2×2 matrix operator independent of space. Thus, the operator is quite different from $\widehat{L}_i$, which is a scalar operator depending on space. With this choice for $\widehat{S}_i$, any spin operator $\widehat{\sigma}_i$ is required to have eigenvalues of ± 1, so that $\widehat{S}_i$ has eigenvalues $\pm\frac{\hbar}{2}$. Substitution of Eq. (9.20) into Eqs. (9.19a) - (9.19c) then leads to

$$\widehat{\sigma}_y\widehat{\sigma}_z - \widehat{\sigma}_z\widehat{\sigma}_y = 2i\widehat{\sigma}_x \tag{9.21a}$$

$$\widehat{\sigma}_z\widehat{\sigma}_x - \widehat{\sigma}_x\widehat{\sigma}_z = 2i\widehat{\sigma}_y \tag{9.21b}$$

$$\widehat{\sigma}_x\widehat{\sigma}_y - \widehat{\sigma}_y\widehat{\sigma}_x = 2i\widehat{\sigma}_z \tag{9.21c}$$

If we reverse the order of indices in the above equations, the sign on the RHS changes. Austrian physicist Wolfgang Ernst Pauli (1900-1958), mentioned earlier in relation to Kronig, treated the spin angular momentum in this manner. In finding a matrix with eigenvalues of ± 1, there are many possible choices for σ_i, satisfying the above relations. But, Pauli's results along these lines were simple and effective and have thus, become a standard. So we will discuss his 2×2 principle spin matrices known as the **Pauli spin matrices**.

Firstly, for $\widehat{\sigma}_z$, Pauli selected the obvious choice deduced from Eq. (9.18):

$$\widehat{\sigma}_z = \begin{bmatrix} 1 & 0 \\ 0 & -1 \end{bmatrix} \tag{9.22}$$

With one down, Pauli obtained the other two matrices by specifying that σ_x and σ_y must have the following form

$$\widehat{\sigma}_x = \begin{bmatrix} 0 & a \\ a^* & 0 \end{bmatrix} \quad \text{and} \quad \widehat{\sigma}_y = \begin{bmatrix} 0 & b \\ b^* & 0 \end{bmatrix}$$

9.4 Principle Spin Matrices and Spinors

a and b are scalars, and a^* and b^* are their complex conjugates, respectively. Substitution of $\widehat{\sigma}_x$ and $\widehat{\sigma}_y$ into Eqs. (9.21a) and (9.21c) leads to the conditions

$$b = -ia \quad \text{and} \quad b^* = ia^*$$

$$ab^* - ba^* = 2i \quad \text{and} \quad a^*b - b^*a = -2i$$

The equations above result in the relation $a^*a = |a|^2 = 1$. Pauli went with $a = 1$, which also implies that $b = -i$. So, we have the following three matrices from Pauli

Principle (Pauli) Spin Matrices

$$\widehat{\sigma}_x = \begin{bmatrix} 0 & 1 \\ 1 & 0 \end{bmatrix} \quad \widehat{\sigma}_y = \begin{bmatrix} 0 & -i \\ i & 0 \end{bmatrix} \quad \text{and} \quad \widehat{\sigma}_z = \begin{bmatrix} 1 & 0 \\ 0 & -1 \end{bmatrix} \quad (9.23)$$

When we square the **principle spin matrices** $\widehat{\sigma}_x$, $\widehat{\sigma}_y$, and $\widehat{\sigma}_z$ we obtain

$$\widehat{\sigma}_x^2 = \begin{bmatrix} 0 & 1 \\ 1 & 0 \end{bmatrix} \cdot \begin{bmatrix} 0 & 1 \\ 1 & 0 \end{bmatrix} = \begin{bmatrix} 1 & 0 \\ 0 & 1 \end{bmatrix} \quad (9.24)$$

$$\widehat{\sigma}_y^2 = \begin{bmatrix} 0 & -i \\ i & 0 \end{bmatrix} \cdot \begin{bmatrix} 0 & -i \\ i & 0 \end{bmatrix} = \begin{bmatrix} 1 & 0 \\ 0 & 1 \end{bmatrix} \quad (9.25)$$

$$\widehat{\sigma}_z^2 = \begin{bmatrix} 1 & 0 \\ 0 & -1 \end{bmatrix} \cdot \begin{bmatrix} 1 & 0 \\ 0 & -1 \end{bmatrix} = \begin{bmatrix} 1 & 0 \\ 0 & 1 \end{bmatrix} \quad (9.26)$$

The square of any of the principle spin matrices is the 2×2 identity matrix $I(2)$, which consequently means that each spin matrix is its own inverse! So, the spin matrices are examples of **involutory matrices**. Let us also consider the products given by

$$\widehat{\sigma}_x \widehat{\sigma}_y = \begin{bmatrix} 0 & 1 \\ 1 & 0 \end{bmatrix} \cdot \begin{bmatrix} 0 & -i \\ i & 0 \end{bmatrix} = \begin{bmatrix} i & 0 \\ 0 & -i \end{bmatrix} = i\widehat{\sigma}_z$$

$$\widehat{\sigma}_y \widehat{\sigma}_x = \begin{bmatrix} 0 & -i \\ i & 0 \end{bmatrix} \cdot \begin{bmatrix} 0 & 1 \\ 1 & 0 \end{bmatrix} = \begin{bmatrix} -i & 0 \\ 0 & i \end{bmatrix} = -i\widehat{\sigma}_z$$

Therefore, we have that

$$\widehat{\sigma}_x \widehat{\sigma}_y + \widehat{\sigma}_y \widehat{\sigma}_x = i\widehat{\sigma}_z - i\widehat{\sigma}_z = 0$$

We obtain a similar result for $\widehat{\sigma}_y\widehat{\sigma}_z$ and $\widehat{\sigma}_z\widehat{\sigma}_x$, where we have

$$\widehat{\sigma}_y\widehat{\sigma}_z = \begin{bmatrix} 0 & -i \\ i & 0 \end{bmatrix} \cdot \begin{bmatrix} 1 & 0 \\ 0 & -1 \end{bmatrix} = \begin{bmatrix} 0 & i \\ i & 0 \end{bmatrix}$$

$$\widehat{\sigma}_y\widehat{\sigma}_x = \begin{bmatrix} 1 & 0 \\ 0 & -1 \end{bmatrix} \cdot \begin{bmatrix} 0 & -i \\ i & 0 \end{bmatrix} = \begin{bmatrix} 0 & -i \\ -i & 0 \end{bmatrix}$$

$$\widehat{\sigma}_z\widehat{\sigma}_x = \begin{bmatrix} 1 & 0 \\ 0 & -1 \end{bmatrix} \cdot \begin{bmatrix} 0 & 1 \\ 1 & 0 \end{bmatrix} = \begin{bmatrix} 0 & 1 \\ -1 & 0 \end{bmatrix} = i\widehat{\sigma}_x$$

$$\widehat{\sigma}_x\widehat{\sigma}_z = \begin{bmatrix} 0 & 1 \\ 1 & 0 \end{bmatrix} \cdot \begin{bmatrix} 1 & 0 \\ 0 & -1 \end{bmatrix} = \begin{bmatrix} 0 & -1 \\ 1 & 0 \end{bmatrix} = -i\widehat{\sigma}_x$$

It follows that

$$\widehat{\sigma}_y\widehat{\sigma}_z + \widehat{\sigma}_z\widehat{\sigma}_y = i\widehat{\sigma}_x - i\widehat{\sigma}_x = 0$$

$$\widehat{\sigma}_z\widehat{\sigma}_x + \widehat{\sigma}_x\widehat{\sigma}_z = i\widehat{\sigma}_y - i\widehat{\sigma}_y = 0$$

Because the principle spin matrices satisfy these relations, they are said to **anti-commute**, or satisfy **anti-commutation**.

Instead adding the products, when we subtract them, we have the relations

$$\widehat{\sigma}_x\widehat{\sigma}_y - \widehat{\sigma}_y\widehat{\sigma}_x = i\widehat{\sigma}_z + i\widehat{\sigma}_z = 2i\widehat{\sigma}_z$$

$$\widehat{\sigma}_y\widehat{\sigma}_z - \widehat{\sigma}_z\widehat{\sigma}_y = i\widehat{\sigma}_x + i\widehat{\sigma}_x = 2i\widehat{\sigma}_x$$

$$\widehat{\sigma}_z\widehat{\sigma}_x - \widehat{\sigma}_x\widehat{\sigma}_z = i\widehat{\sigma}_y + i\widehat{\sigma}_y = i\widehat{\sigma}_y$$

These relations for the spin matrices may be written more compactly in terms of the commutation operator as

Commutation Relations Principle Spin Matrices

$$[\sigma_i, \sigma_j] = 2i\epsilon_{ijk}\sigma_k \tag{9.27}$$

ϵ_{ijk} is the Levi-Civita symbol, which is defined as

$$\epsilon_{ijk} = \begin{cases} +1 & \text{if } ijk = (1,2,3), (2,3,1), or (3,1,2) \\ -1 & \text{if } ijk = (3,2,1), (1,3,2), or (2,1,3) \\ 0 & \text{if } i = j, j = k, \text{ or } i = k \end{cases} \tag{9.28}$$

9.4 Principle Spin Matrices and Spinors

It just helps keep track of the relation between the coordinates, particularly, for a right-handed coordinate system. Using Eqs. (9.24)-(9.26), the $\widehat{S}^2$ operator becomes

$$\widehat{S}^2 = \widehat{S}_x^2 + \widehat{S}_y^2 + \widehat{S}_z^2 = \frac{\hbar^2}{4}\sigma_x^2 + \frac{\hbar^2}{4}\sigma_y^2 + \frac{\hbar^2}{4}\sigma_z^2 = \frac{3}{4}\hbar^2 \mathbf{I} = s(s+1)\hbar^2 \mathbf{I} \quad (9.29)$$

This result only holds for a **spin quantum number** s given by

---
Electron Spin Quantum Number s

$$s = \frac{1}{2} \quad (9.30)$$

Eq. (9.30) is in agreement with our earlier deduction of the spin quantum number, and this was by design. By expressing the operators in matrix form, we *conceal* the two spin magnetic quantum values for m_s into the eigenvalues of these matrix operators. However, there is also the business of the 2×1 vectors that have been consequently introduced.

Since, any of the three spin matrices possess two eigenvalues ± 1, it follows that each $\widehat{\sigma}_i$ must also have 2 corresponding eigenvectors λ_i. The eigenvectors of the principle spin matrices are examples of *spinors*, and they are 2×1 vectors that contain the states of the electron's spin-up and spin-down. Since the spin matrices have unique eigenvalues ± 1, both eigenvectors also span the set of 2×1 vectors. That is, any state vector can be expressed as a linear combination of the spin matrix eigenvectors. So, let us determine the eigenvectors λ_i for the principle spin matrices. For $\widehat{\sigma}_z$, the eigenoperator (or matrix eigenvalue) equation is defined by

$$\widehat{\sigma}_z \lambda_z = \begin{bmatrix} 1 & 0 \\ 0 & -1 \end{bmatrix} \begin{bmatrix} a_1 \\ a_2 \end{bmatrix} = \lambda \begin{bmatrix} a_1 \\ a_2 \end{bmatrix}$$

For $\lambda = 1$, we get

$$\begin{bmatrix} 1 & 0 \\ 0 & -1 \end{bmatrix} \begin{bmatrix} a_1 \\ a_2 \end{bmatrix} = \begin{bmatrix} a_1 \\ a_2 \end{bmatrix} \Rightarrow \begin{bmatrix} a_1 \\ -a_2 \end{bmatrix} = \begin{bmatrix} a_1 \\ a_2 \end{bmatrix}$$

The only number for which $-a_2 = a_2$ is 0, thus $a_2 = 0$, while a_1 is arbitrary. However, to have a normalized vector $|\lambda| = 1$, it is customary to choose for convenience, the value $a_1 = 1$, which leads to the following eigenvector $\lambda_z(\lambda = +1)$

$$\lambda_z(\lambda = +1) = \begin{bmatrix} 1 \\ 0 \end{bmatrix}$$

For the case of $\lambda = -1$, we have

$$\begin{bmatrix} 1 & 0 \\ 0 & -1 \end{bmatrix} \begin{bmatrix} a_1 \\ a_2 \end{bmatrix} = \begin{bmatrix} -a_1 \\ -a_2 \end{bmatrix} \Rightarrow \begin{bmatrix} a_1 \\ -a_2 \end{bmatrix} = \begin{bmatrix} -a_1 \\ -a_2 \end{bmatrix}$$

Now we have $a_1 = 1$, while a_2 is arbitrary. This leads to the following eigenvector $\lambda_z(\lambda = -1)$:

$$\lambda_z(\lambda = -1) = \begin{bmatrix} 0 \\ 1 \end{bmatrix}$$

The pair of eigenvectors also have the property

$$\lambda_z^T(\lambda = +1) \cdot \lambda_z(\lambda = -1) = \begin{bmatrix} 1 & 0 \end{bmatrix} \cdot \begin{bmatrix} 0 \\ 1 \end{bmatrix} = 0$$

λ_z^T is the transpose. This property should be expected as one vector is like the unit vector along x and the other along y. Whenever two vectors form inner products that equal to zero, the eigenvectors are said to be *orthogonal* to each other. This is always true for a matrix with distinct eigenvalues. Thus, since the eigenvalues of spin matrices are ± 1, all three spin matrices will have analogous results, with two *orthogonal* eigenvectors. Carrying out a similar analyses for the other two spin matrices $\widehat{\sigma}_x$ and $\widehat{\sigma}_y$, we find the following (normalized) pairs of eigenvectors:

$$\lambda_x(\lambda = +1) = \frac{1}{\sqrt{2}} \begin{bmatrix} 1 \\ 1 \end{bmatrix} \quad \text{and} \quad \lambda_x(\lambda = -1) = \frac{1}{\sqrt{2}} \begin{bmatrix} 1 \\ -1 \end{bmatrix}$$

For $\widehat{\sigma}_y$, we obtain

$$\lambda_y(\lambda = +1) = \frac{1}{\sqrt{2}} \begin{bmatrix} 1 \\ i \end{bmatrix} \quad \text{and} \quad \lambda_y(\lambda = -1) = \frac{1}{\sqrt{2}} \begin{bmatrix} 1 \\ -i \end{bmatrix}$$

Summarizing, we have that the *principle spin matrices* or *Pauli spin matrices* can be utilized to cast the Schrödinger equation into an alternative two-equation matrix-vector form. This specific choice, enabled by the two distinct values for m_s, leads to the requirement for 2×2 operators in the Schrödinger equation. This form, though not compulsory, turns out to be very convenient. These results can also be generalized such that the spin matrix is along any arbitrary axis, not necessarily a principle axis such as $x, y,$ and z. We'll develop this more general spin matrix in the next section.

9.5 Generalized Spin Matrix For Arbitrary Field Direction

We have only demonstrated how to represent the spin along any of the three principle axes x, y, and z, where it was assumed the magnetic field acts along one of these directions. In a spatially 3D domain, a magnetic field can be along any direction. Here, we show that the results we have obtained for the principle spin matrices $\widehat{\sigma}_i$ also hold for any direction. Let us consider that we have a field of unit amplitude $b = \mathbf{B}/|\mathbf{B}|$ along any direction given by

$$\mathbf{b} = \begin{bmatrix} \sin\theta\cos\varphi \\ \sin\theta\sin\varphi \\ \cos\theta \end{bmatrix} \qquad (9.31)$$

Then, we can write the dot product part of the spin Hamiltonian as

$$\mathbf{S}\cdot\mathbf{b} = \frac{\hbar}{2}\left[b_x\widehat{\sigma}_x + b_y\widehat{\sigma}_y + b_z\widehat{\sigma}_z\right] =$$

$$\frac{\hbar}{2}\left(\begin{bmatrix} 0 & 1 \\ 1 & 0 \end{bmatrix}\sin\theta\cos\varphi + \begin{bmatrix} 0 & -i \\ i & 0 \end{bmatrix}\sin\theta\sin\varphi + \begin{bmatrix} 1 & 0 \\ 0 & -1 \end{bmatrix}\cos\theta\right) \qquad (9.32)$$

By adding the components into a single 2×2 matrix, we obtain

$$\mathbf{S}\cdot\mathbf{b} = \frac{\hbar}{2}\begin{bmatrix} \cos\theta & \sin\theta\cos\varphi - i\sin\theta\sin\varphi \\ \sin\theta\cos\varphi + i\sin\theta\sin\varphi & -\cos\theta \end{bmatrix} \qquad (9.33)$$

Factoring the $\sin\theta$ term in the off-diagonal elements and using the relation $e^{i\theta} = \cos\theta + i\sin\theta$, Eq. (9.33) becomes

$$\mathbf{S}\cdot\mathbf{b} = \frac{\hbar}{2}\begin{bmatrix} \cos\theta & \sin\theta e^{-i\varphi} \\ \sin\theta e^{i\varphi} & -\cos\theta \end{bmatrix} = \frac{\hbar}{2}\widehat{\sigma}_{\theta,\varphi} \qquad (9.34)$$

Thus, we find a spin matrix for any orientation of magnetic field, which is also a 2×2 spin operator $\widehat{\sigma}_{\theta,\varphi}$, defined by

General 2×2 Spin Operator

$$\widehat{\sigma}_{\theta,\varphi} = \begin{bmatrix} \cos\theta & \sin\theta e^{-i\varphi} \\ \sin\theta e^{i\varphi} & -\cos\theta \end{bmatrix} \qquad (9.35)$$

If we use the corresponding polar angles (θ, φ) defining the principle axes (e.g. $x \to (\theta = \pi/2, \varphi = 0)$), we regain the set of Pauli's spin matrices. We leave it as an exercise for the reader to show that the eigenvalues of the matrix in Eq. (9.35) $\widehat{\sigma}_{\theta,\varphi}$ are described by the condition $\lambda^2 = 1$. The only real solutions to this condition are $\lambda = \pm 1$. Thus, we have established that for any direction of **b**, the spin matrix always maintains the same eigenvalues as along the principle axes. We are not constrained to x, y, and z, for the orientation of the magnetic field, though it is often done for convenience.

We must also determine the eigenvectors λ of $\widehat{\sigma}_{\theta,\varphi}$. To find the eigenvector for $\lambda = +1$, we can write

$$\begin{bmatrix} \cos\theta & \sin\theta e^{-i\varphi} \\ \sin\theta e^{i\varphi} & -\cos\theta \end{bmatrix} \begin{bmatrix} a \\ b \end{bmatrix} = +1 \begin{bmatrix} a \\ b \end{bmatrix} \quad (9.36)$$

Note that when we seek eigenvectors, there are infinitely many choices by virtue of their magnitudes. This is because we can multiply Eq. (9.36) by any scalar and the equation remains the same. What is unique, however, is the direction of the vector (or relative magnitude of the components). Hence, we may choose a, then find the resulting value of b that satisfies the eigenvector equation, as in Eq. (9.36). The result is then normalized so that $|\lambda| = 1$. So, let us choose $a = \cos(\theta/2)e^{-i(\varphi/2)}$. Substitution of a into the first of our equations leads to

$$\cos(\theta/2)\cos(\theta)e^{-i(\varphi/2)} + b\sin\theta e^{-i\varphi} = \cos(\theta/2)e^{-i(\varphi/2)}$$

Rearranging this, we have

$$\cos(\theta/2)e^{-i(\varphi/2)} - \cos(\theta/2)\cos(\theta)e^{-i(\varphi/2)} = b\sin\theta e^{-i\varphi}$$

Now, let's express the $\cos\theta$ and $\sin\theta$ in terms of $\theta/2$ by making use of the trigonometric identities $\sin 2x = 2\sin x \cos x$ and $\cos 2x = \cos^2 x - \sin^2 x$. Using these and simplifying the terms, we arrive at

$$e^{-i(\varphi/2)}\left[2\sin^2(\theta/2)\right] = b\sin(\theta/2)e^{-i\varphi}$$

Therefore, b is given by

$$b = \sin(\theta/2)e^{i(\varphi/2)} \quad (9.37)$$

With this result, $|\lambda| = a^*a + b^*b = 1$, hence our choice for a. Carrying out a similar process for the eigenvalue $\lambda = -1$, we have that both eigenvectors for $\sigma_{\theta,\varphi}$ are given by

9.5 Generalized Spin Matrix

> **General 2×1 Spin Operator Eigenvectors**
>
> $$\lambda_{+1} = \begin{bmatrix} \cos\frac{\theta}{2} e^{-i\frac{\varphi}{2}} \\ \sin\frac{\theta}{2} e^{i\frac{\varphi}{2}} \end{bmatrix} \quad \text{and} \quad \lambda_{-1} = \begin{bmatrix} \sin\frac{\theta}{2} e^{-i\frac{\varphi}{2}} \\ -\cos\frac{\theta}{2} e^{i\frac{\varphi}{2}} \end{bmatrix} \qquad (9.38)$$

The spin matrix eigenvectors given by Eq. (9.38) are important in dealing with equations describing *spinors*. Along with the principle spin matrices, or Pauli spin matrices, $\widehat{\sigma}_x$, $\widehat{\sigma}_y$, and $\widehat{\sigma}_z$, they describe the components along the respective axes. To see this, let us compute the expectation value of the x component, given by

$$\langle x \rangle = \lambda_{+1}^{*T} \widehat{\sigma}_x \lambda_{+1} = \lambda_{+1}^{\dagger} \widehat{\sigma}_x \lambda_{+1} \qquad (9.39)$$

The superscript symbol † is shorthand notation for *both* complex conjugating and transposing. Carrying this out, we have

$$\langle x \rangle = \lambda_{+1}^{\dagger} \widehat{\sigma}_x \lambda_{+1}$$

$$= \begin{bmatrix} \cos\frac{\theta}{2} e^{i\frac{\varphi}{2}} & \sin\frac{\theta}{2} e^{-i\frac{\varphi}{2}} \end{bmatrix} \begin{bmatrix} 0 & 1 \\ 1 & 0 \end{bmatrix} \begin{bmatrix} \cos\frac{\theta}{2} e^{-i\frac{\varphi}{2}} \\ \sin\frac{\theta}{2} e^{i\frac{\varphi}{2}} \end{bmatrix}$$

$$= \begin{bmatrix} \cos\frac{\theta}{2} e^{i\frac{\varphi}{2}} & \sin\frac{\theta}{2} e^{-i\frac{\varphi}{2}} \end{bmatrix} \begin{bmatrix} \sin\frac{\theta}{2} e^{i\frac{\varphi}{2}} \\ \cos\frac{\theta}{2} e^{-i\frac{\varphi}{2}} \end{bmatrix}$$

$$= \cos\frac{\theta}{2} \sin\frac{\theta}{2} e^{i\varphi} + \cos\frac{\theta}{2} \sin\frac{\theta}{2} e^{-i\varphi}$$

$$= \cos\frac{\theta}{2} \sin\frac{\theta}{2} \left(e^{i\varphi} + e^{-i\varphi} \right)$$

$$= \sin\theta \cos\varphi$$

In the above, we have made use of the identities $\sin 2x = 2\sin x \cos x$ and $\cos x = (e^{ix} + e^{-ix})/2$. But, this is the x component of the normalized magnetic field direction. We leave it as an exercise for the reader to show that $\langle y \rangle = \sin\theta \sin\varphi$ and $\langle z \rangle = \cos\theta$. This tell us that the x component of the expectation value of the spin direction is along the x component of b, and this was deduced using λ_{+1}. We leave it as an exercise for the reader to show that, using the eigenstate λ_{-1}, the result is an anti-parallel vector.

We will see in the next section how the spinors in Eq. (9.38) can be used to provide initial conditions for time dependent spinor differential equations. This enables us to find solutions to the Schröginer equation in terms of spin matrices. It allows us to analyze spin dynamics.

9.6 THE PAULI EQUATION AND SPIN PRECESSION

Since all electrons possess spin angular momentum, the spin matrices that we have discussed can be used to express the Schrödinger equation, generally, in the two equation form. In the beginning, we wrote the scalar form of the Schrödinger equation, where the spin degree of freedom is introduced via

$$\widehat{H}\psi = \left[\widehat{H}_0 + V_M\right]\psi = \left[\widehat{H}_0 - \gamma \widehat{\mathbf{L}} \cdot \mathbf{B} - g\gamma \widehat{\mathbf{S}} \cdot \mathbf{B}\right]\psi \qquad (9.40)$$

$\widehat{H}_0$ represents the *scalar* Hamiltonian operator, including the kinetic energy and nucleus-electron Coulomb interaction. V_M is the additional magnetic potential energy contribution from the coupling of the angular momentum of the electron to the magnetic field B. This is the simplest Hamiltonian form and it is always justified because the energy of any system is a *scalar quantity*. With the spin, we also found in the introduction of the spin degree of freedom, that by taking the magnetic field along the z-axis, we could replace the spin angular momentum operator $\widehat{\mathbf{S}}$ with it's z-component $\widehat{S}_z$, as we did with $\widehat{\mathbf{L}}$ and $\widehat{L}_z$ operators. From this, both m_ℓ and m_s appeared through the respective energy eigenvalues. Writing the Hamiltonian in 2-equation form, we are just exploiting the fact that m_s takes exactly two values $\pm 1/2$. Whenever spin is included, this alternative form can be used incorporating *both* spin eigenstate equations simultaneously by making substitutions for the spin part of Eq. (9.40) using the relation $\widehat{S}_i = (\hbar/2)/\widehat{\sigma}_i$. So, we can write

$$\widehat{H}\begin{bmatrix}\psi_+\\ \psi_-\end{bmatrix} = \left[\widehat{H}_0 \mathbf{I} - \gamma \widehat{\mathbf{L}} \cdot \mathbf{BI} - \frac{g\gamma\hbar}{2}(\widehat{\sigma}_x B_x + \widehat{\sigma}_y B_y + \widehat{\sigma}_z B_z)\right]\begin{bmatrix}\psi_+\\ \psi_-\end{bmatrix} \qquad (9.41)$$

$\widehat{H}$ is now a 2×2 matrix operator, and Eq. (9.41) is, therefore, a pair of two equations describing the two spin eigenstates ψ_+ and ψ_-. In the Hamiltonian, we can distinguish two types of contributions to the Hamiltonian. The first type is independent of the spin angular momentum. This includes the first two terms of Eq. (9.41). The other contribution includes the spin angular momentum operator, or is a function of $\widehat{\mathbf{S}}$.

9.6 The Pauli Equation and Spin Precession

This distinction leads to a generalized representation of the Hamiltonian that includes spin, where we denote any spin-dependent part of the Hamiltonian as $\widehat{H}_s(\boldsymbol{\sigma})$. Thus, spin-dependent 2×2 Hamiltonian can therefore be expressed as

$$\widehat{H}\begin{bmatrix}\psi_+\\\psi_-\end{bmatrix} = \left[\widehat{H}_{\text{nonspin}} \cdot \mathbf{I}(2) + \widehat{H}_s(\boldsymbol{\sigma})\right]\begin{bmatrix}\psi_+\\\psi_-\end{bmatrix} \quad (9.42)$$

Because the spin-independent operators are scalars, they simply multiply by the 2×2 identity matrix $\mathbf{I}(2)$. These scalar operators are the same, regardless of the spin state. For the spin-dependent Hamiltonian, since it depends on the three principle spin matrices $\widehat{\sigma}_x, \widehat{\sigma}_y,$ and $\widehat{\sigma}_z$, we can define the spin (or Pauli) vector of 2×2 spin matrices

$$\widehat{\boldsymbol{\sigma}} = [\widehat{\sigma}_x, \widehat{\sigma}_y, \widehat{\sigma}_z]^T \quad (9.43)$$

For the hydrogen atom, the explicit form is known as the **Pauli equation** and is given by

> **Pauli Equation**
>
> $$i\hbar\frac{\partial}{\partial t}\begin{bmatrix}\psi_+\\\psi_-\end{bmatrix} = \left[\left(-\frac{\hbar^2}{2m^*}\nabla^2 + V(\mathbf{x}) - \gamma \mathbf{L}\cdot\mathbf{B}\right)\mathbf{I}(2) + \mathbf{H}_s(\boldsymbol{\sigma})\right]\begin{bmatrix}\psi_+\\\psi_-\end{bmatrix} \quad (9.44)$$

The *Pauli equation* was formulated by Wolfgang Pauli in 1927. The *Pauli equation*, and the more general form are unique in that it formulates the Schrödinger equation, in a manner inclusive of spin, in terms of the *spin matrices*. ψ_+ and ψ_- are the spin eigenfunctions, known as *spin-up* and *spin-down*, respectively. The 2×1 eigenvector containing the two spin eigenstates is a *spinor*. The caveat of this form is that we still need to determine expectation values in terms of this 2×1 spinor of eigenfunctions. With a scalar eigenfunction, the *expectation value* of a quantity q is given by

$$\langle q \rangle = \frac{\int_{-\infty}^{+\infty}\psi(x)^*q\psi(x)dx}{\int_{-\infty}^{+\infty}\psi(x)^*\psi(x)dx}$$

So, we must identify the analog of this quantity in the spin world of 2×2 operators. We know that any state of a system can be represented as a linear combination of the eigenstates. Thus, with two spin eigenstates, the probability density of the electron spin state is given by a linear combination of the probability densities of the respective spin eigenstates. The probability density of the system then becomes

$$a_1 \psi_+^* \psi_+ + a_2 \psi_-^* \psi_- = \psi_{spin}^{*T} \psi_{spin} \qquad (9.45)$$

ψ_{spin}^{*T} notation represents the *transposed vector of complex conjugates*. For the electron spin, because of the symmetry of the spin eigenstates, evidenced by their eigenvalues being equal in amplitude, Eq. (9.45) can also be expressed as

$$\psi_{spin}^{*T} \psi_{spin} = \psi_{spin}^{\dagger} \psi_{spin} = a(\psi_+^* \psi_+ + \psi_-^* \psi_-) = a \begin{bmatrix} \psi_+^* & \psi_-^* \end{bmatrix} \begin{bmatrix} \psi_+ \\ \psi_- \end{bmatrix} \qquad (9.46)$$

We know that $a = 1/2$, however, we'll see shortly that the amplitude a plays no role in any expectation value. So, the probability density for the electron spin state in terms of the spinor, can be defined as

Electron Spin State Probability Density

$$\psi_{spin}^{\dagger} \psi_{spin} = \begin{bmatrix} \psi_+^* & \psi_-^* \end{bmatrix} \begin{bmatrix} \psi_+ \\ \psi_- \end{bmatrix} \qquad (9.47)$$

Since the spin operators are 2×2 matrices, we can assume that any function of these operators will also be 2×2 operators. The probability density in Eq. (9.47) provides a way to determine the expectation value of any 2×2 operator depending on $\widehat{S}$. We can define an expectation value of any 2×2 operator $\widehat{O}$ as

$$\langle \widehat{O} \rangle = \frac{\psi_{spin}^* \widehat{O} \psi_{spin}}{\psi_{spin}^* \psi_{spin}}$$

Substituting the 2×1 spinors, we then have

Expectation Value Of A Spin-Dependent Operator $\widehat{O}$

9.6 The Pauli Equation and Spin Precession

$$\langle \widehat{O} \rangle = \frac{\begin{bmatrix} \psi_+^* & \psi_-^* \end{bmatrix} \widehat{O} \begin{bmatrix} \psi_+ \\ \psi_- \end{bmatrix}}{\begin{bmatrix} \psi_+^* & \psi_-^* \end{bmatrix} \begin{bmatrix} \psi_+ \\ \psi_- \end{bmatrix}} \tag{9.48}$$

With this definition, even if we had left the amplitude a in Eq. (9.46), it would have dropped out from the expectation value definition anyway.

Therefore, given the spin eigenstates, we can determine the expectation value using the probability density in Eq. (9.48). When we include the time dependence in the Hamiltonian, the expectation values of the spin operator $\widehat{S}_i$ become functions of time t. The expectation values for the components of $\widehat{\mathbf{S}}$ become

$$\langle \widehat{S}_x \rangle = \frac{\begin{bmatrix} \psi_+^* & \psi_-^* \end{bmatrix} \widehat{S}_x \begin{bmatrix} \psi_+ \\ \psi_- \end{bmatrix}}{\begin{bmatrix} \psi_+^* & \psi_-^* \end{bmatrix} \begin{bmatrix} \psi_+ \\ \psi_- \end{bmatrix}} = \frac{\hbar}{2} \frac{\begin{bmatrix} \psi_+^* & \psi_-^* \end{bmatrix} \widehat{\sigma}_x \begin{bmatrix} \psi_+ \\ \psi_- \end{bmatrix}}{\begin{bmatrix} \psi_+^* & \psi_-^* \end{bmatrix} \begin{bmatrix} \psi_+ \\ \psi_- \end{bmatrix}} \tag{9.49a}$$

$$\langle \widehat{S}_y \rangle = \frac{\begin{bmatrix} \psi_+^* & \psi_-^* \end{bmatrix} \widehat{S}_y \begin{bmatrix} \psi_+ \\ \psi_- \end{bmatrix}}{\begin{bmatrix} \psi_+^* & \psi_-^* \end{bmatrix} \begin{bmatrix} \psi_+ \\ \psi_- \end{bmatrix}} = \frac{\hbar}{2} \frac{\begin{bmatrix} \psi_+^* & \psi_-^* \end{bmatrix} \widehat{\sigma}_y \begin{bmatrix} \psi_+ \\ \psi_- \end{bmatrix}}{\begin{bmatrix} \psi_+^* & \psi_-^* \end{bmatrix} \begin{bmatrix} \psi_+ \\ \psi_- \end{bmatrix}} \tag{9.49b}$$

$$\langle \widehat{S}_z \rangle = \frac{\begin{bmatrix} \psi_+^* & \psi_-^* \end{bmatrix} \widehat{S}_z \begin{bmatrix} \psi_+ \\ \psi_- \end{bmatrix}}{\begin{bmatrix} \psi_+^* & \psi_-^* \end{bmatrix} \begin{bmatrix} \psi_+ \\ \psi_- \end{bmatrix}} = \frac{\hbar}{2} \frac{\begin{bmatrix} \psi_+^* & \psi_-^* \end{bmatrix} \widehat{\sigma}_z \begin{bmatrix} \psi_+ \\ \psi_- \end{bmatrix}}{\begin{bmatrix} \psi_+^* & \psi_-^* \end{bmatrix} \begin{bmatrix} \psi_+ \\ \psi_- \end{bmatrix}} \tag{9.49c}$$

Substitution of the principle spin matrices given by Eq. (9.23), the expectation values of the spin angular momentum components become

$$\langle \widehat{S}_x \rangle = \frac{\hbar}{2} \frac{\begin{bmatrix} \psi_+^* & \psi_-^* \end{bmatrix} \begin{bmatrix} 0 & 1 \\ 1 & 0 \end{bmatrix} \begin{bmatrix} \psi_+ \\ \psi_- \end{bmatrix}}{\begin{bmatrix} \psi_+^* & \psi_-^* \end{bmatrix} \begin{bmatrix} \psi_+ \\ \psi_- \end{bmatrix}} = \frac{\hbar}{2} \frac{\psi_+^* \psi_- + \psi_-^* \psi_+}{\psi_+^* \psi_+ + \psi_-^* \psi_-} \tag{9.50a}$$

$$\langle\widehat{S}_y\rangle = \frac{\hbar}{2}\frac{\begin{bmatrix}\psi_+^* & \psi_-^*\end{bmatrix}\begin{bmatrix}0 & -i\\ i & 0\end{bmatrix}\begin{bmatrix}\psi_+\\ \psi_-\end{bmatrix}}{\begin{bmatrix}\psi_+^* & \psi_-^*\end{bmatrix}\begin{bmatrix}\psi_+\\ \psi_-\end{bmatrix}} = \frac{i\hbar}{2}\frac{\psi_-^*\psi_+ - \psi_+^*\psi_-}{\psi_+^*\psi_+ + \psi_-^*\psi_-} \quad (9.50b)$$

$$\langle\widehat{S}_z\rangle = \frac{\hbar}{2}\frac{\begin{bmatrix}\psi_+^* & \psi_-^*\end{bmatrix}\begin{bmatrix}1 & 0\\ 0 & -1\end{bmatrix}\begin{bmatrix}\psi_+\\ \psi_-\end{bmatrix}}{\begin{bmatrix}\psi_+^* & \psi_-^*\end{bmatrix}\begin{bmatrix}\psi_+\\ \psi_-\end{bmatrix}} = \frac{\hbar}{2}\frac{\psi_+^*\psi_+ - \psi_-^*\psi_-}{\psi_+^*\psi_+ + \psi_-^*\psi_-} \quad (9.50c)$$

We can use Eqs. (9.50a)-(9.50c) to determine the dynamic response (or time-dependent) $\langle S_x\rangle(t), \langle S_y\rangle(t)$, and $\langle S_z\rangle(t)$. Let us do this now. We first solve the Pauli equation form of the Schrödinger equation to obtain ψ_+ and ψ_-. Then, in the final step, we use ψ_+ and ψ_- in the above formulae to determine the expectation values of S_x, S_y, and S_z. In the Hamilonian that we shall consider, we ignore any dissipation for now and we only consider the Schrödinger equation with the spin-dependent Hamiltonian terms. Then, we have

$$i\hbar\frac{\partial}{\partial t}\begin{bmatrix}\psi_+\\ \psi_-\end{bmatrix} = \left[-\frac{g\gamma\hbar}{2}\left(\widehat{\sigma}_x B_x + \widehat{\sigma}_y B_y + \widehat{\sigma}_z B_z\right)\right]\begin{bmatrix}\psi_+\\ \psi_-\end{bmatrix} \quad (9.51)$$

Note that we are solving the time-dependent Schrödinger equation because it describes $\psi(t)$. As we had done earlier, taking the magnetic field along the +z direction, Eq. (9.51) becomes

$$i\hbar\frac{\partial}{\partial t}\begin{bmatrix}\psi_+\\ \psi_-\end{bmatrix} = \left[\frac{g|\gamma|\hbar B}{2}\widehat{\sigma}_z\right]\begin{bmatrix}\psi_+\\ \psi_-\end{bmatrix} \quad (9.52)$$

We have replaced B_z with B. Then, with substitution of the spin matrix $\widehat{\sigma}_z$, Eq. (9.52) becomes

$$i\frac{\partial}{\partial t}\begin{bmatrix}\psi_+\\ \psi_-\end{bmatrix} = \frac{g|\gamma|B}{2}\begin{bmatrix}1 & 0\\ 0 & -1\end{bmatrix}\begin{bmatrix}\psi_+\\ \psi_-\end{bmatrix} \quad (9.53)$$

Eq. (9.53) expresses two simple decoupled ordinary differential equations given by

$$\frac{d\psi_+}{dt} = -i\frac{g|\gamma|B}{2}\psi_+ \quad (9.54a)$$

$$\frac{d\psi_-}{dt} = +i\frac{g|\gamma|B}{2}\psi_- \quad (9.54b)$$

9.6 The Pauli Equation and Spin Precession

Before we can integrate Eqs. (9.54a) and (9.54b), we must specify the *initial condition* of the spinor problem, since we are solving a first order differential equation in time. We must specify the value of the *spinor* $[\psi_+(0) \quad \psi_-(0)]^T$ at $t = 0$. Then, we can solve for the components of the spinor with elements $\psi_+(t)$ and $\psi_-(t)$. The most general spinor we can use to specify the initial condition in terms of real spatial variables is given by the 2×1 eigenvectors we obtained for the general 2×2 spin matrix along an arbitrary field direction. These were given in Eq. (9.38), and are functions of the azimuthal angle φ and polar angle θ. Here, θ is relative to the +z axis, and φ is relative to the +x axis, and they are only initial values, *not* functions of time. Since we take B along the +z axis, we use the spinor given by $\lambda = +1$. So, let our initial condition be given by

$$\begin{bmatrix} \psi_+(0) \\ \psi_-(0) \end{bmatrix} = \begin{bmatrix} \cos\frac{\theta}{2} e^{-i\frac{\varphi}{2}} \\ \sin\frac{\theta}{2} e^{i\frac{\varphi}{2}} \end{bmatrix} \tag{9.55}$$

With this initial condition, the solutions to Eqs. (9.54a) and (9.54b) become

$$\psi_+(t) = \psi_+(0) e^{-\frac{ig|\gamma|B}{2} t} \tag{9.56a}$$

$$\psi_-(t) = \psi_-(0) e^{+\frac{ig|\gamma|B}{2} t} \tag{9.56b}$$

The initial conditions $\psi_+(0)$ and $\psi_-(0)$ are given by Eq. (9.55). Using the solutions for $\psi_+(t)$ and $\psi_-(t)$, the expectation values of $\widehat{S}_x$, $\widehat{S}_y$, and $\widehat{S}_z$ can be determined with Eqs. (9.50a) - (9.50c). Starting with $\widehat{S}_x$, we have

$$\langle \widehat{S}_x \rangle = \frac{\hbar}{2} \frac{\begin{bmatrix} \psi_+^* & \psi_-^* \end{bmatrix} \begin{bmatrix} 0 & 1 \\ 1 & 0 \end{bmatrix} \begin{bmatrix} \psi_+ \\ \psi_- \end{bmatrix}}{\begin{bmatrix} \psi_+^* & \psi_-^* \end{bmatrix} \begin{bmatrix} \psi_+ \\ \psi_- \end{bmatrix}} \tag{9.57}$$

$$= \frac{\hbar}{2} \frac{\psi_+^*(0)\psi_-(0) e^{ig|\gamma|Bt} + \psi_-^*(0)\psi_+(0) e^{-ig|\gamma|Bt}}{\psi_+^*(0)\psi_+(0) + \psi_-^*(0)\psi_-(0)} \tag{9.58}$$

Using the components of the initial spinor, the denominator works out to be $\cos^2\theta/2 + \sin^2\theta/2 = 1$. Then, evaluating the numerator, we have

$$\langle S_x \rangle = \frac{\hbar}{2} \sin\left(\frac{\theta}{2}\right) \cos\left(\frac{\theta}{2}\right) \cdot \left(e^{i\varphi} e^{+ig|\gamma|Bt} + e^{-i\varphi} e^{-ig|\gamma|Bt} \right) \tag{9.59}$$

To further simplify the above result, we can use the trigonometric identity $\sin x \cdot \cos x = \sin(2x)/2$. Then Eq. (9.59) becomes

$$\langle \widehat{S}_x \rangle = \frac{\hbar}{2}\sin\theta \left[\frac{e^{+i(g|\gamma|Bt+\varphi)} + e^{-i(g|\gamma|Bt+\varphi)}}{2}\right] = \frac{\hbar}{2}\sin\theta\cos(g|\gamma|Bt+\varphi) \tag{9.60}$$

Carrying out a similar analysis for $\widehat{S}_y$ leads to

$$\langle \widehat{S}_y \rangle = \frac{\hbar}{2}\frac{\begin{bmatrix}\psi_+^* & \psi_-^*\end{bmatrix}\begin{bmatrix}0 & -i\\ i & 0\end{bmatrix}\begin{bmatrix}\psi_+\\ \psi_-\end{bmatrix}}{\begin{bmatrix}\psi_+^* & \psi_-^*\end{bmatrix}\begin{bmatrix}\psi_+\\ \psi_-\end{bmatrix}} \tag{9.61}$$

$$= \frac{i\hbar}{2}\frac{\psi_-^*(0)\psi_+(0)e^{-ig|\gamma|Bt} - \psi_+^*(0)\psi_-(0)e^{+ig|\gamma|Bt}}{\psi_+^*(0)\psi_+(0) + \psi_-^*(0)\psi_-(0)} \tag{9.62}$$

Using the same trigonometric identity as we used for $\widehat{S}_x$, along with the fact that the denominator is unity, we have

$$\langle \widehat{S}_y \rangle = \frac{\hbar}{2}\sin\theta\left[\frac{e^{i\varphi}e^{+ig|\gamma|Bt} - e^{-i\varphi}e^{-ig|\gamma|Bt}}{2i}\right] \tag{9.63}$$

$$= \frac{\hbar}{2}\sin\theta\sin(g|\gamma|Bt+\varphi) \tag{9.64}$$

We have also used the relation $-i = 1/i$. Finally, for $\widehat{S}_z$, the expectation value becomes

$$\langle \widehat{S}_z \rangle = \frac{\hbar}{2}\frac{\begin{bmatrix}\psi_+^* & \psi_-^*\end{bmatrix}\begin{bmatrix}1 & 0\\ 0 & -1\end{bmatrix}\begin{bmatrix}\psi_+\\ \psi_-\end{bmatrix}}{\begin{bmatrix}\psi_+^* & \psi_-^*\end{bmatrix}\begin{bmatrix}\psi_+\\ \psi_-\end{bmatrix}} \tag{9.65}$$

$$= \frac{\hbar}{2}\frac{\psi_+^*(0)\psi_+(0) - \psi_-(0)\psi_-^*(0)}{\psi_+^*(0)\psi_+(0) + \psi_-^*(0)\psi_-(0)} \tag{9.66}$$

$$= \frac{\hbar}{2}\cos\theta \tag{9.67}$$

We have also used the trigonometric identity $\cos(2x) = \cos^2 x - \sin^2 x$. So, we have found that the z component of the spin angular momentum is constant, or independent of t since θ is only the initial angle. Only $\langle \widehat{S}_x \rangle$

9.6 The Pauli Equation and Spin Precession

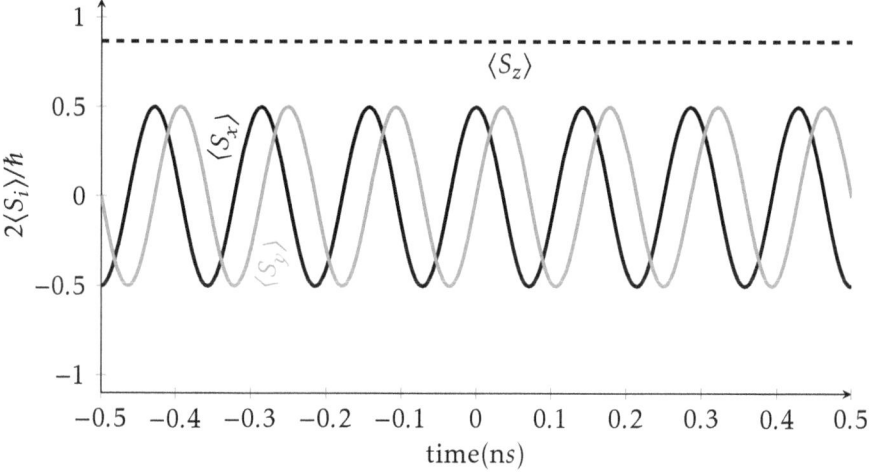

Figure 9.3: Three components of ⟨**S**⟩ as a function of time. The components perpendicular to B precess around the magnetic field along +z.

and $\langle \widehat{S_y} \rangle$ are time-dependent. The two components vary sinusoidally, exactly 90° out of phase with each other. This particular motion associated with the spin angular momentum is known as **precession**. The frequency of rotation appearing in two of the three equations is the rate of rotation about the magnetic field B. It is known as the **Larmor precession frequency**, defined as

---

Spin Precessional Frequency

$$\omega_p = g|\gamma|B \qquad (9.68)$$

The word *precession* describes a rotational vector (the spin) that is additionally rotating about another vector (the magnetic field). Fig. 9.4 illustrates precessional motion about a magnetic field B along the +z axis.

Figs. 9.3 and 9.4 show ⟨**S**⟩ as a function of time. In this example, $B = 0.25$ Tesla (along +z) and $g = 2$, thus the period of precession $T_p = 2\pi/\omega_p$ for an electron spin is only a fraction of a nanosecond ($T_p \approx 0.14$ ns). Fig. 9.4 illustrates the path of the tip of ⟨**S**⟩ while rotating along the surface of a sphere, tracing a circle parallel to the $x - y$ plane. The spin vector, generally, tends to rotate about a magnetic field vector, particularly in

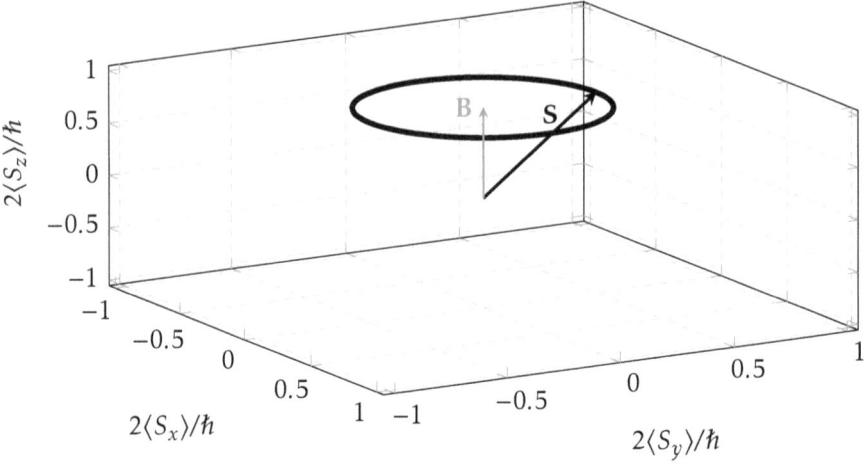

Figure 9.4: 3D view of precessional motion of the spin vector $2\langle \mathbf{S}\rangle/\hbar$. The illustrated ring is the path swept out by the tip of the vector. In 3D, the rotation is on the upper surface of a unit sphere.

a small magnetic field. However, in a sufficiently large field the spin prefers to fully align *opposite* to the magnetic field. To describe this motion, we must introduce dissipation into the Hamiltonian. We'll derive the dissipation and discuss this topic in detail in Chap. 11. Next, we'll introduce the spin density matrix (simplest example of a density matrix), which is also useful for obtaining information from the quantum mechanical system.

9.7 THE SPIN DENSITY MATRIX

The 2×1 spinor vector ψ can be used to construct a 2×2 matrix operator containing all the pertinent information for the spin. Since the state of the electron spin degree of freedom is described by a 2×1 spinor, let us define a 2×2 matrix operator, which we shall call the **spin density matrix** for an electron, given by

$$\widehat{\rho} = \frac{1}{|\psi|^2}\psi\otimes\psi = \frac{1}{|\psi|^2}\psi\psi^\dagger = \frac{1}{\psi_+^*\psi_+ + \psi_-^*\psi_-}\begin{bmatrix}\psi_+\\ \psi_-\end{bmatrix}\begin{bmatrix}\psi_+^* & \psi_-^*\end{bmatrix} \quad (9.69)$$

$\otimes$ denotes the *outer product*, which takes an $n\times 1$ vector and maps it to a $n\times n$ matrix. $\psi^\dagger$ is the transposed vector of complex conjugates of ψ, being a 1×2 vector. Evaluating the *outer product* in Eq. (9.69), we have the *spin density matrix* given by

9.7 The Spin Density Matrix

Electron Spin Density Matrix $\widehat{\rho}$

$$\widehat{\rho} = \frac{1}{|\psi_+|^2 + |\psi_-|^2} \begin{bmatrix} \psi_+\psi_+^* & \psi_+\psi_-^* \\ \psi_-\psi_+^* & \psi_-\psi_-^* \end{bmatrix} \tag{9.70}$$

Let's explore how the 2×2 *density matrix* defined by Eq. (9.70) can be used to obtain expectation values for *any* 2×2 matrix $\widehat{M}$. First, recall the definition of the *trace* of a matrix, denoted by $Tr(\widehat{M})$, is given by

$$Tr(\widehat{M}) = Tr\left(\begin{bmatrix} m_{11} & m_{12} \\ m_{21} & m_{22} \end{bmatrix}\right) = m_{11} + m_{22} \tag{9.71}$$

The trace of a matrix (only defined for a square matrix) is the sum of the diagonal elements of the matrix. Now, let's determine the following:

$$Tr(\widehat{\rho}\widehat{M}) = \frac{1}{|\psi_+|^2 + |\psi_-|^2} Tr\left(\begin{bmatrix} \psi_+\psi_+^* & \psi_+\psi_-^* \\ \psi_-\psi_+^* & \psi_-\psi_-^* \end{bmatrix}\begin{bmatrix} m_{11} & m_{12} \\ m_{21} & m_{22} \end{bmatrix}\right) \tag{9.72}$$

$$= \frac{1}{|\psi_+|^2 + |\psi_-|^2} Tr\begin{bmatrix} \psi_+\psi_+^* m_{11} + \psi_+\psi_-^* m_{21} & \psi_+\psi_+^* m_{12} + \psi_+\psi_-^* m_{22} \\ \psi_-\psi_+^* m_{11} + \psi_-\psi_-^* m_{21} & \psi_-\psi_+^* m_{12} + \psi_-\psi_-^* m_{22} \end{bmatrix} \tag{9.73}$$

$$= \frac{\psi_+\psi_+^* m_{11} + \psi_+\psi_-^* m_{21} + \psi_-\psi_+^* m_{12} + \psi_-\psi_-^* m_{22}}{|\psi_+|^2 + |\psi_-|^2} \tag{9.74}$$

Now, let us consider the expectation value of a matrix $\widehat{M}$ using the definition given by Eq. (9.48), where we have

$$\langle \widehat{M} \rangle = \frac{\psi_{spin}^{*T} \widehat{M} \psi_{spin}}{\psi_{spin}^{*T} \psi_{spin}} = \frac{\begin{bmatrix} \psi_+^* & \psi_-^* \end{bmatrix} \widehat{M} \begin{bmatrix} \psi_+ \\ \psi_- \end{bmatrix}}{\begin{bmatrix} \psi_+^* & \psi_-^* \end{bmatrix}\begin{bmatrix} \psi_+ \\ \psi_- \end{bmatrix}}$$

$$= \frac{\begin{bmatrix} \psi_+^* & \psi_-^* \end{bmatrix}\begin{bmatrix} m_{11} & m_{12} \\ m_{21} & m_{22} \end{bmatrix}\begin{bmatrix} \psi_+ \\ \psi_- \end{bmatrix}}{\begin{bmatrix} \psi_+^* & \psi_-^* \end{bmatrix}\begin{bmatrix} \psi_+ \\ \psi_- \end{bmatrix}}$$

$$= \frac{\begin{bmatrix} \psi_+^* & \psi_-^* \end{bmatrix}\begin{bmatrix} m_{11}\psi_+ + m_{12}\psi_- \\ m_{21}\psi_+ + m_{22}\psi_- \end{bmatrix}}{\psi_+^*\psi_+ + \psi_-^*\psi_-}$$

$$= \frac{\psi_+ \psi_+^* m_{11} + \psi_+ \psi_-^* m_{21} + \psi_- \psi_+^* m_{12} + \psi_- \psi_-^* m_{22}}{|\psi_+|^2 + |\psi_-|^2}$$

But, the last line above is identical to Eq. (9.74). So, we have the important result

Expectation Value With Density Matrix $\widehat{\rho}$

$$\langle \widehat{M} \rangle = Tr(\widehat{\rho}\widehat{M}) \qquad (9.75)$$

As an example, if we apply this result to the x-component of the spin matrix $(\hbar/2)\widehat{\sigma}_x$, we have

$$Tr(\widehat{\rho}\widehat{S}_x) = \frac{\hbar}{2} \frac{\psi_+ \psi_+^* m_{11} + \psi_+ \psi_-^* m_{21} + \psi_- \psi_+^* m_{12} + \psi_- \psi_-^* m_{22}}{|\psi_+|^2 + |\psi_-|^2}$$

$$= \frac{\hbar}{2} \frac{\psi_+ \psi_-^* + \psi_- \psi_+^*}{|\psi_+|^2 + |\psi_-|^2}$$

The last line is identical to the direct calculation of $\langle S_x \rangle$, given by Eq. (9.50a). We leave it as an exercise for the reader to check that $\langle S_y \rangle$ and $\langle S_z \rangle$ also lead to analogous results given by Eqs. (9.50b) and (9.50c).

The spin density matrix is the simplest example of a more general **density matrix** $\widehat{\rho}$ containing information for any number of states. For spin, there are only two eigenstates, leading to a 2×2 matrix. However, the density matrix can be constructed to be *any* size. Let the vector of eigenstates ψ be given by

$$\psi = \begin{bmatrix} \psi_1 \\ \psi_2 \\ \psi_3 \\ \vdots \\ \psi_n \end{bmatrix} \qquad (9.76)$$

Then, the more general definition for any size density matrix is becomes

General Density Matrix $\widehat{\rho}$

$$\widehat{\rho} = \frac{1}{|\psi|^2} \psi \otimes \psi \qquad (9.77)$$

The applications demonstrated for the spin-density matrix are unchanged, for example Eq. (9.75) still holds. If we assume a real-valued or Hermitian Hamiltonian, then $|\psi|^2 = 1$, and the density matrix simplifies to

$$\widehat{\rho} = \psi \otimes \psi \qquad (9.78)$$

This is a more common form of the *general density matrix*. We'll see an example of a non-Hermitian Hamiltonian in Chap. 12 that leads to transitions between states, so $|\psi|^2 = |\psi|^2(t) \neq 1$ for all values of time.

9.8 Chapter Summary

In Chap. 9, we have introduced an important contribution to the angular momentum and energy of an electron. It is an *intrinsic angular momentum of the electron, called spin*. The unwitting experimental discovery of spin by Stern and Gerlach, along with other key experiments, later helped to bring about an even greater justification for a belief in this additional form of angular momentum.

The description of spin angular momentum is always consistent with that of orbital angular momentum. From this, it follows that the *spin magnetic moment* is proportional to the spin angular momentum, where the constant of proportionality for spin introduces the g-factor. Our prior knowledge of how the orbital angular momentum is treated mathematically extended naturally to the mathematical treatment of spin angular momentum and spin magnetic moment. This allowed us to express the Hamiltonian operator with the inclusion of spin. We also found a convenient way to treat spin using principle spin matrices, or Pauli spin matrices.

Lastly, we explored spin behavior using the spin matrices and the Hamiltonian, discovering precessional motion in the magnetic field. Table 9.2 summarizes some of the key results from this chapter.

Table 9.2: Chapter 9 Summary Equations.

Name	Equation
S^2 eigenvalue	$S^2 = s(s+1)\hbar^2$
s spin quantum number	$s = \frac{1}{2}$
S_z eigenvalue	$S_z = m_s \hbar$
m_s spin magnetic quantum number	$m_s = \pm\frac{1}{2}$
magnetic moment-to-spin relation	$\widehat{\boldsymbol{\mu}}_S = \frac{ge}{2m_e}\widehat{\mathbf{S}}$
total angular momentum $\widehat{\mathbf{J}}$	$\widehat{\mathbf{J}} = \widehat{\mathbf{L}} + \widehat{\mathbf{S}}$
total magnetic moment $\widehat{\boldsymbol{\mu}}_T$	$\widehat{\boldsymbol{\mu}}_T = \gamma\widehat{\mathbf{L}} + g\gamma\widehat{\mathbf{S}}$
spin Zeeman splitting	$\Delta E_Z = g\|\gamma\|\hbar B$
principle (Pauli) spin matrix $\widehat{\sigma}_x$	$\widehat{\sigma}_x = \begin{bmatrix} 0 & 1 \\ 1 & 0 \end{bmatrix}$
principle (Pauli) spin matrix $\widehat{\sigma}_y$	$\widehat{\sigma}_y = \begin{bmatrix} 0 & -i \\ i & 0 \end{bmatrix}$
principle (Pauli) spin matrix $\widehat{\sigma}_z$	$\widehat{\sigma}_y = \begin{bmatrix} 0 & -i \\ i & 0 \end{bmatrix}$
generalized spin matrix	$\widehat{\sigma}_{\theta,\varphi} = \begin{bmatrix} \cos\theta & \sin\theta e^{-i\varphi} \\ \sin\theta e^{i\varphi} & -\cos\theta \end{bmatrix}$
general eigenspinor	$\lambda_{+1} = \begin{bmatrix} \cos\frac{\theta}{2}e^{-i\frac{\varphi}{2}} \\ \sin\frac{\theta}{2}e^{i\frac{\varphi}{2}} \end{bmatrix}$
general eigenspinor	$\lambda_{-1} = \begin{bmatrix} \sin\frac{\theta}{2}e^{-i\frac{\varphi}{2}} \\ -\cos\frac{\theta}{2}e^{i\frac{\varphi}{2}} \end{bmatrix}$

9.9 Chapter Problems

Problem 9.1 Similar to how $\langle x \rangle$ was derived using Eq. (9.38), show that

$$\langle y \rangle = \lambda_{+1}^{\dagger} \widehat{\sigma}_y \lambda_{+1} = \sin\theta\sin\varphi \quad \text{and} \quad \langle z \rangle = \lambda_{+1}^{\dagger} \widehat{\sigma}_z \lambda_{+1} = \cos\theta$$

Problem 9.2 Show using the eigenstate λ_{-1} given by Eq. (9.38), that the result is the vector:

$$\langle \mathbf{r} \rangle = - \begin{bmatrix} \sin\theta\cos\varphi \\ \sin\theta\sin\varphi \\ \cos\theta \end{bmatrix}$$

Problem 9.3 Find the eigenvalues of the generalized spin matrix given by

$$\widehat{\sigma}_{\theta,\varphi} = \begin{bmatrix} \cos\theta & \sin\theta e^{-i\varphi} \\ \sin\theta e^{i\varphi} & -\cos\theta \end{bmatrix}$$

Show that they satisfy the condition:

$$|\lambda| = 1$$

Problem 9.4 Using the spin density matrix $\widehat{\rho}$ from Eq.(9.70), show that $\langle S_y \rangle$ and $\langle S_z \rangle$ lead to Eqs. (9.50b) and (9.50c).

Problem 9.5 We have shown several properties of spin matrices in this chapter, including the following:

$$\widehat{\sigma}_x^2 = \widehat{\sigma}_y^2 = \widehat{\sigma}_z^2 = \mathbf{I}(2)$$
$$\widehat{\sigma}_x\widehat{\sigma}_y = -\widehat{\sigma}_y\widehat{\sigma}_x$$
$$\widehat{\sigma}_y\widehat{\sigma}_z = -\widehat{\sigma}_z\widehat{\sigma}_y$$
$$\widehat{\sigma}_z\widehat{\sigma}_x = -\widehat{\sigma}_x\widehat{\sigma}_z$$

Show the following additional properties:

$$\widehat{\sigma}_x\widehat{\sigma}_y\widehat{\sigma}_z = i\mathbf{I}(2)$$
$$\det(\sigma_j) = -1 \quad \text{where} \quad j = x, y, z$$

Problem 9.6 Using the spin density matrix $\widehat{\rho}$ from Eq.(9.70), show that

$$Tr(\widehat{\rho S}_y) = \frac{i\hbar}{2} \frac{\psi_-^*\psi_+ - \psi_+^*\psi_-}{|\psi_+|^2 + |\psi_-|^2} \quad \text{and} \quad Tr(\widehat{\rho S}_z) = \frac{\hbar}{2} \frac{\psi_+\psi_+^* - \psi_-\psi_-^*}{|\psi_+|^2 + |\psi_-|^2}$$

What do these quantities represent?

Problem 9.7 Show that if $|\psi_+|^2 + |\psi_-|^2 = 1$, then the spin density matrix $\widehat{\rho}$ from Eq.(9.70), satisfies

$$\rho^2 = \rho$$

Such a matrix is called an *idempotent matrix*.

Problem 9.8 Using the spin density matrix $\widehat{\rho}$ from Eq.(9.70), prove that

$$\frac{d\widehat{\rho}}{dt} = \frac{-i}{\hbar}\left[\widehat{H},\widehat{\rho}\right]$$

How does this compare to Eq. (3.148), known as the *Generalized Ehrenfest theorem*?

Problem 9.9 Prove the following relation by direct evaluation of both sides:

$$(\mathbf{a}\cdot\widehat{\boldsymbol{\sigma}})(\mathbf{b}\cdot\widehat{\boldsymbol{\sigma}}) = (\mathbf{a}\cdot\mathbf{b})\mathbf{I}(2) + i(\mathbf{a}\times\mathbf{b})\cdot\widehat{\boldsymbol{\sigma}}$$

9.10 Suggested Readings & References

[1] B. Friedrich, D. Herschbach, *Stern and Gerlach: How a Bad Cigar Helped Reorient Atomic Physics*, Physics Today pp.53-59 (Dec 2003)
[2] W. Pauli, *Zeitschrift für Physik A Hadrons and Nuclei In Zeitschrift für Physik*, **43**, No. 9-10. (21 September 1927), pp. 601-623, doi:10.1007/bf01397326
[3] Bandyopadhyay and Cahay, *Introduction to Spintronics*, CRC Press (2008)

CHAPTER 10

Multi-electron Systems

Now that we have considered atomic systems with a single electron, the atomic system can be extended to include more than one electron. Adding even just one more electron to an atom raises the level of complexity considerably. We consider some of the consequences in this chapter. Treating multi-electron systems in quantum mechanics ultimately reveals to us, the rules of engagement between electrons and atoms, evident from transitions between the various states of multi-electron systems. In one of the most general cases for a system of electrons, we can consider the result of having multi-electrons in conditions where it is impossible to contrive an experiment that allows us to tell them apart. Recall that this concept of *indistinguishability* was encountered by Max Planck and others when he treated the *blackbody radiation problem*. As it turned out for an ensemble of photons, it leads to important consequences for electronic systems, as well. Let's find out how.

10.1 Consequences of Electron Indistinguishability

In considering atoms with more than one electron, there is no known mark of distinction when observing the behavior of electrons themselves.

Therefore, a question arises around the implications of the *indistinguishability* of electrons, and other subatomic particles. What implications might this have on wave functions described by the Schrödinger's equation? It is helpful to entertain this question when starting to examine systems with more than one electron. For atoms with two or more electrons having electron-electron interactions, there is no *exact* solution of the Schrödinger equation. The implications of this are important, as this implies that in order to discuss any theoretical analysis for multi-electron systems, *approximations are more often necessary*. As we go, we shall point out these approximations, so it should be kept in mind with any of the analytical results discussed.

Any two electrons that are indistinguishable are defined as follows: swapping the electrons in position should not alter the probabilities of the overall wave function $\psi(\mathbf{r}_1, \mathbf{r}_2)$. $\mathbf{r}_1$ is the coordinate for electron 1, and $\mathbf{r}_2$ is that for electron 2. Let us denote this swapping of electrons by an interchange or swapping operator O_{sw}. We can quantify these two statements as

$$|\psi(\mathbf{r}_1, \mathbf{r}_2)|^2 = |\psi(\mathbf{r}_2, \mathbf{r}_1)|^2 \tag{10.1}$$

$$O_{sw}(\psi(\mathbf{r}_1, \mathbf{r}_2)) = \psi(\mathbf{r}_2, \mathbf{r}_1) \tag{10.2}$$

Since $\psi(\mathbf{r}_1, \mathbf{r}_2)$ is generally complex, the first relation implies

$$\psi(\mathbf{r}_1, \mathbf{r}_2) = \psi(\mathbf{r}_2, \mathbf{r}_1) e^{i\phi} \tag{10.3}$$

By comparing Eq. (10.2) with Eq. (10.3), it follows that

$$O_{sw}(\psi(\mathbf{r}_1, \mathbf{r}_2)) = \psi(\mathbf{r}_2, \mathbf{r}_1) e^{-i\phi} \tag{10.4}$$

If the swapping operator is applied twice to $\psi(\mathbf{r}_1, \mathbf{r}_2)$, we get that

$$O_{sw}[O_{sw}(\psi(\mathbf{r}_1, \mathbf{r}_2))] = O_{sw}\left[\psi(\mathbf{r}_1, \mathbf{r}_2) e^{-i\phi}\right] = \psi(\mathbf{r}_1, \mathbf{r}_2) e^{-2i\phi}$$

But, swapping twice simply means we have taken electrons back to the original configuration. So, it follows that

$$O_{sw}^2(\psi(\mathbf{r}_1, \mathbf{r}_2)) = \psi(\mathbf{r}_1, \mathbf{r}_2) \tag{10.5}$$

Thus, Eqs. (10.4) and (10.5) lead us to conclude that

$$e^{-2i\phi} = 1 \Rightarrow \phi = 0, \pi \tag{10.6}$$

From Eq. (10.6), we have the fundamental result for wave functions of *indistinguishable* particles, where

10.2 Two-Electron wave functions Reveal Exchange

Electron Indistinguishability

$$\psi(\mathbf{r}_1, \mathbf{r}_2) = \psi(\mathbf{r}_2, \mathbf{r}_1) \quad \text{or} \quad \psi(\mathbf{r}_1, \mathbf{r}_2) = -\psi(\mathbf{r}_2, \mathbf{r}_1) \tag{10.7}$$

The two distinct conditions given by Eq. (10.7) define the *two fundamental types of particles* in physical systems. When the wave functions between swapping electrons are equal, given by the first condition, they are said to be *symmetric* under the operation of swapping or exchanging indistinguishable particles. Such particles are **bosons**. Photons of light are examples of *bosons*. When the wave functions are anti-symmetric under the operation of swapping the two particles, given by the second condition, the particles are said to be **fermions**. Electrons, protons, and neutrons are examples of *fermions*. Both were discussed in regards to thermodynamics in Sec. 2.5.

Given these properties for the *total* wave function, we have different possibilities with the wave function form when we solve the Schrödinger equation. We can assume one form, and see which describes the experimental evidence correctly. Therefore, it is important to emphasize that determination cannot be deduced from theory alone. Historically, the patterns and trends from spectral data of multi-electron systems have served to obtain answers to questions like these. Experiments suggest that *two-electrons systems, for example, are always described by anti-symmetric wave functions*, and this is how we have come to know that they are *fermions*. In this chapter, we'll see an example of this with atomic spectra of helium. However, before we get to that, let us begin to look at the implications of these results for electrons (fermions), as dictated by the Schrödinger equation and the energy of such a system.

10.2 Two-Electron wave functions And Exchange Energy

To appreciate the convenience of the simplified Hamiltonian we will ultimately use, let's first consider what the Hamiltonian looks like for a multi-atom multi-electron system. Each atom is taken as a collection of electrons with a nucleus of charge $Z|e|$. Then, the Hamiltonian for a set of n atoms or nuclei, each having Z electrons, can be written as

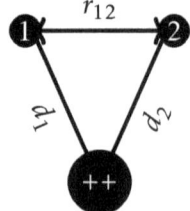

Figure 10.1: Illustration of a helium atom, having a single nucleus and two electrons. The relevant position vectors used in the equations are shown.

$$\widehat{H} = -\sum_{j=1}^{n} \frac{\hbar^2}{2M_j}\nabla_j\psi - \sum_{j=1}^{n}\sum_{k=1}^{Z} \frac{\hbar^2}{2m_k}\nabla_{jk}\psi +$$

$$\left(\sum_{l>l'} \frac{Z_l Z_{l'} e^2}{4\pi\epsilon_0 R_{ll'}} + \sum_{i>i'} \frac{e^2}{4\pi\epsilon_0 r_{ii'}} - \sum_{q>q'} \frac{Z_q e^2}{4\pi\epsilon_0 d_{qq'}}\right)\psi \quad (10.8)$$

M_j is the mass of the jth nucleus; m_k is the mass of the kth electron belonging to the jth atom; Z_l and Z'_l are the atomic numbers of respective atoms l and l'; $R_{ll'}$ denotes the *internuclear distance* between nuclei l and l'; Similarly, $r_{ii'}$ is the interelectronic distance between electrons i and i'; In the last contribution to the potential energy, $d_{qq'}$ is the distance between electron q and nucleus q'. Note that there is but one wave function ψ for this entire system described by Eq. (10.8). Eq. (10.8) simplifies greatly when we consider a single atom with two electrons, such as helium. For a single atom denoted by A, we have the Hamiltonian $\widehat{H}$ given by

$$\widehat{H} = -\frac{\hbar^2}{2M}\nabla_A^2\psi - \frac{\hbar^2}{2m^*}\left[\nabla_1^2 + \nabla_2^2\right]\psi + \left[-\frac{Ze^2}{4\pi\epsilon_0 d_1} + -\frac{Ze^2}{4\pi\epsilon_0 d_2} + \frac{e^2}{4\pi\epsilon_0 r_{12}}\right]\psi$$
(10.9)

∇_A^2 denotes the Laplacian operator (for kinetic energy) applied with respect to the position of nucleus A; ∇_1^2 is the same operator applied with respect to the position of electron 1, while ∇_2^2 is for electron 2. m^* is the *reduced mass* given here by $m^* = m_e m_n/(m_e + m_n)$, where m_n is the mass of the nucleus (c. Chap. 8). Fig 10.1 illustrates the vectors used in Eq. (10.9) and their relations. The next simplification to make for the helium atom is to take the wave function to be the product of a nuclear and electronic wave function, respectively. That is

$$\psi(\mathbf{r}_n, \mathbf{r}_1, \mathbf{r}_2) = \psi_{\text{nucl}} \cdot \psi_{\text{elec}} \quad (10.10)$$

10.2 Two-Electron wave functions Reveal Exchange

This supposed form for ψ can indeed be done, and it means that the dependence on $\mathbf{r}_n$ and the other position variables is separable. By doing this, Eq. (10.9) separates into two separate equations, one depending on $\mathbf{r}_n$ and the other on $\mathbf{r}_1$ and $\mathbf{r}_2$. Recall that this is how we treated the hydrogen atom, and it still works here because we only have a single atom/nucleus. It's not as simple when we extend to two or more atoms, but we'll discuss this point when we consider molecular wave functions later. Note that while the potential terms between the nucleus and electron are attractive, and therefore, negative in sign, the potential function between two electrons is repulsive, and thus, positive in sign. If we take the origin to be at the nucleus, as we did for the hydrogen atom, the Hamiltonian for the electrons become

$$\widehat{H}_{elec} = -\frac{\hbar^2}{2m^*}\left[\nabla_1^2 + \nabla_2^2\right]\psi_{elec} + \left[-\frac{Ze^2}{4\pi\epsilon_0 d_1} - \frac{Ze^2}{4\pi\epsilon_0 d_2} + \frac{e^2}{4\pi\epsilon_0 r_{12}}\right]\psi_{elec}$$

$$= \left(-\frac{\hbar^2}{2m^*}\nabla_1^2 + V(\mathbf{r}_1)\right)\psi_{elec} + \left(-\frac{\hbar^2}{2m^*}\nabla_2^2 + V(\mathbf{r}_2)\right)\psi_{elec}$$

$$\ldots + V(\mathbf{r}_1,\mathbf{r}_2)\psi_{elec}$$

$$= \widehat{H}_1\psi_{elec} + \widehat{H}_2\psi_{elec} + V(\mathbf{r}_1,\mathbf{r}_2)\psi_{elec}$$

The electronic Hamiltonian is composed of the sum of hydrogen-like Hamiltonians $\widehat{H}_1$ and $\widehat{H}_2$, in parentheses, for electron 1 and electron 2, respectively, and an additional interelectron potential term. Because the above *cannot* be separated in the usual way due to the presence of the electron-electron potential term, it has no exact solution. For this reason, in order to proceed with analyzing the Hamiltonian for helium, *approximations must be made* even for the electron Hamiltonian. A common approximation we shall make deals with the explicit form of the wave function ψ_{elec} that allows enforcement of the anti-symmetric property for *fermions* in a simple way. It is motivated by considering Eq. (10.9) *without* the electron-electron interaction, given by

$$\widehat{H} \approx -\frac{\hbar^2}{2m^*}\nabla_1^2\psi + V(\mathbf{r}_1)\psi - \frac{\hbar^2}{2m^*}\nabla_2^2\psi + V(\mathbf{r}_2)\psi \qquad (10.11)$$

In Eq. (10.11), the Hamiltonian *is* separable with respect to the positions of electron 1 and 2. Therefore, in this approximation, ψ_{elec} can be written as a product of functions of $\mathbf{r}_1$ and $\mathbf{r}_2$, or

$$\psi_{elec}(\mathbf{r}_1,\mathbf{r}_2) \simeq \psi_a(\mathbf{r}_1)\psi_b(\mathbf{r}_2) \qquad (10.12)$$

With two electrons, this product takes on more meaning, so let us clarify further the wave function in Eq. (10.12). $\psi_a(\mathbf{r}_1)$ is the wave function associated with the electronic state a housing one of the electrons, presumably electron 1. Similarly, $\psi_b(\mathbf{r}_2)$, is another electronic state denoted by b housing electron 2. Conceptually, its analogous to two people (analogous to two electrons) having two separate homes, homes a and b. In one scenario, person 1 is in home a (state a), and person 2 is in home b (state b). With this analogy, swapping electrons is akin to person 2 and person 1 deciding to *swap* or *exchange* homes. The available homes stay put, but person 1 moves to home b, and person 2 moves to home a. The catch is that if both people are *indistinguishable*, then the neighborhood remains unchanged from this swap, and it goes unnoticed. Thus, the homes represent the two available estates, etc.

Although Eq. (10.12) is a basic solution to the Schrödinger equation, any linear combination of wave functions with this form is also a solution. A slight twist to this is that, with 2 or more electrons, the *indistinguishablity* must be considered explicitly. It was not necessary to consider this when dealing with one electron. However, based on our conclusions from our discussion of indistinguishability of two or more electrons, we require that the total wave function ψ_{elec} be *anti-symmetric* under the swapping of electrons. For now, let us take this as postulate. The anti-symmetry upon swapping electrons means that

$$\psi_{\text{elec}} = \psi(\mathbf{r}_1, \mathbf{r}_2) = -\psi(\mathbf{r}_2, \mathbf{r}_1) \Rightarrow \psi_a(\mathbf{r}_1)\psi_b(\mathbf{r}_2) = -\psi_a(\mathbf{r}_2)\psi_b(\mathbf{r}_1) \tag{10.13}$$

In this approximation for the wave function, there is a problem. If we consider the hydrogen atom wave functions found in Chap. 8, we can immediately find that this condition is, generally, *not* satisfied. For example, take any set of s orbitals, and they will not satisfy the required anti-symmetry condition when exchanging electrons. This necessitates that we take a linear combination of $\psi_a \psi_b$ as functions of the position variables, as a solution, to enforce *indistinguishability* of the electrons. For a two-electron atom, such as helium, for the spatial part of the wave function, we have the following two possible combinations:

$$\psi(\mathbf{r}_1, \mathbf{r}_2) = \psi_a(\mathbf{r}_1)\psi_b(\mathbf{r}_2) \pm \psi_a(\mathbf{r}_2)\psi_b(\mathbf{r}_1) \tag{10.14}$$

When we also include the spin degree of freedom held to the same standards for indistinguishable particles, with spin-up and spin-down eigenfunctions ψ_+ and ψ_- (c. Sec. 9.6), Eq. (10.14) extends to become

10.2 Two-Electron wave functions Reveal Exchange

$$\Psi(\mathbf{r}_1,\mathbf{r}_2,+,-) = \psi(\mathbf{r}_1,\mathbf{r}_2) \cdot \psi_{\text{spin}}(1,2) =$$

$$[\psi_a(\mathbf{r}_1)\psi_b(\mathbf{r}_2) \pm \psi_a(\mathbf{r}_2)\psi_b(\mathbf{r}_1)] \cdot [\psi_+(1)\psi_-(2) \mp \psi_+(2)\psi_-(1)] \quad (10.15)$$

Since the spin eigenfunctions do not depend on spatial variables $\mathbf{r}_1$ and $\mathbf{r}_2$, we denote them with an electron number as the argument to indicate which electron. Because we have a product of two linear combinations, the impact of spin on the energy E_λ is that it affects which form of the spatial wave functions to ensure anti-symmetry for the total wave function. For example, if the spin combination is anti-symmetric, using the − sign, then the spatial function will have to be symmetric, meaning that it takes the + combination so the total wave function remains anti-symmetric. If they both change signs under exchange of electrons, the anti-symmetry condition would not be met, as there would be no net sign change. Thus, for the total wave function to be anti-symmetric upon electron exchange, only one of the two factors in Eq. 10.15 can possess the anti-symmetry property. Hence, we have these two unique possibilities for the two-electron system of helium. For both cases, however, the energy eigenvalue E_λ becomes

$$E_\lambda = \frac{\int \Psi^*(\mathbf{r}_1,\mathbf{r}_2,+,-)\widehat{H}_{\text{elec}}\Psi^*(\mathbf{r}_1,\mathbf{r}_2,+,-)d\mathbf{r}_1 d\mathbf{r}_2}{\int \Psi^*(\mathbf{r}_1,\mathbf{r}_2,+,-)\Psi^*(\mathbf{r}_1,\mathbf{r}_2,+,-)d\mathbf{r}_1 d\mathbf{r}_2} \quad (10.16)$$

Since the spin functions are assumed to be independent of space, it turns out that the electronic energy is determined by the spatial wave functions only, and that the spin degree of freedom influences the energy *only* by dictating which form of symmetry is taken by the linear combination of spatial wave functions. We can see this from substituting Eq. (10.15) into Eq. (10.16), where we get

$$E_\lambda = \frac{\int \psi^*_{\text{spin}}(1,2)\psi^*(\mathbf{r}_1,\mathbf{r}_2)\widehat{H}_{\text{elec}}\psi(\mathbf{r}_1,\mathbf{r}_2)\psi_{\text{spin}}(1,2)d\mathbf{r}_1 d\mathbf{r}_2}{\int \psi^*_{\text{spin}}(1,2)\psi^*(\mathbf{r}_1,\mathbf{r}_2)\psi(\mathbf{r}_1,\mathbf{r}_2)\psi_{\text{spin}}(1,2)d\mathbf{r}_1 d\mathbf{r}_2} \quad (10.17)$$

$$= \frac{\psi^*_{\text{spin}}(1,2)\psi_{\text{spin}}(1,2)\int \psi^*(\mathbf{r}_1,\mathbf{r}_2)\widehat{H}_{\text{elec}}\psi(\mathbf{r}_1,\mathbf{r}_2)d\mathbf{r}_1 d\mathbf{r}_2}{\psi^*_{\text{spin}}(1,2)\psi_{\text{spin}}(1,2)\int \psi^*(\mathbf{r}_1,\mathbf{r}_2)\psi(\mathbf{r}_1,\mathbf{r}_2)d\mathbf{r}_1 d\mathbf{r}_2} \quad (10.18)$$

$$= \frac{\int \psi^*(\mathbf{r}_1,\mathbf{r}_2)\widehat{H}_{\text{elec}}\psi(\mathbf{r}_1,\mathbf{r}_2)d\mathbf{r}_1 d\mathbf{r}_2}{\int \psi^*(\mathbf{r}_1,\mathbf{r}_2)\psi(\mathbf{r}_1,\mathbf{r}_2)d\mathbf{r}_1 d\mathbf{r}_2} \quad (10.19)$$

Therefore, the spatial wave functions are the main contributions that we must consider for the energy of the two-electron single nucleus system.

If we take ψ_a and ψ_b to be normalized and orthogonal eigenfunctions and thus, solutions to the Schrödinger equation, then the following normalized anti-symmetric solution also satisfies the Schrödinger equation (without the electron-electron interaction):

$$\psi(\mathbf{r}_1,\mathbf{r}_2) = \frac{1}{\sqrt{2}}(\psi_a(\mathbf{r}_1)\psi_b(\mathbf{r}_2) - \psi_a(\mathbf{r}_2)\psi_b(\mathbf{r}_1)) \qquad (10.20)$$

Swapping electrons with this wave function leads to

$$\psi(\mathbf{r}_2,\mathbf{r}_1) = \frac{1}{\sqrt{2}}(\psi_a(\mathbf{r}_2)\psi_b(\mathbf{r}_1) - \psi_a(\mathbf{r}_1)\psi_b(\mathbf{r}_2)) = -\psi(\mathbf{r}_1,\mathbf{r}_2)$$

In Chapter problem 10.2, we have you show that this choice of normalized wave function requires that $\int \phi_a^* \phi_a dr = 1$ (normalized) and $\int \phi_a^* \phi_b dr = 0$ (orthogonal). For a real or Hermitian Hamiltonian, solutions to the Schrödinger equation can always be chosen that satisfy these properties.

An anti-symmetric wave function, as given by Eq. (10.15) can also be expressed using the notion of determinants. For the two-electron system, we can write

$$\psi(\mathbf{r}_1,\mathbf{r}_2) = \frac{1}{\sqrt{2}}\begin{vmatrix} \psi_a(\mathbf{r}_1) & \psi_b(\mathbf{r}_1) \\ \psi_a(\mathbf{r}_2) & \psi_b(\mathbf{r}_2) \end{vmatrix} = \frac{1}{\sqrt{2}}(\psi_a(\mathbf{r}_1)\psi_b(\mathbf{r}_2) - \psi_a(\mathbf{r}_2)\psi_b(\mathbf{r}_1)) \qquad (10.21)$$

In this form, each row corresponds to an electron, while each column corresponds to a state. Then, exchanging electrons is effectively done by simply exchanging rows in the matrix, which gives

$$\frac{1}{\sqrt{2}}\begin{vmatrix} \psi_a(\mathbf{r}_2) & \psi_b(\mathbf{r}_2) \\ \psi_a(\mathbf{r}_1) & \psi_b(\mathbf{r}_1) \end{vmatrix} = \frac{1}{\sqrt{2}}(\psi_a(\mathbf{r}_2)\psi_b(\mathbf{r}_1) - \psi_a(\mathbf{r}_1)\psi_b(\mathbf{r}_2)) = -\psi(\mathbf{r}_1,\mathbf{r}_2) \qquad (10.22)$$

This is an application of a well-known mathematical property of matrix determinants. The property states that for any square matrix, *swapping any row or column of the matrix changes the sign of the determinant of that matrix*. This is because when two rows are exchanged, every term in the determinant changes sign, and so the determinant changes sign. This property makes determinants ideal to construct anti-symmetric wave functions that require a sign change upon exchange of electrons. A set of states, usually called *basis functions* are chosen, then inserted

10.2 Two-Electron wave functions Reveal Exchange

into a determinant to generate the total wave function. Because of the generality of this mathematical property, this idea can be generalized to any number n of electrons with an equal number of available states or basis function. Thus, the wave function for the more general case becomes

$$\psi(\mathbf{r}_1, \mathbf{r}_2, \cdots, \mathbf{r}_n) = \frac{1}{\sqrt{n}} \begin{vmatrix} \psi_{a1}(\mathbf{r}_1) & \psi_{a2}(\mathbf{r}_1) & \cdots & \psi_{an}(\mathbf{r}_1) \\ \psi_{a1}(\mathbf{r}_2) & \psi_{a2}(\mathbf{r}_2) & \cdots & \psi_{an}(\mathbf{r}_2) \\ \vdots & \vdots & \ddots & \vdots \\ \psi_{a1}(\mathbf{r}_n) & \psi_{a2}(\mathbf{r}_n) & \cdots & \psi_{an}(\mathbf{r}_n) \end{vmatrix} \quad (10.23)$$

The resulting wave function constructed from this process is known as a **determinantal wave functions**, and this process of constructing wave functions was used by John C. Slater (1900-1926) in 1927. This form turns out to satisfy important properties for fermions. For example, the determinant vanishes for any matrix having two identical or proportional rows. This corresponds to two different electrons being in the same state. Thus, nonzero determinants also ensure no two electrons occupy the same state. For this reason, they are also known as **Slater determinants**, and each basis function used is sometimes called a **Slater orbital**. It is now widely used in numerical computations of quantum mechanics to determined the states of multi-electron systems. One needs to start by wisely choosing a set of *basis functions*, which are the electronic states (like a and b the two electron example) defining each column. They are taken as functions of multiple parameters that are, in turn, used to optimize the system, etc.

Although the Schrödinger equation is *not solvable exactly for systems with two or more electrons*, due to the electron-electron interaction, it is still useful to examine some of the consequences of an anti-symmetric wave function like that given by Eq. (10.20). Let us assume an anti-symmetric form for the eigensolution of the Hamiltonian, however, we shall now include the electron-electron interaction. The energy E_λ of the helium atom becomes

$$E_\lambda = \frac{\int \psi^* \widehat{H} \psi}{\int \psi^* \psi} = -\frac{\hbar^2}{2m^*} \sum_{i=1}^{2} \int \psi^*(\mathbf{r}_1, \mathbf{r}_2) \nabla_i^2 \psi(\mathbf{r}_1, \mathbf{r}_2) d\mathbf{r}_1 d\mathbf{r}_2 \; +$$

$$\sum_{i=1}^{2} \int \psi^*(\mathbf{r}_1, \mathbf{r}_2) V(\mathbf{r}_i) \psi(\mathbf{r}_1, \mathbf{r}_2) d\mathbf{r}_1 d\mathbf{r}_2 \; +$$

$$\int \psi^*(\mathbf{r}_1,\mathbf{r}_2)V(\mathbf{r}_1,\mathbf{r}_2)\psi(\mathbf{r}_1,\mathbf{r}_2)d\mathbf{r}_1 d\mathbf{r}_2 \quad (10.24)$$

In Eq. (10.24), we have taken $\int \psi^*\psi = 1$. Although it looks a bit complicated, the first two integral terms added in the first line are of the same form as what we have seen in dealing with the hydrogen atom. Those are the contributions of the kinetic energy. Likewise, the sum of the integrals on the second line are the *negative* electric potential contributions from the interaction between each electron and the nucleus. These terms will have energy contributions of the form given by Eq. (9.12). So, we have seen these terms before. However, the last term is new, as we have not dealt with it prior to now. It represents contributions to the energy coming from the *electron-electron interaction*. So, let us consider only this last energy term, we shall denote by E_{e-e}, using the anti-symmetric wave function of the form $\psi(\mathbf{r}_1,\mathbf{r}_2) = \psi_a(\mathbf{r}_1)\psi_b(\mathbf{r}_2) - \psi_a(\mathbf{r}_2)\psi_b(\mathbf{r}_1)$. Substituting the Coulomb potential function, we have

$$E_{e-e} = \int \psi^*(\mathbf{r}_1,\mathbf{r}_2)V(\mathbf{r}_1,\mathbf{r}_2)\psi(\mathbf{r}_1,\mathbf{r}_2)d\mathbf{r}_1 d\mathbf{r}_2 \quad (10.25a)$$

$$= \frac{e^2}{4\pi\epsilon_0} \int \frac{\psi^*(\mathbf{r}_1,\mathbf{r}_2)\psi(\mathbf{r}_1,\mathbf{r}_2)}{|\mathbf{r}_2 - \mathbf{r}_1|} d\mathbf{r}_1 d\mathbf{r}_2 \quad (10.25b)$$

Then, substitution of the anti-symmetric wave function leads to

$$E_{e-e} = \frac{e^2}{8\pi\epsilon_0} \times$$

$$\int \frac{\left(\psi_a^*(\mathbf{r}_1)\psi_b^*(\mathbf{r}_2) - \psi_a^*(\mathbf{r}_2)\psi_b^*(\mathbf{r}_1)\right)\left(\psi_a(\mathbf{r}_1)\psi_b(\mathbf{r}_2) - \psi_a(\mathbf{r}_2)\psi_b(\mathbf{r}_1)\right)}{|\mathbf{r}_2 - \mathbf{r}_1|} d\mathbf{r}_1 d\mathbf{r}_2$$

Note that the $1/\sqrt{2}$ from the normalized wave function leads to doubling the coefficient's denominator, hence 4 becomes 8. Multiplying out the terms of the integrand leads to the sum of four terms, as follows:

$$E_{e-e} = \frac{e^2}{8\pi\epsilon_0} \int \frac{\psi_a^*(\mathbf{r}_1)\psi_b^*(\mathbf{r}_2)\psi_a(\mathbf{r}_1)\psi_b(\mathbf{r}_2)}{|\mathbf{r}_2 - \mathbf{r}_1|} d\mathbf{r}_1 d\mathbf{r}_2 \quad -$$

$$\frac{e^2}{8\pi\epsilon_0} \int \frac{\psi_a^*(\mathbf{r}_1)\psi_b^*(\mathbf{r}_2)\psi_a(\mathbf{r}_2)\psi_b(\mathbf{r}_1)}{|\mathbf{r}_2 - \mathbf{r}_1|} d\mathbf{r}_1 d\mathbf{r}_2 \quad -$$

$$\frac{e^2}{8\pi\epsilon_0} \int \frac{\psi_a^*(\mathbf{r}_2)\psi_b^*(\mathbf{r}_1)\psi_a(\mathbf{r}_1)\psi_b(\mathbf{r}_2)}{|\mathbf{r}_2 - \mathbf{r}_1|} d\mathbf{r}_1 d\mathbf{r}_2 \quad +$$

10.2 Two-Electron wave functions Reveal Exchange

$$\frac{e^2}{8\pi\epsilon_0} \int \frac{\psi_a^*(\mathbf{r}_2)\psi_b^*(\mathbf{r}_1)\psi_a(\mathbf{r}_2)\psi_b(\mathbf{r}_1)}{|\mathbf{r}_2-\mathbf{r}_1|} d\mathbf{r}_1 d\mathbf{r}_2 \quad (10.26)$$

The first and fourth terms in Eq. (10.26) are of the same mathematical form, while the second and third terms are, likewise, of the same form, but different from the first and fourth. In the first form, each state's wave function evaluates over the same spatial variable $\mathbf{r}_i$. The two terms become

$$E_{e\text{-}e,1} = \frac{e^2}{8\pi\epsilon_0} \int \frac{\psi_a^*(\mathbf{r}_1)\psi_a(\mathbf{r}_1)\psi_b^*(\mathbf{r}_2)\psi_b(\mathbf{r}_2)}{|\mathbf{r}_2-\mathbf{r}_1|} d\mathbf{r}_1 d\mathbf{r}_2 \; +$$

$$\frac{e^2}{8\pi\epsilon_0} \int \frac{\psi_a^*(\mathbf{r}_2)\psi_a(\mathbf{r}_2)\psi_b^*(\mathbf{r}_1)\psi_b(\mathbf{r}_1)}{|\mathbf{r}_2-\mathbf{r}_1|} d\mathbf{r}_1 d\mathbf{r}_2$$

This sum of terms can be interpreted as a weighted average of *Coulomb interaction* potentials, because each term can be written as

$$C_{12} = \frac{1}{2} \int \psi_a^*(\mathbf{r}_1)\rho(\mathbf{r}_1)\psi_a(\mathbf{r}_1) d\mathbf{r}_1 \quad (10.27)$$

In this form, the average potential ρ from electron 2 is, itself, averaged over the volume occupied by electron 2. The average potential from electron 2, experienced at $\mathbf{r}_1$ is

$$\rho(\mathbf{r}_1) = \frac{e^2}{4\pi\epsilon_0} \int \frac{\psi_b^*(\mathbf{r}_2)\psi_b(\mathbf{r}_2)}{|\mathbf{r}_2-\mathbf{r}_1|} d\mathbf{r}_2 \quad (10.28)$$

Similarly, the second term can be written as

$$C_{21} = \frac{1}{2} \int \psi_a^*(\mathbf{r}_2)\rho(\mathbf{r}_2)\psi_a(\mathbf{r}_2) d\mathbf{r}_2 \quad (10.29)$$

This is the averaging of the mean potential $\rho(\mathbf{r}_2)$ due to electron 1, over the volume occupied by electron 2. $\rho(\mathbf{r}_2)$ is the mean potential from electron 1, experienced at any position of electron 2, and it is given by

$$\rho(\mathbf{r}_2) = \frac{e^2}{4\pi\epsilon_0} \int \frac{\psi_b^*(\mathbf{r}_1)\psi_b(\mathbf{r}_1)}{|\mathbf{r}_2-\mathbf{r}_1|} d\mathbf{r}_1 \quad (10.30)$$

Given the definition of $\rho(r_i)$, it follows that $\rho(r_i) \geq 0$, and since $\psi_k^*\psi_k = |\psi_a|^2 > 0$, the Coulomb integrals are, therefore, always semi-positive definite or $C_{ij} \geq 0$. This is consistent with the expectation that an electron-electron Coulomb interaction is positive, since its a repulsive potential between two like charges. The Coulomb integral contributions therefore *increase* the energy of the system.

The contributions to E_λ from Coulomb interactions become

$$E_{e\text{-}e,1} = C_{12} + C_{21}$$

The second and third terms of the energy summation in Eq. (10.26) are distinct from the Coulomb integral contributions. In these integrals, the same wave function evaluates over two distinct positions. Moreover, the second and third terms are *complex conjugates* of one another, which tells us that the sum of these two integrals is always a real number, because for any complex number $c = a + ib$, $c + c^* = a + ib + a - ib = 2Re(c)$. Therefore, the sum of the second and third terms can be written as a single integral, denoted by J_{ij} and given by

Exchange Integral

$$J_{ij} = Re\left[\frac{e^2}{4\pi\epsilon_0} \int \frac{\psi_a^*(\mathbf{r}_1)\psi_b^*(\mathbf{r}_2)\psi_a(\mathbf{r}_2)\psi_b(\mathbf{r}_1)}{|\mathbf{r}_2 - \mathbf{r}_1|} d\mathbf{r}_1 d\mathbf{r}_2 \right] \quad (10.31)$$

Eq. (10.31) is known as the **exchange integral**, and it is distinct from the Coulomb potential integrals, existing between electron 1 and electron 2. Instead, this single integral accounts for the contributions from interacting electrons. The integrand of Eq. (10.31) is a measure of how much spatial *overlap* exists between the two state's wave functions ψ_a and ψ_b. If ψ_a and ψ_b have no overlap, i.e. if the range of $\mathbf{r}_1$ and $\mathbf{r}_2$ are mutually exclusive spatial domains, then evaluating either wave function will be non-zero for either $\mathbf{r}_1$ or $\mathbf{r}_2$, *but* zero for the other. Thus, there will be no contribution to the energy in this case. It is only when these two wave functions overlap spatially with one another, will the exchange integral be non-zero. This is the hallmark of the exchange integral or *exchange energy* contribution, being more significant when electrons are sufficiently close to each other. This is illustrated in Fig. 10.2. The first case in Fig. 10.2 corresponds to when the two wave functions of each respective electron are far apart, leading to no significant overlap, and thus, zero exchange integral $J \simeq 0$. The other case is when they are sufficiently close to give a nonzero overlap region, which leads to nonzero exchange integral $|J| > 0$. Unlike the Coulomb integral, the analysis has not lead us to require that $J > 0$ or $J < 0$, hence the

10.2 Two-Electron wave functions Reveal Exchange

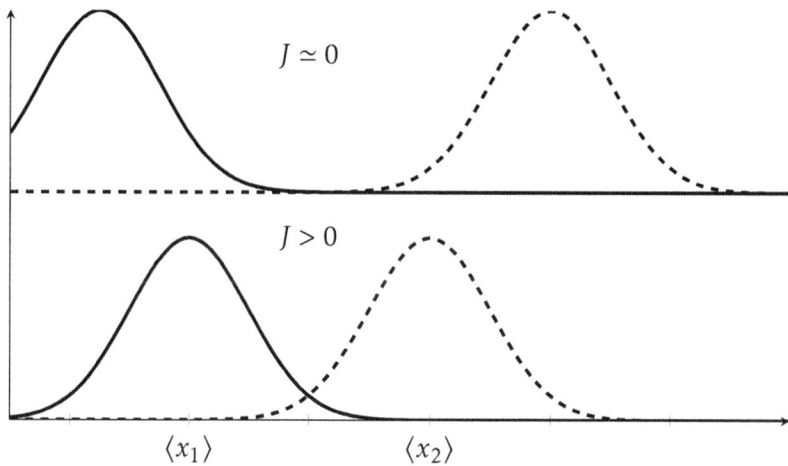

Figure 10.2: Illustration of the two conditions for exchange integral values. The first case corresponds when the two wave functions are far apart, leading to no significant overlap, and thus, zero exchange integral $J \simeq 0$. The other case is when they are sufficiently close to give a nonzero overlap region, which leads to nonzero exchange integral.

exchange integral can be either. When $J > 0$ and the coupled electrons belong to *distinct atoms*, such a material is known as **ferromagnetic**. When $J < 0$ for *two electrons of different atoms*, the material is known as **anti-ferromagnetic**. For the case of helium, the electrons are on a single atom, so these terms to do not apply. The exchange energy between electrons can exist between electrons orbiting about the same nucleus, as well as electrons orbiting about distinct nuclei, of different atoms.

Then, the total of the electron-electron contribution is given by

$$E_{e-e} = C_{12} + C_{21} - J_{12} \tag{10.32}$$

From the derivation of the exchange integral, we find that it has contributions from two identical quantities (real parts of complex conjugates). Therefore, the pairwise *exchange integral* between two electrons is symmetric in that

$$J_{12} = J_{21}$$

For bound electron states, the spatial wave functions are usually real-valued, represented here by ψ_a and ψ_b. Then, it follows that

$$J_{12} = \frac{e^2}{4\pi\epsilon_0} \int \frac{\psi_a(\mathbf{r}_1)\psi_a(\mathbf{r}_2)\psi_b(\mathbf{r}_2)\psi_b(\mathbf{r}_1)}{|\mathbf{r}_2 - \mathbf{r}_1|} d\mathbf{r}_1 d\mathbf{r}_2 \tag{10.33}$$

Because the complex part of the wave function is zero, the case of strictly real wave functions represents a space of upper bounds for the exchange integral contribution. Otherwise, interference phenomena between the wave functions becomes possible, which can lead to destructive interference, reducing the amplitude of Eq. (10.33). Moreover, because of the bound or localized nature of the wave functions of electrons, it is straightforward to see that Eq. (10.33) diminishes with increasing mean distance between electrons, due to the $1/|r_2 - r_1|$ term in the integrand.

The exchange energy turns out to be important for understanding the origin of some types of molecular bonds, including unique bonds between atoms of solid-state matter having spin. This sets up a foundation for the field of *magnetism*. The energy arising from exchange was postulated by Werner Heisenberg c.1926 to explain the origins of ferromagnetism, where electrons *belonging to different atoms* in a solid-state crystal further lower their energy through this peculiar energy contribution. You probably did not even notice we were uncovering the origins ferromagnetism! But, we'll come back to complete this idea, specifically, in Sec. 10.8. Now, let's consider the other case for the anti-symmetric wave function of a two-electron system.

10.3 Symmetric Spatial & Anti-symmetric Spin State

In the previous section, we considered the specific case of *symmetric spin states* which includes *parallel* electron spins. We arrived at our result by first considering the *total* anti-symmetric wave function given by

$$\Psi(\mathbf{r}_1, \mathbf{r}_2, \uparrow, \downarrow) =$$

$$[\psi_a(\mathbf{r}_1)\psi_b(\mathbf{r}_2) \pm \psi_a(\mathbf{r}_2)\psi_b(\mathbf{r}_1)] \cdot [\psi_\uparrow(1)\psi_\downarrow(2) \mp \psi_\uparrow(2)\psi_\downarrow(1)] \quad (10.34)$$

The case that we will now consider, which also yields an anti-symmetric total wave function Ψ, is when we take *anti-symmetric spin states* or anti-parallel spins. In this case, the rightmost factor in Eq. (10.34) takes the minus sign. Then, the resulting *symmetric spatial* wave function ψ becomes

$$\psi(\mathbf{r}_1, \mathbf{r}_2) = \psi_a(\mathbf{r}_1)\psi_b(\mathbf{r}_2) + \psi_a(\mathbf{r}_2)\psi_b(\mathbf{r}_1) \quad (10.35)$$

Recall that when we assume the spin eigenfunctions are independent of space, only the spatial part involving ψ_a and ψ_b will play a role in the

10.3 Symmetric Spatial & Anti-symmetric Spin State

energy E_λ of the helium atom described by Eq. (10.9). Substitution of symmetric ψ given by Eq. (10.35) into the electron-electron potential contribution to E_λ, given by Eq. (10.25b), again leads the sum of four terms now given by

$$E_{\text{e-e}} = \frac{e^2}{8\pi\epsilon_0} \int \frac{\psi_a^*(\mathbf{r}_1)\psi_b^*(\mathbf{r}_2)\psi_a(\mathbf{r}_1)\psi_b(\mathbf{r}_2)}{|\mathbf{r}_2 - \mathbf{r}_1|} d\mathbf{r}_1 d\mathbf{r}_2 \; +$$

$$\frac{e^2}{8\pi\epsilon_0} \int \frac{\psi_a^*(\mathbf{r}_1)\psi_b^*(\mathbf{r}_2)\psi_a(\mathbf{r}_2)\psi_b(\mathbf{r}_1)}{|\mathbf{r}_2 - \mathbf{r}_1|} d\mathbf{r}_1 d\mathbf{r}_2 \; +$$

$$\frac{e^2}{8\pi\epsilon_0} \int \frac{\psi_a^*(\mathbf{r}_2)\psi_b^*(\mathbf{r}_1)\psi_a(\mathbf{r}_1)\psi_b(\mathbf{r}_2)}{|\mathbf{r}_2 - \mathbf{r}_1|} d\mathbf{r}_1 d\mathbf{r}_2 \; +$$

$$\frac{e^2}{8\pi\epsilon_0} \int \frac{\psi_a^*(\mathbf{r}_2)\psi_b^*(\mathbf{r}_1)\psi_a(\mathbf{r}_2)\psi_b(\mathbf{r}_1)}{|\mathbf{r}_2 - \mathbf{r}_1|} d\mathbf{r}_1 d\mathbf{r}_2 \quad (10.36)$$

This result is almost identical to Eq. (10.26) except that the two middle terms are positive in Eq. (10.36), whereas they are negative in Eq. (10.26). The two middle terms make up the *exchange integral* J_{ij} given by Eq. (10.31). Thus, the energy in the case of symmetric spatial+anti-symmetric spin becomes

$$E_{\text{asym spin}} = C_{12} + C_{21} + J_{12} \quad (10.37)$$

C_{ij} are the Coulomb integrals given by the first and last terms of Eq. (10.36). We can determine the difference in energy due solely to different spin configurations. The difference in energy $E_{\text{sym spin}} - E_{\text{asym spin}}$ (all other things remaining the same), between symmetric and anti-symmetric spin states then becomes

$$E_{\text{sym spin}} - E_{\text{asym spin}} = \Delta E_\lambda =$$

$$(C_{12} + C_{21} - J_{12}) - (C_{12} + C_{21} + J_{12}) = -2J_{ij} \quad (10.38)$$

The sign of the energy splitting given by Eq. (10.38) *depends* on the sign of $J_{ij} = J$. Nothing in our analysis restricts the sign of J_{ij}, especially considering that electron eigenfunctions *can* be either positive or negative in sign over the range of space. This statement becomes more true for higher and higher energy states. For the case of helium having two electrons in *similar* orbitals, the wave functions are real and simple enough to approximate the exchange integral given by Eq. (10.33).

10.3.1 Estimation Of J For Electrons In 1s Orbitals

An approximation can be made by taking the two electronic states a and b to be given by s-orbitals, which are defined by $\ell = 0$. They represent the lowest energy states we found for the single electron hydrogen atom. If we take $1s$-orbitals which is the case for the ground state of helium, the normalized single electron orbital state is

$$\psi_{1s}(\mathbf{r}_j) = \left(\frac{Z^3}{a_0^3 \pi}\right)^{1/2} e^{-Zr_j/a_0} \quad \text{where} \quad a_0 = \frac{h^2 \epsilon_0}{m_e \pi e^2} \qquad (10.39)$$

Given that the wave functions generally depend on the energy, self-consistent numerical methods become necessary to compute the energy and wave function simultaneously. However, approximations can be made to get a sense for the energy contribution. The helium atom has been studied extensively and the numerical results that align with experiments with helium reveal that the form of the helium wave function, as well as lithium (Li) and beryllium (Be) wave functions, can take a form of linear combinations of s–orbitals like that given by Eq. (10.39). Let the wave functions ψ_a and $\psi_b = \psi_{1s}$. We'll see that it also leads to the result $J > 0$. Substitution of Eq. (10.39) into Eq. (10.33) *using spherical coordinates* gives

$$J_{12} = \frac{e^2}{4\pi\epsilon_0} \int_V r_1^2 r_2^2 \frac{\psi_{1s}(\mathbf{r}_1)\psi_{1s}(\mathbf{r}_2)\psi_{1s}(\mathbf{r}_2)\psi_{1s}(\mathbf{r}_1)}{|\mathbf{r}_2 - \mathbf{r}_1|} dV$$

$$= \frac{Z^6 e^2}{4\pi^3 a_0^6 \epsilon_0} \int_0^\infty \int_0^\infty \int_0^{2\pi} \int_{\theta=0}^{\pi} r_1^2 r_2^2 \sin\theta \frac{e^{-2Zr_1/a_0} e^{-2Zr_2/a_0}}{|\mathbf{r}_2 - \mathbf{r}_1|} dr_1 dr_2 d\theta d\varphi$$

$$= \frac{Z^6 e^2}{2\pi^2 a_0^6 \epsilon_0} \int_0^\infty \int_0^\infty \int_0^\pi r_1^2 r_2^2 \sin\theta \frac{e^{-2Zr_1/a_0} e^{-2Zr_2/a_0}}{|\mathbf{r}_2 - \mathbf{r}_1|} dr_1 dr_2 d\theta$$

To evaluate the integral, we'll make an additional assumption that the angle between $\mathbf{r}_2$ and $\mathbf{r}_1$ is the polar angle θ. This just means we take $\mathbf{r}_1$ to be along the z axis. Then, we can use the law of cosines to replace $|\mathbf{r}_2 - \mathbf{r}_1|$ with

$$|\mathbf{r}_2 - \mathbf{r}_1| = \left(r_1^2 + r_2^2 - 2r_1 r_2 \cos\theta\right)^{1/2} \qquad (10.40)$$

Since θ in Eq. (10.40) is also the polar angle in our exchange integral, the integral with respect to θ becomes

$$I_\theta = \int_0^\pi \frac{\sin\theta d\theta}{\left(r_1^2 + r_2^2 - 2r_1 r_2 \cos\theta\right)^{1/2}}$$

10.3 Symmetric Spatial & Anti-symmetric Spin State

$$= -\int_{-1}^{1} \frac{dx}{\left(r_1^2 + r_2^2 - 2r_1r_2x\right)^{1/2}}$$

Using the integral formula $\int \left(\sqrt{ax+b}\right)^{-1} dx = 2a^{-1}\sqrt{ax+b}$, we obtain

$$I_\theta = -\frac{1}{2r_1r_2} 2\sqrt{r_1^2 + r_2^2 - 2r_1r_2x}\Big|_{-1}^{1}$$

$$= -\frac{1}{r_1r_2}\left[\sqrt{r_1^2 + r_2^2 - 2r_1r_2} - \sqrt{r_1^2 + r_2^2 + 2r_1r_2}\right]$$

$$= -\frac{1}{r_1r_2}\left(\sqrt{(r_1-r_2)^2} - \sqrt{(r_1+r_2)^2}\right)$$

From the last line, we recognize that there are two solutions implied from the first radical on the RHS being either $\sqrt{(r_1-r_2)^2}$ or $\sqrt{(r_2-r_1)^2}$. If $r_1 > r_2 \to r_1 - r_2$, then we have

$$I_\theta = \frac{2}{r_1} \quad \text{if} \quad r_1 > r_2 \qquad (10.41)$$

But, if $r_2 > r_1 \to r_2 - r_1$, we have

$$I_\theta = \frac{2}{r_2} \quad \text{if} \quad r_2 > r_1 \qquad (10.42)$$

Both conditional results can be used in the evaluation of the exchange integral. This is done by splitting up the integral for r_2 into two parts. With this and substitution, we have

$$J_{12} = \frac{Z^6 e^2}{\pi^2 a_0^6 \epsilon_0} \int_0^\infty dr_1 r_1^2 e^{-2Zr_1/a_0} \times$$

$$\left[\frac{1}{r_1}\int_0^{r_1} r_2^2 e^{-2Zr_2/a_0} dr_2 + \int_{r_1}^\infty r_2 e^{-2Zr_2/a_0} dr_2\right] \qquad (10.43)$$

There are two integral forms we need to solve using the integral relations

$$\int xe^{ax} dx = \left(\frac{x}{a} - \frac{1}{a^2}\right)e^{ax} \qquad (10.44a)$$

$$\int x^2 e^{ax} dx = \left(\frac{x^2}{a} - \frac{2x}{a^2} + \frac{2}{a^3}\right)e^{ax} \qquad (10.44b)$$

With $a = -2Z/a_0$, after some algebra, the exchange integral in Eq. (10.43) becomes

$$J_{12} = \frac{Z^3 e^2}{4\pi^2 a_0^3 \epsilon_0} \int_0^\infty dr_1 r_1^2 e^{-2Zr_1/a_0} \times \left[\frac{1}{r_1} - e^{-2Zr_1/a_0}\left(\frac{1}{r_1} + \frac{Z}{a_0}\right)\right] \quad (10.45)$$

Evaluation of Eq. (10.45) using, again, the integral relations in Eqs. (10.44), one obtains the result

$$J_{12} = \frac{5Ze^2}{32\pi\epsilon_0 a_0} = 34.01 \text{eV} \quad (10.46)$$

This turns out to be a relatively large energy between electrons in the 1s orbital. This energy must be added to the kinetic+potential energy for two electrons, which is negative for bound electrons. Based on our formulation, this is just twice the result for a single electron as we found with the hydrogen atom (c.Eq. 8.50), with $Z = 2$. So we have

$$E_\lambda = E_0 + J_{12} = -2 \times \frac{Z^2 me^4}{32\pi^2 \epsilon_0^2 \hbar^2} + \frac{5Ze^2}{32\pi\epsilon_0 a_0} = -74.83 \text{eV} \quad (10.47)$$

This can be compared to the experimental ground state energy of He, which is $E_\lambda = -79$ eV, which is approximately 5% error.

If $J > 0$, according to Eq. (10.38), in comparing helium atoms in the same state where the only difference is whether spin is symmetric or anti-symmetric, the lower energy state between them should be $E_{\text{sym spin}}$, or the state with parallel spins and anti-symmetric spatial part. This is indeed the experimental observations with helium. Let's take a look at some experimental spectral data of helium, and discuss how it aligns with the above results.

10.4 He Spectra Supports Anti-symmetric Wave Functions

We have seen that atomic spectra of atoms have been useful in the development of quantum mechanics. In practice, without intentional modifications, most elements used in experiments tend to have multiple electrons. The simplest of multi-electron atoms is the element helium, which has two electrons ($Z = 2$) and a nucleus composed of two protons and two neutrons with charge $Z|e|$. Helium spectra turns out to provide

10.4 He Spectra Supports Asymmetry

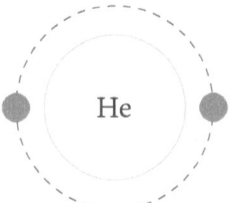

Figure 10.3: Illustration of helium atom, with nucleus, and two orbiting electrons.

supporting evidence of the electron-electron interaction discussed above, being also dependent on the spin-spin interaction. Table 10.1 lists experimental spectral data from helium, showing the ground electron state, the excited electron state, and the observed quantum energies E_λ for both asymmetric and symmetric spin states. Fig. 10.4, illustrates the same states of helium, as a function of the orbital angular momentum quantum number ℓ, to further illustrate the observed transitions.

In the helium spectra, some important features can be observed. If we take the first electron to be in the ground state, or the $1s$ state, then the second electron must have a higher energy (as a fermion). In this scenario, the second electron can have either spin anti-parallel or parallel. These spin-based distinctions for a two-electron system generally define *two sets of two-electron states*. For any multi-electron system, the total spin is given by the sum of the individual electron spins. For two electrons $S = S_1 + S_2$, which is appropriate for a single projection axis for the spin. When the two-electron state has zero total spin angular momentum $S = 0$, it is known as the **parahelium** state. From the analysis above, this type includes the ground state or lowest energy state, which has both electrons in the $1s$ state. This ground state (denoted by 1s) is shown in Fig. 10.4 near $E \sim -79\text{eV}$. For helium having two electrons and four basis or eigenstates, only one state is a *parahelium* state and therefore, it is also known as a **singlet state**.

Table 10.1: Helium experimental states from atomic spectra. The corresponding energies for anti-parallel spin states (parahelium) and parallel spin states (orthohelium) are shown. The difference is always negative (fifth column), in accordance with Eq. (10.38).

Excited Electron	E_λ [eV](asym)	E_λ [eV](sym)	ΔE_λ

Table 10.1

1s	-79.0137	-	-
2s	-58.3625	-59.1581	-0.7956
3s	-56.0579	-56.2605	-0.2026
4s	-55.2491	-55.3634	-0.1143
2p	-57.7604	-58.0144	-0.2540
3p	-55.8915	-55.9717	-0.0802
4p	-55.1928	-55.2361	-0.0433
3d	-55.9043	-55.9048	-0.0005
4d	-55.1855	-55.1880	-0.0025

Fig. 10.4 illustrates these states in relation to one another, and the orbital angular momentum values.

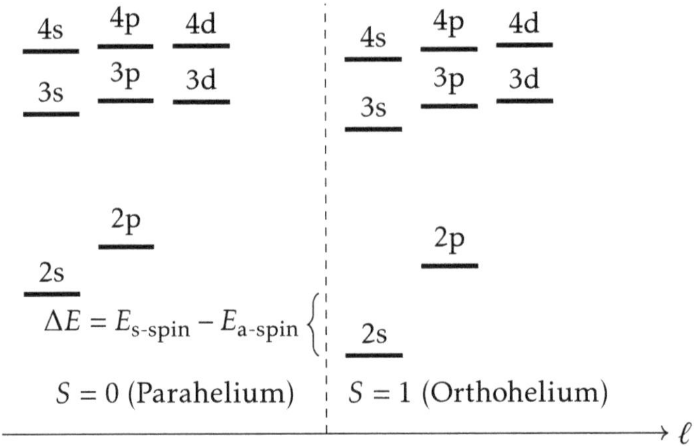

Figure 10.4: Illustration of helium spectra levels as a function of the associated orbital angular momentum state $\ell = 0, 1, 2$ of the excited electron, which is the same as that of the pair if the ground state electron remains in the 1s orbital, which has $\ell = 0$. $E_{\text{s-spin}}$ =symmetric spin state and $E_{\text{a-spin}}$ =anti-symmetric spin state.

If the second electron's spin is parallel to the ground state electron spin, then $S = S_1 + S_2 = 1$, and this set of two-electron states is known as **orthohelium** states. It turns out that the other three eigenstates are of this type. Since three out of the four are in this state, it is called a **triplet state**. We'll discuss these kinds of states more formally later

10.4 He Spectra Supports Asymmetry

in the chapter when we discuss Clebsch-Gordan coefficients. Upon exciting helium to higher states, and separating these two sets of states, the *parahelium* (anti-parallel spins $S = 0$) from the *orthohelium* (parallel spins), as illustrated in Fig. 10.4, one observes that corresponding states have lower energy levels in the orthohelium states compared to parahelium states. For example, the 2s orbital is observed to be lower in energy (more negative) compared to the corresponding 2s orbital in the parahelium states. Similarly, the 2p orthohelium state lies lower in energy compared to the 2p parahelium state. This subtle trend is a telling observation in the helium spectra. Let's think about what this means. Obviously, the two corresponding 2s, 2p, etc. have identical quantum numbers, except m_s! This means that this observed difference not only involves m_s, but involves whether the two electrons have parallel or antiparallel spins. This data is, therefore, some of the first evidence of a **spin-spin interaction**.

Recall that for one electron, there are exactly two possible states. Namely, spin-up and spin-down. With two electrons, there are four possible state configurations for the spins. Lets denote spin-up state with ψ_+ and spin-down with ψ_-, and an example of a two-electron state as $\psi_{\uparrow\uparrow}$. We can express the four states as

$$\psi_{\uparrow\uparrow}, \quad \psi_{\downarrow\downarrow}, \quad \psi_{\uparrow\downarrow}, \quad \text{and} \quad \psi_{\downarrow\uparrow}$$

Keeping the idea of 2×2 matrices, as we used for spin matrices, rather than constructing a vector of four states, we can instead squeeze the four states into a 2×2 matrix of four elements $\boldsymbol{\psi}$, as

$$\boldsymbol{\psi}(x,y,z,t,\sigma) = \begin{bmatrix} \psi_{\uparrow\uparrow} & \psi_{\uparrow\downarrow} \\ \psi_{\downarrow\uparrow} & \psi_{\downarrow\downarrow} \end{bmatrix} \tag{10.48}$$

When we considered Pauli spin matrices for a single spin, with the 2×1 state vectors for ψ_+ and ψ_-, any *spinor* could be expressed as a linear combination of the two eigenvectors of the spin operator σ_i. In this case of two electrons, with a 2×2 state matrix, we therefore, need four matrices to form a basis for $\boldsymbol{\psi}$ (there are four independent elements).

In the helium energy level diagram, one electron is presumed to be in the ground state of a helium atom, the 1s state. An electron in an upper state can have spin antiparallel to the ground state electron (S=0, singlet state, parahelium) or parallel to the ground state electron (S=1, triplet state, orthohelium).

One finds that the *orthohelium* states are lower in energy than the *parahelium* states. The explanation for this is as follows: The parallel spins make the spin part of the wavefunction symmetric. The total wavefunction for the electrons must be anti-symmetric since they are fermions and must obey the Pauli exclusion principle. This forces the space part of the wavefunction to be anti-symmetric. An anti-symmetric spatial wavefunction for the two electrons implies a larger average distance between them than a symmetric function of the same type. This lowers the Coulomb repulsion term, which is positive, between the two electrons sharing a single nucleus. Alternatively, if the electrons are, on the average, further apart, then there will also be less shielding/screening of the nucleus by the other electron. Then, each electron will be more exposed to the nucleus and, thus, more tightly bound and of lower energy. This effect is an example of the so-called the spin-spin interaction and is addressed by Hunds Rule 1. It is part of the ordering of energy levels in multi-electron atoms.

The helium ground state consists of two identical 1s electrons. The energy required to remove one of them is the highest ionization energy of any atom in the periodic table: $\approx$24 eV. The energy required to remove the second electron is 54.4 eV, as would be expected by modeling it after the hydrogen energy levels. The He+ ion is just like a hydrogen atom with two units of charge in the nucleus. Since the hydrogenic energy levels depend upon the square of the nuclear charge, the energy of the remaining helium electron should be just 4x(-13.6 eV) = -54.4 eV as observed. The fact that the second electron is less tightly bound can be interpreted as a shielding effect; the other electron partly shields the second electron from the full charge of the nucleus. Another way to view the energy is to say that the repulsion of the electrons contributes a positive potential energy which partially offsets the negative potential energy contributed by the attractive electric force of the nuclear charge. The description of any electron in a multi-electron atom must find a way to characterize the effect of the other electrons on the energy. If we generalize this idea further, we can begin to understand how the energy of a two-electron system behaves for two-electrons *residing on two different atoms*. There are additional consequences for such a system. This brings us to the topic of molecules, which is the subject of the next section.

10.5 Molecular Wave Functions And Bonds

In the above, we have alluded to two-electron systems where each electron also resides on it's own respective atom. We know that most matter forms by the bonding of atoms with each other, in various forms to create more complex forms of matter. For example, we know air is made of elements such as nitrogen, oxygen, argon, carbon-dioxide, etc. These gases are found in air as diatomic molecules or as N_2, O_2, etc. Just as electrons, protons, and neutrons join to form an atom, they too can join to form more complex structures. Beyond the atom as a unit structure, the next simplest structure turns out to be the *molecule*. A great deal of insight into higher forms of matter can be gleaned from analyzing molecules in quantum mechanics. Specifically, we may ask *why does nature sometimes prefer molecules*? One of the first insightful answers to this question was provided by quantum mechanics. In a similar manner

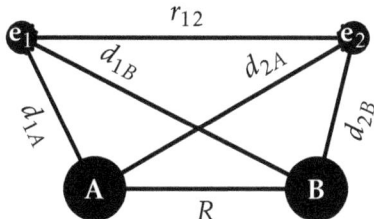

Figure 10.5: Illustration of two single-electron atoms with relevant position vectors used in Eq (10.49).

to how we represented the Hamiltonian for one atom with two electrons for helium, extending it to two atoms A and B with equal masses, each having electron 1 and 2, respectively, the two-nucleus/two-electron Hamiltonian becomes

$$\widehat{H} = -\frac{\hbar^2}{2M}\left[\nabla_A^2 + \nabla_B^2\right]\psi - \frac{\hbar^2}{2m^*}\left[\nabla_1^2 + \nabla_2^2\right]\psi +$$

$$\frac{e^2}{4\pi\epsilon_0}\left[-\frac{1}{d_{1A}} - \frac{1}{d_{1B}} - \frac{1}{d_{2A}} - \frac{1}{d_{2B}} + \frac{1}{r_{12}} + \frac{1}{R}\right]\psi \quad (10.49)$$

As we had with the helium atom, M is the mass of the nucleus; m^* is the reduced mass of the electron with respect to the nucleus; d_{ij} is the distance between electron i and nucleus j. r_{ij} is the interelectron distance between electron i and j; and R is the internuclear separation, which is altogether new for the molecular Hamiltonian. The wave function is a function of the four variables r_A, r_B, r_1, and r_2 (corresponding vectors from the origin). As far as separating Eq. (10.49) into a nuclear and electronic Hamiltonian, because of the presence of several terms nonlinearly coupling r_A, r_B with r_1, r_2, this is not justified. Thus, as we had to do in the case of the helium atom, we must, again, make some approximations to proceed in analyzing this problem. For starters, if we take both nuclei A and B to be *approximately fixed in location*, with only electrons changing position, we can take the first two terms for the kinetic energy of the nuclei to be negligible. The reader should be aware that this assumption is not readily justifiable because although one could argue that $1/M \ll 1/m^*$, we still have the Laplacians of the wave function with respect to r_A and r_B, which may also be large (given the high localization), so the ratios of Laplacian to mass may not possess the same relation. Nonetheless, this approximation gives a useful path to analyze the problem further. It can be considered not a full, but an *approximate separation* of r_A, r_B from r_1, r_2 or nuclei from electrons, and Eq. (10.49) becomes

$$\widehat{H} \simeq -\frac{\hbar^2}{2m^*}\left[\nabla_1^2 + \nabla_2^2\right]\psi + \frac{e^2}{4\pi\epsilon_0}\left[-\frac{1}{d_{1A}} - \frac{1}{d_{1B}} - \frac{1}{d_{2A}} - \frac{1}{d_{2B}} + \frac{1}{r_{12}} + \frac{1}{R}\right]\psi$$
(10.50)

This approximation of ignoring the nuclear kinetic energies in a molecular Hamiltonian is known as the **Born-Oppenheimer approximation**, named after Max Born and J. Robert Oppenheimer, who proposed it around 1927. Keeping r_A and r_B fixed significantly simplifies calculations of the energy. Also, in the limit of $R = |\mathbf{r}_A - \mathbf{r}_B| \to \infty$, the Hamiltonian in Eq. (10.50) simplifies to that of two separate hydrogen atoms, because we then have

$$\lim_{R\to\infty}\widehat{H} = -\frac{\hbar^2}{2m^*}\left[\nabla_1^2 + \nabla_2^2\right]\psi + \frac{e^2}{4\pi\epsilon_0}\left[-\frac{1}{d_{1A}} - \frac{1}{d_{2B}}\right]\psi$$

$$= \left[\widehat{H}_1 + \widehat{H}_2\right]\psi$$

In this limit, all potential terms depending inversely on the distance between the atoms, namely, $1/d_{1B}, 1/d_{2A}, 1/r_{12}$, and $1/R$ all vanish. Because of the nonlinear nature of the Hamiltonian, calculation of the eigenvalues for the H_2 molecule requires more advanced numerical computational techniques. This has been carried out by several research groups by assuming a symmetric spatial trial wave function ψ_T of generalized s–orbitals depending on various parameters. The parametric energy is then effectively minimized with respect to those parameters, as

10.5 Molecular Wave Functions And Bonds

in most common methods of optimization. An example of a rigorously computed ground-state energy for the H_2 molecule is shown in Fig. 10.6. The data shown in Fig. 10.6 has been adapted from the calculations of

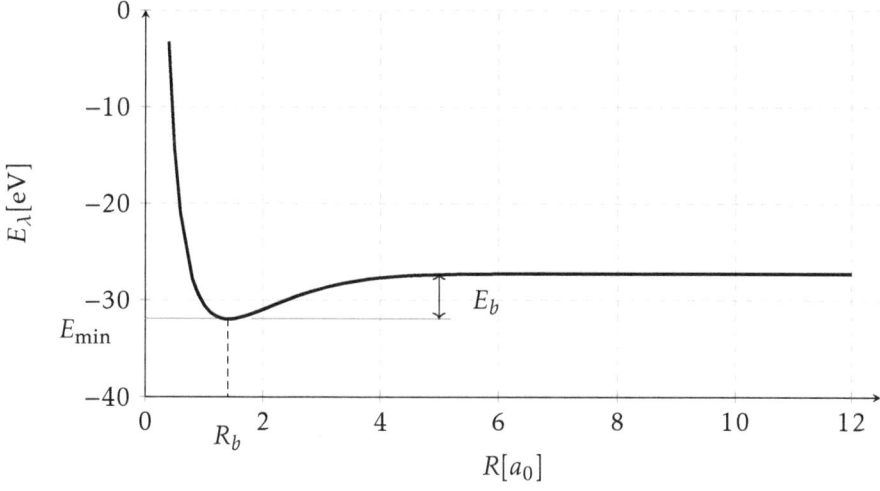

Figure 10.6: Computed ground state energy for H_2 molecule as a function of the internuclear distance R. This data has been adapted from L. Wolniewicz 99 (3), J. Chem. Phys. (1993). Since the ground state energy of a hydroden atom is ≈ -13.6 eV, $\lim_{R \to \infty} E_\lambda(R) = 2 \times -13.6 = -27.2$ eV. The resulting bond length $R_b = 1.4 a_0$ is in excellent agreement with experiments. E_b is the bonding/binding energy of the molecule given by $E_b = 31.94 - 27.2 = 4.74$ eV.

L. Wolniewicz 99 (3), J. Chem. Phys. (1993). The energy has a single minimum value, as a function of the internuclear separation R, implying a unique optimally stable internuclear separation. This optimal nuclear separation defines what is known as the **bond length** of a molecule, and for H_2, it is found to be $R_b = 1.4 a_0 = .74$Å, in good agreement with experimental measurements. Recall that we found the ground state energy of a single hydrogen atom to be ≈ -13.6 eV. The limit of the H_2 molecule in infinite separate should approach two individual hydrogen atoms. And from the calculations, it is seen that $\lim_{R \to \infty} E_\lambda(R) = 2 \times -13.6 = -27.2$ eV, as expected. E_b is the **bonding/binding energy** of the molecule where $E_b = 31.94 - 27.2 = 4.74$ eV. It is defined as the energy difference between two separated atoms and the energy minimum. Another significant observation from solving the Hamiltonian of the hydrogen molecule is that the absolute value of the molecule's energy $E_{min} = 31.94$ is larger than that of two separated hydrogen molecules. Thus, quantum mechanics reveals that there is more energy available from the combining of the two equivalent atoms, than from them remaining separate, both surprising as well as poetic. Hence, the energy is further reduced through the new potential terms of the molecular Hamiltonian.

In this section, we may have gotten a little ahead of ourselves, because the ability to compute the results shown in Fig. 10.6 did not come until the mid-1960s with the availability of computers. However, our understanding of *bond lengths* of molecules started c.1926, through the use of a further simplified hydrogen molecule and Hamiltonian. Around the same time, experimental methods were developed to generate this simpler hydrogen molecule with a missing electron. This simpler molecule turns out to solvable exactly using quantum mechanics. It was this simplified ionic hydrogen molecule that first gave insight into the nature of how molecules form bonds. This simplified molecule is the focus of the next section.

10.6 THE SIMPLEST DIATOMIC MOLECULE: H_2^+

Recall that the two-nucleus/two-electron Hamiltonian is given by

$$\widehat{H} = -\frac{\hbar^2}{2M}\left[\nabla_A^2\psi + \nabla_B^2\right]\psi - \frac{\hbar^2}{2m^*}\left[\nabla_1^2 + \nabla_2^2\right]\psi + \frac{e^2}{4\pi\epsilon_0}\left[-\frac{1}{d_{1A}} - \frac{1}{d_{1B}} - \frac{1}{d_{2A}} - \frac{1}{d_{2B}} + \frac{1}{r_{12}} + \frac{1}{R}\right]\psi$$

M is the mass of the nucleus, taken to be identical for both atoms; m^* is the reduced mass of the electron with respect to the nucleus; d_{ij} is the distance between electron i and nucleus j. r_{ij} is the interelectron distance between electron i and j; and R is the internuclear separation. The last term, which is a combination of attractive and repulsive forces, complicates the Hamiltonian, making an exact solution nontrivial.

The simplest diatomic molecule that can be generated in a laboratory uses hydrogen gas and exposes the gas to electromagnetic radiation and/or electric fields, which serves to free one of the electrons, being in a higher state. This molecule is known as the **hydrogen molecular ion**, denoted by H_2^+. With two nuclei and only one electron, the Hamiltonian simplifies considerably. Using the Born-Oppenheimer approximation, which allows us to ignore the kinetic energy terms of the nuclei, the

10.6 The Simplest Diatomic Molecule: H_2^+

Hamiltonian for H_2^+ becomes

$$\widehat{H} = -\frac{\hbar^2}{2m^*}\nabla^2\psi + \frac{e^2}{4\pi\epsilon_0}\left[-\frac{1}{d_A} - \frac{1}{d_B} + \frac{1}{R}\right]\psi \qquad (10.51)$$

It turns out that this system can be solved exactly given the proper choice of coordinate system, if we assume for any state that R is a parameter independent of d_A, d_B, etc. Then, only the electron energy is needed and the $1/R$ terms is just a parametric offset. The electronic Hamiltonian becomes

$$\widehat{H}_e = -\frac{\hbar^2}{2m^*}\nabla^2\psi + \frac{e^2}{4\pi\epsilon_0}\left[-\frac{1}{d_A} - \frac{1}{d_B}\right]\psi = E\psi \qquad (10.52)$$

The total eigenenergy E_λ is then given by

$$E_\lambda = E + \frac{e^2}{4\pi\epsilon_0 R} \qquad (10.53)$$

In Chap. 8, when deriving the Bohr model for the hydrogen atom, we encountered a unit of length a_0 known as the *Bohr radius*, given by

$$a_0 = \frac{4\pi\hbar^2\epsilon_0}{me^2} \approx 0.52919\text{Å} \qquad (10.54)$$

If we multiply Eq. (10.52) by $m^*/\hbar^2$, it becomes

$$\frac{2m^*}{\hbar^2}\widehat{H}_e = -\frac{1}{2}\nabla^2\psi + \frac{m^*e^2}{4\pi\epsilon_0\hbar^2}\left[-\frac{1}{d_A} - \frac{1}{d_B}\right]\psi \qquad (10.55)$$

$$= -\frac{1}{2}\nabla^2\psi + a_0\left[-\frac{1}{d_A} - \frac{1}{d_B}\right]\psi \qquad (10.56)$$

Then, we may define dimensionless parameters $d_i^* = d_i/a_0$, so we have

$$\frac{2m^*}{\hbar^2}\widehat{H}_e = -\frac{1}{2}\nabla^2\psi + \left[-\frac{1}{d_A^*} - \frac{1}{d_B^*}\right]\psi \qquad (10.57)$$

This form of normalization of the Schrödinger equation is known as **atomic units** (a.u.), and is commonly used. Eq. (10.52) is first expressed in bipolar coordinates (ρ, z, φ), relating to d_A, d_B, and R as follows:

$$d_A^2 = \rho^2 + z^2 \qquad (10.58a)$$

$$d_B^2 = \rho^2 + [R-z]^2 \qquad (10.58b)$$

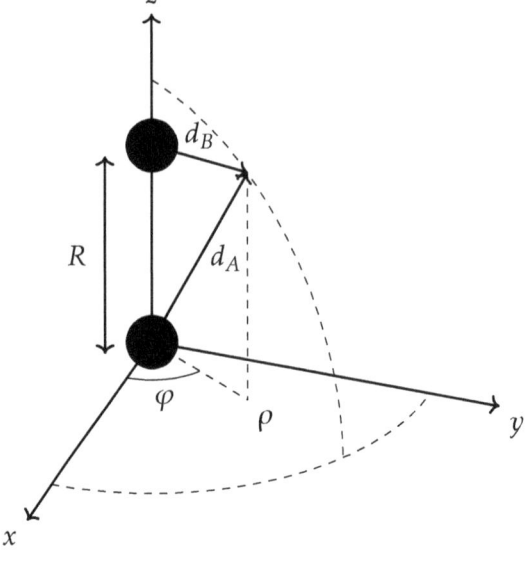

Figure 10.7: Bipolar coordinates for the two nuclei system. The origin is located at nuclei A; the z axis is along the internuclear axis; and the electron location is given by (ρ, z, ϕ).

In the (x, y, z) frame $\mathbf{d}_A = [x, y, z]^T$ and $\mathbf{d}_B = \mathbf{d}_A - \mathbf{R}$. Fig. 10.7 illustrates the bipolar coordinate system used for the H_2^+ molecule. Substitution of Eqs. (10.58a) and (10.58b) into the Hamiltonian of Eq. (10.51), and expressing the Laplacian in corresponding cylindrical coordinates leads to a Hamiltonian in bipolar coordinates (ρ, z, φ). Bipolar coordinates, however, turn out not to be so useful because the Schrödinger equation is still not separable in this form. Around 1927, it was shown that the electronic wave equation (ignoring the $1/R$ term) *separates* in elliptic coordinates. Such a coordinate system transforms Eq. (10.51) into one that is separable in three variables. Fig. 10.8 illustrates the elliptical coordinate system for H_2^+. The origin of the elliptical coordinate system is the center point between both nuclei A and B. The nuclei of the two atoms are located at the foci of the elliptical coordinate system (of both the hyperbola and ellipse). The azimuthal angle φ, around the z axis, is preserved in the transformation. Thus, only ρ and z are replaced. The two variable transformations are defined by

$$\xi = \frac{d_A + d_B}{R} \tag{10.59a}$$

$$\eta = \frac{d_A - d_B}{R} \tag{10.59b}$$

10.6 The Simplest Diatomic Molecule: H_2^+

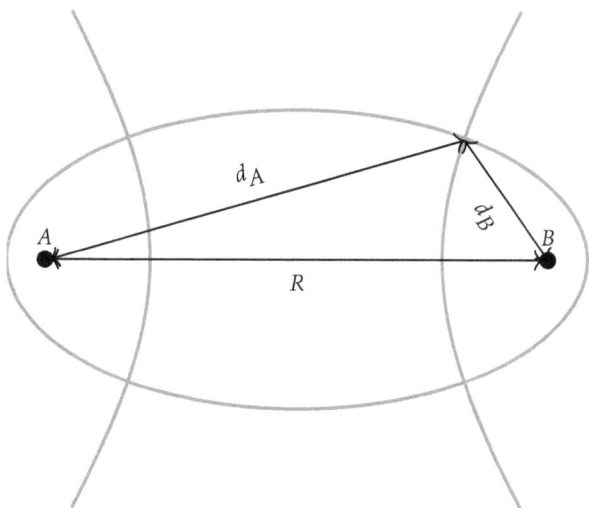

Figure 10.8: Illustration of the elliptical coordinate system that allows the Hamiltonian for the simplified molecule H_2^+ to be solved exactly. Whenever a hyperbola and ellipse have the same foci, they are orthogonal. The H_2^+ nuclei correspond to foci A and B.

After substitution into the Hamiltonian of bipolar coordinates, along with some manipulation of the equation, one obtains

$$\frac{\partial}{\partial \xi}\left[(\xi^2-1)\frac{\partial}{\partial \xi}\right]\psi + \frac{\partial}{\partial \eta}\left[(1-\eta^2)\frac{\partial}{\partial \xi}\right]\psi + 2R\xi + \frac{R^2}{2}E(\xi^2-\eta^2)+$$

$$\left[-\frac{1}{\xi^2-1}+\frac{1}{1-\eta^2}\right]\frac{\partial^2}{\partial \varphi^2}\psi = 0 \quad (10.60)$$

This form of $\widehat{H}$ turns out to be separable in $\xi, \eta,$ and φ. Letting $\psi(\xi,\eta,\varphi) = X(\xi)Y(\eta)\Phi(\varphi)$, the following three eigenvalue equations can be obtained:

$$\left[\frac{\partial}{\partial \xi}(\xi^2-1)\frac{\partial}{\partial \xi}+\left(\frac{ER^2\xi^2}{2}+2R\xi-\frac{m^2}{\xi^2-1}\right)\right]X(\xi) = -\lambda X(\xi)$$

(10.61a)

$$\left[\frac{\partial}{\partial \eta}(1-\eta^2)\frac{\partial}{\partial \eta} + \left(-\frac{ER^2\eta^2}{2}\eta^2 - \frac{m^2}{1-\eta^2}\right)\right]Y(\eta) = \lambda Y(\eta) \quad (10.61\text{b})$$

$$\frac{d^2\Phi}{d\varphi^2} = -m^2\Phi \quad (10.61\text{c})$$

Eq. (10.61c) is familiar because it is identical to the $\widehat{L}_z^2$ eigenvalue equation we obtained in Eq. (7.50). m must be an integer $\ldots, -2, -1, 0, 1, 2, \ldots$ in order to satisfy the boundary condition, or continuity of the wave function $\Phi = 1/\sqrt{2\pi} e^{im\varphi}$ as it revolves about the polar axis. For the other two equations, the solutions can be found by using the *Frobenius method*, but assuming the solutions can be expressed as a linear combination of associated Legendre polynomials $A_\ell^{|m|}$ for Y and Laguerre functions for X. For example, in one of first successful efforts to solve Eq. (10.61b), $Y(\eta)$ takes the form

$$Y(\eta) = \sum_{\ell=|m|}^{\infty} c_\ell A_\ell^{|m|} \quad (10.62)$$

Because the process is relatively involved, we will not solve the equation here. A nice review of the original solution from Hylleraas (1931) can be found in Boyack (2004). We'll only summarize the steps here for $Y(\eta)$. The process is the same for $X(\xi)$. First, substitute Eq. (10.62) into the eigenequation given by Eq. (10.61b). This leads to a recurrence relation for c_ℓ of the form

$$f_{\ell-2}(\ell, |m|, E)c_{\ell-2} + f_\ell(\ell, |m|, E)c_\ell + f_{\ell+2}(\ell, |m|, E)c_{\ell+2} = 0 \quad (10.63)$$

One difficulty with Eq. (10.63) is that we don't know E, even if we march through values of ℓ and m, which is ultimately, one of the aims of the analysis. For all coefficients c_ℓ, Eq. (10.63) helps to define a *trigonal* matrix equation of the form

$$[\mathbf{F}_T]\mathbf{c}_\ell = 0 \quad (10.64)$$

$\mathbf{c}_\ell$ is the vector of coefficients c_ℓ. Because of the form of the recurrence relation, the matrix has diagonal elements corresponding to ℓ, with at most two other nonzero columns for $\ell - 2$ and $\ell + 2$, respectively. Like we found in all previous cases using the Frobenius method, this form of recurrence relation also means that only $c_0, c_2, c_4, etc.$ couple together and likewise for $c_1, c_3, c_5, etc.$ For this reason, there is an even and odd version of Eq. (10.64). Since the matrix $\mathbf{F}_T$ is trigonal, or only has 3

10.6 The Simplest Diatomic Molecule: H_2^+

consecutive elements per row centered on the diagonal, it can be written as

$$[F_T] = [L][U]$$

As we know from linear algebra, such a system can readily be solved. In both cases, since the Associated Legendre functions $A_\ell^{|m|}$ are orthogonal, the coefficients c_ℓ must be nonzero. Thus, for Eq. (10.64) to be valid, the determinant of F_T vanish. This provides the relationship between the eigenvalue λ and the electronic energy E. Therefore, the energy states can be found from the determinants. Then, c_ℓ can be evaluated to express the eigensolution given by Eq. (10.62). The resulting energy as a function of R follows. Fig. 10.9 shows the first two states of H_2^+ as functions of R. The black line in Fig. 10.9 defines the *ground state*

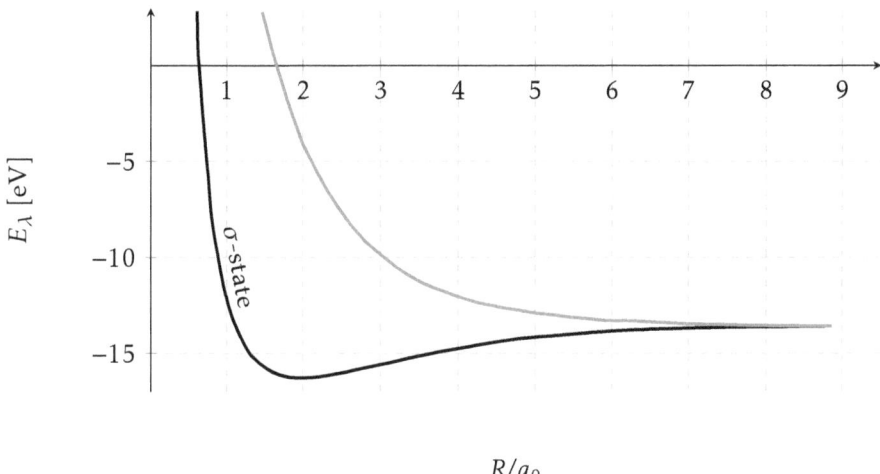

Figure 10.9: The energy E_λ of the first states for the H_2^+ molecule as a function of the internuclear separation R. The ground state evidences a unique minimum $E_{min} = -16.3988$ eV, corresponding to the bond length $R_b = 2a_0$.

of the H_2^+ molecule, as no other state has energy below it. Like we found with H_2, the ground state also has a unique minimum defining the stable ground energy minimum (value from y-axis) $E_{min} = -16.3988$ eV, as well as bond-length $R_b = 2a_0 = 1.0584$Å (value of x-axis). This is the outline for the treatment of the H_2^+, which, even as the simplest molecule, demonstrates a fundamental features of some molecules. In particular, we see the origins of some molecular bonds arising from the balance between electrostatic potential energies of repulsive and attractive types.

436   Chapter 10. Multi-electron Systems

The naming convention for states of molecules parallels the convention for orbitals of atoms. Particularly, the way ℓ values define a set of orbitals s, p, d, f, etc. For molecules, the so-called **molecular orbitals** follow the same convention with the corresponding Greek letters. For example, $\ell = 0$ corresponds to σ orbitals. $\ell = 1$ defines π molecular orbitals, $\ell = 2$ defines δ orbitals, etc. Understanding the physics of molecules improves our ability to measure parameters involved in the physics, such as the bond length. In the next section, we'll discuss a common method used to measure bond length R_b, most applicable to diatomic molecules.

10.7 Vibrational-Rotational Spectra Of Molecules

For over a century, spectral measurements of gas phases have provided a wealth of information relating to the energies of these systems. Quantum mechanics has helped to improve the physical description of these systems and their energies. The two together can, therefore, provide access to parameter values on which the energy depends. One example of this is with the *molecular bond length R_b*, introduced earlier when we discussed the treatment of the hydrogen molecule H_2 and its ion with the Schrödinger equation. We found that solving the Schrödinger equation for the ground state leads to an energy with a unique minimum at R_b, resulting from an energy balance between the electron-electron Coulomb repulsion and attraction. When we treated the lone hydrogen atom, we obtained a radial equation, from which we were able to deduce the eigenenergies of the hydrogen atom. The hydrogen atom was composed of a nucleus and an electron orbiting about it. This picture can be extended to the diatomic molecule by considering one atom to be moving relative to the other atom, which is taken as the origin. This is similar to the bipolar coordinates used to describe H_2^+. Recall that the radial equation describing the radial function $\mathcal{R}e$ we obtained in Sec. 8.3 for the hydrogen atom is given by

$$\frac{1}{r^2}\frac{d}{dr}\left(r^2 \frac{d\mathcal{R}e}{dr}\right) + \frac{2m^*}{\hbar^2}\left[E_r - V(r) - \frac{\ell(\ell+1)\hbar^2}{2m^* r^2}\right]\mathcal{R}e(r) = 0$$

This Hamiltonian leads to the energy with contributions from the orbital angular momentum dependent energy of the electron about the nucleus. This equation can be applied to the diatomic molecule by likening the second atom of the diatomic molecule to the electron of the single hydrogen atom. Then, it can be treated as an energized atom orbiting

10.7 Vibrational-Rotational Spectra Of Molecules

and vibrating relative to another atom. Similar to the potential energy of a harmonic oscillator, in the case of the diatomic molecule, the interatomic potential energy $V(R)$ can be expressed as

$$V(r) = \frac{1}{2}k(R-R_b)^2 \tag{10.65}$$

Instead of d, which was the distance between the electron and nucleus, we use R to denote the internuclear separation variable. k is the stiffness constant which depends on the bonding energy, and it relates to the natural frequency as $\omega = \sqrt{k/m^*}$. m^* is the *reduced mass of the atom relative to the other atom of the diatomic molecule*. For example, for H_2 we have $m^* = m_H m_H/(m_H + m_H) = m_H/2$. From the radial part of the Schrödinger equation, we later defined a function $U = \Re e/r$, then we obtained a modified Hamiltonian given by

$$\frac{d^2 U}{dR^2} + \frac{2m^*}{\hbar^2}\left[E_\lambda - \frac{\ell(\ell+1)\hbar^2}{2m^* R^2} - \frac{1}{2}k(R-R_b)^2\right]U(R) = 0 \tag{10.66}$$

Since we have two atoms in bond, and thus, two or more electrons, we must include both orbital angular momentum *and* spin angular momentum of the molecule. To do this, we replace ℓ in Eq. (10.66) with the total angular momentum quantum number $J = L + S$. We also introduce the variable $\zeta = R - R_b$ so that Eq (10.66) becomes

$$\frac{d^2 U}{d\zeta^2} + \frac{2m^*}{\hbar^2}\left[E_\lambda - \frac{J(J+1)\hbar^2}{2m^*(\zeta+R_b)^2} - \frac{1}{2}k(\zeta)^2\right]U(\zeta) = 0 \tag{10.67}$$

In the next step, an approximation for the total angular momentum term is made in order to simplify the a tricky denominator, having spatial dependence. If we use a binomial series expansion on this term, we can write

$$\frac{1}{(\zeta+R_b)^2} = \frac{1}{R_b^2}\left[1 + \frac{\zeta}{R_b}\right]^{-2} \approx \frac{1}{R_b^2} \tag{10.68}$$

The assumption is valid if $\zeta \ll R_b$. This means that the amount of displacement of the molecular separation is small compared to the equilibrium bond length R_b. Then Eq. (10.67) becomes

$$\frac{d^2 U}{d\zeta^2} + \frac{2m^*}{\hbar^2}\left[E_\lambda - \frac{J(J+1)}{2m^* R_b^2} - \frac{1}{2}k(\zeta)^2\right]U(\zeta) = 0 \tag{10.69}$$

Since the total angular momentum term is now a constant, let us define a shifted energy ε_λ as

$$\varepsilon_\lambda = E_\lambda - \frac{J(J+1)}{2m^* R_b^2} \tag{10.70}$$

We now have

$$\frac{d^2 U}{d\zeta^2} + \frac{2m^*}{\hbar^2}\left[\varepsilon_\lambda - \frac{1}{2}k(\zeta)^2\right]U(\zeta) = 0 \tag{10.71}$$

Eq. (10.71) is identical to the 1D harmonic oscillator equation we obtained, which was solved in Chap. 5 (see Eq. (5.16)). There, we found the eigenenergy to be

$$\varepsilon_\lambda = \left(v + \frac{1}{2}\right)\hbar\omega \tag{10.72}$$

$v = 0, 1, 2, 3, \ldots$ With this solution for ε_λ, solving for the energy E_λ, we have a vibrational-rotational energy eigenvalue given by

Vibrational-Rotational Eigenenergy Of Diatomic Molecule

$$E_\lambda = \underbrace{\frac{J(J+1)}{2m^* R_b^2}}_{\text{rotational}} + \underbrace{\left(v + \frac{1}{2}\right)\hbar\omega}_{\text{vibrational}} \tag{10.73}$$

There are two contributions to the diatomic molecular energy, namely, the rotational energy and vibrational energy. This result can be used along with experimental rotational spectra to extract R_b from the rotational term. It is common to express the energy in units of inverse wavelengths $T = E_\lambda / hc$, where we have

$$T = \frac{E_\lambda}{hc} = \frac{J(J+1)}{2hm^* c R_b^2} + \left(v + \frac{1}{2}\right)\frac{f}{c} = J(J+1)k_r + \left(v + \frac{1}{2}\right)k_v \tag{10.74}$$

k_r is the **rotational wave-number constant**, depending on the bond length; k_v is the **vibrational wave-number constant**. Both are defined as

$$k_r = \frac{1}{2hm^* c R_b^2} \quad \text{and} \quad k_v = \frac{\omega}{2\pi c} \tag{10.75}$$

Both constants have units of inverse length like wave-numbers. These results dictate certain transitional features between the energy states. To see this more clearly, let's look at a transition between the following states:

state 1 : $v = 0, J = J' + 1$

10.7 Vibrational-Rotational Spectra Of Molecules

state 2: $v = 1, J = J'$

With these two states, the change in energy ΔT becomes

$$\Delta T = J'(J'+1)k_r - (J'+1)(J'+2)k_r + \frac{1}{2}k_v - \left(1 + \frac{1}{2}\right)k_v$$

$$= (J'+1)(J'-J'-2)k_r + \frac{1}{2}k_v - \frac{3}{2}k_v$$

$$= -k_v - 2k_r(J'+1)$$

This result means that a change in rotational energy via a change in J corresponds to equally spaced lines of separation $2k_r$, while a change in vibrational energy corresponds to equally spaced spectral lines of separation k_v. Using spectroscopy with diatomic molecules, these features can indeed be observed. Fig. 10.10 shows an example of a pure rotational spectra taken with hydrogen-chloride gas HCl. The peaks in

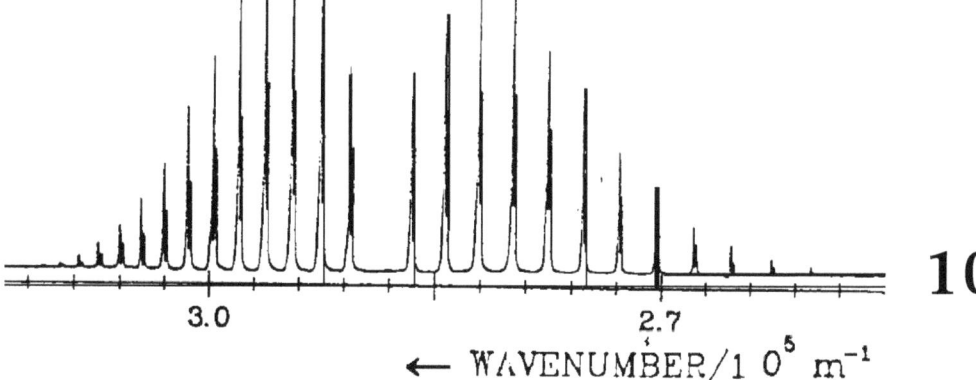

Figure 10.10: Example of an early rotational spectra measurement of HCl, illustrating the period separation between peaks. This provides the means to measure bond length R_b. Data adapted from Ogilvie (1989).

the spectra are also known as *branches*. The two sets of branches shown in Fig. 10.10 correspond to the two isotope species $H^{35}Cl$ and $H^{37}Cl$. Measuring in the region of larger wave-numbers (further to the right on the abscissa) begins to reveal vibrational energy transitions. To measure the bond length of a diatomic molecule, the spacing between the peaks corresponding to the rotational contribution should be used. In Fig. 10.10, the peak separation allows the measurement of k_r. By measuring

k_r^{meas} from the separations in the rotational spectra, the bond length can be found from the relation

$$R_b = \sqrt{\frac{\hbar}{4\pi m^* c k_r^{meas}}} \qquad (10.76)$$

k_r^{meas} is the measured value of k_r, found from spectral measurements. Since c.1889, it's been possible to obtain spectroscopy of molecules. In 1889, the first spectra of this kind within the range of wavelengths between 2.5 μm and 25 μm was done in the laboratory of Swedish physicist Knut Ångström (1857-1910). So, the means to do the measurements predates quantum mechanics, however, quantum mechanics provided a deeper insight into the driving forces behind the spectral measurement observations. The first theoretical interpretation of the vibration-rotational spectra of diatomic molecules with quantum mechanics appeared in 1939 by Chinese physicist T. Y. Wu, from the National Peking University in China in his book *Vibrational Spectra and Structure of Polyatomic Molecules*.

With this, a process has been developed that allows the measurement of molecular bond lengths R_b in diatomic molecules. Now that we have an understanding of how atoms come together to form more complex structures such as molecules, let us return back to the exchange interaction between two electron spins to determine another way to express the energy that is more common in the subject of *magnetism*.

10.8 Exchange Interaction With Spin

Earlier in the chapter, we found that the total anti-symmetric wave function form a pair of electrons is given by Eq. (10.34), and leads to an additional energy contribution given by

$$J_{ij} = Re\left[\frac{e^2}{4\pi\epsilon_0}\int\frac{\psi_a^*(\mathbf{r}_i)\psi_b^*(\mathbf{r}_j)\psi_a(\mathbf{r}_j)\psi_b(\mathbf{r}_i)}{|\mathbf{r}_j - \mathbf{r}_i|}d\mathbf{r}_i d\mathbf{r}_j\right]$$

We also found that the difference in energy $E_{\text{sym spin}} - E_{\text{asym spin}}$ (all other things remaining the same), between symmetric and anti-symmetric spin states is given by

$$E_{\text{sym spin}} - E_{\text{asym spin}} = \Delta E_\lambda = -2J_{ij} \qquad (10.77)$$

10.8 Exchange Interaction With Spin

Here, we will show that the Hamiltonian for $E_{\text{sym spin}}$ and $E_{\text{asym spin}}$ is expressible as a function of the dot product between the two spins $\mathbf{S}_1 \cdot \mathbf{S}_2$. However, we first have to discuss what exactly is the meaning of the quantum mechanical operator $\widehat{\mathbf{S}}_1 \cdot \widehat{\mathbf{S}}_2$. We know from our results with one electron, as in the hydrogen atom, that

$$\widehat{\mathbf{S}}_1^2 = s_1(s_1+1)\hbar^2 \quad \text{and} \quad \widehat{\mathbf{S}}_2^2 = s_2(s_2+1)\hbar^2 \tag{10.78}$$

For two electrons, $\widehat{\mathbf{S}} = \widehat{\mathbf{S}}_1 + \widehat{\mathbf{S}}_2$, thus, it follows that

$$S^2 = \widehat{\mathbf{S}} \cdot \widehat{\mathbf{S}} = (\widehat{\mathbf{S}}_1 + \widehat{\mathbf{S}}_2)^2 = \widehat{\mathbf{S}}_1^2 + \widehat{\mathbf{S}}_2^2 + 2\widehat{\mathbf{S}}_1 \cdot \widehat{\mathbf{S}}_2 \tag{10.79}$$

Eq. (10.79) provides a well-defined operator for $\mathbf{S}_1 \cdot \mathbf{S}_2$, where solving for it, from the above leads to

$$\widehat{\mathbf{S}}_1 \cdot \widehat{\mathbf{S}}_2 = \frac{1}{2}\left[S^2 - (S_1^2 + S_2^2)\right] \tag{10.80}$$

Then, substitution of $S^2 = s(s+1)$, S_1^2, and S_2^2 from Eq. (10.78) (omitting $\hbar^2$) into Eq. (10.80) gives

$$\widehat{\mathbf{S}}_1 \cdot \widehat{\mathbf{S}}_2 = \frac{1}{2}[s(s+1) - (s_1(s_1+1) + s_2(s_2+1))] \tag{10.81}$$

Eq. (10.82) can be evaluated to provide $\mathbf{S}_1 \cdot \mathbf{S}_2$. Let's consider the total spin quantum numbers $s = s_1 + s_2$ for the total system of two electrons. We can determine the values of s for both cases of symmetric spin states $\psi_\uparrow \psi_\downarrow + \psi_\downarrow \psi_\uparrow$ and anti-symmetric spin states $(\psi_\uparrow \psi_\downarrow - \psi_\downarrow \psi_\uparrow)$. While the above two states are only two of representative states for the two electron system, we also know there must be a total of four eigenstates, of which any state for the two electrons can be expressed as a linear combination. Just as we had two eigenstates for one electron, there are four eigenstates for two electrons, eight for three electrons, etc. (2^n in general). The two states above only involve the anti-parallel spin states $\psi_{\uparrow\downarrow}$ and $\psi_{\downarrow\uparrow}$. This means that two more basis states must be included. Along with these two states, the other states are both spins up and both spins down $\psi_{\uparrow\uparrow}$ and $\psi_{\downarrow\downarrow}$, which are symmetric spin states. With these four eigenstates, there is only one anti-symmetric with respect to spin, and the other three are all symmetric. Let's consider all four corresponding s and m_s values. This is summarized in Table 10.2.

Table 10.2: Summary of spin quantum numbers for the two-electron system. s is obtained from $s = \max(m_s)$, just as $\ell = \max(m_\ell)$, etc.

Basis State	Symmetry	m_s	s
$\psi_{\uparrow\uparrow}$	symmetric	$\frac{1}{2} + \frac{1}{2} = 1$	1
$\psi_{\downarrow\downarrow}$	symmetric	$-\frac{1}{2} - \frac{1}{2} = -1$	1
$(\psi_{\uparrow\downarrow} + \psi_{\downarrow\uparrow})/\sqrt{2}$	symmetric	$0 + 0 = 0$	1
$(\psi_{\uparrow\downarrow} - \psi_{\downarrow\uparrow})/\sqrt{2}$	anti-symmetric	$0 - 0 = 0$	0

The first three states correspond to the set of states $m_s = -1, 0, 1$, known as the *triplet state*. The other is a singlet state. There are only two distinct values of s (but four of m_s), which is all we need to determine $\mathbf{S}^2$. When multiple electrons are involved along a single projection axis, m_s is found from first using the appropriate combinations of m_i for each separate spin, and then s is obtained by

Multi-electron Spin Quantum Number

$$S = \max(m_s) \tag{10.82}$$

We'll demonstrate this shortly. The same is true for the multi-electron orbital quantum number **L** as well as **J** = **L**+**S**. In this case, the symmetric spin state has $s = 1$ and thus $m_s = -1, 0, 1$, while the anti-symmetric spin state has a value $s = 0$, and thus $m_s = 0$. Since s has two possible values, so does $s(s+1)$. This leads to the two possible values for $\widehat{\mathbf{S}}_1 \cdot \widehat{\mathbf{S}}_2$. Substitution into Eq. (10.82) leads to

$$\widehat{\mathbf{S}}_1 \cdot \widehat{\mathbf{S}}_2 = \frac{1}{2}\left[1(1+1) - \frac{1}{2}\left(\frac{1}{2}+1\right) - \frac{1}{2}\left(\frac{1}{2}+1\right)\right] = \frac{1}{4} \tag{10.83a}$$

$$\widehat{\mathbf{S}}_1 \cdot \widehat{\mathbf{S}}_2 = \frac{1}{2}\left[0(0+1) - \frac{1}{2}\left(\frac{1}{2}+1\right) - \frac{1}{2}\left(\frac{1}{2}+1\right)\right] = -\frac{3}{4} \tag{10.83b}$$

These binary results along with $E_{\text{sym spin}} - E_{\text{asym spin}} = -2J_{ij}$ can be used to express a new binary Hamiltonian for the two-electron system. Let us express the following Hamiltonian $\widehat{H}_{ij}$

$$\widehat{H}_{ij} = \frac{3}{4}E_{\text{sym}} + \frac{1}{4}E_{\text{asym}} + (E_{\text{sym}} - E_{\text{asym}})\mathbf{S}_i \cdot \mathbf{S}_j \tag{10.84}$$

10.8 Exchange Interaction With Spin

With this Hamiltonian, in the case that we have a symmetric state, where $s = 1$, Eq. (10.84) becomes

$$\widehat{H}_{ij} = \frac{3}{4}E_{sym} + \frac{1}{4}E_{asym} + (E_{sym} - E_{asym})\frac{1}{4}$$

$$= \frac{3}{4}E_{sym} + \frac{1}{4}E_{asym} + \frac{1}{4}E_{sym} - \frac{1}{4}E_{asym}$$

$$= \frac{3}{4}E_{sym} + \frac{1}{4}E_{sym} = E_{sym}$$

Likewise, for the other case of an anti-symmetric spin state ($s = 0$), we obtain

$$\widehat{H}_{ij} = \frac{3}{4}E_{sym} + \frac{1}{4}E_{asym} - (E_{sym} - E_{asym})\frac{3}{4}$$

$$= \frac{3}{4}E_{sym} + \frac{1}{4}E_{asym} - \frac{3}{4}E_{sym} + \frac{3}{4}E_{asym}$$

$$= \frac{3}{4}E_{sym} + \frac{1}{4}E_{sym} = E_{asym}$$

Eq. (10.84) is therefore, a two-spin Hamiltonian that describes the change in energy solely due to pair-wise spin orientation. $\widehat{H}_{ij}(\widehat{\mathbf{S}}_i \cdot \widehat{\mathbf{S}}_j) = E_{sym}$ for symmetric spin states, while $H_{ij}(\widehat{\mathbf{S}}_i \cdot \widehat{\mathbf{S}}_j) = E_{asym}$ for anti-symmetric spin states. Using the relation $E_{sym} - E_{asym} = -2J_{ij}$, the two-spin Hamiltonian becomes

$$\widehat{H}_{ij} = \frac{3}{4}E_{sym} + \frac{1}{4}E_{asym} - 2J_{ij}\widehat{\mathbf{S}}_i \cdot \widehat{\mathbf{S}}_j \quad (10.85)$$

But, we know that $\frac{3}{4}E_{sym} + \frac{1}{4}E_{asym}$ are constants or independent of spin explicitly because they only depend on the Coulomb and exchange integrals obtained earlier. As constants, the variation in energy is due to the last term and so the first two terms are usually disregarded. Therefore, the spin-dependent Hamiltonian becomes

Electron-Pair Exchange Hamiltonian

$$\widehat{H}_{ij} = -2J_{ij}\widehat{\mathbf{S}}_i \cdot \widehat{\mathbf{S}}_j \quad (10.86)$$

Note that $\widehat{\mathbf{S}}_i \cdot \widehat{\mathbf{S}}_j$ is dimensionless, in Eq. (10.86). In terms of spin matrices, Eq. (10.86) becomes

$$\widehat{H}_{ij} = -\frac{1}{2}J_{ij}\widehat{\boldsymbol{\sigma}}_i \cdot \widehat{\boldsymbol{\sigma}}_j \quad (10.87)$$

With a two-spin system, there can be more than one form of exchange coupling. For example, if the energy exists between a pair of electrons in the same orbital of a single atom like the ground state of helium, the exchange energy is known as **intraatomic exchange** coupling, while if the coupling between spins exists between electron spins of distinct atoms, it is known as **inter-atomic** (between atoms) exchange coupling. This form, specifically, is known as the **Heisenberg exchange** as it was postulated by Werner Heisenberg as the origin of *ferromagnetism*, existing between the atoms of certain lattices such as iron (Fe), cobalt (Co), and nickel (Ni).

The above results suggest that for multi-electron systems, hints at some rules that may be followed by energy transitions. The opportunities to study such rules was there with the atomic and molecular spectra. There are indeed some restrictions on the transitions made by electrons. Wolfgang Pauli took a close look at these patterns in atomic spectra, and in 1925, he announced his findings as a *quantum postulate* (meaning that it was not proven, but assumed based on experimental findings). Pauli noticed that no two electrons ever possess the same *four* quantum numbers. Therefore, a unique combination of the four quantum numbers (n, ℓ, m_ℓ, m_s) apparently completely defines the state of an electron. Some time later, it was further generalized to include all half-integer spin particles, or *fermions*. This can be summed up as follows:

Pauli Exclusion Principle: Two or more fermions (particles possessing half-integer spin) *cannot* occupy the same quantum state simultaneously.

The above statement is known as the **Pauli Exclusion Principle**. It is another way of stating the orthogonality between quantum states having distinct quantum number sets, going one step further in that no two electrons will occupy the same state, as *fermions*.

There have been a few occasions in the text where we have mentioned the process of adding angular momentum. In the next section, we shall take up this topic more formally and demonstrate useful applications.

10.9 Multi-electron Total Angular Momentum

The total angular momentum $\widehat{\mathbf{J}}$ is given by the sum of both orbital and spin contributions, or

$$\widehat{\mathbf{J}} = \widehat{\mathbf{L}} + \widehat{\mathbf{S}} \qquad (10.88)$$

For two or more electrons possessing angular momentum $\widehat{\mathbf{J}}_1$ and $\widehat{\mathbf{J}}_2$, respectively, the total angular momentum of the combined system is dictated by further superposition, where we have

$$\widehat{\mathbf{J}} = \widehat{\mathbf{J}}_1 + \widehat{\mathbf{J}}_2 = (\widehat{\mathbf{L}}_1 + \widehat{\mathbf{S}}_1) + (\widehat{\mathbf{L}}_2 + \widehat{\mathbf{S}}_2)$$
$$= (\widehat{\mathbf{L}}_1 + \widehat{\mathbf{L}}_1) + (\widehat{\mathbf{S}}_1 + \widehat{\mathbf{S}}_1) = \widehat{\mathbf{L}} + \widehat{\mathbf{S}}$$

For both $\widehat{\mathbf{J}}_1$ and $\widehat{\mathbf{J}}_2$, we know the commutation relations hold separately, where we have

$$\widehat{\mathbf{J}}_1 = [\widehat{J}_{1x}, \widehat{J}_{1y}, \widehat{J}_{1z}]^T \Rightarrow [\widehat{J}_{1i}, \widehat{J}_{1j}] = i\varepsilon_{ijk}\widehat{J}_{1k}$$

and $\quad \widehat{\mathbf{J}}_2 = [\widehat{J}_{2x}, \widehat{J}_{2y}, \widehat{J}_{2z}]^T \Rightarrow [\widehat{J}_{2i}, \widehat{J}_{2j}] = i\varepsilon_{ijk}\widehat{J}_{2k}$

This property of $\widehat{\mathbf{J}}$ is inherited from $\widehat{\mathbf{L}}$ and $\widehat{\mathbf{S}}$, where it was demonstrated for the former in Chap. 7. It extends to $\widehat{\mathbf{S}}$. ε_{ijk} is the Levi-Cevita symbol keeping track of the order of coordinates, being +1 for right-handedness with distinct coordinates, −1 for the opposite also with distinct coordinates, and 0 otherwise. For multi-electron systems, the total angular momentum $\widehat{\mathbf{J}}$ is also *assumed* to satisfy the same commutation relations as $\widehat{\mathbf{J}}_1$ and $\widehat{\mathbf{J}}_2$. Let's evaluate this relation explicitly for $\widehat{\mathbf{J}}$ for two electrons, to better understand why this must be satisfied. The commutation relation for x and y components of $\widehat{\mathbf{J}}$ is given by

$$[\widehat{J}_x, \widehat{J}_y] = \widehat{J}_x\widehat{J}_y - \widehat{J}_x\widehat{J}_y$$
$$= (\widehat{J}_{1x} + \widehat{J}_{2x})(\widehat{J}_{1y} + \widehat{J}_{2y}) - (\widehat{J}_{1x} + \widehat{J}_{2x})(\widehat{J}_{1y} + \widehat{J}_{2y})$$

Expanding this last line gives

$$[\widehat{J}_x, \widehat{J}_y] = (\widehat{J}_{1x}\widehat{J}_{1y} + \widehat{J}_{2x}\widehat{J}_{1y} + \widehat{J}_{1x}\widehat{J}_{2y} + \widehat{J}_{2x}\widehat{J}_{2y}) - (\widehat{J}_{1y}\widehat{J}_{1x} + \widehat{J}_{1y}\widehat{J}_{2x} + \widehat{J}_{2y}\widehat{J}_{1x} + \widehat{J}_{2y}\widehat{J}_{2x})$$
$$= i\widehat{J}_{1z} + i\widehat{J}_{2z} + (\widehat{J}_{2x}\widehat{J}_{1y} - \widehat{J}_{1y}\widehat{J}_{2x}) + (\widehat{J}_{1x}\widehat{J}_{2y} - \widehat{J}_{2y}\widehat{J}_{1x})$$

In the above, we have used the commutation relations for $\widehat{J}_1$ and $\widehat{J}_2$, respectively. The only way that the total angular momentum for two electrons can satisfy the commutation relation $[\widehat{J}_x, \widehat{J}_y] = iJ_z = i(\widehat{J}_{1z} + \widehat{J}_{2z})$ is if the following conditions are also satisfied

$$\widehat{J}_{2x}\widehat{J}_{1y} - \widehat{J}_{1y}\widehat{J}_{2x} = [\widehat{J}_{2x}, \widehat{J}_{1y}] = 0 \quad \text{and} \quad \widehat{J}_{1x}\widehat{J}_{2y} - \widehat{J}_{2y}\widehat{J}_{1x} = [\widehat{J}_{1x}, \widehat{J}_{2y}] = 0$$

An analogous assumption is made for the other two components, where in order for

$$[\widehat{J}_y, \widehat{J}_z] = iJ_x \quad \text{and} \quad [\widehat{J}_z, \widehat{J}_x] = iJ_y$$

It must also be true that

$$\widehat{J}_{2y}\widehat{J}_{1z} - \widehat{J}_{1z}\widehat{J}_{2y} = [\widehat{J}_{2y}, \widehat{J}_{1z}] = 0 \quad \text{and} \quad \widehat{J}_{1y}\widehat{J}_{2z} - \widehat{J}_{2z}\widehat{J}_{1y} = [\widehat{J}_{1y}, \widehat{J}_{2z}] = 0$$

$$\widehat{J}_{2z}\widehat{J}_{1x} - \widehat{J}_{1x}\widehat{J}_{2z} = [\widehat{J}_{2z}, \widehat{J}_{1x}] = 0 \quad \text{and} \quad \widehat{J}_{1z}\widehat{J}_{2x} - \widehat{J}_{2x}\widehat{J}_{1z} = [\widehat{J}_{1z}, \widehat{J}_{2x}] = 0$$

The above relations can be generalized as

> **$\widehat{J}$ Commutation Postulate For Multielectron System**
>
> $$[\widehat{J}_{ik}, \widehat{J}_{jn}] = 0 \quad i \neq j = 1,2 \quad k \neq n = x,y,z \quad (10.89)$$

The commutation relations above lead to another commutation relation for $\widehat{J}$ that is satisfied by the single electron angular momentum. Using the multi-electron commutation relations above, we have

$$[\widehat{J}^2, \widehat{J}_x] = (\widehat{J}_x^2 + \widehat{J}_y^2 + \widehat{J}_z^2)\widehat{J}_x - \widehat{J}_x(\widehat{J}_x^2 + \widehat{J}_y^2 + \widehat{J}_z^2)$$

$$= (\widehat{J}_x^3 + \widehat{J}_y^2\widehat{J}_x + \widehat{J}_z^2\widehat{J}_x) - (\widehat{J}_x^3 + \widehat{J}_x\widehat{J}_y^2 + \widehat{J}_x\widehat{J}_z^2)$$

$$= (\widehat{J}_y^2\widehat{J}_x - \widehat{J}_x\widehat{J}_y^2) + (\widehat{J}_z^2\widehat{J}_x - \widehat{J}_x\widehat{J}_z^2)$$

If we use the *commutation postulate* to change the orders of $\widehat{J}_i\widehat{J}_j$ since they commute, then we have

$$[\widehat{J}^2, \widehat{J}_x] = (\widehat{J}_y\widehat{J}_y\widehat{J}_x - \widehat{J}_x\widehat{J}_y\widehat{J}_y) + (\widehat{J}_z\widehat{J}_z\widehat{J}_x - \widehat{J}_x\widehat{J}_z\widehat{J}_z)$$

$$= (\widehat{J}_y\widehat{J}_y\widehat{J}_x - \widehat{J}_y\widehat{J}_x\widehat{J}_y) + (\widehat{J}_z\widehat{J}_z\widehat{J}_x - \widehat{J}_z\widehat{J}_x\widehat{J}_z)$$

$$= \widehat{J}_y(\widehat{J}_y\widehat{J}_x - \widehat{J}_x\widehat{J}_y) + \widehat{J}_z(\widehat{J}_z\widehat{J}_x - \widehat{J}_x\widehat{J}_z)$$

10.9 Multi-electron Total Angular Momentum

$$= \widehat{J}_y\left(-i\widehat{J}_z\right) + \widehat{J}_z\left(i\widehat{J}_y\right) = 0$$

Analogous results follow for $\left[\widehat{J}^2,\widehat{J}_y\right]$ and $\left[\widehat{J}^2,\widehat{J}_z\right]$. So, we find that the satisfaction of the commutation postulation of Eq. (10.89) leads to the following commutation relation:

> **$\widehat{J}^2$ Commutation Relation For Multielectron Systems**
>
> $$\left[\widehat{J}^2,\widehat{J}_i\right] = 0 \quad i = x, y, z \qquad (10.90)$$

With the above relations, we can write the following consistent eigenvalue equations:

$$\widehat{J}^2 \Psi = j(j+1)\Psi \quad \text{and} \quad \widehat{J}_z \Psi = m_j \Psi \qquad (10.91\text{a})$$

$$\widehat{J}_1^2 \Psi_1 = j_1(j_1+1)\Psi_1 \quad \text{and} \quad \widehat{J}_{1z} \Psi_1 = m_{1j} \Psi_1 \qquad (10.91\text{b})$$

$$\widehat{J}_2^2 \Psi_2 = j_2(j_2+1)\Psi_2 \quad \text{and} \quad \widehat{J}_{2z} \Psi_2 = m_{2j} \Psi_2 \qquad (10.91\text{c})$$

For each system, the corresponding total angular momentum $\widehat{J}$ commutes with its z component. The quantum numbers are related by

$$j = j_1 + j_2 \quad \text{and} \quad m_j = m_{1j} + m_{2j} \quad m_j = -j, \cdots, 0, \cdots, j \qquad (10.92\text{a})$$

Unlike in the case of a single electron where $j = 1/2$, for multi-electron systems, j takes on more than one value. If we further assume that the Hamiltonian of the system is *separable* in electron one and two, the total wave function ψ can be written as

$$\psi(j_1, j_2, m_1, m_2) = \psi_1(j_1, m_1) \cdot \psi_2(j_2, m_2)$$

All of this leads to the commutation of the following four operators:

$$\widehat{J}_1^2, \widehat{J}_2^2, \widehat{J}^2, \text{ and } \widehat{J}_z \qquad (10.93)$$

We know that the commuting of two operators is synonymous with simultaneous eigenfunctions. Additionally, there is a wave function we shall denote by $\phi(j_1, j_2, j, m)$ describing this system. But, how can we relate ϕ to the separable form given by Ψ. In the next section, a method to establish the coefficients of proportionality between the eigenstates $\phi(j_1, j_2, j, m)$ and $\psi(j_1, j_2, m_1, m_2)$ is discussed. But, before getting to this, let us consider three examples illustrating the total angular momentum states for a system with two electrons.

Exercise 10.1 Determine the total combination of states for a two-electron system where

$$j_1 = s_1 = \frac{1}{2} \quad j_2 = s_2 = \frac{1}{2} \quad \text{and} \quad m_1 = -\frac{1}{2}, \frac{1}{2} \quad m_2 = -\frac{1}{2}, \frac{1}{2}$$

Starting with $m = m_1 + m_2$, the total combinations become:

$$m = m_1 + m_2 = -1, 0, 0, +1$$

The value of $j = j_1 + j_2$ is obtained by

$$j = \max(m) = 1$$

For an electron, the complete set of states must form a set of distinct states. Thus, the first set of distinct states for the pair of electrons is

$$m = (-1, 0, 1) \quad j = 1 \quad \text{(triplet state)}$$

This set of three distinct states is an example of a *triplet state*. For the remaining state corresponding to $m = 0$, it leads to a value for j given by

$$j = \max(m) = 0$$

so

$$m = 0 \quad j = 0 \quad \text{(singlet state)}$$

This lone state is an example of a *singlet state*. This special combination will arise in two-electron systems where $\ell = 0$. ∎

Exercise 10.2 Determine the total combination of states for a two-electron system where

$$j_1 = \ell_1 = 1 \quad j_2 = s_2 = \frac{1}{2} \quad \text{and} \quad m_1 = -1, 0, 1 \quad m_2 = -\frac{1}{2}, \frac{1}{2}$$

Starting with $m = m_1 + m_2$, the total combinations for $m = m_1 + m_2$ are

$$m = m_1 + m_2 = -1 - \frac{1}{2} = -\frac{3}{2}, -\frac{1}{2}, -\frac{1}{2}, +\frac{1}{2}, +\frac{1}{2}, 1 + \frac{1}{2} = +\frac{3}{2}$$

10.9 Multi-electron Total Angular Momentum

We again have to sets of states. For the first set, the value of $j = j_1 + j_2$ is obtained by

$$j = \max(m) = \frac{3}{2}$$

This leads to the first set of distinct states for the pair of electrons given by

$$m = (-\frac{3}{2}, -\frac{1}{2}, +\frac{1}{2}, +\frac{3}{2}) \Rightarrow j = \frac{3}{2}$$

The remaining states form the set given by

$$m = (-\frac{1}{2}, +\frac{1}{2}) \Rightarrow j = \frac{3}{2}$$

In this example, six states are found for the two-electron system, forming two sets of states corresponding to $j = 3/2$ and $j = 1/2$. ∎

Exercise 10.3 Determine the total combination of states for a two-electron system where

$$j_1 = \ell_1 = 2 \quad j_2 = s_2 = \frac{1}{2}$$

and

$$m_1 = -2, -1, 0, 1, 2 \quad m_2 = -\frac{1}{2}, \frac{1}{2}$$

$m = m_1 + m_2$ leads to the total combinations given by

$$m = m_1 + m_2 = -2 - \frac{1}{2} = -\frac{5}{2}, -\frac{3}{2}, -\frac{3}{2}, -\frac{1}{2}, -\frac{1}{2}, +\frac{1}{2}, +\frac{1}{2}, \frac{3}{2}, \frac{3}{2}, 2 + \frac{1}{2} = \frac{5}{2}$$

We have to sets of states. For the first set, the value of $j = j_1 + j_2$ is obtained by

$$j = \max(m) = \frac{5}{2}$$

This leads to the first set of distinct states for the pair of electrons, given by

$$m = (-\frac{5}{2}, -\frac{3}{2}, -\frac{1}{2}, +\frac{1}{2}, +\frac{3}{2}, +\frac{5}{2}) \Rightarrow j = \frac{5}{2}$$

The remaining states form the set given by

$$m = (-\frac{3}{2}, -\frac{1}{2}, \frac{1}{2}, \frac{3}{2}) \Rightarrow j = \frac{3}{2}$$

Thus, we find a total of ten states for the two-electron system, forming two sets of states corresponding to $j = 5/2$ and $j = 3/2$. ∎

Let's move on to discuss a systematic way to determine the coefficients of proportionality between the total angular momentum eigenstates $\phi(j_1, j_2, j, m)$ and the, often useful, electronic states $\psi(j_1, j_2, m_1, m_2)$.

10.10 Clebsch-Gordan Coefficients

For a two-electron system, there are four general eigenstates ϕ of the Hamiltonian. It turns out that this general wave function ϕ can be related directly to the separable wave function ψ under appropriate assumptions, where

$$\phi(j_1, j_2, j, m) \Leftrightarrow \psi(j_1, j_2, m_1, m_2)$$

This can be done by exploiting some of the angular momentum operators we discussed in Chap. 7. A well-known convention for relating these two functions is based on a formalism developed before quantum mechanics. The resulting coefficients relating the two functions are known as **Clebsch-Gordan coefficients**, named after German mathematicians Rudolf Friedrich Alfred Clebsch (1833-1872) and Paul Albert Gordan (1837-1912). The two encountered an equivalent mathematical problem, which paved the way for what we are about to discuss here. In quantum mechanics, these coefficients are obtained by using the *lowering* and *raising ladder* operators $\widehat{J}_-$ and $\widehat{J}_+$, introduced in Sec. 7.10. Consider a system with two electrons both having $j_i = s_i = 1/2$, as in *Example 1* above. For the two-electron system, the operator $\widehat{J}_-$ is defined as

$$\widehat{J}_- = \widehat{J}_{1-} + \widehat{J}_{2-} \tag{10.94}$$

10.10 Clebsch-Gordan Coefficients

$\widehat{J}_{i-}$ denotes the lowering operator for the ith electron. We can use Eq. (10.94) because the LHS operates on the total angular momentum wave function $\phi(j_1, j_2, j, m)$, while the RHS operates on $\psi(j_1, j_2, m_1, m_2)$. Starting at the *top* or highest energy state of the triplet states for the system, let's relate $\psi(\frac{1}{2}, \frac{1}{2}, \frac{1}{2}, \frac{1}{2})$ and $\phi(\frac{1}{2}, \frac{1}{2}, 1, 1)$. Applying Eq. (10.94) gives

$$\widehat{J}_- \phi\left(\frac{1}{2}, \frac{1}{2}, 1, 1\right) = \left(\widehat{J}_{1-} + \widehat{J}_{2-}\right) \psi\left(\frac{1}{2}, \frac{1}{2}, \frac{1}{2}, \frac{1}{2}\right) \tag{10.95}$$

Starting with the LHS, we can substitute the relation for the lowering operator given by Eq. (7.151), so we have

$$\sqrt{(j+m)(j-m+1)}\phi\left(\frac{1}{2}, \frac{1}{2}, 1, 0\right) = \widehat{J}_{1-}\psi\left(\frac{1}{2}, \frac{1}{2}, \frac{1}{2}, \frac{1}{2}\right) + \widehat{J}_{2-}\psi\left(\frac{1}{2}, \frac{1}{2}, \frac{1}{2}, \frac{1}{2}\right) \tag{10.96}$$

For the LHS, $j, m = 1$, while, for the RHS wave functions ψ, $j = m = 1/2$, since $j_i = s_i$. Thus, we have

$$\sqrt{2}\phi\left(\frac{1}{2}, \frac{1}{2}, 1, 0\right) = \sqrt{1}\psi\left(\frac{1}{2}, \frac{1}{2}, -\frac{1}{2}, \frac{1}{2}\right) + \sqrt{1}\psi\left(\frac{1}{2}, \frac{1}{2}, \frac{1}{2}, -\frac{1}{2}\right) \tag{10.97}$$

This gives the following result for $\phi(\frac{1}{2}, \frac{1}{2}, 1, 0)$:

$$\phi\left(\frac{1}{2}, \frac{1}{2}, 1, 0\right) = \frac{1}{\sqrt{2}}\psi\left(\frac{1}{2}, \frac{1}{2}, -\frac{1}{2}, \frac{1}{2}\right) + \frac{1}{\sqrt{2}}\psi\left(\frac{1}{2}, \frac{1}{2}, \frac{1}{2}, -\frac{1}{2}\right) \tag{10.98}$$

As long as ψ is normalized, then ϕ, as given by Eq. (10.98), will also be normalized. Let us continue to the next lowest level by, again, applying the lowering ladder operator $\widehat{J}_-$. Now, we start with Eq. (10.98), applying the ladder operator again, which gives

$$\widehat{J}_- \phi\left(\frac{1}{2}, \frac{1}{2}, 1, 0\right) = \frac{1}{\sqrt{2}}\left(\widehat{J}_{1-} + \widehat{J}_{2-}\right)\psi\left(\frac{1}{2}, \frac{1}{2}, -\frac{1}{2}, \frac{1}{2}\right) +$$
$$\frac{1}{\sqrt{2}}\left(\widehat{J}_{1-} + \widehat{J}_{2-}\right)\psi\left(\frac{1}{2}, \frac{1}{2}, \frac{1}{2}, -\frac{1}{2}\right) \tag{10.99}$$

$$\sqrt{2}\phi\left(\frac{1}{2}, \frac{1}{2}, 1, -1\right) = \frac{1}{\sqrt{2}}\left(\widehat{J}_{1-} + \widehat{J}_{2-}\right)\psi\left(\frac{1}{2}, \frac{1}{2}, -\frac{1}{2}, \frac{1}{2}\right) +$$
$$\frac{1}{\sqrt{2}}\left(\widehat{J}_{1-} + \widehat{J}_{2-}\right)\psi\left(\frac{1}{2}, \frac{1}{2}, \frac{1}{2}, -\frac{1}{2}\right) \tag{10.100}$$

There is a clever simplification hiding on the RHS. In both terms, the state happens to be the lowest state for one of the two operators, so we have

$$\widehat{J}_{1-}\psi\left(\frac{1}{2},\frac{1}{2},-\frac{1}{2},\frac{1}{2}\right)=\widehat{J}_{2-}\psi\left(\frac{1}{2},\frac{1}{2},\frac{1}{2},-\frac{1}{2}\right)=0$$

This follows because for $\widehat{J}_{1-}$, the third argument $m_1 = -1/2$ and likewise for $\widehat{J}_{2-}$, the fourth argument $m_2 = -1/2$, being the lowest states for each electron respectively. Then, further evaluation leads to

$$\sqrt{2}\phi\left(\frac{1}{2},\frac{1}{2},1,-1\right) = \frac{1}{\sqrt{2}}\widehat{J}_{2-}\psi\left(\frac{1}{2},\frac{1}{2},-\frac{1}{2},\frac{1}{2}\right) + \frac{1}{\sqrt{2}}\widehat{J}_{1-}\psi\left(\frac{1}{2},\frac{1}{2},\frac{1}{2},-\frac{1}{2}\right) \quad (10.101)$$

$$= \frac{\sqrt{1}}{\sqrt{2}}\psi\left(\frac{1}{2},\frac{1}{2},-\frac{1}{2},-\frac{1}{2}\right) + \frac{\sqrt{1}}{\sqrt{2}}\psi\left(\frac{1}{2},\frac{1}{2},-\frac{1}{2},-\frac{1}{2}\right) \quad (10.102)$$

This gives a result for $\phi(\frac{1}{2},\frac{1}{2},1,-1)$ as

$$\phi\left(\frac{1}{2},\frac{1}{2},1,-1\right) = \frac{1}{2}\psi\left(\frac{1}{2},\frac{1}{2},-\frac{1}{2},-\frac{1}{2}\right) + \frac{1}{2}\psi\left(\frac{1}{2},\frac{1}{2},-\frac{1}{2},-\frac{1}{2}\right) \quad (10.103)$$

$$\psi\left(\frac{1}{2},\frac{1}{2},-\frac{1}{2},-\frac{1}{2}\right) \quad (10.104)$$

Therefore, this method has allowed us to start at the *top* of the triplet state, then work our way down to the bottom, relating ϕ and ψ by exploiting operator relations for angular momentum. These are the essential steps for obtaining the Clebsch-Gordan coefficients.

That worked well for the triplet state, but what about starting with the singlet state? Unfortunately, because there is only one state in the set of states for $j = 0$, we cannot make use of the ladder operators. However, this can be extricated by making use of the normalization and orthogonality conditions with a state having the same m value, which in this case is $m = 0$. Therefore, Eq. (10.98) is used for this purpose. Let the state be given by a linear combination of ψ states given by

$$\phi\left(\frac{1}{2},\frac{1}{2},0,0\right) = A\psi\left(\frac{1}{2},\frac{1}{2},-\frac{1}{2},\frac{1}{2}\right) + B\psi\left(\frac{1}{2},\frac{1}{2},\frac{1}{2},-\frac{1}{2}\right) \quad (10.105)$$

Use of the *orthogonality condition* leads to

10.10 Clebsch-Gordan Coefficients

$$0 = \int \phi^*\left(\frac{1}{2},\frac{1}{2},0,0\right)\phi^*\left(\frac{1}{2},\frac{1}{2},1,0\right) =$$

$$\int \left[A^*\psi^*\left(\frac{1}{2},\frac{1}{2},-\frac{1}{2},\frac{1}{2}\right) + B^*\psi^*\left(\frac{1}{2},\frac{1}{2},\frac{1}{2},-\frac{1}{2}\right)\right] \times$$

$$\left[\frac{1}{\sqrt{2}}\psi\left(\frac{1}{2},\frac{1}{2},-\frac{1}{2},\frac{1}{2}\right) + \frac{1}{\sqrt{2}}\psi\left(\frac{1}{2},\frac{1}{2},\frac{1}{2},-\frac{1}{2}\right)\right]dx = 0$$

Assuming that ψ states are normalized and orthogonal, evaluation of the above integral leads to the result

$$0 = \int \phi^*\left(\frac{1}{2},\frac{1}{2},0,0\right)\phi\left(\frac{1}{2},\frac{1}{2},1,0\right)dx = \frac{A^*}{\sqrt{2}} + \frac{B^*}{\sqrt{2}} \Rightarrow B^* = -A^*$$

The absolute value of A (or B) is obtained from the normalization condition, given here by

$$\int \phi^*\left(\frac{1}{2},\frac{1}{2},0,0\right)\phi\left(\frac{1}{2},\frac{1}{2},0,0\right)dx = 1$$

The normalization condition leads to

$$1 = \int \phi^*\left(\frac{1}{2},\frac{1}{2},0,0\right)\phi\left(\frac{1}{2},\frac{1}{2},0,0\right)dx = A^2 + B^2 = 2A^2 = 1 \Rightarrow |A| = \frac{1}{\sqrt{2}}$$

Note that we have used the earlier result $A^* = -B^*$. Therefore, this gives for the singlet state

$$\phi\left(\frac{1}{2},\frac{1}{2},0,0\right) = \frac{1}{\sqrt{2}}\psi\left(\frac{1}{2},\frac{1}{2},-\frac{1}{2},\frac{1}{2}\right) - \frac{1}{\sqrt{2}}\psi\left(\frac{1}{2},\frac{1}{2},\frac{1}{2},\frac{1}{2}\right) \quad (10.106)$$

Because we are dealing with linear relations, the Clebsch-Gordan coefficients can be expressed conveniently in matrix form. For the two-electron case that we have considered, the relation in matrix notation becomes

$$\begin{bmatrix} \phi(\frac{1}{2},\frac{1}{2},1,1) \\ \phi(\frac{1}{2},\frac{1}{2},1,0) \\ \phi(\frac{1}{2},\frac{1}{2},1,-1) \\ \phi(\frac{1}{2},\frac{1}{2},0,0) \end{bmatrix} = \begin{bmatrix} 1 & 0 & 0 & 0 \\ 0 & \frac{1}{\sqrt{2}} & \frac{1}{\sqrt{2}} & 0 \\ 0 & 0 & 0 & 1 \\ 0 & \frac{1}{\sqrt{2}} & -\frac{1}{\sqrt{2}} & 0 \end{bmatrix} \begin{bmatrix} \psi(\frac{1}{2},\frac{1}{2},\frac{1}{2},\frac{1}{2}) \\ \psi(\frac{1}{2},\frac{1}{2},-\frac{1}{2},\frac{1}{2}) \\ \psi(\frac{1}{2},\frac{1}{2},\frac{1}{2},-\frac{1}{2}) \\ \psi(\frac{1}{2},\frac{1}{2},-\frac{1}{2},-\frac{1}{2}) \end{bmatrix} \quad (10.107)$$

Note that these are the same states as given in Table 10.2, however, that list was not derived from anything beyond our intuition. However, using Clebsch-Gordon coefficients we have obtained them systematically. Relating the states to those in Table 10.2, note that

$$\psi\left(\frac{1}{2},\frac{1}{2},\frac{1}{2},\frac{1}{2}\right) = \psi_{\uparrow\uparrow}$$

$$\psi\left(\frac{1}{2},\frac{1}{2},-\frac{1}{2},\frac{1}{2}\right) = \psi_{\downarrow\uparrow}$$

$$\psi\left(\frac{1}{2},\frac{1}{2},\frac{1}{2},-\frac{1}{2}\right) = \psi_{\uparrow\downarrow}$$

$$\psi\left(\frac{1}{2},\frac{1}{2},-\frac{1}{2},-\frac{1}{2}\right) = \psi_{\downarrow\downarrow}$$

Using this method, one can always generate a matrix relating the general states under assumptions of separability to the corresponding purely separable eigenstates of the system. The resulting matrix of Clebsch-Gordan coefficients is an example of a **unitary matrix M** which satisfies the condition given by

---

Unitary Matrix Of Clebsch-Gordon Coefficients

$$\overline{M}M = 1 \qquad (10.108)$$

Moreover, for any two-electron system where one of the electrons has a state $j_i = 1/2$, one can obtain systematically, formulas that provide the C-G coefficents. These formulas are listed in Table 10.3.

Table 10.3: Table of Clebsch-Gordon coefficients for a two-electron system, assuming $j_1 = \frac{1}{2}$.

j	$m_1 = -\frac{1}{2}$	$m_1 = \frac{1}{2}$
$j_2 + \frac{1}{2}$	$\sqrt{\dfrac{j_2 + m + 1}{2j_2 + 1}}$	$\sqrt{\dfrac{j_2 - m + \frac{1}{2}}{2j_2 + 1}}$
$j_2 - \frac{1}{2}$	$-\sqrt{\dfrac{j_2 - m + 1}{2j_2 + 1}}$	$\sqrt{\dfrac{j_2 + m + \frac{1}{2}}{2j_2 + 1}}$

Multi-electron systems not only impact the way we treat the wave functions. It also affects the parameter relating magnetic moment and angular momentum. For multi-electron systems, the g-factor generalizes as a constant of proportionality. In the next section, we will derive this generalized parameter, of importance for some magnetic systems.

10.11 Multi-electron Generalized g-Factor

Here, we will discuss the generalization of the electron's g-factor, into a form that depends on the angular momentum quantum numbers. It is no longer a constant $g \approx 2$. To show this, recall that the total magnetic moment $\widehat{\mu}$ of a multi-electron system is given by both the sum of both the orbital angular and spin angular momentum, where

$$\widehat{\mu} = \widehat{\mu}_L + \widehat{\mu}_S = \gamma \widehat{L} + g\gamma \widehat{S} \qquad (10.109)$$

$\widehat{L}$ and $\widehat{S}$ are the sum of the contributions from each electron in the system. Then, the energy becomes

$$V_M = -\widehat{\mu} \cdot \mathbf{B} = -\left[\gamma \widehat{L} + g\gamma \widehat{S}\right] \cdot \mathbf{B} \qquad (10.110)$$

$$= |\gamma|\hbar B(m_\ell + g m_s) \qquad (10.111)$$

Now, substitute $g = 2$ into the above (this will be shown in Chap. 11), so we have

$$V_M = |\gamma|\hbar B(m_\ell + 2m_s) = |\gamma|\hbar B\left(m_j + m_s\right) = \mu_B m_j B\left(1 + \frac{m_s}{m_j}\right)$$

The term in parentheses acts like the g–factor when there is coupling between a magnetic field and spin angular momentum. Recall that we found that the energy for a single electron spin is $V_M = g\mu_B m_s B$. To keep this form for convenient utility with interpretation of atomic spectra, it is common to express the multi-electron system energies having a g–factor × a magnetic quantum number as a coefficient to $\mu_B B$. To express this form from Eq. (10.111), we can define a **generalized g-factor** g_J given by

> **Generalized g-factor**
>
> $$g_J = 1 + \frac{m_s}{m_j} \qquad (10.112)$$

In the case of a ground state hydrogen electron, where $\ell = 0$ and $m_s = m_j = 1/2$, we get $g_J = g = 2$. However, more generally, the g-factor depends on the quantum state. There is a common alternative form of g_J in Eq. (10.112). It is obtained by rewriting V_M with projections of both vectors onto the total angular momentum $\widehat{J}$ as follows:

$$V_M = |\gamma| \frac{[(\widehat{L} + g\widehat{S}) \cdot \widehat{J}][\widehat{J} \cdot \mathbf{B}]}{J^2}$$

$$= |\gamma| \frac{(\widehat{L} \cdot \widehat{J} + g\widehat{S} \cdot \widehat{J})\widehat{J}_z B}{J^2}$$

$$= |\gamma| \frac{\left[\widehat{L} \cdot (\widehat{L} + \widehat{S}) + g\widehat{S} \cdot (\widehat{L} + \widehat{S})\right] m_j \hbar B}{J^2}$$

$$= \mu_B m_j B \frac{\left[L^2 + (1 + g_e)\widehat{S} \cdot \widehat{L} + gS^2\right]}{J^2}$$

Since $\widehat{J} = \widehat{L} + \widehat{S}$, the vectors form a general triangle, as shown in Fig. 10.11. Then, we can rewrite the cosine of the angle between $\widehat{L}$ and $\widehat{S}$, from the dot product term, using the *law of cosines* for a general triangle whose edges are $a = S, b = L$, and $c = J$, given by

$$c^2 = a^2 + b^2 - 2\mathbf{a} \cdot \mathbf{b} \Rightarrow \mathbf{a} \cdot \mathbf{b} = \frac{a^2 + b^2 - c^2}{2} \qquad (10.113)$$

There is one little caveat, however. The law of cosines relates the interior angle α of the triangle formed by $\widehat{S}$ and $\widehat{L}$. But, the physical angle between these two vectors is different, given by β, where the two are related by $\beta = 180° - \alpha$. This relation is also illustrated in Fig. 10.11. Therefore, it follows that

$$\cos(\alpha) = \cos(180 - \beta) = -\cos(\beta)$$

Applying this to $\widehat{S} \cdot \widehat{L}$ with $c = J, a = L$, and $b = S$, we have

$$V_M = \mu_B m_j B \frac{(L^2 + (1 + g)\widehat{S} \cdot \widehat{L} + gS^2))}{J^2}$$

10.11 Multi-electron Generalized g-Factor

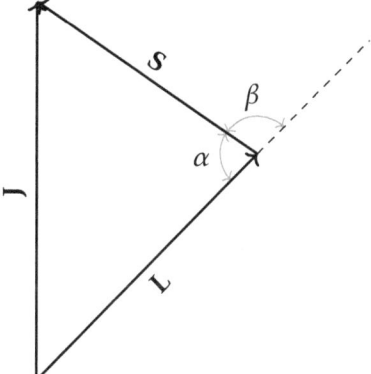

Figure 10.11: Illustration of the triangle vector relations between J, L and S, used to obtain the alternative form of g_J.

$$= \mu_B m_j B \frac{(2L^2 + (1+g)(J^2 - L^2 - S^2) + 2gS^2))}{2J^2}$$

Setting $g = 2$, the magnetic potential V_M becomes

$$V_M = \mu_B m_j B \frac{(2L^2 + 3J^2 - 3L^2 - 3S^2 + 4S^2))}{2J^2}$$

$$= \mu_B m_j B \frac{(3J^2 - L^2 + S^2))}{2J^2}$$

$$= \mu_B m_j B \frac{(3j(j+1) - \ell(\ell+1) + s(s+1)))}{2j(j+1)}$$

$$= g_J \mu_B m_j B$$

From this result, we can express an alternative form for the generalized g-factor denoted by g_J, as

Landé g-factor

$$g_J = \frac{3j(j+1) - \ell(\ell+1) + s(s+1)}{2j(j+1)} \qquad (10.114)$$

The result above is known as the **Landé g-factor**, and it is an alternative form of Eq. (10.112). In the following section, take this further by discussing how this result can be used along with some statistical thermodynamics discussed in Chap. 2, to determine the expectation value of the magnetic moment for any number of decoupled atoms with multiple electrons.

10.12 Spin Expectation Using Boltzmann Distribution

In Sec. 2.5, in our discussion of statistical thermodynamics, we derived the *Boltzmann distribution* for distinguishable fermions. When there is an arrangement of discrete decoupled atoms in a crystal, for example, each atom's distinct location makes all atomic states distinguishable. Because of this, we use the Boltzmann distribution to find the mean or *expectation value* of the magnetic moment from a set of N atoms. We know the energy of a localized electron magnetic moment in a magnetic field is given by

$$V = -\boldsymbol{\mu} \cdot \mathbf{B}$$

The analysis used from this point, follows an approach introduced by French-American physicist Léon Nicolas Brillouin (1889-1969), who was one of the first to intermingle quantum mechanical states with statistical mechanics. Quantum mechanics provides the atomic states, and therefore, the number of quantum states is known, along with a given number of atoms N. We know that when a moment is introduced to a magnetic field, a Zeeman splitting occurs. There appears $2j + 1$ non-degenerate states. This can be included readily using statistical thermodynamics, via the Boltzmann probability obtained in Chap. 2. Using the Boltzmann probability, the total magnetic moment in thermodynamic equilibrium can be expressed in proportion to the expectation or average equilibrium magnetic moment $\mu_{eq} = \langle \mu \rangle$ for an ensemble of N *weakly* interacting magnetic moments. Then, the total magnetic moment is given by

$$M_{eq} = N \langle \mu \rangle \tag{10.115}$$

Note that this equation assumes weak-to-no interactions between the N atoms. Taking B to be along the z axis, the energy becomes

$$V_M = -\boldsymbol{\mu} \cdot \mathbf{B} = -g_J m_J \mu_B B \tag{10.116}$$

g_J is the generalized g-factor discussed in Sec. 10.11. There we found that g_J is a function of m_j and m_s. Furthermore, because the analysis

10.12 Spin Expectation Using Boltzmann Distribution

to be done is simplified when g_J is independent of m_j, let us write Eq. (10.116) as

$$V_M = -\langle g_J \rangle m_j \mu_B B \quad \text{where} \quad \langle g_J \rangle = \sum_{m_j=-j}^{j} \frac{g_J e^{(g_J m_j \mu_B B/k_B T)}}{\sum_{m_j=-j}^{j} e^{(g_J m_j \mu_B B/k_B T)}} \quad (10.117)$$

The expectation value $\langle g_J \rangle$ uses the Boltzmann probability of the corresponding state $E(m_j)$, derived in Chap. 2, and given by Eq. (2.37). In this way, $\langle g_J \rangle$ is independent of m_j. m_j is the total magnetic quantum number where $m_j = -j, -j+1, ..., 0, ..., j-1, j$ for the total angular momentum J. Then, Eq. (10.115) becomes

$$M_{eq} = N\langle \mu \rangle = N \sum_{m_j=-j}^{j} \langle g_J \rangle m_j \mu_B f_B(E_{m_j})$$

$$= N\mu_B \sum_{m_j=-j}^{j} \langle g_J \rangle m_j \frac{e^{-(V_M/k_B T)}}{\sum_{m_j=-j}^{j} e^{-(V_M/k_B T)}}$$

$$= N\mu_B \sum_{m_j=-j}^{j} \langle g_J \rangle m_j \frac{e^{(\langle g_J \rangle m_j \mu_B B/k_B T)}}{\sum_{m_j=-j}^{j} e^{(\langle g_J \rangle m_j \mu_B B/k_B T)}}$$

To simplify notation, let $x = (\langle g_J \rangle \mu_B B)/(k_B T)$. Then, we can write

$$M_{eq} = N \frac{\sum_{m_j=-j}^{j} \langle g_J \rangle m_j \mu_B e^{m_j x}}{\sum_{m_j=-j}^{j} e^{m_j x}} \quad (10.118)$$

Using the relation $d(e^{m_j x})/dx = m_j e^{m_j x}$, Eq. (10.118) becomes

$$M_{eq} = N\langle g_J \rangle \mu_B \frac{\sum_{m=-j}^{j} \frac{d}{dx} e^{mx}}{\sum_{m=-j}^{j} e^{mx}}$$

We have used the fact that μ_B and $\langle g_J \rangle$ does not depend on m_j. Now, let $u = \sum_{m_j=-j}^{j} e^{m_j x}$. Then, we can write

$$M_{eq} = N\langle g_J \rangle \mu_B \frac{1}{u} \frac{du}{dx} = N\langle g_J \rangle \mu_B \frac{d}{dx} \ln u$$

It is useful to write u in the following form

$$u = \sum_{m_j=-j}^{j} e^{m_j x} = e^{-jx} + e^{(-j+1)x} + \cdots + e^{(j-1)x} + e^{jx}$$

$$= e^{-jx}(1 + e^x + e^{2x} + \cdots + e^{2jx})$$

$$= e^{-jx}(1 + r + r^2 + \cdots + r^{2j})$$

If we let $r = e^x$, the summation in parentheses is in the form of a *finite geometric series*, which has the form

$$a + ar + ar^2 + \cdots + ar^n = a\frac{1 - r^{n+1}}{1 - r}$$

In this case, $a = 1$, so u can be written as

$$u = e^{-jx}\frac{1 - r^{2j+1}}{1 - r} = \frac{e^{-jx} - e^{(j+1)x}}{1 - e^x}$$

This leads to the following form for M_{eq}, given by

$$M_{eq} = N\langle g_J\rangle\mu_B \frac{d}{dx}\ln\left[\frac{e^{-jx} - e^{(j+1)x}}{1 - e^x}\right]$$

$$= N\langle g_J\rangle\mu_B \frac{d}{dx}\ln\left[\frac{e^{-(j+\frac{1}{2})x} - e^{(j+\frac{1}{2})x}}{e^{-\frac{x}{2}} - e^{\frac{x}{2}}}\right]$$

$$= N\langle g_J\rangle\mu_B \frac{d}{dx}\ln\left[\frac{\sinh\left(j + \frac{1}{2}\right)x}{\sinh\frac{x}{2}}\right]$$

$$= N\langle g_J\rangle\mu_B \frac{d}{dx}\left[\ln\left(\sinh\left(j + \frac{1}{2}\right)x\right) - \ln\left(\sinh\frac{x}{2}\right)\right]$$

We have multiplied the log argument by $1 = e^{-x/2}/e^{-x/2}$. Using the derivative of the $\ln(\sinh u)$ term which is $du/dx \cdot (\cosh u/\sinh u)$, we obtain the result

$$M_{eq} = N\langle g_J\rangle\mu_B\left[\left(j + \frac{1}{2}\right)\coth\left[\left(j + \frac{1}{2}\right)x\right] - \frac{1}{2}\coth\left(\frac{x}{2}\right)\right]$$

Let us examine the asymptotic limit of $M_{eq}/N\langle g_J\rangle\mu_B$ as $x \to \infty$. Firstly, the hyperbolic cotangent function is defined as

$$\coth x = \frac{e^x + e^{-x}}{e^x - e^{-x}}$$

The limit of this expression is

$$\lim_{x\to\infty}\coth x = \frac{e^\infty + e^{-\infty}}{e^\infty - e^{-\infty}} = \frac{e^\infty}{e^\infty} = 1$$

10.12 Spin Expectation Using Boltzmann Distribution

This means that $M_{eq}/N\langle g_J\rangle\mu_B$ has a corresponding limit given by

$$\lim_{x\to\infty}\frac{M_{eq}}{N\langle g_J\rangle\mu_B} = \left(j+\frac{1}{2}\right)-\frac{1}{2} = j$$

Therefore, the *Brillouin function* can be normalized by defining it as

$$B_J(x) = \frac{1}{J}\left[\left(J+\frac{1}{2}\right)\coth\left(J+\frac{1}{2}\cdot x\right) - \frac{1}{2}\coth\left(\frac{1}{2}\cdot x\right)\right] \quad (10.119)$$

Defining $x = (\langle g_J\rangle\mu_B J B)/(k_B T)$, then B_J can also be written as

Brillouin Function

$$B_J(x) = \frac{1}{J}\left[\left(J+\frac{1}{2}\right)\coth\left(\frac{2J+1}{2J}\cdot x\right) - \frac{1}{2}\coth\left(\frac{1}{2J}\cdot x\right)\right] \quad (10.120)$$

Then, the expectation value of the equilibrium magnetic moment along the B-field axis can be written as

Expectation Value Of Equilibrium Magnetic Moment

$$M_{eq} = N\langle g_J\rangle\mu_B J B_J(x) \quad (10.121)$$

Examples of Brillouin functions are shown in Fig. 10.12 for different values of the total angular momentum quantum number J. The smaller is J, the more shallow B_J becomes. The Brillouin function, thus, contains two parameters that can be extracted from experiments. The slope of the normalized functions $B_J(x)$ is determined by J, while the height is determined by g_J. Then, the saturation levels of $M_{eq}/N\mu_B = g_J B_J(x)$ provide the Landé g-factor g_J. This is illustrated in Fig. 10.13, along with measurement data adapted from W. E. Henry (1952). In Fig. 10.13, each curve saturates at g_J, allowing the determination of this parameter. The data points taken from W. E. Henry (1952), were measured at four different temperatures, on three magnetic ions Cr^{3+}, Fe^{3+}, and Gd^{3+}. The role of the ionic charge is to ionically attract other nonmagnetic charged particles around them. This enables well-separated magnetic ions. Each of these magnetic ions are surrounded by tens of nonmagnetic molecules, which leads to weakened interactions between the magnetic

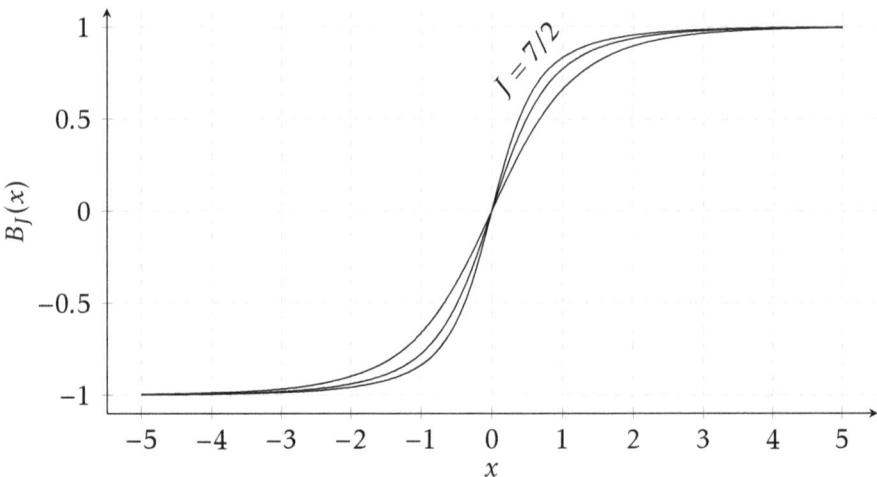

Figure 10.12: Brillouin function for $J = 1/2, 3/2, 5/2$, and $7/2$. The smaller is J, the more shallow B_J transitions.

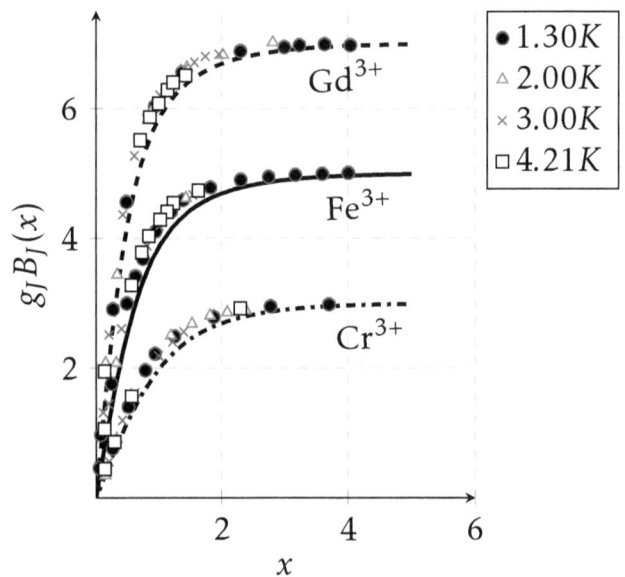

Figure 10.13: Brillouin function for $J = 3/2$, $J = 5/2$, and $J = 7/2$. The data points shown are adapted from the measurement data reported by W. E. Henry, Phys. Rev. 88, 559 (1952), while the lines are Brillouin functions.

10.12 Spin Expectation Using Boltzmann Distribution

Magnetic Ion	J	g_J
Cr^{3+}	$\frac{3}{2}$	3
Fe^{3+}	$\frac{5}{2}$	5
Gd^{3+}	$\frac{7}{2}$	7

Table 10.4: Measured total angular moment quantum number J and the corresponding g-factor g_J.

ions. These details are important because such conditions help to satisfy the assumptions of the Brillouin function derivation. Table 10.4 lists the measured parameters J and g_J for the ions tested by W. E. Henry in 1952. Fe and Cr are in the family of 3d transition elements having their valence electrons in 3d orbitals. This family contains well-known magnetic elements, including cobalt and nickel. However, it may be less obvious that gadolinium (Gd), a rare-earth element with atomic number of 64, has an even stronger net magnetic moment per atom in a magnetic field. The value $g_J = 7$ indicates the 7 unpaired electrons in Gd, in the 4f orbital, that all contribute to the net spin polarization in the isolated atom, coupling to a magnetic field. There is also an additional unpaired electron in the 5d orbital, however, being closer to the valence shell and it plays a role in coupling to other Gd atoms when they are allowed to interact. It does not play a role in the data shown above because the atoms are coated in such a way to minimize interactions. Gd turns out to be a very useful element in magnetic-based technologies, and is widely used in both cutting edge research in magnetism, like influencing reversal of magnetization using only heat, as well as more mature technologies like magnetic resonance spectroscopy (MR).

Because of the form of the potential energy, being proportional to the expectation value of the magnetic moment, this result also leads to a more general form of the **thermodynamic equilibrium magnetic potential**, which becomes

$$\langle V \rangle (B_e, T) = -\langle \boldsymbol{\mu} \rangle \cdot \mathbf{B}_e = -\mu(0) B_e B_J (\langle V \rangle / k_B T) \qquad (10.122)$$

$\mu(0)$ is the magnetic moment at $T = 0$. We have included the Brillouin function to demonstrate that it is obtained in a manner consistent with quantum mechanics by averaging over the quantum states. It is a good example of how quantum mechanics can be used with statistical thermodynamics. The results above are useful for a quantum system with sufficiently weak interactions between atoms or atomic spin carriers.

10.13 Chapter Summary

In this chapter, we have focused on the extension of several concepts in quantum mechanics to systems having more than one electron. We found that electrons, as *fermions*, are described by total anti-symmetric wave functions. This lead to several Coulomb interactions some of which attract (negative energy) while others repel (positive energy). One of these contributions is the *exchange integral*. The exchange integral shows up in the energy of multi-electron systems, as we observed for helium, as well as hydrogen and its ionic molecule. It also lead to an alternative form of energy relating to a dot product between a pair of spins, known as the *Heisenberg exchange*.

We also introduced the *bond length* of molecules, as well as a method to measure this parameter using rotational-vibrational spectroscopy. Additional important properties of angular momentum were discussed, which also lead to the method of determining *Clebsch-Gordan coefficients*, exploiting some properties of ladder operators. The generalized and Landé g-factor for multi-electrons were also introduced and used to obtain equilibrium magnetic moments for weakly interacting atoms. Table 10.5 summarizes some of the key results from this chapter.

Table 10.5: Chapter 10 Summary Equations.

Name	Equation
Coulomb integral	$C_{ij} = \frac{1}{2} \int \psi_a^*(\mathbf{r}_i) \rho(\mathbf{r}_i) \psi_a(\mathbf{r}_i) d\mathbf{r}_i$
Charge density	$\rho(\mathbf{r}_i) = \frac{e^2}{4\pi\epsilon_0} \int \frac{\psi_b^*(\mathbf{r}_j)\psi_b(\mathbf{r}_j)}{|\mathbf{r}_j - \mathbf{r}_i|} d\mathbf{r}_j$
Exchange integral	$J_{ij} = Re\left[\frac{e^2}{4\pi\epsilon_0} \int \frac{\psi_a^*(\mathbf{r}_i)\psi_b^*(\mathbf{r}_j)\psi_a(\mathbf{r}_j)\psi_b(\mathbf{r}_i)}{|\mathbf{r}_j - \mathbf{r}_i|} d\mathbf{r}_i d\mathbf{r}_j \right]$
Commutation postulate	$[\widehat{J}_{ik}, \widehat{J}_{jn}] = 0 \quad i \neq j = 1,2 \quad k \neq n = x,y,z$

10.14 Chapter Problems

Table 10.5

Landé factor	$g_J = \frac{3j(j+1)-\ell(\ell+1)+s(s+1)}{2j(j+1)}$
Brillouin function	$B_J(x) = \frac{1}{J}\left[\left(J+\frac{1}{2}\right)\coth\left(J+\frac{1}{2}\right)x - \frac{1}{2}\coth\left(\frac{1}{2}\right)x\right]$
Moment expectation	$M_{eq} = N\langle g_J\rangle \mu_B J B_J(x)$

10.14 CHAPTER PROBLEMS

Problem 10.1 Consider the two-electron wave function given by $\psi(\mathbf{r}_2,\mathbf{r}_1) = C\psi_{1s}(\mathbf{r}_1)\psi_{1s}(\mathbf{r}_2)(|\uparrow\rangle(1)|\downarrow\rangle(2) - |\uparrow\rangle(2)|\downarrow\rangle(1))$. The spin states here do not depend on space explicitly. Find the normalization constant C using the normalization condition.

Problem 10.2 Show that for a wave function given by $\psi(\mathbf{r}_2,\mathbf{r}_1) = \frac{1}{\sqrt{2}}(\psi_a(\mathbf{r}_2)\psi_b(\mathbf{r}_1) - \psi_a(\mathbf{r}_1)\psi_b(\mathbf{r}_2))$, that it is normalized if and only if $\int \phi_a^*\phi_a dr = 1$ and $\int \phi_a^*\phi_b dr = 0$.

Problem 10.3 Calculate the Landé factor for a helium atom, having two electrons in the s-orbital shell. Then, calculate the same for Lithium. First determine s and ℓ for each electron. Sum to obtain j, then use the formula.

Problem 10.4 In Eq. (10.121), we calculated the expectation value of the magnetic moment for a system with j angular momentum states. Compute the expectation value of the magnetic moment using the Boltzmann distribution for $j = s = 1/2$. Thus, there are only two states.

Problem 10.5 Derive the expectation value distance $\langle |r_2 - r_1|\rangle$ between two electrons in the 1s orbital by evaluating (in spherical coordinates)

$$\langle |r_2 - r_1|\rangle = \int |r_2 - r_1|\psi_{1s}^*(r_1)\psi_{1s}^*(r_2)\psi_{1s}(r_1)\psi_{1s}(r_2)\sin\theta\, dr_1\, dr_2\, d\theta\, d\varphi$$

Hint: Along with Eqs. (10.44), use the integral relation:

$$\int x^3 e^x dx = (x^3 - 3x^2 - 6x + 6)e^x$$

Problem 10.6 Calculate the expectation value $\langle 1/R \rangle$ using the 1s orbital wave function given by Eq. (10.39).

Problem 10.7 Calculate the expectation value $\langle 1/d_A \rangle$ also using the 1s orbital wave function given by Eq. (10.39).

Problem 10.8 Lastly, calculate the expectation value $\langle 1/d_B \rangle$ using the 1s orbital wave function for atoms A and B. given by Eq.

10.15 SUGGESTED READINGS & REFERENCES

[1] N. K. Berezhetskaya, G. S. Voronov, et. al., *Effect Of A Strong Optical-Frequency Electromagnetic Field On The Hydrogen Molecule*, JETP **31**, 3, pp. 403-406 (1970)

[2] B. Duan, Xiao-Yan Gu and Zhong-Qi Ma, Eur. Phys. J. D **19**, 9 (2002)

[3] A. Hutem, and S. Boonodui, J. Math. Chem. **50**, 2086 (2012)

[4] D. R. Lide, *CRC Handbook of chemistry and physics*, 85th Ed. CRCPress, Boca Raton (2004)

[5] J. I. Powell and B. Crosemann, *Quantum Mechanics*, Addison-Wesley, Reading, pp. 456-458 (1961)

[6] J. F. Ogilvie, *Infrared Spectroscopy of Diatomic Molecules-the First Century*, Chin. J. of Phys., **27**, 4(1989)

[7] T. Y. Wu, *Vibrational Spectra and Structure of Polyatomic Molecules* (1939)

[8] W. Pauli Ann. d. Phys., **68** (1922)

[9] K. F. Niessen, *Zur Quantentheorie des Wasserstoffmolekulions*, Dissertation, Utrecht, (1922)

[10] E. A. Hylleraas, *Über die Elektronenterme des Wasserstoffmoleküls*, Zeitschrift für Physik, **71**, pp. 739–763, (1931)

[11] R. M. Boyack, *The Theory of the Hydrogen Molecule Ion, Scalar Beams, and Scattering by Spheroids*, Thesis, (2004)

[12] W. E. Henry, *Spin Paramagnetism of Cr^{+++}, Fe^{+++}, and Gd^{+++} at Liquid Helium Temperatures and in Strong Magnetic Fields*, Phys. Rev. **88**, 559 (1952)

10.15 Suggested Readings & References

CHAPTER 11

Relativistic Quantum Mechanics

When we dealt with the dynamics of the electron spin previously, we used the *Pauli equation*. This is a two-equation form of the *Schrödinger equation*, including the coupling of the spin degree of freedom to a magnetic field B, for example. We introduced spin matrices and saw how they could be used to determine the behavior of the electron spin dynamically, for example, in precessional motion. If a magnetic field were all that electron spins interacted with, we would not need much more than the Schrödinger equation. However, this is not the case. The electron spin *does* interact with energetic aspects of the atom, which we would not find restricting ourselves to the Schrödinger equation. In other words, it is known to be *non-relativistic*, and we'll discuss the meaning of this statement in this chapter. Also, using a *relativistic* quantum mechanical formulation, we will determine the value of the g–factor, which was introduced using the nonrelativistic Schrödinger equation. Up to now, it has been a phenomenological parameter associated with spin angular momentum. It turns out that it can be calculated. An important interaction of special interest, and missing from the *nonrelativistic* Schrödinger equation, of crucial importance in magnetism and spin-physics, is known as the *spin-orbit interaction*. In order for us to unravel where it comes from, we must review some key concepts from the *special theory of relativity*. From there, we can begin

11.1 SPECIAL THEORY OF RELATIVITY & MOMENTUM

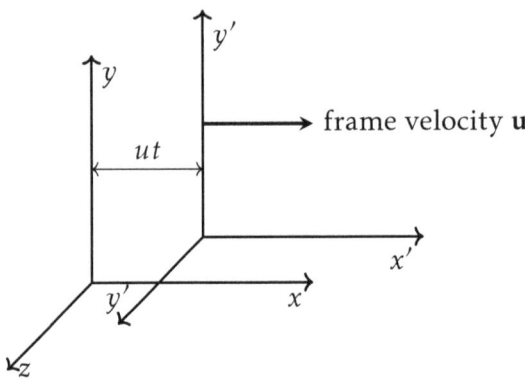

Figure 11.1: Illustration of an inertial reference frame **r**′ moving at uniform velocity **u**, relative to a *stationary* frame **r**.

The concept of *relativity* in the scientific domain goes back towards the end of the nineteenth century. It even predates the work of Albert Einstein who did not work out the special theory of relativity until 1905. It originally referred to the relation between the *position* **r** and time t of two different inertial frames, or their *relative* positions, as shown in Fig. 11.1. One frame **r** is taken to be at rest, and the other frame **r**′ is moving with uniform velocity **u**. Common intuition based on everyday experience suggests we can relate the coordinates between the two frames as

$$\mathbf{r}' = \mathbf{r} - \mathbf{u}t \tag{11.1a}$$

$$\mathbf{r} = \mathbf{r}' + \mathbf{u}t \tag{11.1b}$$

$$t = t' \tag{11.1c}$$

The relations above are known as the **Galilean transformation**, and they are a set of linear equations between position and time coordinates relating one frame to the other, *with the unique feature that time is identical between the two frames.*

11.1 Special Theory Of Relativity & Momentum

 Galilean Frame: A Galilean frame is a physical reference frame, where time is invariant, or identical between all frames. Positions between two distinct frames also follow Eq. (11.1).

Time-differentiation of Eqs. (11.1a) and (11.1b) then gives the following relations for velocities:

$$\mathbf{v'} = \mathbf{v} - \mathbf{u} \tag{11.2a}$$

$$\mathbf{v} = \mathbf{v'} + \mathbf{u} \tag{11.2b}$$

In 1851, a famous set of experiments were conducted by French physicist Hippolyte Fizeau (1819-1896), measuring the speed of light in moving water. The velocity of the light in the water, moving at velocity u, measured in the rest frame, was expected to behave according to Eq. (11.2b). Fizeau found that it did not. Instead, he found that it follows a relation given empirically by

$$v = v' + u\left(1 - \frac{1}{n^2}\right) \tag{11.3}$$

$n = c/v'$ is the *index of refraction* (see Chap. 1). The term in parentheses was regarded as a *drag coefficient* leading to a considerable reduction in the observed velocity of light in moving water. The index of refraction for water $n \sim 1.33$. Thus, substitution of n and using a water flow speed of $u = 1$m/s, with $c = 3 \times 10^8$m/s, in his rest frame, Fizeau found the speed of light to be

$$v = v' + u\left(1 - \frac{1}{n^2}\right)$$
$$= v' + u\left(1 - \frac{1}{1.33^2}\right)$$
$$= v' + 0.434u < v' + u$$

Instead of adding the full speed of the water u, he only needed to add ~43% of it! We'll see shortly that as long as $u \ll c$, the *relativistic* result we shall obtain leads exactly to Fizeau's empirical result in describing the coordinates of light. Note that $v' < v$, as well.

Other well-known experiments at the time also motivated Lorentz. In 1886 Michelson and Morley refined Heinrik Fizeau's experiments confirming his findings, and extending experimental efforts beyond a water medium to try and find evidence of motion of a background *luminiferous aether*, which was a widely believed concept held by many physicists at the time. It was postulated as the medium in which electromagnetic waves travel. The Michelson-Morley experiment on this, however, could not detect such evidence. At the time, physicists were developing models in an effort to explain measurements of light speed in different media, since all experimental results were inconsistent with Eqs. (11.2a) and (11.2b). This lead Lorentz to investigate generalizing the linear transformation equations further. If positions could be different between frames, *depending on the frame velocity*, perhaps the times could be related differently, as well, depending the frame velocity. To generalize Eqs. (11.1a)-(11.1c), we can write x and x' (assume 1D for simplicity) as follows:

$$x' = (x - ut)\gamma_r \qquad (11.4a)$$

$$x = (x' + ut')\gamma_r \qquad (11.4b)$$

In this generalization, $t \neq t'$, unlike in the Galilean transformation. Thus, we have two equations in two unknowns, namely γ_r and t'. Once we find γ_r, we can find the relation between t and t'. γ_r can be found by first computing $x'x$, which gives

$$x'x = \gamma_r^2(x - ut)(x' + ut') = \gamma_r^2(xx' - u^2tt' + xut' - x'ut) \qquad (11.5)$$

Here is where an important leap in all of the theory of relativity enters the picture. *An assumption made by Lorentz was that he took the speed of light as constant in both frames.* Today, this assumption is taken as one of the postulates of relativity.

Postulate of Relativity: The first postulate, or assumption, in the theory of relativity is that *maximum* light speed is invariant in all inertial reference frames.

Then, if x and x' are taken to be the position of the light to be measured, then it follows that

$$x = ct \quad \text{and} \quad x' = ct' \qquad (11.6)$$

Note that the above statement also implicitly assumes synchronization at the initial time $t = t' = 0$. Dividing both sides of Eq. (11.5) by $x'x$, and

11.1 Special Theory Of Relativity & Momentum

substituting Eq. (11.6) then gives

$$1 = \gamma_r^2 \left(1 - \frac{u^2 tt'}{x'x} + \frac{xut'}{x'x} - \frac{x'ut}{x'x}\right)$$

$$= \gamma_r^2 \left(1 - \frac{u^2}{c^2} + \frac{u}{c} - \frac{u}{c}\right)$$

$$= \gamma_r^2 \left(1 - \frac{u^2}{c^2}\right)$$

Solving for γ_r, we have the result

Lorentz Factor

$$\gamma_r = \frac{1}{\sqrt{1 - \frac{u^2}{c^2}}} \quad (11.7)$$

Eq. (11.7) is known as the **Lorentz factor**, and it was obtained by Lorentz c.1895, to try and understand the various experiments involving measurements of light speed. It was motivated not only by Fizeau's experiments, but by the Michelson-Morley experiment, as well as experiments on the astronomical phenomenon known as the *aberration of light*.

Having γ_r, we can determine t' by first multiplying Eq. (11.4a) by x, so we have

$$xx' = c^2 t't = (x^2 - utx)\gamma_r = (c^2 t^2 - utx)\gamma_r \quad (11.8)$$

We have also used Eq. (11.6). Dividing both sides by $c^2 t$, we obtain

$$t' = \left(t - \frac{ux}{c^2}\right)\gamma_r \quad (11.9)$$

A similar procedure for t gives

$$t = \left(t' + \frac{ux'}{c^2}\right)\gamma_r \quad (11.10)$$

We can see that going between the two frames requires only changing the sign of u, along with the respective coordinates for space and time.

This generalized result also contains symmetry, just as with the Galilean transformation. So, we have obtained the following four transformation relations between x, x', t, and t':

$$x' = \frac{x - ut}{\sqrt{1 - \frac{u^2}{c^2}}} \quad \text{and} \quad x = \frac{x' + ut'}{\sqrt{1 - \frac{u^2}{c^2}}} \quad (11.11a)$$

$$t' = \frac{t - \frac{ux}{c^2}}{\sqrt{1 - \frac{u^2}{c^2}}} \quad \text{and} \quad t = \frac{t' + \frac{ux'}{c^2}}{\sqrt{1 - \frac{u^2}{c^2}}} \quad (11.11b)$$

The relations above are known as the **Lorentz transformation**. Although we have only used x for the spatial coordinate, the most general result is obtained by simply $x' \to \mathbf{r}'$ and u with its vector form. The profound implications of the Lorentz relations would not begin to reveal themselves until around ten years later, with the work of Einstein.

Einstein was one of the first to examine closely the implications of the Lorentz transformation on physical systems beyond light, since r and t are usually involved in their physical laws. One of the first variables Einstein considered was the relation between the velocities of the two Lorentzian frames, or the relativistic analog to Eq. (11.2).

To obtain the analog of Eqs. (11.2a) and (11.2b), we need to consider the difference between any two points (t_1, x_1) and (t_2, x_2) in the same reference frame, described by Eqs (11.11a) and (11.11b). Such differences are given by

$$\Delta x' = (\Delta x - u\Delta t)\gamma_r \quad \text{and} \quad \Delta x = (\Delta x + u\Delta t')\gamma_r \quad (11.12a)$$

$$\Delta t' = \left(\Delta t - \frac{u\Delta x}{c^2}\right)\gamma_r \quad \text{and} \quad \Delta t = \left(\Delta t' + \frac{u\Delta x'}{c^2}\right)\gamma_r \quad (11.12b)$$

Taking the ratios $\Delta x'/\Delta t'$ and $\Delta x/\Delta t$ leads to

$$\frac{\Delta x'}{\Delta t'} = \frac{\Delta x - u\Delta t}{\Delta t - \frac{u\Delta x}{c^2}}$$

$$\frac{\Delta x}{\Delta t} = \frac{\Delta x' + u\Delta t'}{\Delta t' + \frac{u\Delta x'}{c^2}}$$

Dividing the numerator and denominator by $\Delta t'$ and Δt, respectively, and taking the limits as $\Delta \to 0$ gives the final result as

$$v' = \frac{v - u}{1 - \frac{uv}{c^2}} \quad (11.13a)$$

11.1 Special Theory Of Relativity & Momentum

$$v = \frac{v' + u}{1 + \frac{uv'}{c^2}} \qquad (11.13b)$$

These relations, obtained by Einstein, are known as the **velocity addition formulae**. By comparing Eq. (11.13a) and (11.13b) to Eqs. (11.2a) and (11.2b), we can see that the relations are the same when $1 \pm uv/c^2 \sim 1$ or $uv/c^2 \ll 1$. The relativistic result given by Eq. (11.13b) turns out to be a generalization of Fizeau's empirical relation in Eq. (11.3). To see this, we can write Eq. (11.13b) as follows:

$$v = \frac{v' + u}{1 + \frac{uv'}{c^2}}$$

$$= \frac{(v' + u)\left(1 - \frac{uv'}{c^2}\right)}{\left(1 + \frac{uv'}{c^2}\right)\left(1 - \frac{uv'}{c^2}\right)}$$

$$= \frac{v' - \frac{uv'^2}{c^2} + u - \frac{u^2 v'}{c^2}}{\left(1 + \frac{uv'}{c^2}\right)\left(1 - \frac{uv'}{c^2}\right)}$$

$$= \frac{v'\left(1 - \frac{u^2 v'^2}{c^4}\right) + u\left(\frac{uv'^3}{c^4} - \frac{v'^2}{c^2} + 1 - \frac{uv'}{c^2}\right)}{\left(1 + \frac{uv'}{c^2}\right)\left(1 - \frac{uv'}{c^2}\right)}$$

$$= v' + u \frac{\left[\frac{uv'}{c^2}\left(1 - \frac{v'^2}{c^2}\right) - \left(1 - \frac{v'^2}{c^2}\right)\right]}{\left(1 + \frac{uv'}{c^2}\right)\left(1 - \frac{uv'}{c^2}\right)}$$

$$= v' + u \frac{\left[1 - \frac{v'^2}{c^2}\right]}{1 + \frac{uv'}{c^2}}$$

Using $v'^2/c^2 = 1/n^2$, the above result reduces to Fizeau's empirical formula in the approximation $1 + uv'/c^2 \sim 1$. Surely, Einstein was onto something by beginning to examine the implications of the Lorentz transformation.

For our aims, we ultimately want to determine how these results affect the energy of a system. To answer this question, we need the relativistic momentum. But to obtain this, we have to discuss one more important concept revealed by Einstein, following from the Lorentz transformation. Einstein and others had noticed that the time variable now *had to be elevated to the status of a coordinate*, meaning that it is now equivalent to x, y, and z as observed in transformations like rotation and translation.

Recall that in the Galilean transformation $t = t'$, which makes it invariant in any Galilean frame. However, in a Lorentzian frame, it is evident that $t \ne t'$, but instead $t = f(t', x')$, just as $x = g(t', x')$, where both f and g are linear operators when u is constant. This seemingly small detail turns out to have enormous implications now that we are not able to treat Galilean time the same in all Lorentzian frames.

Though time is *not invariant in the Lorentzian frame*, perhaps another time-unit quantity might be invariant that can be identified. Einstein found exactly what quantity is conserved in the coordinate system described by a Lorentz transformation. Using four coordinates t, x, y and z, we can form a vector analogous to $\mathbf{r} = [x, y, z]^T$ used in a Galilean frame, by defining a **four-vector S**, also known as the **space-time vector**, given by

$$\mathbf{S} = [ct, x, y, z]^T = [x_0, x_1, x_2, x_3]^T \qquad (11.14)$$

Note that in defining **S**, ct is used instead of t in order to give all coordinates the same physical units. For a vector in a Galilean frame, there is the concept of the *dot product* which is invariant under translations and rotations. These two transformations, in particular, are relevant to describing electron motion. Therefore, by dotting a vector with itself, we know the length of the vector $\mathbf{r}$ is conserved under translations and/or rotations. However, if we take the dot product of **S** with itself, we obtain

$$\mathbf{S} \cdot \mathbf{S} = c^2 t^2 + x^2 + y^2 + z^2$$

Substituting the Lorentz transformation relations for the four coordinates into the above quickly reveals that this quantity is not conserved, or that

$$c^2 t^2 + x^2 + y^2 + z^2 \ne c^2 t'^2 + x'^2 + y'^2 + z'^2$$

But, instead of summing all the terms, let's see what happens if the dot product is defined by a difference between x_0^2 and $x_1^2 + x_2^2 + x_3^2$. For convenience, we will ignore $y = x_2$ and $z = x_3$ coordinates, so we have

$$\mathbf{S} \cdot \mathbf{S} = c^2 t^2 - x^2$$

$$= \left(ct' + \frac{ux'}{c}\right)^2 \gamma_r^2 - (x' + ut')^2 \gamma_r^2$$

$$= \left(c^2 t'^2 + 2t' u x' + \frac{u^2 x'^2}{c^2} - x'^2 - 2x' u t' - u^2 t'^2\right) \gamma_r^2$$

11.1 Special Theory Of Relativity & Momentum

$$= \left[(c^2 - u^2)t'^2 - \left(1 - \frac{u^2}{c^2}\right)x'^2\right]\gamma_r^2$$

$$= \left[\left(1 - \frac{u^2}{c^2}\right)c^2 t'^2 - \left(1 - \frac{u^2}{c^2}\right)x'^2\right]\gamma_r^2$$

$$= c^2 t'^2 - x'^2 = \mathbf{S}' \cdot \mathbf{S}'$$

We have used the definition for the Lorentz factor $\gamma_r^2 = 1/(1 - u^2/c^2)$. This unique definition, using a difference rather than a sum, for the space-time vector inner-product is known as the **space-time interval**, defined as

Invariant Space-Time Interval

$$\mathbf{S} \cdot \mathbf{S} = c^2 t^2 - x^2 - y^2 - z^2 \tag{11.15}$$

This leads to the result that the space-time vector has a vector of differences $\Delta \mathbf{S}$ given by

$$\Delta \mathbf{S} = [c\Delta t, \Delta x, \Delta y, \Delta z]^T \tag{11.16}$$

The significance of this result is that Eq. (11.16) can be used to define *an invariant unit of time Δt^**, analogous to Δt used in a Galilean frame. $\Delta t*$, in a Lorentzian frame, can be defined as

$$\Delta t^* = \frac{1}{c}\sqrt{\Delta \mathbf{S} \cdot \Delta \mathbf{S}} \tag{11.17}$$

$$= \frac{1}{c}\sqrt{c^2 \Delta t^2 - \Delta x^2} \tag{11.18}$$

$$= \Delta t \sqrt{1 - \frac{\Delta x^2}{\Delta t^2}\frac{1}{c^2}} \tag{11.19}$$

The $1/c$ factor is included only to ensure units of time. This quantity of time is known as the **proper time**. Before making use of this result, let's review how the momentum is defined in the Galilean frame. We determine the momentum as

$$p = m_0 \frac{dx}{dt} = m_0 \times \lim_{\Delta t \to 0} \frac{\Delta x}{\Delta t} \tag{11.20}$$

The subtle detail here is that the derivative is done with respect to t, which is invariant in all Galilean frames. However, we know that in the

Lorentzian frame, this is no longer true. Thus, we require a Δ–unit of time that is the same in all Lorentzian frames. That is the significance of Eq. (11.17). Therefore, in the Lorentz frame, the relativistic momentum becomes

$$p = m_0 \times \lim_{\Delta t^* \to 0} \frac{\Delta x}{\Delta t^*} = m_0 \lim_{\Delta t^* \to 0} \frac{\Delta x}{\Delta t \sqrt{1 - \frac{\Delta x^2}{\Delta t^2} \frac{1}{c^2}}}$$

Evaluating this limit leads to our final result given by

Relativistic Momentum

$$\mathbf{p} = \frac{m_0 \mathbf{v}}{\sqrt{1 - \frac{v^2}{c^2}}} \qquad (11.21)$$

Eq. (11.21) is distinguished as the **relativistic momentum**, as opposed to the classical *nonrelativistic* momentum (when $v/c \ll 1$). Thus, Eq. (11.21) approaches $m_0 v$ asymptotically in conditions where $v \ll c$. Now that we have the relativistic momentum, we can determine the relativistic kinetic energy. This will allow us to segue to the quantum mechanics of relativistic energy. This is the subject of the next section.

11.2 Relativistic Energy

When considering a particle traveling at a speed close to the speed of light c, one novel interpretation of the relativistic momentum that arises is that the mass of the particle depends on its speed v since Eq. (11.21) leads to

$$p = \frac{m_0 v}{\sqrt{1 - v^2/c^2}} = m(v)v$$

m_0 is the **rest-mass**, or the mass corresponding to $v = 0$. In practice, these are the conditions in which the mass of most objects are measured. With the relativistic mass m, the corresponding change in relativistic kinetic energy $\overline{E}$ can be found the same way that the non-relativistic kinetic energy is found. It is obtained by using the *work-kinetic energy theorem*, which gives

$$\Delta \overline{K} = \int F \cdot dx = \int \frac{dp}{dt} \cdot dx = \int v \, dp \qquad (11.22)$$

11.2 Relativistic Energy

Using Eq. (11.21) and the relation $dp = d(mv) = dm \cdot v + m \cdot dv$, we then have

$$dp = \frac{m_0}{[1-v^2/c^2]^{1/2}} dv + \frac{m_0 v^2}{c^2[1-v^2/c^2]^{3/2}} dv$$

$$= \frac{m_0[1-v^2/c^2]}{[1-v^2/c^2]^{3/2}} dv + \frac{m_0 v^2}{c^2[1-v^2/c^2]^{3/2}} dv$$

$$= \frac{m_0}{[1-v^2/c^2]^{3/2}} dv$$

Using Eq. (11.22), the change in the relativistic kinetic energy $\Delta \overline{K}$ becomes

$$\Delta \overline{K} = \int v \, dp = \int \frac{m_0 v}{[1-v^2/c^2]^{3/2}} dv = \frac{m_0 c^2}{[1-v^2/c^2]^{1/2}} \bigg|_{v_i}^{v_f} \quad (11.23)$$

Taking the initial velocity to be zero (corresponding to a rest mass), we have our result for the relativistic kinetic energy, given by

$$\Delta \overline{K} = \frac{m_0 c^2}{[1-v^2/c^2]^{1/2}} - m_0 c^2 = mc^2 - m_0 c^2 = \Delta mc^2 \quad (11.24)$$

Eq. (11.24) *forces us to re-evaluate our understanding of mass or inertia,* because this result equates mass with energy. It says that *any change in kinetic energy for a particle arises from a change in mass.* This novel interpretation of mass is one of the hallmarks of the theory of relativity, made known by Einstein. Mass is no longer some trivial constant coefficient of acceleration, but is, itself, a key component to the total energy of a system in motion. From Eq. (11.24), the instantaneous relativistic kinetic energy $\overline{K}$, is evidently given by

Relativistic Kinetic Energy

$$\overline{K} = mc^2 \quad (11.25)$$

Eq. (11.25) appears strange at first, however, we must remember that it contains a non-constant mass, and this is the significance of such a simple equation. Eq. (11.25) can be cast into a more useful form for quantum mechanics if we introduce the energy as a function of the momentum $\bar{\mathbf{p}}$. To do this, we can manipulate the quantity p^2c^2 as follows:

$$p^2c^2 = \frac{m_0^2 v^2 c^2}{1 - v^2/c^2} = \frac{m_0^2 (v^2/c^2) c^4}{1 - v^2/c^2}$$

$$= \frac{m_0^2 (v^2/c^2) c^4 - m_0^2 c^4 + -m_0^2 c^4}{1 - v^2/c^2}$$

$$= \frac{m_0^2 c^4 [v^2/c^2 - 1]}{1 - v^2/c^2} + \frac{m_0^2 c^4}{1 - v^2/c^2}$$

$$= -m_0^2 c^4 + m^2 c^4$$

This leads to the following relation for the *squared relativistic kinetic energy* $\overline{K}^2$, where we have

Square of Relativistic Kinetic Energy

$$\overline{K}^2 = p^2 c^2 + m_0^2 c^4 \qquad (11.26)$$

From Eq. (11.26), the *relativistic kinetic energy* $\overline{K}$ is given by

Relativistic Kinetic Energy-Form 2

$$\overline{K} = \pm \sqrt{p^2 c^2 + m_0^2 c^4} \qquad (11.27)$$

Eq. (11.27) can also be written as

$$\overline{K} = \sqrt{p^2 c^2 + m_0^2 c^4} = \sqrt{m_0^2 + \frac{p^2}{c^2}} \cdot c^2$$

Thus, Eq. (11.27) suggests a definition for the **relativistic mass** m given by

$$m = \sqrt{m_0^2 + \frac{p^2}{c^2}} = m_0 \sqrt{1 + \frac{p^2}{m_0^2 c^2}} \qquad (11.28)$$

Fig. 11.2 plots Eq. (11.28), illustrating how the *relativistic mass* changes with momentum. It is the *relativistic mass* that determines the shape of

11.2 Relativistic Energy

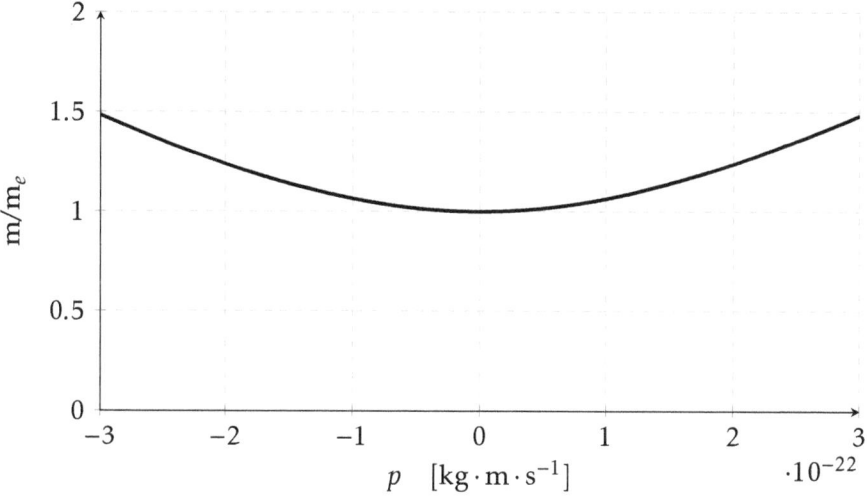

Figure 11.2: The relativistic mass m as a function of momentum p. The rest mass is a minimum mass existing only when $p \approx 0$.

the relativistic kinetic energy curve since $\overline{E} = mc^2$. Moreover, a consequence of the relativistic kinetic energy being squared is that Eq. (11.27) has both a *positive and negative branch*. These are illustrated in Fig. 11.3. Both relativistic energy branches are notably separated by 1e6 eV (1*MeV*), which is staggering for an energy of a single quantum particle compared to any of the energies we have dealt with so far, being on the order of a few eV. Our objective, from here on out, relies on the top branch, where $\overline{K} > 0$. With the relativistic energy and momentum relations, along with the established operators of quantum mechanics, a *relativistic quantum theory* becomes possible. Let's discuss how such a formulation has evolved over the years, starting with the results of Klein and Gordon.

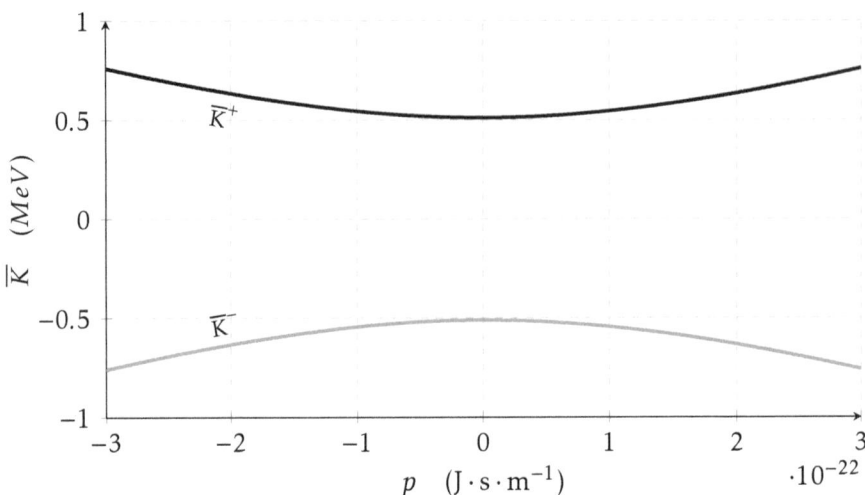

Figure 11.3: Illustration of the relativistic kinetic energy branches as functions of momentum p.

11.3 THE KLEIN-GORDON EQUATION

The Schrödinger equation as we have been using it, is known to be *nonrelativistic*. In particular, it sits near the asymptotic relativistic limit of speeds $v \ll c$ or $v/c \ll 1$, where c is the speed of light in vacuum. In this limit, $m \approx m_0$ and $p \approx = m_0 v = p_0$. For the Schrödinger equation, it means that the energy of a particle moving near the speed of light is *not completely described*. Therefore, some form of corrections are needed to bring it more into alignment with the relativistic theory. It is through the *theory of relativity*, that these corrections may be obtained. This allows us to construct an extended form of the Schrödinger equation that is more consistent with the theory of relativity, though still an arbitrary approximation.

With the relativistic kinetic energy, we just have to add the potential energy V, which is identical to those used in the Schödinger equation. We showed earlier that the *relativistic kinetic energy* $\overline{K}$ of a particle is given by

$$\overline{K}^2 = p^2 c^2 + m_0^2 c^4 \tag{11.29}$$

p is the momentum and m_0 is the rest mass (mass at zero velocity). Adding a potential energy V to the system, the **total relativistic energy** $\overline{E}$ is given by

11.3 The Klein-Gordon Equation

Total Relativistic Energy

$$\overline{E} = \overline{K} + V = \sqrt{p^2 c^2 + m_0^2 c^4} + V \qquad (11.30)$$

Since $\overline{K} = \overline{E} - V \Rightarrow$, use of Eq. (11.29) leads to

$$\overline{K}^2 = \left(\overline{E} - V\right)^2 - p^2 c^2 - m_0^2 c^4 = 0 \qquad (11.31)$$

Having a relativistic energy relation in terms of $\overline{E}$ and p, substitution of the definitions of the quantum mechanical operators for the total energy $i\hbar \partial/\partial t$ and momentum $\widehat{p}_x = -i\hbar \partial/\partial x$ leads to a 1D quantum mechanical version of Eq. (11.31) given by

$$\left[\left(i\hbar \frac{\partial}{\partial t} - V\right)^2 - c^2 \left(-i\hbar \frac{\partial}{\partial x}\right)^2 - m_0^2 c^4\right] \psi = 0 \qquad (11.32)$$

For a free particle with $V = 0$, just as we found in Chap. 3 with the Schrödinger equation, the above relativistic equation results in a plane-wave eigensolution $\psi(x,t)$ given by

$$\psi(x,t) = \psi_0 e^{\frac{i}{\hbar}(\pm \overline{p}x - \overline{E}t)} = \psi_0 e^{i(\pm \overline{k}x - \omega t)}$$

In the 1D case, ψ is traveling along x directions with momentum $\overline{p}_x$. If we also replace the momentum operator p_x with the extended momentum $P = \overline{p}_x + eA_x$ (c. Chap. 9, Sec. 7.9), where A_x is the x-component of the magnetic vector potential **A**, Eq. (11.32) becomes

$$\left[\left(i\hbar \frac{\partial}{\partial t} - V\right)^2 - c^2 \left(eA_x - i\hbar \frac{\partial}{\partial x}\right)^2 - m_0^2 c^4\right] \psi = 0 \qquad (11.33)$$

Extending the 1D equation into its 3D form, Eq. (11.33) becomes

Klein-Gordon Equation

$$\left[\left(i\hbar \frac{\partial}{\partial t} - V\right)^2 - c^2 \sum_{k=1}^{3} \left(eA_{x_k} - i\hbar \frac{\partial}{\partial x_k}\right)^2 - m_0^2 c^4\right] \psi(x,y,z,t) = 0$$

$$(11.34)$$

The summations go through all three Cartesian coordinates $x_k = x, y, z$. Eq. (11.34) is known as the **Klein-Gordon equation**. It was first proposed by Oskar Klein (1894-1977) and Walter Gordon (1893-1939) in 1926. It was one of the first quantum mechanical theories incorporating the theory of relativity. Within a couple of years, however, an extended relativistic quantum theory would be revealed that, to this day, remains one of the most revered theories in quantum mechanics.

In science one tries to tell people, in such a way as to be understood by everyone, something that no one ever knew before.
Paul A. M. Dirac

Dirac would go on to build on the efforts of Klein and Gordon, and we'll discuss how he did that in the next section.

11.4 The Dirac Equation

In comparing the Klein-Gordon equation and the non-relativistic Schrödinger equation, one difference that emerges is that there is a second order operator for time in the Klein-Gordon equation. However, there is a first order operator in time in the Schrödinger equation. The second order derivative in time follows from the first term in Eq. (11.34), leading to

$$\left(\frac{\partial}{\partial t}\right)^2 = \frac{\partial}{\partial t}\frac{\partial}{\partial t} = \frac{\partial^2}{\partial t^2}$$

This is because Klein and Gordon left the equation in terms of the *square* of energy. Consequently, this means that initial conditions for both $\psi(t)$ and $d\psi/dt$ are required, leading to the condition that the integral of the probability density is not uniformly zero. To see this, consider a fundamental particle in a state that must satisfy the normalization condition, where

$$\int \psi^* \psi \, dx = 1$$

Taking the time-derivative of the normalization condition leads to

$$\frac{d}{dt}\int \psi^* \psi \, dx = \int \frac{d\psi^*}{dt}\psi \, dx + \int \psi^* \frac{d\psi}{dt} dx = 0 \qquad (11.35)$$

11.4 The Dirac Equation

However, this result must vanish for the wave function to be *normalizable*. For example, in the Schrödinger equation, since $\psi = \psi(x)e^{iE/\hbar t}$, the sum in Eq. (11.35) adds two terms equal in amplitude, but opposite sign.

But, because a second order equation describes the wave function, the initial condition for $d\psi/dt$ is arbitrary, as is $\psi(0)$. Thus, generally $d\psi/dt \neq 0$ and $d^2\psi/dt^2 \neq 0$. The above normalization condition can be violated by solutions of the Klein-Gordon equation. This is in stark contrast with the capabilities of the Schrödinger equation, in dealing with bounded particles and or bounded states, for example. Moreover, this distinction between the K-G and Schrödinger equations suggest that one does not converge to the other in the relativistic limits. English mathematician and physicist Paul Adrien Maurice Dirac (1902-1984) set out to reconcile this discrepancy for relativistic quantum mechanics. He took an approach that ensures first order operators in time correspond to the energy of the system, (as opposed to the energy squared), yet still obtains convergence to the free particle solution propagating near the speed of light c.

However, before we get to how Dirac aimed to reconcile this dilemma, let us discuss the number of degrees of freedom, or eigenfunctions for the *relativistic* Hamiltonian. The Klein-Gordon equation and Eq. (11.32) have squared energy operators $\overline{E}^2$. Recall that considering a single electron with spin in the Schrödinger equation, we were able to introduce the spin degree of freedom into a matrix-vector form of the Schrödinger equation, because the spin angular momentum has only two possible equilibrium values. Because of this, it allows us to express the Schrödinger equation as a set of two equations for the two spin states $\psi_\uparrow$ and $\psi_\downarrow$. This lead to a 2×2 Hamiltonian operator acting on a 2×1 spinor $[\psi_\uparrow \ \psi_\downarrow]^T$. In the case of relativistic energy, because the operators are squared, this subtly gives rise to an increase in the number of possible states, given by $\pm m$ and $\pm \hbar/2$. Because spin has two eigenstates for all other energy contributions, we go from two simultaneous eigenstates to four simultaneous eigenstates with $\overline{E}^2$. And with four states, the operators of the Hamiltonian necessarily become 4×4 operators.

Let's get to how Dirac approached the reconciliation of the dilemma with first and second order time operators. Earlier, we obtained this equation depending on the relativistic energy

$$\overline{K}^2 - (p^2c^2 + m_0^2c^4) = (\overline{E} - V)^2 - (p^2c^2 + m_0^2c^4) = 0 \tag{11.36}$$

There is a mathematical symmetry in Eq. (11.36), suggesting it may be expressed as

$$\left[\overline{E} - V + \sqrt{p^2c^2 + m_0^2c^4}\right] \cdot \left[\overline{E} - V - \sqrt{p^2c^2 + m_0^2c^4}\right] = 0 \quad (11.37)$$

Factoring in this way reveals that since the RHS equals zero, then so does at least one of these factors. Thus, we only require one of the two terms, *either* the first or second, to obtain the total energy $\overline{E}$. If we only had p^2c^2 under the radical, the problem would be simple. However, we also have the additional constant $m_0^2c^4$. As a work around for this radical term, we seek a *linearization of both radicals*. Then, let us write Eq. (11.37) as

$$\left[\overline{E} - V + (\sigma_1' pc + \sigma_2' m_0 c^2)\right] \cdot \left[\overline{E} - V - (\sigma_1' pc + \sigma_2' m_0 c^2)\right] = 0 \quad (11.38)$$

To facilitate an easier comprehension of the underlying mathematics, let us stay in the space of scalars for now, without introducing higher order operators and spinors. Later, we'll show the analog with the full operators. The linear scalar parameters σ_1' and σ_2' are introduced in order to satisfy the resulting product given in Eq. (11.36). Expanding on the products of the linearized terms in Eq. (11.38) leads to

$$(\overline{E} - V)^2 - (\overline{E} - V)(\sigma_1' pc + \sigma_2' m_0 c^2) +$$
$$(\overline{E} - V)(\sigma_1' pc + \sigma_2' m_0 c^2) - (\sigma_1' pc + \sigma_2' m_0 c^2)^2 = 0 \quad (11.39)$$

The second and third terms are the same, but have opposite signs. Therefore, they cancel each other and we are left only with the first and fourth terms. Expanding the fourth term in Eq. (11.39) gives

$$\left[(\overline{E} - V)^2 - \left((\sigma_1')^2 p^2 c^2 + (\sigma_2')^2 m_0^2 c^4\right) - (\sigma_1' \sigma_2' + \sigma_2' \sigma_1') p m_0 c^3\right]\psi = 0 \quad (11.40)$$

Comparison of Eqs. (11.40) and (11.36) leads to the following conditions for the linear parameters σ_1' and σ_2':

$$(\sigma_1')^2 = (\sigma_2')^2 = 1 \quad (11.41a)$$

$$\sigma_1' \sigma_2' + \sigma_2' \sigma_1' = 0 \quad (11.41b)$$

When we are in the space of scalars, there is no solution satisfying both the above *scalar* equations. However, in the space of 2×2 and 4×4 matrices, this is no longer the case. In higher order spaces, there *are* solutions to the above conditions. This means that Dirac had no choice

11.4 The Dirac Equation

but to include the spin degree of freedom in order to have a possible solution for the linearized energy relation. We have already seen that *all* the principle spin matrices satisfy the conditions in Eq. (11.41). Therefore, it would be convenient to construct 4×4 matrices in terms of the principle spin matrices. Dirac took advantage of this in considering the representation of the linear parameters, which are operators.

Dirac took Eq. (11.38) and generalized it with 4×4 operators denoted by $\widehat{\Sigma}$, replacing the total relativistic energy $\overline{E}$ and momentum with their corresponding operators. This leads to the following two equivalent operator equations

$$\left[i\hbar\frac{\partial}{\partial t} - V + c\widehat{\Sigma}\cdot(\widehat{\mathbf{p}} + e\mathbf{A}) + \widehat{\Sigma}_0 m_0 c^2\right]\psi = 0 \tag{11.42a}$$

$$\left[i\hbar\frac{\partial}{\partial t} - V - c\widehat{\Sigma}\cdot(\widehat{\mathbf{p}} + e\mathbf{A}) - \widehat{\Sigma}_0 m_0 c^2\right]\psi = 0 \tag{11.42b}$$

Unlike the K-G equation, both equations above describe, to first order in time, the total energy $\overline{E}$. Dirac originally used the first of these equations. Both can be rearranged to a similar form as the Schrödinger equation where Eq. (11.42b) becomes

Dirac Equation

$$i\hbar\frac{\partial \psi}{\partial t} = \left[V + c\widehat{\Sigma}\cdot(\widehat{\mathbf{p}} + e\mathbf{A}) + \widehat{\Sigma}_0 m_0 c^2\right]\psi \tag{11.43}$$

The other form is given by

$$i\hbar\frac{\partial \psi}{\partial t} = \left[V - c\widehat{\Sigma}\cdot(\widehat{\mathbf{p}} - e\mathbf{A}) + \widehat{\Sigma}_0 m_0 c^2\right]\psi \tag{11.44}$$

In Eqs. (11.43) and (11.44), ψ is a 4×1 state vector. Eq. (11.43) is known as the **Dirac equation**, and it was derived by Dirac in 1928. With it, he showed it could also account for the duplicity in the hydrogen atomic spectra due to the two spin states along a magnetic field axis. There were two fundamental differences in Dirac's equation, as compared to Schrödinger's equation, however: firstly, the spin matrices appeared *automatically* in constructing the linear operators to satisfy the relativistic energy relation; and secondly, it also enables a relativistic quantum mechanical Hamiltonian. These important attributes both came from the linearization of the operator factors in Eq. (11.37), operating on the 4×1 state vector.

We still need to obtain the conditions satisfying linearization for the 4×4 operators $\widehat{\Sigma}_0$ and $\widehat{\Sigma} = [\widehat{\Sigma}_1, \widehat{\Sigma}_2, \widehat{\Sigma}_3]$, demonstrated by Dirac. For convenience, let us consider the case without any potential energy contributions ($V = \mathbf{A} = 0$). Then, the operators corresponding to the squared energy relation lead to the following product of two linear operators:

$$\left[\left(i\hbar\frac{\partial}{\partial t}\right) + c\sum_{k=1}^{3}\widehat{\Sigma}_k\left(-i\hbar\frac{\partial}{\partial x_k}\right) + \widehat{\Sigma}_0 m_0 c^2\right].$$

$$\left[\left(i\hbar\frac{\partial}{\partial t}\right) - c\sum_{q=1}^{3}\widehat{\Sigma}_q\left(-i\hbar\frac{\partial}{\partial x_q}\right) - \widehat{\Sigma}_0 m_0 c^2\right]\psi = 0 \quad (11.45)$$

Expanding Eq. (11.45), we are left with the following three terms:

$$\left[\left(i\hbar\frac{\partial}{\partial t}\right)^2 - c^2\sum_{k=1}^{3}\sum_{q=1}^{3}\widehat{\Sigma}_k\widehat{\Sigma}_q\left(-i\hbar\frac{\partial}{\partial x_k}\right)\left(-i\hbar\frac{\partial}{\partial x_q}\right) - (\widehat{\Sigma}_0 m_0 c^2)^2\right]\psi = 0$$

(11.46)

All other terms cancel due to the anti-symmetry of the operators. The first and third terms are identical to two of the terms in Eq. (11.32). The second term in Eq. (11.46) is the summation of $3 \times 3 = 9$ terms, which can be regarded as components of a 3×3 matrix. We can separate this summation into two separate sums, one including the diagonal terms ($p = q$), accounting for three of the nine terms, and the other including only off-diagonal terms ($p \neq q$), which includes the remaining six terms. This leads to

$$\left[\left(i\hbar\frac{\partial}{\partial t}\right)^2 - (\widehat{\Sigma}_0 m_0 c^2)^2 - c^2\sum_{k=1}^{3}(\widehat{\Sigma}_k)^2\left(-i\hbar\frac{\partial}{\partial x_k}\right)^2\right] -$$

$$c^2\left[\sum_{k=1,\neq q=1}^{3}\widehat{\Sigma}_k\widehat{\Sigma}_q\left(-i\hbar\frac{\partial}{\partial x_k}\right)\left(-i\hbar\frac{\partial}{\partial x_q}\right) + \widehat{\Sigma}_q\widehat{\Sigma}_k\left(-i\hbar\frac{\partial}{\partial x_q}\right)\left(-i\hbar\frac{\partial}{\partial x_k}\right)\right]\psi = 0$$

(11.47)

We can use the fact that mixed partials are independent of the order of operation, i.e. $\partial/\partial x_1(\partial/\partial x_2) = \partial/\partial x_2(\partial/\partial x_1)$. Then, Eq. (11.46) becomes

11.4 The Dirac Equation

$$\left[\left(i\hbar\frac{\partial}{\partial t}\right)^2 - c^2\sum_{k=1}^{3}(\widehat{\Sigma}_k)^2\left(-i\hbar\frac{\partial}{\partial x_k}\right)^2 - (\widehat{\Sigma}_0 m_0 c^2)^2 - \right.$$

$$\left. c^2\sum_{k=1,q>k}^{3}(\widehat{\Sigma}_k\widehat{\Sigma}_q + \widehat{\Sigma}_q\widehat{\Sigma}_k)\left(-i\hbar\frac{\partial}{\partial x_k}\right)\left(-i\hbar\frac{\partial}{\partial x_q}\right)\right]\psi = 0 \quad (11.48)$$

Comparing Eqs. (11.48) and (11.32), it follows that for both to be equal, the following conditions must be satisfied:

$$\widehat{\Sigma}_0^2 = \mathbf{I}(4) \tag{11.49a}$$

$$\widehat{\Sigma}_k\widehat{\Sigma}_q + \widehat{\Sigma}_q\widehat{\Sigma}_k = 0 \tag{11.49b}$$

$\mathbf{I}(4)$ is the 4×4 identity matrix. Though there are multiple solutions to the above equations, a remarkable observation from the conditions in Eqs. (11.49a) and (11.49b) is that, as mentioned earlier, they are met by the 2×2 spin matrices. Recall that squaring any principle spin matrix, for example, leads to the 2×2 identity matrix $\mathbf{I}(2)$. Additionally, the second condition follows from two properties of the spin matrices we obtained in Chap. 9, namely: the anti-commutation relations for the spin angular momentum, which lead to $\sigma_x\sigma_y - \sigma_x\sigma_y = 2i\sigma_z$, and $\sigma_j\sigma_k = i\varepsilon_{jkl}\sigma_l$, where ε_{jkl} is the Levi-Cevita symbol. This motivated the choices made by Dirac for the 4×4 matrices given in terms of 2×2 principle spin matrices σ_i. The x-component is defined as

$$\widehat{\Sigma}_x = \begin{bmatrix} 0 & 0 & 0 & 1 \\ 0 & 0 & 1 & 0 \\ 0 & 1 & 0 & 0 \\ 1 & 0 & 0 & 0 \end{bmatrix} = \begin{bmatrix} 0 & \widehat{\sigma}_x \\ \widehat{\sigma}_x & 0 \end{bmatrix} \tag{11.50a}$$

The y-component is defined as

$$\widehat{\Sigma}_y = \begin{bmatrix} 0 & 0 & 0 & i \\ 0 & 0 & -i & 0 \\ 0 & i & 0 & 0 \\ -i & 0 & 0 & 0 \end{bmatrix} = \begin{bmatrix} 0 & \widehat{\sigma}_y \\ \widehat{\sigma}_y & 0 \end{bmatrix} \tag{11.50b}$$

The z-component $\widehat{\Sigma}_z$ is defined as

$$\widehat{\Sigma}_z = \begin{bmatrix} 0 & 0 & 1 & 0 \\ 0 & 0 & 0 & -1 \\ 1 & 0 & 0 & 0 \\ 0 & -1 & 0 & 0 \end{bmatrix} = \begin{bmatrix} 0 & \widehat{\sigma}_z \\ \widehat{\sigma}_z & 0 \end{bmatrix} \tag{11.50c}$$

And the last 4×4 matrix from Dirac $\widehat{\Sigma}_0$ is defined as

$$\widehat{\Sigma}_0 = \begin{bmatrix} 1 & 0 & 0 & 0 \\ 0 & 1 & 0 & 0 \\ 0 & 0 & -1 & 0 \\ 0 & 0 & 0 & -1 \end{bmatrix} = \begin{bmatrix} \mathbf{I}(2) & 0 \\ 0 & -\mathbf{I}(2) \end{bmatrix} \quad (11.50d)$$

In Dirac's formulation, the spin matrices arise naturally in the relativistic equations as a result of applying linear operations to enforce equality to the relativistic energy relation. This observation was indeed remarkable at first sight, and for this work, he was awarded the Nobel prize, which he shared with Schr"odinger, in 1933.

And, as was the case with the Schrödinger equation, *if we assume that the potentials are independent of time t*, then we also have that the Dirac equation is separable in space and time. Be aware, however, that this assumption consequently restricts the Dirac equation (and the Schrödinger equation) from being capable of going beyond magnetostatics. This is because we already know, from Maxwell's equations, that if $\mathbf{B} = \nabla \times \mathbf{A}$ varies in time, then $\mathbf{A}$ varies in time, and so does V. But, if we assume $\mathbf{A}$ and V are not varying in time, we can then separate the Dirac equation and write the following two equations:

$$\left[V - c \sum_{k=1}^{3} \widehat{\Sigma}_k \left(eA_k - i\hbar \frac{\partial}{\partial x_k} \right) - \widehat{\Sigma}_0 m_0 c^2 \right] \psi(x, y, z) = \overline{E} \psi(x, y, z) \quad (11.51)$$

$$i\hbar \frac{\partial \psi_t}{\partial t} = \overline{E} \psi_t \quad (11.52)$$

We have solved equations like Eq. (11.52) a few times by now, so we know that in separation, the time-dependent wave function becomes

$$\psi_t(t) = \psi_t(0) e^{-\frac{i\overline{E}}{\hbar} t}$$

With the first equation, if we substitute the Dirac 4×4 spin matrices from above, we can obtain a 4×4 operator equation of the form

$$\begin{bmatrix} A & 0 & C & D \\ 0 & A & D^* & -C \\ C & D & B & 0 \\ D^* & -C & 0 & B \end{bmatrix} \begin{bmatrix} \psi_1 \\ \psi_2 \\ \psi_3 \\ \psi_4 \end{bmatrix} = \overline{E} \begin{bmatrix} \psi_1 \\ \psi_2 \\ \psi_3 \\ \psi_4 \end{bmatrix} \quad (11.53)$$

11.4 The Dirac Equation

The matrix elements, A, B, C, and D are given by

$$A = V - m_0c^2 \tag{11.54}$$

$$B = V + m_0c^2 \tag{11.55}$$

$$C = -c\left[-i\hbar\frac{\partial}{\partial z} + eA_z\right] \tag{11.56}$$

$$D = -c\left[(-i\hbar\frac{\partial}{\partial x} + eA_x) + i(-i\hbar\frac{\partial}{\partial y} + eA_y)\right] \tag{11.57}$$

This equation can also be expressed in terms of the principle spin matrices $\widehat{\sigma}_i$ as follows:

$$\begin{bmatrix} (V - m_0c^2)\mathbf{I}(2) & -c\widehat{\boldsymbol{\sigma}} \cdot (\mathbf{p} + e\mathbf{A}) \\ -c\widehat{\boldsymbol{\sigma}} \cdot (\mathbf{p} + e\mathbf{A}) & (V + m_0c^2)\mathbf{I}(2) \end{bmatrix} \begin{bmatrix} \psi_1 \\ \psi_2 \\ \psi_3 \\ \psi_4 \end{bmatrix} = E \begin{bmatrix} \psi_1 \\ \psi_2 \\ \psi_3 \\ \psi_4 \end{bmatrix} \tag{11.58}$$

Eq. (11.51) can be written as two equations given by

$$\left[V(x,y,z)\mathbf{I}(2) - c\sum_{k=1}^{3}\widehat{\sigma}_k\left(eA_k - i\hbar\frac{\partial}{\partial x_k}\right) + m_0c^2\mathbf{I}(2)\right]\begin{bmatrix}\psi_1 \\ \psi_2\end{bmatrix} = E\begin{bmatrix}\psi_1 \\ \psi_2\end{bmatrix} \tag{11.59}$$

$$\left[V(x,y,z)\mathbf{I}(2) - c\sum_{k=1}^{3}\widehat{\sigma}_k\left(eA_k - i\hbar\frac{\partial}{\partial x_k}\right) - m_0c^2\mathbf{I}(2)\right]\begin{bmatrix}\psi_3 \\ \psi_4\end{bmatrix} = E\begin{bmatrix}\psi_3 \\ \psi_4\end{bmatrix} \tag{11.60}$$

or

$$\left[(V(x,y,z) + m_0c^2)\mathbf{I}(2) - c\sum_{k=1}^{3}\widehat{\sigma}_k\left(eA_k - i\hbar\frac{\partial}{\partial x_k}\right)\right]\begin{bmatrix}\psi_1 \\ \psi_2\end{bmatrix} = E\begin{bmatrix}\psi_1 \\ \psi_2\end{bmatrix} \tag{11.61}$$

$$\left[(V(x,y,z) - m_0c^2)\mathbf{I}(2) - c\sum_{k=1}^{3}\widehat{\sigma}_k\left(eA_k - i\hbar\frac{\partial}{\partial x_k}\right)\right]\begin{bmatrix}\psi_3 \\ \psi_4\end{bmatrix} = E\begin{bmatrix}\psi_3 \\ \psi_4\end{bmatrix} \tag{11.62}$$

The advantage of having the Dirac equation expressed as a pair of two equations, as in Eqs. (11.61) and (11.62), is that it allows us to represent the operator in terms of the 2 × 2 spin matrices.

Now, we may use of the relations above to determine some of the relativistic corrections that can be made to the Schrödinger equation, or energy contributions. Specifically, we are going to uncover what is known as the *spin-orbit interaction* which conveys that just as $\widehat{\mathbf{L}}$ and $\widehat{\mathbf{S}}$ couple to a magnetic field B, they may also couple to each other.

11.5 Spin-Orbit Interaction

The richness of the Dirac equation may not be very obvious from a surface-view of the equation. We can, however, analyze the terms therein to probe the equation more deeply. We do this in the interest of better understanding the origin of some relevant mechanisms to the subject of *magnetism*. In magnetic materials, it is well-known that for materials with net magnetization, the magnetic moments are not equally likely to point in any direction. Instead, they tend to point along a preferred direction (or set of preferred directions), known as *easy-axes*. The origins of this particular phenomenon were originally explained based on electromagnetics, but later more deeply analyzed and explained by quantum mechanics. In this section, we will work out the more general form of this mechanism directly from the Dirac equation, then go on to discuss some of its implications. Recall the following form of the Dirac equation given by

$$\left[V\mathbf{I}(4) + c\widehat{\Sigma}\cdot(\widehat{\mathbf{p}} + e\mathbf{A}) + \widehat{\Sigma}_0 m_0 c^2\right]\psi = \overline{E}\psi \tag{11.63}$$

The eigenvector ψ is a 4×1 vector containing four eigenfunctions ψ_i, or

$$\psi = \begin{bmatrix} \psi_1 \\ \psi_2 \\ \psi_3 \\ \psi_4 \end{bmatrix}$$

Eq. 11.63 can be rearranged to give

$$\left[(\overline{E} - V)\mathbf{I}(4) - c\widehat{\Sigma}\cdot(\widehat{\mathbf{p}} + e\mathbf{A}) - \widehat{\Sigma}_0 m_0 c^2\right]\psi = 0 \tag{11.64}$$

As we found in the previous section, the form of the 4 × 4 matrices involved in the *Dirac equation* leads to two sets of equations for the first

11.5 Spin-Orbit Interaction

two and second two wave functions, respectively, where we have

$$(\bar{E} - V - m_0 c^2)\begin{bmatrix}\psi_1\\\psi_2\end{bmatrix} - c\widehat{\boldsymbol{\sigma}} \cdot (\widehat{\mathbf{p}} + e\mathbf{A})\begin{bmatrix}\psi_3\\\psi_4\end{bmatrix} = 0 \qquad (11.65\text{a})$$

$$-c\widehat{\boldsymbol{\sigma}} \cdot (\widehat{\mathbf{p}} + e\mathbf{A})\begin{bmatrix}\psi_1\\\psi_2\end{bmatrix} + (\bar{E} - V + m_0 c^2)\begin{bmatrix}\psi_3\\\psi_4\end{bmatrix} = 0 \qquad (11.65\text{b})$$

In Eqs. (11.65a) and (11.65b), $\widehat{\boldsymbol{\sigma}}$ is the vector of 2×2 principle spin matrices. We can use Eq. (11.65b) to solve for ψ_3 and ψ_4 and write

$$\begin{bmatrix}\psi_3\\\psi_4\end{bmatrix} = \frac{c}{\bar{E} - V + m_0 c^2}\widehat{\boldsymbol{\sigma}} \cdot (\widehat{\mathbf{p}} + e\mathbf{A})\begin{bmatrix}\psi_1\\\psi_2\end{bmatrix}$$

Substitution of this result back into Eq. (11.65a) leads to

$$(\bar{E} - V - m_0 c^2)\begin{bmatrix}\psi_1\\\psi_2\end{bmatrix} - c\widehat{\boldsymbol{\sigma}} \cdot (\widehat{\mathbf{p}} + e\mathbf{A})\frac{c}{\bar{E} - V + m_0 c^2}\widehat{\boldsymbol{\sigma}} \cdot (\widehat{\mathbf{p}} + e\mathbf{A})\begin{bmatrix}\psi_1\\\psi_2\end{bmatrix} = 0 \qquad (11.66)$$

This *sort-of* leads to an eigenvalue equation for the *shifted* energy $\bar{E}_S = \bar{E} - m_0 c^2$, because Eq. (11.66) may be written as follows:

$$[\widehat{\boldsymbol{\sigma}} \cdot (\widehat{\mathbf{p}} + e\mathbf{A})]\frac{c^2}{\bar{E} - V + m_0 c^2}[\widehat{\boldsymbol{\sigma}} \cdot (\widehat{\mathbf{p}} + e\mathbf{A})]\begin{bmatrix}\psi_1\\\psi_2\end{bmatrix}$$

$$= (\bar{E} - m_0 c^2)\begin{bmatrix}\psi_1\\\psi_2\end{bmatrix} = \bar{E}_S\begin{bmatrix}\psi_1\\\psi_2\end{bmatrix} \qquad (11.67)$$

The sandwiched scalar term on the LHS complicates things because it depends on the potential energy V. To deal with this, an approximation can be made using a **binomial series expansion**, which is the Maclaurin series (around $x = 0$) expansion of the function $f(x) = (1+x)^k$ given by

$$(1+x)^k = 1 + kx + \frac{k(k-1)}{2!}x^2 + \frac{k(k-1)(k-2)}{3!}x^3 + \cdots \frac{k(k-1)\cdots(k-n+1)}{n!}x^n + \cdots$$

To use this, we must write the term involving V in the form $(1+x)^k$. We can do this by rearranging things as follows:

$$\frac{c^2}{\bar{E} + m_0 c^2 - V} = \frac{c^2}{\bar{E}_S - V + 2m_0 c^2}$$

$$= \frac{c^2}{2m_0 c^2}\frac{1}{1 + \frac{\bar{E}_S - V}{2m_0 c^2}} = \frac{1}{2m_0}\left[1 + \frac{\bar{E}_S - V}{2m_0 c^2}\right]^{-1}$$

Chapter 11. Relativistic Quantum Mechanics

Applying the binomial expansion to this result, keeping only the first two terms, with $x = (\overline{E}_S - V)/2m_0c^2$ and $k = -1$, we have

$$\frac{1}{2m_0}\left[1 + \frac{\overline{E}_S - V}{2m_0c^2}\right]^{-1} \approx \frac{1}{2m_0} - \frac{\overline{E}_S - V}{4m_0^2c^2}$$

Substitution of this approximation back into Eq. (11.66) gives

$$(\overline{E}_S - V)\begin{bmatrix}\psi_1\\\psi_2\end{bmatrix} \approx \widehat{\boldsymbol{\sigma}}\cdot(\widehat{\mathbf{p}} + e\mathbf{A})\left[\frac{1}{2m_0} - \frac{\overline{E}_S - V}{4m_0^2c^2}\right]\widehat{\boldsymbol{\sigma}}\cdot(\widehat{\mathbf{p}} + e\mathbf{A})\begin{bmatrix}\psi_1\\\psi_2\end{bmatrix}$$

$$= \frac{[\widehat{\boldsymbol{\sigma}}\cdot(\widehat{\mathbf{p}} + e\mathbf{A})]^2}{2m_0} - \widehat{\boldsymbol{\sigma}}\cdot(\widehat{\mathbf{p}} + e\mathbf{A})\left[\frac{\overline{E}_S - V}{4m_0^2c^2}\right]\widehat{\boldsymbol{\sigma}}\cdot(\widehat{\mathbf{p}} + e\mathbf{A})\begin{bmatrix}\psi_1\\\psi_2\end{bmatrix}$$

The first term can be replaced using an identity relating the vector of principle spin matrices $\widehat{\boldsymbol{\sigma}}$, given by

$$[\widehat{\boldsymbol{\sigma}}\cdot(\widehat{\mathbf{p}} + e\mathbf{A})]^2 = (\widehat{\mathbf{p}} + e\mathbf{A})^2 + e\hbar\widehat{\boldsymbol{\sigma}}\cdot\mathbf{B} \qquad (11.68)$$

B is the magnetic field related to the magnetic vector potential **A** by $\mathbf{B} = \nabla \times \mathbf{A}$. Thus, we now have

$$\overline{E}_S\begin{bmatrix}\psi_1\\\psi_2\end{bmatrix} = \left[V + \frac{(\widehat{\mathbf{p}}+e\mathbf{A})^2}{2m_0} + \left(\frac{e\hbar}{2m_0}\right)\widehat{\boldsymbol{\sigma}}\cdot\mathbf{B} - \widehat{\boldsymbol{\sigma}}\cdot(\widehat{\mathbf{p}}+e\mathbf{A})\left[\frac{\overline{E}_S-V}{4m_0^2c^2}\right]\widehat{\boldsymbol{\sigma}}\cdot(\widehat{\mathbf{p}}+e\mathbf{A})\right]\begin{bmatrix}\psi_1\\\psi_2\end{bmatrix}$$

$$(11.69)$$

This result suggests the value of the g-factor of a single electron. Setting the Zeeman energy term equal to the general relation for the electron, introduced in Eq. (9.2), and used previously in the Schrödinger equation, we find that

$$V_M = -\boldsymbol{\mu}\cdot\mathbf{B} = -g\gamma\widehat{\mathbf{S}}\cdot\mathbf{B} = -g\left(-\frac{e\hbar}{4m_0}\right)\widehat{\boldsymbol{\sigma}}\cdot\mathbf{B} = g\frac{\mu_B}{2}\widehat{\boldsymbol{\sigma}}\cdot\mathbf{B}$$

Recall that $\gamma = -e/2m_e < 0$, for the electron, and so we have a sign change. Given the above relation, the result from the Dirac equation suggests that the value of g, for the electron, is given by

> **Electron g-Factor**
>
> $$g \approx 2 \qquad (11.70)$$

11.5 Spin-Orbit Interaction

With this value for g, it follows that $g\mu_B/2\widehat{\sigma}\cdot\mathbf{B} = \mu_B\widehat{\sigma}\cdot\mathbf{B} = (e\hbar/2m_0)\widehat{\sigma}\cdot\mathbf{B}$, as in Eq. (11.69). Note that g has been obtained from an approximated relativistic Hamiltonian, so it is only as accurate as the Hamiltonian approximation permits. For a more precise prediction, one would have to include more terms in the binomial expansion, etc. Still, Eq. (11.70) turns out to be very close to the experimental value of $g = 2.0023193$. It is remarkable that by carrying out this approximation to obtain some of the lower order terms contained within the Dirac equation, we find a Hamiltonian that contains the Zeeman energy naturally, as well as the g–factor. We have already dealt with the first three terms of Eq. (11.69) in solving the Schrödinger equation in previous chapters. However, in the Schrödinger equation, the Zeeman energy was introduced using the *Principle of Correspondence* based on results from the Maxwell equations. Here, we find that in the relativistic Dirac equation, we begin to uncover energy sources naturally. Since we are familiar with the first three terms, our interest will be the last term of Eq. (11.69), which is unique to the Dirac formulation. To treat this term further, however, let us first get rid of $\overline{E}_S$ by some manipulation involving operators, as follows:

$$(\overline{E}_S - V)\widehat{\sigma}\cdot(\widehat{\mathbf{p}} + e\mathbf{A}) = \widehat{\sigma}\cdot(\widehat{\mathbf{p}} + e\mathbf{A})(\overline{E}_S - V) + [\overline{E}_S - V, \widehat{\sigma}\cdot(\widehat{\mathbf{p}} + e\mathbf{A})]$$

We have used the operator commutation definition $[\widehat{A},\widehat{B}] = \widehat{AB} - \widehat{BA}$ from Eq. (7.97), or $\widehat{AB} = \widehat{BA} + [\widehat{A},\widehat{B}]$. In the first term, we must recognize that $\overline{E}_S - V$ is the *shifted* kinetic energy of the system, for which we already know the operator $\widehat{\mathbf{P}}^2/2m_0$. So, this term becomes

$$(\overline{E}_S - V)\widehat{\sigma}\cdot(\widehat{\mathbf{p}} + e\mathbf{A}) = \widehat{\sigma}\cdot(\widehat{\mathbf{p}} + e\mathbf{A})\frac{(\widehat{\mathbf{p}} + e\mathbf{A})^2}{2m_0} + [\overline{E}_S - V, \widehat{\sigma}\cdot(\widehat{\mathbf{p}} + e\mathbf{A})]$$

For the second term, we can exploit the fact that $\widehat{\sigma}$ is a vector of constant 2×2 matrices, so they commute with $\widehat{\mathbf{P}}$. Then, we can write the following relations:

$$\begin{aligned}[\overline{E}_S - V, \widehat{\sigma}\cdot(\widehat{\mathbf{p}} + e\mathbf{A})] &= (\overline{E}_S - V)\widehat{\sigma}\cdot(\widehat{\mathbf{p}} + e\mathbf{A}) - \widehat{\sigma}\cdot(\widehat{\mathbf{p}} + e\mathbf{A})(\overline{E}_S - V) \\ &= (\overline{E}_S - V)(\widehat{\mathbf{p}} + e\mathbf{A})\cdot\widehat{\sigma} - \widehat{\sigma}\cdot(\widehat{\mathbf{p}} + e\mathbf{A})(\overline{E}_S - V) \\ &= \overline{E}_S(\widehat{\mathbf{p}} + e\mathbf{A})\cdot\widehat{\sigma} - \widehat{\sigma}\cdot(\widehat{\mathbf{p}} + e\mathbf{A})\overline{E}_S - V(\widehat{\mathbf{p}} + e\mathbf{A})\cdot\widehat{\sigma} + \widehat{\sigma}\cdot(\widehat{\mathbf{p}} + e\mathbf{A})V \\ &= \widehat{\sigma}\cdot(\widehat{\mathbf{p}} + e\mathbf{A})V - V(\widehat{\mathbf{p}} + e\mathbf{A})\cdot\widehat{\sigma} \\ &= \widehat{\sigma}\cdot[\widehat{\mathbf{p}} + e\mathbf{A}, V]\end{aligned}$$

We have used the fact that $\overline{E}_S$ is a constant scalar. Using the kinetic energy operator along with this result, we obtain

$$(\bar{E}_S - V)\widehat{\boldsymbol{\sigma}} \cdot (\widehat{\mathbf{p}} + e\mathbf{A}) = \widehat{\boldsymbol{\sigma}} \cdot (\widehat{\mathbf{p}} + e\mathbf{A})\frac{(\widehat{\mathbf{p}} + e\mathbf{A})^2}{2m_0} + \widehat{\boldsymbol{\sigma}} \cdot [\widehat{\mathbf{p}} + e\mathbf{A}, V]$$

Bringing this back into our Hamiltonian term of interest leads to

$$-\widehat{\boldsymbol{\sigma}} \cdot (\widehat{\mathbf{p}} + e\mathbf{A})\left[\frac{\bar{E}_S - V}{4m_0^2 c^2}\right]\widehat{\boldsymbol{\sigma}} \cdot (\widehat{\mathbf{p}} + e\mathbf{A})$$

$$= -\frac{\widehat{\boldsymbol{\sigma}} \cdot (\widehat{\mathbf{p}} + e\mathbf{A})}{4m_0^2 c^2}\left[\widehat{\boldsymbol{\sigma}} \cdot (\widehat{\mathbf{p}} + e\mathbf{A})\frac{(\widehat{\mathbf{p}} + e\mathbf{A})^2}{2m_0} + \widehat{\boldsymbol{\sigma}} \cdot [\widehat{\mathbf{p}} + e\mathbf{A}, V]\right]$$

$$= -[\widehat{\boldsymbol{\sigma}} \cdot (\widehat{\mathbf{p}} + e\mathbf{A})]^2 \frac{(\widehat{\mathbf{p}} + e\mathbf{A})^2}{2m_0} - \widehat{\boldsymbol{\sigma}} \cdot (\widehat{\mathbf{p}} + e\mathbf{A})\widehat{\boldsymbol{\sigma}} \cdot [\widehat{\mathbf{p}} + e\mathbf{A}, V]$$

We also make use of another identity for the spin operators, given by

$$(\mathbf{a} \cdot \widehat{\boldsymbol{\sigma}})(\mathbf{b} \cdot \widehat{\boldsymbol{\sigma}}) = (\mathbf{a} \cdot \mathbf{b}) + i(\mathbf{a} \times \mathbf{b}) \cdot \widehat{\boldsymbol{\sigma}}$$

Applying this to both terms in the last line above, the result becomes

$$-\widehat{\boldsymbol{\sigma}} \cdot (\widehat{\mathbf{p}} + e\mathbf{A})\left[\frac{\bar{E}_S - V}{4m_0^2 c^2}\right]\widehat{\boldsymbol{\sigma}} \cdot (\widehat{\mathbf{p}} + e\mathbf{A})$$

$$= -\frac{(\widehat{\mathbf{p}} + e\mathbf{A})^4}{8m_0^3 c^2} - (\widehat{\mathbf{p}} + e\mathbf{A}) \cdot [\widehat{\mathbf{p}} + e\mathbf{A}, V] - i\widehat{\boldsymbol{\sigma}} \cdot (\widehat{\mathbf{p}} + e\mathbf{A}) \times [\widehat{\mathbf{p}} + e\mathbf{A}, V]$$

Then, using the relation $[\widehat{\mathbf{p}} + e\mathbf{A}, V] = -i\hbar \nabla V$ (c.Eq. 7.128), we have

$$-\widehat{\boldsymbol{\sigma}} \cdot (\widehat{\mathbf{p}} + e\mathbf{A})\left[\frac{\bar{E}_S - V}{4m_0^2 c^2}\right]\widehat{\boldsymbol{\sigma}} \cdot (\widehat{\mathbf{p}} + e\mathbf{A})$$

$$= -\frac{(\widehat{\mathbf{p}} + e\mathbf{A})^4}{8m_0^3 c^2} + \frac{\hbar}{4im_0^2 c^2}(\widehat{\mathbf{p}} + e\mathbf{A}) \cdot \nabla V - \frac{\hbar}{4m_0^2 c^2}\widehat{\boldsymbol{\sigma}} \cdot (\widehat{\mathbf{p}} + e\mathbf{A}) \times \nabla V$$

(11.71)

The first term is regarded as a higher order correction to the ordinary nonrelativistic kinetic energy (e.g. Eq. 11.69). We'll have more to say about this term in Chap. 15, when we discuss *superconductivity*. The second term is new, and of the form $\mathbf{P} \cdot \mathbf{E}$, having a complex coefficient (the operator is still Hermitian). It is the last term that we are after, which gives

11.5 Spin-Orbit Interaction

> **General Spin-Orbit Interaction**
> $$\widehat{H}_{SO} = -\frac{\hbar}{4m_0^2 c^2}\widehat{\boldsymbol{\sigma}}\cdot(\widehat{\mathbf{p}}+e\mathbf{A})\times\nabla V \qquad (11.72)$$

Eq. (11.72) is the **spin-orbit interaction** and, in some solid-state materials having a net magnetic moment, it can be a significant contribution to the energy, giving rise to the magnetic moment having a preferred direction. This preferred direction is known as the *easy axis*. Although we have assumed that the potential gradient is constant, up to now, let's relax this assumption (slightly) for the moment to uncover some of the consequences of this interaction. Let us assume that the potential is in the form of a spherical potential coming from the nucleus having charge Ze. In this case, the electrical potential energy due to the nuclear charge source is given by

$$\nabla V = \nabla\left(\frac{Ze^2}{4\pi\epsilon_0}\frac{1}{r}\right) = -\frac{Ze^2}{4\pi\epsilon_0}\frac{\mathbf{r}}{r^3} = -\frac{Ze^2 C_u}{r^3}\mathbf{r}$$

C_u is a generic constant specific to the *unit* of measurement chosen, where SI units give $C_u = 1/4\pi\epsilon_0$. $\mathbf{r}$ is the relative position vector oriented from the origin located at the nucleus (or source charge) of charge magnitude Ze to the position of the electron. Using the nucleus potential gradient in the spin-orbit interaction, and using the relation $\widehat{\boldsymbol{\sigma}} = 2\widehat{\mathbf{S}}/\hbar$, it follows that

$$\widehat{H}_{SO} = \frac{Ze^2 \hbar C_u}{4m_0^2 c^2}\widehat{\boldsymbol{\sigma}}\cdot(\widehat{\mathbf{p}}+e\mathbf{A})\times\frac{\mathbf{r}}{r^3}$$

Because the momentum operator is followed by a function of space, it generally gives rise to two terms. One term is of order $1/r^2$ and the other is of order $1/r^3$. If we assume $1/r^3 \ll 1/r^2$ (i.e. electron orbits a sufficiently large radius), then the $1/r^2$ dominating term becomes

$$\widehat{H}_{SO} = \frac{Ze^2 \hbar C_u}{4m_0^2 c^2 r^3}\widehat{\boldsymbol{\sigma}}\cdot\widehat{\mathbf{P}}\times\mathbf{r} \qquad (11.73a)$$

$$= -\frac{Ze^2 C_u}{2m_0^2 c^2 r^3}\widehat{\mathbf{S}}\cdot\widehat{\mathbf{L}} \qquad (11.73b)$$

We have used the fact that $\widehat{\mathbf{L}} = \mathbf{r}\times\widehat{\mathbf{P}} = -\widehat{\mathbf{P}}\times\mathbf{r}$. Thus, spin-orbit interaction leads to a **spin-orbit coupling** contribution to the energy. The spin angular momentum can couple to the orbital angular momentum just as both couple to a magnetic field. In SI units, the spin-orbit coupling becomes

Spin-Orbit Coupling

$$\widehat{H}_{SOC} = -\frac{Ze^2}{8\pi\epsilon_0 m_0^2 c^2 r^3} \widehat{S}\cdot\widehat{L} \tag{11.74}$$

For this particular form, an order of magnitude estimate for the spin-orbit coupling can be made by using the fact that both S and $L \approx \hbar$, so the magnitude of this interaction can be estimated as

$$|H_{SOC}|(r) \approx \frac{Ze^2\hbar^2}{8\pi\epsilon_0 m_0^2 c^2 r^3} \tag{11.75}$$

Eq. (11.75) is plotted in Fig. 11.4 for atoms with nuclei charge $Z = 1, 32, 64$. The spin-orbit coupling drops sharply as $1/r^3$, where r is

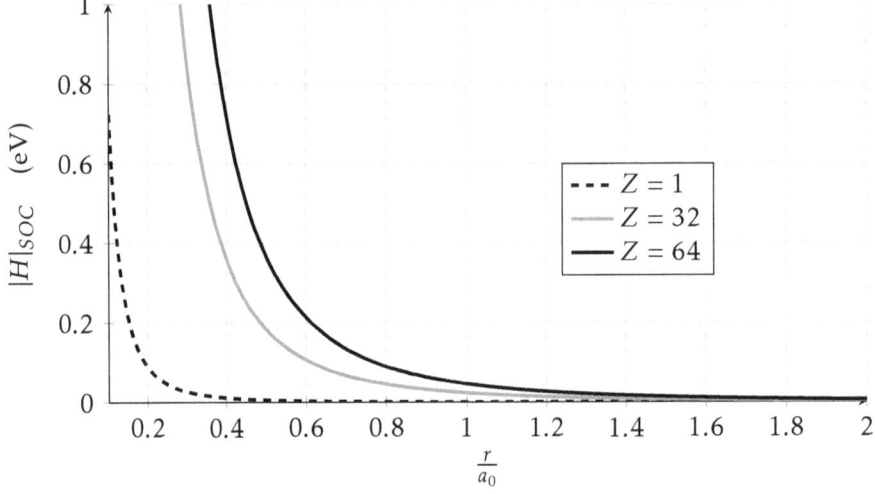

Figure 11.4: Order of magnitude for the *internal* spin-orbit coupling due to a single nucleus. $Z = 1, 32$ and 64, and the x-axis is measured in units of Bohr radius $a_0 = .529\text{Å}$.

the distance between the electron and the positive source of charge associated with $\widehat{L}$. Atoms with larger atomic numbers Z will tend to have stronger coupling, indicating the interaction is electronic in origin.

11.5 Spin-Orbit Interaction

This interaction is sometimes found to be more pronounced in certain crystalline solid-state materials due to contributions from many more atoms. Generally, the more crystalline (as opposed to amorphous or strongly disordered), the stronger can be the spin-orbit coupling. There must be a net of both orbital *and* spin angular momentum. In solid state magnetic crystalline materials that exhibit this behavior, it is known as **magneto-crystalline anisotropy**. The word *anisotropy*, here, means that the there is a definite axis of orientation that will couple to the electron spin, as opposed to an isotropic angular distribution.

This form of spin-orbit interaction leads to a minimum symmetric property known as **bistability**. Because the spin and orbital angular momentum have quantum numbers that are symmetric, e.g. $m_s = \pm 1/2$ and $m_\ell = \pm 0, 1, 2, 3, ..., \ell$, this leads to an interesting consequence for spin-orbit. Normally, without this interaction, a magnetic field splits the energies of the spin into non-degenerate states for spin-dependent contributions, however, this form of interaction can give rise to doubly-degenerate states, quadruply-degenerate, etc. The simplest type is doubly-degenerate which is known as *bistability* because $\widehat{S} \cdot \widehat{L}$ has the same energy (or is degenerate) when both $\widehat{S}$ and $\widehat{L}$ are *up*, as well as when they are both *down*. This can be shown by demonstrating that the quantity $\widehat{S} \cdot \widehat{L}$ depends only on S^2, L^2, and J^2. Let the spin-orbit interaction be given by

$$E_{SL} = C_{s\ell} \widehat{S} \cdot \widehat{L}$$

Fig. 11.5 shows the relationship between $J = L + S$, L, and S. This can be used to relate E_{SL} to quantum numbers s, ℓ, and j. The key to exposing the quantum numbers is to utilize the *law of cosines*, as we did in our derivation of the Landé g–factor in Chap. 10. This can be done by relating properties of the triangle formed by $\widehat{S}, \widehat{L}$, and $\widehat{J}$. Thus, we have

$$E_{SL} = C_{s\ell} S L \cos\beta \qquad (11.76)$$

From Fig. 11.5, the dot product angle β between $\widehat{S}$ and $\widehat{L}$, which is quantized, relates to the angle $\alpha = 180 - \beta$ of the triangle. So, we have

$$E_{SL} = C_{s\ell} S L \cos(180 - \alpha) \qquad (11.77a)$$

$$= C_{s\ell} S L \cos(180) \cdot \cos(\alpha) \qquad (11.77b)$$

$$= -C_{s\ell} S L \cos\alpha \qquad (11.77c)$$

It is $\cos\alpha$ that can be found from the law of cosines applied to the triangle formed by $\widehat{S}, \widehat{L}$, and $\widehat{J}$. Based on Fig. 11.5, we have $\cos\alpha =$

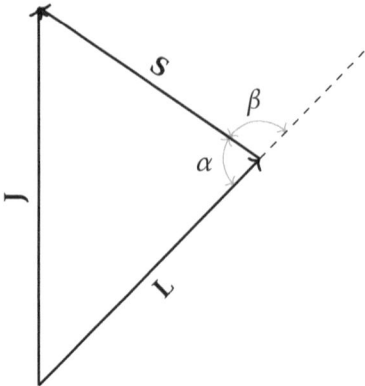

Figure 11.5: Illustration of the triangle vector relations between **J**, **L** and **S**, used to obtain $\cos(\beta)$.

$(J^2 - S^2 - L^2)/(2SL)$, which leads to

$$E_{SL} = C_{s\ell} \cdot \left[\frac{S^2 + L^2 - J^2}{2} \right] \tag{11.77d}$$

$$= C_{s\ell} \cdot \left[\frac{s(s+1) + \ell(\ell+1) - j(j+1)}{2} \right] \tag{11.77e}$$

The above result reveals a condition for the spin-orbit interaction, namely

Condition For Spin-orbit Coupling

$$s(s+1) + \ell(\ell+1) - j(j+1) \neq 0 \tag{11.78}$$

Because the dependence is only on S^2, L^2, and J^2, the above results are unchanged for corresponding symmetric values of m_s and m_ℓ. In other words, there is an inherent symmetry in the spin-orbit interaction of Eq. (11.74). Note that if either $\ell = 0$ or $s = 0$, then spin-orbit coupling vanishes because in the first case, $\ell = 0 \rightarrow j = s$, while in the second case, $s = 0 \rightarrow j = \ell$. For both these cases, $s(s+1) + \ell(\ell+1) - j(j+1) = 0$, so there is no spin-orbit coupling.

A familiar example of *bistability* is widely recognized in hysteresis loops of magnetic materials, as illustrated in Fig. 11.6. It is the *bistability* of spin-orbit coupling that is predominantly responsible for this type

11.5 Spin-Orbit Interaction

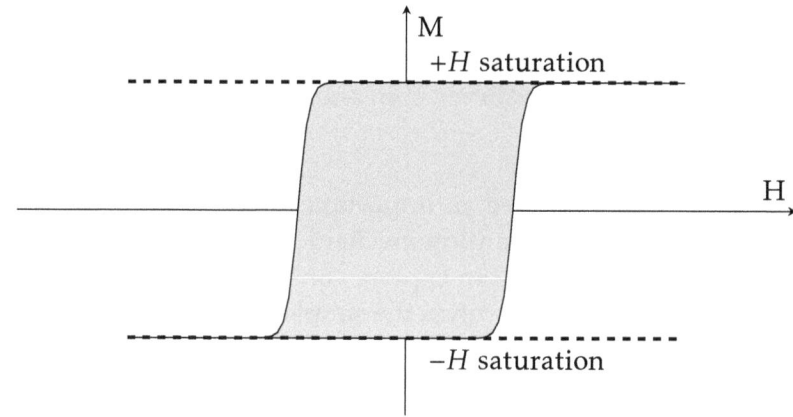

Figure 11.6: Example of a hysteresis loop with bistability resulting from spin-orbit coupling. To achieve reversal of such a material, the magnetic field *must* overcome this effeective spin-orbit field, given approximately by the crossing value along the H axis.

of hysteresis in magnetic materials. An external magnetic field must, therefore, overcome the intrinsic spin-orbit coupling before the electron magnetic moment will align itself with the magnetic field H. When alignment is reached, and all the electron moments in a solid-state material are aligned, the magnetic sample is said to be **magnetically saturated**. And, it acts symmetrically with respect to the external magnetic field $B_e = \mu_0 H$. Therefore, it is the spin-orbit interaction that predominantly leads to preferred directions of electron spins, even in the *absence* of a magnetic field (when $H = 0$), and to additional symmetric degenerate states of spin-up and spin-down.

If we imagine the *spin-orbit coupling* as spin coupling to an *orbital-associated* magnetic field (the electron is moving through an electric field produced from nuclei), then there is no reason why all forms of angular momentum of charged particles should not couple to electric field sources from other angular momentum sources. For example, in Chap. 10, we arrived at an interaction between spins, of the form $E \propto \mathbf{S} \cdot \mathbf{S}$. Now, we have spin-orbit coupling where $E \propto \mathbf{S} \cdot \mathbf{L}$. In general, we can express three types of **angular momentum coupling**, given by

$$E_{SS} = C_{SS} \widehat{\mathbf{S}}_1 \cdot \widehat{\mathbf{S}}_2 \quad \text{(spin-spin interaction)} \tag{11.79a}$$

$$E_{LS} = C_{LS} \widehat{\mathbf{L}} \cdot \widehat{\mathbf{S}} \quad \text{(spin-orbit interaction)} \tag{11.79b}$$

$$E_{LL} = C_{LL} \widehat{\mathbf{L}}_1 \cdot \widehat{\mathbf{L}}_2 \quad \text{(orbit-orbit interaction)} \tag{11.79c}$$

In the cases of the same type of angular momentum, subscripts are needed to denote distinct sources coupling, whereas its more obvious when the coupling exists between spin and orbital angular momentum, e.g. Eq. (11.79b).

Now that we have discussed an important interaction arising from a relativistic treatment of quantum mechanics, let's consider a different kind of problem where we can explore consequences of the extended momentum operator $\widehat{\mathbf{P}}$. It involves the simplest relativistic motion, and it was originally considered by Soviet physicist Lev Davidovich Landau (1908-1968). This is the subject of the next section.

11.6 Cyclotron Orbits And Landau Levels

With the introduction of the magnetic vector potential $\mathbf{A}$ to the momentum $\widehat{\mathbf{P}} = \widehat{\mathbf{p}} + e\mathbf{A}$, the Hamiltonian includes the description of the forces on a charged particle in a static electric and magnetic field, where $\mathbf{B} = \nabla \times \mathbf{A}$ (c. Chap. 9, Sec. 7.9). Consider a situation when an electron is moving, initially, axially along a cylindrical confining structure where the electron begin confined to the cylindrical space, is also exposed to a magnetic field $\mathbf{B}$ pointing along the cylindrical axis (z-axis), as illustrated in Fig. 11.7. Note that the cylindrical structure enforces the boundary conditions on the wave function at the wall. The magnetic field is assumed to be uniform so the magnetic vector potential $\mathbf{A}$ is linear in the spatial variables x, y, and z. Given the relationship between $\mathbf{B}$ and $\mathbf{A}$, we always have more than one choice for $\mathbf{A}$ that can yield a magnetic field B along z. For $B = [0, 0, B_z] = [0, 0, \partial A_y/\partial x - \partial A_x/\partial y]$, the following possibilities all yield a magnetic field along the $z - axis$:

$$\mathbf{A} = [0, B \cdot x, 0]^T \quad \text{(Landau gauge)} \tag{11.80a}$$

$$\mathbf{A} = [-B \cdot y, 0, 0]^T \quad \text{(Landau gauge)} \tag{11.80b}$$

$$\mathbf{A} = \frac{1}{2}[-B \cdot y, B \cdot x, 0]^T \quad \text{(Symmetric gauge)} \tag{11.80c}$$

The first two gauges are called **Landau gauges**, named after Lev Landau, while the last gauge is known as a **symmetric gauge** because it expresses a symmetry in the x and y coordinates. Any gauge chosen from Eq. (11.80) *must yield the same eigenvalues and expectation values*. In this problem, the Hamiltonian includes *only* the kinetic energy, however,

11.6 Cyclotron Orbits And Landau Levels

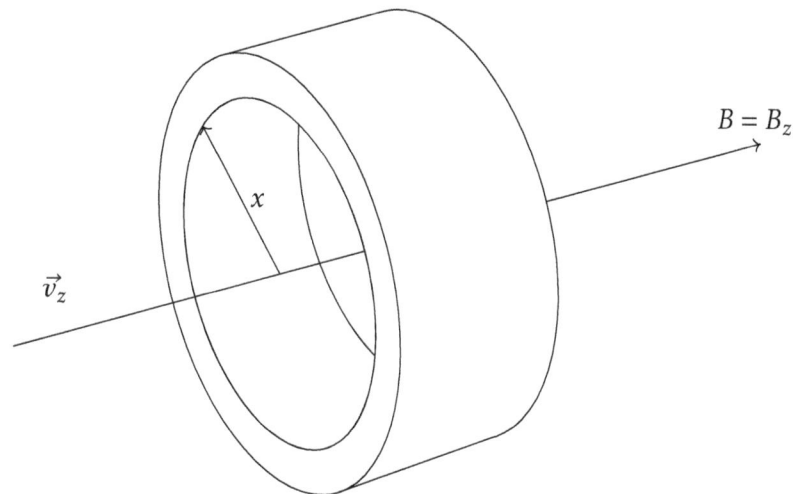

Figure 11.7: Illustration of conditions in the Landau level problem. An electron is traveling along the longitudinal axis, or z axis, while confined within the cylindrical structure.

now we use the extended momentum $\widehat{\mathbf{P}}$ to extend the momentum to include effects of a magnetic field B_z. The Hamiltonian is then given by

$$\widehat{H} = \frac{\widehat{\mathbf{P}}^2}{2m}\psi = \frac{(\widehat{\mathbf{p}}+e\mathbf{A})^2}{2m}\psi = E_\lambda \psi \tag{11.81}$$

Expanding the square of the extended momentum in Eq. (11.81) leads to

$$\frac{1}{2m}\left[\widehat{\mathbf{p}}^2 + e\widehat{\mathbf{p}}\cdot\mathbf{A} + e\mathbf{A}\cdot\widehat{\mathbf{p}} + e^2 A^2\right]\psi = E_\lambda \psi \tag{11.82}$$

$\widehat{\mathbf{p}}^2$ corresponds to $-\hbar^2 \nabla^2 \psi = \nabla \cdot \widehat{\nabla}$. Let us consider the term $e\widehat{\mathbf{p}}\cdot\mathbf{A}$. Since this operates on ψ, we can make use of a vector-calculus identity for the gradient of a vector times a scalar. Applying the identity to the second term of Eq. (11.82) leads to

$$e\widehat{\mathbf{p}}\cdot(\mathbf{A}\psi) = e(\widehat{\mathbf{p}}\cdot\mathbf{A} + \mathbf{A}\cdot(\widehat{\mathbf{p}}\psi))$$

The first term on the RHS is proportional to the divergence of **A**. Because of the form of **A**, this term is identically zero for all gauge forms in Eq. (11.80). Then, we have that

$$\frac{1}{2m}\left[\widehat{\mathbf{p}}^2 + 2e\mathbf{A}\cdot\widehat{\mathbf{p}} + e^2 A^2\right]\psi = E_\lambda \psi \tag{11.83}$$

In the following two subsections, we will solve Eq. (11.83) using a Landau gauge and the symmetric gauge given by Eqs. (11.80), starting with the Landau gauge. We'll see that by adding the magnetic vector potential component $e\mathbf{A}$ to the momentum provides an additional degree of freedom to the motion of the electron which arises in a magnetic field.

11.6.1 Landau Levels Using A Landau Gauge

In the proper form, the Dirac equation, as with the Schrödinger equation, is an eigenvalue equation for the energy operator. In this case, its for the *relativistic* energy $\overline{E}$. For this reason, it is uniquely *Lorentz gauge invariant*, meaning that its results are unchanged from applying a Lorentz transformation to the coordinates. This is also a characteristic of the Maxwell equations. Likewise, the choice of gauge for the magnetic vector potential should lead to the same eigenvalues, although the wave functions can be expressed in different coordinates. This can be illustrated by using two different gauges, first with the *Landau gauge* ($\mathbf{A} = [0, B \cdot x, 0]$), where we have

$$\left[\hat{\mathbf{p}}^2 - 2ie\hbar Bx\frac{\partial \Psi}{\partial y} + e^2 B^2 x^2\right]\Psi = 2mE_\lambda \Psi \tag{11.84}$$

Note that if we had chosen the other Landau gauge in Eq. (11.80b), the result would be essentially the same, where we would only need to swap x and y variables. Expanding the momentum operator in terms of the Laplacian, we have

$$-\hbar^2\left(\frac{\partial^2 \Psi}{\partial x^2} + \frac{\partial^2 \Psi}{\partial y^2} + \frac{\partial^2 \Psi}{\partial z^2}\right) - 2ie\hbar Bx\frac{\partial \Psi}{\partial y} + e^2 B^2 x^2 \Psi = 2mE_\lambda \Psi \tag{11.85}$$

The Hamiltonian $\widehat{H}$ in Eq. (11.85) has a restrictive separable form with $x, y,$ and z coordinates. $\widehat{H}$ only contains an explicit dependence on x. Additionally, one of these terms is coupled to a y− dependence through $\partial \Psi / \partial y$. This conveys that z is separable with respect to x and y. So, the wave function Ψ is separable so that

$$\Psi(x,y,z) = \psi(x,y) \cdot Z(z) \quad \text{and} \quad E_\lambda = E_{xy} + E_z$$

Thus, the Hamiltonian separates into the following equation for z,

$$-\hbar^2 \frac{\partial^2 Z}{\partial z^2} = 2mE_z Z \Rightarrow \frac{\partial^2 Z}{\partial z^2} + k_z^2 Z = 0 \quad \text{where} \quad k_z^2 = \frac{2mE_z}{\hbar^2} \tag{11.86}$$

11.6 Cyclotron Orbits And Landau Levels

The solution to Eq. (11.86) is given by

$$Z(z) = C e^{i k_z z} \qquad (11.87)$$

For the remaining terms of the Hamiltonian that describe both x and y dependence, we have

$$-\hbar^2 \left(\frac{\partial^2 \psi}{\partial x^2} + \frac{\partial^2 \psi}{\partial y^2} \right) - 2 i e \hbar B x \frac{\partial \psi}{\partial y} + e^2 B^2 x^2 \psi = 2 m E_{xy} \psi \qquad (11.88)$$

Let's also assume x and y are separable, so that $\psi(x, y) = X \cdot Y$ leads to

$$\frac{\partial \psi}{\partial y} = X \frac{dY}{dy} \qquad (11.89)$$

If we are able to obtain the exact solution with this assumption, then it is correct. Fortunately, this turns out to be the case. So, let Y be given by

$$Y = C e^{i k_y} \Rightarrow \frac{dY}{dy} = i k_y Y$$

Both the above conditions then lead to the following result for $\psi(x, y)$

$$\frac{\partial \psi}{\partial y} = X \frac{dY}{dy} = X \cdot i k_y \cdot Y = i k_y X Y$$

Substitution of this result back into Eq. (11.88) gives

$$-\hbar^2 \left(Y \frac{d^2 X}{dx^2} + X \frac{d^2 Y}{dy^2} \right) + -2 i e \hbar B k_y x X Y + e^2 B^2 x^2 X Y = 2 m E_{xy} \psi X Y \quad (11.90)$$

The second derivative of $Y(y)$ is just $-k_y^2 Y$, so we also have

$$-\hbar^2 \left(Y \frac{d^2 X}{dx^2} - X Y k_y^2 \right) + 2 e \hbar B k_y x X Y + e^2 B^2 x^2 X Y = 2 m E_{xy} X Y$$

In this form, we see that Y can be factored out of the Hamiltonian, leaving us the following equation for $X(x)$:

$$-\hbar^2 \frac{d^2 X}{dx^2} + \hbar^2 k_y^2 X + 2 e \hbar B k_y x X + e^2 B^2 x^2 X = 2 m E_{xy} X \qquad (11.91)$$

Up to now, we have taken the wave function Ψ to be of the form

$$\Psi(x, y, z) = C_n e^{i(k_y y + k_z z)} X(x) \qquad (11.92)$$

C_n is the normalization constant. $X(x)$ is described by Eq. (11.91), which we can write as follows:

$$\frac{2mE_{xy}}{\hbar^2}X = -\frac{d^2X}{dx^2} + \left(k_y^2 + \frac{2eBk_y}{\hbar}x + \frac{e^2B^2}{\hbar^2}x^2\right)X$$

$$= -\frac{d^2X}{dx^2} + \frac{e^2B^2}{\hbar^2}\left(\frac{\hbar^2 k_y^2}{e^2B^2} + \frac{2\hbar k_y}{eB}x + x^2\right)X$$

$$= -\frac{d^2X}{dx^2} + \frac{e^2B^2}{\hbar^2}q(x)X$$

With the quadratic polynomial $q(x)$ coefficient for $X(x)$ on the RHS, we can use the fact that

$$\frac{\hbar^2 k_y^2}{e^2B^2} + \frac{2\hbar k_y}{eB}x + x^2 = \left(x + \frac{\hbar k_y}{eB}\right)^2 = (x+x_0)^2$$

The shifting point x_0 is then defined by

$$x_0 = \frac{\hbar k_y}{eB} \qquad (11.93)$$

Eq. (11.91) then becomes

$$\frac{d^2X}{dx^2} + \frac{2mE_{xy}}{\hbar^2}X - \frac{e^2B^2}{\hbar^2}(x+x_0)^2 X = \frac{d^2X}{dx^2} + \frac{2m}{\hbar^2}\left(E_{xy} - \frac{1}{2}m\omega_c^2(x+x_0)^2\right)X$$

The last line above is identical in form to the equation we obtained for the harmonic oscillator, in Chap. 5, given by Eq. (5.16). The differences here include that we have a shifted x coordinate and a different definition of the frequency ω_c. This is good news as we can use the previous results for the harmonic oscillator to obtain the eigenvalues found from solving the Hamiltonian. The eigenvalues are then given by

Landau Levels

$$E_{xy} = \left(n + \frac{1}{2}\right)\hbar\omega_c \qquad (11.94)$$

n is an integer $n = 0, 1, 2, ...$ Eq. (11.94) states that energy of the electron is only quantized in the $x-y$ plane, although it propagates continuously along the z axis. The quantized energy levels E_{xy} are known as **Landau levels**, and the frequency, analogous to the harmonic oscillator natural frequency, is called the **cyclotron frequency** ω_c, defined as

11.6 Cyclotron Orbits And Landau Levels

Cyclotron Frequency

$$\omega_c = \frac{eB}{m} \tag{11.95}$$

This analogy to the harmonic oscillator extends to allow us to obtain the wave functions $X(x)$. They are proportional to the harmonic oscillator wave functions, replacing the frequency $\omega \to \omega_c$ and $x \to x + x_0$. $X(x)$ is then given by

$$X(x) = C_n H_n(\zeta) e^{-\frac{1}{2}\zeta^2} \tag{11.96}$$

H_n are Hermite polynomials (see Table 5.4), and ζ is given by

$$\zeta = \alpha(x + x_0) = \sqrt{\frac{m\omega_c}{\hbar}}(x + x_0) \tag{11.97}$$

As with the harmonic oscillator, the normalization constant C_n is given by

$$C_n = \sqrt{\frac{\alpha}{2^n n! \sqrt{\pi}}}$$

In the next subsection, we'll look at the problem with the symmetric gauge to demonstrate that one obtains identical eigenvalues. We'll also illustrate the solutions.

11.6.2 Landau Levels Using The Symmetric Gauge

By finding the solution using the *symmetric gauge*, given by Eq. (11.80c), we should find that the eigenvalues of the Hamiltonian remain unchanged using a different gauge, as well as coordinate system. In using the symmetric gauge, the analysis is a bit more lengthy, but the final result to be same. Illustrating this is one of key objectives of this section. We start with the Hamiltonian we obtained earlier, given by

$$\frac{1}{2m}\left[\widehat{\mathbf{p}}^2 + 2e\mathbf{A}\cdot\widehat{\mathbf{p}} + e^2 A^2\right]\psi = E_\lambda \psi \tag{11.98}$$

Using the momentum operator $\widehat{\mathbf{p}} = -i\hbar\nabla$ along with substitution of the symmetric gauge $\mathbf{A} = (1/2)[-B\cdot y, B\cdot x, 0]^T$, we have

$$E_\lambda \psi = \frac{1}{2m}\left[-\hbar^2 \nabla^2 \psi + 2e\mathbf{A}\cdot\widehat{\mathbf{p}} + \frac{e^2 B^2}{4}(x^2 + y^2)\psi\right] \tag{11.99a}$$

$$= \frac{1}{2m}\left[-\hbar^2\nabla^2\psi - i2\hbar e\mathbf{A}\cdot\nabla\psi + \frac{e^2 B^2}{4}\rho^2\psi\right] \quad (11.99b)$$

The last term appears in a form with a radial coordinate ρ. This suggests we should try a more compatible coordinate system. So, let's use cylindrical coordinates (ρ, φ, z) for the *symmetric gauge*. Therefore, the Laplacian and gradient operators should be in cylindrical coordinates. This gives

$$E_\lambda \psi = -\frac{\hbar^2}{2m}\left(\frac{1}{\rho}\frac{\partial}{\partial\rho}\left(\rho\frac{\partial\psi}{\partial r}\right) + \frac{1}{r^2}\frac{\partial^2\psi}{\partial\varphi^2} + \frac{\partial^2\psi}{\partial z^2}\right)$$

$$-i\frac{\hbar e}{m}\frac{B\rho}{2}\frac{1}{\rho}\frac{\partial\psi}{\partial\varphi} + \frac{e^2 B^2}{8m}\rho^2\psi \quad (11.100)$$

Because the entire equation equals $E_\lambda\psi$, we know the z dependence is separable from the ρ, φ dependence. So, let ψ be given by

$$\psi = F(\rho,\varphi)Z(z) \quad \text{and} \quad E_\lambda = E_{\rho\varphi} + E_z \quad (11.101)$$

Substitution of Eq. (11.101) into Eq. (11.99b), we find that the separation works and for the z dependence, $Z(z)$ is described by

$$-\frac{\hbar^2}{2m}\frac{d^2 Z}{dz^2} = E_z Z(z) \Rightarrow \frac{d^2 Z}{dz^2} + k_z^2 Z = 0 \quad (11.102)$$

As we found from solving Eq. (11.83) using the Landau gauge, the solution to Eq. (11.102) is given by

---
$Z(z)$ Solution

$$Z(z) = Z_0 e^{ik_z z} \quad \text{where} \quad k_z^2 = \frac{2mE_z}{\hbar^2} \quad (11.103)$$

So far, so good. Then, the remaining part of Eq. (11.100) depends only on ρ and φ, and is given by

$$E_{\rho\varphi} = -\frac{\hbar^2}{2m}\left(\frac{1}{\rho}\frac{\partial}{\partial\rho}\left(\rho\frac{\partial F}{\partial\rho}\right) + \frac{1}{\rho^2}\frac{\partial^2 F}{\partial\varphi^2}\right)$$

$$-i\frac{\hbar e}{2m}B\rho\frac{1}{\rho}\frac{\partial F}{\partial\varphi} + \frac{e^2 B^2}{8m}\rho^2 F \quad (11.104)$$

11.6 Cyclotron Orbits And Landau Levels

Continuing with the use of separation of variables, let F be given by

$$F(\rho, \varphi) = R(\rho)\Phi(\varphi) \quad \text{and} \quad E_{\rho\varphi} = E_\rho + E_\varphi$$

To simplify Eq. (11.104) further, it can be shown that if $\Phi(\varphi) = \Phi_0 e^{ik_\varphi(\rho)\varphi}$, then both terms in Eq. (11.104) depending on φ sum to zero for a unique value of k_φ, and thus $E_\varphi = 0$. We'll also determine the value of k_φ. The φ dependence gives

$$-\frac{\hbar^2}{2m}\left(\frac{1}{\rho^2}\frac{\partial^2 \Phi}{\partial \varphi^2}\right) - i\frac{e\hbar B}{2m}\frac{\partial \Phi}{\partial \varphi} = E_\varphi \Phi \tag{11.105}$$

Since differentiation is with respect to φ, let the solution be written as $\Phi(\varphi) = \Phi_0 e^{ik'_\varphi \rho \varphi}$, where $k'_\varphi \neq k_\varphi$. This gives

$$\frac{\partial \Phi}{\partial \varphi} = \Phi_0 i k'_\varphi \rho e^{ik'_\varphi \rho \varphi} = ik'_\varphi \rho \Phi \tag{11.106}$$

$$\frac{\partial^2 \Phi}{\partial \varphi^2} = \Phi_0 (ik'_\varphi \rho)^2 e^{ik'_\varphi \rho \varphi} = -(k'_\varphi)^2 \rho^2 \Phi \tag{11.107}$$

Substitution into the LHS of Eq. (11.105) gives

$$-\frac{\hbar^2}{2m}\frac{1}{\rho^2}\left(-(k'_\varphi)^2 \rho^2 \Phi\right) - i\frac{e\hbar B}{2m}\left(ik'_\varphi \rho \Phi\right) = \frac{\hbar^2}{2m}(k'_\varphi)^2 - \frac{e\hbar B k'_\varphi}{2m} \tag{11.108}$$

Using a value of k'_φ that makes Eq. (15.28) vanish gives

$$0 = \hbar k'_\varphi + eB\rho \Rightarrow k'_\varphi = -\frac{qB}{\hbar}\rho$$

Since the function $\Phi = \Phi(\rho, \varphi) = \Phi_0 e^{ik'_\varphi \rho \varphi}$, we have the following result:

Azimuthal Solution For Landau Levels

$$\Phi(\rho, \varphi) = \Phi_0 e^{-i\left(\frac{e}{\hbar} B \rho^2 \varphi\right)} \tag{11.109}$$

With this result, if we are able to find the exact solution for the remaining part of Eq. (11.100), then the solution will be complete. The remaining terms describing R can be written as

$$E_\rho R = -\frac{\hbar^2}{2m} \frac{1}{\rho} \frac{d}{d\rho}\left(\rho \frac{dR}{d\rho}\right) + \frac{e^2 B^2}{8m} \rho^2 R$$

$$\Rightarrow \frac{d^2 R}{d\rho^2} + \frac{1}{\rho}\frac{dR}{d\rho} + \left(\frac{2mE_\rho}{\hbar^2} - \frac{e^2 B^2}{4\hbar^2}\rho^2\right) R = 0 \quad (11.110)$$

To help solve Eq. (11.110), let's define the variable ζ such that

$$\sqrt{\zeta} = \sqrt{\frac{eB}{\hbar}}\rho = \alpha\rho \Rightarrow \zeta = \alpha^2 \rho^2$$

From the definition of ζ, the following relations follow:

$$d\zeta = 2\alpha^2 \rho \, d\rho \quad \text{or} \quad \frac{d\zeta}{d\rho} = 2\alpha^2 \rho \quad (11.111a)$$

$$\frac{d\Phi}{d\rho} = \frac{d\zeta}{d\rho}\frac{d\Phi}{d\zeta} = 2\alpha^2 \rho \frac{d\Phi}{d\zeta} \quad (11.111b)$$

$$\frac{d^2\Phi}{d\rho^2} = \frac{d}{d\rho}\left(2\alpha^2 \rho \frac{d\Phi}{d\zeta}\right) = 2\alpha^2 \frac{d\Phi}{d\zeta} + 2\alpha^2 \rho \frac{d\zeta}{d\rho}\frac{d}{d\zeta}\left(\frac{d\Phi}{d\zeta}\right) \quad (11.111c)$$

$$= 2\alpha^2 \frac{d\Phi}{d\zeta} + 4\alpha^4 \rho^2 \frac{d^2\Phi}{d\zeta^2} = 2\alpha^2 \frac{d\Phi}{d\zeta} + 4\alpha^2 \zeta \frac{d^2\Phi}{d\zeta^2} \quad (11.111d)$$

Substitution of these relations into Eq. (11.110) leads to the following differential equation in terms of ζ:

$$4\alpha^2 \zeta \frac{d^2\Phi}{d\zeta^2} + 4\alpha^2 \frac{d\Phi}{d\zeta} + \left(\frac{2mE_\Phi}{\hbar^2} - \frac{\alpha^2}{4}\zeta\right)\Phi(\zeta) = 0$$

or

$$4\zeta \frac{d^2\Phi}{d\zeta^2} + 4\frac{d\Phi}{d\zeta} + \left(\frac{2mE_\Phi}{\alpha^2 \hbar^2} - \frac{1}{4}\zeta\right)\Phi(\zeta) = 0 \quad (11.112)$$

We write the wave function Φ in the form $\Phi = \Phi_\infty(\zeta)\Phi_1(\zeta)$, where each term is an asymptotic limit. Then, we consider the asymptotic limit as $\zeta \to \infty$, in which case, Eq. (11.112) becomes

$$4\frac{d^2\Phi_\infty}{d\zeta^2} - \frac{1}{4}\Phi_\infty(\zeta) = 0 \Rightarrow \frac{d^2\Phi_\infty}{d\zeta^2} - \frac{1}{16}\Phi_\infty(\zeta) = 0 \quad (11.113)$$

Eq. (11.113) has a solution given by

$$\Phi_\infty(\zeta) = Ce^{-\frac{1}{4}\zeta} = Ce^{-\frac{e^2 B^2}{4\hbar^2}\rho^2} \quad (11.114)$$

11.6 Cyclotron Orbits And Landau Levels

Having $\Phi_\infty(\zeta)$ in the form of a Gaussian allows us to obtain an equation purely in terms of the other asymptotic limit Φ_1. If $\Phi = \Phi_\infty \Phi_1$, we have the following derivatives of $\Phi(\zeta)$ in terms of Φ_1:

$$\frac{d\Phi}{d\zeta} = \frac{d}{d\zeta}(\Phi_\infty \Phi_1) = \left[\frac{d\Phi_1}{d\zeta} - \frac{1}{4}\Phi_1\right]e^{-\frac{1}{4}\zeta}$$

$$\frac{d^2\Phi}{d\zeta^2} = \left[\left(\frac{d^2\Phi_1}{d\zeta^2} - \frac{1}{4}\frac{d\Phi_1}{d\zeta}\right) + \left(\frac{d\Phi_1}{d\zeta} - \frac{1}{4}\Phi_1\right)\left(-\frac{1}{4}\right)\right]e^{-\frac{1}{4}\zeta}$$

Using these relations in Eq. (11.112), we obtain the following equation for Φ_1:

$$4\zeta\left[\left(\frac{d^2\Phi_1}{d\zeta^2} - \frac{1}{4}\frac{d\Phi_1}{d\zeta}\right) + \left(\frac{d\Phi_1}{d\zeta} - \frac{1}{4}\Phi_1\right)\left(\frac{1}{4}\right)\right]e^{-\frac{1}{4}\zeta}$$

$$+ 4\left[\frac{d\Phi_1}{d\zeta} - \frac{1}{4}\Phi_1\right]e^{-\frac{1}{4}\zeta} + \left[\frac{2mE_\Phi}{\alpha^2\hbar^2} - \frac{1}{4}\zeta\right]\Phi_1 e^{-\frac{1}{4}\zeta} = 0 \quad (11.115)$$

Combining like terms, and noting cancellation of a $\zeta\Phi_1/4 - \zeta\Phi_1/4$ term, then dividing by the common factor $e^{-\zeta/4}$, Eq. (11.115) simplifies to

$$\zeta\frac{d^2\Phi_1}{d\zeta^2} + \left(1 - \frac{1}{2}\zeta\right)\frac{d\Phi_1}{d\zeta} - \left(\frac{1-\Omega}{4}\right)\Phi_1 = 0 \quad (11.116)$$

In the above,

$$\Omega = 2mE_\Phi/\alpha^2\hbar^2 \quad (11.117)$$

Eq. (11.116) is *almost* in the form of the *confluent hypergeometric equation*, which is given by

$$x\frac{d^2y}{dx^2} + (c - x)\frac{dy}{dx} - ay(x) = 0$$

The only difference between them is with the two related terms $c - x \leftrightarrow 1 - (1/2)\zeta$. Unfortunately, for the CHE, we must have 1 as the coefficient for x, or ζ in this case. We can let

$$\zeta^* = \frac{1}{2}\zeta \quad \text{and} \quad \Phi_1^* = \frac{1}{2}\Phi_1$$

Using these relations in Eq. (11.116), we have

$$\zeta^*\frac{d^2\Phi_1^*}{d\zeta^{*2}} + (1 - \zeta^*)\frac{d\Phi_1^*}{d\zeta^*} - \left(\frac{1-\Omega}{2}\right)\Phi_1^* = 0 \quad (11.118)$$

This leads to the exact form of the *confluent hypergeometric equation* that is needed. Thus, the solution to Eq. (11.118) is the **confluent hypergeometric series** (CHS), introduced in Chap. 5. The CHS is given by

$$\text{CHS}(\zeta; a, c) = 1 + \frac{1}{1!}\frac{a}{c}\zeta^* + \frac{1}{2!}\frac{a(a+1)}{c(c+1)}(\zeta^*)^2 +$$

$$\frac{1}{3!}\frac{a(a+1)(a+2)}{c(c+1)(c+2)}(\zeta^*)^3 + \frac{1}{4!}\frac{a(a+1)(a+2)(a+3)}{c(c+1)(c+2)(c+3)}(\zeta^*)^4 +$$

$$\frac{1}{5!}\frac{a(a+1)(a+2)(a+3)(a+4)}{c(c+1)(c+2)(c+3)(c+4)}(\zeta^*)^5 +$$

$$\frac{1}{6!}\frac{a(a+1)(a+2)(a+3)(a+4)(a+5)}{c(c+1)(c+2)(c+3)(c+4)(c+5)}(\zeta^*)^6 + \ldots$$

From Eq. (11.118), the parameters a and c are given by

$$a = \frac{1-\Omega}{2} \tag{11.119a}$$

$$c = 1 \tag{11.119b}$$

Generally, there are two independent solutions, where a and c form one pair, while the other pair is given by $a' = a + c - 1$ and $c' = 2 - c$. However, unlike in the harmonic oscillator problem, here, we have $a = a'$ and $c = c'$. This follows from substitution into Eq. (11.119) since $c = 1$, then $c' = 2 - 1 = 1 = c$ and $a' = a + c - 1 = a$. So, the wave function Φ_1^* is given by the full set of truncated polynomials for identical values of a and c.

Also, recall that in order to satisfy the boundary condition $\Phi_1(\infty) = 0$ for Φ_1 and Φ_1^*, a must be equal to a non-negative integer in order to truncate the CHS to a finite polynomial. This condition leads to

$$\frac{1-\Omega}{2} = -n$$

Substituting $\Omega = 2E_\Phi / \alpha^2 \hbar^2 = 2E_\Phi / \hbar\omega_c$ into the above gives

$$\frac{1 - \frac{2E_\Phi}{\hbar\omega_c}}{2} = -n \Rightarrow \frac{2E_\Phi}{\hbar\omega_c} = 2n + 1$$

Rearranging to solve for E_Φ, we obtain the result

11.6 Cyclotron Orbits And Landau Levels

Landau Levels (Symmetric Guage)

$$E_\Phi = \left(n + \frac{1}{2}\right)\hbar\omega_c \qquad (11.120)$$

Thus, we find that the eigenvalues obtained from using the symmetric gauge are identical to those found using the Landau gauge (Eq. (11.94)), though the paths differed on how we obtained them. Examples of the first four wave functions are illustrated in Table 11.1 for $n = 0, 1, 2$ and 3, $B = 0.5T$ and $1.5T$. The probability density $\rho|\Phi|^2$ is plotted.

(a) $n = 0, B = 0.5T$ \qquad (b) $n = 0, B = 1.5T$

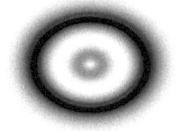

(c) $n = 1, B = 0.5T$ \qquad (d) $n = 1, B = 1.5T$

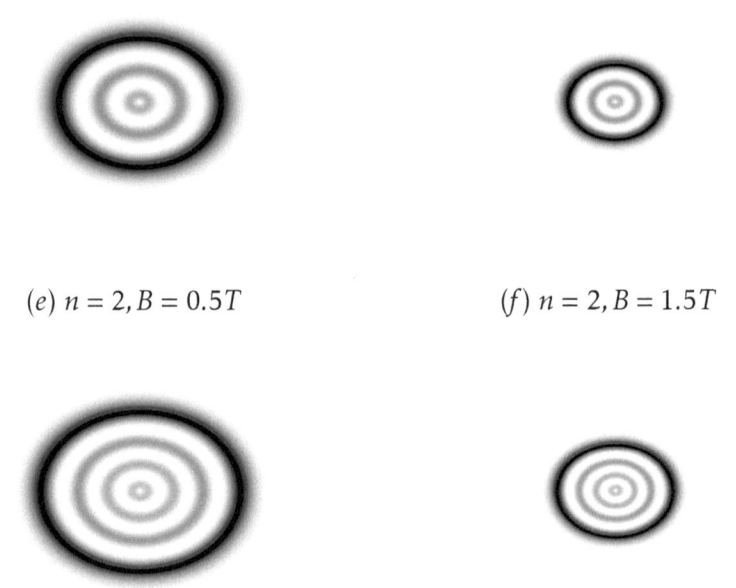

(e) $n = 2, B = 0.5T$ (f) $n = 2, B = 1.5T$

(g) $n = 3, B = 0.5T$ (h) $n = 3, B = 1.5T$

Table 11.1: n–vortices for different values of $n = 0, 1, 2, 3$ and external magnetic field $B = 0.5, 1.5T$. As pure radial functions, they have azimuthal symmetry in the x and y plane. Note that the probability density functions are *not* normalized here. All figures shown are on the same length scale spanning ±200nm, or 400nm.

11

All plots illustrate distributions on the same length scale spanning ±200nm. We see that the electron is predominantly confined to an orbital exterior ring, due to the magnetic field B pointing along the z axis. The orbital motion about the magnetic field inside a confined space is known as a **cyclotron orbit**, and they are *quantized*. The general effect of the magnetic field strength is to further confine the orbital ring radially. These *cyclotron orbits* illustrate the motion of an electron when exposed to a magnetic field, with or without propagation along the z–axis. In the case of propagation, the motion becomes *spiral*. A cyclotron orbit, therefore, is a general spiral of motion of the electron as it moves along the cylindrical axis. k_z can also be zero, which leads to pure circular motion.

When a magnetic field is present, the extended momentum operator must be used in place of the conserved momentum operator $\widehat{\mathbf{p}}$. Without it, this behavior, indicating electrons orbiting in a magnetic field, which is an experimental fact, would not be correctly described. It must also be emphasized that only relativistic equations of motion allow us to uncover this subtle detail relating to the momentum of the electron. Next, we'll move on to the more general relativistic Hamiltonian for the spin-orbit interaction. Specifically, instead of examining the extended momentum in a magnetic field, we'll discuss the spin-orbit interaction, *with* an external magnetic field. These conditions have distinct consequences from the Landau problem.

11.7 SPIN HALL SPIN-ORBIT EFFECTS

In this section, we discuss some important consequences of the spin-orbit interaction. In fact, the particular topic of this section is a very active area of research today, owing to the novel physics that can be exploited to create useful technologies in so many areas including computing, data processing, data storage, sensing, and more. It is a special form of the *spin-orbit interaction* that can be observed in some solid state materials. There is a specific set of conditions giving rise to this spin-orbit interaction, however. For this discussion, we will also need some of the tools we developed in earlier chapters. For example, we'll use results discussed in Chap. 9, where we learned how to determine the time derivative of the expectation value of an operator, known as the *Generalized Generalized Ehrenfest theorem* or *Generalized Ehrenfest theorem*, given by Eqs. (3.147) and (3.148). We'll also require some of the commutation relations developed in Chap. 7. The Generalized Ehrenfest theorem involves commuting the Hamiltonian with the operator whose derivative we are after. With this, we can determine how the expectation value of the extended momentum $\widehat{\mathbf{P}}$ is changed when the Hamiltonian is the *spin-orbit interaction*. We already know the result when only kinetic energy is considered, giving the charge-based Lorentz force. To obtain the time-derivative of the momentum, we must find

$$\frac{d\langle \mathbf{r} \rangle}{dt} = \frac{i}{\hbar}\langle [\widehat{H}_{SO}, \mathbf{r}] \rangle \tag{11.121}$$

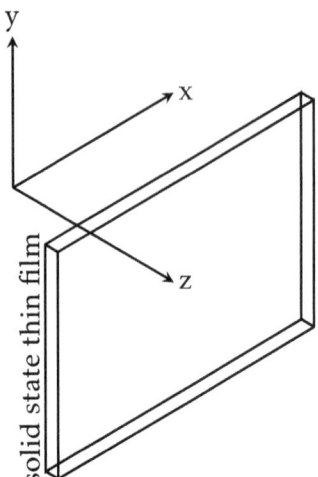

Figure 11.8: A thin film in a defined coordinate system. Solid state thin films of special materials are often used to exploit spin-orbit interactions. In the examples we consider, current flows along the longitudinal axis x.

Omitting the expectation value notation for convenience and substituting the spin-orbit Hamiltonian from Eq. (11.72), we have

$$\frac{d\mathbf{x}}{dt} = \left[-\frac{i}{4m_0^2 c^2}\widehat{\boldsymbol{\sigma}}\cdot(\mathbf{p}+e\mathbf{A})\times\nabla V\right]\mathbf{x} + \mathbf{x}\left[\frac{i}{4m_0^2 c^2}\widehat{\boldsymbol{\sigma}}\cdot(\mathbf{p}+e\mathbf{A})\times\nabla V\right] \quad (11.122)$$

The gradient of the potential energy ∇V is taken as an isolated gradient operation, meaning ∇V is regarded as a spatially uniform vector. It does not operator on the wave function. As an example, ∇V can be proportional to a uniform electric field $\mathbf{E}_0$ driving charges along a wire. Then, using the vector identity $\mathbf{a}\cdot\mathbf{b}\times\mathbf{c} = \mathbf{b}\cdot\mathbf{c}\times\mathbf{a}$, we can rewrite the dot-cross product as

$$\widehat{\boldsymbol{\sigma}}\cdot\widehat{\mathbf{P}}\times\nabla V = \widehat{\mathbf{P}}\cdot(\nabla V\times\widehat{\boldsymbol{\sigma}}) \quad (11.123)$$

This is permitted provided that both ∇V and $\widehat{\boldsymbol{\sigma}}$ are not spatial operators on the wave function. Then, using this result in Eq. (11.122), we have

$$\frac{d\mathbf{x}}{dt} = \frac{i}{4m_0^2 c^2}\left[\mathbf{x}\widehat{\mathbf{P}}\cdot(\nabla V\times\widehat{\boldsymbol{\sigma}}) - \widehat{\mathbf{P}}\cdot(\nabla V\times\widehat{\boldsymbol{\sigma}})\mathbf{x}\right] \quad (11.124)$$

Eq. (11.124) can be evaluated using the result we found in Chap. 7, where we found $[x_i,\widehat{p}_i] = i\hbar$. This relation can be used to expressing Eq. (11.124) as follows:

$$\frac{4m_0^2 c^2}{i}\frac{d\mathbf{x}}{dt} = \left[\mathbf{x}\widehat{\mathbf{P}}\cdot(\nabla V\times\widehat{\boldsymbol{\sigma}}) - \widehat{\mathbf{P}}\cdot(\nabla V\times\widehat{\boldsymbol{\sigma}})\mathbf{x}\right] \quad (11.125a)$$

11.7 Spin Hall Spin-Orbit Effects

$$= \mathbf{x}\left(\widehat{P}_x(\nabla V \times \widehat{\boldsymbol{\sigma}})_x + \widehat{P}_y(\nabla V \times \widehat{\boldsymbol{\sigma}})_y + \widehat{P}_z(\nabla V \times \widehat{\boldsymbol{\sigma}})_z\right) \quad (11.125b)$$

$$- \left((\nabla V \times \widehat{\boldsymbol{\sigma}})_x \widehat{P}_x + (\nabla V \times \widehat{\boldsymbol{\sigma}})_y \widehat{P}_y + (\nabla V \times \widehat{\boldsymbol{\sigma}})_z \widehat{P}_z\right)\mathbf{x} \quad (11.125c)$$

Rearranging the RHS leads to each row of the vector equation having the form:

$$RHS = x_i\left[\left(\widehat{P}_x(\nabla V \times \widehat{\boldsymbol{\sigma}})_x - (\nabla V \times \widehat{\boldsymbol{\sigma}})_x \widehat{P}_x\right) + \left(\widehat{P}_y(\nabla V \times \widehat{\boldsymbol{\sigma}})_y - (\nabla V \times \widehat{\boldsymbol{\sigma}})_y \widehat{P}_y\right) \\ + \left(\widehat{P}_z(\nabla V \times \widehat{\boldsymbol{\sigma}})_z - (\nabla V \times \widehat{\boldsymbol{\sigma}})_z \widehat{P}_z\right)\right] x_i \quad (11.126)$$

In the above, $x_i = x, y, z$. There are three nonzero commutations involving $[x, \widehat{P}_x]$, $[y, \widehat{P}_y]$, and $[z, \widehat{P}_z]$. For all the commutations with different components (i.e. z and $\widehat{P}_x$), they vanish since they commute. The non-commuting components are replaced by their eigenvalue $i\hbar$. This gives the result

$$\frac{4m_0^2 c^2}{i} \frac{d\mathbf{x}}{dt} = \begin{bmatrix} \left(x\widehat{P}_x - \widehat{P}_x x\right)(\nabla V \times \widehat{\boldsymbol{\sigma}})_x \\ \left(y\widehat{P}_y - \widehat{P}_y y\right)(\nabla V \times \widehat{\boldsymbol{\sigma}})_y \\ \left(z\widehat{P}_z - \widehat{P}_z z\right)(\nabla V \times \widehat{\boldsymbol{\sigma}})_z \end{bmatrix} = i\hbar(\nabla V \times \widehat{\boldsymbol{\sigma}}) \quad (11.127)$$

Therefore, we have an additional contribution to the electron's momentum due to spin-orbit interaction. Let us define this additional momentum $\widehat{\mathbf{P}}_{SO} = m_0 d\langle \mathbf{x}\rangle/dt$ (returning the expectation brackets) as

$$\langle \widehat{\mathbf{P}}_{SO}\rangle = m_0 \frac{d\langle \mathbf{x}\rangle}{dt} = -\frac{\hbar}{4m_0 c^2} \nabla V \times \langle \widehat{\boldsymbol{\sigma}}\rangle \quad (11.128)$$

In terms of the spin angular momentum $\widehat{\mathbf{S}}$, the **spin-orbit transverse momentum** becomes

Spin-Orbit Transverse Momentum

$$\langle \widehat{\mathbf{P}}_{SO}\rangle = -\frac{1}{2m_0 c^2}\langle \widehat{\mathbf{S}}\rangle \times \nabla V \quad (11.129)$$

This momentum must be added to the two contributions we have already discussed. Then, the total momentum $\widehat{\mathbf{P}}_T$ with spin-orbit interaction becomes

$$\widehat{\mathbf{P}}_T = \widehat{\mathbf{p}} + e\mathbf{A} + \widehat{\mathbf{P}}_{SO} \qquad (11.130)$$

The first term in Eq. (11.130) is the conserved momentum due to an electric field $\mathbf{E}_0 = \nabla V/e$ responsible for electrical current. The second term corresponds to the ordinary *charge-based Hall effect*. The last term is a *spin-dependent momentum*, and independent of charge. Because the spin-orbit momentum is a function of ∇V, it can also be expressed as a function of the momentum $\widehat{\mathbf{p}}$ using the relation $\nabla V = -\mathbf{F}_e = -\dot{\mathbf{p}}$. Then, we have an alternative form given by

$$\langle \widehat{\mathbf{P}}_{SO} \rangle = \frac{1}{2m_0 c^2} \langle \widehat{\mathbf{S}} \rangle \times \dot{\mathbf{p}} \qquad (11.131)$$

The spin-orbit momentum in Eq. (11.129) is known as **side-jump** and it is the spin-dependent transverse momentum, or velocity $(\widehat{\mathbf{P}}/m_0)$ arising from the spin-orbit interaction with linear momentum $\widehat{\mathbf{p}}$.

The presence of the additional spin-dependent momentum corresponds to a spin-dependent *force* on a charged particle *carrying spin*. We can determine this force by a similar analysis to what was done with the time-derivative of $\langle x \rangle$. As the derivative of the spin-orbit momentum, it defines the *side-jump* force. To obtain this force assuming a *Hermitian Hamiltonian*, we must evaluate the *Generalized Ehrenfest theorem* (c.Sec. 3.10) given by

$$\frac{d\langle \mathbf{P} \rangle}{dt} = \frac{i}{\hbar} \langle [\widehat{H}_{SO}, \mathbf{P}] \rangle \qquad (11.132)$$

Omitting the expectation value brackets (for convenience), we have

$$\frac{d\widehat{\mathbf{P}}_{SO}}{dt} = \left[-\frac{i}{4m_0^2 c^2} \widehat{\mathbf{P}} \cdot (\nabla V \times \widehat{\boldsymbol{\sigma}}) \right] \mathbf{P} + \mathbf{P} \left[\frac{i}{4m_0^2 c^2} \widehat{\mathbf{P}} \cdot (\nabla V \times \widehat{\boldsymbol{\sigma}}) \right] \qquad (11.133)$$

This can be written as

$$\frac{4m_0^2 c^2}{i} \frac{d\widehat{\mathbf{P}}_{SO}}{dt} = \left[\widehat{\mathbf{P}} \mathbf{P} \cdot (\nabla V \times \widehat{\boldsymbol{\sigma}}) - \widehat{\mathbf{P}} \cdot (\nabla V \times \widehat{\boldsymbol{\sigma}}) \mathbf{P} \right]$$

$$= \widehat{\mathbf{P}} \left(\widehat{P}_x \cdot (\nabla V \times \widehat{\boldsymbol{\sigma}})_x + \widehat{P}_y \cdot (\nabla V \times \widehat{\boldsymbol{\sigma}})_y + \widehat{P}_z \cdot (\nabla V \times \widehat{\boldsymbol{\sigma}})_z \right)$$

$$- \left((\nabla V \times \widehat{\boldsymbol{\sigma}})_x \widehat{P}_x + (\nabla V \times \widehat{\boldsymbol{\sigma}})_y \widehat{P}_y + (\nabla V \times \widehat{\boldsymbol{\sigma}})_z \widehat{P}_z \right) \widehat{\mathbf{P}}$$

11.7 Spin Hall Spin-Orbit Effects

The result above is identical in form to Eq. (11.125). Here, instead of using the commutation relation $[x_i, P_i] = i\hbar$, we can use the relevant commutation relation obtained earlier with Eq. (7.120), which is

$$[\widehat{P}_i, \widehat{P}_j] = -ie\hbar\varepsilon_{ijk}B_k \tag{11.134}$$

ε_{ijk} is the Levi-Civita symbol defined in Chap. 7. Substitution of this commutation relation into the above, we end up with all the terms of the form $P_i P_i$ canceling because they commute. Terms of the form $P_i P_j$ utilize Eq. (11.134). Evaluation leads to the following time-derivative of the extended momentum operator:

$$\frac{4m_0^2 c^2}{i}\frac{d\widehat{P}_{SO}}{dt} = -ie\hbar(\nabla V \times \widehat{\boldsymbol{\sigma}}) \times \mathbf{B} \Rightarrow \langle \widehat{\mathbf{F}} \rangle_{SO} = \frac{e\hbar}{4m_0^2 c^2}(\nabla V \times \langle \widehat{\boldsymbol{\sigma}} \rangle) \times \mathbf{B}$$

Using the relation $\mu_B = e\hbar/2m_0$, the spin-dependent force on the electron becomes

Spin-Orbit Force

$$\langle \widehat{\mathbf{F}} \rangle_{SO} = \frac{\mu_B}{2m_0 c^2}(\nabla V \times \langle \widehat{\boldsymbol{\sigma}} \rangle) \times \mathbf{B} \tag{11.135}$$

Eq. (11.135) is *general* for the spin-orbit interaction (with uniform fields), and describes a special force that drives a spin-carrier in a direction normal to the magnetic field **B** and its momentum **P**. This force is a *spin-dependent Lorentz force* because Eq. (11.135) can also be written as

$$\langle \widehat{\mathbf{F}} \rangle_{SO} = -\frac{e}{m_0}\widehat{\mathbf{P}}_{SO} \times \mathbf{B} = -e\langle \mathbf{v}_\sigma \times \mathbf{B} \rangle \tag{11.136}$$

The form of the force in Eq. (11.136) is analogous to the charge-based *Hall effect* which is given by $q\mathbf{v} \times \mathbf{B}$. This result suggests that if there is a contribution to the gradient potential along an axis x of sustained charge motion, with spin-orbit coupling present, the result will be that spins pointing 'up', for example, will develop momentum laterally along y, while forced 'up' (because $(\widehat{\mathbf{x}} \times \pm\widehat{\mathbf{z}}) \times \widehat{\mathbf{x}} = \pm\widehat{\mathbf{z}}$). The spins will end up moving towards the top and bottom surfaces of a finite width sample. The result of this particular force arising from a longitudinal electric field acting on spin-angular momentum is known as the **spin-Hall effect**. Today, the existence of the *spin-Hall* effect, described by Eq. (11.136), is a well-established experimental fact. It has been clearly demonstrated in several experiments to drive the magnetization in thin film devices. One early demonstration of the utility of the *spin-Hall effect* to control the switching of a magnetic layer was carried out by Luqiao Liu *et. al.*, while at Cornell University around 2012. Flowing electrical current in a device composed of a bottom layer of platinum (Pt), a cobalt (Co) layer above the Pt, and an aluminum oxide (AlO_x) layer on top, the spin-Hall effect was demonstrated to reverse the magnetization of the Co layer. The magnetic field was along the same axis of current flow. The resulting force in the Pt *generates a spin current injected vertically* into the Co layer. Fig. 11.9 illustrates a device used to exploit the *spin-Hall effect* to reverse the magnetization of a cobalt layer. In Fig. 11.9, the

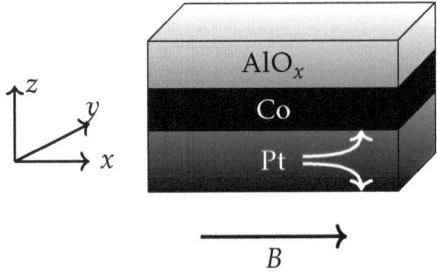

Figure 11.9: Illustration of a thin film device utilizing the *spin-Hall effect*. The platinum layer provides the large spin-orbit coupling, injecting polarized spins (along z) into the Co layer. The force on the spins is proportional to $(\hat{x} \times \pm \hat{z}) \times \hat{x} = \pm \hat{z}$.

bottom platinum layer is essential for the device physics here because it is responsible for the spin-orbit coupling where the overall effect originates. As electrical current flows longitudinally along x ($\sim$ 1mA), the spin-orbit interaction gives rise to spin-orbit momentum $\hat{P}_{SO}$ along the $\pm y$–axis. Then, the magnetic field $B \sim$ 10mT along x causes a *spin-orbit force* along z, which injects spins that are polarized along $\pm z$ into the Co layer. This can generate a special torque known as *spin-transfer torque* on the bottom of the Co layer, causing it to reverse or switch its magnetization direction (also along z) under the right conditions. Note that too large of a magnetic field, however, destroys the spin-Hall effect because beyond a certain magnetic field strength, the z–polarized spins will rotate along the magnetic field axis, perpendicular to z. Without an magnetic field, while this can rotate the magnetization layer, it cannot switch it. In Chap. 12, we'll also derive the so-called *spin-transfer torque* responsible for the reversal of the magnetization in the magnetic layer. To do this, we'll need to introduce *dissipation* into the quantum mechanics picture, which is the subject of the next chapter. Let's move on to discuss another form of spin-orbit coupling which, nowadays, is also the focus of much research. In particular, this next effect is exploited in development of novel types of spin-based electronic transistors. This next spin-orbit effect discussed is known as the *Rashba effect*.

11.8 RASHBA SPIN-ORBIT EFFECT

We saw with the spin-Hall effect that the itinerant electrons initially move along a path due to an external electric field predominantly along the *longitudinal axis* of the device. However, there are more possibilities with the spin-orbit interaction if we introduce additional components to the electric field. The gradient of the potential energy $\nabla V = -e\mathbf{E}$ is still taken to be *nearly uniform*. Otherwise, the Hamiltonian would be time-dependent. One modern example of an application of the *Rashba effect* was first suggested by Supriya Datta and Biswajit Das in 1990, for a device functioning as a voltage modulator. One incarnation of this device is shown in Fig. 11.10. The figure shows a device known as a

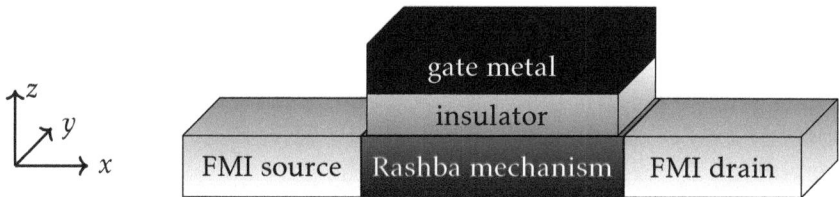

Figure 11.10: Illustration of a three terminal spin-field effect transistor or spin-FET. This device is designed to utilize the Rashba effect in the region just below the gate. The gate provides a perpendicular electric field to manipulate spin current. Polarized spins will precess about the effective Rashba field, varying their angles periodicially with the polarization of the drain source.

spin-field effect transistor or **spin-FET**. The spin-FET is a developing technology (to-date, not yet realized) designed to utilize the *Rashba effect* in the region just below the gate. The gate induces a perpendicular electric field E_g along z. We can use the relation $\mathbf{A} \cdot \mathbf{B} \times \mathbf{C} = \mathbf{C} \cdot \mathbf{A} \times \mathbf{B}$, so the spin-orbit Hamiltonian becomes

$$\widehat{H}_{SO} = -\frac{\hbar}{4m_0^2 c^2}\widehat{\boldsymbol{\sigma}} \cdot \widehat{\mathbf{P}} \times \nabla V = -\frac{\hbar}{4m_0^2 c^2}\nabla V \cdot (\widehat{\boldsymbol{\sigma}} \times \widehat{\mathbf{P}}) \qquad (11.137)$$

From the spin-orbit Hamiltonian, an effective magnetic field per charge carrier can be obtained, and is given by

> **Effective Spin-Orbit Magnetic Field**
>
> $$\mathbf{B}_{SO} = -\frac{1}{2m_0^2 c^2 |\gamma| g} \widehat{P} \times \nabla V \qquad (11.138)$$

Eq. (11.138) is an **effective spin-orbit magnetic field** that is exploited by the spin-FET. We'll have you show the more explicit form applicable to the spin-FET in Chapter problem 11.9. In the spin-FET, another condition (though not essential) is assumed where the flowing or itinerant electrons are *approximately* confined to a two-dimensional 2D plane, such as a current flowing adjacent to an interface between two materials. For reasons discussed in the previous section, this is a reasonable assumption for high spin-orbit interaction, which can cause spin carriers to drift toward the boundaries. The set of electrons confined to this 2D plane is known as a **two-dimensional electron gas** or **2DEG**. Beyond 2D motion, this also implies no interaction between electrons.

Contrary to the spin-Hall device, in the Datta and Das configuration, *there is no external magnetic field*. In the 2D plane of electrons, there exists a general electric field localized and normal to the interface. We denote it by E_I. As mentioned, in a spin-FET, E_I is supplied by the gate voltage E_g, or $E_I = E_g$ (applied on the gate metal in diagram). In the spin-FET device, there are also *ferromagnetic insulators* (FMI) for leads known as *source* and *drain* contacts. A FM insulator is an electrical insulator that, usually, is modified in its fabrication process to possess a net magnetic moment. Well known examples include iron oxide (Fe_3O_4) and chromium oxide (Cr_2O_3). However, generally, higher magnetic moment FMIs are challenging to realize.

In the spin-FET, the so-called Rashba mechanism is used to generate an effective magnetic field $\mathbf{B}_R$ around which the polarized itinerant electron spins precess. The effective Rashba field is proportional to the gate voltage input. By tuning the gate voltage, the precession rate ($\propto \gamma B_R$) varies, and one can generate a modulating voltage/current output across the source and drain through a read mechanism known as *magnetoresistance*. We'll discuss magnetoresistance in some depth in the next chapter. Thus, we'll only discuss the role of the Rashba mechanism to enable an understanding of how a spin-FET device is designed to operate. In the device, the itinerant electrons flow through a

11.8 Rashba Spin-Orbit Effect

FM lead and experience spin-polarization. If the interface of the 2DEG is designed in such a way that an electric field induced from the gate exists normal to the plane of the 2DEG/interface, then, in the 2DEG, we have

$$\nabla V = -eE_g\widehat{\mathbf{z}} \Rightarrow \widehat{H}_{SO} = \frac{\hbar e E_g}{4m_0^2 c^2}\widehat{\mathbf{z}}\cdot(\widehat{\boldsymbol{\sigma}}\times\widehat{\mathbf{P}}) = \frac{\mu_B E_g}{2m_0 c^2}\widehat{\mathbf{z}}\cdot(\widehat{\boldsymbol{\sigma}}\times\widehat{\mathbf{P}}) \quad (11.139)$$

$\widehat{\mathbf{z}}$ is the unit vector normal to the 2DEG plane. If we express $\widehat{\mathbf{P}} = \hbar\mathbf{k}$, we get the more familiar form of the **Rashba Hamiltonian** given by

---
Rashba Hamiltonian

$$\widehat{H}_R = \alpha_R(\widehat{\boldsymbol{\sigma}}\times\mathbf{k})\cdot\widehat{\mathbf{z}} \quad (11.140)$$

α_R is the **Rashba constant** or **Rashba coefficient**, named after the Russian-American physicist Emmanuel I. Rashba (1927-), who, along with Valentin I. Sheka, first discussed the mechanism around 1959. The coefficient is given by

---
Rashba Coefficient

$$\alpha_R = \frac{e\hbar^2 E_g}{4m_0^2 c^2} \quad (11.141)$$

When the conditions in the material essentially set up an electric field in a 3D lattice, still with a net electric field along $\widehat{\mathbf{z}}$ arising from the lattice itself, for example, it is known as the *Dresselhaus effect*. It is named after , who studied the effect of SO coupling on semiconductors, namely those with zincblend structure, the crystal structure of many III-V and II-VI semiconductors such as GaAs, InSb, and CdTe. Ultimately, their Hamiltonians are of the same form, however, their coefficients differ.

As in the case of the spin-Hall effect, the *Rashba effect* can only be detected if there is a resulting force on the spin carriers. It is straightforward to show using the *spin-orbit force*, that in the same conditions generating the spin-Hall effect, adding an interfacial normal electric field *does not result in a Rashba effect*. This is because the spins are polarized along the direction of $\mathbf{E}_I$, which yields *zero* momentum. This matter was lightly debated in the early observations of the spin-Hall effect being used to switch magnetic layers. It was considered an open question as to whether the Rashba effect was also contributing. However, relativistic quantum mechanics does provide a clear answer to the question. The Rashba effect is *not* present in conditions for the spin-Hall, particularly because $\mathbf{E}_I$ and σ are parallel. Alternative arrangements must be used in order to observe the Rashba effect. For example, if there is a gradient potential normal to the plane of motion of the spin-carrier, there can be a net spin polarization laterally in the 2DEG, yielding a force on lateral spins towards the top and bottom surfaces of the device. This scenario of spin-orbit coupling is distinct from the *spin-Hall effect*, both of which can generate spin-current and spin polarization.

Let's now analyze the Rashba Hamiltonian more closely to determine the eigenvalues of this system. The Hamiltonian is given by

$$i\hbar \frac{\partial \psi}{\partial t} = \widehat{\mathbf{H}} \psi = \left[\frac{\widehat{\mathbf{p}}^2}{2m_0} + \widehat{H}_{SO} \right] \psi = \left[\frac{\widehat{\mathbf{p}}^2}{2m_0} + \alpha_R (\widehat{\sigma} \times \widehat{\mathbf{k}}) \cdot \widehat{\mathbf{z}} \right] \psi \quad (11.142)$$

In Eq. (11.142) $\psi = [X_+ \ X_-]^T$. By inspection of the Rashba term, because $\widehat{\sigma}$ and $\widehat{\mathbf{z}}$ are dimensionless, and $\widehat{\mathbf{k}}$ has units of inverse length, we can infer the units of the Rashba constant to be *energy × length* (e.g. J-m). Since the Hamiltonian includes a spin-dependent contribution involving $\widehat{\sigma}$, Eq. (11.142) expresses two equations which can be written in explicit 2 × 2 matrix notation as

$$i\hbar \frac{\partial}{\partial t} \begin{bmatrix} \psi_+ \\ \psi_- \end{bmatrix} = \left[\frac{\hbar^2}{2m_0} \widehat{k}^2 \mathbf{I}(2) + \alpha_R (\widehat{\sigma} \times \widehat{\mathbf{k}}) \cdot \widehat{\mathbf{z}} \right] \begin{bmatrix} \psi_+ \\ \psi_- \end{bmatrix}$$

$$= \left[\frac{\hbar^2}{2m_0} \begin{bmatrix} \widehat{k}_+^2 & 0 \\ 0 & \widehat{k}_-^2 \end{bmatrix} + \alpha_R \begin{bmatrix} \widehat{\sigma}_y \widehat{k}_z - \widehat{\sigma}_y \widehat{k}_z \\ \widehat{\sigma}_z \widehat{k}_x - \widehat{\sigma}_x \widehat{k}_z \\ \widehat{\sigma}_x \widehat{k}_y - \widehat{\sigma}_y \widehat{k}_x \end{bmatrix} \cdot \widehat{\mathbf{z}} \right] \begin{bmatrix} \psi_+ \\ \psi_- \end{bmatrix}$$

$$= \left[\frac{\hbar^2}{2m_0} \begin{bmatrix} \widehat{k}_+^2 & 0 \\ 0 & \widehat{k}_-^2 \end{bmatrix} + \alpha_R \left(\widehat{\sigma}_x \widehat{k}_y - \widehat{\sigma}_y \widehat{k}_x \right) \right] \begin{bmatrix} \psi_+ \\ \psi_- \end{bmatrix}$$

$$= \left[\frac{\hbar^2}{2m_0} \begin{bmatrix} \widehat{k}_+^2 & 0 \\ 0 & \widehat{k}_-^2 \end{bmatrix} + \alpha_R \begin{bmatrix} 0 & \widehat{k}_{y-} + i\widehat{k}_{x-} \\ \widehat{k}_{y+} - i\widehat{k}_{x+} & 0 \end{bmatrix} \right] \begin{bmatrix} \psi_+ \\ \psi_- \end{bmatrix}$$

11.8 Rashba Spin-Orbit Effect

$$= \begin{bmatrix} \frac{\hbar^2 k_+^2}{2m_0} & \alpha_R(\widehat{k}_{y-} + i\widehat{k}_{x-}) \\ \alpha_R(\widehat{k}_{y+} - i\widehat{k}_{x+}) & \frac{\hbar^2 k_-^2}{2m_0} \end{bmatrix} \begin{bmatrix} \psi_+ \\ \psi_- \end{bmatrix}$$

Having no explicit time dependence on the RHS, and no explicit spatial dependence on the LHS, a separable solution in space and time can be tried, where $\psi_+ = X_+(x)T(t)$ and $\psi_- = X_-(x)T(t)$. Substitution of this into the above gives

$$i\hbar \frac{dT}{dt} \begin{bmatrix} X_+ \\ X_- \end{bmatrix} = T \begin{bmatrix} \frac{\hbar^2 k^2}{2m_0} & \alpha_R(\widehat{k}_y + i\widehat{k}_x) \\ \alpha_R(\widehat{k}_y - i\widehat{k}_x) & \frac{\hbar^2 k^2}{2m_0} \end{bmatrix} \begin{bmatrix} X_+ \\ X_- \end{bmatrix} \quad (11.143)$$

We have a pair of equations where the RHS is *coupled* in X_+ and X_-. By dividing each equation by $\psi_\pm = X_\pm T$, we can justify separating the time-dependence from the coupled spin equations. For the time dependence, we get

$$i\hbar \begin{bmatrix} \frac{dT_+}{dt} \\ \frac{dT_-}{dt} \end{bmatrix} = \begin{bmatrix} E_+ T_+ \\ E_- T_- \end{bmatrix} \Rightarrow T_\pm(t) = T_\pm(0) e^{-i\frac{E_\pm}{\hbar}}$$

E_+ is the eigenenergy of the spin-up eigenfunction, and E_- for spin-down. Then, the separated equations depending on space become

$$\begin{bmatrix} \frac{\hbar^2 k^2}{2m_0} & \alpha_R(\widehat{k}_y + i\widehat{k}_x) \\ \alpha_R(\widehat{k}_y - i\widehat{k}_x) & \frac{\hbar^2 k^2}{2m_0} \end{bmatrix} \begin{bmatrix} X_+ \\ X_- \end{bmatrix} = \begin{bmatrix} E_+ & 0 \\ 0 & E_- \end{bmatrix} \begin{bmatrix} X_+ \\ X_- \end{bmatrix} \quad (11.144)$$

Eq. (11.144) is a typical matrix eigenvalue equation which can be written as

$$\begin{bmatrix} \frac{\hbar^2 k_+^2}{2m_0} - E_+ & \alpha_R(\widehat{k}_{y-} + i\widehat{k}_{x-}) \\ \alpha_R(\widehat{k}_{y+} - i\widehat{k}_{x+}) & \frac{\hbar^2 k_-^2}{2m_0} - E_- \end{bmatrix} \begin{bmatrix} X_+ \\ X_- \end{bmatrix} = \begin{bmatrix} 0 \\ 0 \end{bmatrix} \quad (11.145)$$

Since the wave functions are always nontrivial, or nonzero functions, Eq. (11.145) can only be true *if and only if* the determinant of the 2×2 matrix on the LHS vanishes. This gives

$$\left(\frac{\hbar^2 k_+^2}{2m_0} - E_+ \right)\left(\frac{\hbar^2 k_-^2}{2m_0} - E_- \right) - \alpha_R^2 \left(k_{y-} + ik_{x-} \right)\left(k_{y+} - ik_{x+} \right) = 0 \quad (11.146)$$

To simplify notation, let us define the following parameters:

$$\alpha_+^2 = \frac{\hbar^2 k_+^2}{2m_0} \quad \text{and} \quad \alpha_-^2 = \frac{\hbar^2 k_-^2}{2m_0} \quad (11.147a)$$

$$\beta_+^* = \alpha_R(k_{y+} - ik_{x+}) \quad \text{and} \quad \beta_- = \alpha_R(k_{y-} + ik_{x-}) \tag{11.147b}$$

The reader should keep in mind that these operator eigenequations represent differential equations for the wave function spinor $[X_+ \; X_-]^T$. To express Eq. (11.146) in their equivalent differential equations, we just introduce the operator relations $\widehat{k}_i = \widehat{p}_i/\hbar$. When we do this, the pair of differential equations determining the spin-up energy E_+ and spin-down E_- becomes

$$-\frac{\hbar^2}{2m_0}\left(\frac{\partial^2 X_+}{\partial x^2} + \frac{\partial^2 X_+}{\partial y^2}\right) - i\alpha_R\left(\frac{\partial X_-}{\partial y} + i\frac{\partial X_-}{\partial x}\right) = E_+ X_+ \tag{11.148a}$$

$$-\frac{\hbar^2}{2m_0}\left(\frac{\partial^2 X_-}{\partial x^2} + \frac{\partial^2 X_-}{\partial y^2}\right) - i\alpha_R\left(\frac{\partial X_+}{\partial y} - i\frac{\partial X_+}{\partial x}\right) = E_- X_- \tag{11.148b}$$

The matrix notation we will continue to use for convenience. Then, substitution of the relations in Eq. (11.149) into Eq. (11.146) leads to the following results for the energies E_+ and E_-:

$$E_+ = \alpha_+^2 - \frac{\beta_+^* \beta_-}{E_- - \alpha_-^2} \tag{11.149a}$$

$$E_- = \alpha_-^2 - \frac{\beta_+^* \beta_-}{E_+ - \alpha_+^2} \tag{11.149b}$$

The eigenenergies in Eq. (11.149) reveal some unique characteristics which should be highlighted. One important characteristic can be seen by considering highly symmetric conditions in the Rashba picture. This can be done by assuming the following equalities:

$$E_+ = \alpha_+^2 + \Delta E$$

$$E_- = \alpha_-^2 - \Delta E$$

$$k = k_+ = k_-$$

$$\alpha^2 = \alpha_-^2 = \alpha_+^2 = \frac{\hbar^2 k^2}{2m_0}$$

$$\beta = \beta_- = \beta_+$$

ΔE is the **half spin-splitting** because $E_+ - E_- = 2\Delta E$. Substitution of these symmetric conditions into Eq. (11.149) gives

$$\alpha^2 + \Delta E = \alpha^2 - \frac{\beta^* \beta}{\Delta E} \Rightarrow \Delta E^2 = \beta^* \beta \tag{11.150a}$$

11.8 Rashba Spin-Orbit Effect

$$\alpha^2 - \Delta E = \alpha^2 - \frac{\beta^*\beta}{\Delta E} \Rightarrow \Delta E^2 = \beta^*\beta \tag{11.150b}$$

Therefore, $\Delta E = \pm|beta|$, leading to the following eigenenergies:

$$E_+ = \alpha^2 + |\beta| = \frac{\hbar^2 k^2}{2m_0} + \alpha_R \sqrt{k_x^2 + k_y^2} \tag{11.151a}$$

$$E_- = \alpha^2 - |\beta| = \frac{\hbar^2 k^2}{2m_0} - \alpha_R \sqrt{k_x^2 + k_y^2} \tag{11.151b}$$

Eqs. (11.151) tells us that the electron energies are parabolic in k due to the kinetic energy, but this is not surprising. However, the addition of the Rashba interaction leads to a splitting of the energies between spin-up and spin-down electrons. Both parabolas are shifted anti-symmetrically, and thus, the minimum energy momentum points are different for spin-up and spin-down. This momentum at the minimum energy is found from $dE/dk = 0$, which gives

$$k_{\text{min},\pm} = \pm \frac{\alpha_R m_0}{\hbar^2}$$

The split energies $E_+(k)$ and $E_-(k)$ as functions of k are shown in Fig. 11.11. By analyzing the Rashba Hamiltonian along with kinetic energy,

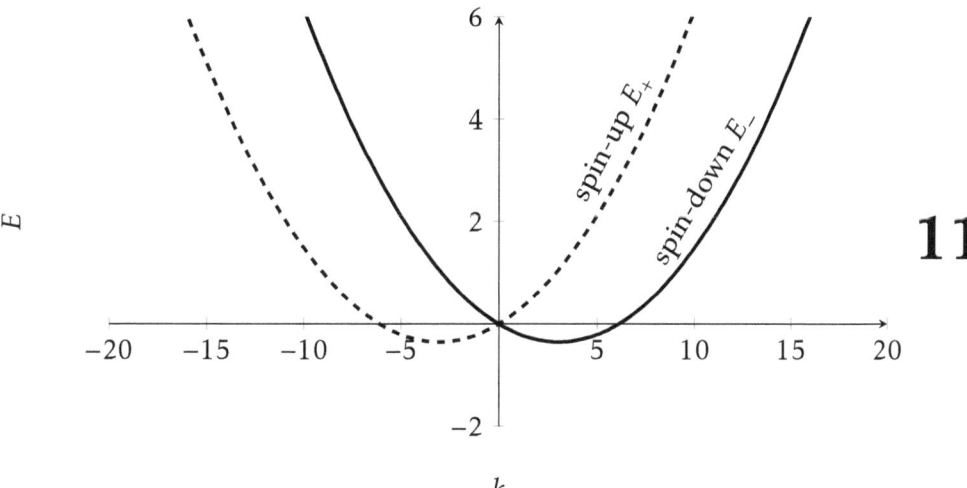

Figure 11.11: Illustration of Rashba splitting due to the Rashba spin-orbit interaction. The spin-up and spin-down momenta corresponding to the minimum energies of each parabola are shifted anti-symmetrically by an amount given by $k_{\min} = \pm \alpha_R m_0/\hbar^2$.

we find distinct effects from the spin-orbit interaction. Firstly, since the electric field is taken along $\hat{z}$, only spins $\hat{\sigma}_x$ and $\hat{\sigma}_y$ affect the energy of the system. This is consistent with the finding that the *spin-orbit force* is zero when $\Delta V \times \boldsymbol{\sigma} \propto \hat{z} \times \hat{z} = 0$. As with other spin-dependent Hamiltonian energy contributions we've discussed (e.g. Zeeman, exchange, and spin Hall), the Rashba effect leads to a spin-splitting where the energies for spin-up and spin-down become distinct or non-degenerate. This is *unlike* in the case of having only kinetic energy. However, we did see a consistent result from adding Zeeman energy, which also leads to a spin-dependent energy splitting.

11.9 Chapter Summary

In this chapter, we have introduced the relativistic quantum mechanics, and analyzed some of the physics resulting from the relativistic forms. To do this properly, we first reviewed the special theory of relativity to derive relativistic momentum and kinetic energy. From there, we developed the various relativistic quantum mechanics, particularly the Dirac equation. The spin-orbit interaction was then derived from the Dirac equation, utilizing a binomial expansion. This result also lead us to the electron g-factor $g = 2$. Then, we considered the effects of the extended momentum $\mathbf{P} = \mathbf{p} + e\mathbf{A}$ on the eigenvalues and eigensolutions in the Landau problem, which is the cyclotron problem. Then, we examined consequences of some specific conditions of the spin-orbit interaction. From this, we found the *spin-Hall* and the *Rashba effects*. Table 11.2 summarizes some of the key results from this chapter.

Table 11.2: Chapter 11 Summary Equations.

Name	Equation
Lorentz factor	$\gamma_r = \frac{1}{\sqrt{1-\frac{u^2}{c^2}}}$
Invariant Space-Time Interval	$\mathbf{S} \cdot \mathbf{S} = c^2 t^2 - x^2 - y^2 - z^2$
Relativistic Momentum	$\mathbf{p} = \frac{m_0 \mathbf{v}}{\sqrt{1-\frac{v^2}{c^2}}}$
Rel. Kinetic Energy	$\overline{K} = \pm\sqrt{p^2 c^2 + m_0^2 c^4}$

11.10 Chapter Problems

Table 11.2

Dirac Eq.	$i\hbar \frac{\partial \psi}{\partial t} = \left[V + c\widehat{\Sigma} \cdot (\widehat{\mathbf{p}} + e\mathbf{A}) + \widehat{\Sigma}_0 m_0 c^2 \right] \psi$
Spin-Orbit	$\widehat{H}_{SO} = -\frac{\hbar}{4m_0^2 c^2} \widehat{\sigma} \cdot (\widehat{\mathbf{p}} + e\mathbf{A}) \times \nabla V$
S-O Momentum	$\langle \widehat{\mathbf{P}}_{SO} \rangle = -\frac{\hbar}{4m_0 c^2} \nabla V \times \langle \widehat{\sigma} \rangle$
S-O Force	$\langle \widehat{\mathbf{F}} \rangle_{SO} = \frac{\mu_B}{2m_0 c^2} (\nabla V \times \langle \widehat{\sigma} \rangle) \times \mathbf{B}$
Rashba Hamiltonian	$\widehat{H}_R = \alpha_R (\widehat{\sigma} \times \mathbf{k}) \cdot \widehat{\mathbf{z}}$

11.10 Chapter Problems

Problem 11.1 Let $\nabla V = q\mathbf{E}_0$, where $\mathbf{E}_0$ in uniform in space. Prove that $\widehat{\sigma} \cdot \widehat{\mathbf{P}} \times \mathbf{E}_0 = \widehat{\mathbf{P}} \cdot (\mathbf{E}_0 \times \widehat{\sigma})$ by evaluating both sides of the relation separately, and show they are equal.

Problem 11.2 Assume the ordinary kinetic energy has the form

$$E_K = \frac{(\widehat{\mathbf{p}} + e\mathbf{A})^2}{2m_0} = an + b$$

We found that the eigenvalue of this operator is quantized, *in contrast* to what is found without the magnetic vector potential $\mathbf{A}$. In this form, there is a one-to-one correspondence between the quantum number n and the eigenenergy. Show that the resulting relativistic correction E_c given by

$$E_c = -\frac{(\widehat{\mathbf{p}} + e\mathbf{A})^4}{8m_0^3 c^2}$$

is no longer one-to-one in the quantum number n. In other words, more than one quantum number yields the same eigenvalue.

Problem 11.3 In the Rashba spin-orbit interaction, the potential gradient is taken to be along the z, typically normal to the thin film interface. In such conditions with electrons flowing initially along x, use the spin-orbit spin-dependent force given by Eq. to show that spin polarization current, due to Rashba, is only possible along z and y, but not x.

Chapter 11. Relativistic Quantum Mechanics

One of these cases leads to spin current and spin-orientation being parallel. This is known as *longitudinal spin current*.

Problem 11.4 It was shown that the spin-orbit momentum is given by

$$\langle \widehat{P}_{SO} \rangle = -\frac{\hbar}{4m_0c^2}(\nabla V \times \langle \widehat{\sigma} \rangle)$$

Use this relation to estimate the magnitude of $\langle \widehat{P}_{SO} \rangle$ per spin for a system with an applied voltage of 5V over a 1cm sample length.

Problem 11.5 Estimate the strength of the spin-orbit force (per spin) for an electron flowing through along 1cm sample, driven by 1 V, in an external magnetic field of 1 Tesla, using the relation

$$\langle \widehat{F} \rangle_{SO} = \frac{\mu_B}{2m_0c^2}(\nabla V \times \langle \widehat{\sigma} \rangle) \times \mathbf{B}$$

Problem 11.6 Show the following property is true for spin matrices by computing the left side and right side:
$[\sigma \cdot (\widehat{p} + e\mathbf{A})]^2 = (\widehat{p} + e\mathbf{A})^2 \mathbf{I}(2) + 2m\mu_B \mathbf{B} \cdot \sigma$

Problem 11.7 Let the number of charge carriers per unit volume be $n = 10^{28} \, \text{m}^{-3}$, with an applied voltage of 1V over a sample 1cm in length L. Using the electrical conductivity of Au, $\sigma = 4.1 \times 10^7 \, \text{S/m}$, first, estimate the average momentum per charge carrier using the relation

$$nev = \frac{nep}{m} = \sigma E \Rightarrow p \approx \frac{\sigma E m}{ne}$$

E is the electric field, which can be estimated by $E \approx V/L$. Use the result to estimate the maximum strength of spin-orbit interaction using the relation

$$SOI \approx \frac{\hbar p E}{4m_0^2 c^2}$$

How does this compare to the ground state energy of an electron in the 1s orbital (use the hydrogen energy).

Problem 11.8 Show that the Dirac 4×4 matrices satisfy the following conditions:

$$\widehat{\Sigma}_0 \widehat{\Sigma}_x = \begin{bmatrix} 0 & \widehat{\sigma}_x \\ -\widehat{\sigma}_x & 0 \end{bmatrix}$$

$$\widehat{\Sigma}_0\widehat{\Sigma}_y = \begin{bmatrix} 0 & \widehat{\sigma}_y \\ -\widehat{\sigma}_y & 0 \end{bmatrix}$$

$$\widehat{\Sigma}_0\widehat{\Sigma}_z = \begin{bmatrix} 0 & \widehat{\sigma}_z \\ -\widehat{\sigma}_z & 0 \end{bmatrix}$$

Problem 11.9 Using the Rashba Hamiltonian $\widehat{H}_R$ (see Eq. 11.140), prove that the effective *Rashba magnetic field* $\mathbf{B}_R$ in a spin-FET, per charge carrier, defined as $-\partial \widehat{H}_R/\partial \boldsymbol{\mu}$, is given by

$$\mathbf{B}_R = \frac{e}{2g|\gamma|m_0^2 c^2} \widehat{\mathbf{P}} \times \mathbf{E}_g$$

$\widehat{\mathbf{P}} = \widehat{\mathbf{p}} + e\mathbf{A}$ is the extended momentum, to include a magnetic field. In the spin-FET, which direction is $\mathbf{B}_R$?

Problem 11.10 Based on your answer to the previous problem, what are the possible orientations of the magnetization vectors in the ferromagnetic insulators of the spin-FET?

11.11 SUGGESTED READINGS & REFERENCES

[1] P. A. M. Dirac, *The Quantum Theory Of The Electron*, **117**, pp. 610-624 (1928)(*note*:available from Proceedings of the Royal Society of London. Series A, Containing Papers of a Mathematical and Physical Character. www.jstor.org).

[2] L. Liu, et. al., *Current-Induced Switching of Perpendicularly Magnetized Magnetic Layers Using Spin Torque from the Spin Hall Effect*, PRL 109, 096602 (2012).

[3] S. Datta and B. Das, *Electronic analog of the electro-optic modulator*, Appl. Phys. Lett. **56**, 665 (1990).

[4] Rashba E I, *Symmetry of energy bands in crystals of wurtzite type: I symmetry of bands disregarding spin-orbit interaction*, Sov. Phys.-Solid State 1, 368–80 (1959).

[5] Rashba E I and Sheka V I, *Symmetry of energy bands in crystals of wurtzite type: II. Symmetry of bands including spin-orbit interaction*, Fiz. Tverd. Tela: Collected Papers 2, 162–76 (1959).

Table 11.3

[6]   Dresselhaus G, Kip A F and Kittel C, *Spin-orbit interaction and the effective masses of holes in germanium*, Phys. Rev. **95**, 568–9 (1954).

[7]   Dresselhaus G, Kip A F and Kittel C, *Cyclotron resonance of electrons and holes in silicon and germanium crystals*, Phys. Rev. **98**, 368–84 (1955).

CHAPTER **12**

Non-Hermitian Operators & Dynamics

Rather than fear apparent limitations, we must celebrate them by embracing the spirit of exploration.
Kwaku Eason

Since the inception of quantum mechanics in the early part of the 20th century, an important capability that was not addressed by Schrödinger and Dirac was the treatment of *energy loss* or *dissipation* in transitions from one eigenstate to another. We are especially interested in loss involving angular momentum-dependent energies, as this ties directly to magnetism. This question was pondered seriously by the author for the first time in conversations with Russian physicist Boris Luk'Yanchuk, while in Singapore. It was also motivated by observation of the inability of quantum mechanics to predict well-known experimentally observed dynamical motion of electron spin in an externally applied magnetic field. In a strong enough magnetic field, an electron's magnetic moment *aligns* itself with a magnetic field, if it is not aligned with it initially. Though, some dynamical equations, atleast in form, pertaining to electron spin have been, none had *not* been obtained from quantum mechanics, beyond precessional motion. *Precession* is only a part of it's motion, which *had* been worked out from quantum mechanics. Summarizing how these difficulties were overcome, this chapter incorporates some of the author's resolutions to the above dilemmas. To demonstrate their correctness, we'll demonstrate several applications to some relevant problems in spin-physics.

We know, from Maxwell, Ampére, Faraday and others that the electromagnetic equations, which are relativistic, include a dynamical wave description. Because of this, it becomes possible to obtain a dynamical description of a loop of current, or a magnetic moment placed in a magnetic field. This will be our starting point. In this chapter, we shall develop this statement more quantitatively, both classically as well as in the domain of quantum mechanics. Note that both Schrödinger and Dirac's formulations contrive energy operators describing *conserved equilibrium* energy states. They still necessarily involve motion, and even acceleration through *orbital* motion, for example. However, the direct evidence of time-dependent transitions between states is indicated by Nature in numerous experimental demonstration.

For example, one experimental technique used to observe electron spin dynamics utilizes an *optical* method known as a *pump-probe technique*. In this technique, a high-powered *laser source* such as a Titanium-Saffire laser is used to emit sequences of ultra-short *pump-pulses* towards a magnetic or *ferrimagnetic* sample. A ferrimagnetic material is composed of two sublattices which are anti-ferromagnetically coupled, but *asymmetrically*, thus, yielding a small net magnetic moment in the sample. The pump-pulse is followed by lower intensity *probe-pulses*, for example, using a time-delay τ. Varying τ allows the construction of the time-response of the average magnetization. In this method, the high-intensity pump-probe thermally excites the sample, which causes the mean magnetization to vanish. In a tilted magnetic field, detecting the optical polarization rotation, e.g. via Kerr rotation of the probe-pulse, one can measure the dynamical motion of a component of the magnetization normal to the sample. This is possible because of two things: (1) short pulses from the laser and (2) excellent control of the time-delay τ, controllable down to femto-seconds (10^{-15}s), and when control loops are also utilized, down even to atto-seconds (10^{-18}s)! This is shorter than the time scale of the motion of the magnetization, which allows for successful temporal resolution to be constructed.

It is also known that magnetization rotations can affect the temperature of a magnetic sample, though only very slightly. This indicates that magnetization rotations involve transfer of energy. Dynamics and dissipation in real magnetic systems is more of a rule than an exception. Fortunately, there *is* a way to determine what some forms of energy dissipation look like using quantum mechanics. In this chapter, dissipation of spin-dependent energy and some consequences are our foci. We begin by revisiting classical torque-energy of the orbital angular momentum of an electron in a magnetic field, but *with a twist*. We will see that this leads to a form of dissipation for momentum-dependent energy. Then, we consider several consequences of the result. Along the way, we'll also revisit some modern topics of spin-physics utilizing the generalized Hamiltonian we'll obtain.

12.1 Lenz' Law Correction For Dynamical Effects

In nearly all the analyses we've done so far in this text, we have *quietly* used a form of potential energy V in the Hamiltonian, that is *Hermitian*, or has real eigenvalues. This also means that time-dependence of the energy is *not described*, as a result of this assumption. With such an assumption, the energy is not changing in time, and thus, cannot describe transitions between any two distinct energy states. From these limited conditions, we can find *equilibrium* and steady states *only*. This has been the outcome of the problems we have dealt with up to now.

Thus, quantum mechanics, as we've discussed it, has said little about physical systems having a *non-Hermitian* Hamiltonian, or a Hamiltonian that has complex eigenvalues. This fact is significant, as we'll find out, because the ability to describe transitions between quantum energy states requires that we step outside of this assumption of a Hermitian Hamiltonian. We treat this extension more formally now, discussing a relevant example of the origins of a non-Hermitian Hamiltonian form, applicable to magnetic materials.

To do this, we first consider a classical problem that inspires the path to obtaining a non-Hermitian Hamiltonian form, first for the orbital magnetic moment $\widehat{\boldsymbol{\mu}}_L$. The resulting non-Hermitian Hamiltonian eigenenergies to be found, will be complex. We'll discuss what this means for the electron states. However, to enable any of this, a subtle, yet important, premise is that in order to obtain a result that is applicable to the *relativistic* Dirac equation, the starting point *must be relativistic*. That is, it must be based on theory that is invariant under the Lorentz transformation (c.Chap. 11). This is why we must begin with the relativistic

Maxwell equations, or with laws that are consistent with the theory of relativity. Specifically, we'll begin with the electromagnetic law known as **Lenz' law**. Around 1833, based on observations of electrical current loop behavior in a magnetic field, the Baltic German physicist Heinrich Friedrich Emil Lenz (1804-1865) observed what he stated as:

The direction of the current induced in a conductor by a changing magnetic field is such that the magnetic field created by the induced current opposes the initial changing magnetic field.
H. Lenz

Today, a common mathematical form of this statement is given by

> **Lenz Law**
>
> $$V_{emf} = -\frac{d\Phi}{dt} \qquad (12.1)$$

In Eq. (12.1), $\Phi = \mathbf{B} \cdot \mathbf{A}_o$ is called the **magnetic flux**, in this case, through the orbital area $\mathbf{A}_o$ of the magnetic moment. If Lenz' law can be applied to electron motion around a loop, because it contains a time-derivative, we'll show that it leads to dynamical effects for the current loop. Let's see how we can apply Eq. (12.1) to an electron in an orbit, giving rise to a magnetic moment $\boldsymbol{\mu}_L$.

For a magnetic moment that tends to align with the magnetic field **B**, seeking equilibrium from an initial condition starting at $\phi = \pi/2$ radians, the magnetic flux Φ only *increases*. In other words, a flux increase of a magnetic moment in a magnetic field is synonymous with an energy decrease. Note that the choice of initial condition $\phi = \pi/2$ is not arbitrary, and we shall have more to say about this shortly. Even if **B** is static and uniform, the effect described by Lenz' law *still* manifests because the electromotive potential V is the time-derivative of the product between the flux area and the magnetic field. As long as the flux area changes, there will be an effect on the motion of the magnetic moment or angular momentum. It is because of the *non-conservative nature of a magnetic field* that we find a nonzero physical electromotive force $\partial V_{emf}/\partial \theta$ (resisting change in motion) around the closed current loop of the magnetic moment $\boldsymbol{\mu}_L$. Thus, the net potential gradient force around the loop is *non-conserving* when a magnetic field is introduced.

12.1 Lenz' Law Correction For Dynamical Effects

Using Eq. (12.1) (Lenz' Law), we can describe dynamical effects on the motion of $\boldsymbol{\mu}_L$ in a magnetic field **B**. In Chap. 9, recall that we obtained *precessional motion* of a magnetic moment, but this is not the whole story. Since we take **B** to be uniform and constant in time, what counts is the amount of area (using its unit normal vector **n**) that is parallel to **B**. Of course, this changes as the loop undergoes rotation. The relevant flux orbital area is given by $A_o \cos\phi$. When **B** and $\boldsymbol{\mu}_L$ are parallel (as in the final equilibrium condition), the quantity $\mathbf{n} \cdot \mathbf{B}$ is maximum and the induced emf is *most* negative. However, when they are perpendicular (as in the initial condition $\varphi = \pi/2$), the flux Φ vanishes because there is no component of the area normal vector **n** parallel to **B**. This initial condition is also illustrated in Fig. 12.1. Therefore, as the loop rotates,

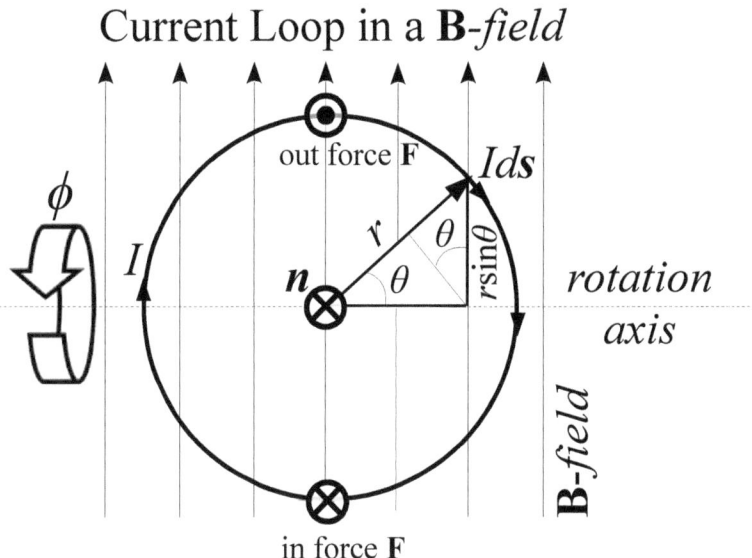

Figure 12.1: Illustration of current I_0 circulation in a closed circular loop, in a uniform magnetic field **B**. The Lorentz force will begin to act on the loop, imparting a torque τ or moment on the loop.

the *additional induced current* I is dependent on the rotation angle ϕ. Lenz' law helps to quantify this variation where we have

$$V_{\text{emf}} = -\frac{d}{dt}(BA_o \cos\phi)$$

$$= BA_o\sin\phi \cdot \dot{\phi}$$

$$= -BA_o\sin\phi \cdot |\dot{\phi}|$$

$\dot{\phi} = d\phi/dt$ is the *rotational rate* of $\boldsymbol{\mu}_L$ towards the magnetic field. For the rotation rate in a static field, $\dot{\phi} \leq 0$ for a magnetic moment transitioning from misalignment to alignment. It follows that for any angle between $\phi = \pi/2$ and $\phi = 0$, $V_{\text{emf}} < 0$ and $\Delta V_{\text{emf}} = V_{\text{emf}}(\phi) - V_{\text{emf}}(\phi_i) = -(BA_o\sin\phi\dot{\phi}) - 0 < 0$. With this dynamical effect present, we find that the system is *loosing or dissipating* emf.

Additionally, in any nonzero magnetic field, since $|V_{\text{emf}}| > 0$ for all angles $\phi > 0$, there is a definite *nonzero* effective **orbital resistance** we can define by $|V_{\text{emf}}| \equiv \Delta I R_o$. ΔI is the *induced* current generated within the current loop arising from the changing flux during rotation. The *orbital resistance* is associated with the circumference of the orbital area A_o. Since $V_{\text{emf}} < 0$, it follows that $\Delta I < 0$, which is a reduction in the loop current. The variation of the induced current ΔI due to rotation of the moment in a **B**-field (taken along the positive direction) becomes

$$\Delta I = -\frac{BA_o}{R_o}|\dot{\phi}|\sin\phi = \Delta I_0 \sin\phi \tag{12.2}$$

The amplitudes of the **induced current** ΔI_0 and corresponding magnetic moment $\Delta \mu_L$ become

$$\Delta I_0 = -\frac{BA_o|\dot{\phi}|}{R_o} \quad \text{(units of Ampéres=Amps)}$$

$$\Delta \mu_L = \Delta I_0 A_o = -\frac{BA_o^2|\dot{\phi}|}{R_o} \quad \text{(units of Amps} \cdot \text{m}^2\text{)}$$

However, we are interested in the total instantaneous magnetic moment and resulting torque. They are then given by

$$I = I_0 + \Delta I \tag{12.3a}$$

$$\mu_T = I_0 A_0 + \Delta I A_0 \tag{12.3b}$$

The net torque based on the instantaneous current I including dynamical effects becomes

$$\tau = A_o I B \sin\phi = A_o \left(I_0 - \frac{BA_o|\dot{\phi}|}{R_o}\sin\phi\right) B\sin\phi$$

12.1 Lenz' Law Correction For Dynamical Effects

$$= A_o I_0 B\sin\phi - A_o \left[\frac{BA_o|\dot\phi|}{R_o}\right] B\sin\phi\sin\phi$$

The units of the term in brackets on the RHS must have the same units of I_0. Let us define a *dimensionless dynamical coefficient* $\alpha(t)$ by

$$\alpha(\phi(t)) = A_o \left[\frac{BA_o|\dot\phi|}{R_o}\right]\frac{1}{\mu_L} \quad (12.4)$$

$$= \frac{|\Delta\mu_d|}{\mu_L} \quad (12.5)$$

Then, in terms of the dimensionless dynamical coefficient α, the torque equation becomes

$$\tau = \mu_L B\sin\phi - \alpha(\phi)\mu_L B\sin^2\phi \quad (12.6)$$

$$= \mu_L B\sin\phi - \frac{\alpha(\phi)}{\mu_L}\mu_L^2 B\sin^2\phi \quad (12.7)$$

α also generally depends on the angle ϕ, the magnetic field B, and magnetic moment μ_L. Note that α is also proportional to the maximum flux amplitude of the loop $\Phi_0 = BA_o$, suggesting that the larger the orbital loop flux, whether via larger orbitals or a stronger magnetic field, the larger the dimensionless damping coefficient. From this, we have the corresponding total torque $\boldsymbol\tau$ in vector form in terms of $\boldsymbol\mu_L$, given by

Torque Equation With Lenz' Law Correction

$$\boldsymbol\tau = \boldsymbol\mu_L \times \mathbf{B} - \frac{\alpha(\phi)}{\mu_L}\boldsymbol\mu_L \times \mathbf{B} \times \boldsymbol\mu_L \quad (12.8)$$

A unique consequence dictated by Lenz' law is that the additional induced current ΔI is generated in the same current loop of I_0, as though the orbiting charged particle motion is impeded physically, to change the current. This leads to a variation in the magnetic moment vector $\Delta\boldsymbol\mu$ that is collinear with $\boldsymbol\mu_L$.

What can be done for quantum mechanics, with this result? Eq. (12.8) allows the determination of the corresponding total magnetic potential energy V_M, given by $\int \tau \cdot d\phi$. Evaluating this integral using Eq. (12.8), we obtain

$$V_M = \int_{\pi/2}^{\phi} \tau(\phi)d\phi = \int_{\pi/2}^{\phi} \mu_L B \sin\phi - \mu_L B \int_{\pi/2}^{\phi} \alpha(\phi)\sin^2\phi \, d\phi \quad (12.9)$$

As mentioned, the dimensionless coefficient $\alpha(\phi)$ is, generally, not constant with time. It changes with ϕ since $\alpha \propto \dot{\phi}$. In order to proceed with the integral in Eq. (12.9), some mathematical simplification is necessary. Fortunately, it can be done without any loss in accuracy to the resulting magnetic potential energy. This is done by using a mathematical theorem known as the *generalized mean-value theorem* which states that for the definite integral of the product of any two integrable functions f_1 and f_2, we have

$$\int_a^b f_1(x)f_2(x)dx = \overline{f_1(x^*)} \int_a^b f_2(x)dx \quad \text{where} \quad a \leq x^* \leq b$$

$\overline{f_1(x^*)}$ represents the average or mean value over the limits of the definite integral. In this case, they are $\phi = \pi/2$ to $\phi = 0$. In this way, we have allowed *no approximation errors* in the evaluation of the integral. The function in the integrand that gets averaged here is the function $\dot{\phi}$. Then, the magnetic potential energy V_M becomes

$$V_M = \int_{\pi/2}^{\phi} \tau(\phi)d\phi = \int_{\pi/2}^{\phi} \mu_L B \sin\phi - \overline{\alpha}\mu_L B \int_{\pi/2}^{\phi} \sin^2\phi \, d\phi \quad (12.10)$$

The second integral on the RHS can be evaluated using $\int \sin^2 x \, dx = \sin(2x)/4 - x/2$ and $\sin(2x) = 2\sin x \cos x$. Eq. (12.10) then becomes

$$V_M = -\mu_L B \cos\phi - \frac{1}{2}\overline{\alpha}\mu_L B \sin\phi \cos\phi + \frac{1}{2}\overline{\alpha}\mu_L B\left(\phi - \frac{\pi}{2}\right) \quad (12.11)$$

The vector representation of the magnetic potential energy V_M is given by

Orbital Magnetic Potential Energy

$$V_M(\mathbf{B}) = -\boldsymbol{\mu}_L \cdot \mathbf{B} - \frac{1}{2}\frac{\overline{\alpha}}{\mu_L}\boldsymbol{\mu}_L \times \mathbf{B} \cdot \boldsymbol{\mu}_L + \frac{1}{2}\overline{\alpha}\mu_L B\left(\phi - \frac{\pi}{2}\right) \quad (12.12)$$

Note that the convention for choosing the sign in the vector form is to take the sign that minimizes the energy with $\boldsymbol{\mu}$ and $\mathbf{B}$ parallel, as

12.1 Lenz' Law Correction For Dynamical Effects

opposed to anti-parallel. The energy is minimized, or is most negative when there is alignment between $\boldsymbol{\mu}_L$ and **B**.

From this analysis, we find that for charges circulating in a loop constituting a magnetic moment $\boldsymbol{\mu}_L$, if the loop is exposed to a magnetic field **B**, there will be a resulting torque $\boldsymbol{\tau}$ on the loop with *two* contributions. The total torque $\boldsymbol{\tau}(\mathbf{B})$ as well as the energy variations $V_M(\mathbf{B})$ are both determined, to an extent, by the strength of μ_L and B. It is important to keep in mind that we have been able to obtain these results using relativistic equations provided by Maxwell's equations. This is what makes the result useful for quantum mechanics, as well. Fig. 12.2 illustrates examples of the potential energy $V(\phi)$ given by Eq. (12.12) as a function of the inclusive angle ϕ, varying the external magnetic field amplitude B, using $\overline{\alpha} = 0.01$.

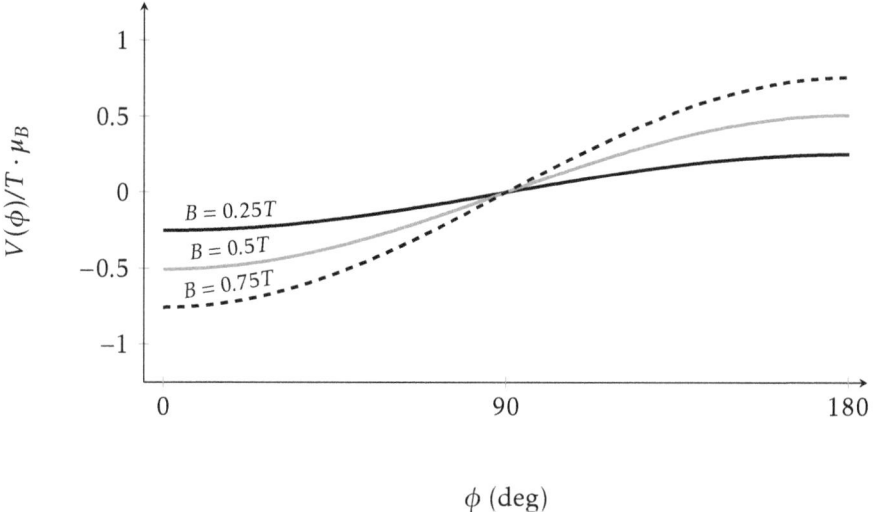

Figure 12.2: Potential energy V per Tesla-Bohr magneton as a function of external magnetic field amplitude B and inclusive angle ϕ. The magnetic field values used are $B = 0.25, 0.5$, and 0.75 Tesla. The energy minimum for all field amplitudes is at $\phi = 0$, corresponding to alignment.

The dominating term in Eq. (12.12) is the first term, or the $-\cos\phi$ term. This is because the other terms multiply by $\overline{\alpha} \ll 1$, so the curve is nearly a $-\cos\phi$ plot. Based on this, it could be tempting to ignore both these terms altogether. However, we shall find out the second term, in particular, is crucial to enable the complete dynamical description of the magnetic moment in a magnetic field. For all cases, the energy minimum in the range of ϕ shown, for all field amplitudes occurs at

$\phi = 0$, or when $\boldsymbol{\mu}$ and $\mathbf{B}$ are aligned parallel.

The above results can be used to determine the corresponding quantum mechanical Hamiltonian that results from the second term of the potential energy of Eq. (12.12). Note that because the third term is just a scalar operator, it has no effect on any of the expectation values that we will consider. In the next section, let's find out where the operator $\boldsymbol{\mu}_L \times \mathbf{B} \cdot \boldsymbol{\mu}_L$ leads.

12.2 Non-Hermition Hamiltonian Including Transient Effects

In the previous section, we found that for an orbital magnetic moment $\boldsymbol{\mu}_L$, there arises an additional time-dependent mechanism, electromagnetic in origin, that gives rise to additional energy of the magnetic moment in a magnetic field B. That result was classical. Now, we want to determine the corresponding quantum mechanical Hamiltonian that is consistent with this classical result. We, therefore, apply the *Principle of Correspondence*. We found that the *dissipated* potential energy V_D is given by

$$V_D = -\frac{\overline{\alpha}}{2\mu_L} \boldsymbol{\mu}_L \times \mathbf{B} \cdot \boldsymbol{\mu}_L \qquad (12.13)$$

We are going to show that Eq. (12.13) leads to an example of a *non-Hermitian operator*. For this, we need to obtain the eigenvalues of this Hamiltonian. Replacing the magnetic moment with $\boldsymbol{\mu}_L = -|\gamma|\mathbf{L}$, the first of the two terms becomes

$$V_D = -\frac{\overline{\alpha}\gamma^2}{2\gamma L} \mathbf{L} \times \mathbf{B} \cdot \mathbf{L} = \frac{\overline{\alpha}|\gamma|}{2L} \mathbf{L} \times \mathbf{B} \cdot \mathbf{L}$$

Since we know the quantum mechanical operator for $\widehat{\mathbf{L}}$ (c.7), the quantum mechanical Hamiltonian eigenvalue equation is given by

$$\widehat{H}_I \psi = \left[\frac{\overline{\alpha}|\gamma|}{2L} \widehat{\mathbf{L}} \times \mathbf{B} \cdot \widehat{\mathbf{L}}\right] \psi = E_d \psi \qquad (12.14)$$

The Hamiltonian in Eq. (12.14) is well-defined since we know the operators involved. B is the magnetic field, and $\widehat{\mathbf{L}}$ is the *angular momentum*

12.2 Non-Hermition Hamiltonian Including Transient Effects

operator. Let us assume that **B** is along the $+z$ direction. Evaluating the cross product term in Eq. (12.14), $\widehat{\mathbf{L}} \times \mathbf{B}$ gives

$$\widehat{\mathbf{L}} \times \mathbf{B} = \begin{bmatrix} i & j & k \\ \widehat{L}_x & \widehat{L}_y & \widehat{L}_z \\ 0 & 0 & B_z \end{bmatrix} = \begin{bmatrix} \widehat{L}_y B_z - \widehat{L}_z B_y \\ -(\widehat{L}_x B_z - \widehat{L}_z B_x) \\ \widehat{L}_x B_y - \widehat{L}_y B_x \end{bmatrix} = \begin{bmatrix} \widehat{L}_y B_z \\ -\widehat{L}_x B_z \\ 0 \end{bmatrix}$$

Dotting this result with $\widehat{\mathbf{L}}$, we have

$$\widehat{\mathbf{L}} \times \mathbf{B} \cdot \widehat{\mathbf{L}} = \widehat{L}_y \widehat{L}_x B_z - \widehat{L}_x \widehat{L}_y B_z$$

$$= -B(\widehat{L}_x \widehat{L}_y - \widehat{L}_y \widehat{L}_x)$$

We have used the fact that B is taken along the $+z$ axis, or $B_z = B$. The term in parentheses can be replaced using the commutation relations for the angular momentum obtained in Chap. 7, given by

$$\widehat{L}_x \widehat{L}_y - \widehat{L}_y \widehat{L}_x = i\hbar \widehat{L}_z$$

This leads to the following result

$$\widehat{H}_I \psi = -i \left[\frac{\overline{\alpha} |\gamma| \hbar B}{2L} \widehat{L}_z \right] \psi$$

$$= i \frac{\overline{\alpha}}{2} \widehat{\boldsymbol{\mu}}_L \cdot \mathbf{B} \psi$$

$$= i\alpha^* V$$

$V = -\boldsymbol{\mu}_L \cdot \mathbf{B}$ is the *ordinary* magnetic energy, without the additional transient contribution from V_D. We have also replaced the classical dimensionless coefficient $\overline{\alpha}$ with a **microscopic dimensionless coefficient** α^* given by

$$\alpha^* = \frac{\overline{\alpha}}{2} \quad (12.15)$$

Therefore, the dissipated energy reduces to a form given by a purely imaginary scalar $-i\alpha^*$ times the *Hermitian Hamiltonian term*. Adding $\widehat{H}_d$ to the original Hamiltonian, the total Hamiltonian $\widehat{H}_T$ becomes

$$\widehat{H}_T \psi = \left[-\boldsymbol{\mu}_L \cdot \mathbf{B} + i\alpha^* \boldsymbol{\mu}_L \cdot \mathbf{B} \right] \psi = (1 - i\alpha^*)(-\boldsymbol{\mu}_L \cdot \mathbf{B}) \psi \quad (12.16)$$

Then, the complex eigenenergy E_λ for the orbital angular momentum becomes

Non-Hermitian Eigenvalues Of Total Hamiltonian

$$E_\lambda = (1 - i\alpha^*)|\gamma|Bm_\ell \hbar \qquad (12.17)$$

When the dissipation-related Hamiltonian $\widehat{H}_I$ is included, we obtain generally complex scalar eigenvalues for the total Hamiltonian, rather than real. Since V_M corresponds to a Hermitian Hamiltonian, and α^* is real, Eq. (12.16) is, therefore, a *non-Hermitian Hamiltonian*.

So, what does this mean? A complex eigenvalue is not in contradiction with anything we have discussed so far. Up to now, a complex potential energy in the Hamiltonian has simply been *beyond the scope of the assumptions made in the Hamiltonian*. However, when we introduce this type of non-Hermitian Hamiltonian, we no longer satisfy the assumption of a real-valued potential energy. For this reason, we've generalized several of the relations we've discussed in order to accommodate a non-Hermitian Hamiltonian. We'll see some examples illustrating this point, later in this chapter.

Let us take a look at the consequence of a complex energy eigenvalue on the resulting wave functions. Eq. (12.17) means that the time-dependent factor of the wave function ψ becomes

$$\psi_\ell(t) = \psi_\ell(0) e^{-iE_\lambda t/\hbar} \qquad (12.18a)$$

$$= \psi_\ell(0) e^{-i(1-i\alpha^*)|\gamma|B\hbar m_\ell/\hbar t} \qquad (12.18b)$$

$$= \psi_\ell(0) e^{-i|\gamma|Bm_\ell t} e^{\alpha^*|\gamma|Bm_\ell t} \qquad (12.18c)$$

$$= \psi_\ell(0) e^{-\alpha^* \mathrm{Re}(E_\lambda) t} e^{-i \mathrm{Re}(E_\lambda) t} \qquad (12.18d)$$

$\psi(0) = \psi_\ell(x, 0)$ is the initial condition, which is the *ordinary Hermitian eigenenergy* solution. This result says that the time-dependent wave function ψ has a real-valued amplitude which either decays or amplifies, depending on the state. Since the wave function ψ_ℓ corresponds to an eigenstate, the general solution will be a linear combination of such states given

General Non-Hermitian Eigensolution

$$\Psi_\ell(t) = \sum_{m_\ell} C_{m_\ell} e^{-\alpha^* Re[E_\lambda(m_\ell)]t} e^{-i Re[E_\lambda(m_\ell)]t} \qquad (12.19)$$

Adding the dynamical effects described by Lenz' law leads to a *non-Hermitian* Hamiltonian. Since the dimensionless coefficient only appears in the real time-dependent factor, α^* is only accessible via measurements of dynamic motion of μ_L, for example, in a magnetic field B. In other words, it cannot be gleaned from measurements of equilibrium states. It turns out that the above results can be generalized to other systems, and we'll demonstrate this throughout this chapter. The generalization, specifically, is the focus of the next section.

For spin-dependent energy, then, let us define for any Hermitian Hamiltonian $\widehat{H}_R$, a corresponding **angular momentum conjugate dissipative operator** given by

Angular Momentum Conjugate Dissipative Operator

$$\widehat{H}_I \psi = -i\alpha \widehat{H}_R \psi \qquad (12.20)$$

The expectation value of any operator $\widehat{O}$ is still given by

$$\langle \widehat{O} \rangle = \frac{\int \Psi^\dagger \widehat{O} \Psi dx}{\int \Psi^\dagger \Psi dx}$$

In the next few sections, we'll be applying *conjugate dissipative operators* in several problems. But first, let's revisit the *continuity equation*. We first discussed the continuity equation in Sec. 4.6, however, assuming a Hermitian Hamiltonian. Now that we have a non-Hermitian Hamiltonian, this topic needs to be revisited.

12.3 Dissipative Continuity Equation

We have now seen an example of a non-Hermitian Hamiltonian operator. Here, we are concerned with what will be the continuity equation for such a system? We'll obtain the answer to this question here starting with the Schrödinger equation, given by

$$i\hbar \frac{\partial \psi}{\partial t} = -\frac{\hbar^2}{2m}\nabla^2\psi + (1 - i\alpha)V\psi \tag{12.21a}$$

The complex conjugate of Eq. (12.21a) becomes

$$-i\hbar \frac{\partial \psi^*}{\partial t} = -\frac{\hbar^2}{2m}\nabla^2\psi^* + (1 + i\alpha)V\psi^* \tag{12.21b}$$

Now, multiply Eq. (12.21a) by ψ^*, and Eq. (12.21b) by ψ, we obtain

$$i\hbar\psi^* \frac{\partial \psi}{\partial t} = -\frac{\hbar^2}{2m}\psi^*\nabla^2\psi + (1 - i\alpha)V\psi^*\psi \tag{12.22a}$$

$$-i\hbar\psi \frac{\partial \psi^*}{\partial t} = -\frac{\hbar^2}{2m}\psi\nabla^2\psi^* + (1 + i\alpha)V\psi\psi^* \tag{12.22b}$$

Taking the difference between the two equations leads to the following:

$$i\hbar\left(\psi^* \frac{\partial \psi}{\partial t} + \psi \frac{\partial \psi^*}{\partial t}\right) = \frac{\hbar^2}{2m}\left(\psi^*\nabla^2\psi - \psi\nabla^2\psi^*\right) - i2\alpha V\psi^*\psi \tag{12.23}$$

We still define the probability density $\rho(x,t)$ as

$$\rho(\mathbf{r}) \equiv \psi \cdot \psi^* = \psi^* \cdot \psi = |\psi|^2(\mathbf{r}) \tag{12.24}$$

Then, dividing both sides by $i\hbar$, Eq. (12.23) becomes

$$\frac{\partial \rho}{\partial t} + \frac{i\hbar}{2m}\left(\psi\nabla^2\psi^* - \psi^*\nabla^2\psi\right) = -\frac{2\alpha}{\hbar}V\rho \tag{12.25}$$

Recall that for the term in parentheses on the LHS, we showed in Chap. 4 that it can be expressed as a divergence (c.Sec. 4.6). Using this result, we have

$$\frac{\partial \rho}{\partial t} + \frac{i\hbar}{2m}\nabla \cdot (\psi\nabla\psi^* - \psi^*\nabla\psi) = -\frac{2\alpha}{\hbar}V\rho \tag{12.26}$$

From this result, the *current density vector operator* $\widehat{\mathbf{J}}$ (defined through its divergence) is the same and given by

$$\widehat{\mathbf{J}} = \frac{i\hbar}{2m}(\psi\nabla\psi^* - \psi^*\nabla\psi) \tag{12.27}$$

We then have a **dissipative continuity equation** given by

12.4 Dynamical Spin Angular Momentum

Dissipative Continuity Equation

$$\frac{\partial \rho}{\partial t} = -\left(\nabla \cdot \mathbf{J} + \frac{2\alpha}{\hbar} V \rho\right) \qquad (12.28)$$

The *dissipative continuity equation* simplifies to the conserved form in Eq. (4.95), when $\alpha = 0$. For a dissipative system whose wave function is given by ψ, Eq. (12.28) should be used. Although we have used the word "dissipative" here, which usually refers to loss of energy, *energy can flow into the system*, depending on the sign of the potential energy V. We'll see an example of this when we discuss spin transport of electrons conducting through a magnet. Let's now move on to explore one of the first examples we'll consider applying a generalized Hamiltonian, or *conjugate dissipative Hamiltonian* on the motion of electron spin subjected to an external magnetic field.

12.4 GENERALIZED TRANSIENT SPIN ANGULAR MOMENTUM

For localized electrons, we return back to the Zeeman energy which involves a magnetic field **B** coupling to the total angular momentum **L** + **S**. As we did in Sec. 9.6 when we derived *spin precessional motion*, we will restrict ourselves to the spin contribution **S**, starting with the Zeeman Hamiltonian $\widehat{H}_Z$ using the conjugate dissipation introduced in the previous section. Taking the magnetic field along the z direction and using spin matrices, we have

$$i\hbar \frac{\partial \psi}{\partial t} = (1 - i\alpha)\widehat{H}_Z \psi \qquad (12.29\text{a})$$

$$= (1 - i\alpha)\left(-\boldsymbol{\mu}_S \cdot \mathbf{B}\right)\psi \qquad (12.29\text{b})$$

$$= \left[\frac{g|\gamma|\hbar B}{2}\widehat{\sigma}_z - \frac{i\alpha g|\gamma|\hbar B}{2}\widehat{\sigma}_z\right]\psi \qquad (12.29\text{c})$$

Using the g-factor for a single electron given by Eq. (11.70) where $g \approx 2$, Eq. (12.29c) becomes

$$i\hbar \frac{\partial \psi}{\partial t} = [\hbar|\gamma|B\widehat{\sigma}_z - i\alpha\hbar|\gamma|B\widehat{\sigma}_z]\psi \qquad (12.30)$$

Since we are dealing with the electron *spin*, we have a spinor state $\psi = [\psi_+ \ \psi_-]^T$. Thus, Eq. (12.30) expresses two equations describing ψ_+ and ψ_- which can be written

$$\frac{d\psi_+}{dt} = -(i|\gamma|B + \alpha|\gamma|B)\psi_+ \tag{12.31}$$

$$\frac{d\psi_-}{dt} = (\alpha|\gamma|B + i|\gamma|B)\psi_- \tag{12.32}$$

The above equations have solutions for ψ_+ and ψ_- given by

$$\psi_+ = \psi_+(0)e^{-(i|\gamma|B+\alpha|\gamma|B)t} = \psi_+(0)e^{-\alpha|\gamma|Bt}e^{-i|\gamma|Bt} \tag{12.33a}$$

$$\psi_- = \psi_-(0)e^{(i|\gamma|B+\alpha|\gamma|B)t} = \psi_-(0)e^{\alpha|\gamma|Bt}e^{i|\gamma|Bt} \tag{12.33b}$$

Because we have taken the magnetic field along +z, the appropriate spinor initial condition is given by Eq. (9.38), so we have

$$\psi_+(0) = \cos\frac{\theta}{2}e^{-i\frac{\varphi}{2}} \tag{12.34a}$$

$$\psi_-(0) = \sin\frac{\theta}{2}e^{i\frac{\varphi}{2}} \tag{12.34b}$$

From Eqs. (12.33), we see the spin eigensolutions ψ_+ and ψ_- are products of a real exponential and complex exponential term. We can find the expectation value of the components of the intrinsic angular momentum $\widehat{S}_x$, $\widehat{S}_y$, and $\widehat{S}_z$ using Eqs. (12.33) and Eqs. (9.50a)-(9.50c). Then, $\widehat{S}_x$ becomes

$$\langle \widehat{S}_x \rangle = \frac{\hbar}{2} \frac{\begin{bmatrix}\psi_+^* & \psi_-^*\end{bmatrix}\widehat{\sigma}_x\begin{bmatrix}\psi_+ \\ \psi_-\end{bmatrix}}{\begin{bmatrix}\psi_+^* & \psi_-^*\end{bmatrix}\begin{bmatrix}\psi_+ \\ \psi_-\end{bmatrix}} = \frac{\hbar}{2}\frac{\psi_+^*\psi_- + \psi_-^*\psi_+}{\psi_+^*\psi_+ + \psi_-^*\psi_-} \tag{12.35}$$

The evaluations are a little lengthy, but straightforward. Substitution of the initial condition given by Eqs. (12.34a)-(12.34b), and the solutions for the eigenfunctions given by Eqs. (12.33a)-(12.33b), $\langle \widehat{S}_x \rangle$ becomes

$$\langle \widehat{S}_x \rangle = \frac{\hbar}{2} \frac{\cos\frac{\theta}{2}\sin\frac{\theta}{2}\left(e^{i\omega_p t}e^{i\varphi} + e^{-i\omega_p t}e^{-i\varphi}\right)}{\left(\cos^2\frac{\theta}{2}e^{-\alpha\omega_p t} + \sin^2\frac{\theta}{2}e^{\alpha\omega_p t}\right)}$$

$$= \frac{\hbar}{2} \frac{\cos\frac{\theta}{2}\sin\frac{\theta}{2}\left(e^{i\omega_p t}e^{i\varphi} + e^{-i\omega_p t}e^{-i\varphi}\right)}{\left(\left[\frac{1+\cos\theta}{2}\right]e^{-\alpha\omega_p t} + \left[\frac{1-\cos\theta}{2}\right]e^{\alpha\omega_p t}\right)}$$

12.4 Dynamical Spin Angular Momentum

$$= \frac{\hbar}{2} \frac{\cos\frac{\theta}{2}\sin\frac{\theta}{2}(e^{i\omega_p t}e^{i\varphi} + e^{-i\omega_p t}e^{-i\varphi})}{\frac{(e^{-\alpha\omega_p t}+e^{\alpha\omega_p t})}{2} + \cos\theta\frac{(e^{-\alpha\omega_p t}-e^{\alpha\omega_p t})}{2}}$$

$$= \frac{\hbar}{2} \frac{(\sin\theta/2)(e^{i\omega_p t}e^{i\varphi} + e^{-i\omega_p t}e^{-i\varphi})}{\frac{(e^{-\alpha\omega_p t}+e^{\alpha\omega_p t})}{2} + \cos\theta\frac{(e^{-\alpha\omega_p t}-e^{\alpha\omega_p t})}{2}}$$

$$= \frac{\hbar}{2} \frac{\sin\theta\cos(\omega_p t + \varphi)}{\cosh(\alpha\omega_p t) - \cos\theta\sinh(\alpha\omega_p t)}$$

In the above, we have used the relation $\omega_p = 2|\gamma|B = g|\gamma|B$ and some useful trigonometric identities given by

$$\cos x = \frac{e^{ix}+e^{-ix}}{2} \quad \cos^2\frac{\theta}{2} = \frac{1+\cos\theta}{2} \quad \sin^2\frac{\theta}{2} = \frac{1-\cos\theta}{2}$$

$$\cosh x = \frac{e^x+e^{-x}}{2} \quad \sinh x = \frac{e^x-e^{-x}}{2} \quad \sin x \cos x = \frac{\sin(2x)}{2}$$

Since the denominator of $\langle\widehat{S}_x\rangle$ is identical for all three expectation values $\langle S_x\rangle$, $\langle S_y\rangle$, and $\langle S_z\rangle$, we won't repeat the denominator for the other two components. We will just use the result above. Taking the initial condition to be that of the canonical problem of the magnetic moment discussed in Chap. 7, $\theta = \pi/2$ while φ is arbitrary, we have the following result for $\langle S_x\rangle$.

$$\langle\widehat{S}_x\rangle = \frac{\hbar}{2}\frac{\cos(\omega_p t + \varphi)}{\cosh(\alpha\omega_p t)} \tag{12.36}$$

Fig. 12.3 illustrates examples of $\langle\widehat{S}_x\rangle$ for three different parameter values $\alpha = 0.01$, $\alpha = 0.03$, and $\alpha = 0.05$. A similar analysis can be done for $\langle S_y\rangle$ using the relation

$$\langle\widehat{S}_y\rangle = \frac{\hbar}{2}\frac{\begin{bmatrix}\psi_+^* & \psi_-^*\end{bmatrix}\begin{bmatrix}0 & -i\\ i & 0\end{bmatrix}\begin{bmatrix}\psi_+\\ \psi_-\end{bmatrix}}{\begin{bmatrix}\psi_+^* & \psi_-^*\end{bmatrix}\begin{bmatrix}\psi_+\\ \psi_-\end{bmatrix}} = \frac{i\hbar}{2}\frac{\psi_-^*\psi_+ - \psi_+^*\psi_-}{\psi_+^*\psi_+ + \psi_-^*\psi_-} \tag{12.37}$$

Substitution of the wave functions from Eqs. (12.33a)-(12.33b), into Eq. (12.37) leads to

$$\langle\widehat{S}_y\rangle = \frac{i\hbar}{2}\frac{\cos\frac{\theta}{2}\sin\frac{\theta}{2}\left(e^{-i\omega_p t}e^{-i\varphi} - e^{i2\omega_p t}e^{i\varphi}\right)}{\frac{(e^{-\alpha\omega_p t}+e^{\alpha\omega_p t})}{2} + \cos\theta\frac{(e^{-\alpha\omega_p t}-e^{\alpha\omega_p t})}{2}}$$

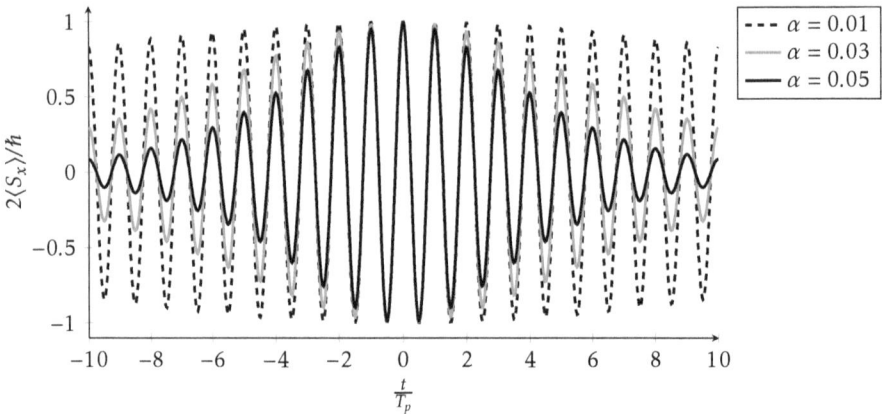

Figure 12.3: Plot of the x-component of the electron spin momentum $\langle S_x \rangle(t)$. $\langle S_x \rangle$ oscillates with frequency $\omega_p = 2|\gamma|B = 2\pi/T_p$, with damping leading to decaying away from zero. The larger the damping α, the faster the damping of the oscillations.

$$= \frac{\hbar}{2} \frac{\sin\theta \sin(\omega_p t + \varphi)}{\cosh(\alpha\omega_p t) - \cos\theta \sinh(\alpha\omega_p t)}$$

We have used the relation $-i = 1/i$ along with some of the trigonometric identities given above. Substitution of the initial boundary condition $\theta = \pi/2$, we obtain the following result for $\langle \widehat{S}_y \rangle$:

$$\langle \widehat{S}_y \rangle = \frac{\hbar}{2} \frac{\sin(\omega_p t + \varphi)}{\cosh(\alpha\omega_p t)} \tag{12.38}$$

Fig. 12.4 illustrates $\langle \widehat{S}_y \rangle$ for the same parameters $\alpha = .01, .03,$ and $.05$: Lastly, for $\langle \widehat{S}_z \rangle$, we have

$$\langle \widehat{S}_z \rangle = \frac{\hbar}{2} \frac{\begin{bmatrix}\psi_+^* & \psi_-^*\end{bmatrix}\begin{bmatrix}1 & 0 \\ 0 & -1\end{bmatrix}\begin{bmatrix}\psi_+ \\ \psi_-\end{bmatrix}}{\begin{bmatrix}\psi_+^* & \psi_-^*\end{bmatrix}\begin{bmatrix}\psi_+ \\ \psi_-\end{bmatrix}} = \frac{\hbar}{2} \frac{\psi_+^*\psi_+ - \psi_-^*\psi_-}{\psi_+^*\psi_+ + \psi_-^*\psi_-}$$

Substitution of the wave functions from Eqs. (12.33) gives

$$\langle \widehat{S}_z \rangle = \frac{\hbar}{2} \frac{\left(e^{-\alpha\omega_p t} - e^{\alpha\omega_p t}\right) + \cos\theta\left(e^{-\alpha\omega_p t} + e^{\alpha\omega_p t}\right)}{\left(e^{-\alpha\omega_p t} + e^{\alpha\omega_p t}\right) + \cos\theta\left(e^{-\alpha\omega_p t} - e^{\alpha\omega_p t}\right)}$$

$$= \frac{\hbar}{2} \frac{-\sinh(\alpha\omega_p t) + \cos\theta\cosh(\alpha\omega_p t)}{\cosh(\alpha\omega_p t) - \cos\theta\sinh(\alpha\omega_p t)}$$

12.4 Dynamical Spin Angular Momentum

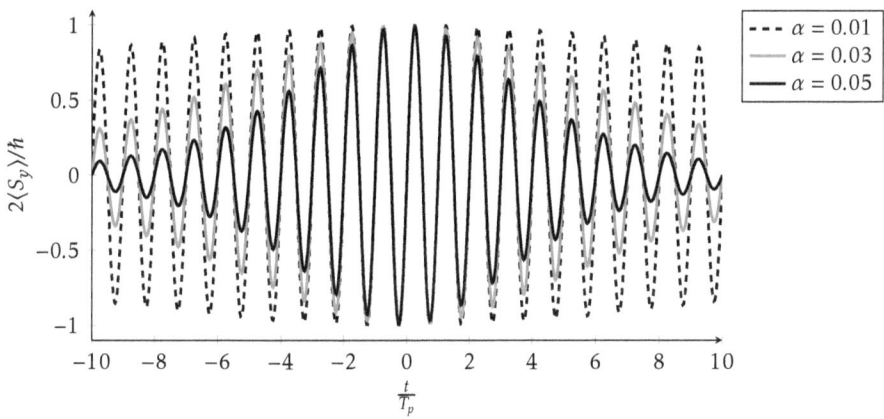

Figure 12.4: Plot of the y-component of the electron spin momentum. Like $\langle \widehat{S}_x \rangle$, $\langle \widehat{S}_y \rangle$ oscillates with frequency $\omega_p = 2\pi/T_p$, with larger dissipation parameter α leading to more rapidly decaying oscillations.

Substituting the initial spinor condition, we have the following result for $\langle S_z \rangle$:

$$\langle \widehat{S}_z \rangle = -\frac{\hbar}{2}\frac{\sinh(\alpha\omega_p t)}{\cosh(\alpha\omega_p t)} = -\frac{\hbar}{2}\tanh(\alpha\omega_p t) \qquad (12.39)$$

From this component specifically, we obtain a time-scale for **spin reversal** τ_S for $\widehat{\mathbf{S}}$ given by

> **Electron Spin-Reversal Time**
>
> $$\tau_S = \frac{1}{\alpha\omega_p} \qquad (12.40)$$

Fig. 12.5 illustrates $\langle \widehat{S}_z \rangle(t)$ for $\alpha = 0.01, 0.03, 0.05$. The component of the spin angular momentum along the z–axis (or field axis) aligns itself *anti-parallel* to the magnetic field. This is consistent with the classical result for orbital angular momentum, we found in Chap. 7. Therefore, we have the following dynamical response of the electron spin magnetic moment in a magnetic field:

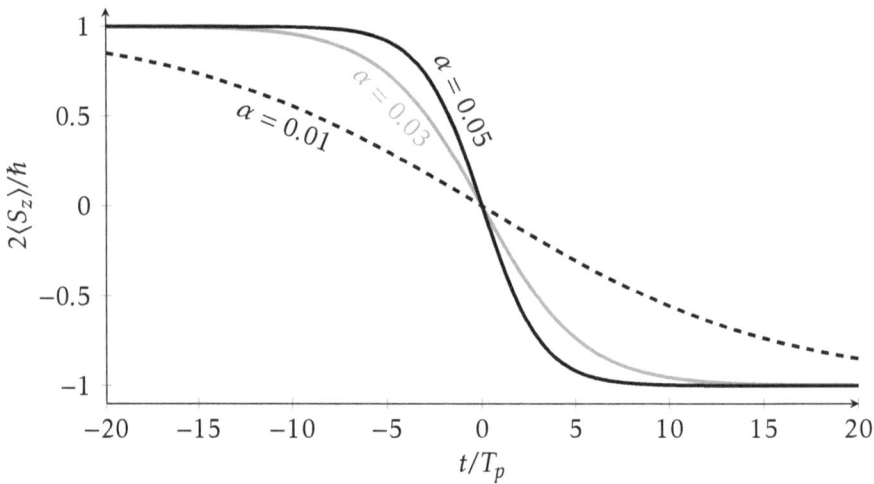

Figure 12.5: Plot of the z-component of the electron spin momentum, which is the component along the field axis, here along +z. This component reverses from alignment with B to being anti-aligned, as expected for the angular momentum. We also see that the larger the parameter α, the faster the transition.

Spin Angular Momentum Expectation Values

$$\langle \widehat{S}_x \rangle = \frac{\hbar}{2} \frac{\cos(\omega_p t + \varphi)}{\cosh(\alpha \omega_p t)}, \langle \widehat{S}_y \rangle = \frac{\hbar}{2} \frac{\sin(\omega_p t + \varphi)}{\cosh(\alpha \omega_p t)}, \langle \widehat{S}_z \rangle = -\frac{\hbar}{2} \tanh(\alpha \omega_p t)$$

(12.41)

The two components $\langle \widehat{S}_x \rangle$ and $\langle \widehat{S}_y \rangle$, perpendicular to the field, oscillate at a frequency $\omega_p = g|\gamma|B$, which is the **electron spin precessional frequency** given by Eq. (9.68). It can also be shown that the results above conserve the total spin angular momentum, in time t. To see this, we consider the length of $\langle S \rangle(t)$, while undergoing a dynamic transition in a magnetic field. The quantity $\langle \widehat{S} \rangle^2$ is given by

$$\langle \widehat{S} \rangle^2 = \langle \widehat{S}_x \rangle^2 + \langle \widehat{S}_y \rangle^2 + \langle \widehat{S}_z \rangle^2 =$$

$$\left[\frac{\hbar}{2} \frac{\cos(\omega_p t + \varphi)}{\cosh(\alpha \omega_p t)}\right]^2 + \left[\frac{\hbar}{2} \frac{\sin(\omega_p t + \varphi)}{\cosh(\alpha \omega_p t)}\right]^2 + \left[-\frac{\hbar}{2} \tanh(\alpha \omega_p t)\right]^2 =$$

12.4 Dynamical Spin Angular Momentum

$$\frac{\hbar^2}{4}\left[\frac{\cos^2(\omega_p t + \varphi) + \sin^2(\omega_p t + \varphi) + \sinh^2(\alpha\omega_p t)}{\cosh^2(\alpha\omega_p t)}\right] =$$

$$\frac{\hbar^2}{4}\left[\frac{1 + \sinh^2(\alpha\omega_p t)}{\cosh^2(\alpha\omega_p t)}\right] = \frac{\hbar^2}{4}\left[\frac{\cosh^2(\alpha\omega_p t)}{\cosh^2(\alpha\omega_p t)}\right] = \frac{\hbar^2}{4} \quad (12.42)$$

Eq. (12.42) tells us that the length of the expectation value of **S** is conserved as it relaxes over time, where

$$|\langle\widehat{\mathbf{S}}\rangle(t)| = \frac{\hbar}{2} \quad (12.43)$$

Our result asserts that the spin angular momentum is conserved as it rotates away from **B**. The solutions for the electron spin magnetic moment $\boldsymbol{\mu}_S$ as a function of time, are then given by

---

$\mu_x(t)$ Expectation Value

$$\langle\mu_x\rangle(t) = -\frac{g|\gamma|\hbar}{2}\frac{\cos(\omega_p t + \varphi)}{\cosh(\alpha\omega_p t)} \quad (12.44)$$

$\mu_y(t)$ Expectation Value

$$\langle\mu_y\rangle(t) = -\frac{g|\gamma|\hbar}{2}\frac{\sin(\omega_p t + \varphi)}{\cosh(\alpha\omega_p t)} \quad (12.45)$$

$\mu_z(t)$ Expectation Value

$$\langle\mu_z\rangle(t) = \frac{g|\gamma|\hbar}{2}\tanh(\alpha\omega_p t) \quad (12.46)$$

Using the relations $g \approx 2$ and $\mu_B = |\gamma|\hbar$, it also follows that

$$|\boldsymbol{\mu}(t)| = \mu(t) = \frac{g|\gamma|\hbar}{2} = \frac{g\mu_B}{2} \approx \mu_B \quad (12.47)$$

Therefore, the generalized Hamiltonian, which includes dissipation, leads to a dynamical solution for the magnetic moment, which maintain a fixed length at the Bohr magneton μ_B, while the spin angular momentum is fixed at $\hbar/2$. Let us examine what kind of time-dependent differential equation is satisfied by the components of magnetic moment given by Eqs. (12.44), (12.45), and (12.46). The above results turn out to be exact solutions to a well-known *nonlinear* differential equation which we'll obtain in the next section.

12.5 Dynamical Equation From G² Ehrenfest Theorem

The equation of motion satisfied by the time-dependent spin or momentum components can be obtained using some of the quantum mechanical tools we have developed in this text. Specifically, we will apply the Generalized-Generalized Ehrenfest Theorem or G² Ehrenfest Theorem we obtained in Eq. (3.147). We only need the Hamiltonian to do this. Recall that the Schrödinger equation with its dissipative conjugate contribution is given by

$$i\hbar \frac{\partial \psi}{\partial t} = \left[\frac{g|\gamma|\hbar}{2} \widehat{\sigma} \cdot \mathbf{B} - i\alpha \frac{g|\gamma|\hbar}{2} \widehat{\sigma} \cdot \mathbf{B} \right] \psi = \widehat{H}_\sigma \psi \qquad (12.48)$$

Eq. (3.147) can be used along with the Hamiltonian above to find the resulting total time-derivative of an electron spin operator. In Chap. 9, we obtained the following general result for the time-derivative of the expectation value of an operator

$$\frac{d\langle \widehat{\mu}_S \rangle}{dt} = \frac{i}{\hbar} \left[\langle \widehat{H}^\dagger \widehat{\mu}_S \rangle - \langle \widehat{\mu}_S \widehat{H} \rangle \right] - \frac{i}{\hbar} \langle \widehat{\mu}_S \rangle \left[\langle \widehat{H}^\dagger \rangle - \langle \widehat{H} \rangle \right] + \left\langle \frac{\partial \widehat{\mu}_S}{\partial t} \right\rangle \qquad (12.49)$$

The partial time-derivative term on the RHS is taken to be zero since the operator does not contain an explicit time dependence. Because the Hamiltonian is the sum of two terms, we'll treat each of the terms separately, then combine the results in the end. The first term of the Hamiltonian $\widehat{H}_Z$ is *Hermitian*. Therefore, Eq. (12.49) simplifies to

$$\frac{d\widehat{\mu}_S}{dt} = \frac{i}{\hbar} \left[\widehat{H}_Z \widehat{\mu}_S - \widehat{\mu}_S \widehat{H}_Z \right] = \frac{i}{\hbar} \left[\widehat{H}_Z, \widehat{\mu}_S \right] \qquad (12.50)$$

Substitution of $\widehat{\mu}_S = (-g|\gamma|\hbar/2)\widehat{\sigma}$ into the above, as well as the spin matrices, we have

$$\frac{d\widehat{\mu}_S}{dt} = -\frac{ig^2|\gamma|^2\hbar}{4} \left[(\widehat{\sigma} \cdot \mathbf{B})\widehat{\sigma} - \widehat{\sigma}(\widehat{\sigma} \cdot \mathbf{B}) \right]$$

12.5 Dynamical Equation From G² Ehrenfest Theorem

Expanding the dot products to better see where this leads, we have

$$\frac{d\widehat{\boldsymbol{\mu}}_S}{dt} = -\frac{ig^2|\gamma|^2\hbar}{4}\left[(\widehat{\sigma}_x B_x + \widehat{\sigma}_y B_y + \widehat{\sigma}_z B_z)\widehat{\boldsymbol{\sigma}} - \widehat{\boldsymbol{\sigma}}(\widehat{\sigma}_x B_x + \widehat{\sigma}_y B_y + \widehat{\sigma}_z B_z)\right]$$

$$= -\frac{ig^2|\gamma|^2\hbar}{4}\begin{bmatrix}(\widehat{\sigma}_y\widehat{\sigma}_x - \widehat{\sigma}_x\widehat{\sigma}_y)B_y + (\widehat{\sigma}_z\widehat{\sigma}_x - \widehat{\sigma}_x\widehat{\sigma}_z)B_z \\ (\widehat{\sigma}_x\widehat{\sigma}_y - \widehat{\sigma}_y\widehat{\sigma}_x)B_x + (\widehat{\sigma}_z\widehat{\sigma}_y - \widehat{\sigma}_y\widehat{\sigma}_z)B_z \\ (\widehat{\sigma}_x\widehat{\sigma}_z - \widehat{\sigma}_z\widehat{\sigma}_x)B_x + (\widehat{\sigma}_y\widehat{\sigma}_z - \widehat{\sigma}_z\widehat{\sigma}_y)B_z\end{bmatrix}$$

We can make use of the cyclic permutation relations for the spin matrices obtained in Eqs. (9.21a)-(9.21c), where $\widehat{\sigma}_i\widehat{\sigma}_j = i\varepsilon_{ijk}\widehat{\sigma}_k$. With this, the above leads to the familiar *precessional* motion, because we have

$$\frac{d\widehat{\boldsymbol{\mu}}_S}{dt} = -\frac{ig^2|\gamma|^2\hbar}{4}\begin{bmatrix}-2i\widehat{\sigma}_z B_y + 2i\widehat{\sigma}_y B_z \\ 2i\widehat{\sigma}_z B_x - 2i\widehat{\sigma}_x B_z \\ -2i\widehat{\sigma}_y B_x + 2i\widehat{\sigma}_x B_z\end{bmatrix} = \frac{g^2|\gamma|^2\hbar}{2}\widehat{\boldsymbol{\sigma}}\times\mathbf{B} = -g|\gamma|\widehat{\boldsymbol{\mu}}_S\times\mathbf{B}$$

This only accounts for one of the two contributions to the dynamic equation for $\widehat{\boldsymbol{\mu}}_S$. Next, we determine the contribution from the dissipative part of the Hamiltonian using Eq. (3.147). There is no simplification here because *the dissipative contribution to the Hamiltonian is non-Hermitian*. Therefore, the first part of the time-derivative due to the dissipative contribution becomes

$$\frac{d\widehat{\boldsymbol{\mu}}_S}{dt} = \frac{i}{\hbar}\left[\widehat{H}_\sigma^\dagger\widehat{\boldsymbol{\mu}}_S - \widehat{\boldsymbol{\mu}}_S\widehat{H}_\sigma\right] = \frac{i}{\hbar}\left[i\alpha\widehat{H}_Z\widehat{\boldsymbol{\mu}}_S + i\alpha\widehat{\boldsymbol{\mu}}_S\widehat{H}_Z\right]$$

$$= \frac{-\alpha}{\hbar}\left(\widehat{H}_Z\widehat{\boldsymbol{\mu}}_S + \widehat{\boldsymbol{\mu}}_S\widehat{H}_Z\right) = \frac{\alpha g^2|\gamma|^2\hbar}{4}(\widehat{\boldsymbol{\sigma}}\cdot\mathbf{B}\widehat{\boldsymbol{\sigma}} + \widehat{\boldsymbol{\sigma}}\widehat{\boldsymbol{\sigma}}\cdot\mathbf{B})$$

$$= \frac{\alpha g^2|\gamma|^2\hbar}{4}\begin{bmatrix}2\widehat{\sigma}_x^2 B_x + (\widehat{\sigma}_y\widehat{\sigma}_x + \widehat{\sigma}_x\widehat{\sigma}_y)B_y + (\widehat{\sigma}_z\widehat{\sigma}_x + \widehat{\sigma}_x\widehat{\sigma}_z)B_z \\ 2\widehat{\sigma}_y^2 B_y + (\widehat{\sigma}_x\widehat{\sigma}_y + \widehat{\sigma}_y\widehat{\sigma}_x)B_x + (\widehat{\sigma}_z\widehat{\sigma}_y + \widehat{\sigma}_y\widehat{\sigma}_z)B_z \\ 2\widehat{\sigma}_z^2 B_z + (\widehat{\sigma}_x\widehat{\sigma}_z + \widehat{\sigma}_z\widehat{\sigma}_x)B_x + (\widehat{\sigma}_y\widehat{\sigma}_z + \widehat{\sigma}_z\widehat{\sigma}_y)B_z\end{bmatrix}$$

Using the relations $\widehat{\sigma}_i^2 = \mathbf{I}(2)$ and the spin matrix anti-commutation relation $\widehat{\sigma}_i\widehat{\sigma}_j + \widehat{\sigma}_j\widehat{\sigma}_i = 0$, the above becomes

$$\frac{d\widehat{\boldsymbol{\mu}}_S}{dt} = \frac{\alpha g^2|\gamma|^2\hbar}{4}\begin{bmatrix}2B_x\\2B_y\\2B_z\end{bmatrix} = \frac{\alpha g^2|\gamma|^2\hbar}{2}\mathbf{B} \qquad (12.51)$$

For the second part of the dissipative contribution, we have

$$\frac{-i}{\hbar}\langle\boldsymbol{\mu}_S\rangle\left[\widehat{H}^\dagger - \widehat{H}\right] = \frac{-i}{\hbar}\langle\boldsymbol{\mu}_S\rangle\left[\frac{i\alpha g|\gamma|\hbar}{2} + \frac{i\alpha g|\gamma|\hbar}{2}\right](\boldsymbol{\sigma}\cdot\mathbf{B}) \qquad (12.52)$$

$$= \frac{2\alpha}{\hbar}\langle\boldsymbol{\mu}_S\rangle\left[\frac{g|\gamma|\hbar}{2}\right](\boldsymbol{\sigma}\cdot\mathbf{B}) \qquad (12.53)$$

$$= -\frac{2\alpha}{\hbar}\left(\boldsymbol{\mu}_S\cdot\mathbf{B}\right)\boldsymbol{\mu}_S \qquad (12.54)$$

To tie our results together, we rewrite the magnetic field **B**. We can make use of a vector identity which states that with a 3D unit vector **u** ($|\mathbf{u}| = 1$), any vector **B** can be decomposed with the unit vector as

$$\mathbf{B} = \frac{1}{\mu_S^2}\left[(\boldsymbol{\mu}_S\cdot\mathbf{B})\boldsymbol{\mu}_S - \boldsymbol{\mu}_S\times\boldsymbol{\mu}_S\times\mathbf{B}\right] \qquad (12.55)$$

In Chapter problem 12.9, we'll also have you prove Eq. (12.55). The idea is illustrated in Fig. 12.6. For a single electron, the saturation moment

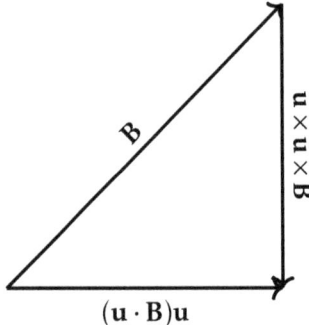

Figure 12.6: Illustration of the vector relations between **B**, **u**×**u**×**B**, and (**u**·**B**)**u**.

μ_S along the axis of the field is given by

$$|\boldsymbol{\mu}_S| = \mu_S = \frac{g|\gamma|\hbar}{2} \qquad (12.56)$$

12.5 Dynamical Equation From G² Ehrenfest Theorem

Substitution of these relations into Eqs. (12.51) gives

$$\mathbf{B} = \frac{2\alpha}{\hbar}\left[(\boldsymbol{\mu}_S \cdot \mathbf{B})\boldsymbol{\mu}_S - \boldsymbol{\mu}_S \times \boldsymbol{\mu}_S \times \mathbf{B}\right] \qquad (12.57)$$

The first term in Eq. (12.57) cancels the result from Eq. (12.54). Then, we are left with the following result for the time-derivative of the operator $\boldsymbol{\mu}_S$:

$$\frac{d\langle\boldsymbol{\mu}_S\rangle}{dt} = -g|\gamma|\boldsymbol{\mu}_S \times \mathbf{B} - \frac{2\alpha}{\hbar}\boldsymbol{\mu}_S \times \boldsymbol{\mu}_S \times \mathbf{B}$$

$$= -g|\gamma|\boldsymbol{\mu}_S \times \mathbf{B} - \frac{\alpha g|\gamma|}{\mu_S}\boldsymbol{\mu}_S \times \boldsymbol{\mu}_S \times \mathbf{B}$$

Therefore, the components of the conserved electron magnetic moment $\boldsymbol{\mu}$ given by Eqs. (12.44), (12.45), and (12.46) satisfy an equation of motion given by

---

Electron Magnetic Moment Dynamical Equation

$$\frac{d\langle\boldsymbol{\mu}\rangle}{dt} = \underbrace{-g|\gamma|\ \boldsymbol{\mu} \times \mathbf{B}}_{\text{precession}} - \underbrace{\frac{\alpha g|\gamma|}{\mu}\ \boldsymbol{\mu} \times \boldsymbol{\mu} \times \mathbf{B}}_{\text{transverse damping}} \qquad (12.58)$$

The first term in Eq. (12.58) describes what is known as *precessional* motion. It is the magnetic moment's rotational motion in the plane perpendicular to the axis of the magnetic field. We saw an example of pure precessional motion in Chap. 9, Sec. 9.6. The second term is known as **transverse damping** and corresponds to the rotation of the conserved moment vector $\boldsymbol{\mu}$ towards the magnetic field **B**. Equations of the form given by Eq. (12.58) are known as **Landau-Lifshitz** type equations. The name originates from the two Soviet physicists who, in 1935, proposed, specifically, the *transverse damping* term as a phenomenological contribution to add to the precessional motion.

That Eqs. (12.44), (12.45), and (12.46) are exact solutions to Eq. (12.58) can be verified directly, which we shall show next. To obtain the differential equation directly, let us assume the magnetic field B is along the z-direction. Then, $\boldsymbol{\mu} \times \mathbf{B}$ gives

$$\boldsymbol{\mu} \times \mathbf{B} = \begin{vmatrix} i & j & k \\ \mu_x & \mu_y & \mu_z \\ 0 & 0 & B_z \end{vmatrix} = \begin{bmatrix} (\mu_y B_z - \mu_z B_y) \\ -(\mu_x B_z - \mu_z B_x) \\ (\mu_x B_y - \mu_y B_x) \end{bmatrix} = \begin{bmatrix} \mu_y B_z \\ -\mu_x B_z \\ 0 \end{bmatrix}$$

Chapter 12. Non-Hermitian Operators & Dynamics

Next, let's use this result to obtain $\boldsymbol{\mu} \times \boldsymbol{\mu} \times \mathbf{B}$. For this, we have

$$\boldsymbol{\mu} \times \boldsymbol{\mu} \times \mathbf{B} = \begin{bmatrix} i & j & k \\ \mu_x & \mu_y & \mu_z \\ \mu_y B_z & -\mu_x B_z & 0 \end{bmatrix} = \begin{bmatrix} \mu_z \mu_x B_z \\ \mu_z \mu_y B_z \\ -\mu_x^2 B_z - \mu_y^2 B_z \end{bmatrix}$$

From this, it follows that

$$-g|\gamma|\boldsymbol{\mu} \times \mathbf{B} - \frac{\alpha g|\gamma|}{\mu} \boldsymbol{\mu} \times \boldsymbol{\mu} \times \mathbf{B} = \begin{bmatrix} -g|\gamma|\mu_y B_z - \frac{\alpha g|\gamma|}{\mu} \mu_z \mu_x B_z \\ g|\gamma|\mu_x B_z - \frac{\alpha g|\gamma|}{\mu} \mu_z \mu_y B_z \\ \frac{\alpha g|\gamma|}{\mu}(\mu_x^2 B_z + \mu_y^2 B_z) \end{bmatrix} \quad (12.59)$$

Next, since our components are ratios, we determine the three components $d\mu_x/dt$, $d\mu_y/dt$, and $d\mu_z/dt$ using the quotient rule given by

$$\frac{d}{dt}\left(\frac{N(t)}{D(t)}\right) = \frac{\frac{dN}{dt} \cdot D - N \cdot \frac{dD}{dt}}{D^2}$$

Starting with $\frac{d\mu_x}{dt}$, we have

$$\frac{d\mu_x}{dt} = -\frac{d}{dt}\left[\mu \frac{\cos(2\omega_p t + \varphi)}{\cosh(\alpha\omega_p t)}\right]$$

$$= -\mu \left[\frac{-\sin(\omega_p t + \varphi)\cosh(\alpha\omega_p t)(\omega_p) - \cos(\omega_p t + \varphi)\cosh(\alpha\omega_p t)(\alpha\omega_p)}{\cosh^2(2\alpha\omega_p t)}\right]$$

$$= -\mu \frac{-\sin(\omega_p t + \varphi)(\omega_p)}{\cosh(\alpha\omega_p t)} - \mu \frac{-\cos(\omega_p t + \varphi)\sinh(2\alpha\omega_p t)(\alpha\omega_p)}{\cosh^2(\alpha\omega_p t)}$$

$$= -g|\gamma|\frac{-\mu\sin(\omega_p t + \varphi)B_z}{\cosh(\alpha\omega_p t)} - g\alpha|\gamma|\frac{-\mu\cos(\omega_p t + \varphi)\sinh(\alpha\omega_p t)}{\cosh^2(\alpha\omega_p t)}$$

$$= -g|\gamma|\mu_y B_z - \frac{g\alpha|\gamma|}{\mu}\mu_z \mu_x B_z$$

We have used the relation $\omega_p = g|\gamma|B$. This result is identical to the x-component of Eq. (12.59). For $\frac{d\mu_y}{dt}$, we have

12.5 Dynamical Equation From G² Ehrenfest Theorem

$$\frac{d\mu_y}{dt} = -\frac{d}{dt}\left[\mu\frac{\sin(\omega_p t + \varphi)}{\cosh(\alpha\omega_p t)}\right]$$

$$= -\mu\left[\frac{\cos(\omega_p t + \varphi)\cosh(\alpha\omega_p t)(\omega_p) - \sin(\omega_p t + \varphi)\sinh(\alpha\omega_p t)(\alpha\omega_p)}{\cosh^2(\alpha\omega_p t)}\right]$$

$$= -\mu\frac{\cos(\omega_p t + \varphi)(\omega_p)}{\cosh(\alpha\omega_p t)} - \mu\frac{-\sin(\omega_p t + \varphi)\sinh(\alpha\omega_p t)(\alpha\omega_p)}{\cosh^2(\alpha\omega_p t)}$$

$$= g|\gamma|\frac{-\mu\cos(\omega_p t + \varphi)B_z}{\cosh(\alpha\omega_p t)} - g\alpha|\gamma|\frac{-\mu\sin(\omega_p t + \varphi)\sinh(\alpha\omega_p t)}{\cosh^2(\alpha\omega_p t)}$$

$$= g|\gamma|\mu_x B_z - \frac{g\alpha|\gamma|}{\mu}\mu_z\mu_y B_z$$

Likewise, this result is identical to the y-component of Eq. (12.59). And lastly, for $\frac{d\mu_z}{dt}$, we have

$$\frac{d\mu_z}{dt} = \frac{d}{dt}\left[\mu\frac{\sinh(\alpha\omega_p t)}{\cosh(\alpha\omega_p t)}\right]$$

$$= \mu\left[\frac{\cosh^2(\alpha\omega_p t)(\alpha\omega_p) - \sinh^2(\alpha\omega_p t)(\alpha\omega_p)}{\cosh^2(\alpha\omega_p t)}\right]$$

$$= \mu(\alpha\omega_p) - \mu\frac{\sinh^2(\alpha\omega_p t)(\alpha\omega_p)}{\cosh^2(\alpha\omega_p t)}$$

$$= \frac{g\alpha|\gamma|B_z}{\mu}\left(\mu^2 - \mu_z^2\right)$$

$$= \frac{g\alpha|\gamma|B_z}{\mu}(\mu_x^2 + \mu_y^2)$$

As we found with x and y components, this result is identical to the z-component of Eq. (12.59). Therefore, it is identical to Eq. (12.58), and the expectation values of $\widehat{\boldsymbol{\mu}}_S$ apparently satisfy the Landau-Lifshitz type of equation. Figs. 12.7 and 12.8 illustrate solutions to Eq. (12.58) obtained by numerical integration, for two different magnetic values of the dissipation parameter α, where Fig. 12.8 has twice the damping coefficient as in Fig. 12.7.

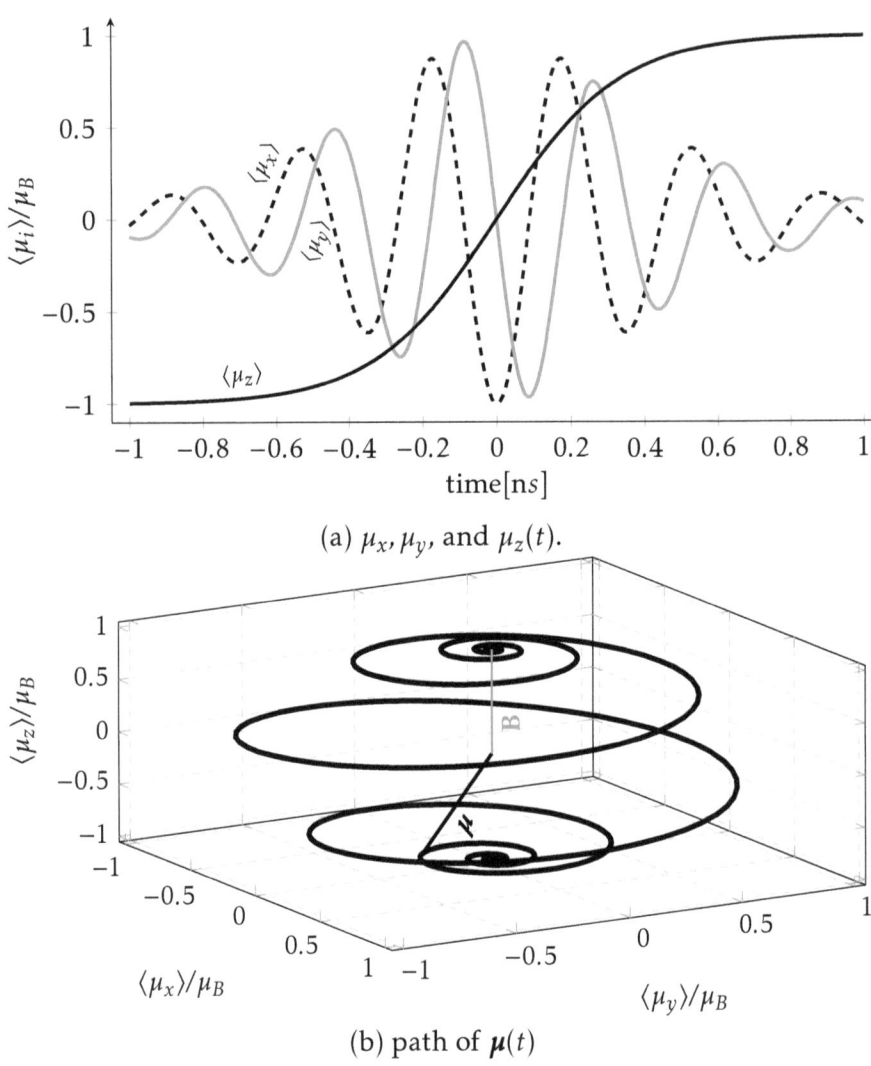

(a) μ_x, μ_y, and $\mu_z(t)$.

(b) path of $\boldsymbol{\mu}(t)$

Figure 12.7: Reversal of magnetic moment $\boldsymbol{\mu}$ in a magnetic field. The gray vertical vector is the magnetic field. The $\boldsymbol{\mu}$ vector's tip traces out the path shown.

If the parameter α is increased, we would find a shorter spin-relaxation time τ_S, and faster reversal in a magnetic field. This also leads to less *ringing* around the sphere during reversal (shown in (b)). An example with increased α by two times that in Fig. 12.7 is illustrated below.

Both figures demonstrate the motion of a *conserved* electron magnetic moment, as given by Eqs. (12.44),(12.45), and (12.46). Thus, we find

12.5 Dynamical Equation From G² Ehrenfest Theorem

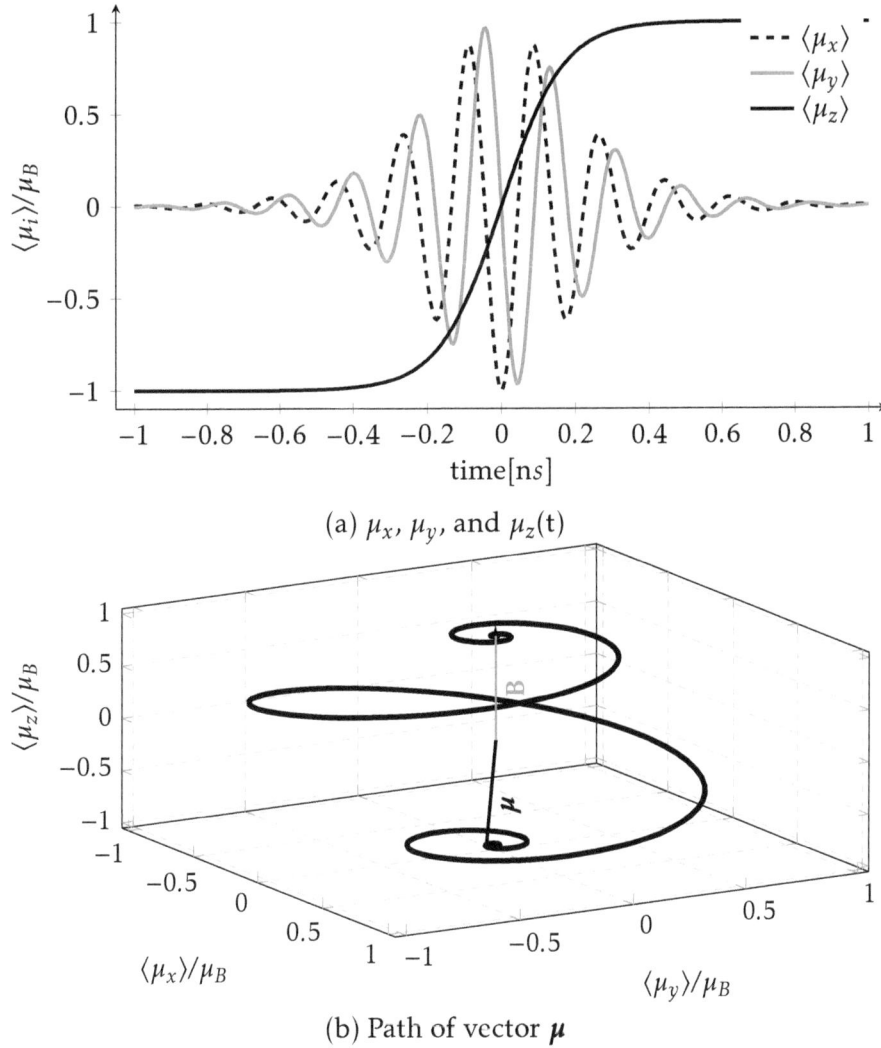

(a) μ_x, μ_y, and $\mu_z(t)$

(b) Path of vector μ

Figure 12.8: Another example of the reversal of a magnetic moment μ in a magnetic field, here using 2× the dissipation parameter α in Fig. 12.7.

that the addition of the dissipative conjugate Hamiltonian to the Zeeman energy of the electron spin goes beyond just precessional motion to a dynamical equation that leads to a reversal of spin angular momentum or magnetic moment. The result here applies to localized electrons. But, what happens with electrons that are moving from atom to atom, if they also have a similar type of non-Hermitian Hamiltonian? By adding kinetic energy to a spin-dependent potential energy, we can answer this question. We examine this scenario, particularly, with exchange energy, rather than the Zeeman energy. This topic has relevance in some important technological applications and is discussed in the next section.

12.6 Asymmetric Spin Flow Of Itinerant Electrons

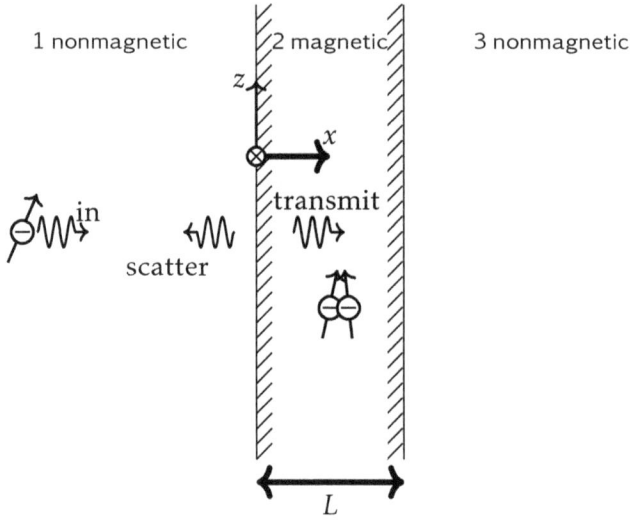

Figure 12.9: Itinerant electron spin moving through a magnetic layer. As we'll see, in certain conditions, some electrons can back-scatter or reflect, while others enter the magnetic material more readily.

The Hamiltonian including dissipation of spin-dependent energy allows us to probe the spin-dependent motion of an itinerant electron, as well. When an electron conducts or flows through a *normal* metal (nonmagnetic) or material, there are generally no spin-dependent energy contributions. This leads to equally probable conditions for both spin-up and spin-down states. However, what about when the electron flows through a conductive *magnetic* metal? In this case, spin-dependent contributions are present in the material, and thus, in the Hamiltonian describing such a system. Because a magnetic layer is being considered, the spin-dependent contribution is the *exchange interaction* potential obtained in Eq. (10.86). Experiments carried out in the 1980s involving flowing electricity through a series of magnetic/nonmagnetic multilayers began to suggest notable distinctions in the behavior of the conduction properties. For example, the resistivity depended on the orientation of the magnetization. Recall that even the Hall effect demonstrates that mag-

12.6 Asymmetric Spin Flow Of Itinerant Electrons

netic fields, for example, coupling to electrons can influence electrical properties. In this case, the field coupled to the charge. There is indeed more to this, and we'll demonstrate that the physics of the conductivity across a magnetic/nonmagnetic junction as illustrated in Fig. 12.9 is also described by quantum mechanics. For the spin transport equations we will obtain later, *we must include dissipation of the spin-dependent energy.*

Therefore, in some parts of this chapter, we focus on the consequences of an itinerant electron experiencing spin-dependent exchange, while moving between a nonmagnetic region (region 1) into a magnetic metallic region (region 2), and eventually back into a nonmagnetic region (region 3), as illustrated in Fig. 12.9. However, in any electrical transport, the physics can be less than straight-forward. This is because there is, in a sense, a balance of the idea of a moving particle (like a propagating wave) with that of a bound electron. The concept is important in the discussion of spin transport, so we'll review concepts in electrical conduction in solid state materials.

12.6.1 Conduction Or Itinerant Electrons

Because the device we are discussing typically has an external voltage source applied to drive the electrons across the layers of the device, its useful to frame the Schrödinger equation and quantum mechanical concepts in terms of accessible or measured parameters relating to electrical conduction. To do this, we must recall that in a solid state material, an electron is ordinarily bound to an atom, resulting from an overall negative eigenenergy E_{bound}. However, *conducting electrons posses an energy above the bound state eigenenergy* ($E_\lambda > E_{bound}$). Moreover, the additional externally driven energy is *directional*. The role of an external electric field, or voltage, is to raise the energy of an electron sufficiently to transform a bound electron, which has $E_{bound} < 0$ into an *itinerant electron*. The description of the energy without an externally applied voltage using the general Schrödinger equation is given by

$$i\hbar \frac{\partial \Psi}{\partial t} = \left[-\frac{\hbar^2}{2m} \frac{\partial^2}{\partial x^2} + V \right] \Psi \tag{12.60}$$

For an electron in an atom with two or more electrons belonging to the atom, the potential $V = V_{n-e} + V_{e-e}$ includes the electric potentials given by

$$V_{n-e}(r) = -\frac{Ze^2}{4\pi\varepsilon_0|\mathbf{r}_1 - \mathbf{r}_2|} \quad \text{and} \quad V_{e-e}(r_i) = \sum_{j=1}^{Z-1}\frac{e^2}{4\pi\varepsilon_0|\mathbf{r}_i - \mathbf{r}_j|} \quad (12.61)$$

As simple as it looks, for a single electron with no electron-electron interactions ($V_{e-e} = 0$), we found this to be a fairly sophisticated problem to solve when we treated the hydrogen atom in Chap. 8. In the absence of a magnetic field, the bound state *eigenenergy* for the hydrogen electron was found to be

$$E_{bound} = -\frac{Z^2 m^* e^4}{32\pi^2\hbar^2\varepsilon_0^2}\frac{1}{n^2} = -13.6\left(\frac{Z}{n}\right)^2 \quad (12.62)$$

n is the *principal quantum number*. For a multi-electron system, we must add V_{e-e} which acts as a *positive* potential energy between the electrons in the atom. V_{e-e} always raises the energy of the electrons. From Eq. (12.62), the larger the principal quantum number n, the smaller is $|E_{bound}|$ and the closer to zero is the negative bound energy. $(Z/n)^2$ ultimately dictates how high the energy will be for this contribution. For example, $Z/n = 79/6 = 13.2$ for gold (Au), while $Z/n = 29/4 = 7.25$ for copper (Cu). This suggests the bound state energy contribution for a valence electron in Cu would be more susceptible to an external electric field than those of Au. In general, the occupied electronic states having the highest energy (closest to zero, but negative in sign) are most susceptible to an external electric field E_{ext}, particularly when they are excited to levels that are in proximity energetically.

Recall that the highest occupied energy state is known as the **Fermi level** at $T = 0K$ (c.Sec. 2.5.1). All the filled states below this level fill what is called the **Fermi sea**. Note that the Fermi level applies to a 1D distribution of energies depending on a single wave vector component $k = k_x$. In 2D, the Fermi level becomes a **Fermi contour**, while in 3D, it becomes a **Fermi surface**, depending of $k = [k_x, k_y, k_z]^T$. The highest shell (which is determined by n) below the *Fermi level* that houses the highest energy electrons is known as the **valence shell**. More generally, just the set of occupied states in this shell form the **valence band**. The remaining nearest *available states* above the valence band form the **conduction band**. In the language of *bands*, valence and conduction states can occupy the same shell, as is the case in metals. When an external electrical potential (i.e. voltage) U_{ext} is applied, there is an associated

12.6 Asymmetric Spin Flow Of Itinerant Electrons

external electric field $\mathbf{E}_{ext}$ generated by the source that acts on the electrons. The energy change due to the electric field is equal to the work done on the electron by E_{ext}. This is given by

$$V_{ext} = -eE_{ext}\delta x > 0 \qquad (12.63)$$

This external energy contribution is *always positive* for the following reasons: Assuming E_{ext} is along the x axis, for an electron having charge $q = -e$, the electric field points along the $-x$ direction. But the force is qE_{ext}, which is positive, or along $+x$ for a negative charge. Thus, the motion is along $+x$ and the work is a positive energy contribution. Reversing the polarity of the external voltage changes the sign of both E_{ext} and δx, which preserves the *positive* energy. Thus, with an externally applied voltage, the electronic energy is *increased* or made *less* negative. This is also consistent with the form qU_{ext}, where a *voltage drop* for the electron charge also yields a positive energy. So, the excited energy becomes

$$E_\lambda = qU_{ext} + E_{bound} + V_{e-e} \qquad (12.64)$$

Generally, electrically conducting electrons, defined as **itinerant electrons**, *are not free electrons*. Instead, they are in a class of their own. In fact, freeing an electron entirely from an atom is akin to providing enough energy for an object to escape the earth's atmosphere. It requires a larger amount of energy for this, compared to hopping around on the earth, as we do in an aircraft. Essentially, electrical conduction is the result of electrons being enabled to hop from atom to atom. This is the meaning of the word *itinerant* (hopping). The hopping process requires less energy because its less work being done since its over a shorter distance. Freeing an electron requires moving the electron a distance of infinity, whereas with hopping, the electron only needs to move a distance on the order of the lattice constant, still finite compared to freeing the electron. The electric field not only raises the electronic energy, but also *kicks* the electron in a definite direction opposite the E–field axis ($F = qE$). When the electron *itinerates* to a nearby atom, there are two possibilities:

1. It is expected that the electron can displace another electron in an arrival atom if its energy conditions are favorable to do so. However, as the electron is excited into itineration, there must be some available states below the maximum electron energy for the electron to occupy, in the arrival atom.

element	valence	available states	no. avail. states
Au (P-shell)	$6s^1$	6s, 6p, 6d, 6f, 6g, 6h	72-1=71
Ag (O-shell)	$5s^1$	5s, 5p, 5d, 5f, 5g	50-1=49
Cu (N-shell)	$4s^1$	4s, 4p, 4d, 4f	32-1=31

2. Contrary and also possible, the electron cannot displace another electron due to unfavorable energy conditions. This occurs when there is no proximate energy state to be occupied by the itinerant electron. This occurs when the available energy states are at a level sufficiently higher than the itinerant electronic energy. In this case, the electron scatters and conduction ceases locally.

These two possibilities dictate that the process of electrical conduction is partly determined by the energies in the vicinity of the valence electron energy levels, or near the Fermi edges (edge = general term for the level, contour, or surface). Thus, conduction properties depend on information relating to the Fermi surface as well as the distribution of available energy states for the conducting material.

Good conductors have a single electron in the lowest possible energy of the highest shell. This leaves the largest number of available proximate energy states within the same shell. This is the case for the top three electrical conductors, which are silver (Ag), copper (Cu), and gold (Au), as summarized in the above table. As mentioned, the proximate available higher energy states (beyond the Fermi edge) form the *conduction band*. Though itinerant electrons generally still have a negative energy, they are energized just enough to *itinerate* from atom to atom, still more or less bound to the lattice of atoms. Therefore, we can only think of the (negative) electronic energy relative to a certain energy point. Only relative to this energy point, can the electron energy be considered *positive*. This *reference point* is, of course, the Fermi edge ϵ_F (1D).

Additionally, within a lattice, which is a periodic arrangement of atoms, because of the combined effects from electron wave scattering at atoms, conditions are reached where some wave numbers corresponding to standing waves that can form, results in no conduction for some wave number/energy pairs. This further removes some $(\mathbf{k}, E)$ points from the distribution of probable energies. These regions of no energy, created from these standing waves introduces what are called **energy gaps** into

12.6 Asymmetric Spin Flow Of Itinerant Electrons

the energy distribution. These details must be computed for the specific lattice to determine the so-called *band structure*. The net result is a further restriction of electron hopping by blocking more potentially available states. This is because for these states given by $E_j(k_j)$, $f(k_j) = 0$. A general prescription can be outlined for how to determine a current density $\widehat{J} = g(E_{ext})$, which depends on the external field E_{ext}. We'll use an effective **conduction wave vector** k_{cond} along the direction of the electric field defined as

$$\frac{\hbar^2 k_{cond}^2}{2m} = \sum_j \left[E_\lambda(E_{ext}) - E_j(k_j) \right] f(k_j) \tag{12.65}$$

$E_\lambda(E_{ext})$ is the eigenenergy from Eq. (12.64), which depends on the external electric field. The sum is taken over the distribution of probable states which is provided by the *band structure* calculations. **k** is varied over 1, 2, or 3D and the expectation of the corresponding energies is computed. In Eq. (12.65), for a large available energy $E_j > E_\lambda$ in the conduction band, the sum adds a negative component to k_{cond}^2, while for small available energies below $E_j < E_\lambda$, there is a positive contribution. When $k_{cond}^2 < 0$, the conduction wave function becomes evanescent (e.g. an insulator), while for $k_{cond}^2 > 0$, a propagating wave continues in the medium. This provides a mean conduction wave vector k_{cond}, which is a function of the external electric field. Thus, as a *quasi-free electron* using the conduction wave vector k_{cond}, the general eigenfunction becomes

$$\psi_{cond}(x) = A e^{ik_{cond}x} + B e^{-ik_{cond}x} \tag{12.66}$$

However, it is the most general eigenfunction for the quasi-free particle which is needed, which is given by (c. Sec. 3.7)

$$\Psi(x,t) = \frac{1}{\sqrt{2\pi\hbar}} \int_{-\infty}^{+\infty} F(p_{cond}) e^{-\frac{i}{\hbar}\left(p_{cond}x - \frac{p_{cond}^2}{2m}t\right)} dp_{cond} \tag{12.67}$$

$F(p_{cond})$ is normally the Fourier transform of the eigenfunction given by Eq. (12.66) and $p_{cond} = \hbar k_{cond}$. Since the largest contributions come from the states in the vicinity of the Fermi surface, this integral can be computed predominantly near the Fermi surface, where *available* states exist and the contributions from the band structure are largest ($f(k_j)$ are largest). The Fourier transform is given by

$$F(p) = \frac{1}{\sqrt{2\pi\hbar}} \int_{-\infty}^{+\infty} \psi(x') e^{-\frac{ip}{\hbar}x'} dx' \tag{12.68}$$

When we introduced $F(p_{cond})$, we also found it to be the probability amplitude of the momentum distribution. This information is determined by the band structure calculations. This yields the appropriate general wave function Ψ for the conduction electron, which can be used to determine the current density $\widehat{J}(E_{ext})$ as

$$\widehat{J}(E_{ext}) = \frac{i\hbar}{2m}\left(\Psi \frac{\partial \Psi^*}{\partial x} - \Psi^* \frac{\partial \Psi}{\partial x}\right)$$

This computation is generally nontrivial and requires implementation of numerical algorithms to obtain results. *For our purposes, the significant details to be emphasized are the meaning of the conduction wave vector k_{cond} along the direction of E_{ext}, and that the effective kinetic energy is proportional to the energy conditions of the electron.* The latter point is especially relevant to the discussion of the effects on electron spin. Note that metals, generally, have small energy gaps in their band structure and even overlap of both conduction and valence bands. At the opposite end of the spectrum, insulators have larger gaps so that electrons generally scatter more strongly. In examples like Cu, Au, and Ag, *the single electron in the valence shell are also spin-down*, since spin-down electrons occupy orbitals before spin-up. This also means that good conductors (with less than half filled valence orbitals) tend to utilize spin-down electrons. It is this subtle detail that gets most exposed when the electron enters a magnetic material. Let's see how.

12.6.2 Spin Splitting In Magnetic Conductors

In treating an itinerant electron entering a magnetic layer, the Hamiltonian can be expressed in the usual way, with the caveat that the meaning of the wave number $k = k_{cond}$ and kinetic energy is *different* from that of a free particle. Instead, it follows from the above discussion on *conduction*. Otherwise, the analysis is straightforward in appearance. We assume the localized spins in the magnetic layer are *fixed* in their orientation, taken to be along the +z axis. In a magnetic layer, the spin of the itinerant electron can couple to that of a localized electron spin via a form of electron *exchange coupling*, whose coefficient is denoted by an effective exchange integral constant J (c.Chap. 10). Since itinerant electrons in metals predominantly come from s–orbitals, and because magnetic materials contain net spin polarization from localized electrons predominantly in d–orbitals, the exchange integral parameter is often denoted by $J = J_{sd}$. Starting with the 1D Schrödinger equation for an electron with kinetic

12.6 Asymmetric Spin Flow Of Itinerant Electrons

energy having dissipative spin-potential energy in the magnet, we have

$$i\hbar \frac{\partial \Psi}{\partial t} = \left[-\frac{\hbar^2}{2m} \frac{\partial^2}{\partial x^2} + (1 - i\alpha) V(\widehat{\sigma}) \right] \Psi \tag{12.69}$$

The magnetic potential energy represented by V in Eq. (12.69) is replaced by the exchange interaction between the itinerant and localized electrons. The general form of the exchange interaction was derived earlier and is given by Eq. (10.86). $V(\widehat{\sigma})$ then becomes

$$V(\sigma) = \widehat{H}_{ij}(\sigma) = -\frac{1}{2} J \widehat{\sigma}_i \cdot \widehat{\sigma}_j = -\frac{1}{2} J_{sd} \widehat{\sigma}_i \cdot \widehat{\sigma}_j \tag{12.70}$$

In Eq. (12.69), the LHS contains only variations in time while the RHS has variations in space. Note that the exchange Hamiltonian itself is independent of both time and space. This suggests we can use a separable form letting Ψ be

$$\Psi(t, x) = \mathbf{T}(t) \mathbf{X}(x) \tag{12.71}$$

Since we have a spin-dependent Hamiltonian, both **T** and **X** are *not* scalar functions. **T** is a temporal diagonal 2×2 matrix, and **X** is a spatio-temporal spinor of size 2×1. The spinor contains both wave functions for spin-up and spin-down, respectively. Substitution of Eq. (12.71) into Eq. (12.69), along with substitution of the exchange Hamiltonian from Eq. (12.70) leads to

$$i\hbar \mathbf{X} \frac{d\mathbf{T}}{dt} = -\frac{\hbar^2}{2m} \mathbf{T} \frac{d^2 \mathbf{X}}{dx^2} - \frac{1}{2} J_{sd} (1 - i\alpha) \widehat{\sigma}_2 \cdot \widehat{\sigma} \mathbf{T}(t) \mathbf{X}(x)$$

We can express the *exchange energy in Zeeman energy-like form* by introducing the magnetic moment for the localized spin polarization $\widehat{\sigma}_P$. Then, we can write

$$i\hbar \mathbf{X} \frac{d\mathbf{T}}{dt} = -\frac{\hbar^2}{2m} \mathbf{T} \frac{d^2 \mathbf{X}}{dx^2} - \frac{J_{sd}}{2\mu_P} (1 - i\alpha) \boldsymbol{\mu}_P \cdot \widehat{\sigma} \mathbf{T}(t) \mathbf{X}(x) \tag{12.72}$$

The **effective exchange field** B^*_{sd}, having units of a magnetic field (e.g. Tesla), is defined as

Effective Exchange Field

$$B^*_{sd} = \frac{J_{sd}}{2\mu_P} \tag{12.73}$$

Eq. (12.72) becomes

$$i\hbar \frac{d\mathbf{T}}{dt}\mathbf{X} = -\frac{\hbar^2}{2m}\mathbf{T}\frac{d^2\mathbf{X}}{dx^2} - B_{sd}^*(1-i\alpha)\boldsymbol{\mu}_P \cdot \widehat{\boldsymbol{\sigma}}\mathbf{T}\mathbf{X} \qquad (12.74)$$

Since $\mathbf{X} = X_\pm$ is a 2×1 spinor, we know there is a corresponding equation for each spin-state. Recall that the spin polarization $\boldsymbol{\mu}_P$ of the magnetic layer is taken to be fixed along the $+z$ axis. Then, the two independent equations for $\Psi_+ = X_+T_+$ and $\Psi_- = X_-T_-$ using $\widehat{\sigma}_z$ become

$$i\hbar X_+ \frac{dT_+}{dt} = -\frac{\hbar^2}{2m}T_+\frac{d^2X_+}{dx^2} - B_{sd}^*\mu_P(1-i\alpha)T_+(t)X_+(x) \qquad (12.75a)$$

$$i\hbar X_- \frac{dT_-}{dt} = -\frac{\hbar^2}{2m}T_-\frac{d^2X_-}{dx^2} + B_{sd}^*\mu_P(1-i\alpha)T_-(t)X_-(x) \qquad (12.75b)$$

Dividing the first equation by $\Psi_+ = T_+(t)X_+(x)$ and the second equation by $\Psi_- = T_-(t)X_-(x)$, we get

$$i\hbar \frac{1}{T_+}\frac{dT_+}{dt} = -\frac{\hbar^2}{2m}\frac{1}{X_+}\frac{d^2X_+}{dx^2} - B_{sd}^*\mu_P(1-i\alpha)$$

$$i\hbar \frac{1}{T_-}\frac{dT_-}{dt} = -\frac{\hbar^2}{2m}\frac{1}{X_-}\frac{d^2X_-}{dx^2} + B_{sd}^*\mu_P(1-i\alpha)$$

The LHS is purely a function of time t while the RHS is purely a function of position x. Since t and x are independent variables, the equations can only be true if both sides equal to the respective eigenenergy constant $E_{\lambda\pm}$ for the electron. This leads to the following four equations:

$$\frac{dT_+}{dt} = -\frac{i}{\hbar}E_{\lambda+}T_+ \qquad (12.76a)$$

$$\frac{dT_-}{dt} = -\frac{i}{\hbar}E_{\lambda-}T_- \qquad (12.76b)$$

$$-\frac{\hbar^2}{2m}\frac{d^2X_+}{dx^2} - B_{sd}^*\mu_P(1-i\alpha)X_+ = E_{\lambda+}X_+ = (E_{R+} + iE_I)X_+ \qquad (12.76c)$$

$$-\frac{\hbar^2}{2m}\frac{d^2X_-}{dx^2} + B_{sd}^*\mu_P(1-i\alpha)X_- = E_{\lambda-}X_- = (E_{R-} - iE_I)X_+ \qquad (12.76d)$$

From Eqs. (12.76a) and (12.76b), the time-dependent parts of the solution are given by

$$T_\pm(t) = T_\pm(0)e^{-i\frac{E_{\lambda\pm}}{\hbar}t} \qquad (12.77)$$

12.6 Asymmetric Spin Flow Of Itinerant Electrons

Note that the eigenenergies $E_{\lambda\pm} = E_{R\pm} + iE_{I\pm}$ are different for each spin state. The real components are given by

$$E_{R+} = E_{K,\text{cond}} - J_{sd}/2 \quad \text{spin up} \tag{12.78}$$

$$E_{R-} = E_{K,\text{cond}} + J_{sd}/2 \quad \text{spin down} \tag{12.79}$$

$E_{K,\text{cond}}$ is the *conduction* kinetic energy term described by Eq. (12.65). Likewise, the imaginary energy components are different for each spin state and are given by

$$E_{I+} = +\alpha B^*_{sd}\mu_P = \alpha J_{sd}/2 \tag{12.80}$$

$$E_{I-} = -\alpha B^*_{sd}\mu_P = -\alpha J_{sd}/2 \tag{12.81}$$

For the Hamiltonians describing space, the complex terms in the last two equations given by Eqs. (12.76c) and (12.76d), describing $X_\pm$, drop out of the left and right hand sides. Eqs. (12.76c) and (12.76d) then become

$$-\frac{\hbar^2}{2m}\frac{d^2 X_+}{dx^2} - B^*_{sd}\mu_P X_+ = \frac{\hbar^2 k_+^2}{2m} X_+ = E_{R+} X_+ \tag{12.82a}$$

$$-\frac{\hbar^2}{2m}\frac{d^2 X_-}{dx^2} + B^*_{sd}\mu_P X_- = \frac{\hbar^2 k_-^2}{2m} X_- = E_{R-} X_- \tag{12.82b}$$

The LHS is just the sum of the conduction kinetic energy $E_{K,\text{cond}}$ and the exchange energy. We can write the wave numbers as

$$k_\uparrow^2 = \frac{2m}{\hbar^2}\left(E_{K,\text{cond}} - B^*_{sd}\mu_P\right) = \frac{2m E_{K,\text{cond}}}{\hbar^2}\left(1 - \frac{J_{sd}}{2E_{K,\text{cond}}}\right) \tag{12.83a}$$

$$k_-^2 = \frac{2m}{\hbar^2}\left(E_{K,\text{cond}} + B^*_{sd}\mu_P\right) = \frac{2m E_{K,\text{cond}}}{\hbar^2}\left(1 + \frac{J_{sd}}{2E_{K,\text{cond}}}\right) \tag{12.83b}$$

This means the spin-dependent wave vectors in a magnetic conductor differ for spin-up and spin-down because of the exchange energy in a magnetic layer. The conduction energy, therefore, has a spin-dependent splitting between spin-up and spin-down. $k_\pm^2$ must be considered carefully because of the different possibilities given by $E_{K,\text{cond}}$. Recall that it can be either positive (e.g. a good conductor) or negative (e.g. a strong insulator). Let us consider the consequences on Eqs. (12.83a) and (12.83b), for these two extreme cases.

Assume $E_{K,\text{cond}} > 0$ and $J_{sd} > 0$. Then, we can consider the ratio given by

$$\frac{k_+^2}{k_-^2} = \frac{1 - \frac{J_{sd}}{2E_{K,\text{cond}}}}{1 + \frac{J_{sd}}{2E_{K,\text{cond}}}} < 1 \tag{12.84}$$

For this ratio, it follows that

$$k_+^2 < k_-^2 \tag{12.85}$$

In this case, *only* k_+^2 has the possibility of being negative. Therefore, spin-down electrons conduct more efficiently through a magnetic conductor. Now, let's consider the insulator case where $E_{K,\text{cond}} < 0$. Then, it follows that

$$k_+^2 = \frac{-2m|E_{K,\text{cond}}|}{\hbar^2}\left(1 + \frac{J_{sd}}{2|E_{K,\text{cond}}|}\right) \tag{12.86}$$

$$= -|k_{\text{cond}}^2|\left(1 + \frac{J_{sd}}{2|E_{K,\text{cond}}|}\right) \tag{12.87}$$

This is a *negative-definite quantity*. Therefore, the spin-up evanescent state resulting from $k_{\text{cond}}^2 < 0$ becomes more pronounced in scattering, and thus, less conductive in the magnetic layer, and more sharply evanescent at the interface. For spin-down, we have

$$k_-^2 = \frac{-2m|E_{K,\text{cond}}|}{\hbar^2}\left(1 - \frac{J_{sd}}{2|E_{K,\text{cond}}|}\right) \tag{12.88}$$

$$= -|k_{\text{cond}}^2|\left(1 - \frac{J_{sd}}{2|E_{K,\text{cond}}|}\right) \tag{12.89}$$

For the spin-down state, if $J_{sd} > 2|E_{\text{cond}}|$, contrary to spin-up, it follows that $k_-^2 > 0$, since the term in parentheses becomes negative. In general, it follows (again) that

$$k_+^2 < k_-^2 \tag{12.90}$$

Since this result follows from *both* conditions, we have a general result given by Ineq. (12.90).

> ® *Asymmetric spin conduction*: When electrons conduct through a magnetic material, the spin-down state conducts more efficiently than spin-up state.

12.6 Asymmetric Spin Flow Of Itinerant Electrons

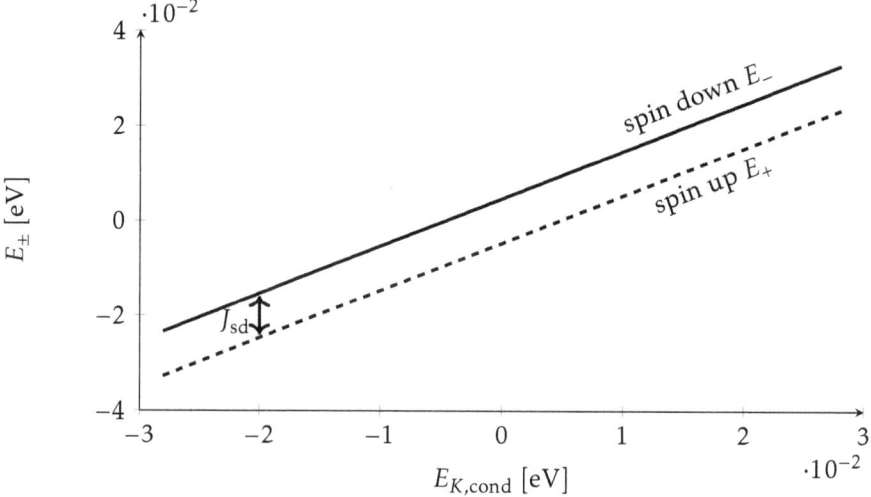

Figure 12.10: Spin-up and spin-down energies depending on the conduction wave number $k^2 = k_{\text{cond}}^2$. An exchange value of $J_{sd} = 1.5 \times 10^{-21}$ J $\approx 1 \times 10^{-2}$ eV is used.

Fig. 12.10 plots E_+ and E_- as functions of k_{cond}. The two spin-dependent energies are proportional to conduction kinetic energy $E_{K,\text{cond}}$, with a splitting or energy difference given by $E_- - E_+ = J_{sd}$. The corresponding plots of $|k_+|$ and $|k_-|$ are shown in Fig. 12.11. In the limit of $E_{K,\text{cond}} \to \pm\infty$,

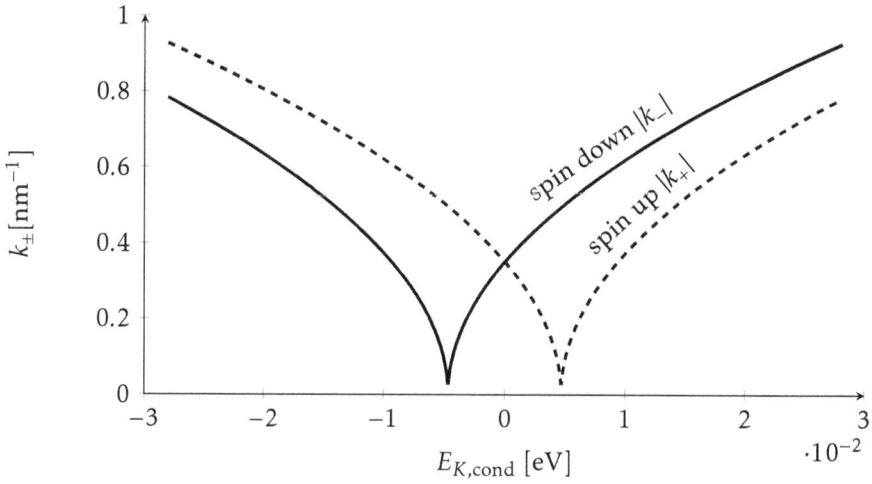

Figure 12.11: Corresponding wave-numbers for spin-up and spin-down depending on the ordinary conduction kinetic energy $E_{K,\text{cond}}^2$.

note that both curves for $k_\pm$ converge to one another. With both wave numbers, the general solutions are given by

$$X_+ = A_+ e^{ik_+ x} + B_+ e^{-ik_+ x} \tag{12.91a}$$

$$X_- = A_- e^{-ik_- x} + B_- e^{-ik_- x} \tag{12.91b}$$

$A_\pm$ and $B_\pm$ are amplitudes determined from boundary conditions at the interfaces between the normal metal and the magnet. Since we also have $T(t)$, we have the complete spinor wave functions for electronic transport in the magnetic layer, given by

$$\Psi_+ = \left(A_+ e^{ik_+ x} + B_+ e^{-ik_+ x}\right) T_+(0) e^{-iE_{R+} t/\hbar} e^{E_{I+} t/\hbar} \tag{12.92a}$$

$$\Psi_- = \left(A_- e^{ik_- x} + B_- e^{-ik_- x}\right) T_-(0) e^{-iE_{R-} t/\hbar} e^{-E_{I-} t/\hbar} \tag{12.92b}$$

We have found that *asymmetric scattering* results from electron transport in a magnetic layer. It arises from the exchange interaction between the spins. Spin-down electrons generally conduct more favorably than spin-up. In the following subsections, we'll consider some consequences of this asymmetric transport. But first, let's take a look at an estimation for the value of J_{sd}, which we have assumed above.

12.6.3 Coarse Estimation of Exchange

We may estimate the order of magnitude of both J_{sd} as well as $k_+ = 1/\delta$. To do this, we can use a relation which we have not derived here, but is well known. Specifically, in a magnetic material with *body-centered cubic crystalline structure*, the exchange stiffness parameter A_X is related to the *localized* interatomic exchange J (an upper bound for J_{sd}) by the relation

$$A_X = \frac{2JS^2}{a}$$

In the above, S is the normalized spin magnitude, which is on the order of 1. For cobalt, for example, the lattice constant is $a \approx 0.286$ nm and $A_X \approx 10^{-11} J/m$. This leads to the following for $J \geq J_{sd}$:

$$A_X = \frac{2JS^2}{a} \Rightarrow J = .5aA_x \approx 1.43 \times 10^{-21} \text{Joules}$$

Using $J_{sd} = J$, we can also estimate a decay constant $\delta_+ = 1/k_+$, say for spin-up electrons, in conditions where $k_+^2 < 0$. Substitution of J_{sd} for a

12.6 Asymmetric Spin Flow Of Itinerant Electrons

conduction kinetic energy of 1.5meV gives

$$\frac{1}{\delta_+^2} = |k_+^2| = \left|\frac{2mE_{K,\text{cond}}}{\hbar^2}\left(1 - \frac{J_{sd}}{2E_{K,\text{cond}}}\right)\right| \Rightarrow \delta_+ \approx 5.2 \times 10^{-9} m$$

Thus, the length scale of the decaying of spins is on the order of nanometers. It is on this length scale that these spin-dependent effects are exploited.

12.6.4 Spin Polarization Of Itinerant Electrons

If the wave functions are known, the expectation value of the *itinerant* spin polarization (magnetic moment) along the direction of the localized spin polarization μ_p (+z) of the magnetic layer can be determined. It should be found at the *exit* of the device (region 3), where the reflection amplitude is zero, and therefore, $\psi = E'_\pm e^{ik_3 x}$. The expectation value $\langle \widehat{\mu_z} \rangle$ is given by

$$\langle \widehat{\mu_z} \rangle = \frac{g\mu_B}{2} \frac{\int_V \begin{bmatrix}\psi_+^* & \psi_-^*\end{bmatrix}\begin{bmatrix}1 & 0 \\ 0 & -1\end{bmatrix}\begin{bmatrix}\psi_+ \\ \psi_-\end{bmatrix} dV}{\int_V \begin{bmatrix}\psi_+^* & \psi_-^*\end{bmatrix}\begin{bmatrix}\psi_+ \\ \psi_-\end{bmatrix} dV} \tag{12.93}$$

The integration is carried out over the volume of region 3, which is nonmagnetic. Then, using the solution in region 3 for $\psi_\pm$, evaluation of Eq. (12.93) leads to

$$\langle \widehat{\mu_z} \rangle = \frac{g\mu_B}{2} \frac{\int_V \psi_+^* \psi_+ - \psi_-\psi_-^* dV}{\int_V \psi_+^* \psi_+ + \psi_-^* \psi_- dV} \tag{12.94}$$

$$= \frac{g\mu_B}{2} \frac{\int_V |E'_+|^2 - |E'_-|^2 dV}{\int_V |E'_+|^2 + |E'_-|^2 dV} \tag{12.95}$$

$$= \frac{g\mu_B}{2} \frac{|E'_+|^2 - |E'_-|^2}{|E'_+|^2 + |E'_-|^2} \tag{12.96}$$

The amplitudes $E'_\pm$ depend on the wave numbers k_1, k_2, k_3 for regions 1, 2, and 3 respectively, as well as the thickness of the magnetic layer. It is k_2 that corresponds to the magnetic layer (see Fig. 12.9). For $E'_\pm$, we can take advantage of the mathematical similarities between this problem and the rectangular potential barrier discussed in Chap. 6. In

Sec. 6.3, the most general expression for $E' = E'_\pm$ for an electron through a rectangular potential barrier was found to be

$$E'_\pm = \frac{4k_1 k_\pm e^{-ik_3 L}}{(k_1 + k_\pm)(k_\pm + k_3)e^{-ik_\pm L} + (k_1 - k_\pm)(k_\pm - k_3)e^{ik_\pm L}} \quad (12.97)$$

The spin polarization ranges from $\langle \mu_z \rangle = 0$, when $E_+ = E_-$ to $\langle \mu_z \rangle = g\mu_B/2 \approx \mu_B$. Fig. 12.12 illustrates two different examples of $\langle \widehat{\mu_z} \rangle$ using Eq. (12.96). $\langle \mu_z \rangle$ is also known as the **spin polarization** P. The spin

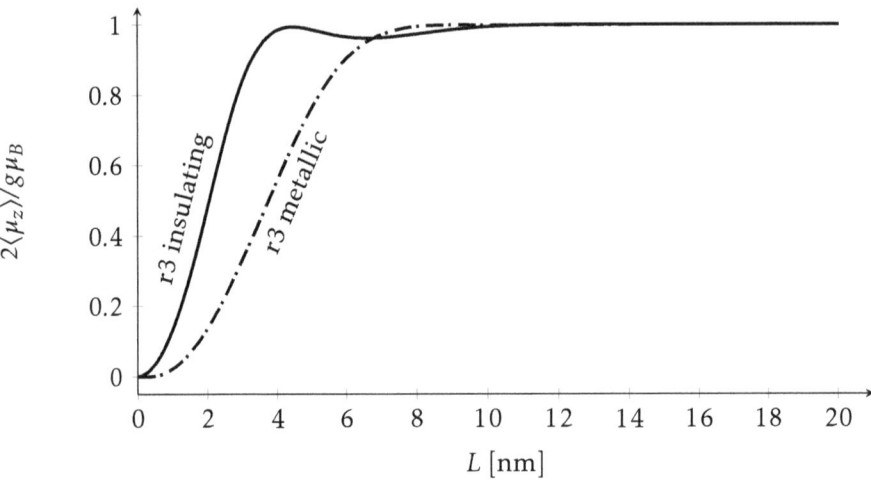

Figure 12.12: Itinerant spin polarization given by Eq. (12.96) as function of the magnetic layer thickness L. Solid curve uses insulating conditions in region 3, where $k_3 = ik'_3$, while the other uses metallic.

polarization is positive and as Fig. 12.12 demonstrates, increases with increasing magnetic layer thickness up to a point of saturation. Between both curves shown, the wave number in region 3 differs. In one case, k_3 is set as a *real* wave number, representing metallic conditions in region 3. The other case uses an imaginary wave number $k_3 = ik'_3$, which is electrical insulating conditions. One finds that insulating conditions can create larger spin polarization. The itinerant spin polarization is therefore also sensitive to the subsequent layer after the magnetic layer.

The spin polarization given by Eq. (12.96) can also be expressed in another form using spin-dependent transmission probabilities T_+ and T_-. Ordinarily, without spin dependent energies, the transmission probability for the potential barrier is given by

$$T = \frac{J_T}{J_I} = \frac{k_3}{k_1}|E'|^2 \quad (12.98)$$

12.6 Asymmetric Spin Flow Of Itinerant Electrons

In this case, however, there are two distinct transmission probabilities corresponding to spin up $T_+ = \frac{k_3}{k_1}|E'_+|^2$ and spin-down $T_- = \frac{k_3}{k_1}|E'_-|^2$. Using this relation, the **spin polarization** δ_P defined as

Itinerant Electron Spin Polarization

$$P = \frac{g\mu_B}{2}\frac{T_+ - T_-}{T_+ + T_-} \tag{12.99}$$

Table 12.1 lists some measured spin polarization values for bulk (thick) magnetic metals taken from Meservey and P.M. Tedrow 1994.

Table 12.1: Measured itinerant spin polarizations P for some magnetic materials. From Table 2, pg. 204 of R. Meservey and P.M. Tedrow 1994 [7].

Material	bulk spin polarization P
Fe	0.40
Co	0.35
Ni	0.23
Gd	0.14
Ho	0.075
Dy	0.070
Tb	0.0655
Er	0.055
Tm	0.027

In determining the spin polarization, we saw that *the adjacent layer to the magnetic layer can influence the extent of the spin polarization*. This fact leads to a novel application of the asymmetric spin transport through a magnetic layer, since there is preference to spin orientation. The application we are alluding to, which is widely used today, is known as *magnetoresistance*. It is the topic of the next section.

12.7 Introduction To Magnetoresistance

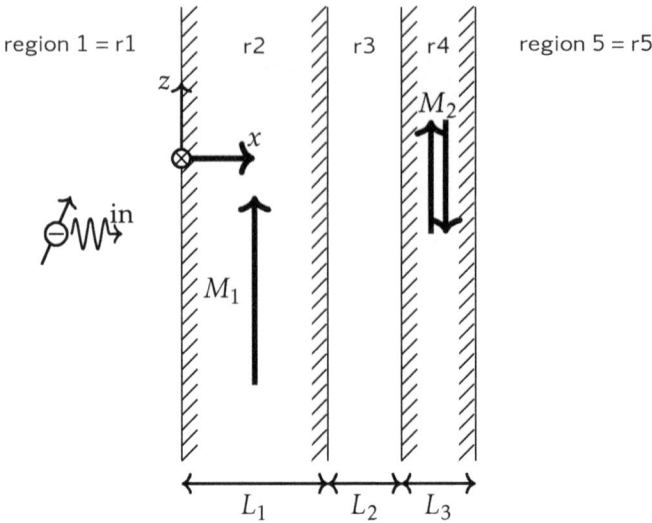

Figure 12.13: Illustration of a magnetoresistance (MR) device, with two magnetic layers, a nonmagnetic layer between them, and two contacts.

Given asymmetric transport of electrons conducting in a magnetic layer, the different wave numbers k_+ and k_- for spin-up and spin-down leads to different electrical currents. This also means there will be differences in the electrical resistances/conductivities for both spin states. This effect can be exploited in a device to provide information on the magnetic layers. For such a device, we need to extend the ideas by considering *two magnetic layers sandwiching a nonmagnetic layer*, as illustrated in Fig. 12.13. One of two magnetic layers is designed so that the magnetization can be controlled to be in one of two possible states (represented as r4). In this device, there are a total of five layers that are taken into consideration. They consist of *electrical contacts* on the extremities of the device (regions 1 and 5), which are nonmagnetic. Between the contacts, there are three layers forming a magnetic|nonmagnetic|magnetic structure. For this device, we can carry out an analysis analogous to what was done in the rectangular barrier problem solved in Sec. 6.3. The general form for the wave functions describing a quasi-free conducting electron is given by

$$X = Ae^{ikx} + Be^{-ikx} \tag{12.100}$$

The first term is the forward propagating wave component and the second term is the reflected wave component, and therefore, backwards

12.7 Introduction To Magnetoresistance

propagating component. Only one of the contacts on the far end has no reflecting wave component, namely, at the exit in r5. Thus, for the five layers, the respective wave functions can be written as

$$X_1(x) = Ae^{ikx} + Be^{-ikx} \tag{12.101a}$$

$$X_2(x) = Ce^{ikx} + De^{-ikx} \tag{12.101b}$$

$$X_3(x) = Ee^{ikx} + Fe^{-ikx} \tag{12.101c}$$

$$X_4(x) = Ge^{ikx} + He^{-ikx} \tag{12.101d}$$

$$X_5(x) = Ie^{ikx} \tag{12.101e}$$

The wave functions may be used to determine the transmission probability T of the structure shown in Fig. 12.13. T is given by the final exiting wave function or $X_5(x)$. The wave component amplitudes are obtained using *boundary conditions*. We set corresponding wave functions and their first derivatives equal at the interfaces between adjacent layers. This includes $x = 0, L_1, L_1 + L_2$, and $L_1 + L_2 + L_3$. In the above equations, there are nine unknown amplitudes given by A, B, C, D, E, F, G, H and I. However, we can express eight equations at the four interfaces at which two equations are satisfied. To enable unique solutions, we divide all the equations by the incoming wave amplitude A (to reduce the number of unknowns from 9 to 8). This gives

$$1 + B' = C' + D' \tag{12.102a}$$

$$1 - B' = \frac{k_2}{k_1}[C' - D] \tag{12.102b}$$

$$C'e^{ik_2 L_1} + D'e^{-ik_2 L_1} = E'e^{ik_3 L_1} + F'e^{-ik_3 L_1} \tag{12.102c}$$

$$C'e^{ik_2 L_1} - D'e^{-ik_2 L_1} = \frac{k_3}{k_2}\left[E'e^{ik_3 L_1} - F'e^{-ik_3 L_1}\right] \tag{12.102d}$$

$$E'e^{ik_3(L_1+L_2)} + F'e^{-ik_3(L_1+L_2)} = G'e^{ik_4(L_1+L_2)} + H'e^{-ik_4 L_1} \tag{12.102e}$$

$$E'e^{ik_3(L_1+L_2)} - F'e^{-ik_3(L_1+L_2)} = \frac{k_4}{k_3}\left[G'e^{ik_4(L_1+L_2)} - H'e^{-ik_4(L_1+L_2)}\right] \tag{12.102f}$$

$$G'e^{ik_4(L_1+L_2+L_3)} + H'e^{-ik_4(L_1+L_2+L_3)} = I'e^{ik_5(L_1+L_2+L_3)} \tag{12.102g}$$

$$G'e^{ik_4(L_1+L_2+L_3)} - H'e^{-ik_4(L_1+L_2+L_3)} = \frac{k_5}{k_4}I'e^{ik_4(L_1+L_2)} \tag{12.102h}$$

The prime denotes division by A. We can make certain combinations with the above equations to obtain the relations that provide the amplitudes $B', C', D', E', F', G', H'$, and I'. We'll use the following combinations:

$$
\begin{aligned}
(12.102\text{g}) + (12.102\text{h}) &\Rightarrow \quad G' \\
(12.102\text{g}) - (12.102\text{h}) &\Rightarrow \quad H' \\
(12.102\text{e}) + (12.102\text{f}) &\Rightarrow \quad E' \\
(12.102\text{e}) - (12.102\text{f}) &\Rightarrow \quad F' \\
(12.102\text{c}) + (12.102\text{d}) &\Rightarrow \quad C' \\
(12.102\text{c}) - (12.102\text{d}) &\Rightarrow \quad D' \\
(12.102\text{a}) + (12.102\text{b}) &\Rightarrow \quad I' \\
(12.102\text{a}) - (12.102\text{b}) &\Rightarrow \quad B'
\end{aligned}
$$

From the above combinations, the solutions are found to be

$$G' = \frac{1}{2}\left(1 + \frac{k_5}{k_4}\right)I'e^{ik_5(L_1+L_2+L_3)}e^{-ik_4(L_1+L_2+L_3)} \tag{12.103a}$$

$$H' = \frac{1}{2}\left(1 - \frac{k_5}{k_4}\right)I'e^{ik_5(L_1+L_2+L_3)}e^{ik_4(L_1+L_2+L_3)} \tag{12.103b}$$

$$E' = \frac{1}{2}\left[\left(1 + \frac{k_4}{k_3}\right)G'e^{ik_4(L_1+L_2)} + \left(1 - \frac{k_4}{k_3}\right)H'e^{-ik_4(L_1+L_2)}\right]e^{-ik_3(L_1+L_2)} \tag{12.103c}$$

$$F' = \frac{1}{2}\left[\left(1 - \frac{k_4}{k_3}\right)G'e^{ik_4(L_1+L_2)} + \left(1 + \frac{k_4}{k_3}\right)H'e^{-ik_4(L_1+L_2)}\right]e^{ik_3(L_1+L_2)} \tag{12.103d}$$

$$C' = \frac{1}{2}\left[\left(1 + \frac{k_3}{k_2}\right)E'e^{ik_3L_1} + \left(1 - \frac{k_3}{k_2}\right)F'e^{-ik_3L_1}\right]e^{-ik_2L_1} \tag{12.103e}$$

$$D' = \frac{1}{2}\left[\left(1 - \frac{k_3}{k_2}\right)E'e^{ik_3L_1} + \left(1 + \frac{k_3}{k_2}\right)F'e^{-ik_3L_1}\right]e^{ik_2L_1} \tag{12.103f}$$

12.7 Introduction To Magnetoresistance

$$B' = \frac{1}{2}\left[\left(1 - \frac{k_2}{k_1}\right)C' + \left(1 + \frac{k_2}{k_1}\right)D'\right] \quad (12.103g)$$

The above solutions can be used to obtain I', which provides the transmission probability T given by $T = hk_5|I'|^2/m$. Note that because there are two spin states, there are two transmission probabilities, corresponding to spin-up and spin-down. Thus, we have

$$T_+ = hk_5|I'_+|^2/m \quad \text{and} \quad T_- = hk_5|I'_-|^2/m \quad (12.104)$$

Although I' is the relative amplitude of the wave function in region 5, which is nonmagnetic, because it depends on the amplitudes of all the previous layers, *there is propagation of the spin splitting behavior* into region 5. Because of this, the amplitude I' has a corresponding I'_+ and I'_-. Then, in region 5, there are also spin-up and spin-down wave functions $X_+(x) = I'_+ e^{ik_5 x}$ and $X_-(x) = I'_- e^{ik_5 x}$, respectively.

In an experiment with an MR device, the voltage and/or current are measured across the contacts (regions 1 and 5) of the device. If an external voltage is applied, then the total current (proportional to the current density) can be found. For this case, we need the expectation value of the total current density $\langle J \rangle$ through the device as a function of the magnetic layer orientations. The expectation value of the total current density across the device is then given by

$$\langle J \rangle = \frac{J_+|X'_+|^2 + J_-|X'_-|^2}{|X'_+|^2 + |X'_-|^2} = \frac{J_+|I'_+|^2 + J_-|I'_-|^2}{|I'_+|^2 + |I'_-|^2} \quad (12.105)$$

For the *magnetoresistance*, $\langle J \rangle$ is computed for the two particular cases of interest: (1) parallel magnetization layers $(M_1 \parallel M_2) \to \langle J_p \rangle$ and (2) anti-parallel magnetization layers $(M_1 \parallel -M_2) \to \langle J_{ap} \rangle$. These currents can be used to define a **dimensionless magnetoresistance fraction** (MR fraction) as

$$\text{MR} = \frac{\langle J_p \rangle - \langle J_{ap} \rangle}{\langle J_{ap} \rangle} = \frac{\langle G_p \rangle - \langle G_{ap} \rangle}{\langle G_{ap} \rangle} = \frac{\langle R_{ap} \rangle - \langle R_p \rangle}{\langle R_p \rangle} \quad (12.106)$$

$G = 1/R$ is the bulk conductance across the device and R is the corresponding resistance. It is often multiplied by 100 to give what is known as the **magnetoresistance ratio** (MR ratio) defined as

> **MR Ratio**
>
> $$\text{MR} = \frac{\langle R_{ap} \rangle - \langle R_p \rangle}{\langle R_p \rangle} \times 100 \qquad (12.107)$$

We have the freedom to vary the properties in region 3, located between both magnetic layers, to observe the effect on the *magnetoresistance*. The properties of such a sandwiched region turn out to be very important. One parameter of the layer which we shall vary is the conduction eigenenergy, which we'll denote by $E^3_{K,\text{cond}}$. Fig. 12.14 illustrates the calculated *magnetoresistance fraction* (c.Eq. 12.106) as a function of $E^3_{K,\text{cond}}$. Note that for conduction electrons in a real device, the energies tend

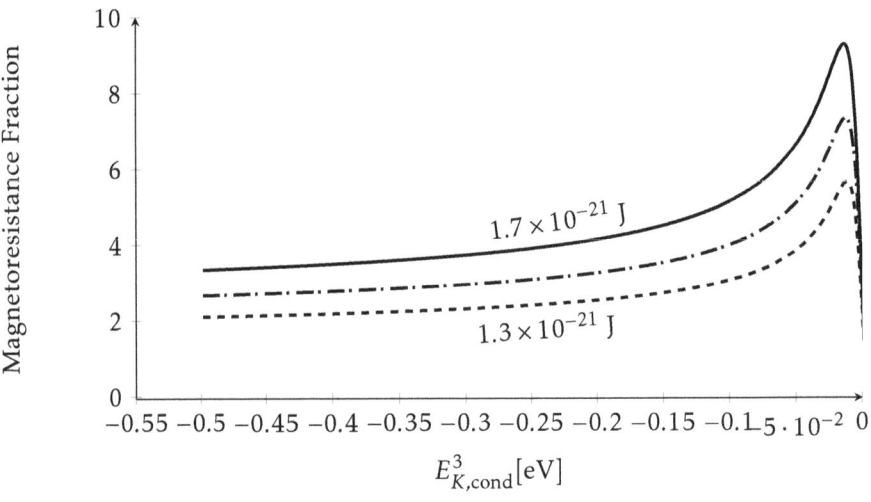

Figure 12.14: Computed magnetoresistance (MR) as a function of the eigenenergy of region 3 $E^3_{K,\text{cond}}$, or the layer between M_1 and M_2, for different exchange $J_{sd} = 1.3, 1.5$ (middle), and 1.7×10^{-21} J. $L_1 = 10$nm and $L_3 = 2$nm. $E^3_{K,\text{cond}}$ is determined by using Eq. (12.64), which includes an external energy $qU_{ext} = .5$meV

to larger negative values (with smaller absolute values since electrons are still bound to the lattice) to smaller energies with larger absolute values. This is because we did *not* express the eigenenergy relative to the Fermi level ϵ_F. Therefore, in this range, the model suggests that as $E^3_{K,\text{cond}}$ becomes more negative, the larger the MR fraction. Closer to the origin represents *more metallic conditions*, while further negative energy

12.7 Introduction To Magnetoresistance

corresponds to more *insulating conditions*. Therefore, the spacer layer becomes an **insulating barrier**. In region 3, the wave number is given by

$$k_3^2 = \frac{2mE_{K,cond}^3}{\hbar^2}$$

As the eigenenergy becomes more negative, then $k_3^2 \ll 0$. Then, the wave function in region 3 becomes more evanescent because $k_3 = ik_3' \rightarrow \psi = e^{ik_3 x} = e^{-k_3' x}$.

We can also consider how the thickness L_2 of the spacer layer affects spin-transport. Fig. 12.15 illustrates the *magnetoresistance* (also in fraction form given by Eq. 12.106), as a function of region three layer thickness L_2. Also from Fig. 12.15, stronger insulating conditions are

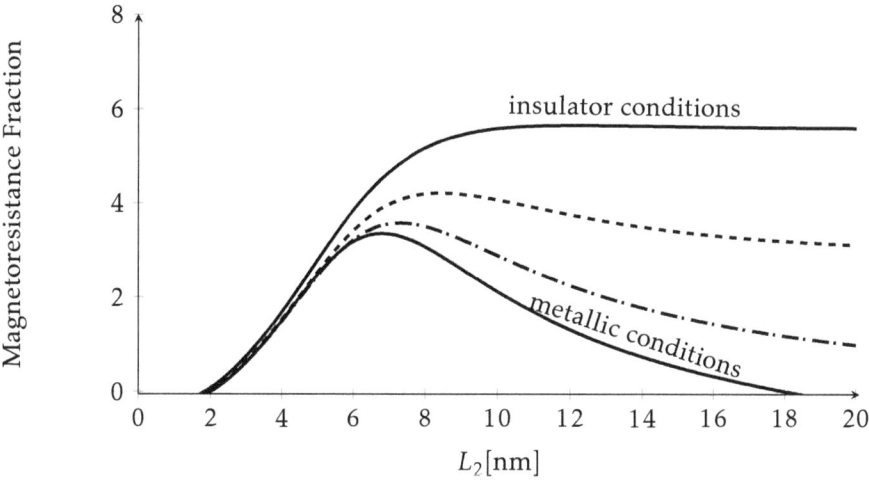

Figure 12.15: Computed magnetoresistance as a function of region 3 thickness, labeled as L_2.

seen to lead to larger MR fractions over the whole range of thicknesses. Metallic conditions tend to rise and then decrease with thickness, while insulating conditions tend to monotonically increase with thickness.

Generally, the more insulating, the higher the MR. The origin of this from this model are related to the asymmetric behavior in the current density between the parallel and anti-parallel states. The reader is encouraged to implement this model in software and explore the solutions. Most of the computed results shown in this text were done using

Python computing language. But, any scientific programming language of choice should do. Because these conditions correspond most to an **electrical insulating barrier** layer, it gives rise to stronger *tunneling* (c.Chap.6). The *magnetoresistance ratio* in this case is also known as **tunneling magnetoresistance ratio** or TMR. A device using an insulating layer between the magnetic layers is known as a **TMR device** or a **magnetic tunnel junction** (MTJ).

One of the first insulating barrier materials investigated in MTJs was alumina or aluminum oxide (Al_2O_3). This compound is more abundant and very stable, and this is one of the reasons it is very widely used in many applications. This property, however, becomes a detriment in this application, because conduction properties cannot be altered easily. The most notable insulating barrier example in an MTJ is magnesia or magnesium oxide MgO. Just as in the case of Al_2O_3, as an insulator, there is generally a sizable gap between the *valence* and *conduction* bands. This imposes an effectively large potential energy that must be overcome by any conduction band electrons (which are supplied by the valence band). However, MgO readily undergoes a unique modification to its valence shell due to an interplay between magnesium Mg, whose valence electrons ordinarily are $3s^2$, and singly oxygen, whose valence electrons are $2p^4$. In MgO, the two electrons in Mg can lower their energies by occupying the two empty 2p orbitals of oxygen (known as charge transfer). This effectively widens the conduction band and has consequences on the conduction properties in these conditions. When this material is sandwiched between two magnetic layers, TMR is enhanced compared to Al_2O_3.

Additionally, in a physical device, similar effects occur at the interfaces, where, if the magnetic layers are, say Fe, then Fe-O forms. Fe has two valence electrons in 4s (recall Mg has two in 3s), so very similar conditions are created at the interfaces. The combination of interface and bulk MgO properties leads to enhanced TMR ratios. Detailed analysis of these special attributes of MgO were first predicted *c.*2000 by William Butler *et. al* while at Oak Ridge National Laboratory in Tennessee, USA. It was revealed by a numerical study of the transport properties of a magnetic tunnel junction with structure Fe|MgO|Fe. Experimental demonstrations verifying this fact with MgO followed in 2004. For the barrier, these details require a relatively complex calculation. The simplest kind has been illustrated here.

12.8 Spin Diffusion Equations

For more general geometries, it can be useful to solve the differential equations describing the physics of asymmetric spin transport. It is also relevant to more kinds of materials than magnetic, for example, in materials with appreciable spin-orbit interaction (see Sec. 11.5), in comparison to unit thermal fluctuation energy $k_B T$. k_B is the Boltzmann constant. One of the first theories rigorously derived for this purpose was done using what is called the Boltzmann transport equation (BTE). It was derived from the BTE in 1993 by French physicists Thierry Valet and Albert Fert. Their theory is known as the *Valet-Fert theory*. Their differential equations in space will, however, be derived here from quantum mechanics along with explicit expressions for the associated parameters. We will obtain the parabolic spatio-temporal spin transport equations, as well.

12.8 SPIN DIFFUSION EQUATIONS

With a splitting in the kinetic energies of the spin-up and spin-down states, the electron motion can be regarded as having two *parallel* currents. One spin-up and the other spin-down. This assumes there is no mixing or correlation between the two states. We found that a ferromagnet essentially *injects* two different *spin currents* into a subsequent nonmagnetic region. In Prob. 6.8 of Chap. 6 problems, the reader is asked to show that for a wave function of the form

$$\psi(x) = A e^{ikx} + B e^{-ikx}$$

The *current density operator* $\widehat{J}$ gives

$$J = \frac{\hbar k}{m}\left(|A|^2 - |B|^2\right)$$

m is the electron mass. Note that this form of wave function is more applicable to metals, where $k^2 > 0$. This means that *the analysis here applies to conducting materials*. From this result, we can see that the currents are independent of space. Consequently, nothing interesting follows if we stick with this wave function. However, when we include the *time-dependent component resulting from the non-Hermitian Hamiltonian*, a deeper description emerges. For this, we must begin with the complete spatio-temporal spin-up and down wave functions given by

$$\Psi_+ = \left(A_+ e^{ik_+ x} + B_+ e^{-ik_+ x}\right) T_+(0) e^{-iE_{R+}t/\hbar} e^{+\alpha J_{sd} t/2\hbar} \qquad (12.108a)$$

$$\Psi_- = \left(A_- e^{ik_- x} + B_- e^{-ik_- x}\right) T_-(0) e^{-iE_{R-}t/\hbar} e^{-\alpha J_{sd} t/2\hbar} \tag{12.108b}$$

$E_{R\pm}$ is the respective real or Hermitian eigenenergy. Using Eqs. (12.108), the modified spin currents then become

$$\widehat{J}_+ = \frac{\hbar k_+}{m} \left(|A_+|^2 - |B_+|^2\right) e^{+\alpha J_{sd} t/\hbar} = K_+ e^{+\alpha J_{sd} t/\hbar} \tag{12.109a}$$

$$\widehat{J}_- = \frac{\hbar k_-}{m} \left(|A_-|^2 - |B_-|^2\right) e^{-\alpha J_{sd} t/2\hbar} = K_- e^{-\alpha J_{sd} t/2\hbar} \tag{12.109b}$$

Since the itinerant electrons have effective motion in an electrical current, through k_+ and k_-, there is a total expectation velocity-like term $\langle v \rangle$ which permits a relation to time t as

$$x = \langle v \rangle t = \frac{\hbar \langle k \rangle}{m} t \Rightarrow t = \frac{m}{\hbar \langle k \rangle} x \tag{12.110}$$

The expectation value $\langle v \rangle$ corresponds to the total effective electrical current. Substitution of t into Eqs. (12.109) gives

$$\widehat{J}_+ = K_+ e^{+\alpha J_{sd} mx/\hbar^2 \langle k \rangle} = K_+ e^{x/\ell_{sf}} \tag{12.111a}$$

$$\widehat{J}_- = K_- e^{-\alpha J_{sd} mx/\hbar^2 \langle k \rangle} = K_- e^{-x/\ell_{sf}} \tag{12.111b}$$

Eqs. (12.111) defines a **spin-diffusion length** ℓ_{sf} as

Spin Diffusion Length

$$\ell_{sf} = \frac{\hbar^2 \langle k \rangle}{\alpha J_{sd} m} \tag{12.112}$$

The *spin diffusion length* ℓ_{sf} is a measure of the length scale over which either spin state decays or amplifies, respectively. It is fundamental to spin-states in general, in any environment with spin-dependent energy. In *ordinary* electro and magnetostatics, the ordinary current density $\mathbf{J}_o$ is given by

$$\mathbf{J}_o = \sigma \mathbf{E}_o \tag{12.113}$$

σ is the electrical conductivity. The electric field is given by

$$\mathbf{E}_o = -\frac{\partial \mu}{\partial x} \quad \text{or in 3D} \quad \mathbf{E}_o = -\nabla \mu$$

12.8 Spin Diffusion Equations

∇ is the conventional gradient operator. Here, μ is the *electric potential* having units of volts, or energy per unit charge. With two distinct spin currents, this relation requires resolving at the level of the electron spin, rather than just at the charge level. For two spin currents, we can write

$$\mathbf{J}_+ = \sigma_+ \mathbf{E}_+ = -\sigma_+ \frac{\partial \mu_+}{\partial x} \qquad (12.114a)$$

$$\mathbf{J}_- = \sigma_- \mathbf{E}_- = -\sigma_- \frac{\partial \mu_-}{\partial x} \qquad (12.114b)$$

$\sigma_\pm$ are the respective spin-dependent conductivities. Note $\mathbf{J}_o \neq \mathbf{J}_+ + \mathbf{J}_-$. These are parallel currents, not currents that superpose. Instead,

$$\mathbf{J}_o = \frac{\mathbf{J}_+ |\Psi_+|^2 + \mathbf{J}_- |\Psi_-|^2}{|\Psi_+|^2 + |\Psi_-|^2} \qquad (12.115)$$

From Eqs. (12.114), we obtain consistent definitions for spin potentials given by

Spin Potential Definitions

$$\mu_+ = -\frac{1}{\sigma_+} \int \mathbf{J}_+(\mathbf{r}) \cdot d\mathbf{r} \quad \text{and} \quad \mu_- = -\frac{1}{\sigma_+} \int \mathbf{J}_-(\mathbf{r}) \cdot d\mathbf{r} \qquad (12.116)$$

Using the above definitions with the solutions for current density $J_\pm$, the spin potentials in 1D become

$$\mu_+ = -\frac{K_+ \ell_{sd}}{\sigma_+} e^{x/\ell_{sd}} + \bar{\mu}_{c+} \qquad (12.117a)$$

$$\mu_- = +\frac{K_- \ell_{sd}}{\sigma_-} e^{-x/\ell_{sd}} + \bar{\mu}_{c-} \qquad (12.117b)$$

$\bar{\mu}_{c\pm}$ are just integration constants. They can be chosen by assuming that at $x = 0$, $\bar{\mu}_+ = \bar{\mu}_-$, which leads to

$$\bar{\mu}_{c+} = \bar{\mu}_{c-} = \bar{\mu}_c \qquad (12.118)$$

Then, we have

$$\mu_+ - \bar{\mu}_c = -\frac{K_+ \ell_{sd}}{\sigma_+} e^{x/\ell_{sd}} \qquad (12.119a)$$

$$\mu_- - \bar{\mu}_c = +\frac{K_-\ell_{sd}}{\sigma_-}e^{-x/\ell_{sd}} \tag{12.119b}$$

By determining $\partial J_\pm/\partial x$ using Eqs. (12.111), we also have

$$\frac{\partial J_+}{\partial x} = \frac{K_+}{\ell_{sd}}e^{x/\ell_{sd}} = -\sigma_+ \frac{\partial^2 \mu_+}{\partial x^2} \tag{12.120a}$$

$$\frac{\partial J_-}{\partial x} = -\frac{K_+}{\ell_{sd}}e^{-x/\ell_{sd}} = -\frac{\partial^2 \mu_-}{\partial x^2} \tag{12.120b}$$

From this result, we have

$$\frac{\partial^2 \mu_+}{\partial x^2} = -\frac{K_+}{\sigma_+ \ell_{sd}}e^{x/\ell_{sd}} \tag{12.121a}$$

$$\frac{\partial^2 \mu_-}{\partial x^2} = \frac{K_-}{\sigma_- \ell_{sd}}e^{-x/\ell_{sd}} \tag{12.121b}$$

If we compare this to Eqs. (12.119), we then have

$$\frac{\partial^2 \mu_+}{\partial x^2} = \frac{\mu_+ - \bar{\mu}_c}{\ell_{sd}^2} \tag{12.122a}$$

$$\frac{\partial^2 \mu_+}{\partial x^2} = \frac{\mu_- - \bar{\mu}_c}{\ell_{sd}^2} \tag{12.122b}$$

The 3D version of this pair of differential equations becomes

Spin Diffusion Equations-Form 1

$$\nabla^2 \mu_+ = \frac{\mu_+ - \bar{\mu}_c}{\ell_{sd}^2} \quad \text{and} \quad \nabla^2 \mu_- = \frac{\mu_- - \bar{\mu}_c}{\ell_{sd}^2} \tag{12.123}$$

These spatial differential equations are also known as the **Valet-Fert equations**, originally obtained from a semi-classical equation known as the *Boltzmann transport equation* or *BTE*. Eqs. (12.123) describe the magnetostatic spin potential distributions only in space.

Eqs. (12.123) can be written another way to give *spatio-temporal equations* for the spin potentials μ_+ and μ_-. Since they alternatively depend on time t, differentiation of both with respect to time gives

$$\frac{\partial \mu_+}{\partial t} = \frac{\mu_+}{\tau} = \frac{\alpha J_{sd}}{\hbar}\mu_+ \tag{12.124a}$$

12.8 Spin Diffusion Equations

$$\frac{\partial \mu_-}{\partial t} = \frac{\mu_-}{\tau} = -\frac{\alpha J_{sd}}{\hbar}\mu_- \tag{12.124b}$$

It follows that

$$\frac{\partial (\mu_+ - \bar{\mu}_c)}{\partial t} = \frac{\mu_+ - \bar{\mu}_c}{\tau} = \frac{\alpha J_{sd}}{\hbar}(\mu_+ - \bar{\mu}_c) \tag{12.125a}$$

$$\frac{\partial (\mu_- \bar{\mu}_c)}{\partial t} = \frac{\mu_- - \bar{\mu}_c}{\tau} = -\frac{\alpha J_{sd}}{\hbar}(\mu_- - \bar{\mu}_c) \tag{12.125b}$$

Using Eqs. (12.122) for substitution on the RHS of Eqs. (12.125), we have

$$\frac{\partial \mu_+}{\partial t} = +\frac{\ell_{\pm}^2}{\tau}\frac{\partial^2 \mu_+}{\partial x^2} \tag{12.126a}$$

$$\frac{\partial \mu_-}{\partial t} = -\frac{\ell_{\pm}^2}{\tau}\frac{\partial^2 \mu_-}{\partial x^2} \tag{12.126b}$$

Eq. (12.126) defines the **spin diffusion constant** D_s having units of m^2/s, given by

$$D_s = \frac{\ell_{sd}^2}{\tau} \tag{12.127}$$

Substitution of ℓ_{sd} and τ gives D_s as

Spin Diffusion Constant

$$D_s = \frac{\hbar^5 \langle k \rangle^2}{\alpha^3 J_{sd}^2 m^2} \tag{12.128}$$

The spatio-temporal spin diffusion equation in 3D become

Spin Diffusion Equations-Form 2

$$\frac{\partial \mu_+}{\partial t} - D_s \nabla^2 \mu_+ = 0 \quad \text{and} \quad \frac{\partial \mu_-}{\partial t} + D_s \nabla^2 \mu_- = 0 \tag{12.129}$$

Notice there is symmetry in the spin diffusion equations, identical in form only with differences in sign arising from the asymmetry in time.

It gets mirrored in space. This completes the objective of this section, obtaining the spatial and spatio-temporal differential equations that can be used to describe the spin potentials.

Now imagine that after the spins leave the magnetic layer, they then flow into another magnetic layer with a free-to-move or *free magnetic moment*, while the itinerant spin polarization direction is maintained. We may ask what happens to the *localized* spins in the free magnetic layer as a consequence of coupling to the itinerant electrons? This question was first analyzed around 1995 by more than one person. We'll examine this question, as well, using the tools we have developed in this chapter. This is the subject of the next section.

12.9 SPIN TRANSFER TORQUE ON LOCALIZED ELECTRONS

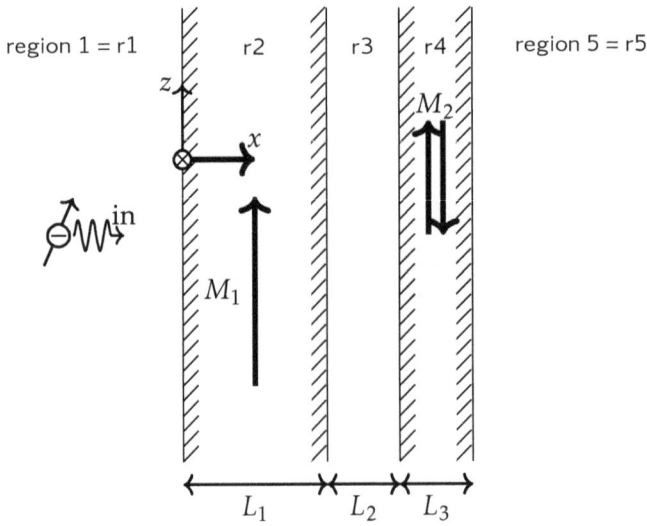

Figure 12.16: Illustration of polarized spin current polarized by M_1, then subsequently injected into a second magnetic layer M_2.

In this section, we examine the potential of what can be done using spin polarized current that passes through a *free* magnetic layer. Since the electrons in the free layer are localized to the atom, we are interested in the resulting dynamics of the localized spins, consequent from interactions with the polarized itinerant spins, interacting through exchange. We must again include the associated dissipation (of the localized spin motion) operator. The itinerant electrons coupling to the localized spin

12.9 Spin Transfer Torque On Localized Electrons

are first spin-polarized by magnetic layer one (region 2), then injected into a nonmagnetic material, which presumably does not alter the spin polarization. The nonmagnetic layer after the spin polarizing layer is also necessary to decouple the two magnetic layers from one another. Otherwise, if one moves, the other may be affected. The objective is to independently manipulate the magnetization of the second magnetic layer M_2. This idea is illustrated in Fig. 12.16.

We first obtain the resulting equation of motion for the localized electrons using quantum mechanical tools. To do this, we *must* use the *Generalized-Generalized Ehrenfest Theorem* or the G^2 Ehrenfest Theorem derived in Sec. 3.10. Thus, we need the Hamiltonian $\widehat{H}$ for the *localized electrons* in M_2, given by

$$\widehat{H}\psi = -(1-i\alpha)\left[\frac{J_{sd}}{2}\widehat{\sigma}\cdot\langle\sigma_p\rangle\right]\psi \tag{12.130}$$

ψ is the 2×1 spinor containing spin-up and spin-down states. α is the dissipation parameter, and J_{sd} is the inter-electron exchange parameter between the itinerant and localized electrons. The quantity $\langle\sigma_p\rangle$ is the expectation value of the itinerant spin polarization. Based on the analysis from Sec. 12.6, we known it depends on the properties (e.g. J_{sd}) of M_1, as well as the electron kinetic energy, etc. In the best case, the itinerant spin polarization is saturated to maximum alignment with M_1. However, generally, the polarization factor P (the dimensionless expectation value of $\langle\mu_p\rangle$) should be used to account for non-saturation of the itinerant spins with M_1. Therefore, to express Eq. (12.130) in terms of the orientation of layer M_1, let us write

$$\widehat{H}\psi = -(1-i\alpha)\left[\frac{J_{sd}P}{2}\widehat{\sigma}\cdot\langle\sigma_p\rangle\right]\psi$$

The fixed spin polarization σ_p is still parallel to magnetic layer M_1. We can use Eq. (3.147) along with Eq. (12.130) to determine an equation of motion for an electron experiencing a polarized itinerant electron spin. Fortunately, because we've encountered an analogous problem already, this time we don't have to work so hard. We can take advantage of the fact that the form of the Hamiltonian can be cast in the same form as the Zeeman Hamiltonian used in Sec. 12.5. So much so, by just redefining the Hamiltonian in terms of an effective exchange magnetic field B_{sd}, the analysis carried out for the exchange energy is identical to that of the Zeeman energy with dissipation. This is because both Hamiltonians are dot products of the spin with a presumed fixed vector in space. In the case of Zeeman energy, the field is the external magnetic field B. Now, we have an internal effective magnetic field B_{sd}. Recall that the Zeeman energy with dissipation, is described by

$$\widehat{H}_Z = -(1 - i\alpha)\boldsymbol{\mu} \cdot \mathbf{B} = \frac{g|\gamma|\hbar B}{2}(1 - i\alpha)\boldsymbol{\sigma} \cdot \mathbf{z}_B$$

$\mathbf{z}_B$ is the unit vector along the direction of the magnetic field. Define the **exchange field** having units of a magnetic field as

---
Exchange Field

$$B_{sd} = -J_{sd}\langle p\rangle/g|\gamma|\hbar \qquad (12.131)$$

Note the negative sign is required to cast the Hamiltonian in identical form to the Zeeman Hamiltionian. Then, we don't have to be concerned about signs, and the analysis will be the same. Eq. (12.130) then becomes

$$\widehat{H} = -(1 - i\alpha)\left[\frac{J_{sd}P}{2}\boldsymbol{\sigma} \cdot \boldsymbol{\sigma}_p\right] = \frac{g|\gamma|\hbar B_{sd}}{2}(1 - i\alpha)\boldsymbol{\sigma} \cdot \boldsymbol{\sigma}_p \qquad (12.132)$$

One key difference between Zeeman and exchange Hamiltonians is the *sign of the Hamiltonians*. They have opposite signs, reflected in B_{sd}.

Zeeman vs Exchange Hamiltonians: Though they have identical form, they also have opposite signs for the same vector orientations. This leads to torque terms with opposite sign to Zeeman effects.

Otherwise, the analyses between the two cases is identical. The time-derivative of the expectation value of $\boldsymbol{\mu}_S$ is determined from use of the relation

$$\frac{d\langle\widehat{\boldsymbol{\mu}}_S\rangle}{dt} = \frac{i}{\hbar}\left[\langle\widehat{H}^\dagger\widehat{\boldsymbol{\mu}}_S\rangle - \langle\widehat{\boldsymbol{\mu}}_S\widehat{H}\rangle\right] - \frac{i}{\hbar}\langle\widehat{\boldsymbol{\mu}}_S\rangle\left[\langle\widehat{H}^\dagger\rangle - \langle\widehat{H}\rangle\right] + \left\langle\frac{\partial\widehat{\boldsymbol{\mu}}_S}{\partial t}\right\rangle \qquad (12.133)$$

Since the last term on the RHS is zero (no explicit time dependence in the operator), the Hamiltonian is the sum of only two terms. We consider each of the terms separately then add the results together. The

12.9 Spin Transfer Torque On Localized Electrons

first term of the Hamiltonian $\widehat{H}$ is *Hermitian*. Then, only the first term on the RHS of Eq. (12.133) is nonzero, which leads to

$$\frac{d\widehat{\boldsymbol{\mu}}_S}{dt} = \frac{i}{\hbar}\left[\widehat{H}\widehat{\boldsymbol{\mu}}_S - \widehat{\boldsymbol{\mu}}_S\widehat{H}\right] = \frac{i}{\hbar}\left[\widehat{H},\widehat{\boldsymbol{\mu}}_S\right]$$

Substitution of $\widehat{\boldsymbol{\mu}}_S = (-g|\gamma|\hbar/2)\widehat{\boldsymbol{\sigma}}$ into the above, as well as the spin matrices, we have

$$\frac{d\widehat{\boldsymbol{\mu}}_S}{dt} = -\frac{ig^2|\gamma|^2\hbar}{4}\left[(\widehat{\boldsymbol{\sigma}}\cdot\mathbf{B}_{sd})\widehat{\boldsymbol{\sigma}} - \widehat{\boldsymbol{\sigma}}(\widehat{\boldsymbol{\sigma}}\cdot\mathbf{B}_{sd})\right]$$

Expanding the dot products to better see where this leads, we have

$$\frac{d\widehat{\boldsymbol{\mu}}_S}{dt} = -\frac{ig^2|\gamma|^2\hbar}{4}\left[(\widehat{\sigma}_x B_x + \widehat{\sigma}_y B_y + \widehat{\sigma}_z B_z)\widehat{\boldsymbol{\sigma}} - \widehat{\boldsymbol{\sigma}}(\widehat{\sigma}_x B_x + \widehat{\sigma}_y B_y + \widehat{\sigma}_z B_z)\right]$$

$$= -\frac{ig^2|\gamma|^2\hbar}{4}\begin{bmatrix}(\widehat{\sigma}_y\widehat{\sigma}_x - \widehat{\sigma}_x\widehat{\sigma}_y)B_y + (\widehat{\sigma}_z\widehat{\sigma}_x - \widehat{\sigma}_x\widehat{\sigma}_z)B_z\\ (\widehat{\sigma}_x\widehat{\sigma}_y - \widehat{\sigma}_y\widehat{\sigma}_x)B_x + (\widehat{\sigma}_z\widehat{\sigma}_y - \widehat{\sigma}_y\widehat{\sigma}_z)B_z\\ (\widehat{\sigma}_x\widehat{\sigma}_z - \widehat{\sigma}_z\widehat{\sigma}_x)B_x + (\widehat{\sigma}_y\widehat{\sigma}_z - \widehat{\sigma}_z\widehat{\sigma}_y)B_z\end{bmatrix}$$

Using the cyclic permutation relations for the spin matrices obtained in Eqs. (9.21a)-(9.21c), where $\widehat{\sigma}_i\widehat{\sigma}_j = i\varepsilon_{ijk}\widehat{\sigma}_k$, the above gives

$$\frac{d\widehat{\boldsymbol{\mu}}_S}{dt} = -\frac{ig^2|\gamma|^2\hbar}{4}\begin{bmatrix}-2i\widehat{\sigma}_z B_y + 2i\widehat{\sigma}_y B_z\\ 2i\widehat{\sigma}_z B_x - 2i\widehat{\sigma}_x B_z\\ -2i\widehat{\sigma}_y B_x + 2i\widehat{\sigma}_x B_z\end{bmatrix} = \frac{g^2|\gamma|^2\hbar}{2}\widehat{\boldsymbol{\sigma}}\times\mathbf{B}_{sd} = -g|\gamma|\widehat{\boldsymbol{\mu}}_S\times\mathbf{B}_{sd}$$

This result only accounts for one of the two contributions to the dynamic equation for $\widehat{\boldsymbol{\mu}}_S$. Now, we determine the contribution from the dissipative part of the Hamiltonian. Then, the first part of the time-derivative for the dissipative contribution becomes

$$\frac{d\widehat{\boldsymbol{\mu}}_S}{dt} = \frac{i}{\hbar}\left[i\alpha\widehat{H}\widehat{\boldsymbol{\mu}}_S + i\alpha\widehat{\boldsymbol{\mu}}_S\widehat{H}\right]$$

$$= \frac{-\alpha}{\hbar}\left(\widehat{H}\widehat{\boldsymbol{\mu}}_S + \widehat{\boldsymbol{\mu}}_S\widehat{H}\right) = \frac{\alpha g^2|\gamma|^2\hbar}{4}\left(\widehat{\boldsymbol{\sigma}}\cdot\mathbf{B}_{sd}\widehat{\boldsymbol{\sigma}} + \widehat{\boldsymbol{\sigma}}\widehat{\boldsymbol{\sigma}}\cdot\mathbf{B}_{sd}\right)$$

$$= \frac{\alpha g^2|\gamma|^2\hbar}{4}\begin{bmatrix}2\widehat{\sigma}_x^2 B_x + (\widehat{\sigma}_y\widehat{\sigma}_x + \widehat{\sigma}_x\widehat{\sigma}_y)B_y + (\widehat{\sigma}_z\widehat{\sigma}_x + \widehat{\sigma}_x\widehat{\sigma}_z)B_z\\ 2\widehat{\sigma}_y^2 B_y + (\widehat{\sigma}_x\widehat{\sigma}_y + \widehat{\sigma}_y\widehat{\sigma}_x)B_x + (\widehat{\sigma}_z\widehat{\sigma}_y + \widehat{\sigma}_y\widehat{\sigma}_z)B_z\\ 2\widehat{\sigma}_z^2 B_z + (\widehat{\sigma}_x\widehat{\sigma}_z + \widehat{\sigma}_z\widehat{\sigma}_x)B_x + (\widehat{\sigma}_y\widehat{\sigma}_z + \widehat{\sigma}_z\widehat{\sigma}_y)B_z\end{bmatrix}$$

Using the relations $\hat{\sigma}_i^2 = I(2)$ and the spin matrix anti-commutation relation $\hat{\sigma}_i\hat{\sigma}_j + \hat{\sigma}_j\hat{\sigma}_i = 0$, the above becomes

$$\frac{d\widehat{\boldsymbol{\mu}}_S}{dt} = \frac{\alpha g^2|\gamma|^2\hbar}{4}\begin{bmatrix}2B_x\\2B_y\\2B_z\end{bmatrix} = \frac{\alpha g^2|\gamma|^2\hbar}{2}\mathbf{B}_{sd} \tag{12.134}$$

For the second part of the dissipative contribution, we have

$$\frac{-i}{\hbar}\langle\boldsymbol{\mu}_S\rangle[\widehat{H}^\dagger - \widehat{H}] = \frac{-i}{\hbar}\langle\boldsymbol{\mu}_S\rangle\left[\frac{i\alpha g|\gamma|\hbar}{2} + \frac{i\alpha g|\gamma|\hbar}{2}\right](\boldsymbol{\sigma}\cdot\mathbf{B}_{sd}) \tag{12.135}$$

$$= \frac{2\alpha}{\hbar}\langle\boldsymbol{\mu}_S\rangle\left[\frac{g|\gamma|\hbar}{2}\right](\boldsymbol{\sigma}\cdot\mathbf{B}_{sd}) \tag{12.136}$$

$$= -\frac{2\alpha}{\hbar}\left(\boldsymbol{\mu}_S\cdot\mathbf{B}_{sd}\right)\boldsymbol{\mu}_S \tag{12.137}$$

Again leveraging the fact that this problem is identical in form to the problem with Zeeman energy, with a magnetic field $\mathbf{B}_{sd}$, we can use the vector identity from Eq. (12.55) (also see Fig. 12.6). The identity guarantees that any 3D *unit* vector $\mathbf{u}$ (i.e. $|\mathbf{u}| = 1$) can be decomposed in a useful way. With application to $\mathbf{B}_{sd}$, we have

$$\mathbf{B}_{sd} = \frac{1}{\mu_S^2}\left[(\boldsymbol{\mu}_S\cdot\mathbf{B}_{sd})\boldsymbol{\mu}_S - \boldsymbol{\mu}_S\times\boldsymbol{\mu}_S\times\mathbf{B}_{sd}\right] \tag{12.138}$$

The saturation moment μ_S for an electron is given by

$$\mu_S = \frac{g|\gamma|\hbar}{2}$$

Substitution of these relations into Eqs. (12.134) and (12.138) gives

$$\mathbf{B}_{sd} = \frac{2\alpha}{\hbar}\left[(\boldsymbol{\mu}_S\cdot\mathbf{B}_{sd})\boldsymbol{\mu}_S - \boldsymbol{\mu}_S\times\boldsymbol{\mu}_S\times\mathbf{B}_{sd}\right] \tag{12.139}$$

In this form, we can see that the first term in Eq. (12.139) cancels the result found in Eq. (12.137). Then, we are left with the following result for the time-derivative of the operator $\boldsymbol{\mu}_S$:

$$\frac{d\langle\boldsymbol{\mu}_S\rangle}{dt} = -g|\gamma|\boldsymbol{\mu}_S\times\mathbf{B}_{sd} - \frac{2\alpha}{\hbar}\boldsymbol{\mu}_S\times\boldsymbol{\mu}_S\times\mathbf{B}_{sd}$$

$$= -g|\gamma|\boldsymbol{\mu}_S\times\mathbf{B}_{sd} - \frac{\alpha g|\gamma|}{\mu_S}\boldsymbol{\mu}_S\times\boldsymbol{\mu}_S\times\mathbf{B}_{sd}$$

After substitution of the exchange field $\mathbf{B}_{sd} = -J_{sd}\mathbf{P}/g|\gamma|\hbar$, the equation of motion for the localized electrons interacting with itinerant spins is given by

12.9 Spin Transfer Torque On Localized Electrons

Spin-Transfer Torque Equation

$$\frac{d\langle\boldsymbol{\mu}_S\rangle}{dt} = \frac{J_{sd}P}{\hbar}\boldsymbol{\mu}_S \times \boldsymbol{\sigma}_p + \frac{\alpha J_{sd}P}{\hbar\mu_S}\boldsymbol{\mu}_S \times \boldsymbol{\mu}_S \times \boldsymbol{\sigma}_p \qquad (12.140)$$

Because the time derivative of angular momentum $\widehat{S}$ is a torque, and angular momentum is proportional to magnetic moment, Eq. (12.140) is proportional to a torque. Since this torque originates from transferred spin angular momenta from itinerant electrons, it is known as **spin-transfer torque** (SST). The first predictions of *spin-transfer torque* came around 1995 with the works of Luc Berger of Carnegie Melon University, and John Slonczewski of IBM. This branch of spin-physics is very active today because it is currently being used to develop novel types of date storage memory cells known as STT *magnetic random access memory* or *magnetoresistive random access memory* (STT-MRAM) cells. They are useful for memory, even intended to replace some forms of RAM used in computer processors.

There is another reason that it is fortuitous that the Hamiltonians (both Zeeman and exchange) have identical mathematical forms. This also means that since we have already obtained the *exact solutions* to the torque equation of this form (in Sec. 12.4), we also have the solutions to the spin-transfer torque equation given in Eq. (12.140). The magnetic moment component solutions are given by

$\mu_x(t)$ Expectation Value

$$\langle\mu_x\rangle(t) = -\frac{g|\gamma|\hbar}{2}\frac{\cos(\omega_p t + \varphi)}{\cosh(\alpha\omega_p t)} \qquad (12.141)$$

$\mu_y(t)$ Expectation Value

$$\langle\mu_y\rangle(t) = -\frac{g|\gamma|\hbar}{2}\frac{\sin(\omega_p t + \varphi)}{\cosh(\alpha\omega_p t)} \qquad (12.142)$$

$\mu_z(t)$ Expectation Value

$$\langle \mu_z \rangle(t) = \frac{g|\gamma|\hbar}{2}\tanh(\alpha\omega_p t) \qquad (12.143)$$

We can also define the spin transfer torque relaxation time constant τ_S, defined by $\langle \mu_z \rangle(t)$ in Eq. (12.143). τ_S is given by

Spin-transfer Torque Relaxation-Time

$$\tau_S = \frac{\hbar}{\alpha J_{sd} P} \qquad (12.144)$$

The precessional frequency ω_p for the electron spin is given by

$$\omega_p = \frac{J_{sd} P}{\hbar} \qquad (12.145)$$

Precessional motion from *spin-transfer torque* is about the exchange field B_{sd}. Since the field is opposite to an external magnetic field, it is in the *opposite* direction to precessional motion about an external magnetic field, if M_1 and B are along the same direction, say +z. This is a consequence of the opposite signs in their respective Hamiltonians, particularly, for any material with $J_{sd} > 0$. Thus, we have found that when itinerant electrons that have been spin polarized from a magnetic layer get injected into another magnetic layer, they impart a torque onto the localized magnetic moments of the free magnetic layer. In addition to obtaining the dynamical equation of motion, we have also obtained the exact time-dependent solutions for a fixed spin exchange field B_{sd}. Both are only possible by including spin-energy dissipation into the Hamiltonian.

12.10 Chapter Summary

In this chapter, we started by introducing a form of energy dissipation into quantum mechanics based on relativistic results applicable to angular momentum. The resulting operators turn out to be *non-Hermitian*,

12.10 Chapter Summary

possessing complex eigenvalues. We found that they also describe the dynamics of spin-systems, and we used this approach to both derive the Landau-Lifshitz-like dynamical equation, as well as obtain its exact solution, *despite its nonlinear form*. We also discussed spin polarization, asymmetric spin transport in magnetic layers, as well as spin-transfer torque from polarized itinerant spins on localized spins of a magnetic layer. These areas of spin-physics are useful in the development of novel technologies such as *spin-transfer torque magnetoresistive access memory* (STTMRAM), as well as magnetic field sensors used in a number of applications. Table 12.3 summarizes some of the key results from this chapter.

Table 12.3: Chapter 12 Summary Equations.

Name	Equation
Nonhermitian Eigenenergy	$E_\lambda = (1 - i\alpha^*)|\gamma|Bm_\ell \hbar$
Dissipative Hamiltonian	$\widehat{H}_d \psi = -i\alpha \widehat{H}_R \psi$
Dissipative Cont. Equation	$\frac{\partial \rho}{\partial t} + \nabla \cdot \mathbf{J} = -\frac{2\alpha}{\hbar} V\rho$
Spin Expectation x	$\langle \widehat{S}_x \rangle = \frac{\hbar}{2} \frac{\cos(\omega_p t + \varphi)}{\cosh(\alpha \omega_p t)}$
Spin Expectation y	$\langle \widehat{S}_y \rangle = \frac{\hbar}{2} \frac{\sin(\omega_p t + \varphi)}{\cosh(\alpha \omega_p t)}$
Spin Expectation z	$\langle \widehat{S}_z \rangle = -\frac{\hbar}{2} \tanh(\alpha \omega_p t)$
spin diffusion (↑)	$\frac{\partial^2 \mu_+}{\partial x^2} = \frac{\mu_+}{\ell_+^2}$
spin diffusion (↓)	$\frac{\partial^2 \mu_-}{\partial x^2} = \frac{\mu_-}{\ell_-^2}$

12.11 Chapter Problems

Problem 12.1 Based on the torque terms in spin transfer torque, can a design be devised for a spin-transfer torque system, with an external magnetic field, that has zero precessional motion? If so, why is this possible?

Problem 12.2 Similarly, based on the torque terms in spin transfer torque, can a design be devised for a spin-transfer torque system, with an external magnetic field, having pure precessional motion, and no transverse rotation? If so, how can this be done?

Problem 12.3 An dynamical equation of motion was derived and given by Eq. (12.140). We found that

$$\frac{d\langle \boldsymbol{\mu}_S \rangle}{dt} = \frac{J_{sd}\langle p \rangle}{\hbar} \boldsymbol{\mu}_S \times \boldsymbol{\sigma}_p + \frac{\alpha J_{sd}\langle p \rangle}{\hbar \mu_S} \boldsymbol{\mu}_S \times \boldsymbol{\mu}_S \times \boldsymbol{\sigma}_p$$

Use this equation to prove that

$$\frac{\partial}{\partial t}\left(\boldsymbol{\mu}_S \cdot \boldsymbol{\mu}_S\right) = 0$$

It may be helpful to note that

$$\boldsymbol{\mu}_S \cdot \frac{\partial \boldsymbol{\mu}_S}{\partial t} = \frac{1}{2}\frac{\partial}{\partial t}\left(\boldsymbol{\mu}_S \cdot \boldsymbol{\mu}_S\right)$$

What is the meaning of this result for a moment that satisfies this equation of motion?

Problem 12.4 Estimate the *precessional frequency* for an electron spin coupled to an exchange energy of value $J = 1 \times 10^{-21}$ Joules. What is the frequency in GHz? Also determine the effective exchange field $B_{exch} = J/2\mu_B$.

Problem 12.5 Use the identity given by

$$\mathbf{a} \times \mathbf{b} \times \mathbf{c} = (\mathbf{a} \cdot \mathbf{c})\mathbf{b} - (\mathbf{a} \cdot \mathbf{b})\mathbf{c}$$

to show that the dynamical equation of motion

$$\frac{d\langle \boldsymbol{\mu}_S \rangle}{dt} = \frac{J_{sd}\langle p \rangle}{\hbar} \boldsymbol{\mu}_S \times \boldsymbol{\sigma}_p + \frac{\alpha J_{sd}\langle p \rangle}{\hbar \mu_S} \boldsymbol{\mu}_S \times \boldsymbol{\mu}_S \times \boldsymbol{\sigma}_p$$

can also be written as

$$\frac{d\langle\boldsymbol{\mu}_S\rangle}{dt} = \frac{J_{sd}\langle p\rangle(1+\alpha^2)}{\hbar}\boldsymbol{\mu}_S \times \boldsymbol{\sigma}_p + \frac{\alpha}{\mu_S}\boldsymbol{\mu}_S \times \frac{d\langle\boldsymbol{\mu}_S\rangle}{dt}$$

Problem 12.6 Prove that a magnetic potential energy of the form given by Eq. (12.12) leads to the relation given by

$$\frac{\partial V_M}{\partial \boldsymbol{\mu}} = -g|\gamma|\mathbf{B} - \frac{g|\gamma|\alpha}{2\mu}\boldsymbol{\mu}\times\mathbf{B} + \frac{g|\gamma|\alpha B}{\mu}\boldsymbol{\mu}$$

Problem 12.7 Show that the expectation value of the eigenenergy is given by $\langle E_\lambda(t)\rangle = \frac{g|\gamma|\hbar B}{2}\tanh(\alpha\omega_p t)$.

Problem 12.8 Let the wave function $\psi(x,t)$ for a Hamiltonian be given by

$$\psi(x,t) = Ae^{i\left(k_r x - \frac{\alpha V}{\hbar}t\right)}e^{-k_i x + \frac{\alpha V}{\hbar}t}$$

$k = k_r + ik_i$ is the wave number, α is the dissipation parameter, while V is a real valued potential. Use the result obtained in Eq. (12.28) which is the *dissipative continuity equation* given by

$$\frac{\partial \rho}{\partial t} + \nabla \cdot \mathbf{J} = -\frac{2\alpha}{\hbar}V\rho$$

Show that it leads to the following condition which must be satisfied:

$$\hbar^2 k_i k_r = 2\alpha m V$$

Problem 12.9 For a pair of 3D vectors $\mathbf{A}$ and $\mathbf{B}$, prove the following relation

$$\mathbf{B} = \frac{1}{|\mathbf{A}|^2}[(\mathbf{A}\cdot\mathbf{B})\mathbf{A} - \mathbf{A}\times\mathbf{A}\times\mathbf{B}]$$

12.12 Suggested Readings & References

[1] L. Landau and E. Lifshitz, *On the theory of the dispersion of magnetic permeability in ferromagnetic bodies*, Physik. Zeits. Sowjetunion **8**, pp. 153-169 (1935).
[2] W. Butler, et. al. *Spin-dependent tunneling conductance of Fe|MgO|Fe sandwiches*, Phys. Rev. B., **63**, 054416 (2001).
[3] S. Yuasa, et. al. *Giant room-temperature magnetoresistance in single-crystal Fe|MgO|Fe magnetic tunnel junctions*, Nature Mat., **3**, pp. 868-871 (2004).
[4] T. Valet and A. Fert, *Theory of the perpendicular magnetoresistance in magnetic multilayers*, Phys. Rev. B., **63**, 054416 (1993).
[5] L. Berger, *Precession of Conduction-Electron Spins Near an Interface Between Normal and Magnetic Metals*, IEEE T. Mag., **31**, pp. 3871-3873 (1995).
[6] J. Slonczewski, *Current-driven excitation of magnetic multilayers*, **159**, pp. L1-L7, J. Mag. Magn. Mater. (1996).
[7] R. Meservey and P.M. Tedrow, *Spin-polarized electron tunneling*, PHYSICS REPORTS (Review Section of Physics Letters) **238**, No. 4 pp. 173—243 (1994).

Effective Exchange Field

$$B^*_{sd} = \frac{J_{sd}}{2\mu_P} \qquad (12.146)$$

CHAPTER 13

Methods Of Approximation

Most problems described by quantum systems are not solvable by exact analytical means. We've only cherry picked, mostly known problems amenable to analytical methods. For most real world problems, *approximate methods are necessary*. We alluded to this, for example, when we discussed the helium atom and the hydrogen molecule. This chapter will give some introduction to two simpler, but common methods of approximation used in quantum mechanics, thought not limited to it. We'll start with an approximate method known as the JWKB or WKB approximation. For this, we need to discuss a special set of functions known as Airy functions.

13.1 Airy Functions: Prelude To JWKB Approximation

In the next few sections, we discuss a well known *approximate* method for dealing with second order differential equations. The method is based on mathematical functions, at its roots, and is not limited to quantum mechanics. It is known as the **JWKB approximation** or the **WKB approximation** method. This approximation method was first introduced by mathematician Harold Jeffreys (1891-1989) in 1923, and later in 1926, by physicists Gregor Wentzel (1898-1978), Hendrik Anthony Kramers (1894-1952), and Léon Nicolas Brillouin (1889-1969), by making a comparison of the second order differential equation being considered to the *Airy equation*. **Airy's differential equation** is given by

$$\frac{d^2y}{dx^2} - xy = 0 \quad \text{(Airy's Equation)} \tag{13.1}$$

The *Airy differential equation* is named after British physicist George Biddell Airy (1801-1892), though it is unclear why. The historical reference (on geometrical optics) usually supporting this does not appear to contain an equation of the form given by Eq. (13.1), but for minimizing confusion, we'll use this name. If we scale y by a constant s where $\widehat{y} = sy$, then we have

$$\widehat{y} = sy \Rightarrow \frac{d\widehat{y}}{dx} = s\frac{dy}{dx} \quad \text{and} \quad \frac{d^2\widehat{y}}{dx^2} = s\frac{d^2y}{dx^2}$$

After substitution into Eq. (13.1), leads to

$$\frac{d^2\widehat{y}}{dx^2} - x\widehat{y} = 0 \tag{13.2}$$

Similarly, if we were to shift x by an amount x_0 where $\widehat{x} = x - x_0$, we have

$$\widehat{x} = x - x_0 \Rightarrow \frac{dy}{dx} = \frac{dy}{d\widehat{x}}\frac{d\widehat{x}}{dx} = \frac{dy}{d\widehat{x}} \quad \text{and} \quad \frac{d^2y}{d\widehat{x}^2} = \frac{d^2y}{dx^2}$$

Substitution of these relations into Eq. (13.1) leads to

$$\frac{d^2y}{d\widehat{x}^2} - \widehat{x}y = 0 \tag{13.3}$$

Therefore, scaling the solution y or shifting the abscissa x results in an identical differential equation, and thus, if $y(x)$ is a solution, so is $\widehat{y}$ and $y(\widehat{x})$. The exact solutions to the Airy differential equation are known as **Airy functions**. Moreover, because Eq. (13.1) is a second order differential equation, we should expect two independent solutions. The two independent solutions are known as the **Airy function of the first kind** $Ai(x)$ and the **Airy function of the second kind** $Bi(x)$, respectively.

A complete method for obtaining the Airy functions is by using the *method of Frobenius* as we had used in Chap. 5, substituting an infinite series of the form $\sum_{n=0}^{\infty} a_n x^n$ into the differential equation. Though the exact solution is found from this approach, it is not as useful for

13.1 Airy Functions: Prelude To JWKB Approximation

our purposes. More than the Airy functions, we are interested in approximations to them. These approximations are used in the JWKB approximation. Thus, towards our objective, a more useful method to obtain a solution to Eq. (13.3) is by use of Fourier transforms, developed earlier in Chap. 3, Sec. 3.4. Let us begin with definitions for y and its Fourier transform $\bar{y}$, given by

$$y(x) = \int_{-\infty}^{+\infty} e^{ipx} Y(p) dp \qquad (13.4)$$

$$Y(p) = \int_{-\infty}^{+\infty} e^{-ipx} y(x) dx \qquad (13.5)$$

We have omitted a normalization constant $1/\sqrt{2\pi}$, for convenience. Using the definition of $y(x)$ in the Fourier transforms (twice-differentiating Eq. (13.4)), the *Airy differential equation* becomes

$$-p^2 y = xy$$

Now, multiplying y in the above equation by $e^{-ipx} dx$, and integrating leads to

$$-p^2 \int e^{-ipx} y(x) dx = x \int e^{-ipx} y(x) dx \qquad (13.6)$$

Also, from the definition of $Y(p)$ given in Eq. (13.5), it follows that

$$\frac{dY}{dp} = -ix \int e^{-ipx} y(x) dx$$

This can used on the RHS of Eq. (13.6) to obtain a first order differential equation for $Y(p)$, given by

$$-p^2 Y = i \frac{dY}{dp} \qquad (13.7)$$

We can rearrange Eq. (13.7) to the following

$$\frac{dY}{Y} = ip^2 dp \Rightarrow \log Y = \frac{ip^3}{3} + C \qquad (13.8)$$

So we have the following solution for $Y(p)$

$$Y(p) = C e^{\frac{ip^3}{3}} \qquad (13.9)$$

Substitution of this result back into Eq. (13.4)(and re-introducing the normalization factor $1/2\pi$) for $y(x)$, leads to

$$y(x) = \frac{1}{\sqrt{2\pi}} \int_{-\infty}^{+\infty} e^{ipx} e^{ip^3/3} dp = \frac{1}{\sqrt{2\pi}} \int_{-\infty}^{+\infty} e^{i\phi(p)} dp \qquad (13.10)$$

The solution, which is denoted by Ai(x), then becomes

Airy Function Of The First Kind

$$\text{Ai}(x) = \frac{1}{\sqrt{2\pi}} \int_{-\infty}^{+\infty} e^{ipx} e^{ip^3/3} dp \qquad (13.11)$$

Eq. (13.11) is the **Airy Function of the First Kind**, and is also known as the **Airy Integral**. The function $\phi(p)$ is given by

$$\phi(p) = px + \frac{p^3}{3} \qquad (13.12)$$

Both Airy functions are illustrated in Fig. 13.1. Both functions are decay-

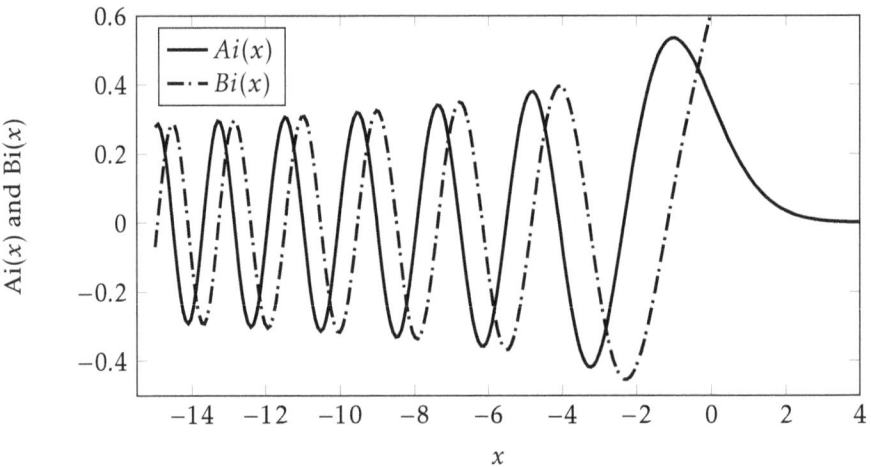

Figure 13.1: Airy functions of the first kind (Ai) and second (Bi). At x=0, the function form changes abruptly, where Ai(x) decays to zero, while Bi(x) diverges to infinity.

ing oscillatory functions to the left, while beyond zero, Ai (illustrated in blue) decays, while Bi diverges. There is also a phase difference between them of 90°. Now let's look at how these functions can be approximated and used in quantum mechanics.

13.2 Approximation To Airy Functions In Region $x < 0$

We saw earlier that any shifted and/or scaled version of the Airy solution is also a solution to the same differential equation. This property can be

13.2 Approximation To Airy Functions In Region $x < 0$

exploited to shift the function to any desired point, say p_0. Let p_0 be in close proximity to regions of zero slope (in phase), where

$$\frac{d\phi}{dp}(p_0) = x + p_0^2 = 0 \Rightarrow p_0 = \pm\sqrt{-x}$$

Let's now consider real solutions for p_0 where $z = -x > 0$ (or $x < 0$), which gives

$$p_0 = \pm\sqrt{z}$$

Note the p_0 is *not arbitrary*, and is assumed to be in a region of little variation, with respect to p. Then, a Taylor series expansion of $\phi(p)$ around p_0 is given by

$$\phi(p) \approx$$

$$\phi(p_0) + \frac{d\phi}{dp}\bigg|_{p=p_0}(p-p_0) + \frac{1}{2}\frac{d^2\phi}{dp^2}\bigg|_{p=p_0}(p-p_0)^2 + \frac{1}{3!}\frac{d^3\phi}{dp^3}\bigg|_{p=p_0}(p-p_0)^3 + 0 \tag{13.13}$$

The Taylor series truncates because $\phi(p)$ is up to third power in p, thus, higher order derivatives vanish. Since p_0 is defined at a location where $d\phi/dp = 0$, ignoring third order and above, the expansion can be approximated by

$$\phi(p) \approx \phi(p_0) + \frac{1}{2}\frac{d^2\phi}{dp^2}\bigg|_{p=p_0}(p-p_0)^2 = \tag{13.14}$$

Because $p_0 = \pm\sqrt{z}$, we have two functions for ϕ, leading to two solutions given by

$$\phi(p_0) = p_0 x + \frac{p_0^3}{3} = -p_0 z + \frac{p_0^3}{3} = \mp z^{\frac{3}{2}} \pm \frac{1}{3}z^{\frac{3}{2}} = \mp\frac{2}{3}z^{\frac{3}{2}}$$

$$\frac{d^2\phi}{dp^2}\bigg|_{p=p_0} = \pm 2\sqrt{z}$$

Substitution of these results into Eq. (13.14), then with that result into the solution for Ai(x), we obtain

$$\sqrt{2\pi} \cdot \text{Ai}(x) = \int_{-\infty}^{+\infty} e^{i\phi(p)}dp \approx \int_{-\infty}^{+\infty} e^{i\left[\mp\frac{2}{3}z^{\frac{3}{2}} + \frac{1}{2}(\pm 2\sqrt{z})(p\mp\sqrt{z})^2\right]}dp$$

$$= e^{\mp i\frac{2}{3}z^{\frac{3}{2}}} \int_{-\infty}^{+\infty} e^{\pm i\sqrt{z}(p\mp\sqrt{z})^2}dp$$

The integral in the last line can be worked out exactly, as it has a well-known form whose integral is $\int_{-\infty}^{+\infty} e^{-ax^2} dx = \sqrt{\pi/a}$. So, we have

$$\sqrt{2\pi} \cdot \text{Ai}(x) = e^{\mp i \frac{2}{3} z^{\frac{3}{2}}} \int_{-\infty}^{+\infty} e^{\pm i \sqrt{z}(p \mp \sqrt{z})^2} dp$$

$$= \sqrt{\frac{\pi}{\mp i \sqrt{z}}} e^{\mp i \frac{2}{3} z^{\frac{3}{2}}}$$

Using the fact that $\sqrt{1/\mp i} = e^{\pm \pi/4}$, our Airy function approximation becomes

$$\sqrt{2\pi} \cdot \text{Ai}(x) \approx \sqrt{\frac{\pi}{\mp i \sqrt{z}}} e^{\mp i \frac{2}{3} z^{\frac{3}{2}}} = \sqrt{\frac{\pi}{\sqrt{z}}} e^{\mp i \left(\frac{2}{3} z^{\frac{3}{2}} - \frac{\pi}{4} \right)}$$

Summing both to obtain the most general solution, we have

$$\sqrt{2\pi} \cdot \text{Ai}(x) \approx \sqrt{\frac{\pi}{\sqrt{z}}} \left[e^{-i\left(\frac{2}{3} z^{\frac{3}{2}} - \frac{\pi}{4}\right)} + e^{i\left(\frac{2}{3} z^{\frac{3}{2}} - \frac{\pi}{4}\right)} \right]$$

$$= 2\sqrt{\frac{\pi}{\sqrt{z}}} \cos\left(\frac{2}{3} z^{\frac{3}{2}} - \frac{\pi}{4} \right)$$

This leads to an approximation of the Airy function of the first kind $\widetilde{\text{Ai}}(x)$, given by

$$\widetilde{\text{Ai}}(x) \approx \sqrt{\frac{2}{\sqrt{-x}}} \cos\left(\frac{2}{3}(-x)^{\frac{3}{2}} - \frac{\pi}{4} \right)$$

or

First Airy Function Approximation ($x < 0$)

$$\widetilde{\text{Ai}}(x) = \sqrt{\frac{2}{\sqrt{-x}}} \sin\left(\frac{2}{3}(-x)^{\frac{3}{2}} + \frac{\pi}{4} \right) \tag{13.15}$$

With Eq. (13.15), which is valid for $x < 0$, we can write the second independent solution to the Airy differential equation, denoted by Bi(x), which is $\pi/2$ out of phase, given by

$$\widetilde{\text{Bi}}(x) \approx \sqrt{\frac{2}{\sqrt{-x}}} \sin\left(\frac{2}{3}(-x)^{\frac{3}{2}} - \frac{\pi}{4} \right)$$

or

13.2 Approximation To Airy Functions In Region $x < 0$

Second Airy Function Approximation ($x < 0$)

$$\widetilde{Bi}(x) = \sqrt{\frac{2}{\sqrt{-x}}} \cos\left(\frac{2}{3}(-x)^{\frac{3}{2}} + \frac{\pi}{4}\right) \tag{13.16}$$

We have used $\cos x = \sin(x + \pi/2)$ and $\sin x = \cos(x + \pi/2)$. Eqs. (13.15) and (13.16) are illustrated in Fig. 13.2, in a region for $x < 0$.

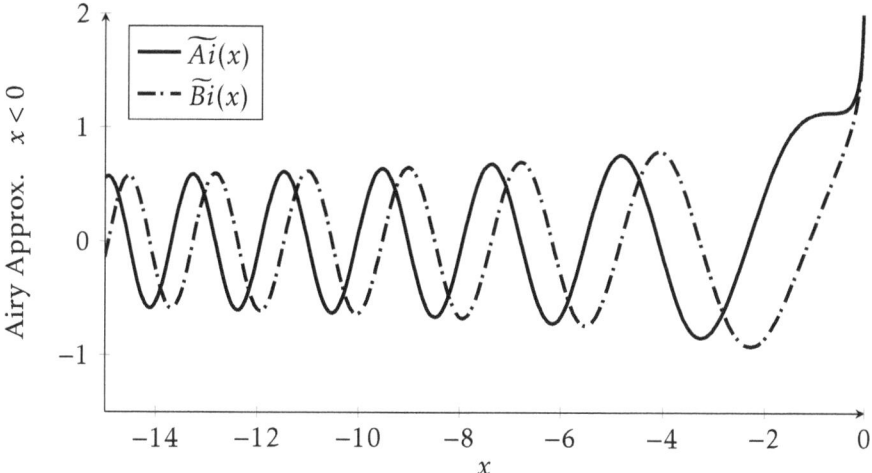

Figure 13.2: Approximations to the Airy functions Ai(x) and Bi(x), in the domain of $x < 0$. Both functions are decaying oscillatory functions in the direction of $x \to -\infty$. The $\sqrt{2/\sqrt{(-x)}}$ forms the envelope of the oscillatory term.

The approximations shown in Fig. 13.2 are good approximations away from $x = 0$. Near this point, the solution starts to diverge rapidly, because of the $\sqrt{2/\sqrt{(-x)}}$ envelope. This point, where the approximations diverge, is called **JWKB turning points**. Even for the exact Airy functions, this corresponds to where teh function behavior radically changes. The *turning points* are significant locations for these approximations, as they define a location for a transition in the behavior of the solution. This will be important in the use of these functions as wave functions, which must vanish at boundaries. Let's now look at the region $x > 0$.

13.3 Approximation To Airy Functions In Region $x > 0$

In this section, we'll determine what an Airy approximation looks like in the domain $x > 0$. We saw that in the domain $x < 0$, the solutions have damped oscillatory behavior, just as the Airy functions do. We will see here that the behavior in the domain $x > 0$ changes significantly, and this is inline with the behavior of the Airy functions. Let's assume a solution of the form

$$y = A(x)e^{-B(x)} \tag{13.17}$$

In this form, both A and B depend on x. The first derivative dy/dx then becomes

$$\frac{dy}{dx} = \frac{dA}{dx}e^{-B} - A\frac{dB}{dx}e^{-B} \tag{13.18}$$

Using this result, the second derivative is given by

$$\frac{d^2y}{dx^2} = \frac{d^2A}{dx^2}e^{-B} - 2\frac{dA}{dx}\frac{dB}{dx}e^{-B} - A\frac{d^2B}{dx^2}e^{-B} + A\left(\frac{dB}{dx}\right)^2 e^{-B}$$

$$= \left[\frac{d^2A}{dx^2} - 2\frac{dA}{dx}\frac{dB}{dx} - A\frac{d^2B}{dx^2} + A\left(\frac{dB}{dx}\right)^2\right]e^{-B} \tag{13.19}$$

Substitution of Eq. (13.19) into the Airy differential equation leads to the following differential equation involving A, but only derivatives of B, given by

$$\frac{d^2y}{dx^2} - xy = 0 \Rightarrow \frac{d^2A}{dx^2} - 2\frac{dA}{dx}\frac{dB}{dx} - A\frac{d^2B}{dx^2} + \left[\left(\frac{dB}{dx}\right)^2 - x\right] = 0 \tag{13.20}$$

Assume that relative to all other terms in Eq. (13.20), that

$$\frac{d^2A}{dx^2} \approx 0 \tag{13.21}$$

This assumption states that the amplitude $A(x)$ of the solution varies more slowly, compared to the phase $B(x)/i$. In the limit that $A \to 0$, the last term is left. Motivated by this asymptotic limit, let

$$\frac{dB}{dx} = \pm\sqrt{x}$$

13.3 Approximation To Airy Functions In Region $x > 0$

This leads to the following result for $B(x)$:

$$B = \int \frac{dB}{dx} dx = \int x^{\frac{1}{2}} dx = \pm \frac{2}{3} x^{\frac{3}{2}} + C \tag{13.22}$$

With this choice, the term in brackets vanishes in Eq. (13.20). We have obtained *two* independent solutions here, which corresponds to both $Ai(x)$ and $Bi(x)$. Next, we use the condition in Eq. (13.21), along with the result from choosing B, so that Eq. (13.20) becomes

$$-2 \frac{dA}{dx} \frac{dB}{dx} - A \frac{d^2 B}{dx^2} = 0 \Rightarrow 2 x^{\frac{1}{2}} \frac{dA}{dx} + \frac{1}{2} x^{-\frac{1}{2}} A = 0$$

We have used the relation $B = x^{1/2} \to dB/dx = (1/2)x^{-1/2}$. This can be rearranged to the following:

$$\frac{dA}{A} = -\frac{1}{4} \frac{dx}{x} \Rightarrow A(x) = C x^{-\frac{1}{4}}$$

Therefore, we have the final approximation in the domain $x > 0$:

> **First Airy Function Approximations $x > 0$**
>
> $$\widetilde{Ai}(x > 0) = C x^{-\frac{1}{4}} e^{-\frac{2}{3} x^{\frac{3}{2}}} \tag{13.23}$$

The approximation for the Airy function of the second kind is

> **First Airy Function Approximations $x > 0$**
>
> $$\widetilde{Bi}(x > 0) = C x^{-\frac{1}{4}} e^{\frac{2}{3} x^{\frac{3}{2}}} \tag{13.24}$$

Thus, one of the approximate solutions diverges as $x \to +\infty$, while the other converges to zero, being dominated by the exponential term. Combining our results for the approximations of the Airy functions almost across the entire domain (except close to $x = 0$), Fig. 13.3 illustrates the approximations on both sides of the *turning point* $x = 0$. Recall that the solution to the Airy equation remains a solution upon shifting the abscissa x to $\widehat{x} = x - x_0$. Then $x = 0$ can be chosen arbitrarily, and our solution simply shifts with it, and the turning point becomes the point

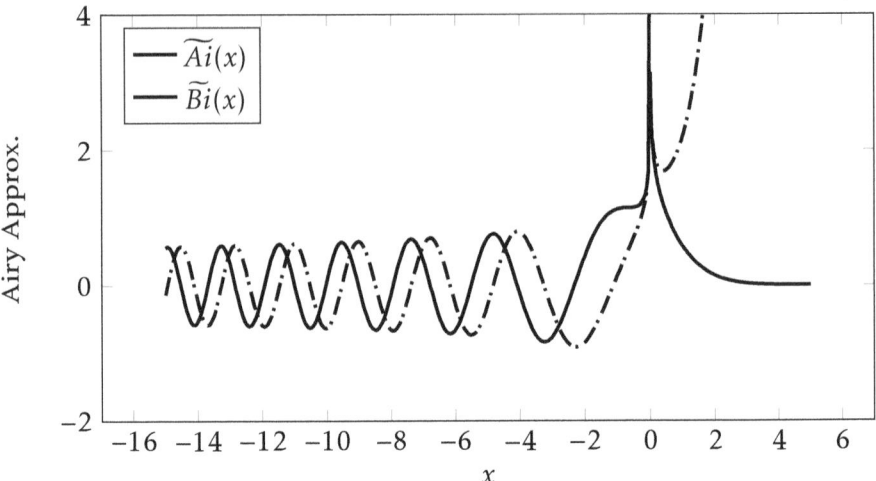

Figure 13.3: Approximations to the Airy functions $\widetilde{Ai}(x)$ and $\widetilde{Bi}(x)$, across both domains of $x < 0$ and $x > 0$. The *turning point* is located at $x = 0$.

$x = x_0$. This property is essential in the use of the JWKB approximation. In this approximation, the solution of the Schrödinger equation is taken to be

$$\psi \approx c_1 \widetilde{Ai}(x) + c_2 \widetilde{Bi}(x) \tag{13.25}$$

In using the JWKB approximation, one first identifies the turning points of the problem. Then, one can contrive the solution based on which side of the turning point is of interest, noting that crossing over the turning point, for $\widetilde{Bi}(x)$ potentially means that the coefficient from Eq. (13.25) may have to be set to zero, since $x \to \infty$ leads to a diverging wave function. In bound-states, the wave function must vanish as $x \to \pm\infty$. This is how the JWKB approximation is applied. When going from one side of the turning point to the other, this process is known as **connecting the formulae**, and leads to the following **connection formulae** using the Airy function approximations $\widetilde{Ai}$ and $\widetilde{Bi}$:

To motivate and connect these results to the JWKB approximation, consider if we take the Airy equation and express it as

$$\frac{d^2y}{dx^2} - xy = 0 \Rightarrow \frac{d^2y}{dx^2} - k^2 y = 0 \tag{13.26}$$

Thus, we allow $k^2 = x \to k = \pm\sqrt{x}$. It follows that

$$\frac{1}{x^{\frac{1}{4}}} = \frac{1}{k^{\frac{1}{2}}} = \frac{1}{\sqrt{k}} \quad \text{and} \quad \frac{2}{3}x^{\frac{3}{2}} = \int x^{\frac{1}{2}} dx = \int k \, dx \tag{13.27}$$

13.4 Zeroth Order JWKB Approximation

$\psi(x < x_0)$	turning point x_0	$\psi(x > x_0)$
$\psi(x) = x^{-\frac{1}{4}} e^{-\frac{2}{3} x^{\frac{3}{2}}}$	$\rightleftarrows$	$\psi(x) = 2(-x)^{-\frac{1}{4}} \sin\left[\frac{2}{3} x^{\frac{3}{2}} + \frac{\pi}{4}\right]$
$\psi(x) = x^{-\frac{1}{4}} e^{+\frac{2}{3} x^{\frac{3}{2}}}$	$\rightleftarrows$	$2(-x)^{-\frac{1}{4}} \cos\left[\frac{2}{3} x^{\frac{3}{2}} + \frac{\pi}{4}\right]$

Table 13.1: Connection formulae table showing the approximation formulae on two sides of a turning point x_0. Because the approximations have singularities at $x = x_0$, we must carefully transition between the two formulae when going from one side of the turning point to the other side.

$\psi(x < x_0)$	turning point x_0	$\psi(x > x_0)$
$\psi(x) = \frac{1}{\sqrt{k}} e^{-\int_x^{x_0} k dx}$	$\rightleftarrows$	$\psi(x) = \frac{2}{\sqrt{k}} \sin\left[\int_x^{x_0} k dx + \frac{\pi}{4}\right]$
$\psi(x) = \frac{1}{\sqrt{k}} e^{+\int_x^{x_0} k dx}$	$\rightleftarrows$	$\frac{2}{\sqrt{k}} \cos\left[\int_x^{x_0} k dx + \frac{\pi}{4}\right]$

Table 13.2: Connection formulae table in terms of parameter k, where $k^2 = x$.

Upon substitution for an arbitrary turning point $x - x_0$, the *connection formulae* becomes as shown in Table 13.2. Therefore, we can expect the JWKB approximation to be more accurate when $k^2 \approx cx$, where c is a constant independent of x. In the next section, we will look at applying these approximations directly to the Schrödinger equation, and later use the results to solve a problem to demonstrate how it can be applied.

13.4 Zeroth Order JWKB Approximation

Let us now focus more on applying the method of JWKB to the time-independent Schrödinger equation, in 1D, given by

$$\frac{d^2 \psi}{dx^2} + \frac{2m}{\hbar^2} [E - V(x)] \psi(x) = 0 \tag{13.28}$$

As we have done previously, let us define the coefficient of the $\psi(x)$ term as

$$k^2(x) = \frac{2m}{\hbar^2} [E - V(x)] \tag{13.29}$$

So we have the following general form of differential equation, that is of interest:

$$\frac{d^2\psi}{dx^2} + k^2(x)\psi(x) = 0 \tag{13.30}$$

The JWKB approximation can be applied to any second order differential equation of the form given by Eq. (13.30). Moreover, any second order differential equation with coefficients that are functions of x only, can be cast into the form of Eq. (13.30). Thus, the JWKB method can be applied to a large class of second order differential equation. Note that the accuracy, however, varies from case to case, depending on the characteristics of the problem. To more clearly understand these limitations, we will discuss the conditions of validity for the JWKB approximation, as well.

Consider the simplest case of this form, when $k(x) = k_0$ is a constant. In this case, we know the solution is in the form of

$$\psi(x) = e^{\pm ikx} \tag{13.31}$$

We can generalize a solution of this form by writing it as

$$\psi(x) = Ae^{i\phi(x)} \tag{13.32}$$

$\phi(x)$ is a phase function of the independent variable x. We will determine its constraints for this approximation shortly. Given the function $\psi(x)$, the derivatives of $\psi(x)$ become

$$\psi(x) = Ae^{i\phi(x)}$$

$$\frac{d\psi}{dx} = i\frac{d\phi}{dx}\psi(x)$$

$$\frac{d^2\psi}{dx^2} = \left[i\frac{d^2\phi}{dx^2} + \left(i\frac{d\phi}{dx}\right)^2\right]\psi(x)$$

All derivatives are proportional to $\psi(x)$. Substitution of these relations into Eq.(13.30) leads to a second order differential equation for $\phi(x)$ given by

$$i\frac{d^2\phi}{dx^2} - \left(\frac{d\phi}{dx}\right)^2 + k^2(x) = 0 \tag{13.33}$$

If $d^2\phi/dx^2$ is sufficiently small such that it can be neglected in Eq. (13.33), we then have

$$\frac{d\phi}{dx} = \pm k(x) \Rightarrow \phi(x) = \pm \int k(x)dx \tag{13.34}$$

Thus, our solution to Eq. (13.33) for ψ becomes

> **Zeroth Order JWKB Solution**
>
> $$\psi_0(x) = Ae^{\pm i \int k(x)dx} \tag{13.35}$$

Eq. (13.35) is known as the **Zeroth Order JWKB approximation**. *The solution is only valid if $d^2\phi/dx^2$ is small in comparison to $(d\phi/dx)^2$*, which means that

$$\left|\frac{d^2\phi}{dx^2}\right| << \left|\frac{d\phi}{dx}\right|^2 \Rightarrow \left|\frac{dk}{dx}\right| << |k|^2 \quad \text{(JWKB approximation condition)} \tag{13.36}$$

We have used the fact that since $\phi = \int k dx$, it follows that the second derivative of ϕ is the first derivative of $k(x)$. When Eq. (13.36) is satisfied, we say that $k(x)$ is *smoothly varying*, much like a straight line with nonzero slope is *smoothly varying*.

13.5 First Order JWKB Approximation

The JWKB approximation can also be extended to be more accurate by redefining the wave function approximation to include more *richness* in its form, building on the zeroth order JWKB approximation $\psi_0(x)$. Let us define, then, the wave function approximation

$$\psi_1 = F(x)\psi_0(x) \tag{13.37}$$

$F(x)$ is a function that we will determine. $\psi_0(x)$ is the zeroth order approximation given by Eq. (13.35). With Eq. (13.37), the first derivative of ψ_1 becomes

$$\frac{d\psi_1}{dx} = \frac{dF}{dx}\psi_0 + F\frac{d\psi_0}{dx} = \left(\frac{dF}{dx} \pm ik(x)F\right)\psi_0 \tag{13.38}$$

In Eq. (13.38), we have used the relation $d\psi_0/dx = \pm ik\psi_0$. The second derivative then becomes

$$\frac{d^2\psi_1}{dx^2} = \frac{d}{dx}\left[\left(\frac{dF}{dx} \pm ikF\right)\psi_0\right]$$

$$= \left(\frac{d^2F}{dx^2} \pm ik\frac{dF}{dx} \pm i\frac{dk}{dx}F\right)\psi_0 + \left(\frac{dF}{dx} \pm ikF\right)\frac{d\psi_0}{dx}$$

$$= \left(\frac{d^2F}{dx^2} \pm ik\frac{dF}{dx} \pm i\frac{dk}{dx}F\right)\psi_0 + \left(\pm ik\frac{dF}{dx} - k^2F\right)\psi_0$$

$$= \left[\frac{d^2F}{dx^2} \pm 2ik\frac{dF}{dx} + \left(\pm i\frac{dk}{dx} - k^2\right)F\right]\psi_0 \quad (13.39)$$

The results from Eq. (13.39) can be used to obtain the following:

$$\frac{d^2\psi_1}{dx^2} + k^2\psi_1$$

$$= \left[\frac{d^2F}{dx^2} \pm 2ik\frac{dF}{dx} + \left(\pm i\frac{dk}{dx} - k^2\right)F\right]\psi_0 + k^2\psi_1$$

$$= \left[\frac{d^2F}{dx^2} \pm 2ik\frac{dF}{dx} \pm i\frac{dk}{dx}F\right]\psi_0 = 0 \quad (13.40)$$

Thus, we have a differential equation for $F(x)$ given by

$$\frac{d^2F}{dx^2} \pm 2ik\frac{dF}{dx} \pm i\frac{dk}{dx}F = 0 \quad (13.41)$$

The key assumption made in the first order approximation is that

$$\left|\frac{d^2F}{dx^2}\right| \ll \left|\pm 2ik\frac{dF}{dx} \pm i\frac{dk}{dx}F\right| \quad \text{(first order JWKB approximation)} \quad (13.42)$$

When this condition is satisfied everywhere, we have that

$$\pm 2ik\frac{dF}{dx} \pm i\frac{dk}{dx}F \approx 0$$

$$\Rightarrow 2k\frac{dF}{dx} = -\frac{dk}{dx}F \Rightarrow \frac{dF}{F} = -\frac{1}{2}\frac{dk}{k}$$

$$\Rightarrow \log F = -\frac{1}{2}\log k = \log k^{-\frac{1}{2}} \Rightarrow F(x) = \frac{A}{\sqrt{k(x)}}$$

Therefore, under our assumption in Eq. (13.42), we obtain the following solution for $\psi_1(x)$:

13.6 The JWKB Eigenvalue Equation

> **First Order JWKB Approximation**
>
> $$\psi_1(x) = \frac{A}{\sqrt{k(x)}} e^{\pm i \int k \, dx} \qquad (13.43)$$

A is a constant, independent of x. These results lead to a well-known eigenvalue equation, which is the subject of the next section.

13.6 THE JWKB EIGENVALUE EQUATION

Now that we have determined approximations for the wave functions that we can use, let us make use of them to obtain an important equation for the energy. We consider a system with two turning points, a and b. The system is taken to be bounded, which means that the energy E must be less than the potential energy V. This leads to three distinguishable regions, which we denote as region I, II, and III, respectively. The turning point a separates region I from II, and b separates region II from III. Using the JWKB approximation, the first two corresponding wave functions for regions ψ_I and ψ_{II} are given by

$$\psi_I = \frac{A}{\sqrt{k}} e^{\left[-\int_x^a k(x) dx\right]} \qquad (13.44a)$$

$$\psi_{II} = \frac{2A}{\sqrt{k}} \sin\left[\int_a^x k(x) dx + \frac{\pi}{4}\right] \qquad (13.44b)$$

The wave function in region II ψ_{II} is the wave function that we must focus on to determine its properties in order than we may connect to region III so that over the entire range, $\psi = \psi_I, \psi_{II}, \psi_{III}$ is square integrable over space, or *normalizable*. In problems that we will consider, it corresponds to the bound state. Before applying the connection formulae, let's rewrite ψ_{II} by rewriting the argument of the sine function in ψ_{II} as

$$\int_a^x k(x) dx + \frac{\pi}{4} = \int_a^b k(x) dx + \int_b^x k(x) dx + \frac{\pi}{4}$$

$$= \int_a^b k(x) dx - \left[\int_x^b k(x) dx + \frac{\pi}{4}\right] + \frac{\pi}{4} + \frac{\pi}{4}$$

$$= \phi + \frac{\pi}{2} - \left[\int_x^b k(x)dx + \frac{\pi}{4}\right]$$

We can also use the relation $\cos x = \sin\left(\frac{\pi}{2} - x\right)$, where we get

$$\sin\left[\int_a^x k(x)dx + \frac{\pi}{4}\right] = \cos\left[\left(\int_x^b k(x)dx + \frac{\pi}{4}\right) - \phi\right]$$

To further rewrite the wave function, we use the trigonometric identity $\cos(\alpha - \beta) = \cos\alpha\cos\beta + \sin\alpha\sin\beta$. This gives

$$\sin\left[\int_a^x k(x)dx + \frac{\pi}{4}\right]$$

$$= \cos\phi\cos\left(\int_x^b k(x)dx + \frac{\pi}{4}\right) + \sin\phi\sin\left(\int_x^b k(x)dx + \frac{\pi}{4}\right)$$

The wave function in region II ψ_{II} becomes

$$\psi_{II} = \frac{2A}{\sqrt{k}}\left[\sin\phi \cdot \sin\left(\int_x^b k(x)dx + \frac{\pi}{4}\right) + \cos\phi \cdot \cos\left(\int_x^b k(x)dx + \frac{\pi}{4}\right)\right] \tag{13.45}$$

With the wave function in this form, let's now consider the connection formulae. Of the four terms in Eq. (13.45), we only apply the connection formaulae to two of them, namely, those with the integrals involving the x coordinate, which goes to ∞. The wave function must be able to connect to region III in a well-behaved manner. If we consider the cosine term, the connection formula given in Tab. 13.2 tells us it must transition to a positive exponential term where

$$\frac{2}{\sqrt{k}}\cos\left[\int_x^{x_0} kdx + \frac{\pi}{4}\right] \Rightarrow \frac{1}{\sqrt{k}}e^{+\int_x^{x_0} kdx}$$

However, in this case, the exponential term diverges because $x > 0$ and $x \to \infty$, and therefore the only way the wave function can be integrable over space is if one of the coefficients vanishes. This leads to an important condition for a wave function given by

$$\cos\phi = 0 \tag{13.46}$$

We know that the cosine function vanishes when ϕ is an odd multiple of $\pi/2$. Given the definition of ϕ, along with this result, we obtain what is known as the **JWKB Eigenvalue Equation**, and it is given by

13.6 The JWKB Eigenvalue Equation

JWKB Eigenvalue Equation

$$\int_a^b k(x)dx = (2n+1)\frac{\pi}{2} \tag{13.47}$$

Note that for the sine term in Eq. (13.45), it connects to an exponentially decaying term, which does not diverge as $x \to \infty$, thus the cosine term is the only point of concern. Next, let's look at an example of how to use Eq. (13.47).

Exercise 13.1 Using the JWKB approximation, compute the energy eigenvalues for a particle in a potential given by $V(x) = \frac{1}{2}m\omega^2 x^2$, using Eq. (13.47).

Solution

With this potential energy function, the wave-number function $k(x)$ becomes

$$k^2(x) = \frac{2m}{\hbar^2}[E - V(x)]$$
$$= \frac{2m}{\hbar^2}\left[E - \frac{1}{2}m\omega^2 x^2\right]$$
$$= \frac{m^2\omega^2}{\hbar^2}\left(x_0^2 - x^2\right)$$

x_0 is the oscillation amplitude we found in Chap. 5 (c. Eq. (5.11)). The amplitude x_0 also gives the location of x where k^2 vanishes. This is where kinetic energy vanishes, or $K = E - V(x) = 0$. Thus, $a = -x_0$ and $b = +x_0$ are the turning points. Using $k(x)$ above in the eiqenvalue equation given by Eq. (13.47), we have

$$\left(n + \frac{1}{2}\right)\pi = \int_a^b k(x)dx$$
$$= \frac{m\omega}{\hbar}\int_{-x_0}^{+x_0}\sqrt{x_0^2 - x^2}\,dx$$

To evaluate this integral, we use substitution where $x = x_0\cos\theta$, which also leads to $dx = -x_0\sin\theta d\theta$. Making this substitution into the integral, we obtain

$$\frac{m\omega}{\hbar}\int_{-x_0}^{+x_0}\sqrt{x_0^2 - x^2}dx = \frac{m\omega x_0^2}{\hbar}\int_\pi^0 \sin\theta(-\sin\theta d\theta)$$

$$= -\frac{m\omega x_0^2}{\hbar}\int_\pi^0 \sin^2\theta d\theta$$

We evaluated this integral in Chap. 5 (c. Eq. (4.63)) using a trigonometric identity, and if we use the result we obtained there, we now have

$$-\frac{m\omega x_0^2}{\hbar}\int_\pi^0 \sin^2\theta d\theta = -\frac{m\omega x_0^2}{2\hbar}\int_\pi^0 (1-\cos 2\theta)d\theta$$

$$= -\frac{m\omega x_0^2}{2\hbar}\left[-\pi - \left(\frac{1}{2}\sin 2\theta\right)\Big|_\pi^0\right]$$

$$= \frac{m\omega x_0^2}{2\hbar} \times \pi$$

Using the relation $x_0^2 = 2E/m\omega^2$, we then have

$$\left(n+\frac{1}{2}\right)\pi = \frac{m\omega x_0^2}{2\hbar}\pi = \frac{m\omega 2E}{2\hbar m\omega^2}\pi$$

$$\Rightarrow E = \left(n+\frac{1}{2}\right)\hbar\omega$$

∎

In this example, you may have noticed that the result obtained is the exact eigenvalue solution to the 1D harmonic oscillator problem. That result was first given by Eq. (5.68). Note, however, that the reason we have obtained the exact solution here is because *k(x) is exactly linear in x* and smoothly varying. This happens to coincide precisely with the assumptions behind the JWKB approximation! In general, exact results should not be expected if k, especially, is not linear in x, or is not smoothly varying. In the next section, we introduce another method of approximation.

13.7 FIRST ORDER PERTURBATION THEORY

Another useful approach to solving quantum mechanical Hamiltonian eigenvalue equations in an approximate form, utilizes the known solution of a *simpler or unperturbed* version of the same Hamiltonian. Then, the difference between the Hamiltonian of interest and the known or unperturbed Hamiltonian and solution can be considered a differential or *perturbation*. Hence, this approach is known as **perturbation theory**. In general, it consists of a set of successive corrections to an unperturbed problem, to a desired order of correction. Let's say that the system we wish to solve is given by

$$\widehat{H}\psi = E\psi \tag{13.48}$$

It must be possible to express Eq. (13.48) in terms of the *unperturbed* Hamiltonian eigenvalue equation that has a known solution. To the known system equation, a perturbation is added. Let this known system be described by

$$\widehat{H}^{(0)}\psi^{(0)} = E^{(0)}\psi^{(0)} \tag{13.49}$$

Perturbation theory provides a method of obtaining an approximate solution, generally, to an arbitrary order n of accuracey. From Eq. (13.48), the Hamiltonian operator, wave function, and eigenvalue are expressed in the form of an 'expansion', where $\widehat{H}\psi$ is expressed by

$$\widehat{H} = c^0 \widehat{H}^{(0)} + c^1 \widehat{H}^{(1)} + c^2 \widehat{H}^{(2)} + O(c^3) \tag{13.50a}$$

$$\psi = c^0 \psi^{(0)} + c^1 \psi^{(1)} + c^2 \psi^{(2)} + O(c^3) \tag{13.50b}$$

$$E = c^0 E^{(0)} + c^1 E^{(1)} + c^2 E^{(2)} + O(c^3) \tag{13.50c}$$

The order of the parameter c^n (which is a constant c raised to the nth power) is used to determine the order of the perturbation correction. The superscripts surrounded by parenthesis are only used for our bookkeeping. Substitution of Eqs. (13.50) into Eq. (13.48) (using $c^0 = 1$) leads to

$$\left[\widehat{H}^{(0)} + c\widehat{H}^{(1)} + c^2\widehat{H}^{(2)} + O(c^3)\right]\left[\psi^{(0)} + c\psi^{(1)} + c^2\psi^{(2)} + O(c^3)\right] =$$
$$\left[E^{(0)} + cE^{(1)} + c^2E^{(2)} + O(c^3)\right)\left(\psi^{(0)} + c\psi^{(1)} + c^2\psi^{(2)} + O(c^3)\right] \tag{13.51}$$

Multiplying out the terms on the LHS leads to nine terms given by

$$\left[\widehat{H}^{(0)} + c\widehat{H}^{(1)} + c^2\widehat{H}^{(2)} + O(c^3)\right]\left(\psi^{(0)} + c\psi^{(1)} + c^2\psi^{(2)} + O(c^3)\right) =$$

$$\widehat{H}^{(0)}\psi^{(0)} + c\widehat{H}^{(0)}\psi^{(1)} + c^2\widehat{H}^{(0)}\psi^{(2)} +$$

$$c\widehat{H}^{(1)}\psi^{(0)} + c^2\widehat{H}^{(1)}\psi^{(1)} + c^3\widehat{H}^{(1)}\psi^{(2)} +$$

$$c^2\widehat{H}^{(2)}\psi^{(0)} + c^3\widehat{H}^{(2)}\psi^{(1)} + c^4\widehat{H}^{(2)}\psi^{(2)} + O(c^3) \quad (13.52)$$

Each term has a coefficient of the form c^n. In any n^{th} order perturbation theory, all terms with coefficients of order higher than c^n are truncated. Continuing with the RHS, similarly expanding the product leads to

$$\left[E^{(0)} + cE^{(1)} + c^2 E^{(2)} + O(c^3)\right]\left[\psi^{(0)} + c\psi^{(1)} + c^2\psi^{(2)} + O(c^3)\right] =$$

$$E^{(0)}\psi^{(0)} + cE^{(0)}\psi^{(1)} + c^2 E^{(0)}\psi^{(2)} +$$

$$cE^{(1)}\psi^{(0)} + c^2 E^{(1)}\psi^{(1)} + c^3 E^{(1)}\psi^{(2)} +$$

$$c^2 E^{(2)}\psi^{(0)} + c^3 E^{(2)}\psi^{(1)} + c^4 E^{(2)}\psi^{(2)} + O(c^3) \quad (13.53)$$

Equating the two results from Eq. (13.52) and (13.53) leads to

$$\left(\widehat{H}^{(0)}\psi^{(0)} - E^{(0)}\psi^{(0)}\right) + c\left(\widehat{H}^{(0)}\psi^{(1)} + \widehat{H}^{(1)}\psi^{(0)} - E^{(0)}\psi^{(1)} - E^{(1)}\psi^{(0)}\right) +$$

$$c^2\left(\widehat{H}^{(0)}\psi^{(2)} + \widehat{H}^{(1)}\psi^{(1)} + \widehat{H}^{(2)}\psi^{(0)} - E^{(0)}\psi^{(2)} - E^{(1)}\psi^{(1)} - E^{(2)}\psi^{(0)}\right) = O(c^3)$$

This is where the order of the *perturbation theory* approximation is invoked. For example, for second order perturbation theory, we truncate all the terms with coefficients c^3 and higher. This means that we assume $O(c^3) = 0$. Under these conditions, since c is an arbitrary parameter, the only way the LHS can equal zero is for all terms in parentheses to be zero. This leads to $n+1$ conditions for the nth order perturbation. For $n = 2$, we have the following conditions:

$$\left(\widehat{H}^{(0)}\psi^{(0)} - E^{(0)}\psi^{(0)}\right) = 0 \quad (13.54a)$$

$$c\left(\widehat{H}^{(0)}\psi^{(1)} + \widehat{H}^{(1)}\psi^{(0)} - E^{(0)}\psi^{(1)} - E^{(1)}\psi^{(0)}\right) = 0 \quad (13.54b)$$

$$c^2\left(\widehat{H}^{(0)}\psi^{(2)} + \widehat{H}^{(1)}\psi^{(1)} + \widehat{H}^{(2)}\psi^{(0)} - E^{(0)}\psi^{(2)} - E^{(1)}\psi^{(1)} - E^{(2)}\psi^{(0)}\right) = 0 \quad (13.54c)$$

13.7 First Order Perturbation Theory

The first of the conditions is expected because it is only restating Eq. (13.49), which the above defines as the zeroth order perturbation. Since this condition is satisfied, let us consider the next term, which defines the **first order perturbation**, given by

$$c\left(\widehat{H}^{(0)}\psi^{(1)} + \widehat{H}^{(1)}\psi^{(0)} - E^{(0)}\psi^{(1)} - E^{(1)}\psi^{(0)}\right) = 0$$

or

$$\widehat{H}^{(0)}\psi^{(1)} + \widehat{H}^{(1)}\psi^{(0)} - E^{(0)}\psi^{(1)} - E^{(1)}\psi^{(0)} = 0 \tag{13.55}$$

Eq.(13.55) can be used to obtain the first order energy perturbation $E^{(1)}$. When we left multiply Eq. (13.55) by $\psi^{(0)}$, then integrate over space, we have

$$0 = \int \psi^{\dagger(0)}\left(\widehat{H}^{(0)} - E^{(0)}\right)\psi^{(1)}dx + \int \psi^{\dagger(0)}\left(\widehat{H}^{(1)}\psi^{(0)} - E^{(1)}\right)\psi^{(0)}dx$$

$$= \int \left(\psi^{\dagger(0)}\widehat{H}^{(0)}\psi^{(1)} - \psi^{\dagger(0)}E^{(0)}\psi^{(1)}\right)dx + \int \psi^{\dagger(0)}\left(\widehat{H}^{(1)}\psi^{(0)} - E^{(1)}\right)\psi^{(0)}dx$$

$$= \int \left(\psi^{\dagger(0)}E^{(0)}\psi^{(1)} - \psi^{\dagger(0)}E^{(0)}\psi^{(1)}\right)dx + \int \psi^{\dagger(0)}\left(\widehat{H}^{(1)}\psi^{(0)} - E^{(1)}\right)\psi^{(0)}dx$$

In the above, we have used Eq. (13.49), which also implies

$$\widehat{H}^{(0)}\psi^{(0)} = E^{(0)}\psi^{(0)} \Rightarrow \psi^{\dagger(0)}\widehat{H}^{(0)} = \psi^{\dagger(0)}E^{(0)}$$

The first term vanishes and if we also assume $\psi^{(0)}$ is normalized, then we obtain the *first order perturbation* result given by

---

First-Order Perturbation Energy $E^{(1)}$

$$\int \psi^{\dagger(0)}\widehat{H}^{(1)}\psi^{(0)}dx = E^{(1)} \tag{13.56}$$

Next, let's look at an example of how to apply first order perturbation theory.

Exercise 13.2 Consider the 1D particle in a box problem, where the potential inside the box is zero, or $V_0 = 0$. Keeping everything else the same, let's add a potential inside the box, of the form

$$\frac{V_0 x}{L}$$

Compute the first order correction to the ground state energy, treating the additional potential as a perturbation.

$$\psi_n^{(0)} = \sqrt{\frac{2}{L}}\sin\frac{n\pi x}{L} \quad \text{and} \quad E_n^{(0)} = \frac{n^2\pi^2\hbar}{2mL} \quad n = 1, 2, 3, \ldots$$

and

$$\widehat{H}^{(1)} = \frac{V_0 x}{L} \quad 0 < x < L$$

Substitution of the Hamiltonian perturbation into Eq. (13.56), we have

$$E^{(1)} = \int \psi^{\dagger(0)} \widehat{H}^{(1)} \psi^{(0)} dx$$

$$= \frac{2V_0}{L^2} \int_0^L x\sin^2\frac{n\pi x}{L} dx$$

$$= \left[\frac{x^2}{4} - \frac{L\sin\left(\frac{2n\pi x}{L}\right)}{4\pi n} - \frac{L^2\cos\left(\frac{2n\pi x}{L}\right)}{8\pi^2 n^2}\right]_0^L$$

$$= \frac{V_0}{2}$$

Therefore, the first order correction becomes

$$E^{(1)} = \frac{V_0}{2}$$

∎

13.8 Chapter Summary

In this chapter, we have introduced some methods of approximation that are often used in quantum mechanics. First, we discussed the JWKB approximation based on Airy functions. We showed how approximations

13.9 Chapter Problems

to the Airy functions can be derived and applied to quantum mechanics. For the JWKB approximation, the potential functions should be smoothly varying.

Lastly, we discussed perturbation theory as another method of approximation that can be used in quantum mechanics. For perturbation theory, the perturbation must be small in comparison to the zero order known energy. Table 13.3 summarizes some of the key formulas from this chapter.

Table 13.3: Chapter 12 Summary Equations.

Name	Equation
Airy Function (First Kind)	$\text{Ai}(x) = \frac{1}{\sqrt{2\pi}} \int_{-\infty}^{+\infty} e^{ipx} e^{ip^3/3} dp$
Airy Function Approximation	$\widetilde{\text{Ai}}(x) = \sqrt{\frac{2}{\sqrt{-x}}} \sin\left(\frac{2}{3}(-x)^{\frac{3}{2}} + \frac{\pi}{4}\right)$
Airy Function Approximation	$\widetilde{\text{Bi}}(x) = \sqrt{\frac{2}{\sqrt{-x}}} \cos\left(\frac{2}{3}(-x)^{\frac{3}{2}} + \frac{\pi}{4}\right)$
Airy Function Approximation	$\widetilde{\text{Bi}}(x>0) = Cx^{-\frac{1}{4}} e^{\frac{2}{3}x^{\frac{3}{2}}}$
First Order JWKB Wave function	$\psi_1(x) = \frac{A}{\sqrt{k(x)}} e^{\pm i \int k dx}$
JWKB Eigenvalue Equation	$\int_a^b k(x) dx = (2n+1)\frac{\pi}{2}$
First-Order Perturbation Energy	$\int \psi^{\dagger(0)} \widehat{H}^{(1)} \psi^{(0)} dx = E^{(1)}$

13.9 Chapter Problems

Problem 13.1 Find the turning points of a JWKB approximation for a system with $k^2(x)$ given by

$$k^2(x) = \frac{2m}{\hbar^2}\left(E_\lambda - \frac{1}{2}m\omega^2 x^2\right)$$

Problem 13.2 Consider a Hamiltonian with kinetic energy and a potential energy defined by

$$V(x) = \begin{cases} \beta x & \text{if } x > 0 \\ \infty & \text{if } x \leq 0 \end{cases}$$

Use the JWKB eigenvalue equation given by Eq. (13.47) to show that

$$E_\lambda = \left(\frac{\hbar^2 \beta^2}{2m}\right)\left[\frac{3\pi}{2}\left(n + \frac{1}{2}\right)\right]^{2/3}$$

Note that the integration should be done from 0 to E_λ/β. What does E_λ/β represent?

It turns out that this result is not a very good approximation to the exact solution. Discuss any contributions to this fact you can see. Consider the assumptions of the JWKB approximation.

Problem 13.3 For the system in the previous problem, write down both the exponentially decaying solution and the trigonometric solution given by Eqs. (13.44a) and (13.44b). First determine $k(x)$ using the turning point E_λ/β.

Problem 13.4 If one assumes the spin remains in a plane containing the itinerant spin and spin of a magnetic polarizer, the *spinor* wave function in the JWKB approximation can be written as

$$\psi_\pm = \begin{bmatrix} \frac{A_+}{\sqrt{k_+}} e^{i\int_0^x k_+ dx} \cos(\theta/2) \\ \frac{A_-}{\sqrt{k_-}} e^{i\int_0^x k_- dx} \sin(\theta/2) \end{bmatrix}$$

$k_\pm$ are the respective wave numbers for spin-up and spin-down, which generally, are *not* equal. This comes from the spatial part due to the JWKB approximation and the spin eigenspinor we derived in Eq. (9.38). Use this spinor to obtain the expectation value of the itinerant spin along the z-axis, or polarizer axis, in this approximation, is given by

$$\langle \mu_z \rangle = \frac{g|\gamma|\hbar}{2} \frac{\frac{|A_+|^2}{k_+}\cos^2\theta/2 - \frac{|A_-|^2}{k_-}\sin^2\theta/2}{\frac{|A_+|^2}{k_+}\cos^2\theta/2 + \frac{|A_-|^2}{k_-}\sin^2\theta/2}$$

Comment on the spin polarization at angles $\theta = 0$ and $\theta = \pi$.

13.10 Suggested Readings & References

Problem 13.5 From the previous problem, show that in the limit of $|A_+|^2 = |A_-|^2$ and $k_+ = k_-$, that

$$\langle \mu_z \rangle = \mu_B \cos\theta$$

Between this result and the spin polarization from the previous problem, compare the two. Which is smaller?

Problem 13.6 Using the spinor in the JWKB approximation, determine the spin-up and spin-down current densities $\widehat{J}_+$ and $\widehat{J}_-$. Use

$$\widehat{J} = \frac{i\hbar}{2m}\left[\psi\frac{\partial\psi^*}{\partial x} - \psi^*\frac{\partial\psi}{\partial x}\right]$$

on the spinor components to determine the respective current densities.

Problem 13.7 Consider a potential given by

$$V_0(x) = \frac{1}{2}kx^2 + \frac{1}{6}\gamma_3 x^3 + \frac{1}{24}\gamma_4 x^4$$

Recall that for the harmonic oscillator, the Hamiltonian $\widehat{H}_0$ is given

$$\widehat{H}_0 = \frac{1}{2}kx^2$$

Show that the first order perturbation correction, using the potential V_0 is given by

$$E_n^{(1)} = \frac{\hbar^2 \gamma_4}{32km}$$

Problem 13.8 Calculate the first-order correction using perturbation theory, to the energy of a particle constrained to move within a region $0 \leq x \leq a$ in the potential

$$V(x) = \begin{cases} V_0 x & \text{if } 0 \leq x \leq \frac{a}{2} \\ V_0(a-x) & \text{if } \frac{a}{2} \leq x \leq a \end{cases}$$

where V_0 is constant.

13.10 Suggested Readings & References

[1] G.B. Airy, *On the intensity of light in the neighbourhood of a caustic* Trans. Cambridge Philos. Soc., **6** (1838) pp. 379–402.

CHAPTER 14

Introduction To BRA & KET Algebra

In this chapter, we introduce a popular and concise notation widely used in quantum mechanics, introduced by Paul. A. M. Dirac. It is both powerful and more concise than a conventional calculus notation with explicit derivatives and integrals. For this reason and more, it is regarded as a *powerfully elegant* notation. It provides an alternative language or notation to express all the operations and eigenstates that we have seen throughout the text. And though it's more concise, it still allows great depth in analyzing problems, as we'll see. This notation has come to be known as *Bra-Ket Algebra*, named for its use of bracket components $\langle$ and $\rangle$. Let's begin with some preliminaries and definitions useful for Bra-Ket algebra.

14.1 Bra-Ket Preliminaries

From any eigenoperator equation, such as the Schrödinger equation, one obtains the solutions known as eigenfunctions ψ, along with their associated eigenvalues. For bound states, it also leads to a given set of quantum numbers and the resulting eigenvalues and eigenfunctions define the eigenstates of the system. The system may be in one of them, or some combination of them. Measurements of the observable states can serve to determine this combination, explicitly. In the Bra-Ket notation, each eigenstate A is represented by a **Ket-vector** $|\cdot\rangle$, expressed as

$$\psi_A : \text{state} A \Rightarrow |A\rangle$$

The name *Ket* is obtained from the fact that *Ket* is the right side of Bra-Ket (as in bracket). That's right, its that simple. Hence, a *Ket-vector points to the right*. Note that it represents a *generalized eigenvector* like the 2×1 spinor or Dirac's 4-state vector, etc. The Ket-vector operates in a linear space, which means that any Ket can also be expressed as a linear combination of other properly chosen *Ket* vectors. For example, consider any two *Ket* vectors $|A\rangle$ and $|B\rangle$ such that their sum is

$$|A\rangle + |B\rangle = |C\rangle$$

Then, $|C\rangle$ is also a *Ket* vector living in the same mathematical space. More generally, the following state is a *Ket* vector in a linear space:

$$c_1|A\rangle + c_2|B\rangle = |D\rangle$$

Coefficients c_1 and c_2 are scalars and the property of a *linear space* is that $|D\rangle$ is also a Ket vector. Additionally, for any Ket vector $|A\rangle$, there also exists a **bra vector** $\langle A|$. In this case, *Bra* is the left side of the word *Bra-Ket*, hence, the *bra vector points to the left*. Given the Bra and Ket vectors, we can write the *inner product* or *scalar product* of any two states using Bra and Ket vectors as

$$\text{inner product} \Rightarrow \langle B|A\rangle$$

Having an *inner product*, we can then define *orthogonal states*. Two states are said to be *orthogonal* if

$$\langle B|A\rangle = 0 \Rightarrow |B\rangle \perp |A\rangle$$

A state $|A\rangle$ is also said to be *normalized* if

$$\langle A|A\rangle = 1 \quad \text{(normalization condition)} \tag{14.1}$$

Eq. (14.1) is the usual *normalization condition*, which we have discussed in the previous chapters. As a scalar product is generally complex. Then, it must also have a complex conjugate. Using Bra-Ket notation, the complex conjugate, denoted by $\overline{\langle A|B\rangle}$ is given by

14.1 Bra-Ket Preliminaries

> **Complex-Conjugate Using Bra-Ket Notation**
>
> $$\overline{\langle A|B\rangle} = \langle B|A\rangle \tag{14.2}$$

It follows from the complex conjugate definition, that if $|B\rangle = |A\rangle$, then

$$\overline{\langle A|A\rangle} = \langle A|A\rangle \Rightarrow \langle A|A\rangle \text{ is real}$$

For a nontrivial state $|B\rangle$, if $\langle A|B\rangle = 0$ for any Bra vector $\langle A|$, then $|B\rangle$ must be the **Null Ket vector**. Let's consider some explicit examples of *Bras* and *Kets*.

Consider the following *Ket* vectors:

$$|A\rangle = \begin{bmatrix} a \\ b \end{bmatrix} \quad \text{and} \quad |B\rangle = \begin{bmatrix} c \\ d \end{bmatrix}$$

a, b, c, and d are all scalars, and thus, both Kets are 2×1 vectors. Their corresponding Bra vectors, are then given by

$$\langle A| = \begin{bmatrix} a^* & b^* \end{bmatrix} \quad \text{and} \quad \langle B| = \begin{bmatrix} c^* & d^* \end{bmatrix}$$

Here a^*, b^*, c^*, and d^* are complex conjugates of a, b, c, and d, respectively. Thus, for a Ket vector, the Bra vector is the *transpose of the complex conjugates*.

Adding these two Kets (2×1 vectors) leads to another Ket (2×1 vectors), as we have

$$|A\rangle + |B\rangle = \begin{bmatrix} a+c \\ b+d \end{bmatrix}$$

Likewise, adding the Bra vectors, which are 1×2 vectors, we have

$$\langle A| + \langle B| = \begin{bmatrix} a^* + c^* & b^* + d^* \end{bmatrix}$$

Computing $\langle A|A\rangle$ becomes

$$\langle A|A\rangle = \begin{bmatrix} a^* & b^* \end{bmatrix} \begin{bmatrix} a \\ b \end{bmatrix} = a^*a + b^*b = |a|^2 + |b|^2 \geq 0$$

Note that because $|a| \geq 0$ and $|b| \geq 0$, then $\langle A|A\rangle = 0$ *if and only if* $|a| = |b| = 0$. For this reason, the inner product can only vanish if one of the states is the *Null Ket*. Otherwise, we have $\langle A|A\rangle > 0$.

Let us now take $a = b = 1$. Then

$$\langle A|A\rangle = 1 + 1 = 2$$

To normalize the Ket $|A\rangle$, we would have to scale $|A\rangle$ by $1/\sqrt{2}$, because then we have

$$\langle A|A\rangle = \frac{1}{\sqrt{2}}\begin{bmatrix}a^* & b^*\end{bmatrix}\frac{1}{\sqrt{2}}\begin{bmatrix}a \\ b\end{bmatrix} = \frac{1}{2}(1+1) = 1$$

This defines a *normalized Ket* $|A'\rangle$ given by

$$|A'\rangle = \frac{1}{\sqrt{2}}\begin{bmatrix}1 \\ 1\end{bmatrix}$$

$\langle A'|A'\rangle = 1$ satisfies the normalization condition in Eq. (14.1). Let the normalized Ket vector $|B'\rangle$ be given by

$$|B'\rangle = \frac{1}{\sqrt{2}}\begin{bmatrix}1 \\ -1\end{bmatrix}$$

Then $\langle B'|A'\rangle$ becomes

$$\langle B'|A'\rangle = \frac{1}{\sqrt{2}}\begin{bmatrix}1 & -1\end{bmatrix}\frac{1}{\sqrt{2}}\begin{bmatrix}1 \\ 1\end{bmatrix} = \frac{1}{2}(1-1) = 0$$

In this case, these two vectors are said to be *orthogonal* to each other. This is easy to visualize in this example, because $|A'\rangle$ is a 45° vector, in the first quadrant of a 2D plane, while $|B'\rangle$ lies at −45° in the fourth quadrant of a 2D plane, i.e. they are 90° apart. This is illustrated in Fig. 14.1. Let us also take a similar set of vectors, involving complex numbers.

$$|C'\rangle = \frac{1}{\sqrt{2}}\begin{bmatrix}1 \\ -i\end{bmatrix} \quad \text{and} \quad |D'\rangle = \frac{1}{\sqrt{2}}\begin{bmatrix}1 \\ +i\end{bmatrix}$$

Testing for orthogonality, we have

$$\langle D'|C'\rangle = \frac{1}{2}\begin{bmatrix}1 & -i\end{bmatrix}\cdot\begin{bmatrix}1 \\ -i\end{bmatrix} = \frac{1}{2}[1\cdot 1 + -i\cdot(-i)] = 0$$

This set of vectors is analogous to $|A'\rangle$ and $|B'\rangle$. They form vectors that are 90° apart in the *complex plane*, as shown in Fig. 14.2. Note that, in this case, the Bra vector of $|D'\rangle$ and the Ket vector $|C'\rangle$ contain the same elements, however, they do not equate, because a Bra can never equal a

14.1 Bra-Ket Preliminaries

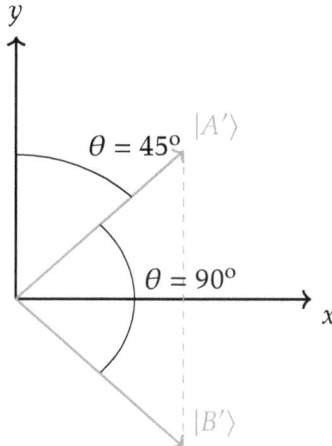

Figure 14.1: Illustration of orthogonal states $|A'\rangle$, represented by Kets. In the state space, the two states form a 2D coordinate system that can describe any state that is a combination of the two eigenstates. These are the conditions for eigenstates which are orthogonal states.

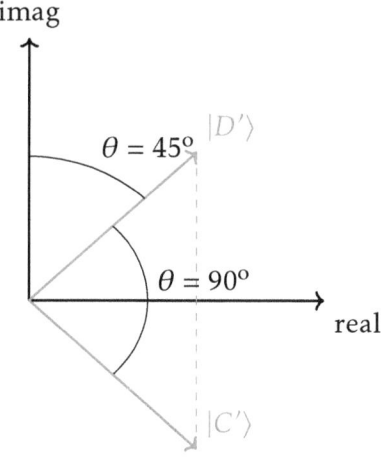

Figure 14.2: Illustration of orthogonal states $|A'\rangle$, represented by Kets. In the state space, the two states form a 2D coordinate system that can describe any state that is a combination of the two eigenstates. These are the conditions for eigenstates obtained, which are orthogonal states.

Ket. They are defined as distinct constructs. For this reason, the vector space of Bras and Kets is said to be a **dual vector space**.

All the examples above contain *fixed* numbers, or constants. They do *not* depend on a variable. These concepts, however, extend to variable elements, as well. An important difference that arises when the Ket vector depends on space **r** is that the inner product then integrates over all of space. For example, let us consider the following states

$$|E\rangle = \psi(x) \quad \text{and} \quad |F'\rangle = \beta(x)$$

Then, the inner product $\langle E|F'\rangle$ becomes

$$\langle E|F'\rangle \equiv \int_{x=-\infty}^{+\infty} \psi^*(x)\beta(x)dx$$

Then, the *orthogonality* condition becomes

$$\langle E|F'\rangle \equiv \int_{x=-\infty}^{+\infty} \psi^*(x)\beta(x)dx = 0$$

Thus, any two states satisfying this condition are said to be orthogonal. This is restating what we have already learned prior to introducing Bra and Ket algebra. However, Bra-Ket notation provides an alternative, more concise notation for expressing this condition. Likewise, the *normalization condition* becomes

$$\langle E|E\rangle \equiv \int_{x=-\infty}^{+\infty} \psi^*(x)\psi(x)dx = 1$$

The concepts extend to 3D, where

$$\langle E|F'\rangle \equiv \int_{z=-\infty}^{+\infty}\int_{y=-\infty}^{+\infty}\int_{x=-\infty}^{+\infty} \psi^*(x,y,z)\beta(x,y,z)dxdydz = 0$$

and

$$\langle E|E\rangle \equiv \int_{z=-\infty}^{+\infty}\int_{y=-\infty}^{+\infty}\int_{x=-\infty}^{+\infty} \psi^*(x,y,z)\psi(x,y,z)dxdydz = 1$$

The conciseness of the Bra-Ket notation is more evident, the higher the dimension or the more degrees of freedom a system has. Let us now discuss the next step of applying operators to Bras and Kets.

14.2 Operators With Bras and Kets

In Chap. 4, we introduced the idea of a Hermitian operator. Now, we will use Bra-Ket algebra as a tool to explore properties of operators, including

14.2 Operators With Bras and Kets

Hermitian or self-adjoint operators. Let's consider a self-adjoint operator where $\overline{\Omega} = \Omega$. And let the eigenoperator equation satisfied by Ω be

$$\Omega|A_n\rangle = \lambda_n|A_n\rangle \tag{14.3}$$

For now, we do not assume anything about λ_n. However, our goal is to show that if $\overline{\Omega} = \Omega$, then it follows that λ_n *must* be real. First, let $|C_n\rangle = \Omega|A_n\rangle$. Our results in the previous sections revealed that this leads to $\langle C_n| = \langle A_n|\overline{\Omega}$. Since Ω is self-adjoint, then we also have

$$\langle A_n|\overline{\Omega} = \langle A_n|\Omega = \lambda_n^*\langle A_n| \tag{14.4}$$

In Eq. (14.4), we have used the eigenoperator equation. Then, right-multiplying by $|A_n\rangle$, we have

$$\langle A_n|\Omega|A_n\rangle = \lambda_n^*\langle A_n|A_n\rangle \tag{14.5}$$

However, if we left-multiply Eq. (14.3) by $\langle A_n|$, we get the following:

$$\langle A_n|\Omega|A_n\rangle = \langle A_n|\lambda_n|A_n\rangle = \lambda_n\langle A_n|A_n\rangle \tag{14.6}$$

But, Eq. (14.5) and Eq. (14.6) are the same, both giving $\langle A_n|\Omega|A_n\rangle$. Then, it follows that

$$\lambda_n^*\langle A_n|A_n\rangle = \lambda_n\langle A_n|A_n\rangle$$

If we assume that $|A_n\rangle$ is not equal to the *null Ket* (this is always assumed for any eigenstate), then it follows that

$$\lambda_n^* = \lambda_n \tag{14.7}$$

The only numbers for which Eq. (14.7) is true are real numbers. Hence, λ_n is real. So, *a self-adjoint or Hermitian operator always has real eigenvalues*. This is why it is also called a *real operator*. Recall that we proved this first in Chap. 4, using conventional mathematical notation. Here, the same result has been obtained using Bra-Ket algebra.

14.2.1 Orthogonal Eigenstates

Let us go further into some consequences for a *self-adjoint operator* using Bra-Ket notation. Consider two distinct eigenstates m and n defined by

$$\Omega|A_n\rangle = \lambda_n|A_n\rangle \tag{14.8a}$$

$$\Omega|A_m\rangle = \lambda_m|A_m\rangle \tag{14.8b}$$

Since Ω is a self-adjoint operator, then both λ_n and λ_m are real, and $\lambda_n \neq \lambda_m$. Now, let $|C_n\rangle = \Omega|A_n\rangle$. Multiplying Eq. (14.8b) by $\langle A_n|$, we have

$$\langle A_n|\Omega|A_m\rangle = \langle A_n|\lambda_m|A_m\rangle = \lambda_m\langle A_n|A_m\rangle \tag{14.9}$$

Additionally, from the definition of $|C_n\rangle$, it follows that

$$\langle C_n| = \langle A_n|\overline{\Omega} = \langle A_n|\Omega = \lambda_n\langle A_n| \tag{14.10}$$

Right-multiplying Eq. (14.10) by $|A_m\rangle$ leads to

$$\langle C_n|A_m\rangle = \lambda_n\langle A_n|A_m\rangle \tag{14.11}$$

Subtracting Eq. (14.11) from Eq. (14.9), we have

$$(\lambda_m - \lambda_n)\langle A_n|A_m\rangle = 0 \tag{14.12}$$

Since we assume $\lambda_m \neq \lambda_n$, then we have the result

$$\langle A_n|A_m\rangle = 0 \tag{14.13}$$

This is the definition of **orthogonal eigenstates**. Thus, for any two distinct states of a self-adjoint operator having distinct eigenvalues, they are orthogonal eigenstates.

14.3 Linear Harmonic Oscillator Revisited

To illustrate a useful application and utility of Bra-Ket algebra, we will revisit the problem of the linear harmonic oscillator (LHO), discussed in Chap. 5. The LHO problem can be solved using Bra-Ket algebra. However, before we do this, lets revisit two useful anti-symmetric operators, originally introduced by P. A. M. Dirac. Recall that the Dirac equation, discussed in Chap. 11, also recognized symmetry in the relativistic energy $\overline{E}$, and exploited the symmetry to devise a first order operator for $\overline{E}$, given by Eqs. (11.43) and (11.44). This result also illustrated how an operator can mirror operator symmetry in the eigenvalues.

Recall an earlier result we found for $[x,\widehat{p}_x]$ (c.Eq. 7.99), where we determined

$$[x,\widehat{p}_x] = x\widehat{p}_x - \widehat{p}_x x = i\hbar \tag{14.14}$$

14.3 Linear Harmonic Oscillator Revisited

x and $\widehat{p}_x$ do not commute. In our first encounter with the LHO problem (1D), we found the Hamiltonian $\widehat{H}$ to be

$$\widehat{H} = \frac{\widehat{p}_x^2}{2m^*} + \frac{1}{2}m^*\omega^2 x^2 \tag{14.15}$$

In Eq. (14.15), m^* is the *reduced mass* (c.Eq. (8.22)), ω is the harmonic oscillator frequency. Just as we observed with relativistic energy, which lead to the Dirac equation, the Hamiltonian in Eq. (14.15) also possesses symmetry of the form $a^2 + b^2 = (a+ib)(a-ib) = c \cdot c*$, where $c*$ denotes the complex conjugate of $c = a + ib$. Then, let us define the following *two* 1D anti-symmetric operators:

$$\widehat{a} \equiv \frac{m^*\omega x + i\widehat{p}_x}{\sqrt{2m^*\hbar\omega}} \quad \text{and} \quad \widehat{a}^\dagger \equiv \frac{m^*\omega x - i\widehat{p}_x}{\sqrt{2m^*\hbar\omega}} \tag{14.16}$$

Note that $\widehat{a}^\dagger$ represents the *adjoint* of $\widehat{a}$, and they are *not equal*. Thus, $\widehat{a}$ is an example of an operator that is *not self-adjoint*. Consider the operator $\hbar\omega\widehat{a}\widehat{a}^\dagger$ given by

$$\hbar\omega\widehat{a}\widehat{a}^\dagger = \left[\frac{m^*\omega x + i\widehat{p}_x}{\sqrt{2m^*\hbar\omega}}\right]\left[\frac{m^*\omega x - i\widehat{p}_x}{\sqrt{2m^*\hbar\omega}}\right]$$

$$= \frac{1}{2}m^*\omega^2 x^2 + \frac{\widehat{p}_x^2}{2m^*} - i\frac{\omega}{2}(x\widehat{p}_x - \widehat{p}_x x)$$

$$= \frac{1}{2}m^*\omega^2 x^2 + \frac{\widehat{p}_x^2}{2m^*} + \frac{\hbar\omega}{2} = \widehat{H} + \frac{\hbar\omega}{2}$$

In obtaining the above result, we have used Eq. (14.14). Then, it follows that $\widehat{H}$, the Hamiltonian for the LHO, is given by

$$\widehat{H} = \hbar\omega\left(\widehat{a}\widehat{a}^\dagger - \frac{1}{2}\right) \tag{14.17}$$

Similarly, for $\hbar\omega\widehat{a}^\dagger\widehat{a}$, we have

$$\hbar\omega\widehat{a}^\dagger\widehat{a} = \left[\frac{m^*\omega x - ip}{\sqrt{2m^*\hbar\omega}}\right]\left[\frac{m^*\omega x + ip}{\sqrt{2m^*\hbar\omega}}\right]$$

$$= \frac{1}{2}m^*\omega^2 x^2 + \frac{\widehat{p}_x^2}{2m^*} + i\frac{\omega}{2}(xp_x - \widehat{p}_x x)$$

$$= \frac{1}{2}m^*\omega^2 x^2 + \frac{\widehat{p}_x^2}{2m^*} - \frac{\hbar\omega}{2} = \widehat{H} - \frac{\hbar\omega}{2}$$

From this result, we have

$$\widehat{H} = \hbar\omega\left(\widehat{a}^\dagger\widehat{a} + \frac{1}{2}\right) \tag{14.18}$$

This result, along with what we already know about the harmonic oscillator, leads to a relation to an operator for the LHO quantum number n. Recall that for the harmonic oscillator, the energy eigenvalue is $\hbar\omega(n + 1/2)$. Comparing to Eq. (14.18), we find an operator for the quantum number or $\widehat{n}$ is given by

---

Operator For LHO Quantum Number n

$$\widehat{n} = \widehat{a}^\dagger\widehat{a} \tag{14.19}$$

Right-multiplying $\hbar\omega\widehat{a}\widehat{a}^\dagger$ by $\widehat{a}$, and left-multiplying $\hbar\omega\widehat{a}^\dagger\widehat{a}$ by $\widehat{a}$ must lead to equal quantities where we have

$$\hbar\omega\widehat{a}\widehat{a}^\dagger\widehat{a} = \widehat{H}\widehat{a} + \frac{\hbar\omega}{2}\widehat{a}$$

$$\hbar\omega\widehat{a}\widehat{a}^\dagger\widehat{a} = \widehat{a}\widehat{H} - \frac{\hbar\omega}{2}\widehat{a}$$

Setting these two equations equal leads to

$$\widehat{H}\widehat{a} = \widehat{a}\widehat{H} - \hbar\omega\widehat{a} \Rightarrow \left[\widehat{a}, \widehat{H}\right] = \hbar\omega\widehat{a} \tag{14.20}$$

Applying Eq. (14.20) to any *eigenket* (or eigenstate) of $\widehat{H}$ where $\widehat{H}|A\rangle = \lambda|A\rangle$, we have

$$\widehat{H}\widehat{a}|A\rangle = \widehat{a}\widehat{H}|A\rangle - \hbar\omega\widehat{a}|A\rangle = \lambda(\widehat{a}|A\rangle) - \hbar\omega(\widehat{a}|A\rangle) \tag{14.21}$$

Eq. (14.21) leads to two important consequences. To explore the first of these, let us define the Ket $|B\rangle = \widehat{a}|A\rangle$. Then, we have that

$$\widehat{H}|B\rangle = \widehat{H}\widehat{a}|A\rangle = \lambda|B\rangle - \hbar\omega|B\rangle = (\lambda - \hbar\omega)|B\rangle \tag{14.22}$$

This process can be continued indefinitely. So, let us apply Eq. (14.20) to a vector $|C\rangle = \widehat{a}|B\rangle$. We then have

$$\widehat{H}|C\rangle = \widehat{H}\widehat{a}|B\rangle = \widehat{a}\widehat{H}|B\rangle - \hbar\omega\widehat{a}|B\rangle$$

$$= (\lambda - \hbar\omega)\widehat{a}|B\rangle - \hbar\omega\widehat{a}|B\rangle$$

14.3 Linear Harmonic Oscillator Revisited

$$= (\lambda - 2\hbar\omega)|C\rangle = \widehat{H}\widehat{a}^2|A\rangle$$

Note that we have used Eq. (14.22) to substitute for $\widehat{H}|B\rangle$. Continuing this systematic process, we obtain a generalized result given by

$$\widehat{H}\widehat{a}^n|A\rangle = (\lambda - n\hbar\omega)|A\rangle \tag{14.23}$$

Note also that, $|A\rangle \equiv \widehat{a}^0|A\rangle$, $|B\rangle = \widehat{a}|A\rangle$, $|C\rangle = \widehat{a}|B\rangle = \widehat{a}^2|A\rangle$, etc. It is important to mention that $|A\rangle$ is a *non-null eigenket* of the LHO Hamiltonian $\widehat{H}$. This property can be used to obtain the second property we are after, which ultimately allows one to obtain the LHO eigenvalues λ.

The second consequence can be worked out as follows: First, for $|B\rangle = \widehat{a}|A\rangle$, let us write

$$\hbar\omega\langle B|B\rangle = \hbar\omega\langle A|\widehat{a}^\dagger \widehat{a}|A\rangle = \langle A|\hbar\omega \widehat{a}^\dagger \widehat{a}|A\rangle \tag{14.24}$$

$$= \langle A|\widehat{H} - \frac{\hbar\omega}{2}|A\rangle \tag{14.25}$$

$$= (\lambda - \frac{\hbar\omega}{2})\langle A|A\rangle \tag{14.26}$$

In the above, we have used our earlier result $\left(\widehat{H} - \frac{\hbar\omega}{2}\right)|A\rangle = (\lambda - \frac{\hbar\omega}{2})|A\rangle$. Since $|A\rangle$ as a *non-null eigenket* of $\widehat{H}$, then $\langle A|A\rangle > 0$. Therefore, the eigenvalues of $\widehat{H}$ must satisfy the condition

$$\lambda = \frac{\hbar\omega}{2} \Rightarrow \langle B|B\rangle = 0 \tag{14.27}$$

We know from the first part, this corresponds to the ground state of the LHO. Thus, another ground state condition becomes

$$B\rangle = \widehat{a}|A\rangle = 0 \quad \text{LHO-Ground-state condition} \tag{14.28}$$

We'll come back to this condition later, when we determine the explicit wave functions from Eq. (14.28). Now, let's continue the process with $|C\rangle = \widehat{a}|B\rangle$:

$$\hbar\omega\langle C|C\rangle = \hbar\omega\langle B|\widehat{a}^\dagger \widehat{a}|B\rangle = \langle B|\hbar\omega \widehat{a}^\dagger \widehat{a}|B\rangle$$

$$= \langle B|\widehat{H} - \frac{\hbar\omega}{2}|B\rangle$$

$$= \langle B|\widehat{H}|B\rangle - \frac{\hbar\omega}{2}|B\rangle$$

$$= (\lambda - \hbar\omega)\langle B|B\rangle - \frac{\hbar\omega}{2}\langle B|B\rangle$$

$$= \left(\lambda - \frac{3\hbar\omega}{2}\right)\langle B|B\rangle$$

From this result, we have that

$$\lambda = \frac{3\hbar\omega}{2} \Rightarrow C\rangle = \widehat{a}^2|A\rangle = 0 \tag{14.29}$$

Applying this process to $|D\rangle = \widehat{a}|C\rangle$, we have

$$\hbar\omega\langle D|D\rangle = \hbar\omega\langle C|\widehat{a}^\dagger \widehat{a}|C\rangle = \langle C|\hbar\omega\widehat{a}^\dagger \widehat{a}|C\rangle$$

$$= \langle C|\widehat{H} - \frac{\hbar\omega}{2}|C\rangle$$

$$= \langle C|H|C\rangle - \frac{\hbar\omega}{2}|C\rangle$$

$$= (\lambda - 2\hbar\omega)\langle C|C\rangle - \frac{\hbar\omega}{2}\langle C|C\rangle$$

$$= (\lambda - \frac{5\hbar\omega}{2})\langle C|C\rangle$$

Hopefully, the pattern is becoming clear. Summarizing the results above, we have

$$\hbar\omega\langle B|B\rangle = 0 = (\lambda - (1 + \frac{1}{2})\frac{\hbar\omega}{2})\langle A|A\rangle$$

$$\hbar\omega\langle C|C\rangle = 0 = (\lambda - \frac{3}{2}\hbar\omega)\langle B|B\rangle$$

$$\hbar\omega\langle D|D\rangle = 0 = (\lambda - \frac{5}{2}\hbar\omega)\langle C|C\rangle$$

$$\vdots$$

Thus, the LHS vanishes, and because $\langle A|A\rangle > 0$, we have the following result for λ:

$$\lambda = \frac{\hbar\omega}{2}$$

$$\lambda = \frac{3}{2}\hbar\omega$$

$$\lambda = \frac{5}{2}\hbar\omega$$

14.4 Annihilation and Creation Operators Revisited

⋮

This result for λ can be expressed, generally, as

Linear Harmonic Oscillator Eigenvalues

$$\lambda = \left(n + \frac{1}{2}\right)\hbar\omega \qquad (14.30)$$

Eq. (14.30) is identical to Eq. (5.68), obtained in Chap. 5, however we have obtained this result utilizing *Bra-Ket algebra*. The approach that has been taken here is also known as the **Dirac Method**. From this, the lowest energy state, or *ground state*, is given from the condition

$$\hat{a}|A\rangle = 0 \quad \text{(ground state condition)} \qquad (14.31)$$

This can be used to determine the ground state wave function.

14.4 ANNIHILATION AND CREATION OPERATORS REVISITED

The operators $\hat{a}$ and $\hat{a}^\dagger$ have special properties based on the analysis we have conducted. We describe them here using the Bra-Ket notation. Let's begin by recalling an earlier result where we obtained

$$\widehat{H}\hat{a}|A\rangle = \lambda(\hat{a}|A\rangle) - \hbar\omega(\hat{a}|A\rangle) = (\lambda - \hbar\omega)\hat{a}|A\rangle \qquad (14.32)$$

$|A\rangle$ is the *eigenket*, representing the eigenstates of the harmonic oscillator described by $\widehat{H}$. For the harmonic oscillator problem, we have

$$\widehat{H}|A\rangle = \lambda|A\rangle = \left(n + \frac{1}{2}\right)\hbar\omega|A\rangle = \left(n + \frac{1}{2}\right)\hbar\omega|n\rangle$$

Representing $|A\rangle$ by $|n\rangle$ reminds us that each eigenket is associated with a specific integer n. It follows that $\hat{a}$ has a direct impact on this integer, and therefore, an impact on the eigenstate because

$$\widehat{H}\hat{a}|n\rangle = \left[\left(n + \frac{1}{2}\right)\hbar\omega - \hbar\omega\right]\hat{a}|n\rangle$$

$$= \left(n - \frac{1}{2}\right)\hbar\omega\hat{a}|n\rangle = \left[(n-1) + \frac{1}{2}\right]\hbar\omega\hat{a}|n\rangle$$

This means that $\hat{a}$ operating on eigenket $|n\rangle$ leads to a eigenket that is proportional to the eigenket $|n-1\rangle$, or

$$\hat{a}|n\rangle = d_n|n-1\rangle$$

Because of this property, $\hat{a}$ is also known as an **annihilation operator** or **destruction operator**. In order to find the proportionality constant d_n, let Ket $B = \hat{a}|n\rangle$. Then, we know from Eq. (14.24) that

$$\hbar\omega\langle B|B\rangle = \langle n|\hbar\omega\hat{a}\hat{a}^\dagger|n\rangle = \langle n|\widehat{H}|n\rangle - \frac{1}{2}\hbar\omega\langle n|n\rangle$$

$$= \left(n + \frac{1}{2}\right)\hbar\omega - \frac{1}{2}\hbar\omega$$

$$= n\hbar\omega$$

We can also use this relation:

$$\hbar\omega\langle B|B\rangle = \hbar\omega|d_n|^2\langle n-1|n-1\rangle = \hbar\omega|d_n|^2$$

We have taken the eigenkets of $\widehat{H}$ to be normalized. From this, we have

$$\hbar\omega|d_n|^2 = n\hbar\omega \Rightarrow d_n = \sqrt{n} \tag{14.33}$$

Note that since we only have $|d_n|^2$ to be real, d_n is within a phase factor (it can be complex, for example).

A similar analysis can be done for $\hat{a}^\dagger$, starting with the relation

$$\widehat{H}\hat{a}^\dagger|A\rangle = \lambda(\hat{a}^\dagger|A\rangle) + \hbar\omega(\hat{a}^\dagger|A\rangle) = (\lambda + \hbar\omega)\hat{a}^\dagger|A\rangle \tag{14.34}$$

It follows that

$$\widehat{H}\hat{a}^\dagger|n\rangle = \left((n+\frac{1}{2})\hbar\omega + \hbar\omega\right)\hat{a}^\dagger|n\rangle$$

$$= (n+\frac{3}{2})\hbar\omega\hat{a}^\dagger|n\rangle = \left([n+1] + \frac{1}{2}\right)\hbar\omega\hat{a}^\dagger|n\rangle$$

Therefore, $\hat{a}^\dagger$ operating on eigenket $|n\rangle$ leads to a eigenket that is proportional to the eigenket $|n+1\rangle$, or

$$\hat{a}^\dagger|n\rangle = c_n|n+1\rangle$$

And because of this unique property, $\hat{a}^\dagger$ is also known as a **creation operator** or **ladder operator**. We can also determine c_n by letting Ket $B = \hat{a}^\dagger|n\rangle$. Then we have

$$\hbar\omega\langle B|B\rangle = \langle n|\hbar\omega\hat{a}^\dagger\hat{a}|n\rangle = \langle n|\widehat{H}|n\rangle + \frac{1}{2}\hbar\omega\langle n|n\rangle$$

14.4 Annihilation and Creation Operators Revisited

$$= \left(n + \frac{1}{2}\right)\hbar\omega + \frac{1}{2}\hbar\omega$$

$$= (n+1)\hbar\omega$$

And since

$$\hbar\omega\langle B|B\rangle = \hbar\omega|c_n|^2 \langle n+1|n+1\rangle = \hbar\omega|c_n|^2$$

We have taken the eigenkets of $\widehat{H}$ to be normalized. From this, we have

$$\hbar\omega|c_n|^2 = (n+1)\hbar\omega \Rightarrow c_n = \sqrt{n+1} \tag{14.35}$$

Thus, we have that operators $\widehat{a}$ and $\widehat{a}^\dagger$ are, alternatively, given by

---
Annihilation and Creation Operators

$$\widehat{a}|n\rangle = \sqrt{n}|n-1\rangle \quad \text{and} \quad \widehat{a}^\dagger|n\rangle = \sqrt{n+1}|n+1\rangle \tag{14.36}$$

This form of the operators is useful in calculating expectation quantities. Recall that our previous definitions for determining expectation values of x, x^2, and $\widehat{p}$ are by evaluating

$$\langle x \rangle = \frac{\int \psi^* x \psi \, dx}{\int \psi^* \psi \, dx} = \int \psi^* x \psi \, dx \tag{14.37a}$$

$$\langle x^2 \rangle = \frac{\int \psi^* x^2 \psi \, dx}{\int \psi^* \psi \, dx} = \int \psi^* x^2 \psi \, dx \tag{14.37b}$$

$$\langle \widehat{p} \rangle = \frac{\int \psi^* \widehat{p} \psi \, dx}{\int \psi^* \psi \, dx} = \int \psi^* \left[-i\hbar \frac{\partial}{\partial x}\right] \psi \, dx \tag{14.37c}$$

The last form assumes the wave functions to be normalized so that $\int \psi^* \psi \, dx = 1$. Now that we have the annihilation and creation operators, these quantities can be computed with them, rather than using the relations in Eqs. (14.37a) - (14.37c). We do not even require the wave functions explicitly. Using the operators $\widehat{a}$ and $\overline{a}$, we can relate them to x as follows:

$$\widehat{a} + \widehat{a}^\dagger = \frac{m^*\omega x + ip}{\sqrt{2m^*\hbar\omega}} + \frac{m^*\omega x - ip}{\sqrt{2m^*\hbar\omega}} = \frac{2m^*\omega}{\sqrt{2m^*\hbar\omega}} x = \sqrt{\frac{2m^*\omega}{\hbar}} x$$

Solving for x gives

$$x = \sqrt{\frac{\hbar}{2m^*\omega}}\left(\widehat{a}+\widehat{a}^\dagger\right) \quad (14.38)$$

Eq. (14.38) can be used, along with Eq. (14.36), to compute the expectation value of x or $\langle x \rangle$ as

$$\langle x \rangle = \langle n|x|n \rangle = \sqrt{\frac{\hbar}{2m^*\omega}}(\langle n|\widehat{a}|n \rangle + \langle n|\widehat{a}^\dagger|n \rangle)$$

$$= \sqrt{\frac{\hbar}{2m^*\omega}}(\langle n|\sqrt{n}|n-1 \rangle + \langle n|\sqrt{n+1}|n+1 \rangle)$$

$$= \sqrt{\frac{\hbar}{2m^*\omega}}(\sqrt{n}\langle n|n-1 \rangle + \sqrt{n+1}\langle n|n+1 \rangle)$$

$$= 0$$

We have used the fact that the eigenstates are *orthogonal*, which is the case for any Hermitian Hamiltonian, as we have for the harmonic oscillator. Carrying out a similar calculation for $\langle x^2 \rangle$, we have

$$\langle x^2 \rangle = \langle n|x^2|n \rangle = \frac{\hbar}{2m^*\omega}\langle n|(\widehat{a}+\text{orthogonal})(\widehat{a}+\widehat{a}^\dagger)|n \rangle$$

$$= \frac{\hbar}{2m^*\omega}\langle n|(\widehat{a}\cdot\widehat{a}+\widehat{a}\cdot\widehat{a}^\dagger+\widehat{a}^\dagger\cdot\widehat{a}+\widehat{a}^\dagger\cdot\widehat{a}^\dagger)|n \rangle$$

$$= \frac{\hbar}{2m^*\omega}(\langle n|\widehat{a}\cdot\widehat{a}|n \rangle + \langle n|\widehat{a}\cdot\widehat{a}^\dagger|n \rangle + \langle n|\widehat{a}^\dagger\cdot\widehat{a}|n \rangle + \langle n|\widehat{a}^\dagger\cdot\widehat{a}^\dagger|n \rangle)$$

We have the sum of four terms. Let's consider each term one by one, starting with $\widehat{a}\cdot\widehat{a}$ and $\widehat{a}^\dagger\cdot\widehat{a}^\dagger$, where we have

$$\langle n|\widehat{a}\cdot\widehat{a}|n \rangle = \langle n|\widehat{a}\sqrt{n}|n-1 \rangle$$

$$= \sqrt{n}\langle n|\widehat{a}|n-1 \rangle = \sqrt{n(n-1)}\langle n|n-2 \rangle$$

$$= 0$$

Similarly, for $\widehat{a}^\dagger\cdot\widehat{a}^\dagger$, we get

$$\langle n|\widehat{a}^\dagger\cdot\widehat{a}^\dagger|n \rangle = \langle n|\widehat{a}^\dagger\sqrt{n+1}|n+1 \rangle$$

$$= \sqrt{n+1}\langle n|\widehat{a}|n+1 \rangle = \sqrt{(n+1)(n+2)}\langle n|n+2 \rangle$$

14.4 Annihilation and Creation Operators Revisited

$$= 0$$

For $\hat{a} \cdot \hat{a}^\dagger$ and $\hat{a}^\dagger \cdot \hat{a}$, we obtain *nonzero* results, as we have

$$\langle n|\hat{a} \cdot \hat{a}^\dagger|n\rangle = \langle n|\hat{a}\sqrt{n+1}|n+1\rangle$$
$$= \sqrt{n+1}\langle n|\hat{a}|n+1\rangle = (n+1)\langle n|n\rangle$$
$$= n+1$$

And for $\hat{a}^\dagger \cdot \hat{a}$, we have

$$\langle n|\hat{a}^\dagger \cdot \hat{a}|n\rangle = \langle n|\hat{a}^\dagger \sqrt{n}|n-1\rangle$$
$$= \sqrt{n}\langle n|\hat{a}^\dagger|n-1\rangle = n\langle n|n\rangle$$
$$= n$$

Putting the results together, we obtain $\langle x^2 \rangle$ to be given by

$$\langle x^2 \rangle = \frac{\hbar}{2m^*\omega}[0 + n + (n+1) + 0] = \frac{\hbar}{2m^*\omega}(2n+1) \qquad (14.39)$$

$$= \frac{\hbar}{m^*\omega}\left(n + \frac{1}{2}\right) \qquad (14.40)$$

If we also determine $\langle \hat{p} \rangle$ and $\langle \hat{p}^2 \rangle$, as we have done for $\langle x \rangle$ and $\langle x^2 \rangle$, the uncertainties Δx and Δp can both be determined for the harmonic oscillator. A relationship between $\hat{p}$ and $\hat{a}$, etc. can be found taking the difference between $\hat{a}$ and $\hat{a}^\dagger$, where we have

$$\hat{a} - \hat{a}^\dagger = \frac{m^*\omega x + ip}{\sqrt{2m^*\hbar\omega}} - \frac{m^*\omega x - ip}{\sqrt{2m^*\hbar\omega}} = i\frac{2}{\sqrt{2m^*\hbar\omega}}\hat{p} = i\sqrt{\frac{2}{m^*\hbar\omega}}\hat{p}$$

Solving for $\hat{p}$ gives

$$\hat{p} = -i\sqrt{\frac{m^*\hbar\omega}{2}}(\hat{a} - \hat{a}^\dagger) \qquad (14.41)$$

Using this result to compute $\langle \hat{p} \rangle$, we have

$$\langle \hat{p} \rangle = \langle n|\hat{p}|n\rangle = -i\sqrt{\frac{m^*\hbar\omega}{2}}(\langle n|\hat{a}|n\rangle - \langle n|\hat{a}^\dagger|n\rangle)$$

$$= -i\sqrt{\frac{m^*\hbar\omega}{2}}(\langle n|\sqrt{n}|n-1\rangle + \langle n|\sqrt{n+1}|n+1\rangle)$$

$$= -i\sqrt{\frac{m^*\hbar\omega}{2}}(\sqrt{n}\langle n|n-1\rangle - \sqrt{n+1}\langle n|n+1\rangle)$$

$$= 0$$

Carrying out the same calculation for $\langle \hat{p}^2 \rangle$, we have

$$\langle \hat{p}^2 \rangle = \langle n|\hat{p}^2|n\rangle = -\frac{m^*\hbar\omega}{2}\langle n|(\hat{a}-\hat{a}^\dagger)(\hat{a}-\hat{a}^\dagger)|n\rangle$$

$$= -\frac{m^*\hbar\omega}{2}\langle n|(\hat{a}\cdot\hat{a}-\hat{a}\cdot\hat{a}^\dagger - \hat{a}^\dagger\cdot\hat{a}+\hat{a}^\dagger\cdot\hat{a}^\dagger)|n\rangle$$

$$= -\frac{m^*\hbar\omega}{2}(\langle n|\hat{a}\cdot\hat{a}|n\rangle - \langle n|\hat{a}\cdot\hat{a}^\dagger|n\rangle - \langle n|\hat{a}^\dagger\cdot\hat{a}|n\rangle + \langle n|\hat{a}^\dagger\cdot\hat{a}^\dagger|n\rangle)$$

This contains the same four terms we had in the determination of $\langle x^2 \rangle$. We determined these separately above, so we can use the same results to compute the above, where we have for $\langle \hat{p} \rangle$

$$\langle \hat{p}^2 \rangle = -\frac{m^*\hbar\omega}{2}(0 - n - (n+1) + 0)$$

$$= \frac{m^*\hbar\omega}{2}(2n+1) = m^*\hbar\omega\left(n+\frac{1}{2}\right)$$

Having these four quantities is sufficient to determine the uncertainty Δx as

$$\Delta x = \sqrt{\langle x^2 \rangle - \langle x \rangle^2} = \sqrt{\langle x^2 \rangle} = \sqrt{\frac{\hbar}{m^*\omega}\left(n+\frac{1}{2}\right)}$$

And for $\Delta \hat{p}$, we now obtain

$$\Delta \hat{p} = \sqrt{\langle \hat{p}^2 \rangle - \langle \hat{p} \rangle^2} = \sqrt{\langle x^2 \rangle} = \sqrt{m^*\hbar\omega\left(n+\frac{1}{2}\right)}$$

This allows the uncertainty $\Delta x \Delta \hat{p}$, for the harmonic oscillator to be determined (using Bra-Ket algebra), yielding the result

Harmonic Oscillator Uncertainty Principle

$$\Delta x \Delta \hat{p} = \hbar\left(n+\frac{1}{2}\right) \tag{14.42}$$

14.5 Ground And Above-Ground States of LHO

In this section, we look at how to obtain the explicit wave functions $|n\rangle = \psi_n(x)$ using the operators $\hat{a}$ and/or $\hat{a}^\dagger$. This connects the Bra-Ket algebra directly to the classical mathematical notation used throughout the text. We begin with the ground state of the harmonic oscillator. Let the ground-state be denoted by $\psi_0 = |0\rangle =$. Eq.(??) is the condition leading to the ground state, defined by

Ground State Condition

$$\hat{a}|0\rangle = 0 \tag{14.43}$$

This condition means that

$$\hat{a}|0\rangle = \hat{a}\psi_0 = \frac{1}{\sqrt{2m^*\hbar\omega}}[m^*\omega x + ip]\psi_0 = 0 \tag{14.44}$$

Since the RHS equals zero, we can ignore the constant factor. Then, using the momentum operator for $\hat{p}$, Eq. (14.44) becomes

$$\left[m^*\omega x + i\left(-i\hbar\frac{d}{dx}\right)\right]\psi_0 = 0$$

This leads to

$$\frac{1}{\psi_0}\frac{d\psi_0}{dx} = \frac{m^*\omega}{\hbar}x \Rightarrow \frac{d\psi_0}{\psi_0} = \frac{m^*\omega}{\hbar}xdx \tag{14.45}$$

Upon integration of Eq. (14.45), we have

$$\int \frac{d\psi_0}{\psi_0} = -\frac{m^*\omega}{\hbar}\int xdx \Rightarrow \log\psi_0 = -\frac{m^*\omega}{2\hbar}x^2 + c_i \tag{14.46}$$

c_i is an integration constant. Taking the exponential of both sides, we have the solution given by

$$\psi_0 = C_0 e^{-\frac{m^*\omega}{2\hbar}x^2} = C_0 e^{-\frac{x^2}{2\sigma^2}} \tag{14.47}$$

Introducing a dimensionless quantity $\xi = x/\sigma$, we can write the wave function $\psi_0 = |0\rangle$ as

---

Ground State For LHO

$$|0\rangle = C_0 H_0 e^{-\frac{1}{2}\xi^2} \tag{14.48}$$

H_0 denotes the zeroth order Hermite polynomial. The latter form uses the simple fact that $H_0 = 1$, however, we'll see that this pattern generalizes further. Eq. (14.48) is identical to the ground state obtained in Chap. 5, for the linear harmonic oscillator. Thus, the condition given by Eq. (14.43) defines the ground-state wave function $|0\rangle$. The normalization constant C_0 is found from the normalization condition $\langle 0|0\rangle = \int \psi_0^* \psi_0 dx = 1$, which gives

$$1 = \int \psi_0^* \psi_0 dx = C_0^2 \int_{-\infty}^{+\infty} e^{-\frac{x^2}{2\sigma^2}} e^{-\frac{x^2}{2\sigma^2}} dx \tag{14.49}$$

$$= C_0^2 \int_{-\infty}^{+\infty} e^{-\frac{x^2}{\sigma^2}} dx \tag{14.50}$$

$$= C_0^2 \left(\sigma\sqrt{\pi}\right) = C_0^2 \sqrt{\frac{\pi\hbar}{m^*\omega}} \tag{14.51}$$

The normalization constant C_0 is, therefore, given by

$$C_0 = \left[\frac{m^*\omega}{\pi\hbar}\right]^{\frac{1}{4}} \tag{14.52}$$

Having the ground state $|0\rangle$, along with the creation/annihilation operators in terms of $\widehat{a}$ enables the determination of above-ground states. This is done using the creation operator, where we have

14.5 Ground And Above-Ground States of LHO

$$\hat{a}^\dagger |n\rangle = \sqrt{n+1}|n+1\rangle \Rightarrow \psi_{n+1} = \frac{1}{\sqrt{n+1}}\hat{a}^\dagger \psi_n$$

$$= \frac{1}{\sqrt{n+1}}\left[\frac{m^*\omega x - i\hat{p}}{\sqrt{2m^*\hbar\omega}}\right]\psi_n$$

$$= \frac{1}{\sqrt{2(n+1)}}\left[\sqrt{\frac{m^*\omega}{\hbar}}x - i\frac{1}{\sqrt{m^*\hbar\omega}}\hat{p}\right]\psi_n$$

Replacing the momentum operator $\hat{p} = -i\hbar d/dx$, leads to

$$\psi_{n+1} = \frac{1}{\sqrt{2(n+1)}}\left[\sqrt{\frac{m^*\omega}{\hbar}}x - \sqrt{\frac{\hbar}{m^*\omega}}\frac{d}{dx}\right]\psi_n$$

$$= \frac{1}{\sqrt{2(n+1)}}\left[\frac{x}{\sigma} - \sigma\frac{d}{dx}\right]\psi_n$$

In terms of the dimensionless quantity $\xi = x/\sigma$, we can write the above relation as

Above-Ground State Iterative Equation For LHO

$$\psi_{n+1}(\xi) = \frac{1}{\sqrt{2(n+1)}}\left[\xi - \frac{d}{d\xi}\right]\psi_n \quad n=0,1,2,\ldots \quad (14.53)$$

Having the ground state and the relation given by Eq. (14.53) allows us to determine any state $n+1$, using it's previous state n. For example, starting with the ground state $|0\rangle = \psi_0$ to obtain $|1\rangle = \psi_1$, we have

$$|1\rangle = \psi_1(\xi) = \frac{1}{\sqrt{2(0+1)}}\left[\xi - \frac{d}{d\xi}\right]\psi_0$$

$$= \frac{1}{\sqrt{2}}\left[\xi - \frac{d}{d\xi}\right]C_0 e^{-\frac{1}{2}\xi^2}$$

$$= \frac{C_0}{\sqrt{2}}[\xi + \xi]e^{-\frac{1}{2}\xi^2}$$

$$= C_1[2\xi]e^{-\frac{1}{2}\xi^2}$$

Referring to Table 5.4, which lists the Hermite polynomials H_n, we find that $H_1(\xi) = 2\xi$, so that $|1\rangle$ can be written as

$$|1\rangle = C_1 H_1(\xi) e^{-\frac{1}{2}\xi^2} \tag{14.54}$$

We can write the normalization constant as

$$C_1 = \frac{C_0}{\sqrt{2 \cdot 1}} \tag{14.55}$$

Evaluating $|2\rangle = \psi_2(\xi)$, we have

$$|2\rangle = \frac{1}{\sqrt{2(1+1)}} \left[\xi - \frac{d}{d\xi} \right] \psi_1$$

$$= \frac{1}{2} \left[\xi - \frac{d}{d\xi} \right] C_1 [2\xi] e^{-\frac{1}{2}\xi^2}$$

$$= \frac{C_1}{2} \left[(2\xi^2) - (2 - 2\xi^2) \right] e^{-\frac{1}{2}\xi^2}$$

$$= C_2 [4\xi - 2] e^{-\frac{1}{2}\xi^2}$$

The term in brackets is the 2nd order Hermite polynomial (see Table 5.4), so that $|2\rangle$ is given by

$$|2\rangle = C_2 H_2(\xi) e^{-\frac{1}{2}\xi^2} \tag{14.56}$$

and

$$C_2 = \frac{C_1}{\sqrt{2 \cdot 2}} = \frac{C_0}{\sqrt{2 \cdot 1}} \frac{1}{\sqrt{2 \cdot 2}} = \frac{C_0}{\sqrt{2!2^2}} \tag{14.57}$$

As the last example, evaluation of $|3\rangle = \psi_3$ gives

$$|3\rangle = \frac{1}{\sqrt{2(2+1)}} \left[\xi - \frac{d}{d\xi} \right] \psi_2$$

$$= \frac{C_2}{\sqrt{2 \cdot 3}} \left[(4\xi^3 - 2\xi) - (8\xi + [4\xi^2 - 2](-\xi)) \right] e^{-\frac{1}{2}\xi^2}$$

$$= \frac{C_2}{\sqrt{2 \cdot 3}} \left[8\xi^3 - 12\xi \right] e^{-\frac{1}{2}\xi^2}$$

$$= C_3 H_3(\xi) e^{-\frac{1}{2}\xi^2}$$

where

$$C_3 = \frac{C_2}{\sqrt{2 \cdot 3}} = \frac{C_0}{\sqrt{2!2^2}} \frac{1}{\sqrt{3 \cdot 2}} = \frac{C_0}{\sqrt{3!2^3}}$$

Hopefully, we have established a clear pattern of the wave functions $|n+1\rangle = \psi_{n+1}$. The general form of the solution is given by

1D Linear Harmonic Oscillator Eigenkets

$$|n+1\rangle = C_{n+1} \cdot H_{n+1}(\xi) \cdot e^{-\frac{1}{2}\xi^2} \qquad (14.58)$$

H_{n+1} are *Hermite polynomials* (c. Chap. 5). This result is identical to our earlier result given by Eq. (14.47). The normalization constant C_n is given by

1D Harmonic Oscillator Normalization Factor

$$C_n = \frac{C_0}{\sqrt{n! \cdot 2^n}} \qquad (14.59)$$

C_0 is the *ground state normalization factor* given by Eq. (14.52). Thus, the *Dirac method* has also lead us to the same eigenstates or eigenkets obtained by solving the Schrödinger equation directly. However, this was done *using two first order operators, rather than one second order operator*. We can go on to express the most general form of the solution. This is the objective of the following section.

14.6 COHERENT STATE OF THE OSCILLATOR REVISITED

With the tools developed in the previous sections, we can find the most general solution for the LHO problem using a general Ket $|C\rangle$, described by

$$|C\rangle = \sum_n c_n |n\rangle \qquad (14.60)$$

Because of the *orthogonality* of the eigenkets, any coefficient c_m can be obtained by

$$\langle m|C\rangle = \sum_n c_n \langle m|n\rangle = \sum_n c_n \delta_{mn} = c_m$$

δ_{mn} is the *Kronecker delta function*. Using the result for c_m, it follows that

$$|C\rangle = \sum_n c_n|n\rangle = \sum_n |n\rangle c_n = \sum_n |n\rangle\langle n|C\rangle \qquad (14.61)$$

Eq. (14.61) implies the completeness condition given by

Completeness Condition In Bra-Ket Notation

$$\sum_n |n\rangle\langle n| = 1 \qquad (14.62)$$

The *completeness condition* given by Eq. (14.62) implies that

$$\sum_n |n\rangle\langle n| \quad \text{is a unitary operator} \qquad (14.63)$$

Letting the annihilation operator $\widehat{a}$ operate on the Ket $|C\rangle$, having eigenvalue β, we get

$$\widehat{a}|C\rangle = \sum_{n=0}^{n=\infty} c_n \widehat{a}|n\rangle = \sum_{n=0}^{n=\infty} c_n \sqrt{n}|n-1\rangle = \beta|C\rangle = \beta \sum_{n=0}^{n=\infty} c_n|n\rangle$$

Note that β is generally complex because the operator $\widehat{a}$ is *not Hermitian*. This is clear from the fact that $\widehat{p}$ *is* Hermitian, but $\widehat{a} \propto i\widehat{p}$. Because we have a ground state $|0\rangle$, the very first term in the series involving $|n-1\rangle$ corresponds to $|0-1\rangle = |-1\rangle$, which must be the null-Ket since it corresponds to an energy *below* that of the ground state. So we can express the above as

$$\sum_{n=1}^{n=\infty} c_n \sqrt{n}|n-1\rangle = \beta \sum_{n=0}^{n=\infty} c_n|n\rangle \qquad (14.64)$$

Now, when we write out the terms explicitly from Eq. 14.64. We obtain the following:

14.6 Coherent State Of The Oscillator Revisited

$$c_1\sqrt{1}|0\rangle + c_2\sqrt{2}|1\rangle + c_3\sqrt{3}|2\rangle + c_4\sqrt{4}|3\rangle + c_5\sqrt{5}|4\rangle + \cdots =$$

$$\beta c_0|0\rangle + \beta c_1|1\rangle + \beta c_2|2\rangle + \beta c_3|3\rangle + \beta c_4|4\rangle + \cdots$$

Equating coefficients of the eigenkets gives the following relations:

$$c_1\sqrt{1} = \beta c_0, \quad c_2\sqrt{2} = \beta c_1, \quad c_3\sqrt{3} = \beta c_2, \quad c_4\sqrt{4} = \beta c_3, \quad c_5\sqrt{5} = \beta c_4 \cdots$$

We can solve for each coefficient in terms of c_0 by solving for c_1, then successively substituting from the lowest order term. Doing this, we obtain

$$c_n = \frac{\beta}{\sqrt{n}} c_{n-1} \Rightarrow c_1 = \frac{\beta}{\sqrt{1}} c_0$$

$$c_2 = \frac{\beta}{\sqrt{2}} c_1 = \frac{\beta^2}{\sqrt{1 \cdot 2}} c_0$$

$$c_3 = \frac{\beta}{\sqrt{3}} c_2 = \frac{\beta^3}{\sqrt{1 \cdot 2 \cdot 3}} c_0$$

$$\vdots$$

$$c_n = \frac{\beta^n}{\sqrt{n!}} c_0$$

Since β is complex, we have that c_n is, generally, complex. With the solutions for the coefficients c_n, the eigenket $|C\rangle$ can be written as

$$|C\rangle = c_0 \sum_{n=0}^{\infty} \frac{\beta^n}{\sqrt{n!}} |n\rangle \tag{14.65}$$

We can solve for c_0 in terms of β by imposing the *normalization condition* on $|B\rangle$. This leads to

$$1 = \langle B|B\rangle = c_0^* c_0 \sum_{m=0}^{\infty} \sum_{n=0}^{\infty} \frac{(\beta^*)^m}{\sqrt{m!}} \frac{\beta^n}{\sqrt{n!}} \langle m|n\rangle$$

$$= |c_0|^2 \sum_{m=0}^{\infty} \sum_{n=0}^{\infty} \frac{(\beta^*)^m}{\sqrt{m!}} \frac{\beta^n}{\sqrt{n!}} \delta_{mn}$$

$$= |c_0|^2 \sum_{n=0}^{\infty} \frac{\left(|\beta|^2\right)^n}{n!}$$

Recall that the MacLauren (Taylor) series expansion for e^x is given by

$$e^x = 1 + x + \frac{x^2}{2!} + \cdots = \sum_{n=0}^{\infty} \frac{x^n}{n!}$$

If we use this relation in our normalization condition above, we have the following result for $|c_0|^2$

$$1 = |c_0|^2 \sum_{n=0}^{\infty} \frac{\left(|\beta|^2\right)^n}{n!} = |c_0|^2 e^{|\beta|^2} \Rightarrow |c_0|^2 = e^{-|\beta|^2}$$

or

$$|c_0| = \sqrt{e^{-|\beta|^2}} = e^{-\frac{|\beta|^2}{2}} \tag{14.66}$$

This leads us to the final form of the eigenket $|C\rangle$, which is given by

Coherent State Of The LHO

$$|C\rangle = e^{-\frac{|\beta|^2}{2}} \sum_{n=0}^{\infty} \frac{\beta^n}{\sqrt{n!}} |n\rangle \tag{14.67}$$

We also know the harmonic oscillator has a *time-dependent solution*. For this, we can write the complete time-dependent solution $|\Psi(x,t)\rangle = |C\rangle e^{-i\frac{E}{\hbar}t}$ as

Complete Time-Dependent LHO Solution

$$|\Psi(x,t)\rangle = e^{-\frac{|\beta|^2}{2}} \sum_{n=0}^{\infty} \frac{\beta^n}{\sqrt{n!}} |n\rangle e^{-i\left(n+\frac{1}{2}\right)\omega t} \tag{14.68}$$

Eq. (14.68) makes use of the energy for the linear harmonic oscillator $E_\lambda = (n + 1/2)\hbar\omega$. The *Coherent state*, is therefore, the initial condition of $|\Psi(x,t)\rangle$ because $|\Psi(x,0)\rangle = |C\rangle e^{-i\frac{E}{\hbar} \cdot 0} = |C\rangle$. We can determine the time-dependent expectation value of x or $\langle x\rangle(t)$ using Eq. (14.68). Carrying out this calculation leads to

$$\langle x\rangle(t) = \langle \Psi|x|\Psi\rangle$$

14.6 Coherent State Of The Oscillator Revisited

$$= \sqrt{\frac{\hbar}{2m^*\omega}} \langle \Psi | \hat{a} + \hat{a}^\dagger | \Psi \rangle$$

$$= \sqrt{\frac{\hbar}{2m^*\omega}} e^{-|\beta|^2} \sum_{m=0}^{\infty} \sum_{n=0}^{\infty} \frac{(\beta^*)^m}{\sqrt{m!}} \frac{\beta^n}{\sqrt{n!}} (\langle m | \hat{a} + \hat{a}^\dagger | n \rangle) e^{-i(m-n)\omega t}$$

We can evaluate $\langle m | \hat{a} + \hat{a}^\dagger | n \rangle$ to obtain

$$\langle m | \hat{a} + \hat{a}^\dagger | n \rangle = \langle m | \hat{a} | n \rangle + \langle m | \hat{a}^\dagger | n \rangle$$

$$= \sqrt{n} \langle m | n-1 \rangle + \sqrt{n+1} \langle m | n+1 \rangle$$

$$= \sqrt{n} \delta_{m,n-1} + \sqrt{n+1} \delta_{m,n+1}$$

Substitution of this result into our expectation value expression leads to

$$\langle x \rangle(t) = \sqrt{\frac{\hbar}{2m^*\omega}} e^{-|\beta|^2} \sum_{m=0}^{\infty} \sum_{n=0}^{\infty} \frac{(\beta^*)^m}{\sqrt{m!}} \frac{\beta^n}{\sqrt{n!}} (\langle m | \hat{a} + \hat{a}^\dagger | n \rangle) e^{-i(m-n)\omega t}$$

$$= \sqrt{\frac{\hbar}{2m^*\omega}} e^{-|\beta|^2} \sum_{m=0}^{\infty} \sum_{n=0}^{\infty} \frac{(\beta^*)^m}{\sqrt{m!}} \frac{\beta^n}{\sqrt{n!}} (\sqrt{n} \delta_{m,n-1} + \sqrt{n+1} \delta_{m,n+1}) e^{-i(m-n)\omega t}$$

The Kronecker delta terms dictate the values m must take in our summations. In the first term, we set $m = n - 1$ and in the second term $m = n + 1$, leading to

$$\langle x \rangle(t) = \sqrt{\frac{\hbar}{2m^*\omega}} e^{-|\beta|^2} e^{-i\omega t} \sum_{n-1=0}^{\infty} \frac{(|\beta|)^{2n-1}}{\sqrt{(n-1)! n!}} \sqrt{n} \; +$$

$$\sqrt{\frac{\hbar}{2m^*\omega}} e^{-|\beta|^2} e^{+i\omega t} \sum_{n+1=0}^{\infty} \frac{|\beta|^{2n+1}}{\sqrt{n!(n+1)!}} \sqrt{n+1}$$

We can also use the following relations:

$$\sqrt{\frac{n}{n!}} = \sqrt{\frac{1}{(n-1)!}} \quad \text{and} \quad \sqrt{\frac{n+1}{(n+1)!}} = \sqrt{\frac{1}{n!}}$$

Then, the expectation value $\langle x \rangle$ becomes

$$\langle x \rangle(t) = \sqrt{\frac{\hbar}{2m^*\omega}} e^{-|\beta|^2} \left[e^{-i\omega t} \sum_{n-1=0}^{\infty} \frac{(|\beta|)^{2n-1}}{(n-1)!} + e^{+i\omega t} \sum_{n+1=0}^{\infty} \frac{|\beta|^{2n+1}}{n!} \right]$$

$$= \sqrt{\frac{\hbar}{2m^*\omega}}|\beta|e^{-|\beta|^2}\left[e^{-i\omega t}\sum_{n=1}^{\infty}\frac{(|\beta|)^{2(n-1)}}{(n-1)!} + e^{+i\omega t}\sum_{n=0}^{\infty}\frac{|\beta|^{2n}}{n!}\right]$$

In the form that we now have, the two series in the last line are identical to one another, and as we discussed above, is in the form of an exponential. Therefore, this leads to

$$\langle x \rangle(t) = \sqrt{\frac{\hbar}{2m^*\omega}}|\beta|e^{-|\beta|^2}\left[e^{-i\omega t}e^{|\beta|^2} + e^{+i\omega t}e^{|\beta|^2}\right]$$

$$= \sqrt{\frac{\hbar}{2m^*\omega}}|\beta|\left[e^{-i\omega t} + e^{+i\omega t}\right]$$

$$= \sqrt{\frac{\hbar}{2m^*\omega}}|\beta|[2\cos(\omega t)] = x_0\cos(\omega t)$$

Therefore, the relationship between the eigenvalue β and the initial amplitude x_0 is given by

$$x_0 = 2|\beta|\sqrt{\frac{\hbar}{2m^*\omega}} \tag{14.69}$$

$$= |\beta|\sqrt{\frac{2\hbar}{m^*\omega}} \tag{14.70}$$

Thus, we have the classical result if the eigenvalue β is given by

Eigenvalue Of Initial State

$$|\beta| = x_0\sqrt{\frac{m^*\omega}{2\hbar}} \tag{14.71}$$

The eigenvalue given by Eq. (14.71) is identical to the eigenvalue we obtained for the coherent state, derived in Chap. 5, and given by Eq. (5.109). Therefore, *the eigenvalues are the same for coherent state and initial condition of the most general state*. Thus, the two functions are one and the same. Also, $|\beta|^2$ becomes

$$x_0^2 = 4|\beta|^2\frac{\hbar}{2m^*\omega} = 2|\beta|^2\frac{\hbar}{m^*\omega} \Rightarrow |\beta|^2 = \frac{1}{2}x_0^2\frac{m^*\omega}{\hbar} \tag{14.72}$$

A similar analysis can also be carried out for $\langle p \rangle$ using, instead, the relation

$$\widehat{p} = i\sqrt{\frac{m^*\hbar\omega}{2}}(\widehat{a}^\dagger - \widehat{a})$$

We leave this as an exercise for the reader. This completes the chapter on *Bra-Ket algebra*. In the next section, we'll summarize this chapter's content.

14.7 CHAPTER SUMMARY

In this chapter, we have introduced a popular notation for quantum mechanics, known as the *Bra-Ket algebra*. The notation was originally introduced by Dirac. We have linked the classical notation to the Bra-Ket algebra to demonstrate how it can also be used to analyze quantum mechanical problems. It not only provides an improved degree of conciseness, but it proves to have far reaching capabilities in analysis, as well.

We revisited the harmonic oscillator problem to demonstrate the use of the Bra-Ket algebra to solve this problem. The annihilation and creation operators were revisited, and used to obtain the solutions and expectation values for the harmonic oscillator. Table 14.1 summarizes some of the key formulas from this chapter.

Table 14.1: Chapter 14 Summary Equations.

Name	Equation
Dirac symmetry operator I	$\widehat{a} = \frac{m^*\omega x + ip}{\sqrt{2m^*\hbar\omega}}$
Dirac symmetry operator II	$\widehat{a}^\dagger = \frac{m^*\omega x - ip}{\sqrt{2m^*\hbar\omega}}$
Oscillator Eigenenergies	$E_\lambda = \left(n + \frac{1}{2}\right)\hbar\omega$
Annihilation operator	$\widehat{a}|n\rangle = \sqrt{n}|n-1\rangle$
Creation operator	$\widehat{a}^\dagger|n\rangle = \sqrt{n+1}|n+1\rangle$

Table 14.1

Oscillator Uncertainty	$\Delta x \Delta \hat{p} = \hbar\left(n + \frac{1}{2}\right)$
Above-Ground State	$\|n+1\rangle = \frac{1}{\sqrt{2(n+1)}}\left[\xi - \frac{d}{d\xi}\right]\|n\rangle$
Coherent State	$\|C\rangle = e^{-\frac{\|\beta\|^2}{2}} \sum_{n=0}^{\infty} \frac{\beta^n}{\sqrt{n!}}\|n\rangle$
General LHO State	$\|\Psi(x,t)\rangle = e^{-\frac{\|\beta\|^2}{2}} \sum_{n=0}^{\infty} \frac{\beta^n}{\sqrt{n!}}\|n\rangle e^{-i\left(n+\frac{1}{2}\right)\omega t}$
Coherent State Eigenvalue	$\|\beta\| = x_0 \sqrt{\frac{m^*\omega}{2\hbar}}$

14.8 Chapter Problems

Problem 14.1 The first order operators introduced by Dirac were given by

$$\hat{a} = \frac{m^*\omega x + ip}{\sqrt{2m^*\hbar\omega}} \quad \text{and} \quad \hat{a}^\dagger = \frac{m^*\omega x - ip}{\sqrt{2m^*\hbar\omega}}$$

In the Bra-Ket notation, we found

$$\hat{a}|n\rangle = \sqrt{n}|n-1\rangle \quad \text{and} \quad \hat{a}^\dagger|n\rangle = \sqrt{n+1}|n+1\rangle$$

Use these two relations strictly in Bra-Ket form to prove that

$$\hat{a}^\dagger\hat{a}|n\rangle = n|n\rangle$$

Problem 14.2 Use the above eigenrelations for $\hat{a}$ and $\hat{a}^\dagger$ to show that reversing the order of the operators gives

$$\widetilde{\hat{a}\hat{a}^\dagger}|n\rangle = (n+1)|n\rangle$$

Problem 14.3 Use Eq. (14.53) to show that the harmonic oscillator state $|2\rangle$ can be expressed in terms of the ground state $|0\rangle$ as

$$|2\rangle = \frac{1}{2\sqrt{2}}\left[\frac{d^2|0\rangle}{d\xi^2} - (1+2\xi) + \xi^2|0\rangle\right]$$

14.8 Chapter Problems

Problem 14.4 Use the relation

$$\widehat{p} = i\sqrt{\frac{m^*\hbar\omega}{2}}(\widehat{a}^\dagger - \widehat{a})$$

to show that $\langle \widehat{p}^4 \rangle$ is given by

$$\langle \widehat{p}^4 \rangle = (m^*\hbar\omega)^2\left(n^2 + n + \frac{1}{4}\right)$$

Start with the following relation and use the properties of the operators $\widehat{a}$ and $\widehat{a}^\dagger$.

$$\langle n|\left(\frac{m^*\hbar\omega}{2}\right)^2(\widehat{a}^\dagger - \widehat{a})^4|n\rangle$$

Problem 14.5 With eigenvalues of the form found for $\langle \widehat{p}^4 \rangle$, discuss the conditions of one-to-one-ness with quantum number n for a value of $\langle \widehat{p}^4 \rangle$.?

Problem 14.6 Use the relation

$$\widehat{p} = i\sqrt{\frac{m^*\hbar\omega}{2}}(\widehat{a}^\dagger - \widehat{a})$$

along with the general time-dependent harmonic oscillator state to show that $\langle p \rangle(t)$ is given by

$$\langle \widehat{p} \rangle(t) = -m^*\omega x_0 \sin\omega t$$

CHAPTER 15

Fundamentals Of Superconductivity

Superconductivity is a set of phenomena that appears in unique materials such that in certain conditions, they demonstrate extraordinary characteristics, both *electrically* and *magnetically*. In earlier times following its discovery, less information was available to aid in understanding this unique phenomenon. For example, there was no concept of electrical *holes* at the time of the discovery, let alone quantum mechanics, which turn out to be a crucial concept for superconductors. Of course, we now know much more about the phenomena today with so much more data available, and we can exploit this fact in analyzing superconductors. As part of *this* discussion, we utilize the concept of holes as well as a *spin-dependent* potential energy to introduce a probable interpretation of some of the key observations in superconductors. We utilize the important role of spin and angular momentum, which has known temperature dependent behavior based on angular momentum states. We apply these concepts aiming to provide sufficiently deep insight into the physics of superconductors. It also has the quality that it is straightforward for the reader to understand. We use this approach to focus on the fundamental characteristics of superconductors. In addition to the potential properties we assume, we'll also examine the effect of higher order corrections dictated from the Dirac equation, derived in Chapter 11.

Chapter 15. Fundamentals Of Superconductivity

Interestingly, the history of *superconductivity* is *slightly* controversial. Some of the historical details are a bit scattered, and this has lead to a few different stories around the discovery of superconductivity. For example, in 1996 an article was published by a Dutch physicist named Jacobus de Nobel, who worked in the same lab where the discovery took place, only years later. de Nobel was a researcher in the same cryogenic laboratory. In 1996, he shared the story that he purportedly had heard from researcher Garrit Flim (who was involved in the discovery) on his arriving to the lab as a young man in 1931, years after the 1911 discovery.

As the story goes, in a bit of an experimental frenzy, they were surprised to all of sudden be measuring zero resistance in mercury (Hg). Repeated trials continued to indicate zero resistance at the low liquid helium temperatures. Everyone thought that a short circuit was responsible, so they replaced parts to troubleshoot. Again and again, the resistance was *zero*. Fortunately, no short circuits could be found. They continued to carry out the experiment. Here is where serendipity arrived. The story has it that a student named Gilles Holst from the instrument-makers school (at the same university of the lab) was made responsible for watching the readings of the pressure meter connected to the experimental apparatus. The helium vapor pressure in the cryostat had to be a bit lower than atmospheric pressure, so air could rush into any small leaks, freeze, and seal them. However, during one experimental run, Holst apparently nodded off. The pressure slowly increased along with the temperature. As the gauge passed near to 4.2 Kelvin, Holst noticed the galvanometer reading suddenly jump as the resistance appeared. According to de Nobel's account, Holst had unwittingly witnessed the transition at which Hg went from a superconducting state (still unsure what it was) to its normal conductive behavior. They would later go on to confirm and better understand what had been observed.

Some additional details of the discoveries around superconductivity are also found in a set of household lab notebooks with challenging legibility, now housed at the Boerhaave Museum, in Leiden, Netherlands. These notebooks are famously dubbed (according to their labels) as *notebooks 56 and 57*. On the cover of the first notebook (number 56), the labeled dates are from 1909-1910, while the second notebook begins with an entry of October 28, 1911. The difficulty with this is that the first public report on superconductivity supposedly given to the Royal Netherlands Academy Of Arts and Sciences appeared in April

1911, at a time juxtapositioned between the spans of both notebooks. To this day, any notes corresponding to this period have not been found. However, there is a sort of saving grace in this scenario. In the first notebook that supports the earlier date of April 1911, although the year was not written, there is a statement which reads *Kwik Nagenoeg Nul*. In Dutch, this translates to *mercury almost null*. This is not believed to be describing a depletion of Hg or anything like that, but instead, describing the observation of the resistance of Hg with temperature falling to almost zero, or measuring practically zero in very low temperatures. Unfortunately, no experimental data exists from the April time-frame. It was a November 1911 presentation at the first Solvay conference in Brussels, Belgium, which first reported plotted data of resistance with temperature, demonstrating for the first time publicly, the phenomenon of *superconductivity*. Based on notes from *notebook number 56*, we can summarize what historians have pieced together. We'll do this in the next section because there is more detail to the story. Then we'll move on to discuss the physics of what was discovered, that is, *superconductivity as well as perfect diamagnetism*. To begin further grasping the discovery of superconductivity, we'll have to go to the Netherlands, and other parts of Europe around the early part of the twentieth century. Perhaps then, we can begin to unravel what became the dawn of a new branch in low finite temperature physics and quantum mechanics. While the temperatures were very low, *thermodynamics* still played a central role in enabling the temperature dependent phenomena that had been observed to be predicted.

15.1 Kwik Nagenoeg Nul

The *discovery of superconductivity* took place at the University of Leiden in the Netherlands. It involved a team of indispensable researchers working in a laboratory lead by a very well-known Dutch physicist named Heike Kamerlingh Onnes (1853-1926). Onnes was the Physics Chair, and a member of the Royal Academy of Sciences of Amsterdam. His team included a masterful glass blower named Oskar Kesselring, who made the crucial double walled capillary tubes that would support drawing down the pressure in the region between the walls, for maximum insulation and temperature control. There was also a talented research student named Gilles Holst (the same one that purportedly nodded off) whose actions and ideas would prove invaluable in the course of there discovery. Holst was also known to have operated the wheatstone bridge circuitry on the galvanometer used for measuring the current, which was in a separate room from the cryogenic system. Not only was this measurement necessary to obtain the resistance, but the wheatstone bridge was important in the ability to control measurement resolution. By May of 1911, its believed that they had a demonstrated resolution of $\approx 30nV$, which impacted how small of a resistance could be measured. Holst also contributed other useful ideas, for example, to solidify the mercury (Hg) used in the experiments by cooling them with liquid nitrogen. This reduced the amount of technical issues during the cooling process, which was the most challenging. There was also the cryogenic lab manager Gerrit Flim (who conveyed the above-mentioned story to de Nobel), who was the lead in getting the cooling systems ready for experiments. There was also Cornelis Dorsman, who was instrumental in controlling and measuring the temperatures in the experiments, which involved calibration procedures, and so on. None of these responsibilities were trivial.

Before the discovery of superconductivity, several significant achievements had already taken place, relevant to this discovery. Let's discuss some of them. In 1877 French physicist Louis P. Cailletet and Swiss scientist Raoul P. Pictet both independently succeeded in liquefying oxygen and nitrogen gases. But, it was in very small quantities, limiting the ability to use the liquids in an application. By 1898, a Scottish-British physicist named James Dewar had found a way to *liquify and even collect several gases in liquid state like oxygen and hydrogen*. To be able to store these liquified gases, Dewar utilized a special double walled flask that enabled unprecedented insulation surrounding the cooled liquified gas. This achievement enabled the tubes used in Onnes' lab. This allowed the cooled liquids to *stay* liquids, and better avoid evaporation. They had become essentially storable. Kesselring would later (after Onnes' death) take a flight to England with one of these *in hand*, to demonstrate what are known as *persisting currents*. These are currents induced with an external magnetic field in the superconducting state. By 1906, Onnes' lab figured out how to liquify hydrogen. This turned out to be an important step in Onnes' lab's eventual liquifaction of helium at $T = 4.15K$, which, with this accomplishment following in 1908, outpaced Dewar who also tried. For this, Onnes was awarded the Nobel prize in 1913.

Up until this point, it was only possible to go down to about $14K$ in any experiment, which is the temperature at which H_2 gas becomes *solid phase* (from a liquid phase). It becomes *fully* liquid at about $20K$. Down

15.1 Kwik Nagenoeg Nul

to 14K, it was possible to observe some unique features in the electrical resistance in some materials. However, at best, one observed only a departure from linear variations, higher order in temperature, but still continuous. With this achievement of helium liquifaction, they were in a position to do experiments at even lower temperatures.

They could now investigate the low temperature dependence of the electrical resistance of liquid mercury (Hg) using a specially designed cooling cycle known as the *Hampson–Linde cycle*. There were several reasons why they chose Hg. One of the most significant reasons was that they were already skilled in purifying liquid Hg using distillation. Their lab had already demonstrated, and was aware of how impurities can lead to a constant residual resistance at lower temperatures. But, they were interested in fundamental questions aimed at understanding how small could resistance become with reducing temperatures, without a residual. At the time, there were at least two diametrically opposed hypotheses. One hypothesis was that the resistance would continuously decrease to zero at $T = 0K$. The other postulated that the resistance would go to a finite minimum, then rise to infinity at $T = 0K$. Resolving dilemmas such as this was part of the plan.

According to *notebook 56*, the April experiment apparently began around 7am. Onnes himself, did not arrive to the lab until around 11:20am, after the liquid helium circulation began. In this experiment, they were cooling from a higher temperature down to a lower temperature. Once the temperature got down to 4.3K, they started to reduce the pressure of the helium, which further lowered temperature down to about 1.8K. In doing this, they were exploiting the so-called *Joule-Thomson effect* (exploited earlier by Dewar). This effect is when the temperature of a gas is reduced in constant volume as it expands through a valve, which lowers the pressure (think $PV = nRT$). In the vicinity of 2.2K, they noticed that boiling of the helium had ceased, and had turned into visible surface evaporation. We now know that they were witnessing another transition known as the *superfluid transition in helium*, which takes place at 2.2K. A superfluid behaves like a fluid, but with zero viscosity. And just like that, they had successfully cooled down to temperatures well below 14K. For the October experiment, which provided the first reported measurement data, the data was collected while increasing the temperature. Onnes and his research team had discovered and confirmed the first material such that when cooled to a low enough temperature, conducts electricity *without* ohmic losses. In other words, electrons could apparently conduct without being scattered by atoms. Moreover, in addition to superconductivity, they had also stumbled upon an entirely new type of phase transition when they observed the superfluid transition near 2.2K. The historical resistance plot presented in the October 1911 meeting is shown in Fig. 15.1. Figs. 15.2 and 15.3

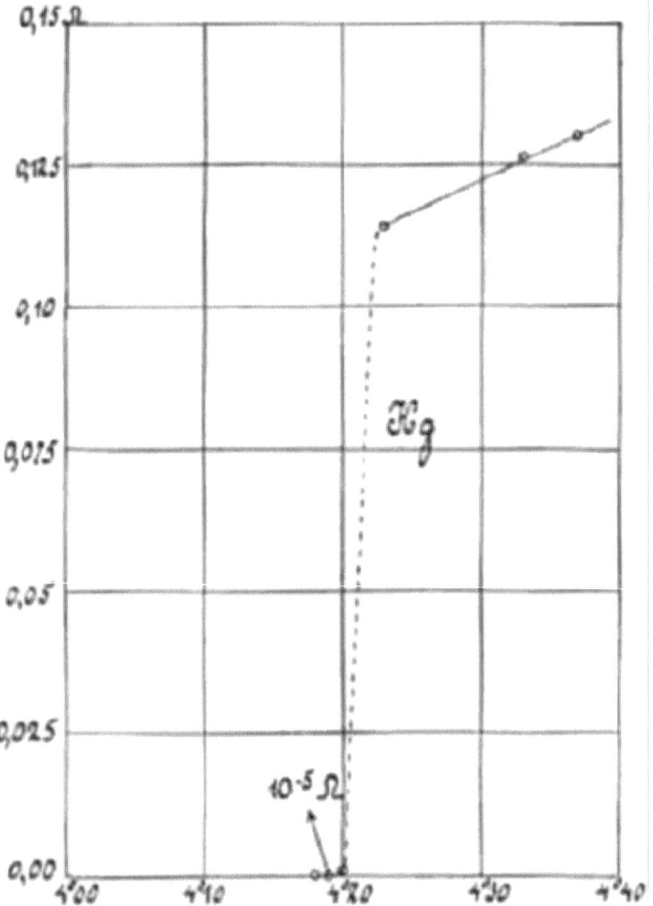

Figure 15.1: The historical plot reported by Onnes' team after their discovery of superconductivity in mercury (Hg) in 1911. The x–axis shows temperature in Kelvin, while the y–axis measures the resistance in Ohms.

illustrates typical features observed in measuring resistivity of electrical conductors as a function of temperature for a normal conductor and superconductor, respectively. For a normal conductor, the resistivity $\rho = RA/\ell$ as a function of temperature can be expressed as

$$\rho(T) = \rho_0 + \rho_1(T) \tag{15.1}$$

15.1 Kwik Nagenoeg Nul

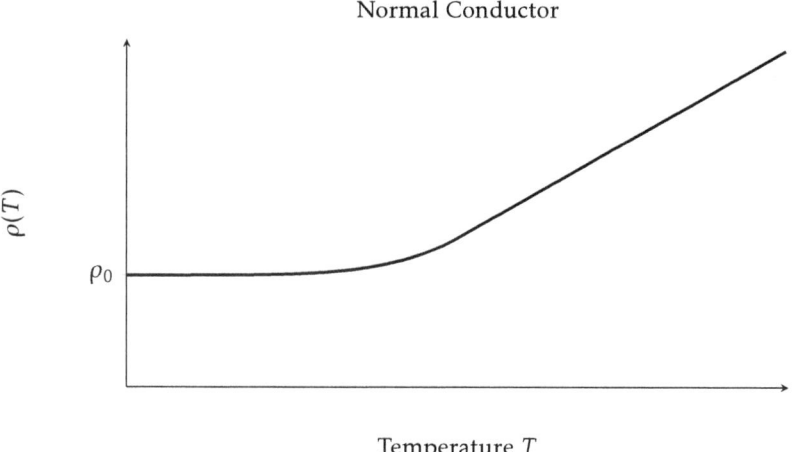

Figure 15.2: Typical temperature dependence of resistivity ρ for a normal conductor.

R is the resistance, which is what is measured. ℓ is the length of the conductor, and A is the cross-sectional area of the conductor. As mentioned above, ρ_0 is the residual resistance due to impurities in the conductor. In the case of a superconductor, as Fig. 15.3 illustrates, there is a notable transition in sufficiently low temperatures. Onnes and his team found that the resistivity sharply drops to practically zero as temperature further approaches zero. For a normal conductor, as the temperature is

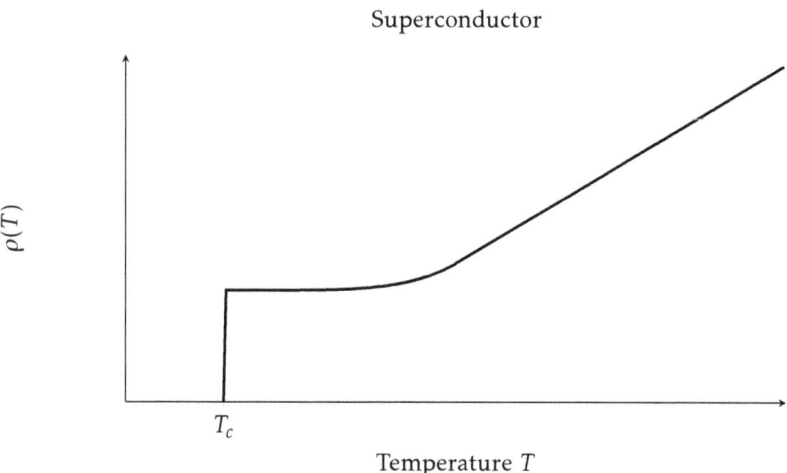

Figure 15.3: Observed temperature dependence of resistivity ρ for a superconductor.

Element	Transition Temperature T_c(K)
V	5.38
Nb	9.50
Tc	7.77
In	3.40
Sn	3.72
La	6.00
Ta	4.48
Hg	4.15
Pb	7.19

Table 15.1: Transition temperatures of some selected superconducting elements.

reduced towards $T = 0$K, the resistivity plateaus to the finite constant value ρ_0 or the *residual resistivity*. This was a significant factor in Onnes using Hg, because it could be purified to a high degree to minimize such contributions. At higher temperatures, a linear $\rho(T)$ relation is typically observed in metals. Between the two extreme regions, there is another region characterized by $\rho \propto T^5$. This transition region, when observed, is described by what is known as the *Bloch-Grüneisen law*.

The temperature at which the superconducting phase transition takes place is known as the **superconducting transition temperature** T_c, or the **critical temperature**. Onnes and his team had also observed that the current they applied to drive the conduction was limited and could not go beyond a certain value, otherwise they would destroy the superconducting state, dashing otherwise high hopes for the possibilities. This upper limit of current became known as the **critical current** I_c. Table 15.1 lists some known *transition temperatures* for some pure element superconductors.

The zero-resistivity characteristic of a superconductor obviously has enormous technological potential, as it means that a superconductor can potentially transmit electrical energy without ohmic loss. One still needs

15.2 The Critical Field Of A Superconductor

to apply an external electric field, but the energy is not wasted ohmically. For extraordinary reasons like this, superconductors remain a very active area of research. They have found their way into groundbreaking technologies such as magnetic resonance imaging (MRI), nuclear magnetic resonance (NMR) imaging machines, and more recently in the development of certain types of quantum computers. In MRI and NMR, superconductors provide large stable magnetic fields required as the background for the resonant field used for imaging. Superconductors are also used in levitation technologies, for example, in high speed magnetically levitated trains, known as *Maglev*. You've gotta ride one, if you haven't yet! Highly sensitive magnetometers (magnetic field detectors) known as superconducting quantum interference devices (SQUIDS) also use superconductors in Josephson junctions, which are bilayers between a dielectric and superconductor utilizing the linear response of the superconductor in an external magnetic field. The list goes on and on.

However, another caveat with superconductors, and it is a big one, is that in order to induce the superconducting phase requires *very low* temperatures. Materials with larger transition temperatures would be more ideal. To reach more in this direction, the awareness of a certain magnetic field effect was also needed, and this effect will be discussed in the next section.

15.2 THE CRITICAL MAGNETIC FIELD OF A SUPERCONDUCTOR

It wasn't until 1933, with the observations of German physicists Fritz Walther Meissner (1882-1974) and Robert Ochsenfeld (1901-1993), that another remarkable property of superconductors would be observed for the first time. They were able to find this additional behavior by exposing the *superconductor* to a uniform external magnetic field B_e and varying the field slowly or quasistatically, while in the superconducting state. They found that the magnetic induction $\mathbf{B}_I$ lines, which ordinarily penetrate through the material, are instead expelled or pushed out of a superconductor. If these lines are only external to the superconductor, then the magnetic induction inside the superconductor must be zero. This idea is illustrated in Fig. 15.4. The only way the magnetic induction can be zero in a material is if the magnetization $\mathbf{M}$ of the superconductor

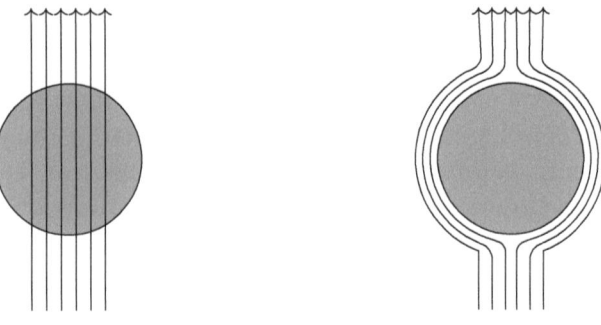

Figure 15.4: Magnetic induction lines on (left) a normal nonmagnetic conductor vs. (right) a superconductor. In the superconductor, the magnetic induction lines are expelled out of the sample. This phenomenon is known as the *Meissner effect*.

M aligns fully opposite to the direction of $\mathbf{B}_e$ so that we have

$$\mathbf{B} = \mathbf{B}_e + \mu_0 \mathbf{M} = 0 \Rightarrow \mu_0 \mathbf{M} = -\mathbf{B}_e \tag{15.2}$$

When the magnetization of a material responds to a magnetic field by pointing *opposite* to the direction of the external magnetic field, the material is said to be **diamagnetic**. This behavior observed by Meissner and Ochsenfeld was not only *diagmagnetic*, but it was the maximum possible form of it, or **superdiamagnetism**, evident because for a diamagnetic material, the maximum relative magnetic susceptibility is given by

$$\chi_r = \frac{\mu_0 \mathbf{M}}{\mathbf{B}_e} = -\frac{\mu_0 \mathbf{B}_e}{\mu_0 \mathbf{B}_e} = -1 \tag{15.3}$$

This was the case for the superconductor and this was a new property that must be considered *in addition* to zero resistivity, found by Onnes in 1911. This observation with an external magnetic field also relates to the amount of electrical current I that can be passed through a superconductor. This is because the current I creates a magnetic field in the superconducting wire. In a circular wire, this field H_I on the surface of the wire is given by

$$\mu_0 H_I = \frac{\mu_0 I}{2\pi R} \tag{15.4}$$

For superconductors, there is greater sensitivity to the field at the surface because, as we'll find, this is where superconducting charge carriers

15.2 The Critical Field Of A Superconductor

tend to concentrate. As mentioned, the superconductor can be driven *out of the superconducting state* when this field exceeds a certain critical magnetic field. More generally, when any magnetic field exceeds a certain value, the superconductor can be driven out of the superconducting state. This magnetic field is known as the **critical magnetic field** B_c. The critical field depends on temperature. This is implied by the fact that without a field, there is a definite T_c as observed by Onnes, however, when the magnetic field is present at some temperature $T < T_c$, it can also be driven out of the superconducting state, i.e. another value of T_c is found with a magnetic field. Experimentally, for several materials including lead (Pb), mercury (Hg), tin (Sn), and indium (In), the B_c is approximately parabolic in temperature, and is often written in the following form:

$$B_c(T) = B_c(0)\left[1 - \frac{T^2}{T_c^2}\right] \tag{15.5}$$

T_c is the transition temperature at zero field. This function is illustrated in Fig. 15.5. Table 15.2 lists both the critical temperature at zero field T_c

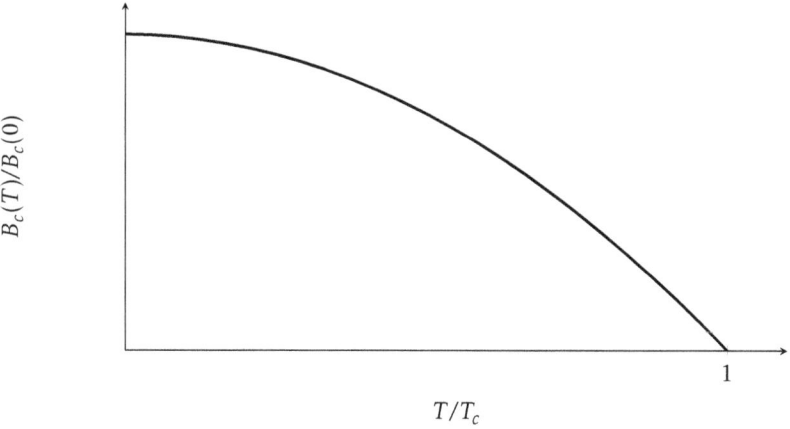

Figure 15.5: Parabolic temperature dependence of the critical field B_c for some single element superconductors.

along with the critical field at zero temperature $B_c(0)$. Since the *critical field* B_c is because of the electrical current flowing in the superconductor, there is also a corresponding average critical current density $J_c = I_c/A$. Using Eq. (15.4), the critical current density is approximately given by

$$J_c = \frac{2\pi r B_c}{\mu_0 A} = \frac{2\pi r B_c}{\mu_0 \pi r^2} = \frac{2B_c}{\mu_0 r} \tag{15.6}$$

Element	T_c[K]	$B_c(0)$[mT]
V	5.38	142.0
Nb	9.50	198.0
Tc	7.77	141.0
In	3.40	29.3
Sn	3.72	30.9
La	6.00	110.0
Ta	4.48	83.0
Hg	4.15	41.2
Pb	7.19	80.3

Table 15.2: Transition temperatures and critical fields of some selected superconducting elements.

Thus, a superconductor has two or three important parameters that characterize it, or determine the ability to transition into/out of the superconducting state.

Superconducting state: The superconducting state depends on two (or three) important parameters, given by the transition temperature T_c and the critical magnetic field B_c (also impacted by the critical current J_c).

With this understanding of the importance of the additional parameter given by the critical field, new opportunities to tune the superconducting state would lead to two distinct types of superconductors. We'll discuss these two types in the next section.

15.3 Type I & Type II Superconductors

In earlier experiments carried out with superconductors of a single element, one that tops the list with the transition temperature as well as critical field value is niobium (Nb). Eventually, the idea to mix elements like this with other materials would lead to further increasing

15.3 Type I & Type II Superconductors

Compound	$T_c(K)$
Nb_3Sn	18.1
Nb_3Ge	23.2
Nb_3Al	17.5
V_3Ga	16.5
V_3Si	17.1
La_3In	10.4
NbTi	10.0

Table 15.3: Transition temperatures and corresponding critical temperatures of some superconducting compounds.

their values, and more. In the 1980s, some compounds were discovered that would demonstrate larger T_c, compared to most single element superconductors. Such compounds demonstrate that superconducting properties *can* be tuned to more favorable critical values, which is an encouraging attribute of superconductors. It hints that better is possible. Table 15.3 lists some compounds that were found to have enlarged transition temperatures. With the single element superconductors, and some of the mixed compounds found later, there was a universal trend with the magnetization dependence on the external magnetic field B_e. Following Eq. (15.2), the expectation is just a linear relation up until the critical field B_c is reached. This is illustrated in Fig. 15.6. This type of behavior, magnetic in origin, defines the so-called **type I superconductor**. It corresponds to the first type observed, e.g. Hg. However, with the mixing of the elements to find additional superconductors, a second extended type of magnetic dependence on B_e was later observed. Rather than the precipitous drop that takes place at the single valued critical field B_c, instead, there is a more gradual exponential decay from a first critical field value B_{c1} down to zero at a second critical field B_{c2}, as illustrated in Fig. 15.7. The behavior shown in Fig. 15.7 defines the **type II superconductor**. This behavior has significant implications because it leads to a superconductor that can carry larger currents without being triggered out of the superconducting state. For this reason, *type II* superconductors are often of more technological importance that *type I*. A list of some type II superconductors is given in Table 15.4. Note that for type II superconductors, up until $B_e = B_{c1}$, the material is in a

Type I Superconductor

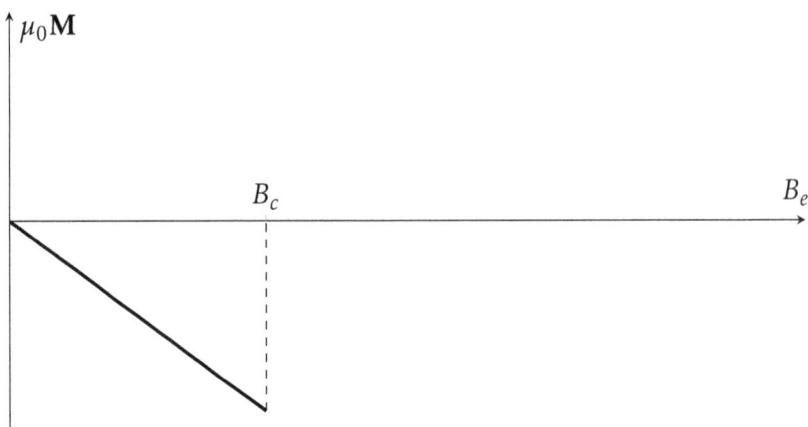

Figure 15.6: Magnetization dependence of the critical field B_c for a Type I superconductor. The slope of the curve is -1.

Type II Superconductor

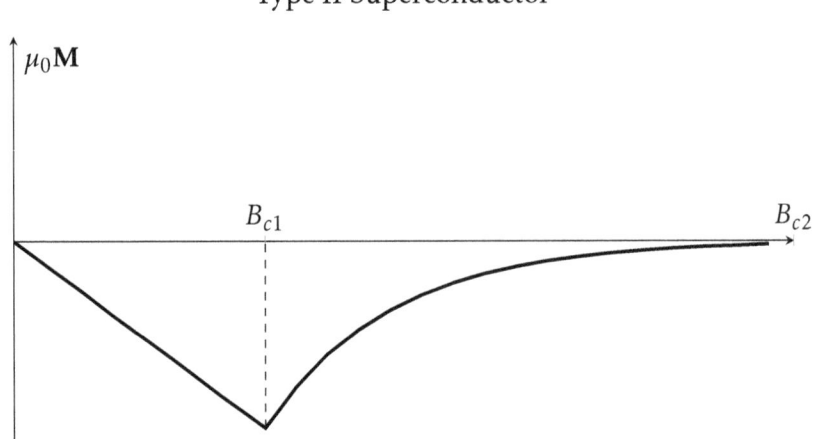

Figure 15.7: Magnetization dependence of the critical field B_c for a Type II superconductor. The behavior indicates two important magnetic field values, B_{c1} and B_{c2}.

pure superconducting state. The region between $B_e = B_{c1}$ and $B_e = B_{c2}$, however, is a *mixed state* of normal + superconducting, going from pure superconducting at B_{c1} to pure mixed beyond B_{c2}.

When superconductivity was discovered experimentally, it would not be long before some theoretical explanations would be attempted. It can

15.3 Type I & Type II Superconductors

Compound	T_c[K]	B_{c2}[Tesla]
Nb_3Sn	18.5	24
Nb_3Ge	21.0	43.9
Nb_3Al	21.0	43.9
NbTi	10.5	12
NbZr	11	9

Table 15.4: Transition temperatures and critical fields of some *type II* superconductors. Because of their magnetic distinction, type II materials are capable of producing very large magnetic fields.

not be overstated enough that this was an incredibly difficult problem for physicists to sort out. Several physicists made serious attempts to offer both macroscopic and microscopic theories to explain the phenomenon of superconductivity. The list of notable physicists who attempted theories for superconductivity includes Einstein, Bohr, Kronig, Landau, Bloch, Brillouin, Born, Feynman, and Heisenberg. We've discussed some of the great contributions from some these physicists already. This list serves as a testament to the level of difficulty of the problem of superconductivity. When the dust finally settled onto paper, only a few had succeeded in offering theories for superconductivity that overwhelmingly agreed with the experimental data that was available, and persisted through the tests of time. One of these theories was carried out by a pair of brothers, German-British physicists named Heinz (1907-1970) and Fritz London (1900-1954). Another was derived by Soviet physicists Vitaly Lazarevich Ginzburg (1916-2009) and Lev Landau (1908-1968). Lastly, the theory developed by American physics John Bardeen (1908-1991), Leon N Cooper (1930-), and John Robert Schrieffer (1931-). These theories didn't clearly connect in the beginning, however, they are consistent. We'll aim to connect them more directly later in the chapter. They cover different aspects of the physics of superconductors. We'll discuss these important theories in the next few sections. Let's start with the work of the London brothers.

15.4 London Equation Leads To Meissner Effect

The London brothers looked to the Maxwell equations to explain some of the physics of superconductors, particularly, the *Meissner effect*, illustrated in Fig. 15.4. Two basic equations were used to arrive at the results of the London brothers. Consider the supercurrent density J_s relation, along with Newton's law including a Coulomb force, given by:

$$\mathbf{J}_s = n_s q_s \mathbf{v}_s \tag{15.7a}$$

$$m_s \frac{d\mathbf{v}_s}{dt} = q_s \mathbf{E} \tag{15.7b}$$

n_s is the mean carrier density of superconducting charge carriers, with units of m^{-3}. q_s is the unit charge on the carrier. $\mathbf{v}_s$ is the mean carrier velocity, and $\mathbf{E}$ is the electric field in the superconductor. This pair of equations describes a *lossless* conductor, as expected in a superconductor. For losses, a dissipation term would have to added to Eq. (15.7b). Then, time differentiation of Eq. (15.7a) leads to

$$\frac{\partial \mathbf{J}_s}{\partial t} = n_s q_s \frac{\partial \mathbf{v}_s}{\partial t} = \frac{n_s q_s^2}{m_s} \mathbf{E}$$

Eq. (15.7b) was used to substitute for $d\mathbf{v}_s/dt$. Taking the curl $\nabla \times$ of both sides then gives

$$\nabla \times \frac{\partial \mathbf{J}_s}{\partial t} = \frac{n_s q_s^2}{m_s} \nabla \times \mathbf{E}$$

Changing the order of the operations on the LHS, and substituting the Maxwell equation $\nabla \times \mathbf{E} = -\partial \mathbf{B}/\partial t$ on the RHS, we have

$$\frac{\partial}{\partial t}(\nabla \times \mathbf{J}_s) = -\frac{n_s q_s^2}{m_s} \frac{\partial \mathbf{B}}{\partial t} = -\frac{n_s q_s^2}{m_s} \frac{\partial}{\partial t}(\nabla \times \mathbf{A})$$

In the above, we have used the relation $\mathbf{B} = \nabla \times \mathbf{A}$. Integration of the above equation leads to a result known as the **London Equation**, given by

London Equation

$$\mathbf{J}_s = -\frac{n_s q_s^2}{m_s} \mathbf{A} \tag{15.8}$$

15.4 London Equation Leads To Meissner Effect

The *London equation* says that the superconducting current density $\mathbf{J}_s$ is proportional to the magnetic vector potential $\mathbf{A}$. What is remarkable about this simple derivation is that it, in fact, leads to the *Meissner effect*. To see this, recall Ampére's law in the form given by

$$\nabla \times \mathbf{B} = \mu_0 \mathbf{J}_s$$

Taking the curl of both sides, and using some of the relations above, gives

$$\nabla \times \nabla \times \mathbf{B} = \mu_0 \nabla \times \mathbf{J}_s = \mu_0 \nabla \times \left(-\frac{n_s q_s^2}{m_s} \mathbf{A} \right)$$

We can replace the LHS using the vector identity given by $\nabla \times \nabla \times \mathbf{B} = \nabla(\nabla \cdot \mathbf{B}) - \nabla^2 \mathbf{B}$. Since the electromagnetic equations dictate that $\nabla \cdot \mathbf{B} = 0$, we have the following result:

$$-\nabla^2 \mathbf{B} = -\left(\frac{\mu_0 n_s q_s^2}{m_s} \mathbf{B} \right) \Rightarrow \nabla^2 \mathbf{B} - k_L^2 \mathbf{B} = 0$$

From previous problems we've analyzed in this book, we know that the solution to this equation is given by

$$\mathbf{B}(\mathbf{r}) = \mathbf{B}_0 e^{-\mathbf{k}_L \cdot \mathbf{r}} \tag{15.9}$$

$\mathbf{B}_0$ denotes the value of the magnetic induction at the boundary of the superconductor. Eq. (15.9) describes a decaying field whose maximum is at the boundary. The wave-number and length scale of the decay are related as

$$k_L^2 = \frac{1}{\delta_L^2} \tag{15.10}$$

This defines the so-called **London penetration depth**, which is the **magnetic induction penetration depth** into the superconductor. It is defined as

Superconductor Magnetic Induction Penetration Depth

$$\delta_L = \sqrt{\frac{m_s}{\mu_0 n_s q_s^2}} \tag{15.11}$$

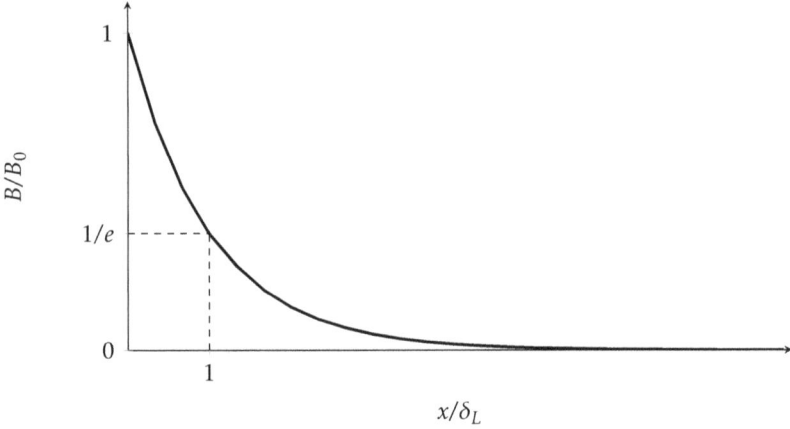

Figure 15.8: Penetration of magnetic induction into a superconductor with depth dimension along x. The location $x = 0$ represents the boundary of the superconductor. At $x/\delta_L = 1$ the induction has reduced to $1/e$ times the peak value on the boundary.

The penetration depth δ_L is, then, the depth into the superconductor where the amplitude of the magnetic induction is $1/e \approx .37$ times the peak value at the boundary. Fig. 15.8 illustrates the magnetic induction penetration depth in a superconductor. In the London equation, and in the penetration depth of the *Meissner effect*, we still have no information on how a superconductor reaches such a state. In fact, the London theory is very general in the way we have presented it. But we know that while in one set of conditions, the material is in a normal state, with the induction field inside the material. Then, changing the conditions, it suddenly gets expelled from the material when the superconducting state is reached. So, what distinguishes this as a relevant theory for superconductors? The answer lies in the carrier concentration n_s. It is this quantity that also changes in the transition. n_s goes from zero to some finite value in the superconducting state.

Another approach taken to explain superconducting behavior, that we'll discuss, uses phenomenological thermodynamical parameters to better understand the behavior of n_s in the vicinity of the transition. Though phenomenological in spirit, the lowest order term in the theory turns out to be formulated in a form *consistent with quantum mechanics*, and this makes it very powerful. This is the *Ginzburg-Landau theory*, and it is the subject of the next section.

15.5 Ginzburg & Landau Theory

In introducing the Ginzburg-Landau theory, because the analysis is somewhat involved, it will be done in three steps. *First*, we'll show that the energy change of a superconductor transitioning from zero field to the critical field is strictly positive. We'll use this result to argue the properties of the free energy density proposed by Ginzburg and Landau. *Second*, we'll derive the Ginzburg-Landau equation. The third and last part of this section is the messiest. It entails a complete derivation of the solution to what is called the *linearized Ginzburg-Landau equation*, which turns out to be a quantum mechanical equation. Therefore, the solutions obtained turn out to be quantum mechanical solutions, and they provide the expected results for a charge carrier in a magnetic field, for example. So, let's begin.

15.5.1 External Field Superconductor Energy

When a material transitions from one phase into another, there is always an associated change in some aspect of the energy of the system. We may consider a change in the thermodynamic energy due to an external magnetic field $\mathbf{B}_e$ applied to a superconductor. Upon varying the field, it can induce the transition in or out of the superconducting state. The potential energy on the conductor magnetization $\mathbf{M}$ is given by

$$V = -\mathbf{M} \cdot \mathbf{B}_e$$

If we consider the transition from the normal conducting state to the superconducting state, the change in the field corresponds to a change in energy. The change in energy is the work done on the system, and is given by

$$W = -\int \mathbf{M} \cdot d\mathbf{B}_e = +\frac{1}{\mu_0} \int_0^{B_c} \mathbf{B}_e \cdot d\mathbf{B}_e = +\frac{B_c^2}{2\mu_0}$$

Then, the increase in the free energy $F_s(B_c) - F_n(0) = \Delta F$ of the superconductor when going from normal to superconducting state is

Isothermal Superconductor Free Energy Increase

$$\Delta F = \frac{B_c^2}{2\mu_0} \tag{15.12}$$

This result ignores any contributions from *paramagnetism*, which can also do work on the magnetization linearly with B. Because the sign is opposite (positive) for paramagnetism, it would weaken this type of effect.

15.5.2 The Ginzburg-Landau Equation

Prior to working on superconductivity, Lev Landau had already proposed a general theory of phase transitions. He took an interest in applying it to superconductors. Eventually, Landau together with Ginzburg would work to apply Landau's theory of phase transitions to provide more insight into the behavior of the probability density, which is a measure of the number of charge carriers per unit volume n_s. This parameter is proportional to the quantum mechanical wave function ψ. Since wave functions are *normalizable*, we can simply write

$$n_s = \psi_s^* \psi_s = |\psi_s|^2 \tag{15.13}$$

This parameter is positive, and in a normal conductor, we expect $n_s \approx 0$, while in the superconducting state, we expect $n_s > 0$. According to Landau's phase transition theory, this parameter is ideal to use as an **order parameter**. This is the first step in applying the phase transition theory, to identify a so-called *order parameter*. The order parameter in the Landau-Ginzburg theory, which is n_s, must generally possess the binary property where:

(1) order parameter $= 0$ for $T > T_c$

(2) order parameter $\neq 0$ for $T < T_c$

T_c is the superconductor phase transition temperature. Any parameter that satisfies the above can be used as an order parameter. Since the wave function ψ_s generally depends on space, the order parameter also does.

We then consider the change in free energy or free energy density, near the transition point. With an order parameter identified, the next step

15.5 Ginzburg & Landau Theory

is to write the difference in the free energy density $\Delta f = f_s - f_n$, as an expansion, along with the additional relevant energies associated with the phase transition. For this, we must include a general kinetic energy term, as well as the added work done on the magnetization by the changing magnetic field. Since we are writing the energy density, the kinetic energy density in terms of the superconductor wave function ψ_s becomes $\psi_s^* \widehat{H}_K \psi_s$, where $\widehat{H}_K$ is the kinetic energy operator. This leads to an energy density given by

$$f_s - f_n = \alpha |\psi_s|^2 + \frac{\beta}{2}|\psi_s|^4 + \psi_s^* \frac{(-i\hbar\nabla - q_s\mathbf{A})^2}{2m_s}\psi_s - \int \mathbf{M} \cdot d\mathbf{B}_e \quad (15.14)$$

Because the last two terms are positive, since they are the kinetic energy of the carrier and the contribution given by Eq. (15.12), which was shown to be positive, the function $\psi_{\min}$ that minimizes Eq. (15.14) also minimizes *only* the first two terms. This can be used to relate the minimizing wave function $\psi_{\min}$ to the parameters α and β. Therefore, minimizing only the first two terms of Eq. (15.14) with respect to ψ_s^* leads to

$$\psi_{\min}(\alpha + \beta|\psi_{\min}|^2) = 0$$

The above equation has two distinct solutions given by

$$\psi_{\min} = 0 \quad \text{and} \quad |\psi_{\min}|^2 = \frac{-\alpha}{\beta} \quad (15.15)$$

Since $|\psi_{\min}|^2 \geq 0$, it must be that α and β have opposite signs. This makes sense for an expansion of this type with even powers. Since $|\psi_{\min}|^{2q} \geq 0$, general convergence is only possible if the coefficients of the expansion have opposite sign. Otherwise, a nearly infinite expansion would diverge. For a favorable transition, $f_s - f_n$ should be negative. Therefore, the leading term, which is the largest in amplitude, is negative, while the next term is positive. Thus, $\alpha < 0$ and $\beta > 0$.

Minimizing *all* the terms of Eq. (15.14) w.r.t. ψ_s^* leads to the well-known **Ginzburg-Landau equation** given by

Ginzburg-Landau Equation

$$\alpha\psi_s + \beta|\psi_s|^2\psi_s + \frac{(-i\hbar\nabla - q_s\mathbf{A})^2}{2m_s}\psi_s = 0 \quad (15.16)$$

Because of the third order term with β, the *Ginzburg-Landau equation* is generally nonlinear. However, it is still useful to look at an approximate form of Eq. (15.16) and study the solutions. It reveals spatial information regarding $n_s = \psi_s^* \psi_s$. To do this, one neglects the nonlinear term $\beta |\psi_s|^2 \psi_s$ and solves the equation.

15.5.3 Exact Solution To Linearized Ginzburg-Landau Equation

Ignoring the nonlinear term in the Ginzburg-Landau equation, and setting $q_s = -e$ gives the following equation to solve:

$$\frac{(-i\hbar\nabla + e\mathbf{A})^2}{2m_s}\psi_s - |\alpha|\psi_s = \frac{1}{2m_s}(\widehat{\mathbf{p}} + e\mathbf{A})^2 - |\alpha|\psi_s = 0 \tag{15.17}$$

We have used the property $\alpha < 0$. To solve Eq. (15.17), we must also specify the magnetic vector potential $\mathbf{A}$, as a function of external magnetic field B_e. Then, we can proceed to obtain the exact solution to Eq. (15.17). Squaring the extended momentum operator gives

$$|\alpha|\psi_s = \frac{1}{2m_s}\left[\widehat{\mathbf{p}}^2 + 2e\mathbf{A}\cdot\widehat{\mathbf{p}} + e^2 A^2\right]\psi_s \tag{15.18}$$

$$= \frac{1}{2m_s}\left[-\hbar^2\nabla^2 - i2e\hbar\mathbf{A}\cdot\nabla + e^2 A^2\right]\psi_s \tag{15.19}$$

Eq. (15.19) gives the immediate interpretation of the Ginzburg-Landau parameter $|\alpha|$. Our equation is in the exact form of the eigenvalue equation for the kinetic energy of a charge carrier in an electric as well as magnetic field. Therefore, *α is a proxy for the kinetic energy operator eigenvalue for a charge carrier in a uniform electric and magnetic field.*

Ginzburg-Landau parameter $|\alpha|$: This parameter can be interpreted, through the linearized-GL equation, as just the eigenvalue of the kinetic energy of a charge carrier in a uniform electric and magnetic field

Eq. (15.19) will be solved in a cylindrical coordinate system (r, φ, z). We'll use the *symmetric gauge* given by $\mathbf{A} = (1/2)[-B_e \cdot y, B_e \cdot x, 0]^T$ describing a uniform magnetic field. Note that $y = r\sin\varphi$ and $x = r\cos\varphi$ in the cylindrical coordinate system. Therefore, since the azimuthal unit vector

15.5 Ginzburg & Landau Theory

in the cylindrical coordinate system is given by $\boldsymbol{\varphi} = -\sin\varphi\hat{\mathbf{x}} + \cos\varphi\hat{\mathbf{y}}$, the magnetic vector potential $\mathbf{A}$ is oriented azimuthally, or circles around the cylindrical z-axis. $\hat{\mathbf{x}}$ and $\hat{\mathbf{y}}$ are unit vectors. Then, we can write $\mathbf{A} = (rB_e)/2\mathbf{e}_\varphi$. Using the ordinary momentum operator $\hat{\mathbf{p}} = -i\hbar\nabla = -i\hbar[\partial/\partial r, (1/r)\partial/\partial\varphi, \partial/\partial z]^T$ along with the *symmetric gauge*, we then have

$$|\alpha|\psi_s = \frac{1}{2m_s}\left[-\hbar^2\nabla^2\psi_s - ier\hbar B_e \mathbf{e}_\varphi \cdot \nabla\psi_s + \frac{e^2 B_e^2}{4}r^2\psi_s\right] \tag{15.20a}$$

$$= \frac{1}{2m_s}\left[-\hbar^2\left(\frac{1}{r}\frac{\partial}{\partial r}\left(r\frac{\partial\psi_s}{\partial r}\right) + \frac{1}{r^2}\frac{\partial^2\psi_s}{\partial\varphi^2} + \frac{\partial^2\psi_s}{\partial z^2}\right) - ie\hbar B_e\frac{\partial\psi_s}{\partial\varphi}\right. \tag{15.20b}$$

$$\left. + \frac{e^2 B_e^2}{4}r^2\psi_s\right] \tag{15.20c}$$

From the form of this equation, the z dependence is easily separable from the r, φ dependence. In fact, the solution is separable in all three variables, so let ψ_s be given by

$$\psi_s = R(r)\Phi(\varphi)Z(z) \quad \text{and} \quad |\alpha| = |\alpha|_r + |\alpha|_\varphi + |\alpha|_z \tag{15.21}$$

Substitution of Eq. (15.21) into Eq. (15.20b), we can separate the equations, particularly, for the z dependence, where $Z(z)$ is described by

$$-\frac{\hbar^2}{2m_s}\frac{d^2 Z}{dz^2} = |\alpha|_z Z(z) \Rightarrow \frac{d^2 Z}{dz^2} + k_z^2 Z = 0 \tag{15.22}$$

The solution to Eq. (15.22) is given by

$Z(z)$ Solution for ψ_s

$$Z(z) = Z_0 e^{ik_z z} \quad \text{where} \quad k_z^2 = \frac{2m_s|\alpha|_z}{\hbar^2} \tag{15.23}$$

The remaining part of Eq. (15.20b) depends on r and φ and gives the following equation for functions $R(r)$ and $\Phi(\varphi)$:

$$\frac{1}{2m_s}\left[-\hbar^2\left(\Phi\frac{1}{r}\frac{\partial}{\partial r}\left(r\frac{\partial R}{\partial r}\right) + R\frac{1}{r^2}\frac{\partial^2\Phi}{\partial\varphi^2}\right) - ie\hbar B_e R\frac{\partial\Phi}{\partial\varphi} + \right.$$

$$\frac{e^2 B_e^2}{4} r^2 R\Phi \Big] = (|\alpha|_r + |\alpha|_\varphi) R\Phi \quad (15.24)$$

Next, let's deal with $\Phi(\varphi)$. It can be shown that if $\Phi(\varphi) = C_\varphi e^{i\kappa_\varphi(r)\varphi}$, then both terms in Eq. (15.24) depending on φ sum to zero, which also means $|\alpha|_\varphi = 0$. We'll show this now, and determine the value of κ_φ. The φ dependence leads to

$$-\frac{\hbar^2}{2m_s}\left(\frac{1}{r^2}\frac{\partial^2 \Phi}{\partial \varphi^2}\right) - i\frac{e\hbar B_e}{2m_s}\frac{\partial \Phi}{\partial \varphi} = |\alpha|_\varphi \Phi \quad (15.25)$$

Now, let's assume a form for the solution. Since we are differentiating with respect to φ, let the solution be written as $\Phi(\varphi) = C_\varphi e^{ik_\varphi r\varphi}$. Note that $\kappa_\varphi \neq k_\varphi$. This leads to

$$\frac{\partial \Phi}{\partial \varphi} = C_\varphi i k_\varphi r e^{ik_\varphi r\varphi} = ik_\varphi r \Phi \quad (15.26)$$

$$\frac{\partial^2 \Phi}{\partial \varphi^2} = C_\varphi (ik_\varphi r)^2 e^{ik_\varphi r\varphi} = -(k_\varphi)^2 r^2 \Phi \quad (15.27)$$

Substitution of these results into the LHS of Eq. (15.25) gives

$$-\frac{\hbar^2}{2m_s}\frac{1}{r^2}\left(-(k_\varphi)^2 r^2 \Phi\right) - i\frac{e\hbar B_e}{2m_s}\left(ik_\varphi r\Phi\right) = \frac{\hbar^2}{2m_s}(k_\varphi)^2 + \frac{e\hbar B_e k_\varphi}{2m_s} \quad (15.28)$$

We want to find the value of k_φ that makes Eq. (15.28) vanish. This leads to

$$0 = \hbar k_\varphi + eB_e r \Rightarrow k_\varphi = -\frac{eB_e r}{\hbar}$$

Since the function $\Phi(\varphi) = \Phi(r,\varphi) = C_\varphi e^{ik_\varphi r\varphi}$, we have the following result:

Azimuthal Solution $\Phi(\varphi)$

$$\Phi(r,\varphi) = C_\varphi e^{-i\frac{e}{\hbar}B_e r^2 \varphi} \quad (15.29)$$

C_φ is the normalization factor, determined from the normalization condition. Evaluating this condition in the cylindrical coordinate gives

$$1 = \int_0^{2\pi} C_\varphi^* C_\varphi e^{i\frac{q_s}{\hbar}B_e r^2 \varphi} e^{-i\frac{q_s}{\hbar}B_e r^2 \varphi} d\varphi$$

$$= |C_\varphi|^2 \int_0^{2\pi} d\varphi$$

$$= |C_\varphi|^2 \cdot 2\pi$$

This gives the **azimuthal normalization factor** as

> **Normalization Factor For $\Phi(\varphi)$**
>
> $$|C_\varphi| = \frac{1}{\sqrt{2\pi}} \qquad (15.30)$$

Let's talk about the magnetic flux that one would measure from such a system.

15.5.4 Magnetic Flux Relation

The solution given by Eq. (15.29) leads to an important property when we consider the following: if the wave function is forming a loop around the magnetic field axis, for the continuity of the wavefunction in the φ direction, one complete loop around must return to the same starting point. With this part of the solution given by $\Phi = C_\varphi e^{ik_\varphi r \varphi}$, the exponent becomes the phase ϕ of the carrier around the loop. Then, the continuity condition for a general number of loops leads to

$$\frac{\Phi}{C_\varphi} = e^{i(k_\varphi 2\pi r n)} = \cos(k_\varphi 2\pi r n) + i\sin(k_\varphi 2\pi r n) = 1 \qquad (15.31)$$

Since the RHS is purely real, it must be that the cosine term gives unity, while the sine term values, or

$$\cos(k_\varphi 2\pi r n) = 1 \qquad (15.32)$$

This is true if and only if the argument is a multiple of 2π. This leads to

$$k_\varphi 2\pi r n = 2\pi n \Rightarrow k_\varphi r = 1 \qquad (15.33)$$

Using $k_\varphi = -eB_e r/\hbar$ and $\hbar = h/2\pi$, we have

$$-\frac{eB_e r^2}{\hbar} = -\frac{eB_e r^2}{h/2\pi} = -\frac{2eB_e \pi r^2}{h} = 1 \qquad (15.34)$$

Recognizing that the flux $\Phi_s = B_e A = B_e \pi r^2$, gives an expression for a **unit of quantized flux** given by

> **Unit Of Quantized Flux**
>
> $$\Phi_s = -\frac{h}{2e} \tag{15.35}$$

Eq. (15.35) is also known as the **flux quantum** and it describes the unit of flux that satisfies the continuity of the wave function. The quantized nature of the flux in superconductors was first measured around 1961 by Deaver and Fairbank, and by Doll and Nabauer. They used cylinders in well-controlled magnetic fields in which the superconducting cylinders were moved cyclically along the axis of the magnetic field. This allowed determination of the fluxes. They observed that the fluxes were indeed quantized by the unit described by Eq. (15.35).

Now that we have $Z(z)$ and $\Phi(\varphi)$, if we are also able to find the exact solution to the remaining radial parts of Eq. (15.20b), then the solution will be complete. Now, down to the radial terms, we have

$$\frac{1}{2m_s}\left[-\hbar^2 \frac{1}{r}\frac{\partial}{\partial r}\left(r\frac{\partial R}{\partial r}\right) + \frac{e^2 B_e^2}{4}r^2 R\right] = |\alpha|_r R \tag{15.36}$$

The derivative term expands leading to a radial equation described by

$$\frac{d^2 R}{dr^2} + \frac{1}{r}\frac{dR}{dr} + \left(\frac{2m_s |\alpha|_r}{\hbar^2} - \frac{e^2 B_e^2}{4\hbar^2} r^2\right) R = 0 \tag{15.37}$$

To solve this equation, we define the variable ζ by the relation

$$\sqrt{\zeta} = \sqrt{\frac{eB_e}{\hbar}} r = vr \Rightarrow \zeta = v^2 r^2 \tag{15.38}$$

From the definition of ζ, we obtain the following relations:

$$d\zeta = 2v^2 r\, dr \quad \text{or} \quad \frac{d\zeta}{dr} = 2v^2 r$$

$$\frac{dR}{dr} = \frac{d\zeta}{dr}\frac{dR}{d\zeta} = 2v^2 r\frac{dR}{d\zeta}$$

$$\frac{d^2 R}{dr^2} = \frac{d}{dr}\left(2v^2 r\frac{dR}{d\zeta}\right) = 2v^2 \frac{dR}{d\zeta} + 2v^2 r\frac{d\zeta}{dr}\frac{d}{d\zeta}\left(\frac{dR}{d\zeta}\right)$$

$$= 2v^2 \frac{dR}{d\zeta} + 4v^4 r^2 \frac{d^2 R}{d\zeta^2} = 2v^2 \frac{dR}{d\zeta} + 4v^2 \zeta \frac{d^2 R}{d\zeta^2}$$

15.5 Ginzburg & Landau Theory

Substitution of these relations into Eq. (15.37) leads to the following differential equation in terms of ζ:

$$4v^2\zeta\frac{d^2R}{d\zeta^2} + 4v^2\frac{dR}{d\zeta} + \left(\frac{2m_s|\alpha|r}{\hbar^2} - \frac{v^2}{4}\zeta\right)R(\zeta) = 0$$

or

$$4\zeta\frac{d^2R}{d\zeta^2} + 4\frac{dR}{d\zeta} + \left(\frac{2m_s|\alpha|r}{v^2\hbar^2} - \frac{1}{4}\zeta\right)R(\zeta) = 0 \quad (15.39)$$

We can try another mathematical trick here by writing the wave function R in the form $R = R_\infty(\zeta)R_1(\zeta)$, where $R_\infty(\zeta)$ is an asymptotic solution we will find. For the asymptotic limit $\zeta \to \infty$, Eq. (15.39) becomes

$$4\frac{d^2R_\infty}{d\zeta^2} - \frac{1}{4}R_\infty(\zeta) = 0 \Rightarrow \frac{d^2R_\infty}{d\zeta^2} - \frac{1}{16}R_\infty(\zeta) = 0 \quad (15.40)$$

Eq. (15.40) has a solution given by

$$R_\infty(\zeta) = C_\infty e^{-\frac{1}{4}\zeta} = C_\infty e^{-\frac{eB_e}{4\hbar}r^2} \quad (15.41)$$

Because $R_\infty(\zeta)$ is in the form of a Gaussian, it allows us to obtain an equation purely in terms of R_1. If $R = R_\infty R_1$, we have the following derivatives of $R(\zeta)$ in terms of R_1:

$$\frac{dR}{d\zeta} = \frac{d}{d\zeta}(R_\infty R_1) = \left[\frac{dR_1}{d\zeta} - \frac{1}{4}R_1\right]e^{-\frac{1}{4}\zeta}$$

$$\frac{d^2R}{d\zeta^2} = \left[\left(\frac{d^2R_1}{d\zeta^2} - \frac{1}{4}\frac{dR_1}{d\zeta}\right) + \left(\frac{dR_1}{d\zeta} - \frac{1}{4}R_1\right)\left(-\frac{1}{4}\right)\right]e^{-\frac{1}{4}\zeta}$$

Using these relations in Eq. (15.39), we obtain the following equation for R_1:

$$4\zeta\left[\left(\frac{d^2R_1}{d\zeta^2} - \frac{1}{4}\frac{dR_1}{d\zeta}\right) + \left(\frac{dR_1}{d\zeta} - \frac{1}{4}R_1\right)\left(-\frac{1}{4}\right)\right]e^{-\frac{1}{4}\zeta}$$

$$+ 4\left[\frac{dR_1}{d\zeta} - \frac{1}{4}R_1\right]e^{-\frac{1}{4}\zeta} + \left[\frac{2m_s|\alpha|r}{v^2\hbar^2} - \frac{1}{4}\zeta\right]R_1e^{-\frac{1}{4}\zeta} = 0 \quad (15.42)$$

We can cancel the Gaussian factors and write all the terms as

$$4\zeta\frac{d^2R_1}{d\zeta^2} - \zeta\frac{dR_1}{d\zeta} - \zeta\frac{dR_1}{d\zeta} + \frac{1}{4}\zeta R_1$$

$$+4\frac{dR_1}{d\zeta} - R_1 + \left[\frac{2m_s|\alpha|_r}{v^2\hbar^2} - \frac{1}{4}\zeta\right]R_1 = 0$$

Combining terms simplifies the above equation to

$$4\zeta\frac{d^2R_1}{d\zeta^2}(4-2\zeta)\frac{dR_1}{d\zeta} - \left[1 - \frac{2m_s|\alpha|_r}{v^2\hbar^2}\right]R_1 = 0$$

or

$$\zeta\frac{d^2R_1}{d\zeta^2}\left(1-\frac{1}{2}\zeta\right)\frac{dR_1}{d\zeta} - \left[\frac{1-\frac{2m_s|\alpha|_r}{v^2\hbar^2}}{4}\right]R_1 = 0 \tag{15.43}$$

This has the form

$$\zeta\frac{d^2R_1}{d\zeta^2} + \left(1-\frac{1}{2}\zeta\right)\frac{dR_1}{d\zeta} - \left(\frac{1-\Omega}{4}\right)R_1 = 0 \tag{15.44}$$

Ω is defined as

$$\Omega = \frac{2m_s|\alpha|_r}{v^2\hbar^2} \tag{15.45}$$

Eq. (15.44) is *almost* in the form of the *confluent hypergeometric equation*(CHE), given by

$$x\frac{d^2y}{dx^2} + (c-x)\frac{dy}{dx} - ay(x) = 0$$

The only difference between them is with the two corresponding terms $c - x \leftrightarrow 1 - (1/2)\zeta$. Unfortunately, for the CHE, we must have 1 as the coefficient for x, or ζ in this case. We can let

$$\zeta^* = \frac{1}{2}\zeta \quad \text{and} \quad R_1^* = \frac{1}{2}R_1$$

Using these relations in Eq. (15.44), we have

$$\zeta^*\frac{d^2R_1^*}{d\zeta^{*2}} + (1-\zeta^*)\frac{dR_1^*}{d\zeta^*} - \left(\frac{1-\Omega}{2}\right)R_1^* = 0 \tag{15.46}$$

This gives the exact form of the *confluent hypergeometric equation*. We know that the solution is given by the **confluent hypergeometric series** (CHS), derived in Chap. 5. The CHS is given by

$$\text{CHS}(\zeta^*;a,c) = {}_1F_1(\zeta^*;a,c) = 1 + \frac{1}{1!}\frac{a}{c}\zeta^* + \frac{1}{2!}\frac{a(a+1)}{c(c+1)}(\zeta^*)^2 +$$

15.5 Ginzburg & Landau Theory

$$\frac{1}{3!}\frac{a(a+1)(a+2)}{c(c+1)(c+2)}(\zeta^*)^3 + \frac{1}{4!}\frac{a(a+1)(a+2)(a+3)}{c(c+1)(c+2)(c+3)}(\zeta^*)^4 +$$

$$\frac{1}{5!}\frac{a(a+1)(a+2)(a+3)(a+4)}{c(c+1)(c+2)(c+3)(c+4)}(\zeta^*)^5 +$$

$$\frac{1}{6!}\frac{a(a+1)(a+2)(a+3)(a+4)(a+5)}{c(c+1)(c+2)(c+3)(c+4)(c+5)}(\zeta^*)^6 + \ldots$$

Thus, we have arrived at the exact solution to the linearized Ginzburg-Landau equation. The complete solution is given by

Linearized Ginzburg-Landau Equation Solution

$$\psi_s(r,z) = C_n e^{-\frac{eB_e}{4\hbar}r^2} e^{i\kappa_\varphi \varphi} e^{ik_z z} {}_1F_1\left(\frac{eB_e}{2\hbar}r^2; -n, 1\right) \quad (15.47)$$

C_n is the normalization constant. The parameters κ_φ and k_z are given by

$$\kappa_\varphi = \frac{-eB_e r^2}{\hbar} \quad \text{and} \quad k_z = \sqrt{\frac{2m_s |\alpha|_z}{\hbar^2}} \quad (15.48)$$

The azimuthal solution $\Phi(\varphi)$ can be written in a more conventional form, also, because we have

$$\kappa_\varphi \varphi = \frac{-eB_e r^2}{\hbar}\varphi = \frac{-eB_e r}{\hbar} r\varphi = k_\varphi \cdot d\ell \quad (15.49)$$

Then, the momentum in the azimuthal direction becomes

$$p_\varphi = \hbar k_\varphi = -eBr \Rightarrow v_\varphi = \frac{p_\varphi}{m_s} = -\frac{eB_e r}{m_s} = -\omega_s r \quad (15.50)$$

The rotational frequency ω_s, as we will see shortly is, in fact, that for the charge carrier in a superconductor. Note that the wavenumber k_z has units of inverse length, while κ_φ is dimensionless, since it multiplies by φ in $e^{i\kappa_\varphi \varphi}$. Since $k_\varphi r\varphi$ is dimensionless, it follows that $\xi_0^2 = \frac{\hbar}{eB_e}$ has units of length^{-2}. Recall that ξ_0 showed up as a natural unit to scale the radial dimension when we solved the radial equation. Since k_φ depends on the radial dimension r, an equivalent length scale can also be expressed in a more conventional form. This can be done by defining a more general parameter ξ having units of inverse length, in terms of the energy eigenvalues $|\alpha|_r$ as

Superconducting Coherence Length

$$\xi = \sqrt{\frac{\hbar^2}{2m_s|\alpha|_r}} \tag{15.51}$$

Eq. (15.51) is known as the **superconducting coherence length** (or Ginzburg-Landau coherence length) and it provides a length scale for the variation of the charge density in the radial direction, or in the direction normal to the superconductor surface. Because this dimension is quantized, there will always be variation over finite lengths along this direction.

The exact solutions to the *linearized Gingzburg-Landau equation* turn out to be *quantized radially*, or in the direction normal to the surface. The complete solution describes quantized radial positioning in the motion of the carriers, while traveling both along z and simultaneously *whirling* about the z axis or magnetic field axis. This second motion is along the $\mathbf{e}_\varphi$ azimuthal axis. In other words, Eq. (15.47) describes a *quantized vortex motion* for the charge carrier. This type of solution is known as an *n*-**vortex**. Vortex behavior in superconductors was first predicted by Alexei Abrikosov in 1957 to explain experiments with type II superconductors. After this prediction, the first experimental observation of vortices was made with a lead-indium (PbIn) compound in 1967 using a method known as *bitter patterns*, which was a mature method at the time for observing domain walls in magnetic materials. Around 1989 experiments using low-temperature scanning-tunneling microscopy would show with much more resolution the direct evidence of an array of *n*-vortices in NbSe$_2$. This array of vortices is also known as an **Abrikosov lattice**, found in type II superconductors. Fig. 15.9 shows one of the first examples of direct observation of the Abrikosov lattice made in 1967 by Essmann and Trauble. For this prediction, Abrikosov was awarded the Nobel Prize for physics, later, in 2003. The solutions for ψ_s allow us to visualize n_s and observe the distribution in a cylindrical coordinate system because it is proportional to the probability of finding the charge carrier in the location (r, φ). Fig. 15.5 shows examples of the probability distribution $r^2 \psi_s^* \psi_s$, given by Eq. (15.47), for $n = 1, 5, 10,$ and 15. From the figures in Table 15.5 and 15.6, we can see that the probability densities are mostly concentrated towards an exterior ring, extending more radially as *n* increases. As long as there is a magnetic

15.5 Ginzburg & Landau Theory

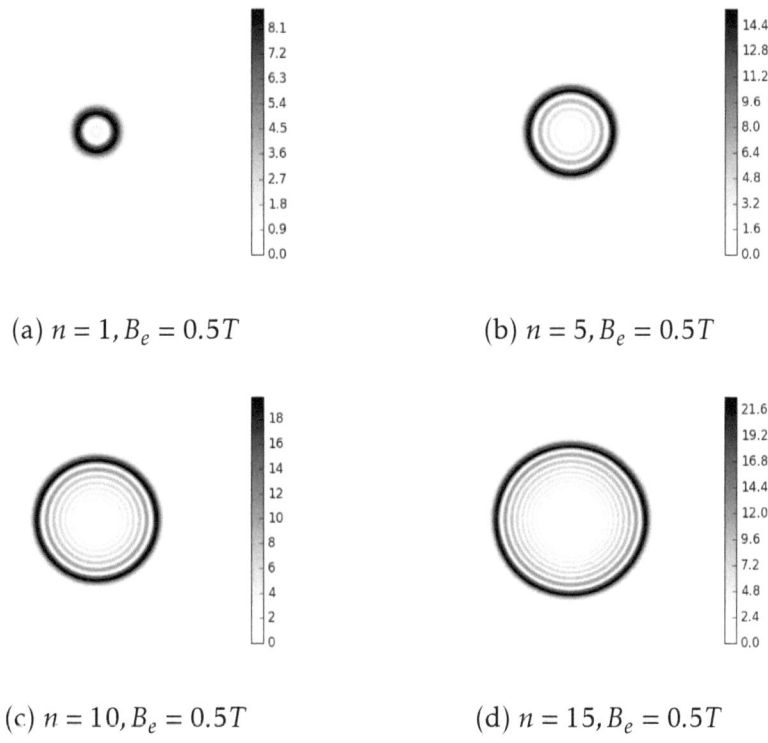

(a) $n = 1, B_e = 0.5T$

(b) $n = 5, B_e = 0.5T$

(c) $n = 10, B_e = 0.5T$

(d) $n = 15, B_e = 0.5T$

Table 15.5: Probability density distributions $r\psi_s^*\psi_s(r,\varphi)$, which are proportional to $n_s(r,\varphi)$ for different quantum numbers n. The field strength used is 0.5T, and all plots are on the same length scale of $1\mu m$ width and height. For larger energy states, the charge carrier pushes more radially outward.

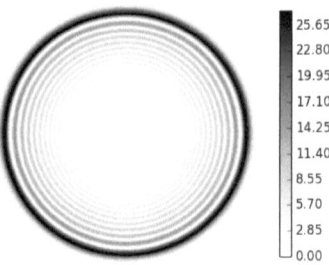

(a) $B_e = 0.5T, n = 25$

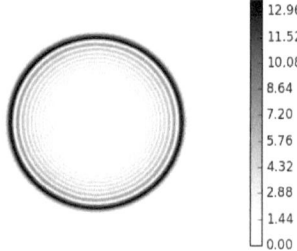

(b) $B_e = 1T, n = 25$

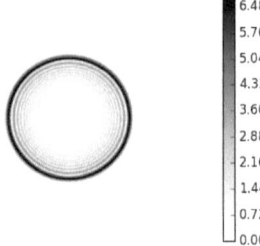

(c) $B_e = 2T, n = 25$

Table 15.6: Probability densities $r\psi_s^*\psi_s(r,\varphi)$ for $n = 25$, varying the external magnetic field strength B_e. More magnetic field compresses the vortex core to smaller diameters. All plots are on the same length scale of $800nm$ width and height.

15.5 Ginzburg & Landau Theory

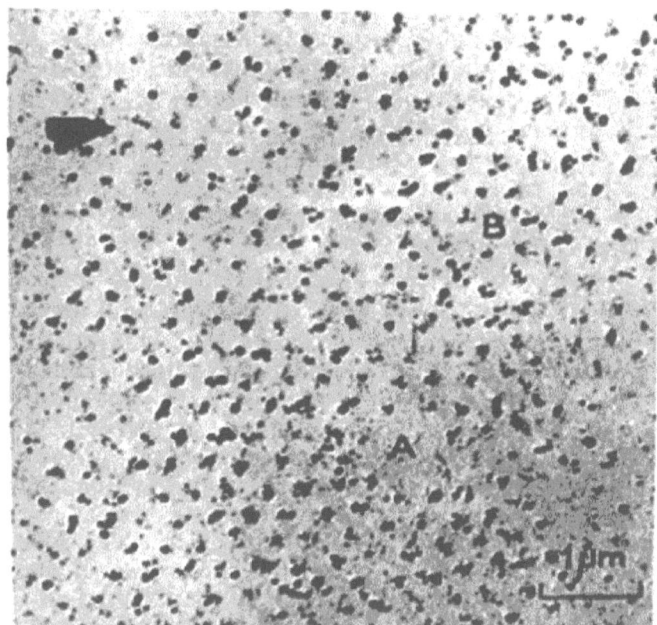

Figure 15.9: One of the first direct observations of the **Abrikosov lattice**, in Pb$_4$In reported in the historic paper from Essmann and Trauble (1967).

field, the state is described by Eq. (15.47). In a large system, there can be multiple vortices as there are multiple charge carriers. In the case of a finite size cylindrical wire, the carriers end up squeezing towards the surface of the wire. Table 15.6 shows that the magnetic field opposes the trend of extending radially. Instead, we find the probability reducing with more and more magnetic field and being more confined towards the center, still maintaining zero probability in the center. These trends have all been observed experimentally in superconductors.

We are also interested in the eigenvalue $|\alpha|_r$, which represents the eigenenergy for the kinetic energy operator, with the extended momentum $\widehat{\mathbf{P}} = \widehat{\mathbf{p}} - q_s \mathbf{A}$. From Eq. (15.46), we can identify the parameters a and c as

$$a = \frac{1-\Omega}{2} \tag{15.52a}$$

$$c = 1 \tag{15.52b}$$

Generally, there are two independent solutions in a second order differential equation, where one solution is determined from a and c, and the other solution is determined by $a' = a + c - 1$ and $c' = 2 - c$ We determined these relations in Chap. 5. However, unlike in the harmonic oscillator problem, uniquely, we have $a = a'$ and $c = c'$. This follows from substitution into Eq. (15.52) since $c = 1$, then $c' = 2 - 1 = 1 = c$ and $a' = a + c - 1 = a$. Therefore, the wave function R_1^* is given by the full set of truncated confluent hypergeometric series $_1F_1$ for identical values of a and c.

In order to satisfy the boundary condition for any wave function, $R_1(\infty) = R_1^* = 0$. As discussed in Chap. 5, this can only happen if a is a non-positive integer. Then, the infinite series CHS is truncated to a finite polynomial. This condition leads to

$$a = \frac{1-\Omega}{2} = -n$$

The parameter Ω, defined in Eq. (15.45), equals $\Omega = 2m_s|\alpha|_r/v^2\hbar^2$. The quantity $q_s B_e/m_s$ has units of an angular frequency. Then, we can define the following frequency

Rotational Frequency ω_s

$$\omega_s = \frac{eB_e}{m_s} \qquad (15.53)$$

Then $\Omega = 2|\alpha|_r/\hbar\omega_s$, using $v = \sqrt{q_s B_e/\hbar}$ (c.Eq. 15.38). Substitution into the above gives

$$\frac{1 - \frac{2|\alpha|_r}{\hbar\omega_s}}{2} = -n \Rightarrow \frac{2|\alpha|_r}{\hbar\omega_s} = 2n + 1$$

Rearranging to solve for $|\alpha|_r$, we have the result

$|\alpha|_r$ Levels

$$|\alpha|_r = \left(n + \frac{1}{2}\right)\hbar\omega_s \qquad (15.54)$$

[15] Note that the wave-number k_z does not have to be along the direction of the magnetic field.

15.5 Ginzburg & Landau Theory

We now have a picture coming into view that starts to explain some of the experimental observations in superconductors. We have found circulating surface currents from solving the *linearized Ginzburg-Landau equation*, which is exactly the kinetic energy quantum mechanical operator eigenvalue equation, with magnetic field effects on moving charge carriers included. Thus, the magnetic energy $E_b = |\alpha|_r$, or

$$E_b = \left(n + \frac{1}{2}\right)\hbar\omega_s = \left(n + \frac{1}{2}\right)\frac{\hbar e B_e}{m_s} \Rightarrow n = n_b = \frac{m_s E_b}{\hbar e B_e} - \frac{1}{2} \quad (15.55)$$

The total number of superconducting charge carriers n_T can be expressed as the sum of the number due to an external magnetic field n_b and any carriers in the absence of B_e. Let's denote this as n_a. Thus, n_T can be written as

$$n_T = n_a + n_b = n_a + \frac{m_s E_b}{\hbar e B_e} - \frac{1}{2} \Rightarrow \frac{n_T}{n_a} = \frac{n_b}{n_a} + 1 \quad (15.56)$$

If the magnetic field only affects the superconducting charge carriers n_b, then Eq. (15.56) gives the relative dependence of the total number of charge carriers for a constant or conserved energy E_b, relative to its corresponding value with $B_e = 0$. Fig. (15.10) plots an example of n_s/n_Δ given by Eq. (15.56). We see that the total number of superconducting

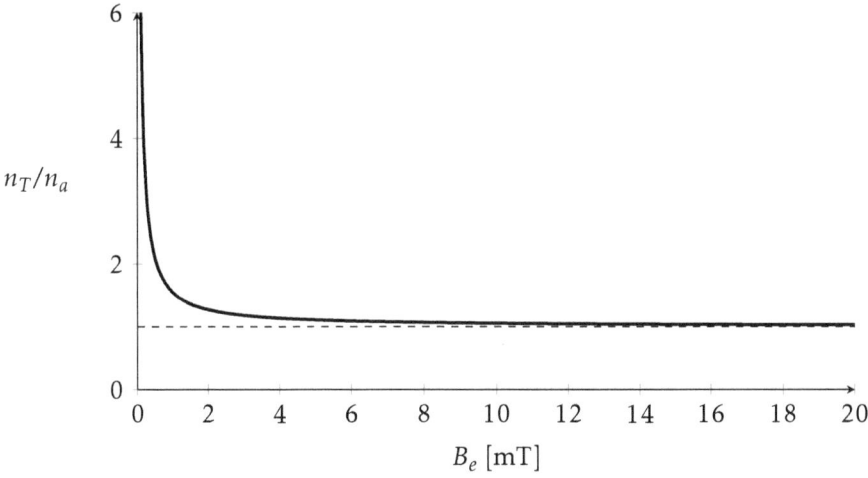

Figure 15.10: Relative total number of charge carriers n_T/n_a as a function of external magnetic field B_e.

charge carriers is maximum without an external field. Introducing a magnetic field reduces the number of carriers relatively sensitively, working to occupy lower energy states. This is done to offset the increase in energy due to B_e. The relative carrier density can also be expressed as a function of the relative energy of a superconductor E_s/Δ, where Δ is the energy of the superconducting carrier without an external field. The energy E_s can be expressed as the sum of the magnetic energy and the nonmagnetic related energy

$$E_\lambda = E_a + E_b \Rightarrow \frac{E_\lambda}{E_a} = \frac{E_b}{E_a} + 1 \tag{15.57}$$

Fig. 15.11 illustrates n_T as a function of the relative energy given by Eq. (15.57). Figs. 15.10 and 15.11 show that the minimum energy state

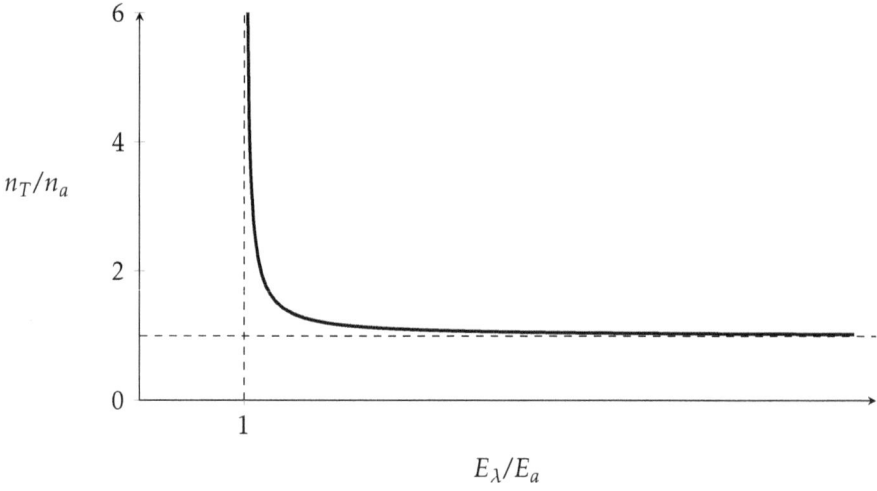

Figure 15.11: Number of carriers as a function of relative energy for a superconductor. E_a is the energy the system has, even in the absence of B_e.

corresponds to the largest number of charge carriers. As the magnetic energy or field increases, the number of carriers gets reduced.

The eigensolutions we've obtained also reveal that the radius of the circulating current must be reduced discretely in the presence of an external magnetic field, in conserving the energy. Therefore, we expect that from zero field to some finite critical field B_c, the surface current predominantly remains confined to the surface. It can be shown that these conditions, with maximum surface current, can lead to *perfect diamagnetic* behavior. This is the subject of the next section.

15.6 DIAMAGNETISM FROM CIRCULATING SURFACE CURRENTS

The presence of a magnetic field leads to circulating charge carriers about the magnetic field axis. According to the solutions we obtained from the *linearized Ginzburg-Landau equation* in the previous section, increasing the external magnetic field reduces the radius of carrier circulation. This suggests that under a sufficiently weak magnetic field, we can take the charge carriers to be on the surface of a cylindrical body with radius r. Loop currents can be described in terms of a magnetic moment $\mu = IA$.

Magnetic moments associated with materials have *three distinct origins*. They are

1. Electron spin-associated angular momentum
2. Electron orbital-associated angular momentum
3. Change in orbital motion of electrons caused by an external magnetic field

Diamagnetism, in particular, arises from the third mechanism in the list.. However, diamagnetism is a distinct form of magnetism. Similar to conventional paramagnetism, it is only present or created when an external field is applied.

Diamagnetic material: A diamagnetic material is defined as any material with negative magnetic susceptibility χ.

For most materials, it causes a very weak magnetic moment that *opposes an externally applied magnetic field*. In the case of weak diamagnetism, it arises from either small numbers of charge carriers participating, circulating in smaller current loops, or because of the presence of other forms of magnetism which tend to align with an external field, and this competes to ultimately reduce the net strength. However, for any structure, if other forms of opposing magnetism are absent, and a sufficiently large number of charge carriers per unit volume circulates at the exterior surface, then diamagnetism can be largest. We are going to show now that circulating currents inherently lead to diagmagnetism in an external magnetic field. This behavior applies to the charge carriers on the surface of a superconductor. One key distinction is in the fact that *diamagnetism* arises from *variations* in the external magnetic field. The variation in the field causes a corresponding *change* in magnetic moment. The magnetic moment is defined by

$$\mu_c = I_s \mathbf{A} \tag{15.58}$$

Recall in Chap. 7, when we derived the current for a single electron in a circular orbit (c.Eq 7.23), we found the current and azimuthal velocity are related by

$$I_s = -\frac{e}{T_s} = -\frac{e|v_\varphi|}{2\pi r} \tag{15.59}$$

Note that use of the negative sign assumes *the charge of a Cooper pair is negative*, or due to electrons. $|v_\varphi|$ is the magnitude of the velocity around the loop $v_\varphi = -r\omega_s$ from Eq. (15.50). The diamagnetic moment relation $\mu_c = I_s A$ then becomes

$$\mu_c = -\frac{e|v_\varphi|r}{2} \tag{15.60}$$

The diamagnetic moment is maximal in amplitude at larger radii. Thus, surface currents correspond to the maximum possible magnetic moment. This leads to a variation relation given by

$$\Delta\mu_c = -\frac{e\Delta v_\varphi r}{2} \tag{15.61}$$

We need to determine the relationship between Δv_φ and ΔB_e. This will allow us to obtain the magnetic susceptibility $\Delta\mu_c/\Delta B_e$. Fortunately, this is straightforward because we already have the result

$$|v_\varphi| = r\omega_s = r\frac{eB_e}{m_s} \Rightarrow \Delta v_\varphi = r\omega_s = r\frac{e\Delta B_e}{m_s} \tag{15.62}$$

Substitution of Eq. (15.62) into Eq. (15.61) gives the result

$$\Delta\mu_c = -\frac{e^2 r^2}{2m_s}\Delta B_e = -\frac{\mu_0 e^2 r^2}{2m_s}\Delta H_e \tag{15.63}$$

H_e is B_e/μ_0 and $\mu_0 = 4\pi 10^{-7} H/m$ is the magnetic permeability in vacuum. Eq. (15.63) contains the **diamagnetic susceptibility** $\chi_c = \Delta\mu_c/\Delta H_e$. So, we have

15.6 Diamagnetism From Circulating Surface Currents

Diamagnetic Susceptibility Of Circulating Currents

$$\chi_c = -\frac{\mu_0 e^2 r^2}{2m_s} < 0 \qquad (15.64)$$

From Eq. (15.60), the unit magnetic moment of a charge carrier in a surface current then becomes

Magnetic Moment Per Charge Carrier Of Circulating Current

$$\mu_c = -\frac{\mu_0 e^2 H_e r^2}{2m_s} \qquad (15.65)$$

Note that the SI units of μ_c are A/m. To express in Tesla, we have

$$\mu_0 \mu_c = -\frac{\mu_0^2 e^2 H_e r^2}{2m_s} \qquad (15.66)$$

This result allows us to express the magnetic induction $\mathbf{B}_I$ in a material, given by

$$\mathbf{B}_I = -\frac{\langle n_s \rangle \mu_0^2 e^2 \mathbf{H}_e r^2}{2m_s} + \mu_0 \mathbf{H}_e = \left(-\frac{\langle n_s \rangle \mu_0 e^2 r^2}{2m_s} + 1\right) \mu_0 \mathbf{H}_e \qquad (15.67)$$

This a **superconductor relative magnetic permeability** of

Superconductor Relative Magnetic Permeability

$$\mu_r = 1 - \frac{\langle n_s \rangle \mu_0 e^2 r^2}{2m_s} \qquad (15.68)$$

$\langle n_s \rangle$ is the number of charge carriers per unit volume. The term in parentheses is defined as the *relative magnetic permeability* $\mu_r^* = \mu_s/\mu_0$, which necessarily satisfies the condition $0 < \mu_r^* < 1$, for any finite amount of circulating current. Weak diamagnetism corresponds to a value just below unity or $\mu_r^* \approx 1$, while strong diamagnetism, or the maximum circulating current, corresponds to the condition $\mu_r^* \approx 0$. This condition implies

$$B_I = \mu_r^* \mu_0 H_e = 0 \qquad (15.69)$$

Let's consider an example that allows us to estimate the number of charge carriers per unit volume in a superconductor. The geometrical parameters we'll use are taken from experiments carried out in 1961, by Deaver and Fairbank on cylinders of tin (Sn) of approximately 0.8cm in length, with a radius of $r = 2.33 \times 10^{-3}$ cm.

> **Exercise 15.1** Use the following condition to estimate the number of charge carriers per unit volume $\langle n_s \rangle$ in a tin superconducting cylinder of radius $r = 2.33 \times 10^{-3}$ cm, using unit charge and mass e, m_e to be that of the electron:
>
> $$-\frac{\langle n_s \rangle \mu_0 e^2 r^2}{2 m_s} + 1 \approx 0$$
>
> Solving for $\langle n_s \rangle$ from the above condition gives
>
> $$\langle n_s \rangle \approx \frac{2 m_s}{\mu_0 e^2 r^2}$$
>
> $$= \frac{2 \cdot \left(9.10938356 \times 10^{-31} \, kg\right)}{4\pi 10^{-7} H \cdot m^{-1} \left(1.60217662 \times 10^{-19} C\right)^2 \left(2.33 \times 10^{-5} m\right)^2}$$
>
> $$= 1.04 \times 10^{23} m^{-3}$$
>
> Therefore, there are on the order of $\langle n_s \rangle \approx 10^{23}$ charge carriers per unit volume in a cylindrical superconductor of size $r = 2.33 \times 10^{-3}$ cm.
>
> ∎

Solving the eigenvalue equation for the extended momentum based kinetic energy operator, or the linearized Ginzburg-Landau equation, allowed us to analyze conditions leading to surface currents, as well as magnetic field effects with these surface currents. This analysis, however, has not involved temperature at all, while we know this is one of the key ingredients in the observations from Onnes' team. Fortunately, this can also be done at the microscopic level. For this, we need to try and understand the origins of the free charge carriers that end up on the surface of a superconductor, and how this can lead to superconductivity. We'll begin answering fundamental questions like this in the next section.

15.7 Correlating Cooper Pairs

The results will complement the above to provide a more complete understanding of superconductors.

15.7 CORRELATING COOPER PAIRS

The fact that measuring $\mu_0 M$ as a function of an external magnetic field B_e yields zero in zero magnetic field indicates that all *the spins of the surface current charge carriers are balanced*, before any field is applied. Moreover, the demonstrated magnetic moment originates from the circulating current, as shown in Sec. 15.6. This is still the superconducting state. Therefore, there must somehow be pairs of equal and opposite spins among the charge carriers, otherwise there would be an additional net magnetic moment tending to align with the magnetic field. The Cooper problem looks at what conditions could lead to such properties. Fig. 15.12 illustrates possible conditions for these so-called pairs that we are interested to describe. Before an electric field is applied to the

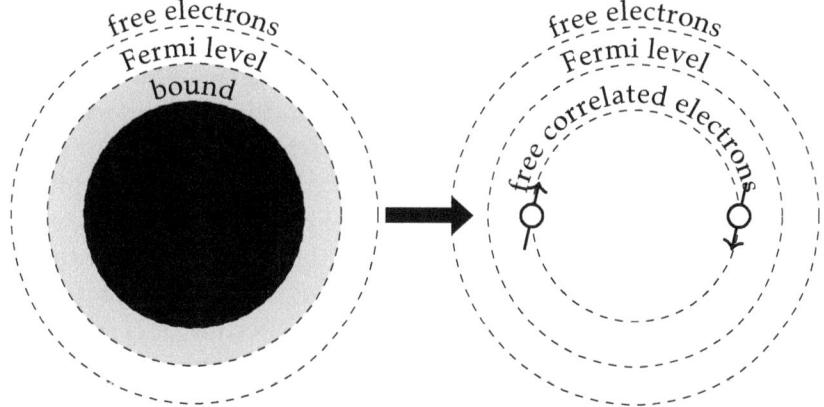

Figure 15.12: Illustration of correlated electrons in a superconductor (left) in originally in bounded states in finite temperature, prior to excitation by an external electric field; and (right) after excitation in a thermally activated potential turned on after cooling below a critical temperature T_c. The energy level for the free electrons on the right is lower than the bound state energy illustrated on the left.

superconductor, the electrons are in bound states below the Fermi level, or in the *valence band*. When an electric field is applied, two important conditions must be met for superconductors. First, the electrons must be excited out of the bound state, into the *conduction band*. There is a catch however. Ordinarily, an electron in the conduction band, once freed from localization, will itinerate or hop to another atom. In conditions for a superconductor, we will discuss at least two energy contributions that arise, *that uniquely lower the energy of the itinerant electron*. Moreover, these mechanisms are symmetric in the spin-states of the electron.

Thus, there is an effective concomitant potential energy that becomes activated, particularly for conducting electrons. There are two important consequences from this. First, free charge carriers can build a population. And second, because these charge carriers are uniquely stablized as free charge carriers in an energy lowering potential, this also leaves open vacancies in the potential electronic states for the atoms. These vacancies are examples of *holes* for the electron. It is the unique combination of these conditions that contributes to *superconductivity*. This idea is illustrated in Fig. 15.12. On the left, electrons are in bound states just below the Fermi level at finite temperature. Once excited by the external electric field, the electrons become uniquely stablized conducting electrons with even potentially lower energy than in their bound state.

In the beginning, one of the biggest intellectual challenges to understanding the origins of superconductivity had to do with how the unique charge carriers are created. What type of conditions support the existence of superconducting charge carriers, which are also free to form *Abrikosov vortices* (see Sec. 15.5), etc. Most of the earlier attempts to explain superconductivity utilized the idea of multiple electrons arranged in some sort of structure, moving through the superconductor together. It was thought that if electrons are involved in superconduction, perhaps if they *band together* somehow, they can become less sensitive to scattering, and other similar mechanisms that contribute to resistivity. This idea, unfortunately, does not jive with the best understanding today, however, it certainly is a reasonable hypothesis. The enhancement in the conducting properties becomes possible because of the unique creation of *stable potential holes* in the superconductor. We'll have more to say about this later. Let's begin by stripping down a relevant Hamiltonian to its simplest form that would allow a peak into how superconductivity might be possible.

The simplest multi-electron problem we may consider is with a *pair* of electrons. Recall that in Chap. 10, we saw an example of a Hamiltonian for two nuclei A and B, with two electrons 1 and 2, given by

15.7 Correlating Cooper Pairs

$$\hat{H} = -\frac{\hbar^2}{2M}\left[\nabla_A^2 + \nabla_B^2\right]\psi_s - \frac{\hbar^2}{2m_s}\left[\nabla_1^2 + \nabla_2^2\right]\psi +$$

$$\frac{e^2}{4\pi\epsilon_0}\left[-\frac{1}{d_{1A}} - \frac{1}{d_{1B}} - \frac{1}{d_{2A}} - \frac{1}{d_{2B}} + \frac{1}{r_{12}} + \frac{1}{R}\right]\psi_s \quad (15.70)$$

M is the sum of both nuclei masses, while m^* is the reduced mass obtained after expressing the system in the coordinates of the center of mass R and the electron relative position r_1 and r_2 (see Sec. 8.3). In the *Born-Oppenheimer approximation*, the kinetic energy contributions from both nuclei are ignored. Another justification for why it can be ignored here is because we are primarily interested in free or conduction electrons, which are not bound to any nuclei. However, their respective attractions to the nuclei are still included. Then, we have

$$\hat{H} = -\frac{\hbar^2}{2m_s}\left[\nabla_1^2 + \nabla_2^2\right]\psi_s + \underbrace{\frac{Ze^2}{4\pi\epsilon_0}\left[-\frac{1}{d_{1A}} - \frac{1}{d_{2A}} + \frac{1}{r_{12}}\right]}_{\text{Electric Potential V}}\psi_s \quad (15.71)$$

We have generalized the atomic number beyond unity by including any number of nuclear protons Z. From Eq. (15.71), because of the combination of signs (electron-nuclei attraction + electron-electron repulsion), the last term in brackets can be positive or negative, depending on the *inter*distances. In Chap. 10, we found from substituting explicit forms for the multi-electron wave function, that it shows up in interactions like the *exchange integral* and the *Coulomb integral*. With the exchange integral, for example, we found that bound electrons on different atoms with opposite spins can lower their energy by overlapping their respective wave functions. This gives rise to the well-known *covalent bond* for anti-aligned spins, yet ferromagnetism for some aligned spin systems. In general, since the Hamiltonian is the eigenoperator for the energy, and since the kinetic energy terms are positive, the energy can only be lowered through some kind of combination of interactions between the electrons with themselves, and between the electrons with nuclei or atoms. Since the Coulomb potential energy is somewhat restrictive because it is a specific electrostatic interaction, it can not provide an accurate account for dynamical effects in conditions extending beyond the simple electrostatic potential. We don't want this restriction. So, let us write Eq. (15.71) more generally as

$$\hat{H} = \left[-\frac{\hbar^2}{2m^*}\nabla_1^2 - \frac{\hbar^2}{2m^*}\nabla_2^2 - V_s\right]\psi_s = E_s\psi_s \quad (15.72)$$

In this form, it is assumed that $V_s > 0$ (since a negative sign is explicitly included), however $E_s \geq 0$. These conditions are reasonable for metals, and recall that *this indicates energy above the Fermi level*. Otherwise, we'd have a bound state if the eigenenergy was sufficiently negative, with the external electric field. Ignoring the interaction potential between the two electrons for the moment, Eq. (15.72) simplifies to

$$\widehat{H}_0 = \left[-\frac{\hbar^2}{2m^*}\nabla_1^2 - \frac{\hbar^2}{2m^*}\nabla_2^2\right]\psi_k = 2\epsilon_k = E_k\psi_k \qquad (15.73)$$

ψ_k is the two-electron eigenfunction without an interaction, and ϵ_k is the single electron kinetic energy. For this to be true, using $\widehat{p} = \hbar k$, it must be that

$$\frac{-\hbar^2 k_1^2}{2m} = \frac{-\hbar^2 k_2^2}{2m}$$

This suggests the first condition for what is known as the **Cooper pair**, where we have

$$V(\mathbf{k}_1) = V(-\mathbf{k}_2) \quad \text{and} \quad V(\widehat{\sigma}_1) = V(-\widehat{\sigma}_2) \qquad (15.74)$$

For the second condition, we can readily identify one mechanism satisfying this condition, namely, the *spin-orbit interaction*. Recall in Sec. 11.5, we derived this interaction given by

$$\widehat{H}_{SO} = -\frac{\hbar}{4m_0^2 c^2}\widehat{\sigma}\cdot(\widehat{\mathbf{p}}+e\mathbf{A})\times\nabla V = -\frac{\hbar}{4m_0^2 c^2}\widehat{\sigma}\cdot\widehat{\mathbf{p}}\times\nabla V \qquad (15.75)$$

Without a magnetic field ($B = 0 \to A = 0$), the symmetry of the spin-orbit suggests that when the first equation of Eq. (15.74) is satisfied, then $\widehat{H}_{SO}$ is conserved for both spins because the pair $(+\mathbf{p},+\boldsymbol{\mu})$ has *equal* spin-orbit energy as $(-\mathbf{p},-\boldsymbol{\mu})$. Therefore

$$V(\widehat{\sigma}_1) = V(-\widehat{\sigma}_2) \qquad (15.76)$$

These are hallmark conditions for a **Cooper pair**, as they can still satisfy the property of fermions having different states, but symmetric in certain energy forms. Spin-orbit like interactions can therefore *lower the energy of free electrons* in such a way that spin-up and spin-down electrons are degenerate and correlated through this interaction, as with the kinetic energy.

Additionally, recall that it was shown in Chap. 11, that due to the *spin-orbit interaction* with an external electric field, *charge carriers can propagate to the surface of the material body.* Therefore, we have the following condition for a so-called *Cooper pair*:

 Cooper pairs: They have equal energies for equal and opposite momentum, or for momenta 180° out of phase. Likewise, for their spin angular momentum. Two electrons moving in opposite directions, as well as two circling in a loop with this phase difference both satisfy the Cooper pair condition, but a magnetic field reduces the symmetry.

Note that when a magnetic field is introduced, the symmetry property is reduced because $\widehat{\mathbf{P}} = \pm\widehat{\mathbf{p}} + e\mathbf{A}$. We know opposite charge and spin don't do the same things in a magnetic field. In addition to having 180° phase difference between the electron pair, another condition assumed concerns the energetic conditions of electrons that are excited just beyond the Fermi level ϵ_F, to within a narrow band above it. The threshold energy is sometimes called the *cutoff energy* of $\epsilon_F + \hbar\omega_D$. Below the Fermi level, particularly when $T = 0$, the states are all occupied, and thus in a bound state. This changes for finite temperature, and we'll treat this case, as well. The Cooper-pair problem considers what is possible for electrons excited just beyond the Fermi level, initially, while also possessing some small potential energy.

Particularly, for a spin-dependent potential energy like the spin-orbit interaction, we can consider the simplest spin-dependent potential V which is a two state system symmetric about some energy point of the system. Next, we'll introduce a way to treat this kind of system.

15.8 SMALL SYMMETRIC SPLITTING WITH SPIN MATRICES

Excitation of charge carriers into the conduction band is taken to be under conditions where there is a energetic mechanism of potential energy which correlates the liberated electron pairs having equal and opposite momenta, particularly linear and spin angular momenta. We have discussed atleast one mechanism with this property, the *spin-orbit interaction*. Recall that it was obtained from the relativistic quantum mechanics in Sec. 11.7. As mentioned, the spin-orbit interaction gives rise to a small splitting with electrons of equal and opposite linear momentum with equal and opposite electron spins. These spins have respective momenta directed in equal and opposite directions towards the exterior boundaries of the material domain. This form of spin-orbit interaction is enabled by an external electric field. While this may not be the only mechanism at play, *spin-orbit interaction* is strongly implicated based on the materials that demonstrate superconductivity. Also, it is one of the only known potentials to imbue electrons with this type of pairwise correlation. However, regardless of the exact mechanism, it's properties can be used to determine the consequences of this type of pairwise correlation. It turns out that quantum mechanics has something to add to the conversation started by Ginzburg and Landau.

In this scenario, one energy state is negative relative to the energy of the system in the absence of this potential. Let us regard this reference energy level as a sort of Fermi level ϵ_F. The higher energy state is positive relative to the Fermi level. We can regard the resulting state of this particular energy as a two state system, equivalent to a spin system with spin-up and spin-down electrons. This can be formalized into a mathematical description. In Chap. 9, we introduced the 2×2 *principle spin matrices* $\widehat{\sigma}_x, \widehat{\sigma}_y$, and $\widehat{\sigma}_z$ (c. Eq. (9.23)) to describe the two states along x, y, and z respectively. It was introduced to describe the states of the electron spin angular momentum, when coupling to a magnetic field. Recall that the Hamiltonian $\widehat{H}_Z$ for the Zeeman energy (spin coupling to an external magnetic field) of the spin angular momentum is given by

$$\widehat{H}_Z = -\widehat{\boldsymbol{\mu}} \cdot \mathbf{B} = \frac{g|\gamma|\hbar B}{2}\sigma_z = \frac{V_B}{2}\widehat{\sigma}_z$$

In this case, the factor of $1/2$ comes from the spin angular momentum relation $\widehat{S}_z = (\hbar/2)\widehat{\sigma}_z$. With spin matrices, we are describing *two* Hamiltonians representing the two respective eigenenergy states of the two-state system. However, here, these two states can corresponds to the two electrons in the Cooper pair. The two eigenenergies E_λ of the system can be written as

$$E_\lambda(\lambda = +1) = +\frac{V_B}{2} \tag{15.77a}$$

$$E_\lambda(\lambda = -1) = -\frac{V_B}{2} \tag{15.77b}$$

Then, the expectation of the energy difference, or splitting between the two states can be expressed in a form ΔE given by

$$\langle \Delta \widehat{H}_Z \rangle = \Delta E_Z = E_{\lambda=-1} - E_{\lambda=+1} - 2 \times \frac{V_B}{2} = -g|\gamma|\hbar B = -V_B \tag{15.78}$$

15.8 Small Symmetric Splitting With Spin Matrices

The energy splitting can be described as being due to a net potential of $-V_B$. This definition is appropriate because it determines the likelihood of the lowest energy state. The absolute energies are *symmetric with respect to the middle of the splitting*. With Zeeman energy, the symmetry point is whatever the energy is without the external magnetic field. When a magnetic field is introduced, the energy is then changed by $-V_B/2$ or $+V_B/2$ depending on the direction of the spin. In the case of Cooper pairs, the symmetry point is taken to be in proximity to the Fermi level ϵ_F. This can be used to describe the pair using spin matrices We apply this to the Cooper pair problem.

A *Cooper pair* is taken to have two possible eigenstates:

$|1\rangle$ representing occupation of the Cooper electron-pair state, and

$|0\rangle$ for *nonoccupation* of the Cooper pair state.

In the latter case, there are two bound electrons having combined negative eigenenergies. Let's describe a *general transition* of a Cooper pair with momentum $\pm k_1, \pm \sigma_1$ to another state $\pm k_2, \pm \sigma_2$. We will denote such a transition as $k_1 \rightarrow k_2$. Before we get to describing a transition using spin matrices, we should review some essential properties of spin matrices that are useful in this type of description. Recall that when we derived the principle spin matrices, or Pauli matrices, in Chap. 9, we also derived the corresponding eigenkets or 2×1 eigenvectors for each 2×2 matrix operator. If $\hat{\sigma}_i$ operates on its own eigenvector $|0\rangle$ or $|1\rangle$, we know the eigenvector remains unchanged. However, what happens if we let $\hat{\sigma}_i$ operate on the eigenvectors of the other two spin operators $\hat{\sigma}_j$ and $\hat{\sigma}_k$. We can find out by looking at an example. We'll do this now. In Chap. 9, we found the eigenvectors for $\hat{\sigma}_z$ are given by

$$|1_z\rangle = \lambda_{+1} = \begin{bmatrix} 1 \\ 0 \end{bmatrix} \quad \text{and} \quad |0_z\rangle = \lambda_{-1} = \begin{bmatrix} 0 \\ 1 \end{bmatrix} \tag{15.79}$$

Since these are the eigenkets of $\hat{\sigma}_z$, let's see what happens when we operate on them using $\hat{\sigma}_x$ and $\hat{\sigma}_y$. With $\hat{\sigma}_x$, we have

$$\hat{\sigma}_x |1_z\rangle = \begin{bmatrix} 0 & 1 \\ 1 & 0 \end{bmatrix} \begin{bmatrix} 1 \\ 0 \end{bmatrix} = \begin{bmatrix} 0 \\ 1 \end{bmatrix} = |0_z\rangle$$

Likewise, when operating on $|0_z\rangle$, we obtain

$$\hat{\sigma}_x |0_z\rangle = \begin{bmatrix} 0 & 1 \\ 1 & 0 \end{bmatrix} \begin{bmatrix} 0 \\ 1 \end{bmatrix} = \begin{bmatrix} 1 \\ 0 \end{bmatrix} = |1_z\rangle$$

We, therefore, have a *toggling of the energy states* when $\hat{\sigma}_x$ operates on the eigenkets of $\hat{\sigma}_z$. Now, letting $\hat{\sigma}_y$ operate on the eigenvectors of $\hat{\sigma}_z$, we obtain

$$\hat{\sigma}_y |1_z\rangle = \begin{bmatrix} 0 & -i \\ i & 0 \end{bmatrix} \begin{bmatrix} 1 \\ 0 \end{bmatrix} = \begin{bmatrix} 0 \\ i \end{bmatrix} = i|0_z\rangle$$

Then, operating on $|0_z\rangle$ leads to

$$\hat{\sigma}_y |0_z\rangle = \begin{bmatrix} 0 & -i \\ i & 0 \end{bmatrix} \begin{bmatrix} 0 \\ 1 \end{bmatrix} = \begin{bmatrix} -i \\ 0 \end{bmatrix} = -i|1_z\rangle$$

Thus, we have that the other two spin matrices, apart from z, toggle the eigenstates of $\hat{\sigma}_z$. But they do so along independent axes. One is along the real axis, and the other is along the imaginary axis. This property allows one to define a linear combination of these to make *two adjoint operators that create and annihilate the energy states* of a binary system. Note that this is a general property of the principle spin matrices because if we use the two eigenkets of $\hat{\sigma}_x$, for example, we can also toggle its eigenstates using $\hat{\sigma}_y$ and $\hat{\sigma}_z$, etc.

> *Toggling Eigenstates With Spin Matrices*: The toggling of spin states can be achieved by choosing the two possible states of the system in the form of eigenkets of one spin matrix $\hat{\sigma}_i$. Then, the other two spin matrices $\hat{\sigma}_j$ and $\hat{\sigma}_k$ can be used to toggle the eigenstates of $\hat{\sigma}_i$.

It is because of these properties of the principle spin matrices that *any symmetric binary or two eigenstate system can use principle spin matrices to represent transitions between symmetric binary states*. Therefore, if we choose the eigenstates of the Cooper pair to be those of $\hat{\sigma}_z$, we can define two general *unitary operators* $\hat{\sigma}^+$ and $\hat{\sigma}^-$ as

$$\hat{\sigma}^+ = \frac{1}{2}(\hat{\sigma}_x + i\hat{\sigma}_y) = \frac{1}{2}\left(\begin{bmatrix} 0 & 1 \\ 1 & 0 \end{bmatrix} + i \begin{bmatrix} 0 & -i \\ i & 0 \end{bmatrix} \right) = \begin{bmatrix} 0 & 1 \\ 0 & 0 \end{bmatrix} \quad (15.80)$$

$$\hat{\sigma}^- = \frac{1}{2}(\hat{\sigma}_x - i\hat{\sigma}_y) = \frac{1}{2}\left(\begin{bmatrix} 0 & 1 \\ 1 & 0 \end{bmatrix} - i \begin{bmatrix} 0 & -i \\ i & 0 \end{bmatrix} \right) = \begin{bmatrix} 0 & 0 \\ 1 & 0 \end{bmatrix} \quad (15.81)$$

Eqs. (15.80) and (15.81) turn out to be **binary creation and annihilation operators**, defined as

15.8 Small Symmetric Splitting With Spin Matrices

Binary Creation (+) And Annihilation (-) Operators

$$\widehat{\sigma}^+ = \frac{1}{2}\left(\widehat{\sigma}_x + i\widehat{\sigma}_y\right) \quad \text{and} \quad \widehat{\sigma}^- = \frac{1}{2}\left(\widehat{\sigma}_x - i\widehat{\sigma}_y\right) \tag{15.82}$$

It is important to point out that the two operators involved Eq. (15.82) *depend on the choice of system eigenkets*. When we apply the *binary creation operator* $\widehat{\sigma}^+$ to the eigenkets of $\widehat{\sigma}_z$, which we have chosen as the representation for the Cooper pair eigenstates (dropping the z subscript from here on out) of a two-state system, we obtain

$$\widehat{\sigma}^+|0\rangle = \frac{1}{2}\left(\widehat{\sigma}_x + i\widehat{\sigma}_y\right)|0\rangle = \begin{bmatrix} 0 & 1 \\ 0 & 0 \end{bmatrix}\begin{bmatrix} 0 \\ 1 \end{bmatrix} = \begin{bmatrix} 1 \\ 0 \end{bmatrix} = |1\rangle \tag{15.83}$$

However, when we apply $\widehat{\sigma}^+$ to $|1\rangle$, we get

$$\widehat{\sigma}^+|1\rangle = \begin{bmatrix} 0 & 1 \\ 0 & 0 \end{bmatrix}\begin{bmatrix} 1 \\ 0 \end{bmatrix} = \begin{bmatrix} 0 \\ 0 \end{bmatrix} \neq |0\rangle \tag{15.84}$$

Instead, one obtains the *nullket*. These are hallmark characteristics of *creation* and *annihilation* operators, which are *ladder operators*. We found ladder operators for the orbital angular momentum problem in Chap. 7, as well as the harmonic oscillator problem in Chap 5. Based on the results from the *linearized Ginzburg-Landau* analysis of Sec. 15.5, the latter case of the harmonic oscillator turns out to be most relevant to Cooper pair problem, but we'll come back to this point shortly. When the maximum energy state is reached in a finite population of states, for all states at the maximum energy state and beyond, the creation operator returns the *nullket*, as there are no energy eigenstates above the maximum energy eigenstate. Likewise, for the annihilation operator with the ground state. For the annihilation operator, we have

$$\widehat{\sigma}^-|1\rangle = \frac{1}{2}\left(\widehat{\sigma}_x - i\widehat{\sigma}_y\right)|1\rangle = \begin{bmatrix} 0 & 0 \\ 1 & 0 \end{bmatrix}\begin{bmatrix} 1 \\ 0 \end{bmatrix} = \begin{bmatrix} 0 \\ 1 \end{bmatrix} = |0\rangle \tag{15.85}$$

But, when we apply $\widehat{\sigma}^-$ to $|0\rangle$, we get

$$\widehat{\sigma}^-|0\rangle = \begin{bmatrix} 0 & 0 \\ 1 & 0 \end{bmatrix}\begin{bmatrix} 0 \\ 1 \end{bmatrix} = \begin{bmatrix} 0 \\ 0 \end{bmatrix} = \text{null Ket} \tag{15.86}$$

Notice also, that

$$\widehat{\sigma}^- = [\widehat{\sigma}^+]^T = [\widehat{\sigma}^+]^\dagger \quad \text{and} \quad \widehat{\sigma}^+ = [\widehat{\sigma}^-]^T = [\widehat{\sigma}^-]^\dagger \tag{15.87}$$

Both operators are respective transposes and because principle spin matrices have real eigenvalues, the creation and annihilation operators are also *adjoints* of one another. For a direct transition, $\widehat{\sigma}^+$ and/or $\widehat{\sigma}^-$ would suffice. However, it is more general to take Cooper pairs to form from indirect interactions, since all the exact mechanisms are not fully understood. Therefore, we want to describe *a transition* of the Cooper pair binary system from $\pm \mathbf{k}_1$ to $\pm \mathbf{k}_2$ as first transitioning from a Cooper pair having kinetic energy $\pm \mathbf{k}_1$ to *not* occupying the Cooper pair state, then from the non-Cooper pair state into the $\pm \mathbf{k}_2$ Cooper pair state.

Because a Cooper pair involves conducting electrons, we have to be careful in our description of the momentum states. This is because we know that *free particles have a continuous spectrum of momenta*. This means that, in writing a momentum state as we are doing here, it is more proper to write the expectation value $\langle k_1 \rangle$ and $\langle k_2 \rangle$. In the following analysis, we shall take care to account for this subtle detail, however, for now, we are writing this terse form for convenience. Since the starting state for a transition is a Cooper pair in, say, state $\pm \mathbf{k}_1$, the spin operator for transitioning to $\pm \mathbf{k}_2$ becomes an ladder operator sequence given by

$$\widehat{O}(k_1 \Rightarrow k_2) = \widehat{\sigma}^+_{k_2} \widehat{\sigma}_{k_1} \tag{15.88}$$

The first spin matrix annihilates the spin-pair state from $\pm \mathbf{k}_1$, then the second spin matrix creates the Cooper pair state $\pm \mathbf{k}_2$. Eq. (15.88) represents a general *indirect transition operator for a single Cooper pair*. The Hamiltonian for a single Cooper pair transition is then given by

$$\widehat{\Delta H} = \widehat{\sigma}^+_{k_2} \widehat{\sigma}_{k_1} \tag{15.89}$$

If there are N_c Cooper pairs that can make such an indirect transition, then we sum all these pair transition energies together. We'll denote the sum as

$$\widehat{O}^N_c(k_1 \Rightarrow k_2) = \sum_{(k_1 \rightarrow k_2)} \widehat{\sigma}^+_{k_2} \widehat{\sigma}_{k_1} \tag{15.90}$$

Introducing the potential energy expectation value (since it resides outside the summation) like we have in Eq. (15.78), we obtain the following Hamiltonian for a collection of N_c **electron Cooper pairs transitions**:

Cooper Pair Transition Energy Hamiltonian

15.8 Small Symmetric Splitting With Spin Matrices

$$\widehat{\Delta H_c} = -\langle V_c \rangle \sum_{(k_1 \to k_2)} \widehat{\sigma_{k2}^+}\widehat{\sigma_{k1}} \tag{15.91}$$

The ground state energy for the collection of Cooper pairs can be obtained by minimizing the energy difference of the system (shifted relative to the Fermi level). Let's first deal with the potential energy contribution given by Eq. (15.91). For a single pair, the general state is a linear combination of both eigenstates, given by

$$\psi_p = u_p|0\rangle + v_p|1\rangle \quad \text{where} \quad u_p^2 + v_p^2 = 1 \tag{15.92}$$

u_p is the *probability amplitude* of the non-occupation of the Cooper pair state, while v_p is that for the occupation of the Cooper pair state. The second condition follows from the normalization condition of the state because we have

$$1 = \langle \psi_p | \psi_p \rangle$$
$$= \left(\langle 0|u_p + \langle 1|v_p\right)\left(u_p|0\rangle + v_p|1\rangle\right)$$
$$= u_p^2\langle 0|0\rangle + u_p v_p\langle 0|1\rangle + v_p u_p\langle 1|0\rangle + v_p^2\langle 1|1\rangle$$

The condition in Eq. (15.92) follows from using the *orthogonality condition* of the eigenstates, where $\langle 0|1\rangle = \langle 1|0\rangle = 0$, along with the normalization condition of the basis eigenstates. This can be verified directly using the eigenstates of $\widehat{\sigma}_z$ given by Eq. (15.79), which are the assumed eigenstate representation for the Cooper pair.

A subtle point here is that because the Hamiltonian is the summation of each Cooper pair's transitional potential energy, and no coupling between any of them is assumed, the eigenstate takes the form of a product of each pair's transitional eigenstates. This defines what is known as the **BCS wave function** ϕ_{BCS} given by

$$\phi_{BCS} = \prod_{p=1}^{N_c} u_p|0\rangle + v_p|1\rangle \tag{15.93}$$

Using the above properties for a single Cooper pair, we can readily evaluate the normalization condition of the BCS state, which becomes

$$\langle \phi_{BCS} | \phi_{BCS} \rangle = \left\langle \prod_{p=1}^{N_c} \langle 0|u_p^\dagger + \langle 1|v_p^\dagger \prod_{q=1}^{N_c} u_q|0\rangle + v_q|1\rangle \right\rangle$$

The above expression is an integral of a product of products. For the left and right products, there are corresponding states, which generally include spatial wave function components. For example, $p = 1, q = 1$, $p = 2, q = 2$, etc. are corresponding states. The spinor parts of the product effectively come out of the expectation value spatial integral (remember the expectation value integrates over wave function volume). Under the integral remains only spatial parts of the wave function, which are also normalizable. The spatial integral parts for each respective state, thus, becomes unity. This ultimately allows us to drop the outer expectation braces. Then, the product of products can be written as a single product given by

$$\langle \phi_{BCS} | \phi_{BCS} \rangle = \left\langle \prod_{p=q=1}^{N_c} \left(\langle 0|u_p^* + \langle 1|v_p^* \right) \left(u_q|0\rangle + v_q|1\rangle \right) \right\rangle \quad (15.94)$$

$$= \prod_{p=1}^{N_c} \left(|u_p|^2 + |v_p|^2 \right) = \prod_{p=1}^{N_c} 1 = 1 \quad (15.95)$$

Under the given assumptions, *the BCS state is normalized*. We can use Eq. (15.91) with Eq. (15.93) to evaluate the expectation value of the potential energy contribution. From substitution of the BCS state, we have

$$\langle \Delta E_V \rangle = \langle \phi_{BCS} | \widehat{\Delta H_c} | \phi_{BCS} \rangle \quad (15.96)$$

$$= -\langle V_c \rangle \prod_{p=1}^{N_c} \langle 0|u_p^* + \langle 1|v_p^* \left(\sum_{(k_1 \to k_2)} \widehat{\sigma_{k_2}^+ \sigma_{k_1}} \right) \prod_{q=1}^{N_c} u_q|0\rangle + v_q|1\rangle$$

(15.97)

The above expression is not as difficult as it appears. This is because we are assisted by the properties of the BCS state. For each of the terms of the sum, the creation-annihilation operator only operates on one of the states in the product, while the remaining bras and corresponding kets come together to yield an inner product of unity, as in the normalization condition of the BCS state. Therefore, $\langle \Delta E_V \rangle$ becomes

15.8 Small Symmetric Splitting With Spin Matrices

$$\langle \Delta E_V \rangle =$$

$$-\langle V_c \rangle \prod_{p-} \langle 0|u_p^\dagger + \langle 1|v_p^\dagger \left(\sum_{(k_1 \to k_2)} \left(\langle 0|u_{k2}^\dagger + \langle 1|v_{k2}^\dagger \right) \widehat{\sigma}_{k2}^+ \widehat{\sigma}_{k1} (u_{k1}|0\rangle + v_{k1}|1\rangle) \right) \times$$

$$\prod_{q-} u_q|0\rangle + v_q|1\rangle = -\langle V_c \rangle \left(\sum_{(k_1 \to k_2)} \left(\langle 0|u_{k2}^\dagger + \langle 1|v_{k2}^\dagger \right) \widehat{\sigma}_{k2}^+ \widehat{\sigma}_{k1} (u_{k1}|0\rangle + v_{k1}|1\rangle) \right)$$

(15.98)

Note that $q-$ and $p-$, above, denote that *the product has had one state removed*, and brought into the summation explicitly. Thus, the products have $N_c - 1$ terms for each contribution to the sum. We can evaluate this directly by using the property $\widehat{\sigma}^+ = [\widehat{\sigma}^-]^\dagger$. This leads to

$$\langle \Delta E_V \rangle = -\langle V_c \rangle \left(\sum_{(k_1 \to k_2)} \left(\langle 0|u_{k2}^\dagger + \langle 1|v_{k2}^\dagger \right) \left[\widehat{\sigma}_{k2}^- \right]^\dagger \widehat{\sigma}_{k1} (u_{k1}|0\rangle + v_{k1}|1\rangle) \right)$$

$$= -\langle V_c \rangle \left(\sum_{(k_1 \to k_2)} \left(0 \cdot u_{k2}^\dagger + \langle 0|v_{k2}^\dagger \right) (u_{k1} \cdot 0 + v_{k1}|0\rangle) \right)$$

$$= -\langle V_c \rangle \sum_{(k_1 \to k_2)} v_{k2}^\dagger v_{k1}$$

The above result says that the energy contribution from each pair is proportional to the product of the probability amplitudes of k_2 and k_1. If either are zero, then all bets are off for the Cooper pair formation transitioning from k_1 to k_2. That makes sense. There is also an alternative and more useful form and interpretation of $\langle \Delta E_V \rangle$. It can be found from, again, exploiting the fact that the BCS state wave function spinors are independent of space, just like spinors of spin angular momentum. Eq. (15.97) can also be written

$$\langle \Delta E_V \rangle = -\langle V_c \rangle \left\langle \prod_{p=1}^{N_c} \langle 0|u_p^\dagger + \langle 1|v_p^\dagger \left(\sum_{(k_1 \to k_2)} \widehat{\sigma}_{k2}^+ \widehat{\sigma}_{k1} \right) \prod_{q=1}^{N_c} u_q|0\rangle + v_q|1\rangle \right\rangle$$

$$= -\langle V_c \rangle \prod_{p=1}^{N_c} \langle 0|u_p^\dagger + \langle 1|v_p^\dagger \prod_{q=1}^{N_c} u_q|0\rangle + v_q|1\rangle \left(\sum_{(k_1 \to k_2)} \langle \widehat{\sigma}_{k2}^+ \widehat{\sigma}_{k1} \rangle \right)$$

We can now connect this result directly to those of Ginzburg, Landau, and the Abrikosov vortex core discussed in Sec. 15.5. There, we found the eigenenergy E_λ of a vortex core given by

$$E_\lambda = \hbar\omega_s\left(n+\frac{1}{2}\right)$$

$\omega_s = q_s B_e/m_s$ is the superconductor charge carrier oscillator frequency. The mathematical form of the linearized Ginzburg-Landau problem turned out to be identical to that of the harmonic oscillator. In the oscillator problem, when described using annihilation $\widehat{a}^\dagger$ and creation $\widehat{a}$ operators, we also found that (c.Eq. 14.19)

$$\langle \widehat{a}^\dagger \widehat{a}\rangle = \langle n \rangle$$

n is an integer that counts the states of the system. Because of the properties of the latter operators, it counts exactly, since once the highest possible energy state is reached, the result is the nullket. Note that $\widehat{a}^\dagger$ is also the *adjoint* of $\widehat{a}$. Using this property of the *creation* and *annihilation* operators, and since the BCS state is normalized, we obtain the result

$$\langle \Delta E_V \rangle = -\langle V_c \rangle \prod_{p=1}^{N_c}\langle 0|u_p^\dagger + \langle 1|v_p^\dagger \prod_{q=1}^{N_c} u_q|0\rangle + v_q|1\rangle \left(\sum_{(k_1 \to k_2)} \langle \widehat{\sigma}_{k2}^+ \widehat{\sigma}_{k1} \rangle\right)$$

$$= -\langle V_c \rangle \left(\sum_{(k_1 \to k_2)} \langle n_{k_1 \to k_2} \rangle\right)$$

$$= -\langle V_c \rangle \langle N_c \rangle$$

$\langle n_{k_1 \to k_2}\rangle$ is the expectation value of the number of Cooper pairs making the transition from k_1 to k_2. Since the expectation value for the ensemble of Cooper pairs can also be written as $\langle \Delta E_V \rangle = \langle N_c \rangle \langle \Delta E_V^{pair}\rangle$, we have that

$$\langle \Delta E_V^{pair}\rangle = \left\langle \frac{\Delta E_V}{N_c}\right\rangle \tag{15.99}$$

This result tells us that the properties of the superconductor are determined, on average, from the mean properties of a Cooper pair. It is sufficient for us to then examine the properties of the average Cooper pair to enable a better understanding of the properties of the ensemble. By considering a potential energy source that is connected to a thermal bath, we can also determine the temperature dependence of the probability of a Cooper pair. This information can then be used to treat the Cooper pair at finite equilibrium temperatures. This will be useful to explain the temperature dependent behaviors observed in superconductors.

15.8 Small Symmetric Splitting With Spin Matrices

15.8.1 Simplest Electronic Potential Connected To A Thermal Bath

To introduce finite temperature into the superconductor problem, let us determine the temperature dependent expectation value of the transition potential energy of a Cooper pair. This can be done readily for *a potential connected to a thermal bath of the superconducting material* using statistical thermodynamics. Particularly, we can use the Boltzmann probability distribution obtained in Chap. 2 (c. Eq. (2.37)). Using the Boltzmann probability distribution, the energy in thermodynamic equilibrium can be expressed in proportion to the expectation or average energy per pair $\langle \epsilon \rangle$. The expectation value of the potential energy changes becomes

$$\frac{\langle \Delta E_V \rangle}{2}(T) = \sum_{j=1}^{2} \Delta V_j f_B(\Delta V_j)$$

$$= \sum_{j=1}^{2} \Delta V_j \frac{e^{-(\Delta V_j)/k_B T}}{\sum_{k=1}^{2} e^{-(\Delta V_j)/k_B T}}$$

$$= \frac{\frac{\langle V_c \rangle}{2} e^{-\langle V_c \rangle/2k_B T} - \frac{\langle V_c \rangle}{2} e^{\langle V_c \rangle/2k_B T}}{e^{-\langle V_c \rangle/2k_B T} + e^{\langle V_c \rangle/2k_B T}}$$

k_B is the Boltzmann constant and T is the *thermodynamic equilibrium temperature*. $\langle \Delta V_j \rangle$ are the corresponding *expected* symmetric eigenenergies (with respect to everything *except temperature*) for the Cooper pair. These are relative to the Fermi level ϵ_F and given by $\pm \langle V_c \rangle/2$, (c.Eq. 15.77). Then, we have

$$\frac{\langle \Delta E_V \rangle}{2} = \frac{\langle V_c \rangle}{2} \frac{e^{-\langle V_c \rangle/2k_B T} - e^{\langle V_c \rangle/k_B T}}{e^{-\langle V_c \rangle/2k_B T} + e^{\langle V_c \rangle/2k_B T}}$$

$$= -\frac{\langle V_c \rangle}{2} \tanh \frac{\langle V_c \rangle}{2k_B T}$$

We know from the symmetry conditions that the net energy splitting for a Cooper pair is twice the above result, or $\langle \Delta E_V \rangle$. This gives the mean temperature dependent Cooper Pair transition energy as

Mean Temperature Dependent Potential Splitting

$$\langle \Delta E_V \rangle(T) = -\langle V_c \rangle(0) \tanh\left(\frac{\langle V_c \rangle}{2k_B T}\right) \qquad (15.100)$$

In this result, $\langle V_c \rangle$ is the expectation value with respect to all dependent variables. We have expressed the coefficient as $\langle V_c \rangle(0)$. It follows from Eq. (15.100) since the expectation value at $T = 0$ is given by

$$\langle \Delta E_V \rangle(0) = -\langle V_c \rangle \tanh\left(\frac{\langle V_c \rangle}{2k_B \cdot 0}\right) = -\langle V_c \rangle(0) \tag{15.101}$$

When we compare this result to Eq. (15.89), the expectation value of a Cooper pair transition then becomes

$$-\langle V_c \rangle(0)\langle \hat{\sigma}^+ \hat{\sigma}^- \rangle(T) = -\langle V_c \rangle(0) v_{k2}^\dagger v_{k1} = -\langle V_c \rangle(0)\tanh\left(\frac{2\epsilon_k - E_s}{2k_B T}\right) \tag{15.102}$$

From this, we also have that

$$v_{k2}^\dagger v_{k1}(T) = \tanh\left(\frac{\langle V_c \rangle}{2k_B T}\right) \tag{15.103}$$

Therefore, the probability of a transition of a Cooper pair also depends on temperature. Fig. 15.13 illustrates examples of the probability of formation for a Cooper pair $v_{k2}^\dagger v_{k1}$ as a function of temperature for mean splittings of $0.5, 0.75$, and $1.0 meV$. From Fig. 15.13, we see that *a Cooper*

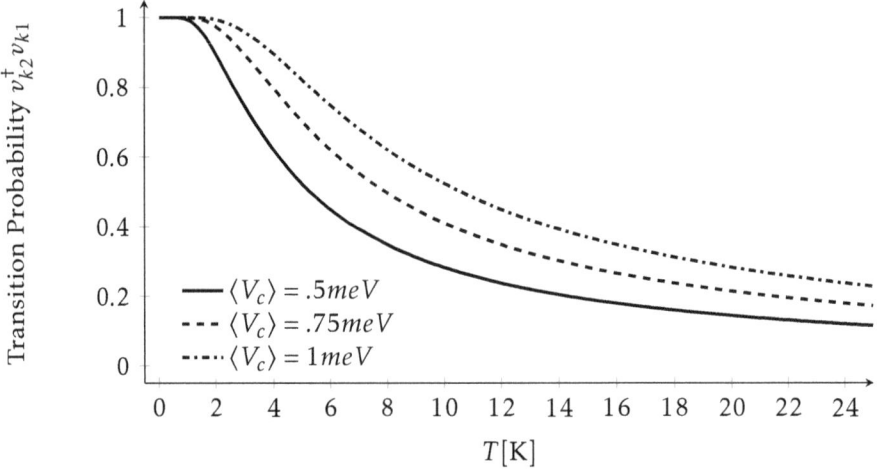

Figure 15.13: Cooper pair transition probability $v_{k2}^\dagger v_{k1}$ as a function of temperature T for mean splittings of $0.5, 0.75$, and 1.0 meV. We see that a Cooper pair has larger finite probability for lower temperatures.

pair formation generally has larger finite probability for lower temperatures, where the extent depends on the potential splitting strength at $T = 0K$. This also has implications on what is called *the correlation energy* of a

15.8 Small Symmetric Splitting With Spin Matrices

Cooper pair. Let's return to the Hamiltonian for the transition energy for a Cooper pair, using the above results. We now have

$$\Delta \widehat{H} \psi_s = [2\epsilon_k + \langle \Delta E_V \rangle] \psi_s = E_s \psi_s$$

This leads to

$$\langle \Delta E_V \rangle \psi_s = [E_s - 2\epsilon_k] \psi_s \Rightarrow \frac{\langle \Delta E_V \rangle}{E_s - 2\epsilon_k} \psi_s = \psi_s \quad (15.104)$$

Based on the assumptions for a Cooper pair, it must be that $E_s - 2\epsilon_k < 0$, for any ϵ_k. Another consequence of the assumptions made in the Cooper problem is that there are *continuous* momentum states (for free electrons), and therefore, continuous energy states within the assumed range ϵ_F to $\epsilon_F + \hbar \omega_D$. We know from our analysis of the free particle problem (c.Eq. 3.107) that the most general form of the solution for ψ_s is written using the *Fourier transform*, and is given by

$$\psi_s(x,t) = \frac{1}{\sqrt{2\pi}} \int_{-\infty}^{+\infty} F_s(k) e^{-i(\mathbf{k}\cdot\mathbf{x} - \epsilon_k t)} d\mathbf{k} \quad (15.105)$$

$F_s(\mathbf{k})$ is the Fourier transform of the wave function ψ_s. It's sufficient and convenient to use the initial condition of Eq. (15.105) (set $t = 0$). This is known as the *coherent state*. Substitution of the coherent state $\psi_s(x,0)$, then left multiplying by $\psi_s^*(t = 0)$ in Eq. (15.104), then integrating leads to

$$\int_{-\infty}^{+\infty} \psi_s^*(x,0) \psi_s(x,0) d\mathbf{x} = \int_{-\infty}^{+\infty} \psi_s^*(x,0) \frac{\langle \Delta E_V \rangle}{E_s - 2\epsilon_k} \psi_s(x,0) d\mathbf{x}$$

For the LHS, if $\psi_s(x,0)$ is normalized such that $\int \psi_s^* \psi_s = 1$, then we have that

$$1 = \frac{1}{\sqrt{2\pi}} \frac{1}{\sqrt{2\pi}} \int \int F_s(k) e^{i\mathbf{k}\cdot\mathbf{x}} d\mathbf{k} \int F_s^*(k') e^{-i\mathbf{k}'\cdot\mathbf{x}} d\mathbf{k}' \frac{\langle \Delta E_V \rangle}{E_s - 2\epsilon_k} d\mathbf{x} \quad (15.106)$$

Note that all integrals are currently from $-\infty$ to $+\infty$. Since we are combining two independent integrals with respect to a wavenumber k, different dummy variables must be used, hence the use of $\mathbf{k}$ and $\mathbf{k}'$. We can regroup the integral terms by writing

$$1 = \frac{1}{2\pi} \int \int \int F_s(\mathbf{k}) F_s^*(\mathbf{k}') e^{i(\mathbf{k}-\mathbf{k}')\cdot\mathbf{x}} \frac{\langle \Delta E_V \rangle}{E_s - 2\epsilon_k} d\mathbf{k}' d\mathbf{k} d\mathbf{x} \quad (15.107)$$

The *integral representation of the Dirac Delta function*, derived earlier and given by Eq. (3.76), can be used to simplify Eq. (15.107). The Dirac

Delta function turns out to be contained in the result above, specifically, in the integral parts with respect to **x** and d**x**. Using the Dirac Delta function, we then have

$$1 = \int\int F_s(\mathbf{k})F_s^*(\mathbf{k}')\delta(\mathbf{k}-\mathbf{k}')\frac{\langle\Delta E_V\rangle}{E_s - 2\epsilon_k}d\mathbf{k}'d\mathbf{k} \tag{15.108}$$

We have reduced three integrals down to two. Now, we'll reduce two integrals down to one by using the property of the Dirac Delta function given by $\int \delta(x-x')f(x')dx' = f(x)$ (c.Eq. 3.77). With this substitution, we have

$$1 = \int |F_s(\mathbf{k})|^2 \frac{\langle\Delta E_V\rangle}{E_s - 2\epsilon_k}d\mathbf{k} \tag{15.109}$$

It is more convenient to express Eq. (15.109) as an integral with respect to the kinetic energy states ϵ_k. This can be done by recognizing that the quantity $|F_s(\mathbf{k})|^2 d\mathbf{k} = dN/N_\Delta$ is the probability distribution of the states depending on the wavevector **k**, within the interval d**k**. N_Δ is the total number of states between the interval of integration. The integral can alternatively be expressed in terms of the *density of states probability distribution* introduced in Sec. 4.3. There, we found

$$f_D d\epsilon_k = D(\epsilon_k)/N_\Delta = [(dN/d\epsilon_k)/N_\Delta]d\epsilon_k = dN/N_\Delta$$

$D(\epsilon_k)$ is the density of states defined here as $D(\epsilon_k) = dN/d\epsilon_k$. Because there is a one-to-one correspondence between each pair $(\mathbf{k},-\mathbf{k})$ and a corresponding kinetic energy ϵ_k, Eq. (15.109) can also be written as

$$1 = \frac{1}{N_\Delta}\int D(\epsilon_k)\frac{\langle\Delta E_V\rangle}{E_s - 2\epsilon_k}d\epsilon_k \tag{15.110}$$

The number of states over the finite integration range N_Δ is given by

$$N_\Delta = \int_{\epsilon_F}^{\epsilon_F + \hbar\omega_D} D(\epsilon_2)d\epsilon_2$$

The conditions of the problem allow us to impose finite limits of integration, instead of the usual limits from $-\infty$ to $+\infty$. The reasoning is as follows: below the Fermi level, the electrons around bounded, and their is no kinetic energy for a Cooper pair state. And due to the finite amound of potential, there is always an upper bound on the kinetic energy a Cooper pair will have. Thus, ϵ_F (at $T = 0K$) can replace the lower limit of the integral since below this value, $|F(p)|^2 = 0$ for the electron pair. An upper bound on ϵ_k is also assumed for the Cooper

15.8 Small Symmetric Splitting With Spin Matrices

pair and it is typically expressed as $\epsilon_F + \hbar\omega_D$, where ω_D is the effective energy frequency associated with the system energy change. It's value depends on the coupling mechanism or mechanisms. Then, we have

$$1 = \int_{\epsilon_F}^{\epsilon_F+\hbar\omega_D} \frac{\langle\Delta E_V\rangle}{E_s - 2\epsilon_k} \frac{D(\epsilon_k)}{N_\Delta} d\epsilon_k = \int_{\epsilon_F}^{\epsilon_F+\hbar\omega_D} \frac{\langle\Delta E_V\rangle}{E_s - 2\epsilon_k} f_D(\epsilon_k) d\epsilon_k \quad (15.111)$$

Using Eq. (15.100), we have

$$1 = -\int_{\epsilon_F}^{\epsilon_F+\hbar\omega_D} \frac{\langle V_c\rangle(0)\tanh\left(\frac{\langle V_c\rangle}{2k_BT}\right)}{E_s - 2\epsilon_k} f_D(\epsilon_k) d\epsilon_k \quad (15.112)$$

Having continuous values over the range of integration, we can pull out the denominator if it is replaced by the expectation value. Then, the other expectation values can also come out of the integral, so we have

$$1 = -\frac{\langle V_c\rangle(0)\tanh\left(\frac{\langle V_c\rangle}{2k_BT}\right)}{E_s - 2\langle\epsilon_k\rangle} \int_{\epsilon_F}^{\epsilon_F+\hbar\omega_D} f_D(\epsilon_k) d\epsilon_k \quad (15.113)$$

Using the result we obtained in Eq. (4.57), the integral on the RHS is unity, so we have

$$1 = -\frac{\langle V_c\rangle(0)\tanh\left(\frac{\langle V_c\rangle}{2k_BT}\right)}{E_s - 2\langle\epsilon_k\rangle} \quad (15.114)$$

The **correlation energy** of a Cooper pair, typically denoted by Δ_c is defined as the energy difference attributed to the potential energy splitting. In other words, $\Delta_c = 2\langle\epsilon_k\rangle - E_s = \langle V_c\rangle$, where E_s is the eigenenergy of the Cooper pair. Then, in terms of the correlation energy, we then have the relation

Cooper Pair Correlation Energy With Temperature

$$\Delta_c(T) = \Delta_c(0)\tanh\left(\frac{\Delta_c(T)}{2k_BT}\right) \quad (15.115)$$

Eq. (15.115) is a nonlinear equation in $\Delta_c(T)$. Therefore, to obtain the temperature dependence of the correlation energy, it must be solved numerically using methods of solving such equations. Fig. 15.23 illustrates the temperature dependence of the Cooper pair correlation energy from

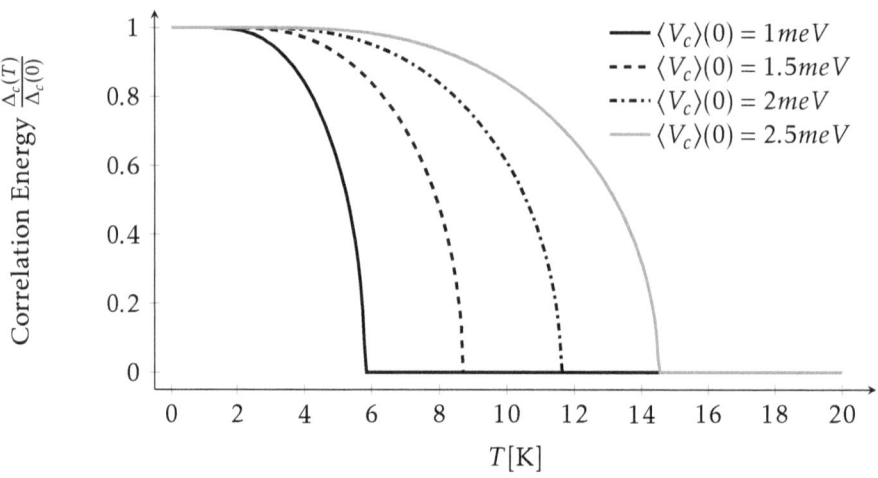

Figure 15.14: Cooper pair correlation energy as a function of temperature T for four different values of $\langle V_c \rangle(0) = 1, 1.5, 2.0,$ and 2.5 meV.

solving Eq. (15.115). Above a certain critical temperature T_c, Cooper pairs have no correlation energy, and therefore, they can no longer form. Fig. 15.14 also illustrates that the critical temperature is proportional to the $T = 0$ correlation energy $\langle V_c \rangle(0)$. For larger $\langle V_c \rangle(0)$, the critical temperature increases in proportion.

 Correlation energy and Critical Temperature: The zero temperature correlation energy $\langle V_c \rangle(0)$ is proportional to the critical temperature T_c.

We can *estimate* this relation of proportionality by using an expansion of the tanh$x$ function near zero, which is the condition for T_c. In the conditions of small x, tanh$x$ can be approximated as

$$\tanh x \approx x - \frac{x^3}{3} \qquad (15.116)$$

Substitution of Eq. (15.116) into Eq. (15.115) gives

$$\Delta_0 \tanh x \approx \Delta_0 \left(\frac{\Delta_c}{2k_B T} - \frac{\Delta_c^3}{24 k_B^3 T^3} \right) = \Delta_c \qquad (15.117)$$

We can divide both sides by $\Delta_c/2k_B T$, which then gives

$$\left(1 - \frac{\Delta_c^2}{12 k_B^2 T^2} \right) = \frac{2 k_B T}{\Delta_0} \qquad (15.118)$$

This can be rearranged to so we now have

$$1 - \frac{2k_B T}{\Delta_0} = \frac{\Delta_c^2}{12 k_B^2 T^2} \tag{15.119}$$

The critical temperature T_c is defined as the finite temperature at which Δ_c vanishes (or becomes negative). Therefore, Eq. (15.119) suggests that T_c satisfies the approximate condition given by

---

Critical Temperature-Correlation Energy Linear Relation

$$\Delta_0 \approx 2 k_B T_c \tag{15.120}$$

This is in agreement with what is illustrated in Fig. 15.14. For example, doubling the zero temperature correlation energy doubles the critical temperature. When normalizing the temperature axis by the critical temperature T_c, all the curves fall onto the same line, from $\langle V_c \rangle (T/T_c = 0)/\langle V_c \rangle (0) = 1$ on the y-axis down to 0 at $T/T_c = 1$. The disability of Cooper pairs to form at T_c and above corresponds to the region in which the material returns to normal conductivity.

15.9 Discontinuities In Electronic Properties Near T_c

Two of the first parameters observed to have discontinuities in their properties, in superconductors, was in the electrical resistance $R(T)$ or resistivity $\rho(T)$ and the heat capacity $C(T)$. Now, we'll examine how the above properties lead to such discontinuities. Let's discuss the resistance first.

15.9.1 Discontinuity In Electrical Resistance

We have enough information to determine the consequences of the above results on the electrical conductivity $\sigma_s = 1/\rho_s$, in the vicinity of the transition temperature T_c. ρ_s is the bulk resistivity (proportional to the measured resistance R). If we assume that over the thermoequilibrium temperature variation, the electric field applied to the superconductor is held constant, then variation with temperature is given by the current density $\widehat{J}_c$. If we are concerned with the conduction along the axis of the cylinder ($z-axis$), then we need to determine the current density along this axis. Then, this component of current density *per charge carrier* becomes

$$\widehat{J}_z = -\frac{ie\hbar}{2m_s}(\psi\nabla\psi^* - \psi^*\nabla\psi)$$

$$= -\frac{ie\hbar}{2m_s}\left[(-ik_z)C_n e^{ik_z z}C_n^* e^{-ik_z z} - ik_z C_n^* e^{-ik_z z}C_n e^{ik_z z}\right]\Phi^*\Phi R^* R$$

$$= \frac{ie\hbar k_z}{2m_s}\left(2|C_n|^2 R^2\right)$$

$$= \frac{e\hbar k_z}{m_s}|C_n|^2 R^2$$

We have used the fact that $\Phi^*\Phi = e^{-ik_\varphi d\ell}e^{ik_\varphi d\ell} = 1$. R is the radial wave function describing the *Abrikosov quantized vortices* obtained in Sec. 15.5. To determine k_z, we must return to the eigenenergy of a charge carrier given by

$$E_s = \epsilon_k + \langle\Delta E_V\rangle$$

$$= \frac{\hbar^2 k_z}{2m_s} - V_c(0)\tanh\left(\frac{V_c}{2k_B T}\right)$$

$$= \frac{\hbar^2 k_z}{2m_s} - \Delta_c(0)\tanh\left(\frac{\Delta_c}{2k_B T}\right)$$

Rearranging, we have

$$\frac{\hbar^2 k_z}{2m_s} = \Delta_c(0)\tanh\left(\frac{\Delta_c}{2k_B T}\right) + E_s \tag{15.121}$$

Because the potential energy is negative and we have free charge carriers, this constitutes a *potential hole* experienced by the carriers, as discussed in Sec. 6.1. This raises the momentum along z, *free of charge*. Solving for k_z gives

$$k_z = \sqrt{\frac{2m_s}{\hbar^2}\left[\Delta_c(0)\tanh\left(\frac{\Delta_c}{2k_B T}\right) + E_s\right]} \tag{15.122}$$

Since $J_c \propto k_z$ times a constant, only vary temperature means that $k_z(T) \propto \sigma$, and thus $\rho(T) \propto 1/k_z(T)$. Using this result, we are interested to see

15.9 Discontinuities In Electronic Properties Near T_c

what happens *very close* to the critical temperature T_c, as the temperature is reduced. In this case, the eigenenergy E_s can be taken to be the kinetic energy without the potential present, *just above* T_c. Let us denote this energy by E_s^+. So, we have

$$k_z = \sqrt{\frac{2m_s}{\hbar^2}\left[\Delta_c(0)\tanh\left(\frac{\Delta_c}{2k_BT}\right) + E_s^+\right]} \qquad (15.123)$$

Then, we can rewrite Eq. (15.122) as

$$k_z = \sqrt{\frac{2m_s}{\hbar^2}\left[\Delta_c(0)\tanh\left(\frac{\Delta_c}{2k_BT}\right) + E_s^+\right]}$$

$$= \frac{\sqrt{2m_s E_s^+}}{\hbar}\sqrt{\frac{\Delta_c(0)}{E_s^+}\tanh\left(\frac{\Delta_c}{2k_BT}\right) + 1}$$

Defining the dimensionless ratio $\hbar k_z/\sqrt{2m_s E_s^+} = k_z(T)/k_z^+(T_c^+)$, we obtain the dimensionless ratio, proportional to the conductivity, as

Discontinuous Conductivity Factor

$$\frac{k_z(T)}{k_z^+} = \sqrt{\frac{\Delta_c(0)}{E_s^+}\tanh\left(\frac{\Delta_c}{2k_BT}\right) + 1} \qquad (15.124)$$

The inverse of Eq. (15.124) is proportional to the temperature dependent resistivity. Fig. 15.15 plots examples of $k_z^+(T_c^+)/k_z(T)$ for different ratios of $\Delta_c(0)/E_s^+ = 1000, 500$, and 100. The sudden appearance of the potential as the temperature is reduced in the vicinity of T_c, leads to a discontinuity in the conductivity/resistivity. Notice, also, that the discontinuity is most pronounced for systems where the kinetic energy just above T_c, denoted by E_s^+ is smallest, suggesting conditions of poor conduction above T_c will express stronger discontinuities.

15.9.2 Discontinuity In Superconductor Heat Capacity

Since the temperature has been introduced into the energy of the system using a potential energy connected to a thermal bath, properties involving energy and temperature can be analyzed, as well. The electronic

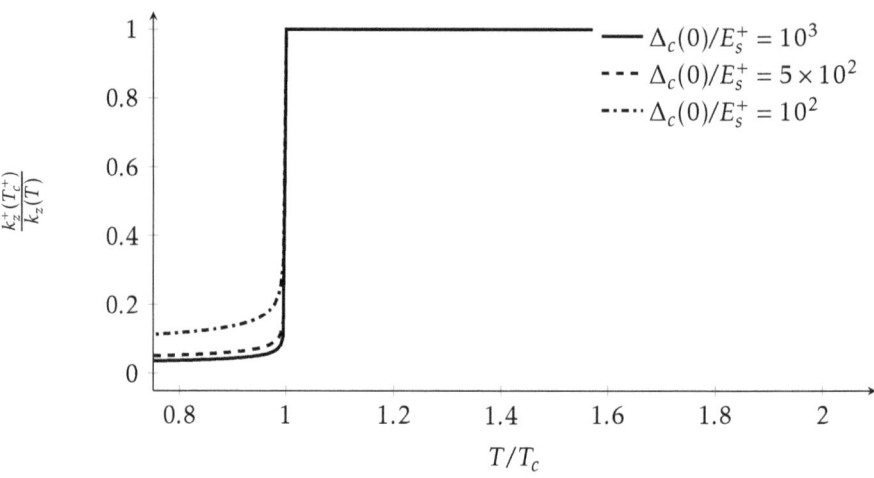

Figure 15.15: Discontinuities in $k_z^+/k_z(T)$ *near* critical temperature T_c, which is proportional to the resistivity, for three ratios $\Delta_c(0)/E_s^+$ 1000, 500, and 100.

heat capacity C_e is a measure of the *amount of change in electronic energy of a system per unit change in temperature*. For a superconductor, the electronic heat capacity can be defined as

$$C_e = \frac{\partial(\langle E_s \rangle + \langle E_n \rangle)}{\partial T} = \frac{\partial}{\partial T}(2\langle \epsilon_k \rangle + \Delta E_V)) + \frac{\langle E_n \rangle}{\partial T} = C_{se} + C_{ne} \quad (15.125)$$

In general, there are two contributions. C_{se} is the heat capacity due to charge carriers contributing to superconductivity, while C_{ne} is the heat capacity due to normal conductivity. For the superconductor energy contribution, we're using the energy for the correlated pair of electrons (we can always divide the result by 2 to express per electron). Based on Eq. (15.100), the total energy expectation value becomes

$$\langle E_s \rangle = 2\langle \epsilon_k \rangle + \langle \Delta E_V \rangle(T) = 2\langle \epsilon_k \rangle - \langle V_c \rangle(0)\tanh\left(\frac{\langle V_c \rangle}{2k_B T}\right) \quad (15.126)$$

We are primarily interested in the contributions from the carriers involved in superconductivity. For this contribution, all the temperature dependence is contained in the potential energy term. It is also responsible for the effective change in the wavenumber k, as we saw with the resistance temperature dependence in the previous subsection. Therefore, Eq. (15.125) becomes

$$C_{se} = \frac{\partial \langle E_s \rangle}{\partial T} = \frac{\partial}{\partial T}\left[-\langle V_c \rangle(0)\tanh\left(\frac{\langle V_c \rangle}{2k_B T}\right)\right]$$

15.9 Discontinuities In Electronic Properties Near T_c

The term in square brackets is just the Cooper pair correlation energy $\Delta_c(T)$ given by Eq. (15.115), so we have

$$C_{se} = \frac{\partial \Delta_c}{\partial T} \tag{15.127}$$

Therefore, to determine C_{se}, we only need to determine $\partial \Delta_c / \partial T$. So, we have

$$C_{se} = \frac{\partial \Delta_c}{\partial T} = \frac{\partial}{\partial T}\left[-\langle V_c \rangle(0)\tanh\left(\frac{\Delta_c}{2k_B T}\right)\right]$$

$$= -\langle V_c \rangle(0)\left[1 - \tanh^2\left(\frac{\Delta_c}{2k_B T}\right)\right]\left[\frac{1}{2k_B T}\frac{\partial \Delta_c}{\partial T} + \frac{\Delta_c}{2k_B}\left(\frac{-1}{T^2}\right)\right]$$

The first term in square brackets on the RHS is the derivative of tanh, while the second term in square brackets on the RHS arises from the temperature dependent argument of the tanh function. This gives

$$\frac{\partial \Delta_c}{\partial T} = -\frac{\langle V_c \rangle(0)}{2k_B T}\left[1 - \tanh^2\left(\frac{\Delta_c}{2k_B T}\right)\right]\frac{\partial \Delta_c}{\partial T} +$$

$$\frac{\langle V_c \rangle(0)}{2k_B T^2}\frac{\Delta_c}{2k_B}\left[1 - \tanh^2\left(\frac{\Delta_c}{2k_B T}\right)\right] \tag{15.128}$$

We have $\partial \Delta_c / \partial T$ on both the LHS and RHS. This is what we are after, as it gives C_e. Rearranging to solve for $\partial \Delta_c / \partial T$, one obtains

$$\frac{\partial \Delta_c}{\partial T} = \frac{\frac{\langle V_c \rangle(0)\Delta_c}{2k_B T^2}\left[1 - \tanh^2\left(\frac{\Delta_c}{2k_B T}\right)\right]}{1 - \frac{\langle V_c \rangle(0)}{2k_B T}\left[1 - \tanh^2\left(\frac{\Delta_c}{2k_B T}\right)\right]}$$

Multiplying the above numerator and denominator by $2k_B T$ gives for the **superconductor electronic heat capacity** $C_{se}(T)$

Electronic Heat Capacity C_{se} Of Superconductor

$$C_{se} = \frac{\frac{\langle V_c \rangle(0)\Delta_c}{T}\left[1 - \tanh^2\left(\frac{\Delta_c}{2k_B T}\right)\right]}{2k_B T - \langle V_c \rangle(0)\left[1 - \tanh^2\left(\frac{\Delta_c}{2k_B T}\right)\right]} \tag{15.129}$$

Eq. (15.129) can be solved simultaneously with Δ_c, given by Eq. (15.115), to obtain both as functions of temperature T. Fig. 15.16 shows the

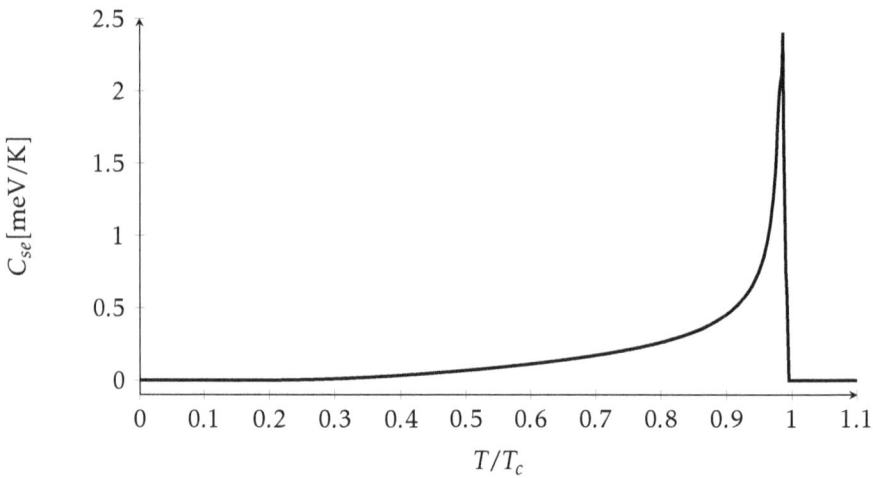

Figure 15.16: Temperature dependence of electronic heat capacity of a superconductor approaching the critical temperature T_c.

solution to Eq. (15.129) for the case of $\langle V_c \rangle(0) = 1.5\text{meV}$. The heat capacity due to superconductor carriers exhibits an exponential rise to a maximum near T_c. What goes up, must come down, so it undergoes are sharp decrease to zero. This means that the normal conduction carriers become the only contributors to C_e after exceeding T_c. Then, C_n becomes the only contribution, which is linear in temperature. With a linear dependence on C_n, the total heat capacity has the general form

$$C_e = C_{se} + \gamma_n T \tag{15.130}$$

γ_n is the proportionality constant for the normal conduction carriers. We see that the sudden appearance of the correlation energy leads to discontinuities in the temperature dependent parameters of superconductor.

Let us now discuss how to bring the above results into alignment with the linearized Ginzburg-Landau results revealed by Abrikosov. This will be the subject of the next section.

15.10 Connecting Cooper, Ginzburg, and Landau

The above results allow us to connect directly to the results we obtained solving the linearized Ginzburg-Landau equation. There we found that

15.10 Connecting Cooper, Ginzburg, and Landau

when a magnetic field is present, there is an additional eigenenergy contribution to the kinetic energy which is positive, or increases the energy. It is given by (c.Eq15.54)

$$|\alpha|_r = E_b = \hbar\omega_s\left(n+\frac{1}{2}\right) = \frac{\hbar e B_e}{m_s}\left(n+\frac{1}{2}\right) \qquad (15.131)$$

Recall that $|\alpha_r|$ is the eigenvalue of the radial part of the *linearized Ginzburg-Landau equation*, which is a form of the relativistic kinetic energy quantum mechanical equation of motion for a charge carrier in a magnetic field. This energy source is taken as separate from the potential energy source ΔE_V. When the magnetic energy is included, the correlation energy becomes

$$\Delta_c(T,B) = 2\langle\epsilon_k\rangle - (E'_s + E_b)$$

$$= \Delta_c(0)\tanh\left(\frac{\Delta_c(T)}{2k_B T}\right) - E_b$$

E'_s is the eigenenergy *without the magnetic field*. Substitution of Eq. (15.131) into the above gives a **magnetic field dependent Cooper pair correlation energy** $\Delta_c(T,B)$ given by

---
Correlation Energy With External Magnetic Field

$$\Delta_c(T,B) = \Delta_c(0,B)\tanh\left(\frac{\Delta_c(T)}{2k_B T}\right) - \frac{\hbar q_s B_e}{m_s}\left(n+\frac{1}{2}\right) \qquad (15.132)$$

Fig. 15.17 shows the temperature dependent correlation energy for several values of external field, computed by solving Eq. (15.132). The correlation energy for a Cooper pair is reduced in an external magnetic field. The entire temperature dependent curve shifts downward by an amount E_b, consequently resulting in a reduction of the critical temperature T_c. Fig. 15.18 shows the change in T_c with an external magnetic field. The external field where the correlation energy *vanishes for all temperatures* corresponds to the *critical magnetic field B_c*. In the above example, $B_c \approx 17.5$ milliTesla. Likewise, the magnetic field at which the correlation energy vanishes changes as a function of thermodynamic equilibrium temperature T. Fig. 15.19 shows an example of B_c as a function of T. Because the curve is shifted downwards, the slope at $T = 0$ remains unchanged, however, the slope at $T = T_c$ is reduced for larger

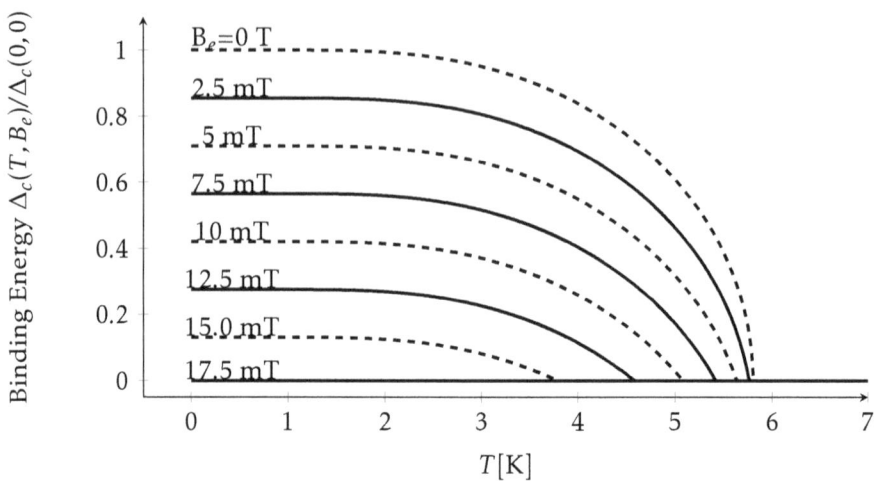

Figure 15.17: Cooper pair correlation energy as a function of temperature T and external magnetic field B_e for $V_0 = 1$meV.

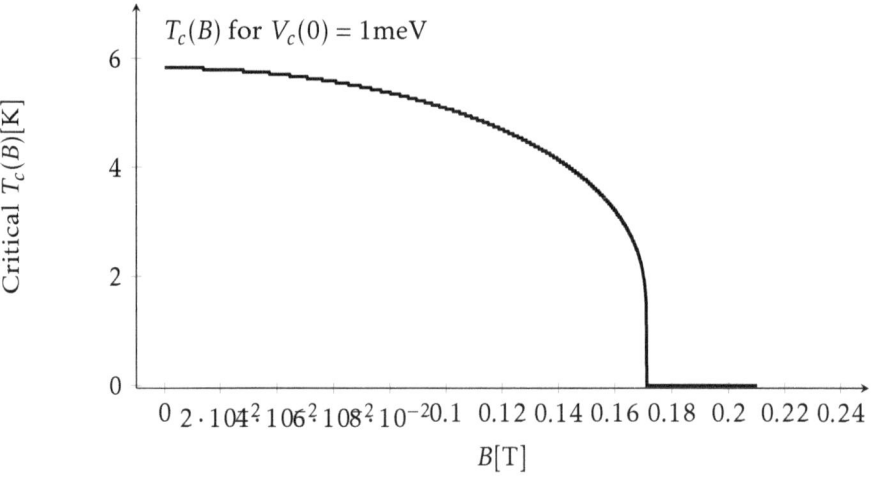

Figure 15.18: Critical temperature T_c as a function of an external magnetic field.

and larger magnetic fields. This is also a characteristic of the approximation using $B_c(T) = B_c(0)(1 - T^2/T_c^2)$. The effect of the magnetic field as described here is maximized in this particular analysis. In the next section, we discuss how higher order relativistic quantum mechanical corrections will tend to reduce and even reverse the sensitivity of the correlation energy to an external magnetic field.

15.11 Higher Order Relativistic Correction

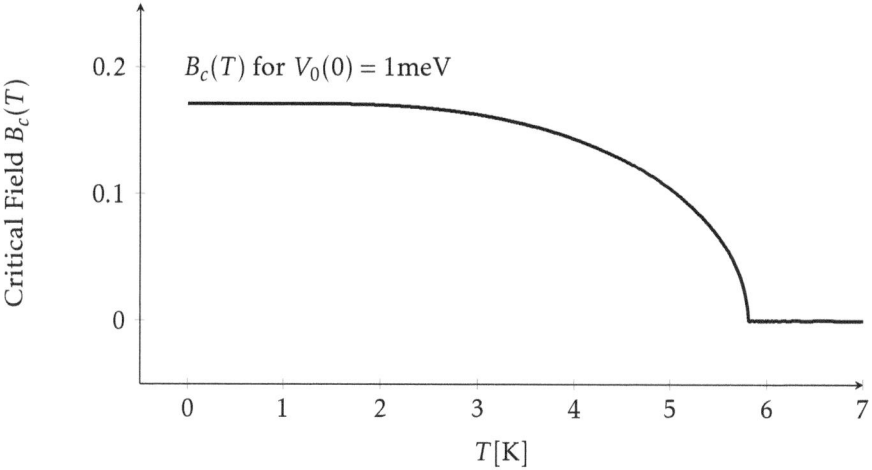

Figure 15.19: Cooper pair critical magnetic field B_c as a function of temperature. Reciprocal to a magnetic field effect on T_c, increasing temperature likewise reduces B_c. This function is often approximated by $B_c(0)(1 - T^2/T_c^2)$.

15.11 HIGHER ORDER RELATIVISTIC CORRECTION

In Sec. 11.5 of Chap. 11, in deriving the spin-orbit interaction, we also derived a few other energy contributions. One of them was the Zeeman energy term (c.Eq. 11.69), which gave us the g-factor. Another one of the energy terms we obtained, which we quietly ignored, had the appearance of a higher order kinetic energy term given by

$$E_c = -\frac{(\widehat{\mathbf{p}} + e\mathbf{A})^4}{8m_0^3 c^2} \tag{15.133}$$

The form of this correction term means that if we know the eigenenegy for the kinetic energy, being proportional to $(\widehat{\mathbf{p}} + e\mathbf{A})^2$, then we also know the eigenenergy for the correction term E_c. Notice the sign of E_c, while the kinetic energy term has a positive sign, is instead, negative. This suggests that the correction will tend to reduce the net kinetic energy. The kinetic energy eigenvalue (only the magnetic contribution) we obtained earlier is given by

$$\frac{(\widehat{\mathbf{p}} + e\mathbf{A})^2}{2m_s}\psi_s = \frac{eB_e \hbar}{m_s}\left(n + \frac{1}{2}\right)\psi_s = E_0 \psi_s \Rightarrow (\widehat{\mathbf{p}} + e\mathbf{A})^2 = 2m_s E_0$$

From this, we have

$$(\widehat{\mathbf{p}} + e\mathbf{A})^4 = 4m_s^2 E_0^2 = 4e^2 \hbar^2 B_e^2 \left(n + \frac{1}{2}\right)^2$$

Substituting this result into Eq. (15.133) gives for the energy correction term

$$E_c = -\frac{e^2\hbar^2 B_e^2}{2m_0^3 c^2}\left(n+\frac{1}{2}\right)^2 \quad (15.134)$$

Then the combined energy due to an external magnetic field becomes $E_b = E_0 + E_c$, or

Relativistically-Corrected Eigenenergy

$$E_b = \frac{e\hbar B_e}{m_0}\left(n+\frac{1}{2}\right) - \frac{e^2\hbar^2 B_e^2}{2m_0^3 c^2}\left(n+\frac{1}{2}\right)^2 \quad (15.135)$$

This result can also be connected to the Ginzburg-Landau theory. Recall the energy density from Eq. (15.14) given by

$$f_s - f_n = \alpha|\psi_s|^2 + \frac{\beta}{2}|\psi_s|^4 + \psi_s^* \frac{(-i\hbar\nabla - q_s\mathbf{A})^2}{2m_s}\psi_s - \int \mathbf{M}\cdot d\mathbf{B}_e$$

f_s is the energy density of a superconductor charge carrier, while f_n the same for the charge carrier in the *normal* conducting state. We must add the BCS mean potential energy V_c, so we have

$$f_s - f_n = \alpha|\psi_s|^2 + \frac{\beta}{2}|\psi_s|^4 + \psi_s^* \frac{(-i\hbar\nabla - q_s\mathbf{A})^2}{2m_s}\psi_s - \int \mathbf{M}\cdot d\mathbf{B}_e - \langle V_c \rangle \quad (15.136)$$

Recall that the sign of α and β are opposite in the Ginzsburg-Landau theory, as with the terms of E_b, given by Eq. (15.135). The higher order β term of the Ginzburg-Landau theory is a proxy for the higher order correction in Eq. (15.135), based on the Dirac equation. Comparing the β term to the correction term of E_b, we find that $\beta = \beta_n$ is given by

Ginzburg-Landau β Coefficient

$$\beta_n = \frac{e^2\hbar^2 B_e^2}{m_0^3 c^2}\left(n+\frac{1}{2}\right)^2 \quad (15.137)$$

There is an important distinction that must be made in this comparison, however. From the higher order correction to E_b based on the Dirac

15.11 Higher Order Relativistic Correction

equation, we know that the energy is given by

$$E_\beta^D = -\int_V \frac{\beta}{2}|\psi_s|^2 dV \tag{15.138}$$

However, the Ginzburg-Landau theory leads to an energy given by

$$E_\beta^{GL} = -\int_V \frac{\beta}{2}|\psi_s|^4 dV \tag{15.139}$$

From this, we find that the interpretation from the Ginzburg-Landau theory has a slight incompatibility with relativistc quantum mechanics because the eigenfunction $|\psi_s|^2$ in 3D has physical units of m^{-3}, while these are the units for $|\psi_s|^4$ in the Ginzburg-Landau theory. Based on this, this argument, the G-L theory cannot fully be reconciled with quantum mechanics. For this reason, we believe that quantum mechanics ultimately has to provide the additional energy contributions we are after.

Ginzburg-Landau Higher Order β Term: The higher order correction can approximately be interpreted as a higher order correction to the kinetic energy of a *free* charge carrier in a superconductor.

One significant consequence of this correction is that it restricts the maximum of the energy increase for the charge carrier in an external magnetic field, particularly for larger magnetic fields. Combined with the BCS potential energy $\langle V_c \rangle < 0$, since this energy is negative, the charge carriers become more stable as free electrons, with this correction being in combination with the holes that are stablized by the BCS potential $\langle V_c \rangle$. Fig. (15.20) illustrates the comparison between the eigenenergy without the correction $E_b = E_0$ and with the correction. From adding the relativistic correction energy term, the linear dependence becomes a non-monotonic energy variation with a single maximum. Figs. 15.21 and 15.22 demonstrate that the effect of the relativistic energy correction is to enhance the critical magnetic field. Note that the differences shown are for a zero temperature potential $\langle V_c \rangle = 1\,\text{meV}$. From Fig. 15.20, we can see that the two curves are close in value in this range, so the differences are relatively minor. However, we must not expect such modest differences for cases of larger potentials V_c, where the differences become much larger. One of the last features of superconductors we will discuss concerns an observable feature in the amount of energy needed

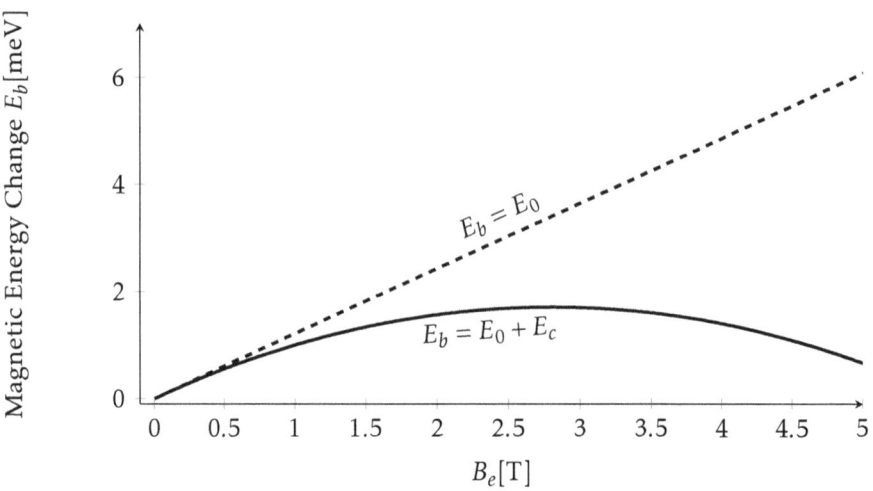

Figure 15.20: Eigenenergy versus the external magnetic field, with and without the relativistic correction. A quantum number of $n = 10$ is used. The correction reduces the energy relative to its value without the correction, in such a way to create a single maximum in the dependence on the magnetic field.

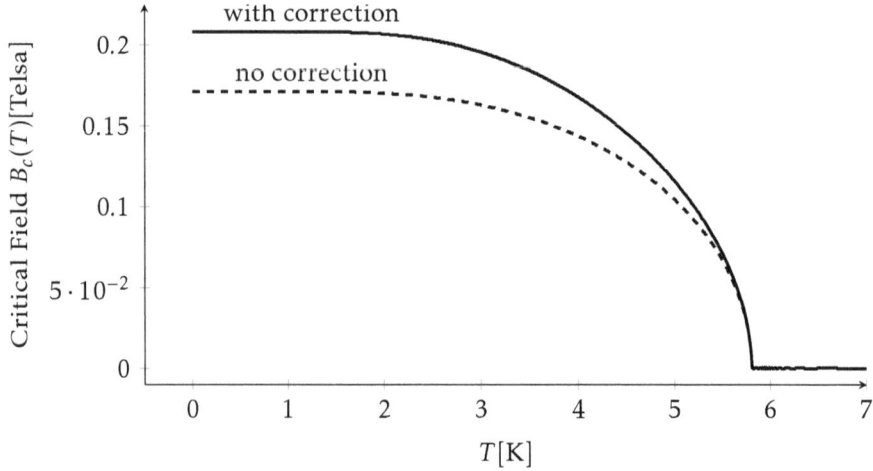

Figure 15.21: Cooper pair critical magnetic field B_c as a function of temperature, without and with the relativistic correction.

in order to excite the so-called Cooper pairs. There is an energy gap that must be overcome when applying the external electric field, even though the temperature is below T_c. This feature can be teased out of a low temperature analysis. This is done in the next section.

15.12 An Energy Gap To Excite Cooper Pairs

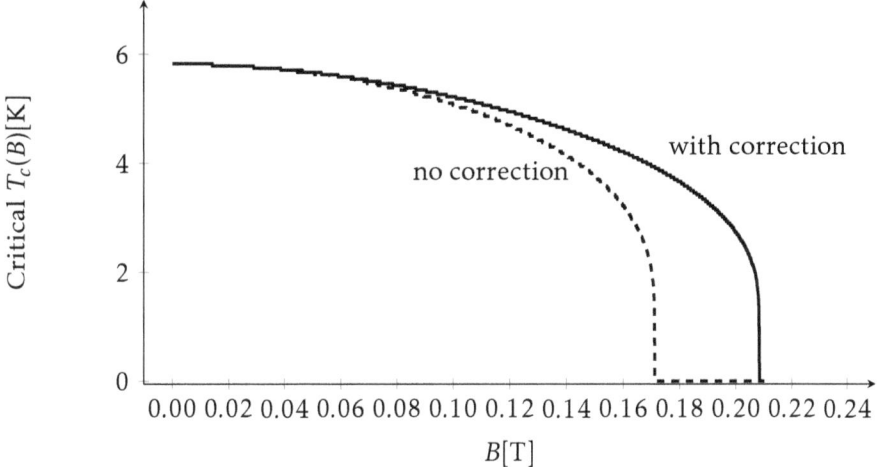

Figure 15.22: Cooper pair critical temperature T_c as a function of an external magnetic field, with and without the relativistic correction.

15.12 AN ENERGY GAP TO EXCITE COOPER PAIRS

To expose the fact that a potential gap exists in exciting the correlating electrons, let's return to the integral relation we obtained earlier, given by

$$1 = -\int_{\epsilon_F}^{\epsilon_F+\hbar\omega_D} \frac{\langle V_c \rangle(0)\tanh\left(\frac{\langle V_c \rangle}{2k_B T}\right)}{E_s - 2\epsilon_k} f_D(\epsilon_k) d\epsilon_k \quad (15.140)$$

For this analysis, we can take the temperature to be $T = 0$, which gives the simplification $\tanh(V_c/2k_B T) = 1$. Then, Eq. (15.140) becomes

$$1 = -\int_{\epsilon_F}^{\epsilon_F+\hbar\omega_D} \frac{\langle V_c \rangle(0)}{E_s - 2\epsilon_k} f_D(\epsilon_k) d\epsilon_k \quad (15.141)$$

Since Eq. (15.141) has a leading negative sign and it has been assumed that $E_s - 2\epsilon_k < 0$ in order to have mobile or free charge carriers, for finite potential, the integral in Eq. (15.141) is positive definite. Note that the potential at $T = 0$ also satisfies $V_s(0) > 0$ (since the energy difference has a negative sign), and the density of states distribution denoted by f_D satisfies $f_D(\epsilon_k) = D(\epsilon_k)/N_\Delta > 0$, where D is the density of states $dN/d\epsilon_k$ and N_Δ is the number of states over the range of integration. Now, let us further assume that *the interval of integration is sufficiently small* so $f_D(\epsilon_k)$ can be regarded as approximately constant over the interval of integration. This translates to two probable assumptions: first, the range of kinetic energies is small, and/or a relatively small correlation potential V_c between Cooper pairs. Then, in this approximation, we can take $f_D(\epsilon_k)V_s(0)$ outside the integral. Then, using the integral formula $\int (ax+b)^{-1} dx = a^{-1}\ln|ax+b|$, the evaluation with respect to ϵ_k becomes

$$-\int_{\epsilon_F}^{\epsilon_F+\hbar\omega_D} \frac{V_s(0)}{E_s - 2\epsilon_k} f_D(\epsilon_k) d\epsilon_k \approx -f_D(\epsilon_F)V_s \int_{\epsilon_F}^{\epsilon_F+\hbar\omega_D} \frac{1}{E_s - 2\epsilon_k} d\epsilon_k$$

$$= -\left(-\frac{1}{2}\right) f_D(\epsilon_F) V_s(0) \ln|E_s - 2\epsilon_k|\Big|_{\epsilon_F}^{\epsilon_F+\hbar\omega_D}$$

$$= \frac{1}{2} f_D(\epsilon_F) V_s(0) \ln\left|\frac{E_s - 2(\epsilon_F + \hbar\omega_D)}{E_s - 2\epsilon_F}\right|$$

$$= \frac{1}{2} f_D(\epsilon_F) V_s(0) \ln\left|1 - \frac{2\hbar\omega_D}{E_s - 2\epsilon_F}\right|$$

Therefore, we have the result

$$\frac{1}{2} f_D(\epsilon_F) V_s(0) \ln\left(1 - \frac{2\hbar\omega_D}{E_s - 2\epsilon_F}\right) = 1 \tag{15.142}$$

Since $E_s - 2\epsilon_k < 0$ for all ϵ_k in the range of integration, we can drop the absolute value bars (since the subtracted terms becomes an added positive term because the denominator is negative). Then, Eq. (15.142) becomes

$$e^{\frac{2}{f_D V_s(0)}} = 1 - \frac{2\hbar\omega_D}{E_s - 2\epsilon_F}$$

Multiplying both sides by $e^{-2/f_D V_s(0)}$, and solving for $E_s - 2\epsilon_F$ gives

$$E_s - 2\epsilon_F = \frac{2\hbar\omega_D e^{-\frac{2}{f_D V_s(0)}}}{e^{-\frac{2}{f_D(\epsilon_F)V_s(0)}} - 1} \tag{15.143}$$

The LHS of Eq. (15.143) is the difference in the energy of the correlated Cooper pair and the kinetic energy of two non-correlated itinerant electrons. Since this quantity is negative, the **correlation energy of the Cooper pair** is $\Delta_c = -(E_s - 2\epsilon_F) > 0$. This gives the result

---

Cooper Pair Correlation Energy At Absolute 0

$$\Delta_c(0) = \frac{2\hbar\omega_D e^{-\frac{2}{f_D(\epsilon_F)V_s}}}{1 - e^{-\frac{2}{f_D(\epsilon_F)V_s}}} \tag{15.144}$$

15.12 An Energy Gap To Excite Cooper Pairs

Eq. (15.144) can be used to make reasonable estimates of the possible correlation energy for a *Cooper pair* over a small range of potential energies. For this, we can use the *density of states* $D(\epsilon_k)$(DOS) relations we obtained for the 3D infinitely deep potential well problem, in Chap. 4. There, we found a relation for the density of states for the case of when fundamental particles are confined to motion inside of a finite sized 3D structure. This can be used along with a reasonable estimate for the Fermi level and Debye frequency ω_D for the example material. The *Debye frequency* denotes in a general way, the energy in frequency ($\Delta E_D = \hbar\omega_D$) responsible for the generation of Cooper pairs. The **Debye frequency** ω_D that should be used here is the phonon lattice modal vibration frequency near the Fermi energy (or Fermi vector k_F), which is proportional to its energy at the temperature of interest. Fig. 15.23 shows correlation energy estimates as a function of the potential energy V_s. The amplitude of the binding energies Δ_c resulting from the

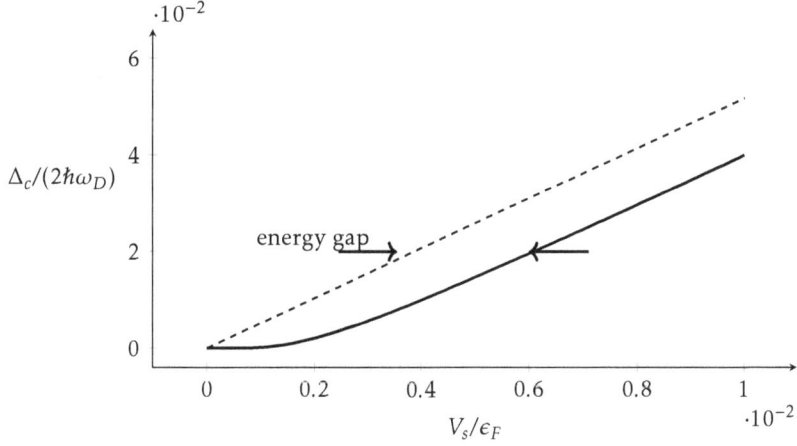

Figure 15.23: Estimated relative Cooper pair correlation energy Δ_c from Eq. (15.144), as a function of potential coupling strength V_s/ϵ_F. The example uses parameters relevant to (Nb). The Debye frequency used is $\omega_D = 2^{13} Hz$ and Fermi energy $\epsilon_F = 5.32 eV$, taken from published experimental data (see references). Note that the value of ω_D was measured for Nb at 300K, which means it overestimates the energies expected at lower temperatures.

conditions introduced by Cooper turn out to be relatively small, being one-hundredth of an eV or smaller. Typical measurements of Δ_c tend to be on the order of meV, indicating very small potentials V_s. However, a small correlation energy is consistent with the fact that this type of binding can only observed at very low temperatures. But that's not all the evidence that supports Cooper's analysis. The other supporting evidence relates to the density of states. The larger the density of states, the stronger the effect, and this is supported experimentally, as well. Most of the superconducting elements are transition metals with relatively large density of states, having d-orbital states near the Fermi energies. For this reason, these materials tend to have larger critical or transition temperatures T_c, or become superconducting at higher temperatures.

15.13 Chapter Summary

In this chapter, we have introduced the topic of *superconductivity*, which is low temperature phenomena which demonstrates a practical loss of resistivity below a critical temperature T_c. A superconductor is also an ideal *diagmagnet*, demonstrating not just a negative susceptibility χ_c, but a maximum of it. We showed how the London equation readily suggests that any conducting material without electrical loss demonstrates ideal diamagnetism, through the fact that the magnetic induction vanishes inside the conductor.

We then discussed the Ginzburg-Landau theory. We worked out the exact solution to the *linearized* form, which is the quantum mechanical eigenoperator equation for kinetic energy of a charge carrier in a magnetic field. We found that surface currents are predicted as part of the solution, and the surface current charge carriers gain energy from an external field. However, the extent of the gain can be reduced by the higher order relativistic correction to the kinetic energy.

The temperature effects were introduced in a unique manner by using well-known properties of other forms of electron-electron coupling potentials such as exchange interaction and spin-orbit coupling. From this, we considered a potential connected to a thermal bath at thermodynamic equilibrium. This implicitly assumes the potential energy is spin-dependent. Then, we examined the effects of temperature on parameters like the correlation energy, resistivity and heat capacity, all demonstrating discontinuities at the critical temperature T_c. Lastly, we also discussed dependence on external magnetic field where we found that a necessary energy gap is required to excite the so-called Cooper pairs. Table 15.7 lists some of the key equations from this Chapter.

15.13 Chapter Summary

Table 15.7: Chapter 15 Summary Equations.

Name	Equation
London Equation	$\mathbf{J}_s = -\frac{n_s q_s^2}{m_s}\mathbf{A}$
B penetration depth	$\delta_L = \sqrt{\frac{m_s}{\mu_0 n_s q_s^2}}$
Ginzburg-Landau Equation	$\alpha \psi_s + \beta |\psi_s|^2 \psi_s + \frac{(-i\hbar\nabla - q_s\mathbf{A})^2}{2m_s}\psi_s = 0$
Flux Quantization Unit	$\Phi_s = -\frac{h}{2e}$
Azimuthal lG-L solution	$\Phi(r,\varphi) = C_\varphi e^{-i\frac{e}{\hbar}B_e r^2 \varphi}$
Energy levels	$|\alpha|_r = \left(n + \frac{1}{2}\right)\hbar\omega_s$
Magnetic moment	$\mu_c = -\frac{\mu_0 e^2 H_e r^2}{2m_s}$
Relative Permeability	$\mu_r = 1 - \frac{\langle n_s\rangle \mu_0 e^2 r^2}{2m_s}$
Binary Creation Operator	$\hat{\sigma}^+ = \frac{1}{2}\left(\hat{\sigma}_x + i\hat{\sigma}_y\right)$
Binary Annihilation Operator	$\hat{\sigma}^- = \frac{1}{2}\left(\hat{\sigma}_x - i\hat{\sigma}_y\right)$
Mean Binary Spin Potential	$\langle \Delta E_V\rangle(T) = -\langle V_c\rangle(0)\tanh\left(\frac{\langle V_c\rangle}{2k_B T}\right)$
Cooper Pair Correlation Energy	$\Delta_c(T) = \Delta_c(0)\tanh\left(\frac{\Delta_c(T)}{2k_B T}\right)$
Discont. Conductivity Factor	$\frac{k_z(T)}{k_z^+} = \sqrt{\frac{\Delta_c(0)}{E_s^+}\tanh\left(\frac{\Delta_c}{2k_B T}\right) + 1}$
Electronic Heat Capacity	$C_{se} = \frac{(\langle V_c\rangle(0)\Delta_c)/T)\left[1-\tanh^2(\Delta_c/2k_B T)\right]}{2k_B T - \langle V_c\rangle(0)\left[1-\tanh^2(\Delta_c/2k_B T)\right]}$
Corrected Magnetic Energy	$E_b = \frac{e\hbar B_e}{m_0}\left(n + \frac{1}{2}\right) - \frac{e^2\hbar^2 B_e^2}{2m_0^3 c^2}\left(n + \frac{1}{2}\right)^2$
Zero Temp Correlation Energy	$\Delta_c(0) = \frac{2\hbar\omega_D e^{-\frac{2}{f_D(\epsilon_F)V_s}}}{1 - e^{-\frac{2}{f_D(\epsilon_F)V_s}}}$

15.14 Chapter Problems

Problem 15.1 Estimate the penetration depth of a magnetic induction field B in a superconductor having $10^{23} m^{-3}$ charge carriers per unit volume. How would introducing an external magnetic field change the penetration depth?

Problem 15.2 Use the result given by Eq. (15.68) or

$$\mu_r = 1 - \frac{\langle n_s \rangle \mu_0 e^2 r^2}{2 m_s}$$

And estimate the *relative* diamagnetic permeability of superconductor using $\langle n_s \rangle = 10^{23} m^{-3}$, and a superconductor radius of 2×10^{-3} cm, and use the electron charge and mass.

Problem 15.3 Using the eigenenergy we obtained in Eq. (15.135), given by

$$E_b = \frac{e\hbar B_e}{m_0}\left(n + \frac{1}{2}\right) - \frac{e^2 \hbar^2 B_e^2}{2 m_0^3 c^2}\left(n + \frac{1}{2}\right)^2$$

Calculate the ground state energy ($n = 0$) of a superconductor n–vortex in a magnetic field of 1 Tesla. Compare the relativistic correction to energy to the energy without the correction.

Problem 15.4 Using the solution for $|\alpha|_r$, show that the ground state $n = 0$ for a superconductor leads to the maximum coherence length of $\xi = \nu = \sqrt{q_s B_e / \hbar}$.

Problem 15.5 Show that the orbital magnetic moment of the n-vortex core is given by $\mu_L^s = \frac{q_s^2 B_e r^2}{2 m_s}$. Hint: Use the definition of magnetic moment and the solution for k_φ.

Problem 15.6 Use the solution for ψ_s and the definition of the current density to show that the current density in the direction φ is given by $J_\varphi = v_\varphi |\psi_s|^2 = r\omega_s |\psi_s|^2$.

Problem 15.7 Using the definitions for $\widehat{\sigma}_{k2}^+$ and $\widehat{\sigma}_{k2}$, show that $-V_c \langle 1|\widehat{\sigma}_{k2}^+ \widehat{\sigma}_{k1}|1\rangle = -V_c \mathbf{I}(2)$, where $\mathbf{I}(2)$ is the 2×2 identity matrix.

Problem 15.8 In Chap. 14, the uncertainty relation was derived for the harmonic oscillator, given by Eq. (14.42), or

$$\Delta x \Delta \widehat{p} = \hbar \left(n + \frac{1}{2} \right)$$

We found that the harmonic oscillator energy is the same form of energy of superconducting circulating currents. Use the momentum relation for a superconductor given by

$$p = eBr$$

Also discuss what the uncertainty principle suggests about Δx, which is a measure of the width of the circulating current loop, particularly, if there is an increase in magnetic field B? Is the result consistent with the conservation of momentum, for an increase in magnetic field?

15.15 Suggested Readings & References

[1] A. A. Abrikosov. *On the magnetic properties of superconductors of the second group.* Soviet Physics JETP, **5**, 1774, (1957).
[2] U. Essmann and H. Trauble. *The direct observation of individual flux lines in type II superconductors,* Phys. Lett., **24a**, pp. 526, (1967).
[3] H. F. Hess *et. al.*, *Scanning-Tunneling-Microscope Observation of the Abrikosov Flux Lattice and the Density of States near and inside a Fluxoid,* **62**, pp. 214-, Phys. Rev. Lett. (1989).
[4] B. S. Deaver, Jr., and W. M. Fairbank. *Experimental Evidence For Quantized Flux In Superconducting Cylinders.* Phys. Rev. Lett., **7**, pp. 43-46, (1961).
[5] M. Holt *et. al.*, *Phonon dispersions in niobium determined by x-ray transmission scattering,* Phys. Rev. B **66**, 064303 (2002).
[6] P. K. Sharma *et. al.*, *Vibration Spectra and Debye Temperatures of Some Transition Metals,* Z. Naturforsch. **26 a**, pp.747-752 (1971).
[7] R. de Bruyn Ouboter, *Heike Kamerlingh Onnes's Discovery of Superconductivity,* Scientific American, pp.98-103 (1997).
[8] D. v Delft and P. Kes, *The Discovery Of Superconductivity,* Physics Today, pp.38-43 (2010).

APPENDIX A

Supplemental: Real Electric Potential

Potential energy is an important and necessary part of the energy of a real physical system. Without it, any description of a physical system can be profoundly altered. However, we have also seen that it is possible to make reasonable approximations in some circumstances, for heuristic purposes. An example was with the infinitely deep potential well problem, or particle in a box. In this appendix, we review one form of potential energy that is essential for the description of any fundamental particle carrying charge q, such as an electron. Though we've used it in several of the problems considered in this text, we've not discussed its origins, and therefore, its limitations. The limitations are important to consider as they restrict any formulation's ability to describe time-dependent phenomena. This understanding is important in the treatment of most physical systems involving electrons. Even before the electron was discovered, there was an understanding that forces exist between charged bodies. The electron turns out to be subject to the same forces. This tells us that any charged particle possesses a potential energy that manifests anytime another charged particle appears in sufficient proximity. When we move towards the description of physical systems such as atoms, this form of potential energy *must* be included. Let us discuss what is known about this special form of electrical potential energy discovered even prior to the electron.

A.1 Potential And Discrete Electric Charges

Consider the scalar function of position **r** given by

$$\phi(\mathbf{r}) = \frac{k}{|\mathbf{r} - \mathbf{r}'|} \tag{A.1}$$

The constant of proportionality k here carries the choice of physical units. For example, in SI units, $k = 1/4\pi\varepsilon_0$. Recall that this is the form used in the hydrogen atom problem (c. Eq. 8.13). Eq. (A.1) is a scalar equation that depends on a single variable **r**, or equivalently, $|\mathbf{r} - \mathbf{r}'|$. For this to be true, **r**' must be constant. This gives one of the *first assumptions* for this form of static potential, namely:

Static potential function: In a static potential, the sources located at **r**' are assumed to be fixed in space.

The potential function is infinite when $\mathbf{r} = \mathbf{r}'$, and zero when **r** is infinitely far away from **r**'. This function shows up in forms of potential such as electrical potential, but also magnetic, as well as gravitational potential. It is a *potential* because whether or not charges at position x are introduced, the potential energy is present as long as the source charges at x' are hanging around. These are useful properties to keep in mind with any potential of the form given by Eq.(A.1). Extending this idea, there can be any number of sources with positions **r**', in which case, an electron sees a combination of peaks described by Eq.(A.1), as illustrated in Fig. A.1. Fig.A.1 illustrates an example of the potential function $\phi(x)$ depending on a single spatial variable x, and parameterized (since x' are held constant) by positions of a 1D line of three charges q_1', q_2', and q_3' at locations x_1', x_2', and x_3', respectively. **r**', or in this case x', denotes the location of a source charge q'. Generally, there can any number of charges contributing to the potential. For N source charges located at $\mathbf{r}_1', \mathbf{r}_2', ..., \mathbf{r}_N'$, respectively, where each potential adds. Then, evaluation of Eq. (A.1) with contributions from all the source charges becomes

$$\phi(\mathbf{r}) = k \sum_{i=1}^{N} \frac{1}{|\mathbf{r} - \mathbf{r}_i'|} \tag{A.2}$$

Since Eq. (A.1) and (A.2) are scalar functions, the gradient operator ∇ can be applied by differentiating with respect to $\mathbf{r} = [x, y, z]^T$. For one

A.1 Potential And Discrete Electric Charges

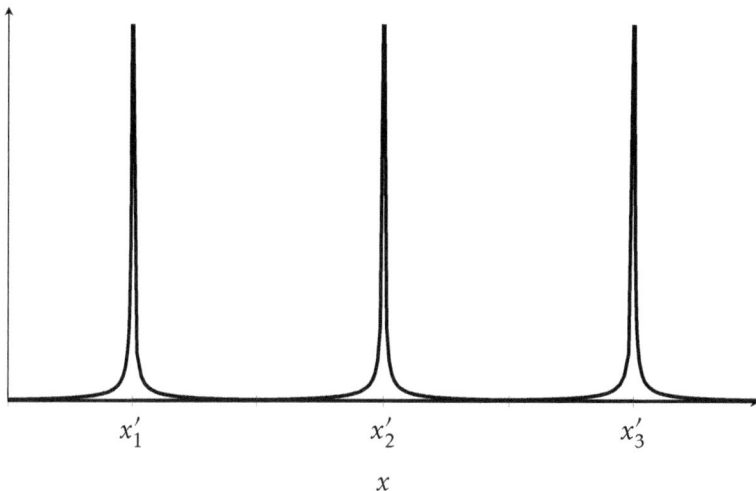

Figure A.1: Example of a potential function $\phi(x)$ function for a line of three discrete charges.

contribution, we have

$$\nabla\left(\frac{k}{|\mathbf{r}-\mathbf{r}'|}\right) = -k\frac{\mathbf{r}-\mathbf{r}'}{|\mathbf{r}-\mathbf{r}'|^3} \tag{A.3}$$

For N charges, this becomes

$$\nabla\left(\sum_{i=1}^{N}\frac{k}{|\mathbf{r}-\mathbf{r}'_i|}\right) = -k\left(\sum_{i=1}^{N}\frac{\mathbf{r}-\mathbf{r}'}{|\mathbf{r}-\mathbf{r}'|^3}\right) \tag{A.4}$$

In Eqs. (A.3) and (A.4), the gradient operator *operates on a scalar function*, and is taken with respect to $\mathbf{r}$, *not* $\mathbf{r}'$. Therefore, while Eq. (A.1) is a scalar function, the gradient operator in Eq. (A.3) results in a *vector* function. This result is related to the *electrostatic force*. Recall the electrostatic force known as the **Coulombic empirical force** (or Coulomb) known to exist between any pair of *static point charges* q and q' given by

$$\mathbf{F}_c(r) = kq(\mathbf{r})q(\mathbf{r}')\frac{\mathbf{r}-\mathbf{r}'}{|\mathbf{r}-\mathbf{r}'|^3}$$

As a vector field, superposition holds for the force due to all the charges at $\mathbf{r}'_1, \mathbf{r}'_2, ..., \mathbf{r}'_N$, respectively. Then, the force generalizes to

$$\mathbf{F}_c(r) = kq(\mathbf{r})q(\mathbf{r}')\sum_{i=1}^{N}\frac{\mathbf{r}-\mathbf{r}'_i}{|\mathbf{r}-\mathbf{r}'_i|^3}$$

A unit vector pointing in the direction of the force can be defined pointing *from* the source charge at $\mathbf{r}'$ *to* the position of evaluation $\mathbf{r}$, defined as

$$\mathbf{e}_r = \frac{\mathbf{r} - \mathbf{r}'}{|\mathbf{r} - \mathbf{r}'|} \tag{A.5}$$

The, we also have

Coulomb Electrostatic Force

$$\mathbf{F}_c(r) = k \frac{q(\mathbf{r})q(\mathbf{r}')}{|\mathbf{r} - \mathbf{r}'|^2} \mathbf{e}_r \tag{A.6}$$

The constant of proportionality k, in SI units, is given by

$$k = \frac{1}{4\pi\varepsilon_0} \tag{A.7}$$

ε_0 is the *permittivity* of free space. The Coulomb force F_c is directed along the vector between charge $q = q(\mathbf{r})$ and $q' = q(\mathbf{r}')$. If we use Eq. (A.3), the Coulomb force at position $\mathbf{r}$, due to charge q' at $\mathbf{r}'$, is given by

$$\mathbf{F}_c = \frac{qq'}{4\pi\varepsilon_0} \frac{\mathbf{r} - \mathbf{r}'}{|\mathbf{r} - \mathbf{r}'|^3} = -\frac{qq'}{4\pi\varepsilon_0} \nabla\left(\frac{1}{|\mathbf{r} - \mathbf{r}'|}\right) \tag{A.8}$$

This extends naturally for any number of source charges, where the general Coulomb force is given by

$$\mathbf{F}_c = q \sum_{i=1}^{N} \frac{q'_i}{4\pi\varepsilon_0} \frac{\mathbf{r} - \mathbf{r}'_i}{|\mathbf{r} - \mathbf{r}'_i|^3} = -q \sum_{i=1}^{N} \frac{q'_i}{4\pi\varepsilon_0} \nabla\left(\frac{1}{|\mathbf{r} - \mathbf{r}'_i|}\right) \tag{A.9}$$

Now, we can define the **electrostatic electric field E** at a point $\mathbf{r}$ as the *force per unit charge*, defined as

$$\mathbf{E}(\mathbf{r}) = \frac{\mathbf{F}_c}{q} \tag{A.10}$$

Substitution of Eq. (A.4) into the Coulomb force gives

$$\mathbf{E} = -\frac{qq'}{4\pi\varepsilon_0} \sum_{i=1}^{N} \nabla\left(\frac{1}{|\mathbf{r} - \mathbf{r}'_i|}\right) \tag{A.11}$$

Using the definition of the electrostatic potential given by Eq. (A.1), we have

A.1 Potential And Discrete Electric Charges

Electrostatic Electric Field

$$E = -\nabla \phi \qquad (A.12)$$

Eq. (A.12) is the familiar definition for the *electrostatic electric field*. In this context, it means the source charges are static, relative to the position r. The above results extend naturally to a countably infinitely number of charges, however, we can express the form slightly differently to allow for additional definitions of parameters, like the charge density. We can also determine the corresponding differential equations describing E, which is of practical significance.

It is useful to imagine a collection of charges in 3D space to be contained within two *disjoint* geometries: *volumes* and *surfaces*. For all physical bodies, they are a collection of volumes and bounding surfaces. Charges exist either within the volume *or* within the bounding surface (confined to the 2D surface), however, not in both. For this continuum approach, it is further assumed that there are necessarily a large number of source charges over a unit differential element of either the volume or surface. We may then speak of (and shall introduce shortly) an average charge density. Such quantities must result in the correct total charge over the unit differential element. For a continuum approach, this also suggests that the spatial variation (on the order of the differentials) of the resulting electric field must be over a length scale much larger than the distances between the point charges, particularly, within the differential elements. If variation distances are relatively large, the charge over the differential element, from the perspective of the whole geometry, acts more like a point charge, and thus an average approximation, yielding the correct total charge, is valid. So, it becomes important to emphasize that there must be a large number of source point charges q', within the differential element, and consequently within the overall structure.

Given such a partition of charges, we now treat separately, volumes and surfaces to obtain electric fields that result from all source charges of a given physical body. First, contributions from charges within the volume are treated, followed by contributions from charges on a corresponding bounding surface.

A.2 CONTINUOUS CHARGE IN VOLUMES

The charge can be represented as a spatially continuous distribution with a variation length scale on the order of a *differential volume element*, $\delta V'$. The charge over a small differential volume $\delta V'$ can be expressed using a continuous *Dirac delta function*, resulting in

$$\delta q' = \frac{1}{\delta V'}\left[\sum_{i=1}^{N}\delta(\mathbf{r}-\mathbf{r}')q_i'(\mathbf{r}')\right]\delta V' = \rho(\mathbf{r}')\delta V' \tag{A.13}$$

Thus, the differential volume charge density $\rho(\mathbf{r}')$ is defined by

Volumetric Charge Density Definition

$$\rho = \frac{1}{\delta V'}\sum_{i=1}^{N}\delta(\mathbf{r}-\mathbf{r}')q_i'(\mathbf{r}') \tag{A.14}$$

From this definition, it follows that

$$\delta q' = \rho(\mathbf{r}')\delta V' \tag{A.15}$$

The use of the Dirac delta function allows the later use of continuous methods, e.g. integration, to be carried out seamlessly. Accounting for the contributions of all $_\delta$ differential element charges $\delta q_j'$, and we have

$$\mathbf{E} = \sum_{j=1}^{N_\delta}\frac{1}{4\pi\varepsilon_0}\frac{\mathbf{r}-\langle\mathbf{r}'\rangle}{|\mathbf{r}-\langle\mathbf{r}'\rangle|^3}\delta q_i'(\mathbf{r}') = \sum_{j=1}^{N_\delta}\frac{1}{4\pi\varepsilon_0}\frac{\mathbf{r}-\langle\mathbf{r}'\rangle}{|\mathbf{r}-\langle\mathbf{r}'\rangle|^3}\rho(\mathbf{r}')\delta V'$$

In the limit, as the number of differential volume elements approaches infinity and thus, $\delta V'$ approaches zero, the electrostatic field due to the volume charges becomes

$$\mathbf{E} = \lim_{N_\delta\to\infty}\sum_{i=1}^{N_\delta}\frac{1}{4\pi\varepsilon_0}\frac{\mathbf{r}-\mathbf{r}'}{|\mathbf{r}-\mathbf{r}'|^3}\rho(\mathbf{r}')\delta V' = \frac{1}{4\pi\varepsilon_0}\int_{V'}\frac{\mathbf{r}-\mathbf{r}'}{|\mathbf{r}-\mathbf{r}'|^3}\rho(\mathbf{r}')\delta V'$$

This result assumes that $\langle\mathbf{r}'\rangle = \mathbf{r}'$, which is true in the limit of a vanishing differential volume. Then, substituting Eq.(A.3) leads to the following for the electric field **E**:

$$\mathbf{E} = -\frac{1}{4\pi\varepsilon_0}\int_{V'}\nabla\left(\frac{1}{|\mathbf{r}-\mathbf{r}'|}\right)\rho(\mathbf{r}')\delta V' \tag{A.16}$$

A.2 Continuous Charge In Volumes

Remember that the gradient in Eq.(A.16) is with respect to **r**, not **r**'. Thus, the gradient operator can be *pulled out* of the integral (i.e. applied after integration), which is evaluated over **r**', so the electric field becomes

$$\mathbf{E} = -\nabla \left(\frac{1}{4\pi\varepsilon_0} \int_{V'} \frac{1}{|\mathbf{r}-\mathbf{r}'|} \rho(\mathbf{r}') \delta V' \right) \tag{A.17}$$

We can see that **E**, again, is the gradient of the *continuous* scalar potential defined here as

Continuous Electrostatic Potential

$$\phi(\mathbf{r}) = \frac{1}{4\pi\varepsilon_0} \int_{V'} \frac{\rho(\mathbf{r}')}{|\mathbf{r}-\mathbf{r}'|} \delta V' \tag{A.18}$$

Thus, an explicit solution to the electric field due to volume charges, is known if one is given the volume source charge density distribution $\rho(\mathbf{r})$.

This summarizes the exact relations for the electrostatic electric potential and its corresponding electric field. The final result we are after is to find out what is the differential equation describing this sort of electric potential ϕ, or it's corresponding electric field **E**. For the result we are after, we'll consider an integral relation given by *Gauss' Theorem* (c.Eq.4.111), also known as the *Divergence theorem*:

$$\int_V \nabla \cdot \mathbf{E} \, dV = \oint_A \mathbf{E} \cdot d\mathbf{A} \tag{A.19}$$

It is applied to a volume V with a bounding surface area A (i.e. the surface area of volume V) containing a single charge q. In order to evaluate Eq. (A.19), we are going to use a bit of a hand-waving argument to justify that the result after integration over one closed surface area is the same for any other closed bounded surface area. A good argument and sort-of-proof can be found in Feynman's Lectures on Physics, Volume II. It is a consequence of the symmetry of the angles around a closed surface. Based on this reasoning, the simplest shape suffices. So, we'll use a sphere whose origin-center is the location of the source charge location **r**', illustrated in Fig. A.2. The integral over the closed and bounding area A, with differential area $dA = r^2 \sin\theta \, d\theta \, d\varphi$ becomes

$$\oint_A \mathbf{E} \cdot d\mathbf{A} = \frac{q}{4\pi\varepsilon_0} \oint \frac{\mathbf{r}-\mathbf{r}'}{|\mathbf{r}-\mathbf{r}'|^3} \cdot d\mathbf{A}$$

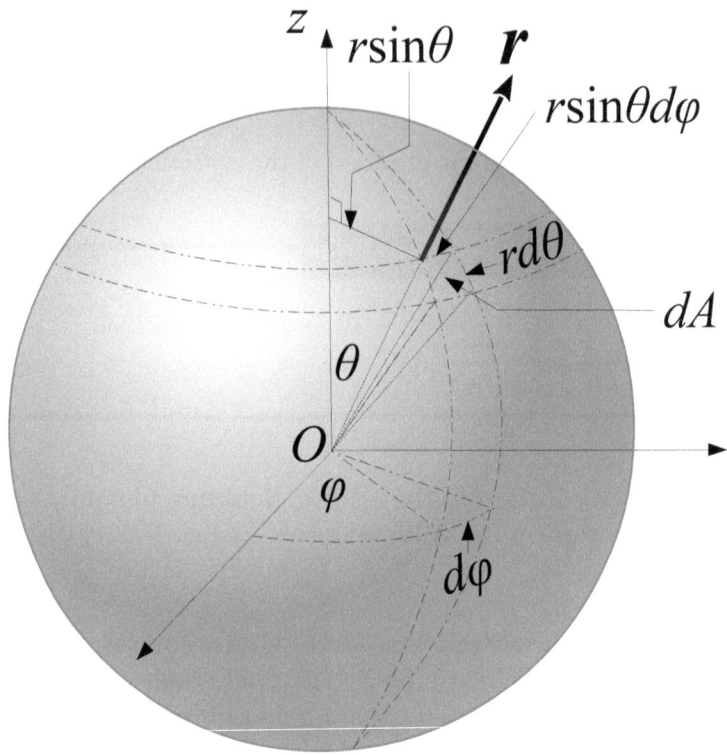

Figure A.2: Illustration of spherical coordinate system and its differential surface.

$$= \frac{q}{4\pi\varepsilon_0} \oint \frac{1}{r^2} \mathbf{e}_r \cdot d\mathbf{A}$$

$$= \frac{q}{4\pi\varepsilon_0} \frac{1}{r^2} \oint \sin\theta r^2 d\theta d\phi = \frac{q}{4\pi\varepsilon_0} \frac{1}{r^2} \times 4\pi r^2 = \frac{q}{\varepsilon_0}$$

Note that in a spherical coordinate system described by (r, θ, ϕ), the unit vector along $\mathbf{r}$ is collinear with the normal vector $\mathbf{n}$ of the area $d\mathbf{A}$. Using the relation $\int \sin\theta d\theta = -\cos\theta$, with the limits of integration sweeping out the surface of a sphere from $\theta = 0$ to π, and $\phi = 0$ to 2π, leads to the above result. Since adding charges just leads to addition of the results using superposition, for N charges, we have that

$$\int_V \nabla \cdot \mathbf{E}\, dV = \oint_A \mathbf{E} \cdot d\mathbf{A} = \frac{\sum_{i=1}^N q_i}{\varepsilon_0} = \frac{1}{\varepsilon_0}\int_V \rho\, dV$$

For this statement to be true anywhere in the volume of charge, the following differential equation that must then be satisfied everywhere in the volume:

> **Gauss' Electrostatic Equation**
>
> $$\nabla \cdot \mathbf{E} = \frac{\rho}{\varepsilon_0} \tag{A.20}$$

This result is the well-known *Gauss' Law*, of electrostatics in the Maxwell equations. Thus, the objective here has been to illustrate that the electric potential ϕ, of the form given by Eq. (A.1), is an *electrostatic potential*, and therefore does not contain time-dependent information. This is an important condition to understand as used in quantum mechanics, because it has limitations with time-dependent descriptions. In fact, we know from the Maxwell equations that time-dependent potential gives rise to time-dependent magnetic fields, and the electrodynamic *electric field* is then described by

$$\mathbf{E}(\mathbf{r}, t) = -\nabla \phi - \frac{\partial \mathbf{A}}{\partial t} \tag{A.21}$$

While ϕ is the electric potential, the relation between it and the potential energy V we have used in the quantum mechanical descriptions is

$$\phi = V/q \tag{A.22}$$

where q is the charge. The electric potential ϕ is, thus, the potential energy per unit charge.

A.3 SUMMARY

In this appendix, we have aimed to discuss in a bit more detail, the origins and characteristics of the electrostatic potential often used in quantum mechanics of atoms and molecules. This form of potential, for example, was used in the hydrogen atom problem. These nuance details are important to consider in a quantum mechanical formulation because it is not time-dependent. This tends to limit the description capability of the quantum mechanical formulation. Particularly, in describing dynamic phenomena. We also learned that we can extend the capability when a magnetostatic field **B** is present, by adding the magnetic vector potential term $e\mathbf{A}$ to the momentum. We called this the *extended momentum*. These still describe systems whose expected velocity amplitude v are constant. Ultimately, to go beyond this, non-Hermitian operators are needed. For spin dependent energy, in particular, a useful approach for extending the dynamical description was introduced in Chap. 11. In modeling and computations of electronic systems exploiting spin, the combination of such details should prove very useful.

About The Author

The author, Kwaku Eason, earned a PhD in engineering from the Georgia Institute of Technology in Atlanta, GA, USA in the area of computational and theoretical electromagnetics and magnetism, applied to the development of magnetometers with significant small length scale features. His thesis *fully* combined micromagnetics with the Maxwell dynamical equations through the use of advanced numerical schemes. Additionally, he earned a Master's and bachelor of science (BS) from the Georgia Institute of Technology, as well as a BS in mathematics from Morehouse College, also in Atlanta, GA, USA. After obtaining his PhD, he moved overseas from the US to Singapore taking a position as a Research Fellow at the Data Storage Institute (DSI), located at the National University of Singapore. There, he developed novel modeling techniques for data storage channel models that could be exploited in developing and designing disk drive components as well as error correction coding strategies for a host of future data storage technologies. At DSI, he later went on to lead the Spintronics and Magnetism team within the Advanced Concepts and Nanotechnology group where his team initiated and began development of technologies (along with disk drive technologies) utilizing spin-transfer torque of both conventional spin polarizers and spin-orbit. He later went to San Jose, Ca, USA to join the Advanced Technology Organization of Western Digital Technology (WD). There, he lead a team of computational modelers in areas spanning quantum mechanics, electromagnetics, micromagnetics, heat transfer, thermodynamics, and mechanics for the development of future disk drive storage technologies. During his time with WD, he was also invited to join the MINT Research Center at the University of Minnesota, co-leading an industry-academia joint project, as well as doing some teaching on topics in spintronics and magnetism within the ECE department. He is an expert in areas of solid state, quantum, electromagnetic, and spin-physics, and creating accurate grounded computational models describing systems based on this physics. He is also the founder of Symphonious Technologies, which focuses on data storage and data processing technology development, exploiting these areas of physics, while also moonlighting as a technical writer.

Index

Symbols

G^2-Ehrenfest theorem 125
2DEG 521

A

Abrikosov lattice 688
addition of angular momentum 445
adjoint 207
Airy Function 604
Airy functions.............. 602
Airy Integral 604
Airys differential equation ... 602
Alexei Abrikosov 688
alpha-particles 28
amber resin 22
Ampére's law 15
Ampere-Maxwell equation ... 15
angular momentum 265
angular momentum conjugate dissipative operator 544
annihilation operator ... 211, 640
anomalous Zeeman effect....368
anti-commutation 384
anti-ferromagnetic 417
associated Legendre funs 288
Augustine Fresnel 14
Avogadros number 45

B

Balmer series 73
band structure 250
barrier reflection probability 246
BCS wave function 709
binomial series expansion ... 492
bistability 498
blackbody radiation 34
Bohr H discrete energy levels 332
Bohr magneton..............273
Bohr radius 333
Bohr's hydrogen model 330
Boltzmann probability distribution 51
Boltzmann's constant 62
bond length 429
bonding energy 429
Born's conditions........... 168
Born-Oppenheimer approximation 428, 701
Bose-Einstein distribution 46, 65
Bose-Einstein thermodynamic probability 47
bosons.................. 46, 407
bound energy states 332
bra vector 628
Bra-Ket algebra.............627
Bra-Ket inner product 628
Bra-Ket scalar product 628
Brackett series 73
branches rotational spectra .. 439

C

canonical magnetic moment problem 268
Cauchy sequence 356
center of mass 335
centripetal force 266
Claudius Ptolemy 5
Clebsch-Gordan coefficients . 450
coherent state 715
commutation 208, 301
commutation postulate 446
Compton scattering 72
Confluent Hypergeometric equation 187, 343
Confluent Hypergeometric function 192, 344
Confluent Hypergeometric Series 192
connecting the formulae 610
connection formulae 610
Continuity equation 169, 174
continuity equation 166
continuous linear operation .. 83
Cooper pair 702
Cooper pair transition Hamiltonian 708
corpuscular model of light 6
correlation energy of Cooper pair 732
Coulombic empirical force .. 741
creation operator 212, 641
critical current 666
critical magnetic field 669
critical temperature 666
Crooke's tube 24
current density 545
current density operator 169
cyclotron frequency 505

cyclotron orbit 513

D

de Broglie relation 74
degeneracy level 47
degenerate states 47, 155
degree of assoc. Legendre fun. 289
density matrix 400
density of states 156
destruction operator 640
destructive interference 13
determinantal wave functions 413
diamagnetic material 695
diamagnetic susceptibility ... 696
diamagnetism 668, 695
diffraction 11, 128
diffuse orbital 351
Dirac delta function 89, 744
Dirac equation 486
Dirac method 639
Dirichlet Integral 99
discharge tubes 328
dissipative continuity equation 545
DOS 156
dual vector space 630

E

easy axes 491
effective exchange field 568
effective spin-orbit magnetic field 521
eigenoperator equations 108
electric displacement field 15
electrical permittivity of vacuum 16
electron gyromagnetic ratio . 273
electron orbital families 351
electron shells 351
electron spin 367
elementary charge 26

entropy 37
equipartition of energy 46
Erwin Schrödinger 73
Euler's equation 8
Euler's theorem 277
evanescent wave function . . . 232
exchange field 591
exchange integral 416
expectation value 116

F

Faraday's law 15
Fermat's Principle 8
Fermi contour 563
Fermi level 50
Fermi sea 563
Fermi surface 563
Fermi-Dirac distribution 49
Fermi-Dirac thermodynamic probability 47
fermions 47, 407
ferromagnetic 417
fine structure 358
finite geometric series 460
first law of thermodynamics . . 36
flux quantum 684
four-vector 475
Fourier Integral Theorem 103
Fourier transform 103
Frobenius method 185
fundamental orbital 351
fundamental quantum numbers 379

G

g-factor 493
Galilean transformation 470
Gauss' theorem 174
Generalized Ehrenfest theorem125
generalized g-factor 456

geometric series 281
George Eugene Uhlenbeck . . . 371
glass blowing 23
gyromagnetic ratio 273

H

H center of mass wave function 350
Hall effect 517
Hamiltonian 8, 88
harmonic oscillator eigenfunctions 203
harmonic oscillator energy . . 196
Heaviside function 92
Heisenberg exchange 444
Hermite polynomial 198
Hermite Polynomials 198
Hermitian operator 166
holes 230, 700
Hooke's law 181
Humphreys series 73
Huygen's Principle 10
hydrogen atom 327
hydrogen atomic spectra 329
hydrogen molecular ion 430

I

Ibn Sahl 5
ideal gas law 45
indicial equation 190, 294
infinitely deep potential well 140
infinitely deep potential well 3D 148
infinitely deep potential well 3D, IDPW 3D 151
integral of action 6
integral representation of Dirac Delta function 102
integration by parts 7
inter-atomic 444

interference 249
intraatomic exchange 444
Inverse Fourier transform ... 103
involutory matrices 383
Isaac Newton 9
isothermal superconductor free energy 677
itinerant electrons 564

J

James Clerk Maxwell 15
Joseph-Louis Lagrange 8

K

Ket-vector 628
Klein-Gordon equation 483

L

ladder operator 641
Lagrangian 8
Landau gauge 501
Landau levels 505
Landau-Lifshitz equation 556
Larmor precession frequency 397
Legendre differential equation 282, 293
Legendre polynomials 281
Leibniz rule 283
Leibniz rule of differentiation 283
Leonhard Euler 6
Levi-Civita symbol 306, 384
lodestones 22
London Equation 674
longitudinal spin current 529
Lord Rayleigh 59
Lorentz factor 472
Lorentz force 268
Lorentz transformation 473
Lortentz radius 371

Louis de Broglie 73
Lyman series 73

M

magnetic moment 267
magnetic permeability in vacuum 16
magnetic potential energy ... 271
magnetic quantum number .. 350
magnetic saturation 500
magnetic tunnel junction 244
magnetoresistance ratio 580
Max Born 168
measurable 107
method of Frobenius ... 185, 293, 344
method of separation of variables 148, 292
MgO 583
microscopic dimensionless coefficient 542
molecular orbitals 436
MR ratio 580

N

n-vortex 688
natural frequency 181
Neils Bohr 330
non-Hermitian Hamiltonian 534, 543
non-Hermitian operators 533
normal distribution 96
normalization condition 167
normalized wave function ... 113
null Ket vector 629
null-space 638
number of complexions 38

O

observable 107

occupation probability distribution 159
orbital d 350
orbital f 350
orbital p 350
orbital s 350
orbital angular momentum .. 263
orbital angular momentum lowering operator 312
orbital angular momentum quantum number 350
orbital angular momentum raising operator 316
orbital magnetic dipole 270
orbital moment torque 270
orbital resistance 537
order of Associated Legendre func. 289
order parameter 678
ordinary Zeeman effect 358
ordinary Zeeman interaction energy 359
orthogonal eigenstates 634
orthogonality condition 162, 164
orthogonality property of Hermite polynomials 203
orthohelium 424
Otto Stern 372

P

parahelium 423
Parseval's Theorem 105
particle in a 3D box 151
Paschen series 73
Pauli Exclusion Principle 444
Pauli spin matrices 382
perturbation theory 619
Pfund series 73
Pierre de Fermat 8
Planck's constant 66
Planck's distribution 65

plum-pudding model 27
Pochhammer symbol 192
potential hill 230
potential hole 230
potential well 250
Poynting vector 31
precession 397
precession frequency 397
precessional frequency 551
principal orbital 351
principal quantum number .. 344
Principle of Correspondence 180
probability amplitude 148
probability density function . 168
Problems 30, 77, 136, 175
proper time 477
Python computing language . 583

Q

quantum tunneling 244

R

radial part of Schrödinger equation for hydrogen atom 340
radial wave function normalization factor 346
radial wave functions 345
radial wave functions for hydrogen 345
radioactive decay 28
Ralph Kronig 368
Rashba coefficient 522
Rashba constant 522
Rashba effect 520
rectangular potential barrier 237
recurrence relation 191
reduced mass 337
reflection probability 235
refraction 5

relativistic kinetic energy....479
relativistic mass............480
relativistic momentum......477
René Descartes................6
rest-mass....................478
rising factorial..............192
Rodrigues' formula..........285
rotational inertia............264
rotational wave-number constant 438
Rudolf Clausius..............37
Rudolph Clausius............38
Rydberg constant........72, 334
Rydberg formula.............72

S

Samuel Abraham Goudsmit . 371
scattering...................231
Schrödinger equation.........88
Schwartz' Inequality........119
separation of variables . 110, 148
sharp orbital................351
side jump...................517
singlet state...........423, 448
Sir James Jeans..............59
Slater determinants.........413
Slater orbital................413
Snell's law...................6
solid angle..................40
space-time interval..........476
space-time vector...........475
special theory of relativity...470
spherical harmonics.........297
spherical harmonics table ... 324
spin.........................368
spin density matrix..........398
spin diffusion constant......588
spin Hall....................514
spin polarization............575
spin quantum number..369, 385

spin-diffusion length........585
spin-down state.............380
spin-FET....................520
spin-field effect transistor ... 520
spin-Hall effect..............518
spin-orbit effective field.....520
spin-orbit interaction...491, 496
spin-orbit transverse momentum 516
spin-spin interaction........425
spin-transfer torque....519, 594
spin-up state................380
spinor.......................381
spintronics..................368
square integrable...........167
steradian.....................41
Stern-Gerlach experiment ... 372
Stirling's approximation .. 53, 64
STT.........................594
superconducting coherence length 688
superconductor heat capacity 723
superconductor mag. permeability...................697
superdiamagnetism.........668
superfluid transition in helium 663
superposition principle.......12
symmetric gauge............501

T

table of electron quantum numbers..................379
thermodynamic probability...38
Thomas Young............12, 13
TMR ratio...................583
total quantum number......344
total relativistic energy......481
transition temperature......666
translational inertia.........264
transmission electron microscopy 75

INDEX

transmission probability 235
transverse damping 556
trigonometric identities 548
triplet 357
triplet state 424, 448
tunneling 244
tunneling magnetoresistance ratio 583
turning points 607
two-dimensional electron gas 521
type I superconductor 671
type II superconductor 671

U

ultraviolet catastrophe 59
uncertainty 117
uncertainty principle 118
unitary matrix 454
unitary operators 706
units of wave function 87
universal gas constant 45

V

valence band 699
valence shell 563
Valet-Fert theory 584
vibrational wave-number constant 438
Virial theorem 332

W

Walter Gerlach 372
wave function interference .. 249
wave-number 19
Wien's blackbody distribution 61
Wilhelm Wien 60
Willebrord Snell 6

Y

Young's interference experiments 12

Z

Zeroth Order JWKB approximation 613

www.ingramcontent.com/pod-product-compliance
Ingram Content Group UK Ltd.
Pitfield, Milton Keynes, MK11 3LW, UK
UKHW041438190426
11946UKWH00021B/31